GENETICS AND NEUROBIOLOGY OF DOWN SYNDROME

GENETICS AND NEUROBIOLOGY OF DOWN SYNDROME

BANI BANDANA GANGULY
MGM New Bombay Hospital, Vashi, Navi Mumbai, India
MGM Institute of Health Sciences, Kamothe, Navi Mumbai, India

ELSEVIER
ACADEMIC PRESS
An imprint of Elsevier

Academic Press is an imprint of Elsevier
125 London Wall, London EC2Y 5AS, United Kingdom
525 B Street, Suite 1650, San Diego, CA 92101, United States
50 Hampshire Street, 5th Floor, Cambridge, MA 02139, United States
The Boulevard, Langford Lane, Kidlington, Oxford OX5 1GB, United Kingdom

Notices

Knowledge and best practice in this field are constantly changing. As new research and experience broaden our understanding, changes in research methods, professional practices, or medical treatment may become necessary.

Practitioners and researchers must always rely on their own experience and knowledge in evaluating and using any information, methods, compounds, or experiments described herein. In using such information or methods they should be mindful of their own safety and the safety of others, including parties for whom they have a professional responsibility.

To the fullest extent of the law, neither the Publisher nor the authors, contributors, or editors, assume any liability for any injury and/or damage to persons or property as a matter of products liability, negligence or otherwise, or from any use or operation of any methods, products, instructions, or ideas contained in the material herein.

ISBN: 978-0-323-90456-8

For information on all Academic Press publications visit our website at https://www.elsevier.com/books-and-journals

Publisher: Nikki P. Levy
Acquisitions Editor: Joslyn Chaiprasert-Paguio
Editorial Project Manager: Samantha Allard
Production Project Manager: Maria Bernard
Cover Designer: Matthew Limbert

Typeset by TNQ Technologies

Contents

An overview of this book

Down syndrome (DS), the most common chromosomal disorder, was first characterized clinically in 1862 by John Haydon Langdon-Down, and genetically in 1959 by Jérôme Lejeune. The epidemiological incidence is estimated at one in 600 to 700 newborns; however, this ratio is heavily influenced by age of the mother. DS is a well-exemplified aneuploidal genetic disorder manifested with the syndrome-specific but variable facial and clinical features and early onset of Alzheimer disease (AD)-associated dementia. The long history of understanding the syndrome in different geo-political boundaries has been narrated in Chapter 1 of this book. One must know the journey of these individuals from institutionalized homes, care-givers' centers to living with parents, and their acceptance by the general public. They require schooling with special arrangements for facilitating improvement of their learning skill. Besides typical syndromic craniofacial appearance, their intellectual disability and delayed motor milestones segregate them in performance from their age-matched peers. Indeed, one extra copy (trisomy) of the human chromosome 21 (HSA21), which harbors ~500 genes, arises *de novo* during prefertilization gametogenesis (meiotic nondisjunction) of the parents (most with full trisomy, ~95%) or postfertilization mitotic nondisjunction (most of mosaic DS). This trisomic HSA21 confers an altogether different facial configuration, cognitive impairment, congenital heart disease (CHD, in ~50 of DS) early onset of AD, and many other clinical complications throughout their life. One must know the risk of parents carrying a Robertsonian translocation between "D" group and HSA21 are at greater risk of reproducing a child with DS, while parents carrying one such translocation between two HSA21s will deliver 100% children with DS. Albeit conceptions with DS have a high rate of pregnancy loss or early death in postnatal life, this autosomal aneuploidy has exhibited extended lifespan due to early intervention of medical therapeutics and increasing parental awareness.

The chapters have been arranged based on the order of developmental sequences of the knowledge bank on the syndrome, which started with identification of trisomy 21 as genetic etiology to expression of trisomic HSA21 genes and interaction with non-HSA21 disomic genes leading to a complex scenario. Indeed, the DS patients with the trisomy of the smallest chromosome have not yet been fully understood for clinical management of their cognitive performance, facial correction, or amelioration of AD at pre- or postnatal age, even with the advent of the current gene editing CRISPR technologies or therapeutic targets. The impact of gene-dosage imbalance is highly critical for DS pathobiology, albeit variations in phenotypic manifestations observed among the DS population. Notably, over 80 phenotypes have been demonstrated in DS, of which the unique craniofacial features, muscle hypotonia, hallux varus, intellectual disability, and risk of AD causing early aging and dementia are constant features for all individuals with DS, while remaining are variable, but affecting majority of the DS individuals (illustrated in Chapter 2). Among the variable characters, atlantoaxial instability, reduced neural density, CHD, transient myeloproliferative disorder (TMPD) with increased risk of acute megakaryocytic leukemia (AMKL), Hirschsprung disease, impairment of hearing and vision, autoimmune disorders, vulnerability to coronavirus disease 2019 (COVID-19) infection, etc. are not observed in all DS births but may develop in some of the individuals during their lifetime. Taken together, DS features vary in quantum in terms of presence/absence/gravity of the features and frequency of DS individuals affected, though all of them carry the extra HSA21. Noticeably, DS population showed a 4-fold increase in hospitalization and a 10-fold higher risk of fatality related to COVID-19 during the SARS-COV-2 pandemic. Nevertheless, DS individuals of the present era live longer (≥60 years) than before (25 years in 80s). Notably, parents' awareness, facility at special schools, health monitoring, and surveillance are enormously contributing to developing professional skills with increased memory functions and improvement in quality of life. Males are less fertile than females due mainly to anatomical defects and inadequate spermatogenesis. Collectively, the gene-dosage imbalance in a complex genetic-biochemical-physiological-cellular micro-environment really makes the disorder highly complex. To unzip the pathogenetic expression on the DS patients, regulation of gene and transcriptional expression at the backdrop of overexpressing nuclear-encoded genome and cytoplasmic mitochondrial dysfunction require expanded knowledge, which are included in the forthcoming chapters.

The triplicated HSA21 contributes to a myriads of DS-specific phenotypes leading to discriminatory intellectual disability and craniofacial dysmorphism besides several systemic complications, including early onset of leukemia and neurodegenerative diseases. Interestingly, the extra HSA21 comes from one of the parents though DS is not a hereditary disorder and mostly influenced by maternal age. Meiotic nondisjunction leads to inclusion of both 21s in 1 cell, which subsequently produces an aneuploid zygote with three copies of HSA21 following fertilization with another apparently normal gamete. The consequences of meiotic nondisjunction in first or second phase and etiology of trisomic HSA21 in DS have been discussed in Chapter 3 with figures and real human karyotypes for understanding the origin of the extra 21. However, there could be myriads of other intrinsic or extrinsic factors to contribute in nondisjunction of chromosomes, which may include paternal age; maternal grandmother's age at the time of mother's birth; exposure to occupational and ambient environment and life-style factors; intrinsic genetic mechanisms; and so on (narrated in Chapter 4). Moreover, DS children are also born to young mothers due mainly to their antenatal complexity, exposure to radiation, chemicals, other hazardous elements, infections, and many other reasons, which could induce meiotic or mitotic nondisjunction. Thus, alongside improvement of life of DS patients, further investigation on the possible causes of aberrant chromosome separation would be important for understanding and effective implementation of preventive modalities. The focus of research on DS has been revolving around understanding the contribution of triplicated HSA21 at a possible co-operating setting of other chromosomes or genes on the phenotypic and clinical manifestation of the syndrome. Nonetheless, in the era of molecular karyotyping and targeted sequencing, conventional karyotyping still stands gold standard for allowing identification of the chromosomal basis of DS, possible consequences of future transmission, and more so, at significantly low cost and short timeframe. Indeed, a basic karyotyping is doable by nations of all economic strata, though the application of conventional cytogenetics is being diminished at developed nations.

The advancement of knowledge on the disease and technological developments has established guidelines toward prenatal and/or preimplantation screening of the trisomic status in a growing fetus during the first-trimester gestational age. Prenatal testing modalities are almost safe and definitive for reproductive choice of termination or continuation of the pregnancies with trisomy 21 or DS. Importantly, presence of the extra chromosome in all (full trisomy in classic DS: 90%) or fraction (mosaicism in variant DS) of cells, but all tissues of a body; thus, it cannot be subtracted and/or cured as of now. Prenatal screening of dual-, triple-, or quadruple markers in the maternal serum combined with ultrasound imaging, and testing of trisomy 21 in chorionic villi (CVS) or amniocytes have long been under practice for prevention of birth with DS or other genetic disorders. However, invasive sampling of CVS or amniotic fluid is gradually being replaced by noninvasive collection of fetal DNA (cell free fetal DNA or cffDNA) in maternal plasma. The compromised life of DS patients with respect to facial appearance, intellectual impairment, and early onset of neurodegeneration, is responsible for societal discrimination. Thus, parents of such children are always at stake to be blamed by spouses/family members and even the child with the syndrome, for either carrying such pregnancies and/or lack of their awareness. Genetic counseling, prior to sampling, testing, and post testing, explains the genetic consequences of the disease, and purpose, procedure, and prospect of prenatal screening and testing. Notably, anticipated issues of prenatal and carrier screening of at-risk pregnancies can be controlled through genetic counseling (discussed in Chapter 5). Additionally, policies of the public health departments can effectively mitigate the burden of economic status of such individuals born with untreatable genetic impairments and curve an equitable niche in the society with that of the general population. Indeed, these individuals can achieve acumen in diversified professions and become financially independent. Taken together, increase in awareness of the life of these individuals and prenatal screening can be exacerbated effectively by the involvement of governmental policies, aid, and ethical aspects.

For therapeutic development and intervention, understanding of the consequences of gene-dosage imbalance appears to be the primary concern of modulating the phenotypic expression in individuals with DS. In particular, molecular characterization of the complex mechanism of antagonistic or synergistic impact of the triplicated HSA21 and disomic genes has resulted in tagging of some of the HSA21 genes in relevance to the genotype–phenotype basis of the syndrome. However, phenotypic variability in DS and contribution of the disrupted genomic homeostasis have yet to be delineated. Consistent with this, a cocktail of over- and underexpressed and compensated genic expression has indicated interaction of genic and nongenic complements, including epigenetic and environmental components. The situation is more complex for mosaic or translocated DS. Nonetheless, research conducted on primates and/or cell lines developed from DS patients for unraveling the DS pathomechanism has identified specific link of triplicated DSCR1, DYRK1A, APP, SOD1, SYNJ1, OLIG2, RCAN1, RUNX1, ETS2, ERG, etc. with DS phenotypes, AD, oxidative stress, production of reactive oxygen species (ROS), cardiac defects, leukemia, and others. However, reduction of trisomy to disomy through cross-breeding of DS models failed to rescue some of the impaired phenotypes. Thus, collective effects of gene-dosage imbalance, allelic variations, and contribution of modifier genes of the global genome and coincidental small effects were thought to confer on DS phenotypes. Moreover, expression of dosage imbalance and altered transcription to proteomic translation render variability in phenotypic manifestation in these individuals (explained in Chapter 6). Notably, assay of the entire smallest autosome, including sequencing could merely contribute to change the scale of the problem of dosage sensitivity on the disorder known for over 6 decades.

Indeed, there is a great deal of variability in phenotypic characteristics and severity of disease expression among individuals with DS, though all of them carry trisomic dosage imbalance in their genome. Moreover, the molecular dissection of the HSA21 genes has been restricted to a limited number of human samples, partial trisomies in particular, for identification of the phenotype-specific genes. Expression of HSA21 genes greatly differs in different tissues at different developmental age. In addition, interplay among the genes contributes significantly against the backdrop of variable background genome and polymorphisms (narrated in Chapter 7), which have been studied in a number of mouse models produced with HSA21 orthologous segmental trisomies alongside human DS. Thus, dissection of the individual alleles responsible for phenotypic expressions is the need of the present situation, which can only be mated out with an intense level of cooperation between different countries for exchange of DS samples of different race and geography and technological collaborations. However, the catch is to comply with the ethical and regulatory issues on cross-border exchange of DNA samples, though the system is already in place for diagnostic purposes.

Despite the contribution of the nuclear genome-encoded trisomic genes, involvement of the extra-nuclear mitochondrial DNA (mtDNA) and synergistic cooperation with upregulated HSA21 genes on neurodegeneration and many other diseases have drawn attraction for understanding the pathogenetic mechanisms of DS phenotypes. Indeed, the 16.5 kb circular mtDNA is the power house of cells for producing ATP, maintaining cellular calcium and redox homeostasis, and cellular energy-production processes. Moreover, the dynamic structure of mtDNA depends on fusion and fission mechanism, and the genetic composition of the mtDNA varies greatly in hundreds to thousands of mitochondria present in each cell, which is higher in more energy consuming cells such as the brain and heart. Indeed, the mitochondrial alterations significantly perturb the brain and cardiac development because of their dependence on higher energy consumption and susceptibility to oxidative stress, starting from the fetal age. The fission of this organelle has been demonstrated to be significantly higher in trisomic fetal heart and brain, and also in DS mouse models, resulting in perturbation in oxidative phosphorylation (OXPHOS) and generating free radicals such as reactive oxygen species (ROS). The results suggest involvement of mtDNA in proliferation and protection of cells from endogenous toxic environment and external stressors. Overexpression of HSA21 genes and imbalance in genomic constitution collectively contributes to mitochondrial dysfunction resulting in disrupted ATP production and respiratory capacity in DS patients (demonstrated in Chapter 8). Striking impairment in OXPHOS and damage in redox homeostasis and mitochondrial dynamics were observed in all types of tissues and organs in DS. Therefore, exploration of the interplay of the trisomic HSA21 genes and mtDNA has become a major concern of drug development for DS, as the context is highly complex and not unraveled fully.

Since trisomic HSA21 underlies the deregulation of genome-wide gene expression contributing to variability of DS phenotypes, disease consequences at cellular and systemic levels are directly set off by their interplay on a background of generalized metabolic perturbations. Moreover, the impact of the ubiquitous OXPHOS and its impairment might affect the cellular functions of every organ of the body. Indeed, mitochondrial dysfunctions play the central role for myriads of inevitable DS phenotypes. Prominently, the mitochondrial abnormalities and altered energy metabolism are responsible for neuronal defects, learning disability, heart defects, type 2 diabetes, immune disorders, early aging, neurodegeneration, AD, AD-related dementia, and so on (described in Chapter 9). Albeit the mitophagic efficacy of the cells for clearing the dead and apoptotic cells, an adaptive response triggered by a longer survival has been demonstrated as the effect of the sub-lethal levels of stress on oxidative metabolism. Impaired scavenging of free radicals may disturb mitophagy. Taken together, the combinatorial expression of chronic energy stress and impaired ROS scavenging on age-related health indices may accelerate development and progression of DS diseases. Therefore, compelling therapeutic amelioration of the mitochondrial defects and subsequent improvement of the OXPHOS function was targeted to improve the quality of life of DS individuals and their families. A number of pharmacologically active compounds, including natural bio-elements, have shown convincing outcome for correction of mitochondrial dysfunctions and dysregulated oxidative metabolism.

Defects in neurogenesis caused by gene-dosage imbalance are responsible for cognitive impairment in people with DS. Morphologically, the DS brains are characterized with reduced size and weight, specifically the cerebellum and the hippocampus, and defective neocortical and brainstem organization (illustrated in Chapter 10). Reduced neuronal density in diverse regions of the brain, prominently in the cortex, hippocampus, and cerebellum, has a gestational onset, and gets aggravated through adulthood. The consequences of neurogenesis have been studied in human autopsies and in DS models. However, the contribution of the reduced brain size or neuronal density to cognitive deficits in DS is not a straightforward mechanism, rather a collective expression of triplicated HSA21 genes and altered proteins contributing to dysregulation of several signaling pathways. Furthermore, perturbation of OXPHOS due to dysfunctional mitochondrial expression and overproduction of ROS disturb the development DS brains. Imaging of different brain regions exhibited reduced functional connections between brain circuit elements resulting in dysfunctional circuitry in DS adults. However, it may not be rational to attribute the cognitive deficits in DS only to reduced brain size or neuronal density. Notably, different aspects of neuronal cell structure, including the size of the basal dendritic arbors, their branching structure, and spine density may vary independently. The neural mechanisms underlying intellectual impairment may include abnormalities in the formation of neuronal networks and/or in information processing aspects of brain plasticity. Additionally, abnormalities in dendritic architecture and subsequent neuronal disruption determine the nature of a neuron's inputs and its role in cortical circuitry, which are distinct in DS subjects and correlate with their cognitive profile to some extent. However, the question remains: if the degree of dendritic alterations correlates with cognitive function, then DS subjects with moderate level of intellectual deficits should exhibit moderate level of dendritic abnormalities, and, if dendritic abnormalities indicate the global cognitive dysfunction, they should be observed in many other syndromes associated with mental retardation. Thus, dendritic alterations should parallel the defects in neuronal maturation and/or decline in cognitive performance. Furthermore, neurons in DS may not achieve the level of structural complexity as that of the normal children or greater dendritic retraction may occur during maturation. In total, age-related consequences of neurodevelopment and implication of dosage-imbalance of genes and proteins might perturb many signaling pathways involved in neurogenesis, which collectively affects the cognitive development in DS, in cooperation with dysfunctional mitochondrial action and deregulated energy production.

The abnormalities of neurotransmission originate (at least in part) during neurodevelopment at gestation and that goes through complex synaptic, physiological, and circuit changes to adulthood, ultimately causing cognitive impairment and other neurological manifestations. Seemingly, the basic wiring of the human nervous system is configured during early development, when the key events such as birth and migration of neurons, synapse formation, and maturation occur with refinement of circuit connections of functional neural circuits with balanced excitatory (glutamatergic)/inhibitory (GABAergic) (E/I) components (discussed in Chapter 11). Seemingly, the aberrant neurotransmission perturbs the synaptic plasticity through alteration of E/I balance, which finally impairs learning and memory in individuals with DS. The γ-amino-butyric acid (GABA) is produced by the GABAergic neurons in mammalian brain during the formation of GABAergic synapses and acts as the major inhibitory neurotransmitter. GABA binds to GABA-receptors (GABA-A and GABA-B), opens a chloride channel, and reduces neuronal excitability in the central nervous system (CNS). The strength of GABAergic synapses is regulated by glutamatergic transmission. Moreover, GABA signaling, in most immature CNS regions, is established before glutamatergic transmission, indicating that GABA is the primary excitatory transmitter during early development of brain. DS-related GABAergic and glutamatergic neurotransmission have been studied mostly in DS mouse models. Indeed, cooperation of depolarizing GABA-A signaling and glutamatergic excitatory alterations appeared to be brain-region specific in trisomic Ts65Dn mice – a DS model. Summarily, the alterations in neurotransmission, especially the impaired GABAergic transmission, represent the collective expression of different molecular mechanisms exerted by gene–dosage imbalance to emphasize the multifaceted consequences of DS brain abnormalities. Nevertheless, as a matter of fact, the unmet identification of effective pharmacological interventions for alleviation of DS-related cognitive deficits constitutes an incredible complex challenge for extraction of the insights about underlying GABA-related impairments in DS models and other neurodevelopmental disorders related to defective GABAergic transmission (such as autism, Fragile X syndrome, epilepsy, and possibly Rett syndrome).

The genesis of the defective GABAergic transmission tracks back into fetal onset of defective neurodevelopment, with possible E/I imbalance and neuronal and plasticity impairments that percolate to adulthood. However, the gravity of the GABAergic inhibitory vs. the glutamatergic excitatory transmission in this irregularity is still less understood. Understanding the abnormal signaling from GABA-A and GABA-B receptors (GABA-AR/BR) and E/I balance on the neuronal defects has demonstrated discordant results in Ts65Dn DS mice. Changes in GABA-AR subunit composition, specific for human DS, were lacking in the Ts65Dn mice, which could be responsible for the failure of therapeutic outcome with a α5-selective antagonist in DS patients. Thus, a low predictive potential of the Ts65Dn mouse model needed to be replaced by use of trisomic iPSCs for pharmaceutical investigation of potential drug candidates. Moreover, creation of the "perfect" replica of trisomic HSA21 in DS models has come out with unintended rearrangements. Thus, future assays on the efficacy of each new pharmacological treatment could focus on more than one DS models both *in vitro* and *in vivo* and that can control the failures, inadequacy, and discrepancies in clinical trials. Overall, although the available data from human studies, while limited, seem not to support the data derived from animal research of increased GABA-mediated transmission due to the overproduction of GABAergic interneurons. Further electrophysiological studies are needed on neurons in DS-derived iPSCs are needed. On the other hand, no study has yet conclusively assessed the occurrence of depolarizing GABA-A signaling or increased GABAB-mediated transmission in human DS neurons.

Alzheimer disease (AD) in DS (DSAD) is an unavoidable clinical event, which is characterized with aggregation of Aβ plaques, tau hyperphosphorylation, and neurofibrillary tangles in the brain that exacerbates age-related dementia in individuals with DS. DSAD population is genetically driven with interactive gene-dosage effects among trisomic HSA21 and non-HSA21 (nontrisomic) genes in ways specific to DSAD, thus somewhat different from sporadic or familial AD. Unlike sporadic and familial AD, DSAD occurs with early beginning at around 30 years of age with expression of AD features and overt dementia by 40 and 50 years, respectively. Overexpression of the triplicated HSA21 genes, APP and BACE2 in particular, and interaction among the trisomic and global disomic genes in DS accelerate overproduction of Aβ-protein via processing, cleavage, and clearance of senile plaques in DS population. Influence of gene-dosage on mitochondrial dysfunction and associated oxidative stress further aggravates the context of DSAD mechanism. In addition, neuroinflammation accentuated by higher dosage of S100β, SOD1, ADAMTS1, ADAMTS5, CXADR, PSEN1/2, and many other genes results in activation of microglial cells and generation of cascade events, including abnormal release of cytokines and chemokines. Therapeutics for sporadic AD and DSAD on normalization of APP or either active vaccination or passive immunization to introduce anti-Aβ antibodies in the DS population and AD patients have presented limited outcome in clinical trials and mouse models of DS. Thus, a keynote address (narrated in Chapter 12) on the genetic consequences of Aβ biology has highlighted further requirement of understanding of DSAD for prevention or therapeutic intervention.

Reunderstanding the genomics, metabolomics, lipidomics, and other (-)omics, and associated mechanistic lines of investigation may crucially define the DSAD pathophysiology. Besides prominent overexpression of APP and chronic oxidative stress, studies of exosomes, TREM2, neurotransmitters, neurofilament light chain (NFL), and other novel biomarkers and modifiable risk factors (such as sleep apnea in DS) would contribute significantly. However, gaps exist in the knowledge of the specific biomarkers of DSAD, including markers of vascular, inflammatory and oxidative stress, neuroexcitation, calcification, etc. across the horizon of the disease. Interestingly, the domain of exosome application, especially brain-derived exosomes (BDEs), remains relatively novel in understanding the disease mechanisms and biomarkers of the course of AD, and their ability to mediate cell-to-cell communication in the processes associated with AD (described in Chapter 12). Furthermore, the mechanisms of bidirectional transport of cargo-carrying exosomes across the blood-brain barrier (BBB), changes in number or size of exosomes secreted, alterations in cargo constituents, and identification of differences in specific cell types may help in reversal of physiological changes in nervous system disorders. Nevertheless, the substantial lacuna in the present understanding of aging process in people with DS influencing early development of DSAD and dementia may continue to persist but are critically important to address. Notably, limited success has been achieved with FDA-approved treatments for AD in the general population. Therefore, exploration of novel therapeutic interventions with potential prophylactic approaches stands fundamentally important for further understanding of aging, dementia, and AD for individuals with and without DS.

The gene dosage of triplicated HSA21 and synergistic or antagonistic association with disomic genes have unequivocally been identified as the key elements of pathomechanisms in DS population. Furthermore, contribution of the noncoding genes of RNAs, termed as microRNAs (miRNAs/miR), have been envisaged with DS phenotypes with constitutive genomic overexpression and the consequent production of haploinsufficiency of their specific proteins. Of the several HSA21-encoded miRNAs, miR-155, miR-125b-2, miR-802, miR-99, and let-7c are overexpressed but lead to underexpression of their specific target proteins, which are characterized for prevention or reversal of DS-related complications. The phenotypic association of miRs is annotated as: miR-155 with neuropathology; miR-155 and miR-802 with cognitive disability; the cluster of miR-99 and let-7c favor CHD; and miR-125b-2 with acute megakaryocytic leukemia (AMKL), while miR-99a-5p, miR-155-5p, and let-7c-5p were markedly overexpressed in trisomic hearts (demonstrated in Chapter 13). Other miRNAs (miR-17, -20a, -26a/b, -101, -106b, and -199b) appeared important for reduction of tau and Aβ damage in DSAD. The miR125b-2 has been considered as an oncogenic miRNA for DS patients, whereas protective role of miR-125b, miR-99, and let-7c on breast, prostrate, and lung cancers might explain the low incidence of solid tumors in DS. In addition, interaction of HSA21-miRNAs with non-HSA21-miRNAs has significant role in DS pathomechanism. Overexpression of miR-99a-5p, miR-155-5p, and let-7c-5p in the heart of trisomic fetuses as well as other miRNAs in several other organs in human DS, DS models, and DS-derived iPSCs have directed development of miRNA-targeted drugs for alleviating the DS features and/or reversal of the consequences of overexpressed miRNAs. Indeed, identification of antagomiRs is increasingly being considered for testing in different systems for deciphering their druggable potential.

Additionally, global hypermethylation was observed in the DS genome, where expression of methylation status varied depending on the tissue of origin and developmental stage. Targeted gene analysis has demonstrated the implications of overexpressed DNMT3L on regulation of tumor suppressor TP53 in DS patients and alteration in neurogenesis (demonstrated in Chapter 13). Interestingly, expression profiling of epigenetic regulation in DS has indicated silencing of pathways and expression of proteins and thus, contributes to pre-existing knowledge of oxidative stress, gliosis, and mitochondrial dysfunction of DS population. Indeed, epigenetic DNA methylation pertaining to OXPHOS, ubiquitination, neurodevelopment, and other pathways raised likelihood of significant effects of DNA methylation on DS features. Noticeably, precise understanding of gene expression in the context of miRNAs and epigenetic regulation is directing identification of small molecule inhibitors as compelling drug targets and consequent therapeutic development for alleviating DS pathology and improvement of life-degrading features, prominently intellectual disability, dementia, and AD. In this line, small molecule inhibitors have yielded promising results, in particular for management of hematopoietic malignancies, and thus, similar target is expected to work for DS leukemogenesis. However, DS patients have many more clinical consequences of gene-dosage effect and consequent expression of miRNAs and epigenetic patterns. Thus, screening of "methylomes" (methylation signatures) contributing to several pathways, and precise understanding of gene-dosage expression of the triplicated HSA21 in the context of miRNAs and epigenetic regulation necessitates further exploration of the independent expression and synergistic or antagonistic interaction in the complex genetic environment. Moreover, the pharmacological mechanism of action of the proposed small molecule inhibitors of miRNAs and epigenetic markers of DS genome would require appropriate test systems targeting specific phenotypic features of DS.

Nevertheless, the collective contribution of all HSA21 genes could not lead to any success, particularly improvement of cognitive skill or deferment of dementia or DSAD. Due to wide variability in clinical expression of trisomy 21, such as presence or absence of CHD, degree of mental retardation (MR), severity of hypotonia, etc. among the DS people, tagging of specific gene to specific phenotype of DS has always been a grave hunt in DS research. Genotype–phenotype association in DS was mostly limited with less number of human DS with partial trisomy. Owing to such limitation of *in vivo* study in humans, the shortcomings led to develop models of DS (DS models) with full HSA21 or trisomic orthologous segments syntenic to HSA21 through chromosomal rearrangements, and recombinant and gene-editing technologies for investigation of gene expression. Mouse chromosomes 10 (MMU10), 16 (MMU16), and

17 (MMU17) have HSA21 orthologous segments. Various mouse models have been generated (described in Chapter 14), of which Ts65Dn model has been extensively considered in multiple aspects of DS research, including aging and age-related complications of human DS. The Ts65Dn mice are trisomic for 60 MMU17 genes and disomic for 16 MMU16 genes orthologous to HSA21, and have remarkable phenotypic similarities with Dp(16)1Yey/+ mice that is exclusively trisomic for the entire MMU16 region syntenic to HSA21. MMU10 and MMU17 orthologous regions might ameliorate some of the expression of the trisomic MMU16 regions. Several targeted DS mouse models were created, initially with segmental trisomy relevant to HSA21 and later plenty of deletion and duplications models, for precise recapitulation of dosage sensitivity responsible for specific phenotypes of human DS. Further, cross-breeding of these strains with mice bearing deletions of chromosomal segments syntenic to HSA21 (such as Ms1Yah and Ms1Rhr) allowed subtraction of the anticipated responsible gene and eventually helped in identification of the sensitive genes associated with specific DS phenotype. Crucially, exploitation of trisomic mouse strains, in combination with YAC or BAC transgenics and gene knockouts with inactivation of the targeted genes, identified responsible dosage-sensitive genes for different DS-specific features and helped in therapeutic development, mostly for improvement of cognition and AD-related dementia. Notably, the advent of chromosome engineering technology enabled creation of trisomic animals (Tc1) with the entire stretches of HSA21. The orthologous mouse regions have become realistic tools and resource models to unravel the dosage-sensitive genes of HSA21. Interestingly, breeding of the trisomic mice with knockout mice for some alleles has enabled bringing trisomic mice back to disomic state for some targeted genes, which allowed study of expression of genes in disomic as well as trisomic condition, and also facilitated understanding the rescue of the trisomic effect in the cross-bred mice. However, some of the important and crucial points to be taken into account in studies of DS research, which includes, at least: the expression trisomic genes against the genetic makeup of different mouse models, which may act differently, and individuals with DS are not inbred. Thus, the expression observed in a non-inbred mouse strain or in different genetic backgrounds might be more rational for studying the relevance to the human DS than the effects reported in inbred mouse models. Therefore, association of a single gene and/or a single phenotype shall be considered critical for different DS models. Indeed, target is underway to evaluate all mouse genes and their contribution in phenotypic development through manipulation of gene numbers, their structure, and expression for drawing an appropriate interpretation of the consequences of trisomic status and possible interference of or interaction with the disomic part of the genome.

The extra HSA21 in DS has been explored in patients with full and partial trisomies. The implication of trisomic gene-dosage and co-operating effect of disomic genes has been reported in myriads of studies in DS patients, including DS autopsies, DS-derived cell lines, and DS-derived organoids of different cell/tissue types. Extensive use of DS-derived iPSC lines and embryonic cell lines (ESCs), including neural progenitor cells (NPCs) in DS research has become beneficial for identification and testing of druggable candidates. However, the gene expression in these cell lines of different tissue lineages does not replicate the age-related consequences. Thus, organoids developed from DS-derived iPSCs have drawn attraction in DS-research (discussed in Chapter 14). The cerebellar organoids are increasingly being used for elucidating the perturbation in neurogenesis, which are reminiscent to DS patients. Indeed, CRISPR/Cas9 and CRISPR-mediated rearrangement (CRISMERE) techniques are facilitating creation of DS-derived organoids for DS research. Importantly, more and more studies are essential to unravel the expression of genes independently or in cooperation with others, which may lead to revision of today's knowledge of genotype–phenotype association in DS individuals.

Although knowledge on contribution of genes has been expanded through creation of DS mouse models and derivation of iPSCs and trisomic organoids from DS patients, still no therapeutic candidate has been identified for complete correction of a DS-specific clinical feature. Genes of mitochondrial biogenesis were considered as compelling therapeutic targets. The effective therapeutic activators for the PGC-1α and/or the downstream PPAR proteins, such as use of Metformin and Resveratrol for induction of AMPK and SIRT1, respectively, have been tested for neurodegenerative diseases, including AD in mouse models in which mitochondrial alterations play the central role (discussed in Chapters 8 and 9). PPARs, AMP-activated kinases (AMPKs), and the NAD-dependent de-acetylase SIRT1 can control the PGC-1α activity. The posttranslational modifications of direct phosphorylation can promote the PGC-1α dependent induction at the PGC-1α promoter level by AMPK. Additionally, stimulation of PGC-1α activity through deacetylation by SIRT1, might induce mitochondrial biogenesis. Interestingly, studies involving Metformin treatment exhibited multiple positive outcome on the mitochondrial activity in human DS cells, and treatment with Resveratrol and Epigallocatechin3-gallate (EGCG) overturned the severe impairment of mitochondrial bioenergetics and biogenesis in hippocampal progenitor cells of Ts65Dn. Additionally, thiazolidinediones, pioglitazones, and bezafibrates, which selectively stimulate PPARs of PGC-1α pathway, were also recommended for testing as pharmacologicals to reinstate the mitochondrial function in DS. Such targets on new therapeutic candidates can improve the mitochondria-mediated neurological perspectives of DS and thus, prevent associated pathologic manifestations, including AD, diabetes, and hypertrophic cardiopathy. However, therapeutic approaches have not been routinely undertaken with the drugs available for modulating the activity or to control the overall mitochondrial dysfunction observed in DS patients. The association of the miRNA, in particular miR-155, with respiratory chain function and ROS production was considered to propose the potential roles of mitochondrial miRNAs (mitomiR) as novel therapeutic targets for DS patients. However, further studies on the transient or durable regulatory effects and delivery mechanisms of the mitomiRs to mitochondria may establish the therapeutic potential of miRNAs.

Therapeutic interventions aimed at reducing oxidative stress and improving mitochondrial function starting from prenatal stage could be more beneficial to prevent neurobehavioral phenotypes. However, long-time treatment is recommended, since short-time prenatal or neonatal intervention appeared ineffective in adulthood. A number of natural bioactive compounds, such as polyphenol extract from fruits or plants, are suggested for therapeutic applications in DS due to long-term retention of their safety and efficacy. Efficacy has also been tested from combination therapy such as omega-3 fatty acids with polyphenols or other pharmacologic agents. Nevertheless, prior testing is essential before considering application on humans. Importantly, the challenge for improving mitochondrial function may ameliorate cognitive impairment and many other DS-clinical phenotypes, including AD. Several promising druggable candidates have already presented convincing amount of hope following testing on DS mouse models, DS-derived trisomic cells lines, and also on human DS. In principle, each of the triplicated genes that perturb the mitochondrial biogenesis, and neuronal cells, should be targeted for their contribution to OXPHOS dysfunction and neurogenesis. The prenatal effect of the proposed drugs on heart and brain development can be performed in DS animal models *in vivo* but that involves more time, effort, and resources for maintenance. Interestingly, few of the proposed drugs have been tested in pregnant women based on adequate preclinical evidence and appeared to be safe, which have been approved for the treatment of other pathologies in DS.

Therapeutically, a number of potential compounds have shown improvement in learning in the Ts65Dn mice of DS models. These mice presented improvement in hippocampal-based learning and LTP deficits from chronic treatment with picrotoxin or pentylenetetrazole and also after discontinuation of treatment, via reduction in GABA-mediated inhibition in the hippocampus development. The treatment with noncompetitive N-methyl-D-aspartic acid receptor (NMDAR) antagonist, Memantine, has also improved learning in this strain. Memantine has been demonstrated to partially inhibit the opening of the NMDAR and is proposed to mitigate the effect of trisomic RCAN1 on the function of the receptor. Further studies and clinical trials would be important for investigation of the potential of many more druggable candidates with a view to improving the cognition in people with DS and development of new targeted therapeutics for alleviating the consequences of gene dosage on DS phenotypes.

The most noticeable challenge in the pharmacologic identification of small molecule inhibitors is their unpredictable specificity of action in the cellular microenvironments both *in vivo* and *in vitro*. Thus, future outreach must take such hurdles into accounts of targeted drug developments. Indeed, studies on preclinical DS model have achieved limited success while investigating the efficacy and safety of the niche of small molecules. Moreover, seeking the answers of insightful questions by taking care of the overlooked factors might precisely establish the translational and clinical breakthroughs related to development of DS therapeutics for management of clinical consequences of trisomic HSA21. Since CRISPR/Cas9-mediated gene editing in humans has ethical and societal restrictions, identification and testing of more natural products (such as extract of green tea with EGCG) can reduce the function of overexpressing genes.

Inactivation of DYRK1A was targeted with introduction of the XIST (noncoding X-inactivation gene) into the DYRK1A locus on one of three HSA21 in trisomic iPSCs. Silencing of most of the protein coding genes of HSA21 by using this noncoding RNA transcribed by the XIST was achieved without affecting the expression of other genes located elsewhere in the genome. Thus, silencing of the extra HSA21 in DS would be a remarkable postnatal therapy for the individuals with DS and *in utero* embryonic manipulation of trisomic HSA21 and other full chromosome aneuploidies. Indeed, such audacious goal of silencing protein coding genes could be exercised for correction of anomalous genes and/or chromosomes therapeutically; however, that remains as a therapeutic dream so far. Notably, the RNA product of the noncoding XIST gene accumulates across the interphase over the inactive X-chromosome (the second X-chromosome in females) and silences transcription of most of the genes through numerous heterochromatin modification processes. Nevertheless, inactivation of the extra 21 could rescue the DS phenotypes has yet to be tested in DS models *in vivo*. Moreover, theoretical use of this approach in preimplantation embryos and/or embryonic inactivation or correction of overexpressing genes by CRISPR/Cas9 gene-editing technology is still subject to ethical debates, since there could be significant side effects, influenced potentially by such gene manipulation techniques. A new strategy of DOX-inducible XIST after neurogenesis has been suggested more useful than at the pluripotent stem cell stage. Ideally, inactivation of the entire HSA21 may shed some insight on the possibilities of cure and/or control of some of the features of DS individuals; however, turning on the XIST expression may not block all genes present on HSA21 because only handful of genes may escape CpG methylation. Thus, further elucidation of the concept of making "chromosome 21 Barr body" by heterochromatization and its clinical application would be extremely important for "silencing HSA21" or use of "chromosome therapy" in medical practice. Integrative approaches of epigenetic chromosome correction beyond the early stages of embryo development would be ideal because time would be insufficient between the diagnosis and intervention at early embryonic stages. Moreover, the influence of disomic genes, nongenetic or epigenetic factors, gene–environment interaction, and the fundamental problem of interindividual genetic variation also need to be addressed.

Summarily, therapeutic aspects of DS have been attempted with inhibitors of GABAergic and glutamatergic systems; kinase inhibitors of DYRK1A; activators of SHH-pathway; activators of NMDA receptors, serotonergic signaling, neurogenesis and dendritic spines; degeneration of cholinergic system; inactivation of the extra 21; and so on in DS models *in vivo* and DS-iPSCs and organoids *in vitro*. Future therapeutic targets may consider fetal and/or embryonic stage for evaluating neurogenesis and neuronal migration to the neocortex. Neurogenesis and neuronal migration in the hippocampal dentate gyrus (DG) and cerebellum continue after birth, while hippocampal neurogenesis continues throughout life at slow rate. Thus, most of the therapeutic testing in mouse models has focused on hippocampal circuit–related performance in memory and learning. In practice, trisomy 21 cannot be removed or neutralized with any kind of medical intervention, and supportive treatment for mental retardation or stimulation of cognitive skill could not yet be established in practice because stimulation of brain matter for cognitive performance has resulted in seizures and other side effects. Cosmetic reconstruction of the facial look of DS by plastic surgery or partial glossectomy for tongue reduction has sometimes been advocated and performed on children with DS to correct their mongoloid face or improve their speech. The pattern of cognitive impairment and learning in DS is largely caused by late maturity of prefrontal cortex, hippocampus, and cerebellum, which are augmented by dementia caused by AD. Thus, developmental abnormalities of the brain matter, synaptic plasticity, and dendritic spines related to memory deficits need better resolution in suitable DS models. In the recent era, molecular mechanisms have recognized several key molecules of HSA21 causing defects in spine morphology leading to aberrant synaptic vesicle trafficking process and synaptic transmission. Thus, identification of better assay systems *in vitro* and *in vivo* will be the guiding factor for testing newer therapeutics of specific synaptic dysfunctions in DS.

In conclusion, the studies on DS have heralded with information over decades, on several aspects of DS features; however, the complex interactive expression of gene-transcription-translation-function and regulation of the entire mechanism could not be resolved yet. There is a continual urge for better dissection of the genotype–phenotype association in different DS models for ultimate recapitulation in human DS. Nevertheless, the models with different trisomic segments, cell lines, and organoids have significantly contributed in understanding gene-dosage consequences of DS and also, for identification and testing of targeted therapeutic candidates for dysregulating the aberrant expressions and reversal of DS pathomechanism or gene expression. Indeed, *in vitro* inactivation of HSA21 and single or combinatorial therapies are being tested on DS models, and some are considered for clinical trials. The target is to understand the genetic consequences of dosage imbalance on expression of DS pathophysiology at all ages and therapeutic developments accordingly. Importantly, the significant variability in copy number observed in human genomic studies needs a synergistic corroboration for molecular dissection of implications of the dosage-sensitive genes and causative pathways involved in gene expression at the backdrop of trisomic HSA21. Therefore, deciphering the genotype–phenotype association remains a challenging endeavor for understanding the insights into euploid and aneuploid biology of DS. Indeed, complex interaction and allelic variation in HSA21 and non-HSA21 genes, variation in individual genetic background, and genetic modifiers create puzzles for unraveling the variability in obligatory phenotypes in fully trisomic DS.

Nevertheless, development of therapeutics based on the knowledge of the dosage-sensitive genes is another challenge for this complex disorder. Importantly, a number of novel drugs have already been formulated for enhancement of learning and memory in people with DS. A range of therapeutics is in clinical trials for ameliorating the DS-associated AD. Indeed, advancement in technological development for genetic manipulation and creation of genetically modified animals for testing the expression of genes of choice has enormously contributed in the field of DS research and that is heralded with information on contribution of genes for DS-specific phenotypes. Mouse models produced with chromosomal regions syntenic to HSA21 in parallel with transgenic and/or gene-knockout models have enabled unprecedented opportunities of understanding the pathomechanism in DS patients, extracted crucial information on genotype–phenotype association, and identified therapeutic molecules for pharmaceutical developments, specifically for addressing cognitive impairment and proneness to AD.

Studies on monitoring the changes in overall health, cognitive function, development of dementia and AD, mortality, and so on would be essential for development of policies and therapeutic interventions with a view to promoting healthy aging in DS population. The social and medical infrastructure is not sufficiently prepared for providing care to sporadic or familial AD; thus, clinical management of DSAD remains a key challenge. Establishment of a consensus research framework and development of a data-driven assessment battery for DSAD may advance clinical research and clinical trials to identify recruitment priorities, and power trials appropriately. In addition, establishment of coordinated research and medical care centers-of-excellence within the framework of existing DS and/or AD canters may contribute in better understanding of the progression and phenoconversion of DSAD, and accelerate development of effective treatments.

Importantly, a precise standard definition of phenotype is essential. Also, formulation of standardized protocols would be important, which can be universally applied by multiple centers on sufficiently large number of samples for a meaningful genotype–phenotype analysis. Availability of carefully designed and curetted bio-bank of detailed phenotypic description along with DNA and tissue samples of participating DS individuals would be useful. Pooling of such large-size datasets derived from individual candidate gene and genome-wide analyses will change much to the present understanding of DS. Undoubtedly, the current progress in DS research has facilitated unparalleled comparison of data collected from DS population and myriads of DS models for testing genetic associations, exploration of biological mechanisms, and therapeutic trials.

Several international research consortia and collaborations, including the Horizon 21 Genetics Consortium (funded by J-L Institute), AD Biological correlates in DS (NIH, PI Granholm), Inflammation and NGF Dysfunction in the Evolution of AD Pathology in DS (multicenter), Improving AD care in adults with DS (submitted to GBHI and Alzheimer's Association Pilot Award), Horizon 21 Cognitive Project, HEROES consortium funded by the JPND in Europe, the NIH Alzheimer's Biomarker Consortium for

Down syndrome (ABC-DS), and the Crnic Institute's Human Trisome Project (NCT02864108) are advancing the understanding of DSAD. In addition, ABC-DS, LonDownS Consortium, Down Syndrome Biomarker Initiative (DSBI), and DABNI have demonstrated critical data on the natural mechanism of AD in DS and enabled clinical trials. Moreover, there is a growing need for the autopsy study of brain tissues from well characterized DSAD patients of different ages for study by the international research community. The initial network, including 10 participating universities in the United States and Europe, and the Down Syndrome Biobank Consortium (DSBC) is developing the standardized protocols for harmonized procurement of DS tissues from different sites of the DS brains.

Development of clinical trial networks and infrastructure for multicenter collaborations using harmonized protocols has been considered as another important priority with several efforts and currently underway, including the LuMIND IDSC Foundation, Down Syndrome Clinical Trials Network (DS-CTN), the NIH-funded Alzheimer's Clinical Trial Consortium-Down Syndrome (ACTC-DS), and the Longitudinal Investigation for Enhancing Down Syndrome Research (LIFE-DSR) study. Expanded development and utilization of brain banking, collaborative data sharing, and evaluation across studies using pharmacological and non-pharmacological intervention as well as combination therapy approaches are also essential for prevention of dementia in the DS population. Taken together, someday will come as today for offering effective treatment of DS and to give a normal life to this population.

Regarding societal and demographic aspects, developed nations have already changed their attitude toward DS and advocate for the rights of equal opportunities for these people. In India, the Self Advocates Forum of India (SAFI) is meant for the people born with some kind of disabilities; in fact, the "disability" has been defined as "specially able or differently able" in the present era. Through SAFI, these people voice their opinion of rights and attract the attention of the regulators. Nowadays, there are several organizations to provide training in sports, dance, music, computer work, and other vocational training and education round the year for individuals with intellectual disabilities including DS. Two-teacher system of education is followed in some countries, wherein the DS children attend the same class along with other normal children and a second teacher fosters DS children separately with special attention and techniques. The syndrome is honored with celebration of the "World Down Syndrome Day," which was first held on March 21, 2006, indicating trisomy (third month) 21 (date). The concept was first proclaimed by the European Down Syndrome Association during their European Congress held in 2005 in Palma de Mallorca. The fact is that the DS is not a curable disorder; therefore, there is a need of established direction of early onset of occupational and speech therapy and offering education in special schools accompanying with vocational education, which are enabling individuals with DS not only to achieve strength and sustain in real life, but also compete with their peers in several reality shows.

Preface

Down syndrome (DS) is the most common chromosomal disorder caused by trisomy 21 (HSA21) and is largely influenced by the maternal age (≥35 years). The syndrome cannot be cured as of now but prevented through prenatal screening and testing. The search on the contribution of the triplet set of HSA21 genes on the clinical features of the syndrome, particularly cognitive skill and early onset of aging and Alzheimer disease (AD), has always been a primary concern for improving the life of individuals suffering from this syndrome. However, variable expression of the trisomic HSA21 on different individuals with the syndrome could not label the genes with the syndromic features. The therapeutic approach for improvement of the cognitive sense has limited success due to severe side effects, including seizures. Several transgenic and gene knockout mouse models, stem cell lines, and trisomic organoids were developed for precise recognition of genes contributing to specific phenotypes. Targeted "chromosome therapy" following heterochromatinization of the extra chromosome using noncoding RNA transcribed from the XIST gene was successful in trisomic induced pluripotent stem cells *in vitro*; however, application of this technique in medical practice is far away. Collectively, a clear understanding of the clinical impact of genetic-epigenetic-environmental interaction in the presence of the extra HSA21 has yet to be clearly reached.

More than 60 years have passed since identification of trisomy 21 as the heart of this syndrome, and more than 160 years have passed after clinical characterization of the syndrome. Screening and preventive practices have been established with extensive awareness and participation of the at-risk population. Yet, the epidemiological index of the syndrome is estimated at one in 600–700 live births, and ~90% of pregnancies with this trisomy die *in utero*. Despite accelerated acceptance of these children, societal discrimination and stigmatization pose serious psychological impact on the children as well as their families. Nevertheless, an early initiation of supportive medical intervention, specialized education, and vocational training in special schools is enabling them to live a better life with an increased life-span and to walk alongside their peers with normal genetic makeup in various walks of life.

This book aims to discuss the history of the syndrome; phenotypic features and health consequences; genetic counseling, prenatal screening and testing and associated ethical issues, societal and demographic aspects; genetic etiology and underlying mechanism of nondisjunction, and *de novo* and hereditary factors on acquisition of the extra HSA21; HSA21-encoded genes and the impact of their overexpression; dosage-sensitive genes and genotype–phenotype association; mitochondrial dysfunction, oxidative stress and associated diseases of the syndrome; genetic consequences of neurogenesis and cognitive disability in Down syndrome; glutamatergic and GABAergic mechanisms and therapeutic targets; consequences of Alzheimer disease; microRNAs and epigenetic signatures; *in vivo* and *in vitro* models of DS research; and future directions of therapeutic developments. Such information is available in text books on medical genetics, research articles, and research magazines in fragmented sections; however, this book attempts to comprehensively present all major aspects of the syndrome in a holistic manner. The book will surely be attractive to the medical practitioners and professionals engaged in teaching and education, geneticists, to a wider circle of biologists, public health educators and regulators, and Parent-Patient Groups across the globe. In a nutshell, this book will occupy a coveted place in every book shelf of all those who participate in the health of each family and collective health of our society.

Diagnosis catered to thousands of children born with Down syndrome and thousands of prenatal diagnoses with full, partial, and translocated HSA21 in trisomy, and counseling offered to their parents has prompted me to learn more about the impact of the smallest human chromosome 21. This compiled knowledge has been reshaped into this book. Indeed, this book has gathered copious information on the gene–dosage imbalance and interaction of the trisomic and disomic genes at the backdrop of an individual's own genetic makeup, allelic variation, and genetic modifiers, along with the impact of noncoding RNAs, and epigenetic and environmental factors on the life of individuals afflicted with the syndrome. Besides the effect of the dosage imbalance of nuclear genome, the role of cytoplasmic mitochondrial DNA (mtDNA) on impaired energy metabolism contributing to syndromic heart and brain defects is significant. In total, motivation accrued from decades of diagnostic work has left enormous temptation for accumulating all aberrant chromosome configurations and linking them to the clinical histories of patients of different ages, which is under preparation. Many years ago, it was my father's wish for me to apply my knowledge for the benefit of the common people that prompted me for a transition from fundamental research in the field of radiation and chemical mutagenesis to diagnostic research of hereditary and acquired genetic diseases. Put together, I hope my contribution in diagnostic research, and the labor put into this book will make my parents happy in their heavenly abode.

I do express my sincere gratitude to the management of Mahatma Gandhi Mission Trust, Mr. Kamal kishor Kadam, Chairman and Prof. Dr. Sudhir Kadam, Director of MGM Hospitals, Navi Mumbai, India, in particular, for extending their all-round support for this book. Special thanks are due to Prof. Dr. Nitin N. Kadam, Medical Director of MGM New Bombay Hospital, Vashi, Navi Mumbai, and Pro-Vice Chancellor of the MGM Institute of Health Sciences, Navi Mumbai, India, for his continual impetus for writing a book on Down syndrome, and kindly extending the required financial and administrative supports. I would also like to acknowledge the support and encouragement extended by Prof. Dr. S. Dalvi, the Vice Chancellor of the MGM Institute of Health Sciences, Navi Mumbai, India. Contribution of Ms. Swati Bhilare is also acknowledged for her support in literature search on several aspects of Down syndrome. The members of the editorial team at Elsevier Inc., USA—Melina Tucker, Samantha Allard, Joslyn T. Chaiprasert-Paguio, Maria Bernad and others are specially remembered and thanked for their immense contribution to broadening the scope of this book, and their spirited support through its entire construction. Sincere acknowledgments are also due to my family—my sons Dr. Shameek Ganguly (Robotics Researcher, California) and Dr. Shouvik Mandal (Assistant Professor Conservative in Dentistry, India), and my daughters-in-law Debapriya Pal (Sr. Software Engineer, Microsoft, California) and Dr. Alivia Mandal (Sr. Clinical Researcher, India) for their continuous encouragement, guidance and valuable comments, in spite of their busy schedules. I cannot thank my husband Mr. Prodosh Ganguly (Consultant in Banking Regulations) enough who has been right by my side through this entire effort. Lastly, I express my sincere gratitude to all those patients who shared their stories with me during genetic counseling or otherwise—it is their stories and questions that have propelled me to learn incessantly about genetic disorders, and gain a deeper understanding that I can share with the society.

I dedicate this book to my beloved late parents Prahlad Chandra Ghosh and Satadal Basini Ghosh, who always wanted me to utilize my education and skill for the good of mankind, and also to my late brother Prof. Bijay Krishna Ghosh for his unwavering faith in me, continuous inspiration, and his deep interest in the compilation of this book.

Bani Bandana Ganguly, Ph.D., FICMCH

January 27, 2022

CHAPTER

1

Down syndrome: history and overview

Bani Bandana Ganguly, Nitin N. Kadam

1. History

English physician John Haydon Langdon Down first characterized Down syndrome (DS) in 1862 on a small group of children who had common physical features and were different from normal children with a distinct form of mental disability. The description was widely published in a report in 1866 (Down, 1866). Due to shared physical and facial similarities (epicanthic folds) of DS with those of Blumenbach's Mongolian race, JHL Down used the term "mongoloid," which was derived from the prevailing ethnic theory (Conor, 1999). Some of the prominent characteristics in infants with DS includes decreased muscle tone, flat face, upslanting eyes, irregular-shaped ears, ability to extend joints beyond the usual, large space between the big toe and its neighboring toe, large tongue relative to the mouth, intellectual disability, etc (Fig. 1.1).

Because of Dr. J. L. Down's contribution to research on DS more than 150 years ago, the syndrome is much known today, which was named after him. However, there was a controversy that the first description of DS was made in 1838 by Jean Etienne Dominique Esquirol (1772–840), a founder of modern alienism, followed by another clinical description made by Édouard Séguin (1812–80) in 1844 (Who named it: http://www.whonamedit.com/synd.cfm/322.html). Even several years before, psychiatrist Andrew Levitas and geneticist Cheryl Reid identified a 16th-century Flemish Nativity painting in which one angelic figure appears distinctly different from other individuals in the painting with an appearance of DS (Fig. 1.1).

Historically, most individuals with DS were killed, abandoned, or ostracized from the society. In the 20th century, it was common for individuals with DS to be institutionalized and was denied appropriate treatment for the associated medical complications such as heart disorders, vision defects, intestinal problems, and so on. Many children with the syndrome, therefore, used to die during infancy or early adulthood. By the 20th century, DS had become the most recognizable form of mental retardation. With the rise of the eugenics movement, the forced sterilization of individuals with DS and comparable degree of disability was introduced to 33 of the then 48 US states and several other countries. The ultimate expression of this type of public policy was "Action T-4" in Nazi Germany, a program of systematic murder. The sterilization program killed large proportions of people with DS until court challenges, scientific advances, and protests came from the general public, which led to its discontinuation or repeal of such sterilization program during the decades after World War II. Although at that point in history (middle of

Genetics and Neurobiology of Down Syndrome
https://doi.org/10.1016/B978-0-323-90456-8.00008-9

FIGURE 1.1 The most known personalities: (A) Down syndrome diagnosis at the Adoration of the Christ Child. Available in Public domain: http://www.downsyndromeprenataltesting.com/down-syndrome-diagnosis-at-the-adoration-of-the-christ-child/. 16th-century Flemish Nativity painting of Down syndrome by unknown Flemish painter (1515) showing typical characteristics of Down syndrome in the angel next to Mary and also in one of the shepherds (Metropolitan Museum of Art, New York). (B) Dr. Down; (C) Kay McGee (Founder of National Association for Down Syndrome, USA) and his wife; and (D) their daughter Tricia with Down syndrome (from left). *Reproduced from: (A) Levitas and Reid (2003). (B–D): National Association of Down Syndrome. https://www.nads.org/about-us/history-of-nads/. Accessed June 25, 2021.*

20th century), the cause of DS was not understood and the syndrome was noticed in all races with rarity of recurrence. It was presumed that combination of several heritable genetic factors, older maternal age, birth injuries, and injury during pregnancy cause DS and illness with other physical and mental disabilities (Warkany, 1971).

With the discovery of karyotyping techniques in the 1950s, it became possible to identify abnormalities of the chromosomal number or shape. In 1959, Jérôme Lejeune discovered that DS results from genetic consequences with an extra chromosome (Lejeune et al., 1959) (Table 1.1). The extra chromosome was subsequently labeled as the 21 and the genetic condition of DS was known as trisomy 21. However, Lejuene's discovery was disputed when the Scientific Council of the French Federation of Human Genetics awarded Marthe Gautier with the Grand Prize for this discovery in 2014.

As time went by, many other people contributed to the study of DS as well. In 1930, Brewster

TABLE 1.1 Timeline of developments and activities to support Down syndrome.[a]

Year	The key person behind	Development
Several years ago	Psychiatrist Andrew Levitas and geneticist Cheryl Reid	Identified a 16th-century Flemish Nativity painting The Adoration of the Christ Child in which one angelic figure appears distinctly different from other individuals in the painting with an appearance of Down syndrome
1838	Jean Etienne Dominique Esquirol (1772–840), a founder of modern alienism	First described Down syndrome
1844	Édouard Séguin (1812–80)	Clinical description of the syndrome and cretinism
1862	British physician, John Langdon Down	Characterized Mongolism as a different form of mental disability
1866	British physician, John Langdon Down	Detailed description of Down syndrome
1959	French Pediatrician/Geneticist Professor Jérôme Jean Louis Marie Lejeune (1926–94)	Down syndrome is caused by genetic factor and has an extra 21st chromosome
1960	Kay McGee	Foundation of Mongoloid Development Council (MDC) in Chicago
1961	Scientists and People's Republic of Mongolia	Requested to Lancet for changing the name from mongoloid to Down syndrome
1965	WHO recommended the change	Down syndrome universally
1972	Kay McGee	MDC changed to National Association for Down's Syndrome (NADS)
1973		National Down Syndrome Congress (DSC) was founded
~1973		National Down Syndrome Society (NDSS)
1979	NADS	Parent's training program
1975	NIH Convened a conference	Unifying abnormal characteristics and drop of possessive form "Down's"
1975	NADS	Education for All Handicapped Children Act (Public Law 94–142)
		First local support group was formed
1979	NADS	First parent support training program was held
1980	NADS	Hospital education program was developed
1982	NADS	Temporary foster care program was developed
1984	NADS	First group of public speakers was trained
1987	NADS	Trained parents to become education facilitators
1989	NADS	Developed a fellowship University Affiliated Program (UAP), University of Illinois at Chicago
1990	NADS	*A Baby First* was first published
1991	NADS	Commitment by Lutheran General Hospital to start a clinic for adults with Down syndrome

(*Continued*)

TABLE 1.1 Timeline of developments and activities to support Down syndrome.[a]—cont'd

Year	The key person behind	Development
1992	NADS	The Adult Down Syndrome Center (ADSC) was opened
1996	NADS	NADS website was developed
1998	NADS	First behavior retreat was held
1990s		DS was characterized with additional disorders such as Autism or Attention Deficit Hyperactivity Disorder (ADHD)
2005	NADS	Program for Hispanic families was formalized
2010	The 35th anniversary of the passage of Public Law 94-142	Disabilities Education Act (IDEA)
2014	Lejeune's claim on discovery of trisomy 21 disputed	French Federation of Human Genetics unanimously awarded his colleague Marthe Gautier for this discovery

[a] *Reproduced from NADS and other sources available in public domain.*

and Cannon were the first to report an association between DS and acute leukemia, and later Schunk and Lehman were the first to report an association between Down's descriptions of the syndrome and transient leukemia in 1954. In 1959, Lejeune was the first to associate DS features with trisomy 21 genetics. So, it all developed into a chain of research, but it was John Langdon Down who established the first solid base for the study of the disorder. DS patients are also affected with other conditions, including congenital heart disease, Alzheimer disease, and early aging.

In 1961, 19 geneticists wrote to the editor of *The Lancet* suggesting that Mongolian idiocy has "misleading connotations," "an embarrassing term," and "an insult to the Mongolian race," and thus, should be changed (Gordon et al., 1961). *The Lancet* supported 'Down syndrome.' The World Health Organization (WHO) officially dropped references to mongolism in 1965 following a request by the Peoples Republic of Mongolia delegate (Howard-Jones, 1979). However, almost 50 years later, the term "mongolism" still appears in leading medical texts such as General and Systematic Pathology, fourth Edition, 2004, edited by Professor Sir James Underwood, and also in medical diagnostic reports.

In 1975, the US National Institutes of Health (NIH) convened a conference to standardize the nomenclature of malformations. They recommended eliminating the possessive form: "The possessive use of an eponym should be discontinued, since the author neither had nor owned the disorder (A Planning Meeting, 1974)". Although both the possessive and nonpossessive forms are used in the general population, 'Down syndrome' is the accepted term among the professionals in the United States, Canada, and other countries; however, 'Down's syndrome' is still used in the United Kingdom and other regions. In other parts of the world, today DS is popularly known as just "Downs" (Leshin, 2003).

2. Incidence

The term "incidence" of DS means the annual diagnosis rate, or the number of new cases of DS diagnosed each year, and the word "prevalence" usually means the estimated population of people who are managing the syndrome at any given time (i.e., people with DS). Nevertheless, epidemiological incidence of DS is the highest among all the genetic disorders involving abnormal chromosomes. Worldwide the incidence of DS birth is estimated at 1 in 700 to 1000 newborns. This statistics is heavily influenced by the age of the mother as the chance of giving birth of a child with this condition increases as woman gets older (Mégarbané et al., 2009). Other biological and environmental factors may also play important roles, and thus, the women of all ages can have a child with DS. Notably, the population of DS has increased over the years. Each year, approximately 3000 to 5000 children are born with this chromosomal disorder, and it is believed there are about 250,000 families in the United States alone who are affected with the syndrome (http://www.who.int/genomics/public/geneticdiseases/en/index1.html#). A cross-sectional study conducted by Shin et al. (2009) during 1979–2003 from 10 population-based birth defects registries revealed an increase in DS birth by 31.1%, from 9.0 to 11.8 per 10,000 live births, wherein the pooled prevalence of DS among children and adolescents was lower among the non-Hispanic black individuals and other racial/ethnic groups compared with non-Hispanic white individuals (Fig. 1.2). Such varying estimates according to region, race/ethnicity, and gender suggest possible variation in the prevalence at birth or in survival rates of DS. A marked increase of DS birth has been reported during 2004–06 to 14.47 per 10,000 live births (Parker et al., 2010). Incidence rate for DS in the United States is approximately 1 in 800 or 0.12%.

Importantly, the frequency of DS is much higher at conception, and 25% of all miscarriages are contributed by DS. Approximately 75% and 50% of DS pregnancies identified during the first and second trimester, respectively, are aborted spontaneously before term (Morris et al., 1999;

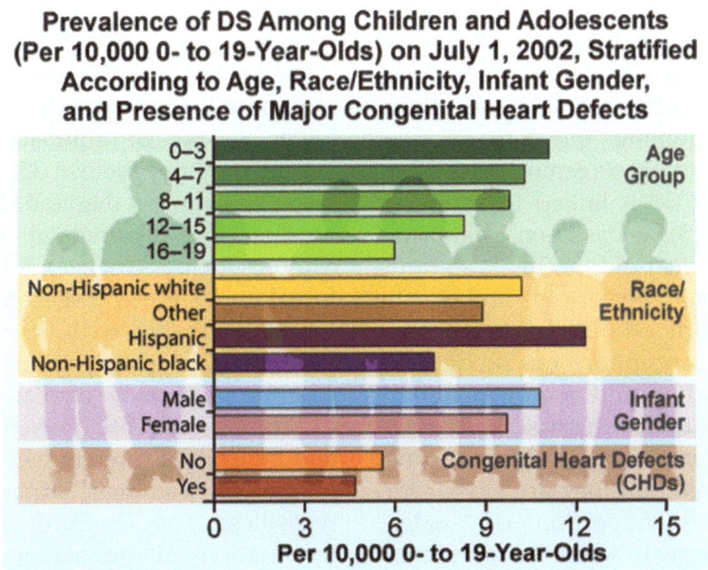

FIGURE 1.2 Trend of increasing birth of Down syndrome: Incidence of DS in children and adolescents and in different ethnic groups estimated in 10 regions of the United States. Available in Public domain: http://developmentalcognitivedisabilities.weebly.com/statistics.html Accessed April 6 2022. *Credit: Shin et al. (2009).*

Spencer, 2001). However, 25% of all children born with the syndrome will go through the regular 9 months of pregnancy and survive till birth. DS occurs in all human populations, and analogous effects have been found in other species such as chimpanzees and mice. Recently, researchers have created transgenic mice with most of human chromosome 21 (in addition to the normal mouse chromosomes).

In Australia, the rate of DS is 1 in 1100–1700 mainly because of a high termination rate. However, overall rate of DS birth has increased from 253% in 2003 to 276% in 2009 (Shin et al., 2009). In the United Kingdom, DS has risen by 71% in the past 20 years, from 1075 in 1989–90 to 1843 in 2007–08 (http://www.nhs.uk/news/2009/10October/Pages/Down-syndrome-termination-rate-screening.aspx). European Union of 15 member states have shown rise of DS birth with over 7000 affected pregnancies every year with dramatic rise of elderly mother from 8% to 14% or more (Dolk et al., 2005). Interestingly, some regions show two-fold increase of birth with DS owing to consideration of termination of pregnancy as illegal, such as Ireland and Malta. Therefore, EUROCAT network has presented a variable incidence of DS in the European countries combining the influences of maternal age and prenatal screening practice.

The prevalence of DS is linked to advanced maternal age (Fig. 1.3). It has been established that the risk for a 40-year-old mother is 16 times greater than that for a 25-year-old mother. The trend for later motherhood due to late marriage or electively late conception on account of personal career building and professional considerations has resulted in increase in birth of children with DS. Some couples simply do not want to have children until they are older. During 1995–1999, the proportion of "older" mothers increased from 10% to 25% in different regions in Europe, and the total prevalence (including termination of pregnancy) of DS increased from 1 to 3 per 1000 births (Dolk et al., 2005). Another fact is that meiotic

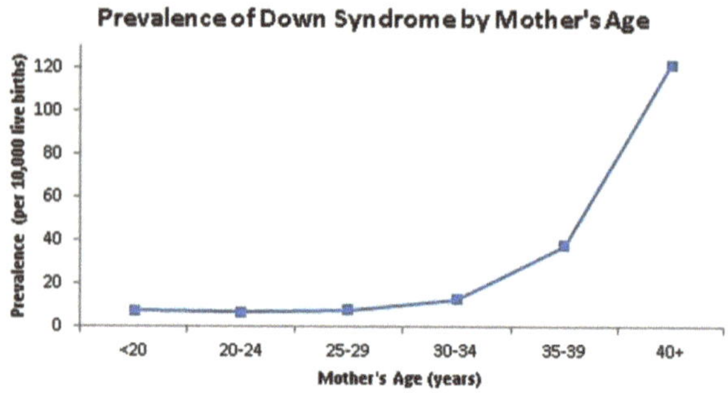

FIGURE 1.3 Risk of pregnancies with Down syndrome is directly linked to maternal age. *Reproduced from: Data and Statistics on Down Syndrome. Centers for Disease Control and Prevention. https://www.cdc.gov/ncbddd/birthdefects/downsyndrome/data.html. Accessed June 25, 2021.*

nondisjunction is higher in females contributing to 80%–90% maternal disjunction, whereas paternal nondisjunction contributes to 10%–20%. However, maternal age augments the chance of meiotic error resulting in gametes with trisomy 21.

Nevertheless, despite the large rise in pregnancies with DS, the increased uptake of the prenatal screening practice has resulted in decrease in the number of babies born with DS, due to terminations of pregnancy. On the other hand, lower uptake of antenatal screening among young mothers (below 35 years), as their age are at a lower risk than older women, influenced birth of DS. The antenatal screening and testing can provide personalized information to parents for taking their own decisions on how to proceed with a pregnancy with DS: whether to keep the pregnancy, or give it up for adoption, or terminate the pregnancy, or planning for the welfare of a child with DS. Therefore, the importance and benefits of screening compared to risk and harms of prenatal sampling are manifold, regardless of age. In Europe, the extent to which termination of pregnancy was concentrated among older mothers varied between regions from 0% in Ireland, Malta, and other regions where termination of pregnancy was illegal, to less than 50% in 14 further regions of EUROCAT, to 77% in Paris (Dolk et al., 2005). Thus, prenatal screening did not control the overall rise of DS in European countries during 1995–99.

Prevalence of DS in different countries has been calculated as extrapolations of various prevalence or incidence rates against the populations of a particular country or region. The statistics used for prevalence/incidence of DS are typically based on the United States, the United Kingdom, Canadian, or Australian prevalence or incidence statistics, which are extrapolated using only the population of the other country. Thus, this extrapolation calculation does not take into account of any genetic, cultural, environmental, social, racial, or other differences across the various countries and regions and does not use data sources or statistics about any country other than its population. Hence, these incidences of extrapolations for DS are only estimates, which only gives a general indication as to the actual prevalence or incidence of DS in that region and may have very limited relevance to the actual incidence of DS in any region. Hence, warning is inserted to view the extrapolated incidence of DS in different countries.

3. Sex ratio

The prevalence of DS is variable in different sex. Among live births, males are higher with DS than females, although the discrepancy is less severe among fetuses suggesting differential in utero survival between the sexes (Lary and Paulozzi, 2001; Iselius and Lindsten, 1986). Some phenotypic features are associated with differences of sex. Males are more affected with cognitive and language impairment (Aoki et al., 2018; Startin et al., 2020; Morris et al., 2014), while congenital heart defects are more prevalent in females with DS (Takano et al., 2019; Santoro et al., 2018).

4. Recurrence

DS is solely caused by trisomy 21, and 95% of the children born with the syndrome are diagnosed at birth. Trisomy 21 occurs through nondisjunction or mis-segregation of chromosomes during meiosis or the cell division for gamete formation (gametogenesis). The frequency of meiotic nondisjunction has been reported as maternal disjunction (80%–90%) and paternal disjunction (10%–20%), with ~1% chance of recurrence. The disorder is rarely passed down from the parents or occurs as a hereditary disorder. The age of the expecting mother increases the risk of occurrence. For women between the ages of 35 and 40 years, the risk of having a baby with DS increases to 1 in 350. For women at the age of 40, the risk increases to 1 in 100. For women at the age of 45 years, the risk increases to a stunning 1 in 30. The basic trisomy 21 means trisomy of the whole chromosome 21; also known as 'free trisomy 21.' Approximately, 95% of DS are born with whole of the extra 21, whereas, remaining DS can have partial trisomy due to translocations.

Partial trisomies that involve only a segment of the chromosome 21 are less frequent than whole chromosome trisomies, and these may occur through translocations. Partial, but with almost the whole of the 21q, occurs through Robertsonian translocation involving one 21 and any other acrocentric chromosome, most frequently the chromosome 14. Fifty percent of the people born with translocation are scattered, and another 50 percent are due to balanced translocation present in one parent; thus, it is hereditary factor. To this day, however, the reoccurrence rate is unknown. Only ~5% of DS cases may be hereditary.

Sometimes, trisomy of the whole chromosome 21 is present in a fraction of the body's cells, in association with the normal chromosomal pattern of 46,XX/XY in some other fraction. Therefore, a mixture of normal cells and trisomic cells constitute the genome of an individual with DS. This condition is called as mosaic DS and occurs with a lower incidence compared to those having the constitutive trisomy in all cells and tissues. These cases may have milder form of cognitive disability and other facial features; however, they appear with recognizable features of the syndrome. Such mosaicism is caused most likely by mitotic nondisjunction and that occurs after fertilization during embryonic growth.

After the birth of a child with DS, the risk of DS in a future pregnancy is determined by reviewing the chromosome reports of the parents and the affected child and by considering maternal age at the time of a future pregnancy. If the chromosome report of the child indicates trisomy 21 (free or whole chromosome), the future risk is relatively low, but when the syndrome is due to a Robertsonian translocation involving 21, the risk may be significantly higher. Then, reviewing the blood chromosome report of the parents stands mandatory. Indeed, a parent with a Robertsonian translocation involving two 21s will give birth with DS in all pregnancies. The positive family history also raises the suspicion for an inherited unbalanced translocation.

5. What other names do people use for Down syndrome?

Since the time of the first description of the DS in 1866, the physicians and the living society started noticing birth with the syndrome. Interestingly, similarity of the phenotypic features of the syndrome with the Mongoloid population described the disorder as "mongoloid" or "mongol" in short. Surprisingly, some continent of the globe understood that the birth of children with the syndrome indicates "Mongol" or "blessings" of the God. Notably, semi- or illiterate people in India thought that their child is born with God's blessings and has brought happiness for the family. The protest from the Mongolian race and the practice of using eponym in describing such disorders have led to name the syndrome as "Down syndrome." The chronological development of the naming has been presented in Table 1.2. However, the syndrome is still represented differently by different users in scientific journals, medical practice, and medical books, such as:

Down's syndrome: Indicates John Langdon Down's contribution on the description;

Down syndrome: Deleted the "'s" because John Langdon Down neither had nor owned the disorder;

Trisomy 21: Presence of three copies of the chromosome 21;

47,XX,+21: A female karyotype with the syndrome with whole chromosome 21 in 100% cells of the body;

47,XY,+21: A male karyotype with the syndrome with whole chromosome 21 in 100% cells of the body;

Translocation DS: DS with trisomy 21, when 21 is translocated with other chromosome (often with chromosome 14). Translocation DS can also have partial trisomy of 21.

Mongoloid disorder: Presence of the facial features of the Mongolian race.

6. Can Down syndrome be inherited?

Most cases of DS are not inherited. When the condition is caused by free trisomy 21, the chromosomal abnormality occurs as a random event during the formation of reproductive cells or gametes due nondisjunction resulting in a clone with abnormal number of the chromosome 21. An egg or sperm cell with an extra copy of chromosome 21, upon fertilization with a normal gamete having 23 chromosomes, contstitute the genetic makeup of a child with 47 chromosomes. This is a constitutive abnormality for its presence (extra chromosome 21) in each of the body's cells. Therefore, DS with free trisomy 21 is not an inherited disorder. Mosaic DS is also not

TABLE 1.2 The transition of naming the Down syndrome.

Year	Naming or other issues	Further suggestion	Individual or association	Outcome	References
1960s	Langdon-Down's anomaly or Down's syndrome	To abandon "Mongolism" or mongoloid disorder because the word indicates a descriptive expression of the mongol-like features.	Scientific community	The name was changed to Down's syndrome	Lancet (1961, i: 775)
1966	The possessive or nonpossessive, hyphenated or nonhyphenated expression of "Langdon (–) Down('s) syndrome"	Commemorate the centennial anniversary of Down's paper. Identification of Down's syndrome with a racial group - the Mongols appeared inappropriate	A group of specialists met in London	"Mongolid" persisted and used by the physicians. "Mongol" was used in the United Kingdom.	Wolstenholme (1967)
Beyond 1960s	The Russians used Down syndrome for 50–60 years and even call the patients as "Downs."	Use of "trisomy 21" or "translocation-type trisomy 21" was preferred because these were neither eponymous nor euphemistic.			
1962	A delegation from Mongolia requested WHO to drop the objectionable terms "mongolism" and "mongol"	It was honored by WHO in 1965	Mongolian population	Although honored by WHO, the trend continued in scientific journals and textbooks	Medical History (1979, 23:104)
1966	Trisomy 21 or G	"trisomy G" was used before identification and numbering the chromosome as 21	International Standing Committee on Human Cytogenetic Nomenclature. Lisa G Shaffer and Tommerup (2005)	"Trisomy 21" is preferred because of 21's involvement	In all descriptions of the Down syndrome
1974	Debate on the possessiveness using "'s" to express "Down's syndrome" continued	To discontinue the possessive form of an eponym, since the author neither had nor owned the disorder	National Institute of Health. USA	The disorder was named as "Down syndrome" ("syndrome" is not capitalized). However, the United Kingdom and Europe delayed the practice. Notably, many text books still used "Down's syndrome"	Medical History (1979, 23:104)

Source: Information avialable at public domain. Modified from: Leshin (2003).

inherited. This mostly occurs as a random event during postfertilization embryonic growth or early stage of fetal development. As a result, some of the body's cells will have the normal set of 46 chromosomes, and other cells will have three copies of chromosome 21 or trisomy 21. Therefore, trisomy 21 occurs de novo due to meiotic or mitotic nondisjunction or missegregation.

DS with Robertsonian or unbalanced translocation can be inherited. One of the parents may carry a translocation involving 21. Although such carriers will not have signs of DS, they are at an increased risk of having meiotic nondisjunction and affecting children with DS. Interestingly, parents carrying a Robertsonian translocation between two copies of chromosome 21 will obviously have all children (100% pregnancies) with DS (Fig. 1.4). These DS children with translocated 21 are affected with two abnormalities, such as the Robertosonian translocation involving two 21s as the structural abnormality and trisomy 21 as the numerical alteration, along with a copy of free 21. The incidence of parents carrying a balanced translocation involving one 21 increases the risk of DS birth with most likely partial trisomy of 21.

Such incidences of parents being carrier of translocation increase the risk of meiotic error and reproduce DS, and thus, a hereditary factor is involved, though the parents do not have DS. Therefore, if a DS child has translocation, a chromosome study on both parents must be performed for preventing further birth with DS. Genetic counseling and discussion of testing of other at-risk family members should also be considered.

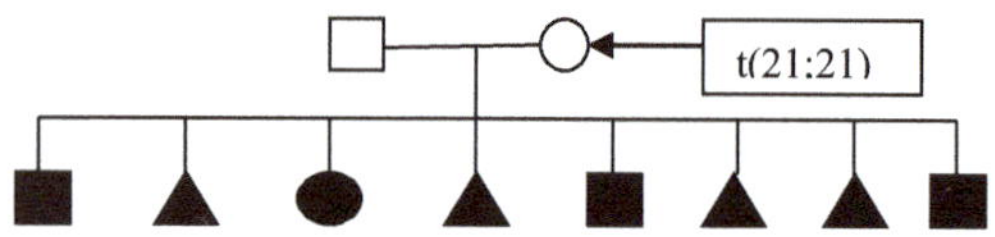

FIGURE 1.4 The mother is carrying a Robertsonian translocation involving both 21s with 45,XX,t(21;21) karyotypic pattern. She had altogether eight pregnancies: four live births (two males and two females) and four spontaneous abortions. All children of four live births had features of DS with 46,XX,+21,t(21;21) and 46,XY,+21,t(21;21) karyotype of the female and male children, respectively. Interestingly, karyotyping of all four abortus also showed trisomy 21 following the fluorescence in situ hybridization (FISH) test. The solid geometric shapes indicate presence of trisomy 21, the Robertsonian translocation and DS in live children and abortuses. *Source: Ganguly, B.B. (unpublished). Variable Down syndrome and risk of reference. Manuscript under preparation*

7. Secular trends and racial and ethnic variation

JL Down characterized the appearance of the people with DS as greatly resembling people of the Mongoloid race in 1866. The facial appearance of European children with DS was interpreted as the result of degeneration and that reflected upon the racial notion. It was stated that the "Caucasian race" was superior to the "Mongoloid race" in intellect.

Variation in prevalence of DS by race/ethnicity has been more consistently reported. However, contradictory reports have been published in favor of or against the effect of secular trends on the prevalence of DS (O'leary et al., 1996, CDCP, 1994; Baird and Sadovnick, 1988). It has been established that race or ethnicity has significant effect on prevalence of DS. Higher incidence was found among Hispanic infants than non-Hispanic whites, even after adjustment of the maternal age, whereas non-Hispanic black women have the lowest observed prevalence (CDCP, 1994; Canfield et al., 2006; Agopian et al., 2012; Kucik et al., 2013). The Hispanics had the highest frequency of birth with DS compared with Asians, whites, Native Americans, and African-Americans (Chavez et al., 1988). Other studies found increased risk in Vietnamese mothers when compared with Californian mothers (Shaw et al., 2008). The underlying factor behind such racial or ethnic differences on the prevalence of DS could be influenced by the level of awareness, facility available for or use of prenatal diagnostic services, and underdiagnosis of the defects at birth. Racial composition of women who use prenatal screening services varied from the racial composition of the US population (Meaney et al., 1993), though racial difference in usage was not found in other studies (Naber et al., 1987). Use of prenatal diagnostic services and elective abortion significantly reduced the DS birth among white women but not among other races of Atlanta (Krivchenia et al., 1993); however, that was not supported by a Los Angeles study (Wilson et al., 1992). Moreover, the biological factors including maternal age, in association with social factors such as attitudes toward elective termination or access to care, might contribute to the racial/ethnic differences to enhance the birth of DS (Jackson et al., 2014).

Some of the religious groups have reservation on terminating the pregnancies with trisomy 21 and/or participating in screening and testing for DS. It has been experienced that one couple of the Christian community of India had gracefully continued their pregnancy after confirmation of DS with an explanation (personal experience of the author): "it is God's choice to give us the child with this condition; hence, we have to deliver and take care of the child, though with Down syndrome."

References

A Planning Meeting, 1974. Classification and nomenclature of morphological defects (discussion). The Lancet 1974 305 (7905), 513. https://doi.org/10.1016/S0140-6736(75) 92847-0 (1975).

Agopian, A.J., Marengo, L.K., Mitchell, L.E., 2012. Predictors of trisomy 21 in the offspring of older and younger women. Birth Defect. Res. A Clin. Mol. Teratol. 94 (1), 31–35. https://doi.org/10.1002/bdra.22870.

Aoki, S., et al., 2018. Developmental trend of children with Down's syndrome - how do sex and neonatal conditions influence their developmental patterns? Brain & Dev. 40, 181–187.

Baird, P.A., Sadovnick, A.D., 1988. Life expectancy in Down syndrome adults. Lancet 2 (8624), 1354–1356. https://doi.org/10.1016/s0140-6736(88)90881-1, 10.

Canfield, M.A., Honein, M.A., Yuskiv, N., Xing, J., Mai, C.T., Collins, J.S., Devine, O., Petrini, J., Ramadhani, T.A., Hobbs, C.A., Kirby, R.S., 2006. National estimates and race/ethnic-specific variation of selected birth defects in the United States, 1999–2001. Birth Defect. Res. A Clin. Mol. Teratol. 76 (11), 747–756. https://doi.org/10.1002/bdra.20294. PMID: 17051527.

Centers for Disease Control and Prevention (CDCP), August 1994. Down syndrome prevalence at birth–United States, 1983–1990. MMWR Morb. Mortal. Wkly. Rep. 43 (33), 617–622, 26.

Chavez, G.F., Cordero, J.F., Becerra, J.E., 1988. Leading major congenital malformations among minority groups in the United States, 1981–1986. MMWR CDC Surveill. Summ. 37, 17–24.

Conor, W.O., 1999. John Langdon down: the man and the message. Down Syndr. Res. Pract. 6 (1), 19–24. https://doi.org/10.3104/perspectives.94.

Dolk, H., Loane, M., Garne, E., De Walle, H., Queisser-Luft, A., De Vigan, C., Addor, M.C., Gener, B., Haeusler, M., Jordan, H., Tucker, D., Stoll, C., Feijoo, M., Lillis, D., Bianchi, F., 2005. Trends and geographic inequalities in the prevalence of Down syndrome in Europe, 1980–1999. Rev. Epidemiol. Sante Publ. 53 (2), 2S87-95.

Down, J.L.H., 1866. Observations on an ethnic classification of idiots. Clin. Lect. Rep. 3, 259–262. London Hospital.

Gordon, A., Benda, C.E., Böök, J.A., Carter, C.O., Ford, C.E., Chu, E.H.Y., Hanhart, E., George, J., Langdon-Down, W., Lejeune, J., Nishimura, H., Oster, J., Penrose, L.S., Polani, P.E., Potter, E.L., Stern, C., Turpin, R., Warkany, J., Yannet, H., 1961. Mongolism (correspondence). The Lancet 1961 1 (7180), 775.

Howard-Jones, N., 1979. On the diagnostic term Down's disease. Med. Hist. 23 (1), 102–104.

Iselius, L., Lindsten, J., 1986. Changes in the incidence of Down syndrome in Sweden during 1968–1982. Prog. Clin. Biol. Res. 209B, 345–350.

Jackson, J.M., Crider, K.S., Cragan, J.D., Rasmussen, S.A., Olney, R.S., 2014. Frequency of prenatal cytogenetic diagnosis and pregnancy outcomes by maternal race-ethnicity, and the effect on the prevalence of trisomy 21, Metropolitan Atlanta, 1996–2005. Am. J. Med. Genet. 1 64A (1), 70–76. https://doi.org/10.1002/ajmg.a.36247.

Krivchenia, E., Huether, C.A., Edmonds, L.D., et al., 1993. Comparative epidemiology of Down syndrome in two United States populations, 1970–1989. Am. J. Epidemiol. 137, 815–828.

Kucik, J.E., Shin, M., Siffel, C., Marengo, L., Correa A for the Congenital Anomaly Multistate Prevalence and Survival Collaborative, 2013. Trends in survival among children with Down syndrome in 10 regions of the United States. Pediatrics 131 (1), e27–e36. https://doi.org/10.1542/peds.2012-1616.

Lancet, 1961, i:775.

Lary, J.M., Paulozzi, L.J., 2001. Sex differences in the prevalence of human birth defects: a population-based study. Teratology 64 (5), 237–251.
Lejeune, J., Gautier, M., Turpin, R., 1959. Etude des chromosomes somatiques de neuf enfants mongoliens. Comptes Rendus Hebd. Seances Acad. Sci. 248 (11), 1721–1722.
Leshin, L., 2003. What's in a Name, 2011.
Levitas, A.S., Reid, C.S., February 1, 2003. An angel with Down syndrome in a 16th century Flemish Nativity painting. Am. J. Med. Genet. 116A (4), 399–405.
Meaney, F.J., Riggle, S.M., Cunningham, G.C., Stern, K.S., Davis, J.G., 1993. Prenatal diagnostic services: toward a national data base. Clin. Obstet. Gynecol. 36, 510–520.
Mégarbané, A., Ravel, A., Mircher, C., Sturtz, F., Grattau, Y., et al., 2009. The 50th anniversary of the discovery of trisomy 21: the past, present and future of research and treatment of Down syndrome. Genet. Med. 11 (9), 611–616.
Medical History, vol. 23, 1979, p. 104.
Wolstenholme, G.E. (Ed.), 1967. Mongolism (Ciba Foundation Study Group No 25). Little Brown and Co.
Morris, J.K., Wald, N.J., Watt, H.C., 1999. Fetal loss in Down syndrome pregnancies. Prenat. Diagn. HC 19, 142–145.
Morris, J.K., et al., 2014. Major congenital anomalies in babies born with Down syndrome: a EUROCAT population-based registry study. Am. J. Med. Genet. 164A, 2979–2986.
Naber, J.M., Huether, C.A., Goodwin, B.A., 1987. Temporal changes in Ohio amniocentesis utilization during the first twelve years (1972–1983), and frequency of chromosome abnormalities observed. Prenat. Diagn. 7, 51–65.
O'Leary, P., Bower, C., Murch, A., et al., 1996. The impact of antenatal screening for Down syndrome in Western Australia: 1980–1994. Aust. N. Z. J. Obstet. Gynaecol. 36, 385–388.
Parker, S.E.1, Mai, C.T., Canfield, M.A., Rickard, R., Wang, Y., Meyer, R.E., Anderson, P., Mason, C.A., Collins, J.S., Kirby, R.S., Correa, A., National Birth Defects Prevention Network, 2010. Updated National Birth Prevalence estimates for selected birth defects in the United States, 2004–2006. Birth Defects Res. A Clin. Mol. Teratol. 88 (12), 1008–1016. https://doi.org/10.1002/bdra.20735.
Santoro, M., et al., 2018. Sex differences for major congenital heart defects in Down syndrome: a population based study. Eur. J. Med. Genet. 61, 546–550.
Shaffer, L.G., Tommerup, N., 2005. International Standing Committee on Human Cytogenetic Nomenclature.
Shaw, S.W., Hsu, J.J., Lee, C.N., et al., 2008. First and second trimester Downs' syndrome screening: current strategies and clinical guidelines. Taiwan. J. Obstet. Gynecol. 47 (2), 157–162.
Shin, M., Besser, L.M., Kucik, J.E., Lu, C., Siffel, C., Correa, A., 2009. The congenital anomaly multistate prevalence and survival (CAMPS) collaborative. Prevalence of down syndrome among children and adolescents in 10 regions of the United States. Pediatrics 124 (6), 1565–1571. https://doi.org/10.1542/peds.2009-0745.
Spencer, K., 2001. What is the true fetal loss rate in pregnancies affected by trisomy 21 and how does this influence whether first trimester detection rates are superior to those in the second trimester? Prenat. Diagn. 21, 788–789.
Startin, C.M., et al., 2020. Health comorbidities and cognitive abilities across the lifespan in Down syndrome. J. Neurodev. Disord. 12, 4.
Takano, T., et al., 2019. Sex differences in congenital heart disease in Down syndrome: study data from medical records and questionnaires in a region of Japan. BMJ Paediatr. 3, e000414.
Warkany, J., 1971. Congenital Malformations. Year Book Medical Publishers, Inc, Chicago, ISBN 0-8151-9098-0, pp. 313–314.
Wilson, M.G., Chan, L.S., Herbert, W.S., 1992. Birth prevalence of Down syndrome in a predominantly Latino population: a 15-year study. Teratology 45, 285–292.

CHAPTER

2

Implications of trisomy 21 on congenital features and health aspects

Bani Bandana Ganguly, Nitin N. Kadam

1. Introduction

The eponym of Down syndrome (DS) has been derived from John Langdon Down, who first described the clinical aspects of the syndrome in 1866 (Down, 1866). The syndrome is the most common genomic disorder of trisomic aneuploidy of *Homo sapiens* chromosome 21 (HSA21) causing intellectual disability as a major problem throughout life. There are several other DS-specific phenotypes involving manifestations of multiple bodily systems, especially, the musculoskeletal, neurological, and cardiovascular systems. Individuals with DS commonly appear with short stature, muscle hypotonia, atlantoaxial instability, reduced neuronal density, cerebellar hypoplasia, intellectual disability, and congenital heart defects (CHDs; particularly atrioventricular septal defects (ASDs)) (Table 2.1) (Antonarakis et al., 2020; Rodrigues et al., 2019). Patients with DS are also at higher risk for developing certain health conditions, such as hypothyroidism, autoimmune diseases, obstructive sleep apnea, epilepsy, hearing and vision problems, hematological disorders (including leukemia), recurrent infections, anxiety disorders, and early-onset of Alzheimer disease (AD) (Antonarakis et al., 2020). Interestingly, most of solid tumors are less frequent in individuals with DS than their peers in the general population (Hasle et al., 2000).

Other phenotypic features of DS are characterized by mongoloid craniofacial profiles, congenital defects, including heart defects and gastro-enteric Hirschsprung disease, immune disorders, increased susceptibility to infection developing common cold and fever, lymphoid and myeloid leukemia, diabetes mellitus, and obesity (Van Cleve and Cohen 2006; Van Cleve et al., 2006). Most of the DS individuals are surviving till the age of 60 years with increasing awareness and early onset of specialty medical care. However, development of Alzheimer disease (AD) and early aging are of critical concern at their advanced age. Approximately, 10%–25% of individuals with DS show signs of AD, even before 50 years age, and 50%–70% with the clinical symptoms at 60 years of age, which is then also coupled with dementia and reduced life expectancy (Bittles and Glasson, 2004). Therefore, DS has been grouped with the progeroid and neurodegenerative diseases (Head et al., 2016). Also, individuals with DS commonly suffer from obstructive sleep apnea,

Genetics and Neurobiology of Down Syndrome
https://doi.org/10.1016/B978-0-323-90456-8.00009-0

TABLE 2.1 Phenotypic and systemic features of Down syndrome.[a]

Involvement of organs	Systems affected	Features	Obligatory/ variable	Diseases
Brain	Neurodevelopmental	Intellectual disability Developmental delay Language disorders Cerebellar hypoplasia	Obligatory (variable extent)	Alzheimer disease Epilepsy Dementia
	Psychiatric	Anxiety and depression Behavioral disturbance	Obligatory (variable extent)	Autism, ADHD
	Major neuroimaging findings	Overall reduced brain volume; progressive brain atrophy; basal ganglia calcifications; malformations of the corpus callosum; Stroke or hemorrhage (complications of Moyamoya syndrome)		
Head and neck	Craniofacial	Small low-set ears Epicanthic folds Hypertelorism Protruding tongue Flat nasal bridge Flat occiput Small mouth and ears Upslanting palpebral fissures	Obligatory (variable extent)	Typical mongoloid face of DS
	Major imaging findings	Reduction in head size/microcephaly; brachycephaly without craniosynostosis; platybasia; macroglossia; stenotic external auditory canal; ossicular chain abnormalities; hypoplasia/aplasia of the bony island of the lateral semicircular canal; semicircular canal dehiscence; stenosis of the cochlear nerve canal; stenosis of the internal auditory canal		
Ear	Sensory	Conductive and sensorineural hearing loss	Variable	Hearing loss
Eye	Vision	Refractive errors Cataracts Keratoconus Amblyopia	Obligatory (variable extent)	Poor vision
Heart	Cardiovascular	Congenital heart defects (atrioventricular septal defects: AVSD)	Variable	Moyamoya syndrome Aberrant subclavian artery
Lungs	Respiratory	Obstructive sleep apnea	Obligatory (variable extent)	Recurrent infections

TABLE 2.1 Phenotypic and systemic features of Down syndrome.[a]—cont'd

Involvement of organs	Systems affected	Features	Obligatory/ variable	Diseases
		Respiratory tract infections	Variable	Increased risk of infection, such as H1N1, COVID-19
Bone/muscle	Musculoskeletal	Short stature	Obligatory (variable extent)	Weakness, neck pain
		Short fingers		Spasticity, spinal cord injury
		Hypotonia		Gait difficulties
		Altantoaxial instability		Hyperreflexia
		Clinodactyly		
		Increased joint laxity		
	Imaging findings of spine	Craniocervical instability; flattened surface of the occipital condyles; bifid anterior or posterior C1 arches; atlanto-occipital assimilation; congenital Os odontoideum "Mickey Mouse" pelvis		
Intestine	Gastrointestinal	Gut abnormalities	Variable	Hirschsprung disease
		Duodenal atresia		Severe constipation
		Annular pancreas		Gastro-esophageal reflux disease
		Imperforate anus		
Blood	Hematology	Acute megakaryocytic leukemia (AMKL)	Variable	Acute lymphoid leukemia (ALL)
		Transient myeloproliferative disorder (TMPD)		Acute myeloid leukemia (AML)
		Anemia		
Others/global genetic expression	Autoimmune	Thyroid disease	Variable	Hypothyroidism
		Celiac disease		
		Alopecia		
		Psoriasis		
		Type 1 diabetes mellitus		
		Obesity		
	Dysregulated functions	Immune dysfunction	Obligatory (variable extent)	Increased rate of infections
		Bowel dysfunction		
	Reproductive	Male infertility	Variable (not clearly known)	Infertility, increased risk of DS and other syndromes

Obligatory (variable extent), features are present in all DS with variable degree of expression; Variable, not present in all DS individuals. ADHD, attention-deficit/hyperactivity disorder; COVID-19, coronavirus disease 2019; DS, Down syndrome.

[a] *Modified from Antonarakis et al. (2020), Rodrigues et al. (2019).*

and the instability of the atlanto-axial joint may lead to spinal cord compression and development of the neurologic symptoms.

The discovery of trisomy of HSA21 causing DS-specific phenotypes and systemic complications was established as an important landmark for the development of genetic medicine. Following understanding the chromosomal pattern in DS in 1959 (LeJeune et al., 1959), subsequent developments included mouse model of DS for genotype-phenotype research and documentation of the complete sequencing of nucleotides of the long arm of HSA21 (21q) by a multinational consortium of investigators (Hattori et al., 2000). That was a landmark report on the sequencing of the 21q. Eventually, substantial progress has occurred over the last 2 decades for understanding the molecular pathophysiology of the different phenotypic manifestations of DS, which is currently considered as an immune disorder caused by dysregulation of gene expression. Parallel understanding of the prenatal screening methods and genetic pathomechanisms due to gene–dosage imbalance are contributing to the management of the different symptoms with a view to improving the quality of life of these individuals.

Individuals with DS display a complex spectrum of ≥80 variable clinical features, which could be contributed by various genetic mechanisms in a complex genetic environment with imbalance of gene dosage (Epstein et al., 1991). The physical characteristics, including an almond shape to the eyes caused by oblique eye fissures with epicanthic skin folds on the inner corner of the eyes, up-slanting palpebral fissures, muscle hypotonia (poor muscle tone), a flat face with flat nasal bridge, cerebellar hypoplasia, midface skeletal retrusion, hypertelorism, irregular-shaped/low set ears, a single transverse palmar fold (a single instead of a double crease across one or both palms, also called the Simian crease), a large protruding tongue (due to small oral cavity and an enlarged tongue near the tonsils), a short neck, shorter limbs, white spots on the iris known as Brushfield spots, excessive joint or ligamental laxity including atlanto-axial instability, CHDs, a larger than normal gap between large toe and second toe known as sandal gap or hallux varus, clinodactyly/a single flexion furrow of the fifth finger, and a higher number of ulnar loop dermatoglyphs. Most individuals with classical DS (100% with free +21) have delayed global and cognitive development, and mental retardation in the mild (IQ 50–70) to moderate (IQ 35–50) or severe range (CDC-birth defects, Down syndrome-NCBDDD. CDC.gov.2013-11-06; Bull and the Committee on Genetics, 2011; Papavassiliou et al., 2015); however, individuals having mosaic DS (presence of cells with +21 and normal cell lineages) have IQ typically 10–30 points higher. In addition, individuals with DS can have serious abnormalities affecting many of physiological systems, in particular, histopathologic development of AD by the fourth decade of life. They also may have a broad head and a very round face. In general, DS is associated with several diseases, including metabolic syndrome, dyslipidemia, and hyperinsulinemia, and the incidence of such complications are much higher than the general population (Figs. 2.1 and 2.2) (Ordóiez-Munoz et al., 2005; Buonuomo et al., 2016).

The effect of trisomy 21 on structures and functions of different cells, including cascading effect of several systems and pathways, produces clinically observed phenotypic end points throughout development of DS individuals. Typical systemic complications, including thyroid dysfunction, CHD with septal defects (ASD/VSD) (~50%), failure to thrive, obesity, premature senility, hearing and speech impairment, cataract, Hirschsprung disease, AD, skeletal problems, obstructive sleep apnea, etc., have been reported in 40%–45% individuals with DS (Epstein et al., 1991; Bull and Committee on Genetics, 2011). Many of the common physical features of the syndrome also appear in people with a standard set of 46 chromosomes.

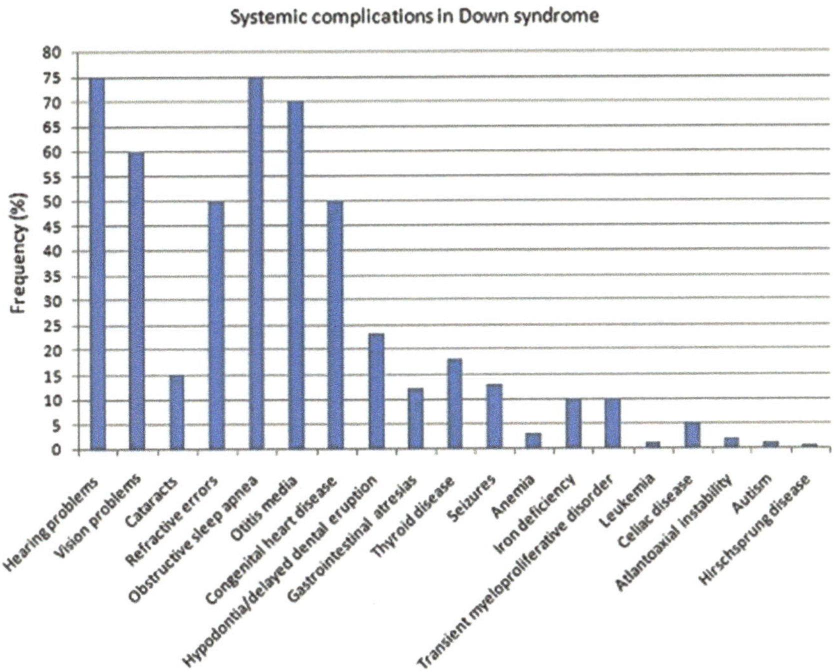

FIGURE 2.1 Relative incidence of phenotypes and systemic complications in DS. *The figures are created based on the data available in Bull and Committee on Genetics (2011).*

Some common but serious complications are described in detail.

Till the present era, enormous challenges remain to elucidate genetic association of each of the phenotypic components of the syndrome and developing therapeutic candidates based on the precise understanding of the complex biological mechanisms. Furthermore, development of animal models and cell lines to decipher the contribution of >200 triplicated protein-coding genes of HSA21, the effects of the triplicated non-coding genes, and the direct or indirect effects of the disomic genes pose extremely challenging tasks for understanding the contribution of the smallest chromosome of the human genome toward the structural and physiological ontology of the syndrome. In this section, we discuss the prevalence of the DS phenotypes and their impact on health condition of individuals with the syndrome, management of the phenotypic features, and medical conditions for rendering an improved life to individuals with DS. Current understanding of genetics and pathophysiology resulting from the extra HSA21 has been emphasized mainly for the prominent and prevalent characteristics of DS so that integration of individuals with DS into the society should eliminate their discrimination from their peers in all corners of the globe.

2. Phenotypic features and systemic complications of Down syndrome

2.1 Craniofacial malformations

Besides varying degrees of intellectual capacity, delayed speech and learning development, ophthalmological problems, hypothyroidism, epilepsy, transient myeloproliferative disorders,

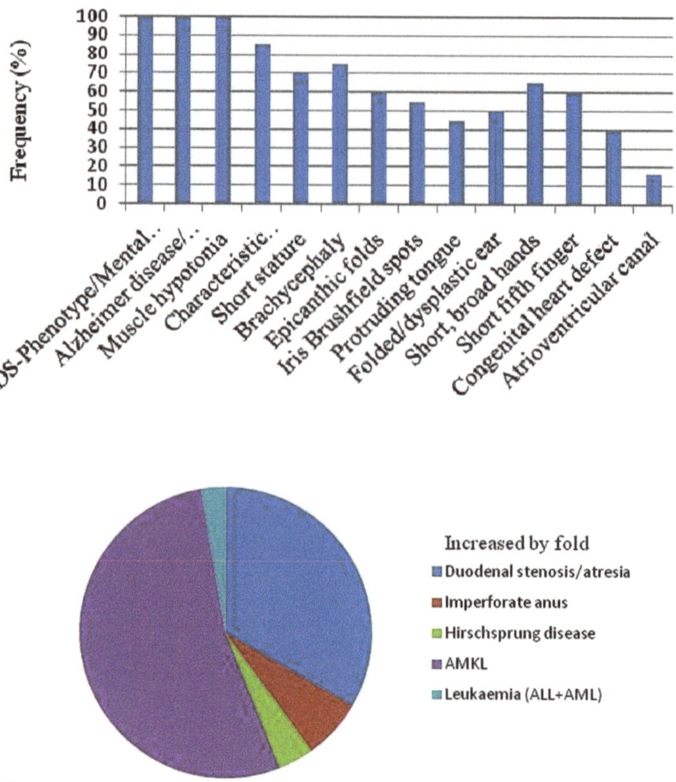

a.

b.

FIGURE 2.2 Presence of craniofacial expression and other features in individuals with Down syndrome. (A) Phenotypes are expressed in percentage; (B) fold of increased risk of the diseases. *The figures are created based on the data available in Antonarakis (2004).*

and many other variable features, the typical facial characteristics of the individuals with DS are demonstrated with reduced facial dimensions, small low set ears, epicanthic folds, midfacial hypoplasia, malocclusion with posterior cross-bite and anterior open bite, and advanced position of the tongue and macroglossia, flat and depressed nasal bridge, upturned nose, flat occiput, small mouth, upslanting palpebral fissures, and protruding tongue. Their craniofacial characteristics often include reduced skull size (microcephaly) accompanying with abnormal calvarial widening in the transverse diameter, flattened occiput (posterior portion of the head), brachycephaly (shortened fronto-back diameter of the skull), variable tooth morphogenesis and eruption patterns, small midface, and reduced size of the maxilla and the mandible (Richtsmeier et al., 2002). In brachycephaly, the ratio of biparietal diameter (BPD) to occipitofrontal diameter (OFD) (corresponding to the cephalic index) is increased and that can be identified visually and may also be evaluated in neuroimaging studies (Rodrigues et al., 2019). The sutures typically remain patent in DS, though brachycephaly is caused by craniosynostosis involving the coronal and lambdoid sutures limiting anteroposterior growth of the skull. Some of these features, namely the ones involving the superior airway, can increase the risk of obstructive sleep apnea in children with DS, which affects about 30%–75% of DS individuals (Radhakrishnan and Towbin, 2014).

Their slower craniofacial growth rates discriminate them from their peers (Megarbane et al., 2009). Notably, the internal architecture of the facial configuration and genetic consequences on how trisomy 21 perturbs the developmental stability to configure facial morphology of DS remains limited. The cephalometric characteristics of DS individuals revealed that the anterior cranial base length (SN) was significantly shorter in comparison with healthy populations detected from the analysis of lateral teleradiographs of their head (Vicente et al., 2020). Notably, the smaller brain size in DS individuals was linked to the lesser development of the anterior cranial base (Silva and Valladares-Neto, 2013). However, the growth of the cranial base was described to be autonomous, as compared to the cranial vault (Enlow, 1990).

There is versatile information on the age-related growth deficit of the cranial base as reported at prenatal age or 4 to 8 years (Alió et al., 2008). The posterior canal base was significantly reduced in size in DS individuals, which grows till 18 years in healthy individuals reported from histological study of autopsies. This growth gradually decreases in individuals with DS up to 15 years of age with cessation of growth of the spheno-occipital synchondrosis at an early age. Collectively, the anterior and posterior cranial bases are smaller indicating a significant decrease in the total cranial base length compared to the general population (Alió et al., 2008). Conversely, the cranial base angle was significantly larger in DS individuals than healthy control subjects. An angle superior to 143 degrees defines platybasia, which can be measured using a line joining the nasion with the center of the pituitary fossa and a line joining the anterior border of the foramen magnum and the center of the pituitary fossa. Collectively, facial feature of DS is associated with flattening of the skull base or platybasia and that is reflected by an increase of the skull base angle (Rodrigues et al., 2019).

The delay in intrasphenoidal synchondrosis fusion in postnatal life is suggested to be essential for the flattening mechanism of the cranial base. The radiographic data collected from individuals with DS exhibited a delay in fusion, as the synchondrosis remains without fusion between the ages of 1 and 7 years, which is fully obliterated by the end of the first year of life in the general population. A meta-analysis showed a significant difference in measurement between DS cases and control subjects in the 5–22 years age range (Vicente et al., 2020); however, the difference was variable in different reports and not clinically significant (Almeida et al., 2017).

The anteroposterior position of the maxilla appeared smaller with a significantly shorter angle of the sagittal position of subspinale relative to cranial base maxilla in individuals with DS. Maxillary length (CoA) was significantly shorter in DS individuals than control groups and that corroborates the existence of maxillary hypoplasia in the sagittal plane (Vicente et al., 2020; Klingel et al., 2017).

The mandible in DS individuals was not clinically relevant. The larger cranial base angle observed in this group would lead to less symphysis projection, which was more acute in control subjects contributing to a more anterior mandibular position (Almeida et al., 2017).

Skeletal class III malocclusions were reported in DS individuals based on the maxillary and mandibular anteroposterior position. Also significantly smaller anterior and posterior face heights were described in these individuals where maxillary hypoplasia in the vertical plane could lead to developmental deficiency in the facial middle third (Alió et al., 2011). The decrease in anterior face heights was more pronounced in the 5–22 years age range. Notably, the individuals with DS have a tendency to keep the mouth half open in repose, which causes frequent infection of the upper airway. This tendency toward the brachyfacial pattern is most likely contributed by the mandibular autorotation in maximum intercuspation due to maxillary vertical hypoplasia. Taken together, the individuals with DS present a cranial base of reduced length, which is more flattened than in healthy subjects. In addition, the maxilla is of reduced size in the sagittal plane, and there is a tendency toward skeletal class III malocclusion. DS individuals also have reduced facial height in both posterior and anterior regions (Vicente et al., 2020).

Dental agenesis is a commonly recognized phenotypic characteristic of DS patients along with a higher prevalence of angle class III malocclusion, posterior cross bite, and anterior open bite (Doriguêtto et al., 2019). DS patients often present delays of both deciduous and permanent dentition and disorders in the eruption sequence. The dental agenesis (excluding third molars) is present in approximately 54.6%–58.5% of DS subjects (Palaska and Antonarakis, 2016). Also, microdontia and taurodontism are fairly common in these individuals (Abeleira et al., 2014).

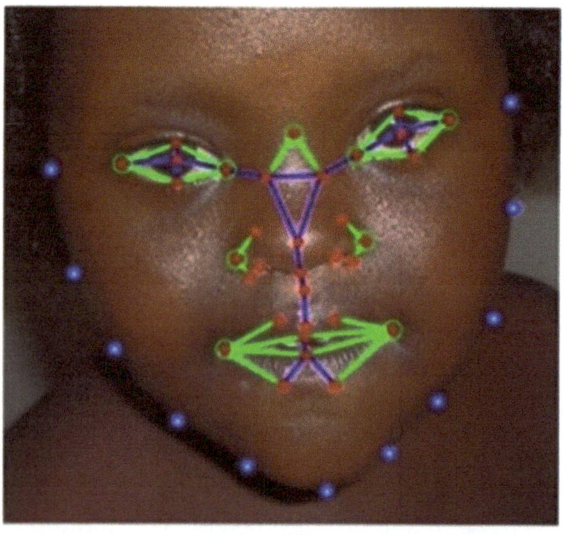

FIGURE 2.3 Facial landmarks of a Down syndrome (DS) patient. *Red spots* indicate the inner facial landmarks, and *blue spots* represent the external landmarks. The calculated distances are presented as *blue lines*, and the calculated angles are indicated by *green circles*. *Reproduced with permission from Agbolade et al. (2020).*

Imbalance in gene dosage in DS underlies the craniofacial dysmorphology, but which genes of HSA21 contributing to which craniofacial characteristic in DS are yet to be precisely assigned. Besides genetic analysis of craniofacial malformations, algorithm of machine learning of facial dysmorphism has demonstrated the patterns of recognition of individuals with DS (Fig. 2.3) (Agbolade et al., 2020).

2.2 Learning, memory, brain development, and function

All individuals with DS exhibit some form of learning and memory impairment, which varies greatly in terms of severity (Lott and Dierssen, 2010). Cognitive development can be detected in infants with DS and slow down in first 2 years of age due to delayed brain myelination. Learning deficits between birth and 11 years (school age) is characterized with difficulty in maintenance of the existing skills, while visuospatial processing in adolescent years exhibits more difficulty in working memory and short-term verbal memory relative to controls with other forms of mental retardation (Fildler et al., 2009). Language and communication development is affected by hearing impairment but improved with mental age. The linguistic strength is markedly influenced by pragmatics or social language at school-aged or adolescent children.

Studies of mouse models have indicated that defects in neurogenesis, synaptogenesis, synaptic transmission, and cell signaling pathways could all contribute to this phenotype, potentially by causing excessive inhibitory neurotransmission (Belichenko et al., 2007; Chakrabarti et al., 2007). Indeed, regular application of gamma-aminobutyric acid (GABA) antagonists to mitigate the effects of inhibitory neurons could reverse the cognitive deficit in DS mouse models (Fernandez et al., 2007). Studies on individuals with partial trisomy of HSA21 demonstrated that multiple regions of the chromosome 21 contribute to the cognitive deficits, suggesting involvement of several genes and pathways for this phenotype (Korbel et al., 2009; Lyle et al., 2009). DS with partial trisomy of HSA21 showed that the Down syndrome Critical Region (DSCR) was not required in three copies to manifest intellectual disability, though DSCR may partially contribute to the phenotype and that was confirmed in mouse model (Korbel et al., 2009; Lyle et al., 2009; Olson et al., 2007). Nevertheless, genes located in the DSCR, including DYRK1A, SIM2, DSCAM, and KCNJ6 are expressed in the developing and adult nervous system (Hammerle et al., 2008).

2.3 Alzheimer disease

Alzheimer disease (AD) is an inevitable cause of early-onset of dementia in individuals with DS (DSAD). AD is a neurodegenerative disorder caused by overproduction of APP protein and accumulation of Aβ and formation of neurofibrillary tangles in the brain (Wiseman et al., 2015). Triplication of APP encoded on HSA21 is responsible for AD in individuals with DS. Clinical symptoms of early onset of AD (EOAD) appear within 35–40 years in 40% of DS, and late onset of AD (LOAD) occurs in 77% cases after 50 years of age, and overt AD is expressed in 100% cases at age over 60 years (Strydom et al., 2010; Antonarakis et al., 2020). The cognitive decline is a major symptom of AD in individuals with DS, who are cognitively impaired (background cognition) from birth. Thus, understanding the progressive decline in cognitive phenotype may help in an early clinical intervention (Firth et al., 2018). Since memory and attention deficits are congenital features of DS, recognition of AD onset goes unnoticed until severe behavioral changes appear in these individuals (Ballard et al., 2016; Lott and Head 2019). The AD-related symptoms are nearly similar in sporadic and DSAD. Severity of AD symptoms and dementia is occasionally augmented by the development of seizures and neurological signs such as myoclonus indicating disease progression and final culmination into death (Hithersay et al., 2019; De Simone et al., 2010).

Besides gene–dosage effect of APP and cooperation of other disomic genes, allelic expression of APOE gene (mapped onto HSA19), in particular the APOE ε4 allele, has been reported to enhance the risk of further decline of the background cognition in AD leading to death (Lott and Head, 2019; Startin et al., 2019). A routine baseline screening of cognitive and adaptive functioning age is recommended for all individuals with DS at 30 years of age, to help in drawing an accurate diagnosis and future monitoring and surveillance of AD (Bayen et al., 2018). Practically, interventional management should pay importance in early detection for introducing the necessary supportive measures on time. Therapeutic use of acetylcholinesterase inhibitors has indicated positive outcomes; however, reduced heart rate and other adverse effects were evidenced in some individuals (Eady et al., 2018). An appropriate and adequate support should be extended to individuals with DSAD and dementia, keeping the fact in mind that their needs may increase with the progress of the disease. Arrangement of an adequate home accommodation facilitating their living within their families will be ideal than transferring them to other institutional accommodations, which may be opted only when such circumstances and needs arise (Dodd et al., 2018).

2.4 Epilepsy

Epilepsy is a variable phenotype in individuals with DS, still reported in 8%–13% of DS children (reviewed in Antonarakis et al., 2020). Interestingly, two distinct categories of epilepsy of bimodal age of onset, including one peak appears before 3 years of age (infant onset; ~40%; also known as West syndrome) and the other one occurs after 30 years of age (adult onset) are demonstrated in DS patients (Roizen and Patterson, 2003). Infant onset is the most frequently observed form of infantile spasms and is associated with psychomotor regression and hypsarrhythmia in ~6%–32% of infantile seizures (an abnormal pattern of activity between convulsions) detected on electroencephalography (Sanmaneechai et al., 2013). Tonic-clonic or myoclonic seizures are reported in ~40% of individuals with DS in their first 3 decades. Epilepsy of adults (>40 years) with DS (~84%) is linked to the development of AD and is related to the rapid decline in their cognitive functions (De Simone et al., 2010). Clinically, epilepsy is managed with the use of antiepileptic drugs for reducing the seizures effectively; however, management depends on age of onset and etiology (Bayen et al., 2018; Gholipour et al., 2017). DS children with infantile spasm do respond better to antiepileptic drugs compared to other kids with the same problem, and therefore, early intervention and treatment improve the developmental outcome in children with DS. DS children also develop Lennox-Gastaut syndrome exhibiting reflex seizures along with an increased incidence of EEG abnormalities (Ferlazzo et al., 2009).

2.5 Immune systems

Individuals with DS are more prone to infections, especially of the respiratory tract, suggesting susceptibility of this group of individuals to pathogens due mainly to their compromised immune status (Chan et al., 2017; Grut et al., 2017). Patients with infections should be treated promptly to reduce the risk of cross-infection with other pathogens (Bull and Committee on Genetics, 2011). Interestingly, immunization of the DS children did not report any contraindication, and thus, the standard schedule and protocol of immunization could be followed for these children.

Individuals with DS are also at higher predisposition for developing autoimmune diseases compared to the general population, such as type 1 diabetes mellitus, alopecia areata, Addison disease, celiac disease, and so on (Giménez-Barcons et al., 2014). Celiac disease is highly prevalent in DS population; thus, testing of immunoglobulin A antibodies to tissue transglutaminase 2 is recommended for recognition of celiac disease. Opportunistic screening of asymptomatic individuals may appear unnecessary and has not proved cost-effective (Roizen and Patterson, 2003; Swigonski et al., 2006).

2.6 SARS CoV-2 infection

Age, obesity, deregulated immune system, and many other phenotypes of >80 clinical features of DS facilitate respiratory syncytial virus to cause severe bronchitis, which was further supported by higher rate of hospitalization, endotracheal intubation, and death during H1N1 pandemic (Perez-Padilla et al., 2010). Lymphopenia was observed in two DS individuals detected at the time hospital admission (within few days of suspected infection). However, variability in severity of the infection was noticed between two patients; one required continuous positive airway pressure (CRAP) support and mechanical ventilation, while other one did not (Vita et al., 2021). Notably, owing to the presence of genetic background, individuals with DS are demonstrated with immune dysfunction, CHDs, and pulmonary pathology manifested with chronic cold and cough, given its prevalence, albeit likely to be an unconfirmed risk factor for severe COVID-19 infection (Illouz et al., 2021). Of 8.26 million COVID-19-infected adults in a study cohort of the United Kingdom, 0.05% was trisomic Down syndrome (no interaction with age, sex, or BMI of DS) who exhibited pneumonitis (25.0% DS vs. 14.4% non-DS) and other complications (35.3% DS vs. 65.3% non-DS) leading to a fatality rate of 39.7% (20.3% in non-DS) due to COVID-19 infection (Clift et al., 2021). Taken together, a 4-fold increased risk for hospitalization and a 10-fold increased risk of death related to COVID-19 were demonstrated in adults with DS. However, pathomechanism of the severity of COVID-19 infection in DS is poorly understood. Nevertheless, likeliness of overexpression of TMPRSS2 gene located on 21q22.3 that encodes transmembrane protease serine 2 could partially explain the severity in COVID-19-infected DS patients (Schuler et al., 2021). Age-related expression of TMPRSS2 has been demonstrated in ciliated and secretory cells (Sungnak et al., 2020), suggesting that the cellular spectrum infected by SARS-CoV-2 virus in the lung epithelium and developmental regulation of TMPRSS2 may underlie the relative protection of infants and children from severe respiratory illness. Additionally, interaction of the viral spike receptor binding domain (RBD) of the surface glycoprotein of the virus with the ACE2 protein (SARS-CoV-2 uses the ACE2 (angiotensin converting enzyme 2) receptor to invade the host cell) could be mediated by TMPRSS2 receptor–dependent interaction and that is implicated in the activation of SPIKE protein (Hoffmann et al., 2020). Evaluation of TMPRSS2 in individuals with DS may explain the underlying mechanisms of COVID-19 and may further allow the development of targeted therapeutics.

The prognosis of SARS-CoV-2 infection in DS could be worse in the presence of associated cardiac malformations and anatomical differences in the upper respiratory tract (a smaller trachea, enlarged adenoids/tonsils, glossoptosis, upper airway narrowing, and tracheal bronchus airway malacia) contributed by gene–dosage effect in DS (Colvin and Yeager, 2017). Increased necessity of oxygen supplementation (up to 15 L/min) coupling with CARP supplementation, though variable among COVID-19 infected DS adults, may indicate a marked increase in inflammatory cytokine markers, in particular, type I and type III interferons (IFNs) (Kaouini et al., 2021; Vita et al., 2021). Activation of Toll-like receptors and subsequent induction of IFN antiviral response resulted in a cytokine storm with worsening of liver pathology and culminating into death in an animal model carrying triplicated IFN receptors (Tuttle et al., 2020). https://dx.doi.org/10.1101/2020.04.07.024455 DS individuals aged ≥40 years exhibited 3-fold higher risk of severe outcome than the general population, though comorbidities were not considered for drawing such inference (Hüls et al., 2021). Additional risk factors for COVID-related hospitalization were male gender, obesity (due to poor metabolic rate), intellectual disability, diabetes, and CHD; of these, male gender and AD/dementia contributed more for mortality.

2.7 Hematology

Hematological anomalies, including neutrophilia, thrombocytopenia, and polycythemia are seen in newborn with DS, but these are mild and get resolved in first few (~3) weeks of life. Indeed, DS children are severely predisposed (>200-fold) to hematological malignancies compared with the children without DS. Transient myeloproliferative disorder (TMPD) is reported in ~20% of newborns with DS in association with peripheral megakaryoblasts, thrombocytopenia, variable leukocytosis, anemia, hepatosplenomegaly, skin rash, etc., before 3 months of age (Gamis and Smith, 2012). TMPD is practically a benign form of clonal proliferation and resolves spontaneously within 3–4 months of life without initiating any treatment; however, it can predispose to leukemia by 5 years of age in 20%–30% DS children along with risk of other illnesses in later life (Bull and Committee on Genetics, 2011). Thus, TMPD is also referred to as transient leukemia of DS. Owing to such high incidence of TMPD, a blood count and film are suggested to be analyzed within 3 days of birth to enable the identification and monitoring of this disorder at early stage (Tunstall et al., 2018). TMPD may also occur in the fetus and that may lead to spontaneous abortion.

Risk of acute lymphoid leukemia (ALL) is ~20 times and acute myeloid leukemia (AML) is ~50 times higher in DS, with an ALL-AML ratio of 1.7 (Hitzler and Zipurskey, 2005; Lana-Elola et al., 2011; Mateos et al., 2015). Leukemia in DS children constitutes ~2% of ALL and ~10% of all AML of children. Incidence of acute megakaryocytic leukemia (AMKL) is ~500 times higher in DS than the general population without DS. The susceptibility of DS to acute leukemia was first reported in 1930 by Brewster and Cannon, while the risk of TMPD was reported in 1954 by Schunk and Lehman (Ganguly and Kadam, 2016).

Besides triplicated genes and their dosage imbalance, an inextricable link between loss of functions in GATA1 (mapped onto X chromosome) and trisomy 21 has been reported for development of neonatal leukemia in ~30% of newborns with DS. In contrast to hematologic malignancies, malignancies of solid tumors are less common in this syndrome, which could possibly be explained by the presence of increased number of tumor suppressor genes harbored on the HSA21 such as ETS2 (Lana-Elola et al., 2011; Yang and Reeves, 2011) and/or decreased angiogenesis (Reynolds et al., 2010). Exome-sequencing outlined the "genetic identity card" of leukemia by a group of geneticists in Geneva, who described that RAS-opathy is the underlying mechanism in tumorigenesis of 1/3rd of ALL cases with DS (Gamis and Smith, 2012). Complex rearrangements with 38 chromosomes were reported in bone marrow of a 2-year-old DS child having acute erythroleukemia (FAB: M6), where trisomy 21 could not be recognized following conventional cytogenetics due to complexity of rearrangements (Fig. 2.4) (Ganguly et al., 2017). Mutation in JAK2 (Janus Kinase 2) has been demonstrated in ~30% of ALL of DS children (Kearney et al., 2009).

2.8 Congenital heart defects

Congenital heart defects (CHD) are prevalent in children with DS. Approximately, 50% of the individuals born with DS were presented with CHD (most commonly with AVSD in ~42% of DS), and of the AVSD, ventricular septal defect (VSD in ~22% of DS) and atrial septal defect (ASD in ~16% of DS) were frequently demonstrated (Bergstrom et al., 2016). CHDs affect the quality of life of these individuals and increase the rate of morbidity and mortality for the individuals with DS in the first 2 years of life.

Children with DS are also affected with secundum atrial defect (10%), isolated PDA (4%), and tetralogy of Fallot (6%). Notably, ~30% individuals with DS are affected with more than one cardiac defect. Interestingly, the incidence of the specific type of CHD varies with age, ethnic origin, and geographical variation. VSD is higher in Asia and secundum-type ASD is prevalent in Latin America. Nevertheless, clinical management and treatment outcome of CHD are the same as that of the general population (Morales-Demori, 2017).

Screening through a fetal echocardiography during pregnancy and postnatal cardiac examination are recommended. Further echocardiography examination within the first month after birth, and annual screening throughout life should be performed to minimize or eliminate the risk of acquired valve disease and heart failure. Surgical repair of CHD and eventual rate of mortality in children with DS are almost similar to that of the general population (Russell et al., 2018). Interestingly, a lower incidence of hypertension is demonstrated in individuals with DS (Draheim et al., 2002). MicroRNA (miRNA), especially the triplicated miR-155, was hypothesized to control hypertension by targeting one

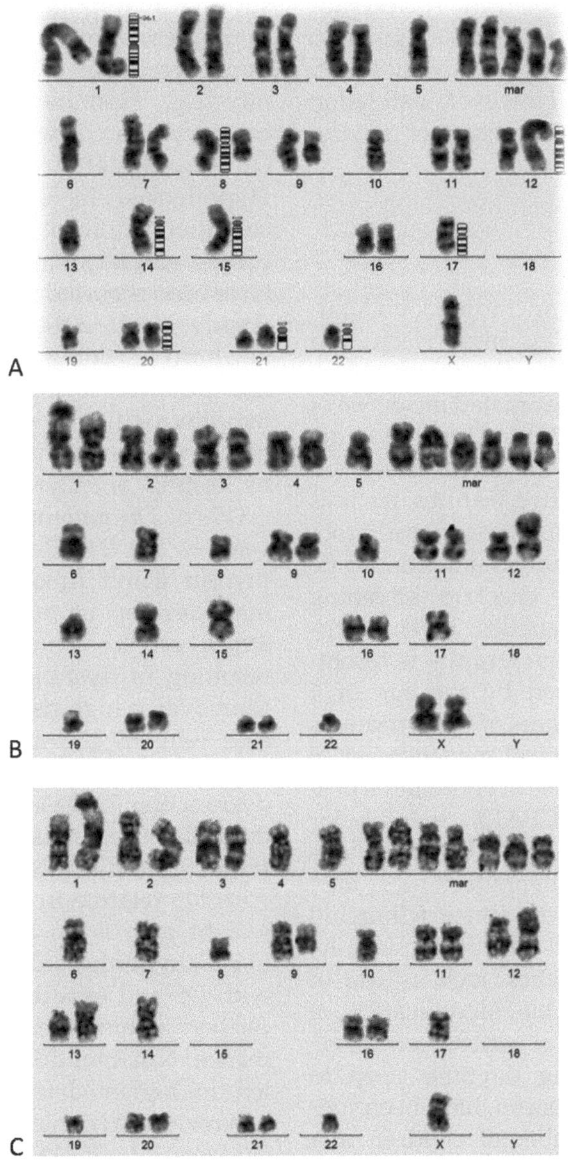

FIGURE 2.4 Description on chromosomal rearrangements and clonal abnormalities in a Down syndrome patient developed acute myeloid leukemia A. Der(1p),-5,-6,der(8)?t(8;21),del(9q),-10,der(12p),-13,-14?t(13;14),-15?t(14;15),-17,i(17q),-18x2,-19,der(21),-22,-X,+marx5 B. Der(1p),-5,-6,-7,-8,der(8)?t(8;21),del(9q),-10,der(12p),-13,-14?t(13;14),-15?t(14;15),-17,i(17q),-18x2,-19,-22,+marx6 C. Der(1p),-4,-5,-6,-7,-8,der(8)?t(8;21),del(9q),-10,der(12p),-13,-14?t(13;14),-15?t(14;15),-17,i(17q),-18x2,-19,der(21),-22,-X,+marx7. *Reproduced from: Ganguly et al., (2016).*

allele of the type-1 angiotensin II receptor (AGTR1) gene, resulting in its downregulation and lowering the risk of hypertension in DS (Sethupathy et al., 2007). However, validation of this hypothesis necessitates further studies on the topic.

2.9 Sleep apnea

Obstructive sleep apnea, including loud snoring, heavy breathing, restless night sleep, and daytime drowsiness, is frequently reported in individuals with DS with an estimated incidence of 54%–90% (Capone et al., 2018; Churchill et al., 2012). These symptoms are often presented with abnormal neurocognitive features, such as cognitive decline, irritability, behavioral problems, depression, and paranoia, and should be included in every health checkup screening (Bull and Committee on Genetics, 2001). Practically, an overnight polysomnography is recommended for all DS children by the age of 4 years, despite the consequences of the symptoms (Esbensen et al., 2018). Clinical management of sleep apnea, including the use of continuous positive airway pressure (CPAP), mandibular advancement devices, and surgical interventions for tonsillectomy and adenoidectomy, may reduce the intensity of sleep apnea (Churchill et al., 2012; Venekamp et al., 2015). Ideally, further alternative approaches such as use of oxymeter at home to enable identification of at-risk individuals (Hill et al., 2018), control on weight gain, and avoiding daytime sleep to help in getting a deep night sleep, have been suggested to reduce the number of children that require full diagnostic studies. Individuals with DS are at increased risk for obesity and that increases with increasing age.

2.10 Thyroid dysfunction

Individuals with DS are at increased risk for dysfunction of the thyroid gland, including subclinical hypothyroidism (24%; more likely to progress to overt hypothyroidism), congenital hypothyroidism (1%), and thyroid autoimmunity (e.g., Hashimoto's disease or Grave's disease) compared with the normal population (Antonarakis et al., 2020; Whooten et al., 2018). Hypothyroidism is seen as the most common condition in almost 1/3rd of those with the syndrome, and its abnormalities in thyroid function have been reported in >50% of neonates with DS (Purdy et al., 2014; Iughetti et al., 2015). The newborn screening program in several countries has indicated a higher incidence of congenital hypothyroidism in children born with DS than others. Acquired hypothyroidism is established by testing antithyroid autoantibodies in 13%–34% of DS patients and that increases after 8 years of life (Hawli et al., 2009). Irregularity of thyroid gland function may be congenital or may develop at infancy or childhood due to attack on the gland by the immune system resulting in dysfunction of the thyroid gland. Moreover, a high risk of developing thyroid disease remains throughout life, and the risk of developing autoimmune-related thyroid dysfunction increases with age. Hypo- or hyperthyroidism *per se* is an auto-immune disorder. Hyperthyroidism is infrequent in individuals with DS relative to hypothyroidism. Moreover, weight gain is an accompanying symptom of hypo-/hyperthyroidism for the individuals with DS and experimental animals, which may further deteriorate their intellectual performance. Notably, differences in thyroid function led to higher deterioration of cognition and behavior in trisomic Ts65Dn mice of DS models (Vidal et al., 2021) because hypothyroidism is associated with fatigue and attentional and memory dysfunctions. Obesity may augment the risk of thyroid problem in DS.

Children with DS may display hypothyroidism at the subclinical level with elevated thyroid-stimulating hormone (TSH) and normal thyroxine levels. Therefore, regular blood screening, including TSH and thyroxine (T4), would be important for any age group, though a clinical diagnosis can be difficult at birth or infancy. In particular, the levels of TSH and T4 should be screened at birth, and at 6 and 12 months of age, and follow-up screening should be performed annually for TSH surveillance (Bull and Committee on Genetics, 2011).

Treatment with bexarotene for AD in trisomic Ts65Dn mouse model, increased the levels of TSH and reduced the levels of T4 compared to their vehicle-treated littermates indicating bexarotene-induced primary hypothyroidism in these animals (Vidal et al., 2021). A contrast association of bexarotene with central hypothyroidism has been demonstrated in the vast majority of studies (Makita et al., 2019).

Children with DS have been reported with significant delay in puberty in both genders (Hawli et al., 2009). The girls with DS experience delay in menarche or adrenarche due to primary hypogonadism, while the DS boys can exhibit cryptorchidism, micropenis, small/undescended testes, low sperm count, and scanty axillary and pubic hairs. Ambiguous genitalia is also prevalent in children born with DS.

Children with DS are prone to type 1 diabetes mellitus, and thus, the insulin-like growth factor might contribute to delay in skeletal maturation giving a short stature to DS individuals.

2.11 Gastrointestinal complications

Individuals with DS are affected with a number of structural and functional disorders of the gastrointestinal (GI) tract, where structural defects can be detected anywhere in the tract from the mouth to anus.

DS children have increased risk of Hirschsprung disease (~2%) due to the absence of the normal myenteric ganglion nerve cells that control the function of parts of the colon and suffer from severe constipation. The disease can be detected within a few days after birth. Notably, ~12% of patients with Hirschsprung disease have DS (Holmes, 2014). In particular, the neural cells of the lower intestine fail to migrate to the distal segment of the rectum leading to an aganglionic segment, which lacks a normal peristalsis, and that results in failure to a normal defecation reflex and a functional obstruction (Amiel et al., 2008). These children usually display the signs and symptoms of intestinal obstruction, failure to pass meconium, enterocolitis, and bilious vomiting as the relevant signs and symptoms of Hirschsprung disease.

Other congenital gut anomalies in DS frequently include duodenal and small bowel atresia or stenosis, annular pancreas, imperforate anus, gastroesophageal reflux disease (GERD), and also celiac disease. Celiac disease occurs in ~5% of children born with DS. Incidence of duodenal stenosis and imperforate anus, identified in the neonatal period with signs of bilious vomiting, are 260 and 33-fold higher in DS compared to normal population (Berrocal et al., 1999). In the absence of treatment, the condition may cause severe dehydration and electrolyte imbalance. Imperforate anus with malformed rectum may increase the risk of a group of other specific anomalies called as VACTERL (vertebral anomalies, anal atresia, cardiovascular anomalies, tracheoesophageal fistula, esophageal atresia, renal, and limb defects). Triplication of the DSCAM gene, mapped onto HAS21 and expressed in neural crest, is attributed to the enteric nervous system.

Besides, DS children are often presented with chronic constipation and intermittent diarrhea. Screening of celiac disease would help in diagnosis of the disease and suggest a gluten-free diet throughout life (Wallace, 2007). These conditions of the GI tract may dampen the metabolic rate in DS children affecting the growth parameters such as height, weight, and head circumference, which are smaller than age and sex-matched normal children. Notably, the DS adults appear with short stature (men: ~5′ and women: ~4.5′).

2.12 Neurology

Clinically, DS is a neuro-developmental disorder manifested with moderate to severe degree of learning and short-term memory deficits that lead to delay in global development on acquisition of speech, expressive language, learning and motor skills, verbal working, and episodic memory, which collectively renders mild to severe level of cognitive impairment and mental retardation with an IQ range of 25 and 50 (Kazemi et al., 2016). The manifestation of neurologic consequences, including decreased muscle tone or hypotonia, delayed developmental milestones (sitting, walking, and talking), and epilepsy in DS children occur early in life. Notably, their cognitive profile displays their capability in visual learning. Noticeably, owing to wide variation in cognitive function (IQ, language, attention, memory, and functional abilities) in individuals with DS, cognitive deficit is considered a variable characteristic of DS features, for which trisomic HSA21 might not be the only contributing factor. Individuals with DS also appear with inconsistent behavior and symptoms of attention deficit hyperactivity disorder (ADHD) in ~6% of cases. Notably, ~10%–25% DS adults develop autism spectrum disorder (ASD) as a behavioral anomaly, which aggravates in 50%–70% cases with overt clinical presentation of ASD—many a time associated with dementia (Dykens, 2007; Head et al., 2016). Therefore, appropriate screening of ADHD/ASD for DS individuals would be helpful for their parents. Interestingly, clinical presentation of ADHD or ASD in DS patients may differ from the general population; thus, identification of the problem and counseling by the specialists may help in appropriate interventional assessments. Nevertheless, standard treatment protocol for ADHD may not extract the outcome or effect from application of stimulant medication for DS patients to that extent observed in individuals without having intellectual disability. Intellectual disability in DS may interfere in such situation.

Neurodegeneration in DS individuals leading to early aging and death suggested grouping of DS as a progeroid and neurodegenerative disease (Head et al., 2016). In human, elevation of oxidative stress and production of reactive oxygen species (ROS) perturb the mitochondrial respiratory system and stimulate apoptosis and gliosis that collectively promotes neuronal damage. Such neuronal losses during neurogenesis and neuronal degeneration contribute to dementia, early aging, and AD in DS adults. In addition, synergistic interaction of HSA21 genes, such as APP and S100B, increase hydrogen peroxide (H_2O_2) levels and decrease membrane potential in the mitochondria of human DS neuroprogenitor cells (NPC) and contribute to patho-mechanism of neurodegeneration.

2.13 Mental health

Individuals with DS often exhibit depression and anxiety disorders, most likely due to their behavioral and mental health problems compared to their peers in the general population (Capone et al., 2018; Tassé et al., 2016). The adolescents and young adults may develop depression and experience sharp decline in skills and attitude of independent performance compared to their previous levels of functioning that they achieved through practice and/or special education. This situation is described as Down syndrome disintegrative disorder, which seems to occur through exposure to emotional stressors (Mircher et al., 2017), but the mechanism of such decline and definitive treatment for this regressive presentation are still unelucidated. Nevertheless, the extent of psychological stressors such as transitions in accommodation, school or care-giving arrangements, bereavement, emotional deprivation, etc., should be avoided to the maximum extent (Nevill and Benson, 2018). The condition of their intellectual deficits, difficult communication, behavioral swings, and many other atypical symptoms may not help in appropriate understanding and diagnosis of their mental health status. Therefore, standard clinical practice of behavioral interventions, psychotropic medication, and psychological therapy should be followed towards treatment for mental illness of individuals with DS.

2.14 Atlantoaxial instability

In individuals with DS, ligamentous laxity increases the risk of misalignment of the first and second cervical vertebrae, medically termed as atlantoaxial subluxation—reported in ~1%–2% of children with DS (Brockmeyer, 1999). Recognition of the problem may prohibit these children from participation in sports, which impose some increased risk of spinal cord injury. Parents should be careful about this feature in their DS children. For management of this condition, symptoms, including neck pain, weakness, spasticity, gait difficulties, and hyperreflexia, should be examined medically and using cervical spine radiography for monitoring the ligamental laxity (Bull and Committee on Genetics, 2011).

Children with DS are at an increased risk of hypotonia (reduced muscle mass). Increased ligamentous laxity may cause retardation of gross motor skills and can result in joint dislocation. Inadequate intake and malabsorption of vitamin D cause deficiency of vitamin D in these patients. Use of antiepileptic drugs may additionally affect intake of vitamin D in these children. These factors and decreased bone mass increase the risk of fractures in children with DS.

2.15 Ophthalmology and otolaryngology

Ophthalmic (ocular and orbital) disorders are common in individuals with DS affecting almost 50%–100% of these patients which include strabismus (two eyes do not move in tandem: 5%–47%); amblyopia (10%–26%); nystagmus (3%–33%) with lesser visual acuities; refractive errors (~80%) and astigmatism (10%–40%); upward slanting palpebral fissures (60%–80%) and large epicanthal folds (30%–90%); blepharitis (5%–30%); lacrimal duct obstruction (4%–40%); keratoconus (develops at early adulthood; 0.5%–15%); Brushfield spots in iris (38%–90%; differ in various ethnic populations) (Fig. 2.5); congenital and developmental cataracts (opacity of the lens) (4%–40%); anomalous retina (13%–38%); optic nerve hypoplasia; glaucoma (increased eye pressures; 0%–1%); and so on. Recognition of the problems of vision and use of treatment with corrective aids (supportive spectacles, lenses) can help DS individuals, and if left untreated, the ocular anomalies can significantly affect the lives of these patients.

Ophthalmological examination for vision management should be performed at regular intervals for DS patients, in particular, at birth and regularly throughout life, ideally every 1–2 years (Bull and Committee on Genetics, 2011; Merrick and Koslowe, 2001). Cataract surgery is frequently required for individuals with DS.

Patients with DS often encounter with ear, nose, and throat problems. The anatomical structure of their ears predisposes them to hearing impairment with a high prevalence of hearing loss in 38%–78% of cases (Clark et al., 2017). Conductive hearing loss (CHL) is the most

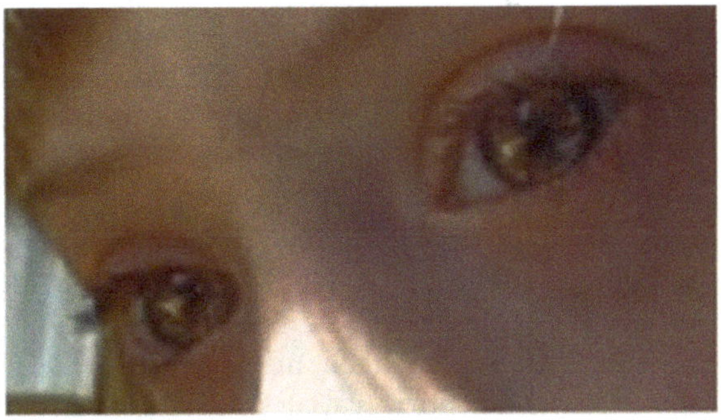

FIGURE 2.5 Brushfield spots in iris of a child with DS. The figure has been obtained from public domain.

common subtype affecting 50% with hearing impairment – generally arising from sequelae of chronic otitis media (infections of the middle ear; has a higher incidence in DS) and effusion because of impaction of cerumen and eardrum perforation (Austeng et al., 2013). Occasionally, CHL can be secondary to stenotic external auditory canal, dehiscence of the facial nerve canal, or ossicular chain abnormalities, which may either be related to chronic infections (including erosion of the long process of the incus, erosion of the manubrium of the malleus, and erosion of the stapes) or attributed to congenital deformities, such as malformation of the stapes. These findings should be considered and actively searched for with appropriate evaluation by neuroimaging in DS children having a persistent CHL though maximal management of middle ear infections is recommended (Rodrigues et al., 2019). Postnatally, baseline hearing tests should be performed using brainstem auditory acoustic response, then every 6 months until school age, and thereafter once every year (Bull and Committee on Genetics, 2011).

Sensorineural hearing loss (SNHL) appears increasingly common because of the structural malformation in the inner ear in ~75% of the DS cases. Abnormality in the bony island of the lateral semi-circular canal (LSCC) is the most common anomaly observed in about 50% of DS patients (Radhakrishnan and Towbin, 2014). Sometimes, the bone island appears totally absent, giving a shape (Rodrigues et al., 2019), and its absence and small size constitute a spectrum of LSCC dysplasia. This LSCC dysplasia is postulated to result from the late formation in embryogenesis of the LSCC and occurs with higher incidence than that of the superior or posterior semicircular canals. In addition, semicircular canal dehiscence is also associated in ~9% of DS patients as a cause of SNHL.

Stenosis of the cochlear nerve canal and/or tight internal auditory canal (IAC) is reported in ~20%–25% of individuals with DS, which can be associated with aplasia or hypoplasia of the cochlear nerve (Radhakrishnan and Towbin, 2014). The malformation of the membranous labyrinth in patients with SNHL is proposed to inhibit the normal production of nerve growth factor, and that is crucial for the normal growth of the vestibule-cochlear nerve and may result in aplasia or hypoplasia of the cochlear nerve. In addition, many other cochlear anomalies have been demonstrated, such as arrest of IAC formation in absence of the stimulus from the cochlear nerve, a short cochlea as well as the Mondini anomaly, enlarged vestibule with normal semi-circular canals, and enlarged vestibular aqueduct in DS patients (Intrapiromkul et al., 2012).

A routine screening and an early assessment by computed tomography (CT) and/or magnetic resonance imaging (MRI) studies of the temporal bone would be helpful to initiate effective treatment that can reduce the risk of hearing loss and prevent long-term sequelae and minimize impairment in life quality (American Academy of Pediatrics Committee on Genetics, 2001). Indeed, amplification with hearing aids, and/or with cochlear implants, can improve the communication skills of DS patients with SNHL (Intrapiromkul et al., 2012). The pressure equalization tubes can also be useful for these individuals.

Delay in speech development is an inherent problem in individuals with DS, which is further compounded by hearing loss. Identification and early intervention of speech therapy, and use of communication aids and sign language, can be helpful for treatment of hearing and speech impairment. The use of specially developed language program, such as Makaton, can also be beneficial (Roizen and Patterson, 2003).

3. Life-span of individuals with Down syndrome

Effects of trisomy 21 vary greatly among people with DS, depending on the type of trisomic

21 (free, mosaic, translocated, partial trisomy), genetic history, environmental factors, and pure random chance. Although people with DS had a shortened life expectancy in past, today's DS patients are living longer than before. In 1929, average life expectancy of DS children was 9, which increased to 25 years by 1980. Indeed, that has increased to 60 years or beyond in the present day with the discoveries of antibiotics, open heart surgery, and even fetal management of many of the clinical complications pertaining to this syndrome (Antonarakis et al., 2020). In the past, untreated CHD, leukemias, hypothyroidism, and infections contributed to early mortality, whereas the causes of death in the present era are more likely due to complications of AD. Early start of medical intervention with special therapeutic approaches, particularly for neurodegenerative complications, is improving their life-span and quality of life. Since the syndrome can be identified during pregnancy or at birth, expecting parents may educate themselves for handling children with DS; however, many of the parents wish to terminate such pregnancies, if detected at an early gestational age.

4. Attention to be paid for health monitoring of Down syndrome

4.1 At birth

Initial examination of newborns may notice some of the physical signs, which are more commonly found in children with DS. Evaluation of a robust red reflex can help to identify congenital cataracts for neonates born with DS by a pediatrician shortly after birth. In cases with congenital cataract, surgical intervention may be useful. Movement of the eyes can identify strabismus; constipation can be associated with Hirschsprung disease, and feeding problem of any extent should prompt intense understanding to ensure adequate input and nutrition.

At birth, an ultrasound of the heart (echocardiogram) can take care of the CHD, while a complete blood count can identify pre-existing leukemia. A hearing test using brainstem auditory evoked responses (BAERS) testing can detect any hearing deficits and further management. The thyroid function should also be included in a newborn screening program of the metabolic diseases. The requirement of monitoring or intervention of associated eye disease increases with age, from 38% of children at <1 year age to 80% at 5–7 year age of the DS children (Table 2.2) (Bull and the Committee on Genetics 2011; Roizen and Patterson, 2003).

4.2 At childhood and adult age

As the DS children grow, their progress should be plotted on a growth chart in order to detect deviations from the expected growth and should be compared with special growth charts available to monitor development of DS children. Thyroid monitoring should be performed at 6-month interval for initial 2 years and, thereafter, on yearly basis. Evaluation of the ears for infection as well as objective hearing tests should be performed at every visit of health checkup. The possibility of atlanto-axial instability should be checked by obtaining X-ray of the neck at the age of 3–4 years. Screening of obstructive sleep apnea should be performed as the child grows on an yearly basis. Comprehensive guidelines for the care of children with DS are available (American Academy of Pediatrics Committee on Genetics, 2001).

Ophthalmic examination should be performed for formal evaluation of refractive errors and its correction with glasses at least every 2 years with subjective vision assessments. The American Academy of Pediatrics provides the

TABLE 2.2 Suggested health surveillance programs for children born with Down syndrome.

Surveillance modalities	Prenatal	Birth–1 m	1 m–1 y	1–5 y	5–13 y	13–21 y
Counseling regarding prenatal screening test and imaging results	Once					
Plan for delivery	Once					
Referral to geneticist	Once					
Parent-to-parent contact, support groups, current books and pamphlets	Once	Do, if not done before	Once	Do, if not done before		
Physical exam for evidence of trisomy 21		Once				
Chromosomal analysis to confirm dx	Once	Do, if not done before	Do, if not done before	Do, if not done before		
Discuss risk of recurrence of Down syndrome	Once	Do, if not done before	Do, if not done before	Do, if not done before		
Echocardiogram		Once				
Radiographic swallowing assessment if marked hypotonia, slow feeding, choking with feeds, recurrent or persistent respiratory sx, FTT		Once	Do, if not done before			
Eye exam for cataracts		Once	Once	Yearly	Yearly	Yearly
Newborn hearing screen and follow-up		Once				
Hx and PE assessment for duodenal or anorectal atresia		Once				
Reassure parents delayed and irregular dental eruption (hypodontia are common)			Do, if not done before			
If constipation, evaluate for limited diet or fluids, hypotonia, hypothyroidism, GI malformation, Hirschsprung		Repeat at any visit				
CBC to R/O transient myeloproliferative disorder, polycythemia		Once				
Hb annually; CRP and ferritin or CHr if possible risk iron deficiency or Hb < 11 g.				Repeat annually		
Hemoglobin						Repeat annually
TSH (may be part of newborn screening)		Once	Repeat at 6 and 12 m	Repeat annually		

TABLE 2.2 Suggested health surveillance programs for children born with Down syndrome.—cont'd

Surveillance modalities	Prenatal	Birth–1 m	1 m–1 y	1–5 y	5–13 y	13–21 y
Discuss risk of respiratory infection		Once				
If cardiac surgery or hypotonic: Evaluate apnea, bradycardia, or oxygen desaturation in car seat before discharge		Once				
Discuss complementary and alternative therapies		Repeat at all health maint. visits				
Discuss cervical spine positioning, especially for anesthesia or surgical or radiologic procedures		Repeat at all health maint. visits				
Review signs and symptoms of myopathy		Repeat at all health maint. visits				
If myopathic signs or symptoms: Obtain neutral position spine films and, if normal, obtain flexion and extension films and refer to pediatric neurosurgeon or orthopedic surgeon with expertise in evaluating and treating atlantoaxial instability		Repeat at any visit				
Instruct to contact physician for change in gait, change in use of arms or hands, change in bowel or bladder function, neck pain, head tilt, torticollis, or new-onset weakness				Repeat biennially		
Advise risk of some contact sports, trampolines				Repeat at all health maint. visits		
Audiology evaluation at 6 month			Once			
If normal hearing established, behavioral audiogram and tympanometry until bilateral ear specific testing possible. Refer child with abnormal hearing to ot				Repeat every 6 months		
If normal ear-specific hearing established, behavioral audiogram				Repeat annually		
Assess for obstructive sleep apnea sx			Once	Repeat at all health maint. visits		
Sleep study by age 4 years				Once		
Ophthalmology referral to assess for strabismus, cataracts, and nystagmus			Once			
Refer to pediatric ophthalmologist or ophthalmologist with experience of Down syndrome				Repeat annually	Repeat every 2 year	Repeat every 3 year
If congenital heart disease, monitor for signs and sx of congestive heart failure		Repeat at all visits				
Assess the emotional status of parents and intrafamilial relationships		Repeat at all health maint. visits				
Check for sx of celiac disease; if sx present, obtain tissue transglutaminase IgA and quantitative IgA				Repeat at all health maint. visits		
Early intervention: Physical, occupational, and speech therapy				Repeat at all health maint. visits		

(Continued)

TABLE 2.2 Suggested health surveillance programs for children born with Down syndrome.—cont'd

Surveillance modalities	Prenatal	Birth–1 m	1 m–1 y	1–5 y	5–13 y	13–21 y
At 30 months, discuss transition to preschool and development of IEP				Once		
Discuss behavioral and social progress				Repeat at all health maint. visits		
Discuss self-help skills, ADHD, OCD, wandering off, transition to middle school					Repeat at health maint. visits	
If chronic cardiac or pulmonary disease, 23-valent pneumococcal vaccine at age >2 years				Once		
Reassure regarding delayed and irregular dental eruption				Once		
Establish optimal dietary and physical exercise patterns				Repeat at health maint. visits		
Discuss dermatologic issues with parents					Once	
Discuss physical and psychosocial changes through puberty, need for gynecologic care in the pubescent female					Once	Once
Facilitate transition: Guardianship, financial planning, behavioral problems, school placement, vocational training, independence with hygiene and self-care, group homes, work settings						Repeat at health maint. visits
Discuss sexual development and behaviors, contraception, sexually transmitted diseases, recurrence risk for offspring						Repeat at health maint. visits

ADHD, attention-deficit/hyperactivity disorder; CBC, complete blood count; CHr, reticulocyte hemoglobin; dx, diagnosis; FTT, failure to thrive; GI, gastrointestinal; Hb, hemoglobin; Hx, history; IEP, Individualized Education Plan; IgA, immunoglobulin A; Maint. maintenance; OCD, obsessive compulsive disorder; ot, occupational therapy; PE, physician examination; R/O, rule out; sx, symptoms.
Adapted from: Bull and the Committee on Genetics (2011).

following recommendations for visits to a pediatric ophthalmologist for the DS children:

1. Evaluation of strabismus, cataracts, and nystagmus along with visual acuity testing within the first 6 months of life;
2. Annual follow-up until age 5 years with evaluation for refractive error, strabismus, or other conditions, which could result in amblyopia; and
3. Biannual follow up from 5 to 13 years age to evaluate the onset of new ophthalmic disorders.

Refractive error is treated with prescription glasses. Use of bifocals due to reduced accommodative ability may be important for optimal near vision to promote learning potential. Treatment of visual acuity testing can be measured in preverbal DS children using Teller, Keeler, and Cardiff acuity cards with 100% success rates. Optic nerve edema, papilledema, or idiopathic intracranial hypertension should be suspected in these children with elevated optic disks, which may represent transient benign findings. These children should be monitored for signs of primary congenital glaucoma, and if present, medical and surgical interventions are required.

All individuals with DS have some degree of cognitive impairment, and most of them have moderate to severe degree of mental retardation. Therefore, an early intervention from childhood should occur, ideally shortly after birth, to help coordinate and plan effective strategies for learning and developmental progress (Table 2.2). Early intervention may include physiotherapy, speech, and occupational therapy that help in each of the four main areas of development such as gross motor and fine motor skills, language, social development, and self-help skills. Early intervention programs should be individualized to help and monitor each child in every area of development. Vocational education and training can help them to learn wide spectrum of activities and develop skill in some of the areas, which can make them able for earning their living at adult age and when necessary (Bull and the Committee on Genetics, 2011). Some of them have achieved professional achievements such as Jimmie Brewer, a DS model in the United States, and others (discussion on socio-cultural aspects of DS is included in Chapter 5). Although many of the physical limitations cannot be overcome due to defect in their genetic makeup, a conducive family environment, proper care, and early start of education at special school (meant for such children) will improve the quality of their life.

5. Fertility of Down syndrome

The implication of the trisomic HSA21 on the life and health of the individuals with DS is many-fold. The concern of reproductive ability and outcome is subject to data available in this regard. Given the situation of meiotic gametogenesis, production of normal gamete with 46, XX/XY genome is expected in 50:50 ratio in DS individuals having free trisomy. Birth of chromosomally normal children has been demonstrated in ~60% of the pregnancies of a female DS; however, 25% of these had some form of DS-like phenotype (Sheridan et al., 1989). Nevertheless, such reports are not only scanty but also noninformative about the growth, development, and life of such children born to DS parent. Taking the situation of complex genotype–phenotype expression in DS individuals and contribution of the disomic genes, their meiotic recombination is most likely to yield a recombinant gamete against the backdrop of genomic aneuploidy, which is expected to be different from the normal gamete formed in a normal parent. Moreover, regardless of the risk of trisomy 21, meiotic nondisjunction of other autosome or sex chromosome cannot be ruled out. Co-occurrence of XXY and trisomy 21 has been witnessed by the authors.

Majority of the reports question about the reproductive potential of individuals with DS. The females with DS have ovaries producing a lower number of small follicles, which has a higher risk of atresia and can cause sub-fertility (Hojager et al., 1978). On the other hand, males with the syndrome usually appear well masculinized but are reported to be infertile. Cryptorchidism, small penis, and small testicular size have indicated occasional (56%) or no (44%) incidence of spermatogenesis. Pubic and/or facial hair growth was mostly absent in males with DS. Histological evaluation of the testes revealed significantly reduced spermatogenesis or oligospermia with diminished testicular size. Taken together, infertility in the males with DS is most likely related to their sexual impotence and inability to produce sufficient gametes. Interestingly, incidence of male DS fathered apparently normal child has been documented, though the test was not mentioned to exclude the possibility of paternal low-degree mosaicism, which might favor a fruitful reproductive outcome (Hsiang et al., 1987; Sheridan et al., 1989; Pradhan et al., 2006). Interestingly, many of the male DS patients exhibited normal pubertal development with normal penile length, testicular volume, and serum testosterone, LH, and FSH levels within the normal range. However, many of them revealed hormonal deficits, morphological alterations of the gonads, and abnormal spermatogenesis, but the definite cause of the insufficient and inadequate spermatogenesis remains to be deciphered (Stefanidis et al., 2011).

Nevertheless, both males and females with mosaic DS can have a better reproductive outcome, albeit such reports are rare. Moreover, it is obvious that the extra chromosome 21 has a direct and/or indirect detrimental effect on the reproductive capacity of the affected male or female patient.

Intellectual disability in individuals with DS might contribute to their less awareness about reproduction, sexual activity and fertility. Myriads of other DS phenotypes such as autoimmune disorders, anxiety, obesity, psychological, and social factors related to the mental retardation, etc. collectively contribute to infertility in males with normal spermatogenesis and reproductive hormones. Given the fact that DS patients have increased life expectancy, it is imperative to counsel the parents about the reproductive ability and possible outcome in their DS children. DS males and females should also be educated about contraceptive measures. Based on the fact that 95% of pregnancies with trisomy 21 end up with spontaneous abortion at first-trimester gestation, it is apparent that DS individuals may experience a higher rate of gonadal and reproductive failure. However, improvement in medical care and systematic provision of specific support since childhood improves their cognitive function and autonomy at the adult stage. Importantly, DS patients and their families are now seeking equal rights as that of normal population. Hence, the presumption of infertility in DS people is an old concept drawn based on old studies. Utilization of assisted reproductive technology may change the scenario of infertility and address the avenue of fertility, without entering into highly controversial ethical considerations related to parenthood for people with DS (Parizot et al., 2019). Moreover, care has to be taken in cases of both parents having DS (even with free 21) because both will produce 50% gamete with chromosomal aneuploidy and following fertilization, a complex outcome will be derived, including tetrasomy of 21 (25%), trisomy 21 (50%) and apparently normal (25%). Mosaicism, translocation, and/or partial trisomy may draw a more complex configuration of the pregnancy outcome of the partners with DS.

References

Abeleira, M.T., et al., 2014. Dimensions of central incisors, canines, and first molars in subjects with Down syndrome measured on cone-beam computed tomographs. Am. J. Orthod. Dentofacial Orthop. 146, 765–775. https://doi.org/10.1016/j.ajodo .2014.08.016.

Agbolade, O., Nazri, A., Yaakob, R., Ghani, A.A., Cheah, Y.K., 2020. Down syndrome face recognition: a review. Symmetry 12, 1182. https://doi.org/10.3390/sym12071182.

Alió, J.J., Lorenzo, J., Iglesias, C., 2008. Cranial base growth in patients with Down syndrome: a longitudinal study. Am. J. Orthod. Dentofacial Orthop. 133, 729–737. https://doi.org/10.1016/j.ajodo .2006.03.036.

Alió, J., Lorenzo, J., Iglesias, M.C., Manso, F.J., Ramírez, E.M., 2011. Longitudinal maxillary growth in Down syndrome patients. Angle Orthod. 81, 253–259. https://doi.org/10.2319/04051 0-189.1.

Almeida, K.C.M., Raveli, T.B., Vieira, C.I.V., Santos-Pinto, A.D., Raveli, D.B., 2017. Influence of the cranial base flexion on Class I, II and III malocclusions: a systematic review. Dental. Press. J. Orthod. 22, 56–66. https://doi.org/10.1590/2177-6709.22.5.056-066.oar.

American Academy of Pediatrics Committee on Genetics, 2001. Health supervision for children with Down syndrome. Paediatrics 107 (2), 442–449.

Amiel, J., Sproat-Emison, E., Garcia-Barcelo, M., Lantieri, F., Burzynski, G., Borrego, S., Pelet, A., Arnold, S., Miao, X., Griseri, P., Brooks, A.S., Antinolo, G., de Pontual, L., Clement-Ziza, M., Munnich, A., Kashuk, C., West, K., Wong, K.K., Lyonnet, S., Chakravarti, A., Tam, P.K., Ceccherini, I., Hofstra, R.M., Fernandez, R., Hirschsprung Disease Consortium, January 2008. Hirschsprung disease, associated syndromes and genetics: a review. J. Med. Genet. 45 (1), 1–14.

Antonarakis, S.E., 2004. Comparison of human chromosome 21 conserved nongenic sequences (CNGs) with the mouse and dog genomes shows that their selective constraint is independent of their genic environment. Genome Res. 14 (5), 852–859.

Antonarakis, S.E., Skotko, B.G., Rafii, M.S., Strydom, A., Pape, S.E., Bianchi, D.W., Sherman, S.L., Reeves, R.H., 2020. Down syndrome. Nat. Rev. Dis. Prim. 6 (1), 9. https://doi.org/10.1038/s41572-019-0143-7.

Austeng, M.E., et al., 2013. Otitis media with effusion in children with in Down syndrome. Int. J. Pediatr. Otorhinolaryngol. 77, 1329–1332.

Ballard, C., Mobley, W., Hardy, J., Williams, G., Corbett, A., 2016. Dementia in Down's syndrome. Lancet Neurol. 15, 622–636.

Bayen, E., Possin, K.L., Chen, Y., Cleret de Langavant, L., Yaffe, K., 2018. Prevalence of aging, dementia, and multimorbidity in older adults with Down syndrome. JAMA Neurol. 75, 1399–1406.

Belichenko, P.V., Kleschevnikov, A.M., Salehi, A., Epstein, C.J., Mobley, W.C., 2007. Synaptic and cognitive abnormalities in mouse models of Down syndrome: exploring genotype-phenotype relationships. J. Comp. Neurol. 504, 329–345.

Bergstrom, S., et al., 2016. Trends in congenital heart defects in infants with Down syndrome. Pediatrics 138, e20160123.

Berrocal, T., Lamas, M., Gutiérrez, J., 1999. Congenital anomalies of the small intestine, colon, and rectum. Radiographics Radiol Bras 19, 1219–1236.

Bittles, A.H., Glasson, E.J., 2004. Clinical, social, and ethical implications of changing life expectancy in Down syndrome. Dev. Med. Child Neurol. 46, 282–286.

Brockmeyer, D., 1999. Down syndrome and craniovertebral instability. Topic review and treatment recommendations. Pediatr. Neurosurg. 31, 71–77.

Bull, M.J., Committee on Genetics, 2011. Health supervision for children with Down syndrome. Pediatrics 128, 393–406. https://doi.org/10.1542/peds.2011-1605.

Buonuomo, P.S., Bartuli, A., Mastrogiorgio, G., Vittucci, A., Di Camillo, C., Bianchi, S., et al., 2016. Lipid profiles in a large cohort of Italian children with Down syndrome. Eur. J. Med. Genet. 59, 392–395.

Capone, G., et al., 2018. Co- occurring medical conditions in adults with Down syndrome: a systematic review toward the development of health care guidelines. Am. J. Med. Genet. 176, 116–133.

Chakrabarti, L., Galdzicki, Z., Haydar, T.F., 2007. Defects in embryonic neurogenesis and initial synapse formation in the forebrain of the Ts65Dn mouse model of Down syndrome. J. Neurosci. 27, 11483–11495.

Chan, M., et al., 2017. The burden of respiratory syncytial virus (RSV) associated acute lower respiratory infections in children with Down syndrome: a systematic review and meta- analysis. J. Glob. Health 7, 020413.

Churchill, S.S., Kieckhefer, G.M., Landis, C.A., Ward, T.M., 2012. Sleep measurement and monitoring in children with Down syndrome: a review of the literature, 1960–2010. Sleep Med. Rev. 16, 477–488.

Clark, C.M., Patel, H.H., Kanekar, S.G., Isildak, H., 2017. Enlarged vestibular aqueducts and other inner-ear abnormalities in patients with Down syndrome. J. Laryngol. Otol. 131 (4), 298–302.

Clift, A.K., Coupland, C.A.C., Keogh, R.H., Hemingway, H., Hippisley-Cox, J., 2021. COVID-19 mortality risk in down syndrome: results from a cohort study of 8 million adults. Ann. Intern. Med. 174 (4), 572–576. https://doi.org/10.7326/M20-4986.

Colvin, K.L., Yeager, M.E., 2017. What people with Down syndrome can teach us about cardiopulmonary disease. Eur. Respir. Rev. 26, 143.

De Simone, R., Puig, X.S., Gelisse, P., Crespel, A., Genton, P., 2010. Senile myoclonic epilepsy: delineation of a common condition associated with Alzheimer's disease in Down syndrome. Seizure 19, 383–389.

Dodd, K., et al., 2018. Consensus statement of the international summit on intellectual disability and Dementia related to post- diagnostic support. Aging Ment. Health 22, 1406–1415.

Doriguêtto, P.V.T., et al., 2019. Malocclusion in children and adolescents with Down syndrome: a systematic review and meta-analysis. Int. J. Paediatr. Dent. 29, 524–541. https://doi.org/10.1111/ipd.12491.

Down, J.L.H., 1866. Observations on an ethnic classification of idiots. Lond. Hosp. Rep. 3, 259–262.

Draheim, C.C., McCubbin, J.A., Williams, D.P., 2002. Differences in cardiovascular disease risk between nondiabetic adults with mental retardation with and without Down syndrome. Am. J. Ment. Retard. 107, 201–211.

Dykens, E.M., 2007. Psychiatric and behavioral disorders in persons with Down syndrome. Ment. Retard. Dev. Disabil. Res. Rev. 13, 272–278.

Eady, N., et al., 2018. Impact of cholinesterase inhibitors or memantine on survival in adults with Down syndrome

and dementia: clinical cohort study. Br. J. Psychiatry 212, 155–160.
Enlow, D.H., 1990. Facial Growth, third ed. WB Saunders Company, Philadelphia.
Epstein, C.J., Korenberg, J.R., Anneren, G., Antonarakis, S.E., et al., 1991. Protocols to establish genotype-phenotype correlations in Down syndrome. Am. J. Hum. Genet. 49, 207–235.
Esbensen, A.J., Hoffman, E.K., Stansberry, E., Shaffer, R., 2018. Convergent validity of actigraphy with polysomnography and parent reports when measuring sleep in children with Down syndrome. J. Intellect. Disabil. Res. 62, 281–291.
Ferlazzo, E., Adjien, C.K., Guerrini, R., Calarese, T., Crespel, A., Elia, M., Striano, P., Gelisse, P., Bramanti, P., di Bella, P., Genton, P., 2009. Lennox-Gastaut syndrome with late-onset and prominent reflex seizures in trisomy 21 patients. Epilepsia 50 (6), 1587–1595.
Fernandez, F., Morishita, W., Zuniga, E., Nguyen, J., Blank, M., Malenka, R.C., Garner, C.C., 2007. Pharmacotherapy for cognitive impairment in a mouse model of Down syndrome. Nat. Neurosci. 10, 411–413.
Fildler, D., Most, D., Philofsky, A., 2009. The Down syndrome behavioral phenotype: taking a developmental approach. Down Syndrome Res. Pract. 12 (3), 11. https://eric.ed.gov/?id=EJ850564. (Accessed 21 June 2021).
Firth, N.C., et al., 2018. Aging related cognitive changes associated with Alzheimer's disease in Down syndrome. Ann. Clin. Transl. Neurol. 5, 741–751.
Gamis, A.S., Smith, F.O., 2012. Transient myeloproliferative disorder in children with Down syndrome: clarity to this enigmatic disorder. Br. J. Haematol. 159, 277–287.
Ganguly, B.B., Kadam, N.N., October 21, 2016. Down syndrome: from the age of characterization to the era of curative approach. Nucleus. https://doi.org/10.1007/s13237-016-0187-y. Published online.
Ganguly, B.B., Kadam, N.N., Mondal, P.K., 2017. Complexity of chromosomal rearrangements in Down syndrome leukemia. J. Cancer Res. Therapeut. 13, 381–383.
Gholipour, T., Mitchell, S., Sarkis, R.A., Chemali, Z., 2017. The clinical and neurobehavioral course of Down syndrome and dementia with or without new- onset epilepsy. Epilepsy Behav. 68, 11–16.
Giménez- Barcons, M., et al., 2014. Autoimmune predisposition in Down syndrome may result from a partial central tolerance failure due to insufficient intrathymic expression of AIRE and peripheral antigens. J. Immunol. 193, 3872–3879.
Grut, V., Söderström, L., Naumburg, E., 2017. National cohort study showed that infants with Down's syndrome faced a high risk of hospitalisation for the respiratory syncytial virus. Acta Paediatr. 106, 1519–1524.
Hammerle, B., Elizalde, C., Tejedor, F.J., 2008. The spatio-temporal and subcellular expression of the candidate Down syndrome gene Mnb/Dyrk1A in the developing mouse brain suggests distinct sequential roles in neuronal development. Eur. J. Neurosci. 27, 1061–1074.
Hasle, H., Clemmensen, I.H., Mikkelsen, M., 2000. Risks of leukaemia and solid tumours in individuals with Down's syndrome. Lancet 355, 165–169.
Hattori, M., et al., 2000. The DNA sequence of human chromosome 21. Nature 405, 311–319.
Hawli, Y., Nasrallah, M., El-Hajj Fuleihan, G., June 2009. Endocrine and musculoskeletal abnormalities in patients with Down syndrome. Nat. Rev. Endocrinol. 5 (6), 327–334.
Head, E., Lott, I.T., Wilcock, D.M., Lemere, C.A., 2016. Aging in Down syndrome and the development of Alzheimer's disease neuropathology. Curr. Alzheimer Res. 13, 18–29.
Hill, C.M., et al., 2018. Home oximetry to screen for obstructive sleep apnoea in Down syndrome. Arch. Dis. Child. 103, 962–967.
Hithersay, R., et al., 2019. Association of dementia with mortality among adults with Down syndrome older than 35 years. JAMA Neurol. 76, 152–160.
Hitzler, J.K., Zipursky, A., 2005. Origins of leukaemia in children with Down syndrome. Nat. Rev. Cancer 5, 11–20.
Hoffmann, M., Kleine-Weber, H., Schroeder, S., Krüger, N., Herrler, T., Erichsen, S., et al., 2020. SARS-CoV-2 cell entry depends on ACE2 and TMPRSS2 and is blocked by a clinically proven protease inhibitor. Cell 181 (2), 271–280 e8.
Hojager, B., Peters, H., Byskov, A.G., Faber, M., 1978. Follicular development in ovaries of children with Down's syndrome. Acta Paediatr. Scand. 67, 637–643.
Holmes, G., 2014 Winter. Gastrointestinal disorders in Down syndrome. Gastroenterol. Hepatol. Bed. Bench 7 (1), 6–8.
Hsiang, Y.H., Berkovitz, G.D., Bland, G.L., Migeon, C.J., Warren, A.C., 1987. Gonadal function in patients with Down syndrome. Am. J. Med. Genet. 27, 449–458.
Hüls, A., Costa, A.C.S., Dierssen, M., et al., 2021. Medical vulnerability of individuals with Down syndrome top severe COVID-19-data from the trisomy 21 Research Society and the UK ISARIC4C survey. EClin. Med. 33, 100769. https://doi.org/10.1016/j-eclim.2021.100769.
Illouz, T., Biragyn, A., Frenkel-Morgenstern, M., et al., 2021. Specific susceptibility to COVID-19 in adults with Down syndrome. NeuroMolecular Med. 23 (4), 561–571. https://doi.org/10.1007/s12017-021-08651-5.
Intrapiromkul, J., Aygun, N., Tunkel, D.E., Carone, M., Yousem, D.M., 2012. Inner ear anomalies seen on CT images in people with Down syndrome. Pediatr. Radiol. 42 (12), 1449–1455.
Iughetti, L., Lucaccioni, L., Fugetto, F., Mason, A., Predieri, B., 2015. Thyroid function in Down syndrome. Expet Rev. Endocrinol. Metabol. 10, 525–532.

Kaouini, A.E., Rhalete, A.E., Aabdi, M., et al., 2021. COVID-19 pneumonia in Down syndrome patients: about 2 cases. Ann. Med. & Surg. 65, 102324.

Kazemi, M., Salehi, M., Kheirollahi, M., 2016. Down syndrome: current status, challenges and future perspectives. Int. J. Mol. Cell. Med. 5, 125–133.

Kearney, L., Gonzalez De Castro, D., Yeung, J., Procter, J., Horsley, S.W., Eguchi-Ishimae, M., Bateman, C.M., Anderson, K., Chaplin, T., Young, B.D., Harrison, C.J., Kempski, H., So, C.W., Ford, A.M., Greaves, M., January 15, 2009. Specific JAK2 mutation (JAK2R683) and multiple gene deletions in Down syndrome acute lymphoblastic leukemia. Blood 113 (3), 646–648.

Klingel, D., Hohoff, A., Kwiecien, R., Wiechmann, D., Stamm, T., 2017. Growth of the hard palate in infants with Down syndrome compared with healthy infants—a retrospective case control study. PLoS One 12, e0182728. https://doi.org/10.1371/journ al.pone.0182728.eCollection.

Korbel, J.O., Tirosh-Wagner, T., Urban, A.E., Chen, X.N., Kasowski, M., Dai, L., Grubert, F., Erdman, C., Gao, M.C., Lange, K., et al., 2009. The genetic architecture of Down syndrome phenotypes revealed by high-resolution analysis of human segmental trisomies. Proc. Natl. Acad. Sci. U.S.A. 106, 12031–12036.

Lana-Elola, E., Watson-Scales, S.D., Fisher, E.M.C., Tybulewicz, V.L.J., 2011. Down syndrome: searching for the genetic culprits. Dis. Models & Mechan. 4, 586–595. https://doi.org/10.1242/dmm.008078.

LeJeune, J., Gautier, M., Turpin, R., 1959. Study of somatic chromosomes from 9 mongoloid children [French]. C. R. Hebd. Seances Acad. Sci. 248, 1721–1722.

Lott, I.T., Dierssen, M., 2010. Cognitive deficits and associated neurological complications in individuals with Down's syndrome. Lancet Neurol. 9, 623–633.

Lott, I.T., Head, E., 2019. Dementia in Down syndrome: unique insights for Alzheimer disease research. Nat. Rev. Neurol. 15, 135–147.

Lyle, R., Bena, F., Gagos, S., Gehrig, C., Lopez, G., Schinzel, A., Lespinasse, J., Bottani, A., Dahoun, S., Taine, L., et al., 2009. Genotype-phenotype correlations in Down syndrome identified by array CGH in 30 cases of partial trisomy and partial monosomy chromosome 21. Eur. J. Hum. Genet. 17, 454–466.

Makita, N., Manaka, K., Sato, J., Mitani, K., Nangaku, M., Iiri, T., 2019. Bexarotene-induced hypothyroidism: characteristics and therapeutic strategies. Clin. Endocrinol. 91, 195–200.

Mateos, M.K., Barbaric, D., Byatt, S.-A., Sutton, R., Marshall, G.M., 2015. Down syndrome and leukemia: insights into leukemogenesis and translational targets. Transl. Pediatr. 4, 76–92.

Mégarbané, A., Ravel, A., Mircher, C., Sturtz, F., Grattau, Y., et al., 2009. The 50th anniversary of the discovery of trisomy 21: the past, present and future of research and treatment of Down syndrome. Genet. Med. 11 (9), 611–616.

Merrick, J., Koslowe, K., July 2001. Refractive errors and visual anomalies in Down syndrome. Downs. Syndr. Res. Pract. 6 (3), 131–133.

Mircher, C., et al., 2017. Acute regression in young people with Down syndrome. Brain Sci. 7, 57.

Morales- Demori, R., 2017. Congenital heart disease and cardiac procedural outcomes in patients with trisomy 21 and Turner syndrome. Congenit. Heart Dis. 12, 820–827.

Nevill, R.E., Benson, B.A., 2018. Risk factors for challenging behaviour and psychopathology in adults with Down syndrome. J. Intellect. Disabil. Res. 62, 941–951.

Olson, L.E., Roper, R.J., Sengstaken, C.L., Peterson, E.A., Aquino, V., Galdzicki, Z., Siarey, R., Pletnikov, M., Moran, T.H., Reeves, R.H., 2007. Trisomy for the Down syndrome 'critical region' is necessary but not sufficient for brain phenotypes of trisomic mice. Hum. Mol. Genet. 16, 774–782.

Ordóiez-Munoz, F.J., Rosety-Rodríguez, M., Rosety-Rodríguez, J.M., Rosety-Plaza, M., 2005. Anthropometrical measurements as predictor of serum lipid profile in adolescents with Down syndrome. Rev. Invest. Clin. 57, 691–694.

Palaska, P.K., Antonarakis, G.S., 2016. Prevalence and patterns of permanent tooth agenesis in individuals with Down syndrome: a meta-analysis. Eur. J. Oral Sci. 124, 317–328. https://doi.org/10.1111/eos.12282.

Papavassiliou, P., Charalsawadi, C., Rafferty, K., Jackson-Cook, C., 2015. Mosaicism for trisomy 21: a review. Am. J. Med. Genet. 167A, 26–39.

Parizot, E., Dard, R., Janel, N., Vialard, F., 2019. Down syndrome and infertility: what support should we provide? J. Assist. Reprod. Genet. 36, 1063–1067.

Perez-Padilla, R., Fernandez, R., Garcia-Sancho, C., et al., 2010. Pandemic (H1N1) 2009 virus and Down syndrome patients. Emerg. Infect. Dis. 16, 1312–1314.

Pradhan, M., Dalal, A., Khan, F., Agrawal, S., 2006. Fertility in men with Down syndrome: a case report. Fertil. Steril. 86, 1765–e1–3.

Purdy, I.B., Singh, N., Brown, W.L., Vangala, S., Devaskar, U.P., 2014. Revisiting early hypothyroidism screening in infants with Down syndrome. J. Perinatol. 34, 936–940.

Radhakrishnan, R., Towbin, A.J., 2014. Imaging findings in Down syndrome. Pediatr. Radiol. 44 (5), 506–521.

Reynolds, L.E., Watson, A.R., Baker, M., Jones, T.A., D'Amico, G., Robinson, S.D., et al., 2010. Tumor angiogenesis is reduced in the Tc1mouse model of Down's syndrome. Nature 465, 813–817.

Richtsmeier, J.T., Zumwalt, A., Carlson, E.J., Epstein, C.J., Reeves, R.H., 2002. Craniofacial phenotypes in segmentally trisomic mouse models for Down syndrome. Am. J. Med. Genet. 107, 317–324.

Rodrigues, M., Joana Nunes, J., Figueiredo, S., Martins de Campos, A., Geraldo, A.F., 2019. Neuroimaging assessment in Down syndrome: a pictorial review. Insights Imag. 10, 52. https://doi.org/10.1186/s13244-019-0729-3.

Roizen, N.J., Patterson, D., 2003. Down's syndrome. Lancet 361, 1281–1289.

Russell, M.W., Chung, W.K., Kaltman, J.R., Miller, T.A., 2018. Advances in the understanding of the genetic determinants of congenital heart disease and their impact on clinical outcomes. J. Am. Heart Assoc. 7, e006906.

Sanmaneechai, O., et al., 2013. Treatment outcomes of West syndrome in infants with Down syndrome. Pediatr. Neurol. 48, 42–47.

Schuler, B., Habermann, C., Plosa, E.J., Taylor, C.J., Jetter, C., Negretti, N.M., et al., 2021. Age-determined expression of priming protease TMPRSS2 and localization of SARS-CoV-2 in lung epithelium. J. Clin. Invest. 131 (1), e140766. https://doi.org/10.1172/JCI140766.

Sethupathy, P., Borel, C., Gagnebin, M., Grant, G.R., Deutsch, S., Elton, T.S., et al., 2007. Human microRNA-155 on chromosome 21 differentially interacts with its polymorphic target in the AGTR1 3′ untranslated region: a mechanism for functional single-nucleotide polymorphisms related to phenotypes. Am. J. Hum. Genet. 81, 405–413.

Sheridan, R., Llerena Jr., J., Matkins, S., Debenham, P., Cawood, A., Bobrow, M., 1989. Fertility in a male with trisomy 21. J. Med. Genet. 26, 294–298.

Silva, J., Valladares-Neto, J., 2013. Craniofacial morphological differences between Down syndrome and maxillary deficiency children. Eur. J. Orthod. 35, 124–130. https://doi.org/10.1093/ejo/cjr105.

Startin, C.M., et al., 2019. Cognitive markers of preclinical and prodromal Alzheimer's disease in Down syndrome. Alzheimers Dement 15, 245–257.

Stefanidis, K., Belitsos, P., Fotinos, A., Makris, N., Loutradis, D., Antsaklis, A., 2011. Causes of infertility in men with Down syndrome. Andrologia 43 (5), 353–357. https://doi.org/10.1111/j.1439-0272.2010.01043.x.

Strydom, A., et al., 2010. Dementia in older adults with intellectual disabilities—epidemiology, presentation, and diagnosis. J. Pol. Pract. Intellect. Disabil. 7, 96–110.

Sungnak, W., Huang, N., Bécavin, C., Berg, M., Queen, R., Litvinukova, M., et al., 2020. SARS-CoV-2 entry factors are highly expressed in nasal epithelial cells together with innate immune genes. Nat. Med. 26 (5), 681–687.

Swigonski, N.L., Kuhlenschmidt, H.L., Bull, M.J., Corkins, M.R., Downs, S.M., 2006. Screening for celiac disease in asymptomatic children with Down syndrome: cost- effectiveness of preventing lymphoma. Pediatrics 118, 594–602.

Tassé, M.J., et al., 2016. Psychiatric conditions prevalent among adults with Down syndrome. J. Pol. Pract. Intellect. Disabil. 13, 173–180.

Tunstall, O., et al., 2018. Guidelines for the investigation and management of transient leukaemia of Down syndrome. Br. J. Haematol. 182, 200–211.

Tuttle, K.D., Minter, R., Waugh, K.A., et al., 2020. JAK1 inhibition blocks lethal sterile immune responses: implications for COVID-19 therapy. bioRxiv. https://doi.org/10.1101/2020.04.07.024455.

Van Cleve, S.N., Cohen, W.I., 2006. Part I: clinical practice guidelines for children with Down syndrome from birth to 12 years. J. Pediatr. Health Care Off. Publ. Natl. Assoc. Pediatr. Nurse Assoc. Pract 20, 47–54. https://doi.org/10.1016/j. pedhc.2005.10.004.

Van Cleve, S.N., Cannon, S., Cohen, W.I., 2006. Part II: clinical practice guidelines for adolescents and young adults with Down syndrome: 12–21 years. J. Pediatr. Health Care 20 (3), 198–205. https://doi.org/10.1016/j.pedhc.2006.02.006.

Venekamp, R.P., et al., 2015. Tonsillectomy or adenotonsillectomy versus non- surgical management for obstructive sleep- disordered breathing in children. Cochrane Database Syst. Rev. 10, CD011165.

Vicente, A., Bravo-González, L.-A., López-Romero, A., Muñoz, C.S., Sánchez-Meca, J., 2020. Craniofacial morphology in down syndrome: a systematic review and meta-analysis. Sci. Rep. 10, 19895. https://doi.org/10.1038/s41598-020-76984-5.

Vidal, V., Puente, A., García-Cerro, S., Unzueta, M.T.G., Rueda, N., Riancho, J., Martínez-Cué, C., 2021. Bexarotene impairs cognition and produces hypothyroidism in a mouse model of down syndrome and Alzheimer's disease. Front. Pharmacol. 12, 613211. https://doi.org/10.3389/fphar.2021.613211.

Vita, S., Di Bari, V., Corpolongo, A., et al., 2021. Down syndrome patients with COVID-19 pneumonia: a high-risk category for unfavorable outcome. Int. J. Infect. Dis. 103, 607–610.

Wallace, R.A., March 2007. Clinical audit of gastrointestinal conditions occurring among adults with Down syndrome attending a specialist clinic. J. Intellect. Dev. Disabil. 32 (1), 45–50.

Whooten, R., Schmitt, J., Schwartz, A., 2018. Endocrine manifestations of Down syndrome. Curr. Opin. Endocrinol. Diabetes Obes. 25, 61–66.

Wiseman, F.K., et al., 2015. A genetic cause of Alzheimer disease: mechanistic insights from Down syndrome. Nat. Rev. Neurosci. 16, 564–574.

Yang, A., Reeves, R.H., 2011. Increased survival following tumorigenesis in Ts65Dn mice that model Down syndrome. Cancer Res. 71, 3573–3581.

CHAPTER

3

Chromosomal etiology of Down syndrome: genesis of trisomy 21

1. Introduction

Trisomies were categorized as complete or whole chromosome trisomy (also known as free trisomy), partial trisomy, microtrisomies, and trisomy of single genes or single functional genomic elements occuring through triplication of full set of genes or segmental triplication (Antonarakis et al., 2004). Freely segregating full trisomies arise through meiotic or mitotic nondisjunction mechanism and appear in ~0.3%–0.5% live births with Down syndrome (DS) caused by trisomy of chromosome 21. DS accounts for the most prevalent autosomal disorder with an epidemiological incidence of 1 in 700 live births and the first clinical syndrome characterized by chromosomal abnormality (trisomy 21, also represented as T21 or HSA21) (Parker et al., 2010). Pregnancies with trisomic HSA21 contribute to a significant proportion of spontaneous abortions, which is estimated at 1 in 43 of such abortion (Hsu, 1998). The history of understanding chromosomal consequences of origin of trisomy 21 in DS has been described in the following table (Table 3.1): As the time went by, many other studies were conducted on DS; however, Lejeune et al. was the first to link the features of DS with trisomy 21 (Lejeune et al., 1959).

DS is really not an inherited disorder, but more of a genetic abnormality or of the chromosome makeup of the cells. A typical human karyotype is designated as 46,XX or 46,XY, representing 46 chromosomes with 22 homologous pairs of autosomes and one pair of sex chromosomes with an XX homologous arrangement for females and XY arrangement for males. The karyotypic anomaly of DS is characterized by the presence of an extra copy of the 21st chromosome, either in whole (free trisomy 21 or +21; 95% cases) in 100% cells of a body (constitutive) or mixed with normal diploid cells in mosaicism (2%–3%) or in part due to translocation of 21 with other acrocentric chromosomes of D-/G-group (2%–3%). Full trisomies of HSA21 are strongly associated with increasing maternal age, while ~9% of full trisomy is caused by paternal meiotic error. So the disorder is rarely passed down from parents, but there are myriads of negative factors to cause DS. The extra chromosomal material of HSA21 can render the characteristic phenotypic and clinical features of DS and disrupts the normal course of development in life. Interestingly, the effects of the trisomic HSA21 vary greatly among DS people, depending on the frequency of the extra copy, genetic history, environmental factors, and socio-environmental impact on expression of

Genetics and Neurobiology of Down Syndrome
https://doi.org/10.1016/B978-0-323-90456-8.00004-1

TABLE 3.1 Chronological understanding of the chromosomal anomaly in Down syndrome.

Year	Knowledge gained
1932	Waardenburg and Davenport suspected that Down syndrome might be caused by chromosomal alteration
1950	Penrose conducted a study on testicular sample of Down syndrome; however, could not conclude the matter of aneuploidy
1956	Human karyotyping technique was established
1959	Lejeune et al. and Jacobs et al. confirmed trisomy 21 in Down syndrome
1960	Fraccaro et al. Identified translocation in a child with Down syndrome inherited from the father
1960	Harnden et al. Identified co-existence of Klinefelter syndrome with Down syndrome
1961	Mosaicism: Presence of Trisomy 21 and normal cells in Down syndrome was demonstrated
1961	Down syndrome or trisomy 21 anomaly replaced the term "mongolism"
1989	Characterization of Down syndrome critical region (DSCR) was done
2000	Sequencing of chromosome 21 was done by Hattori et al.

Adapted From: Mégarbané et al. (2009) and information available in literature.

the genetic imbalance. Moreover, genomic dosage imbalance is frequently reported in cancer of the somatic cells, where tetrasomies, trisomies, and monosomies truly contribute to progression of cancer (Ganguly et al., 2016).

2. Chromosome 21

The human chromosome 21 (HSA21 of *Homo sapiens*), the smallest autosomal chromosome, is one of the two acrocentric chromosomes (21 and 22) of the "G"-group of human karyogram and smaller in size compared to other three pairs of acrocentric chromosomes of "D"-group, such as 13, 14, and 15. The acrocentric chromosomes in the human genome do not have a recognizable short ("p")-arm, and satellites appear as p-arm in a condensed form (Fig. 3.1). Satellites are connected to the centromere through the satellite-stalk, known as the Nucleolus Organizer Regions (NOR), which is made up of ribosomal RNA. The p-arm of the HSA21 encodes largely repetitive DNA, and it is likely that it makes little contribution to the phenotypes of DS. The long ("q")-arm of 21 is prominent and contains functional genes. HSA21 was the second human chromosome to be sequenced completely (Hattori et al., 2001), the first being chromosome 22, and since then, it has been extensively annotated for better understanding of the genotype–phenotype consequences in DS (Gupta et al., 2016; Harrow et al., 2014).

The HSA21q contains majority of the genes annotated with the syndrome. Practically, the gene-rich and gene-poor regions of 21q correlate to G + C content. If G + C correlates to <43%, called L-isochore, it indicates poor gene content (1 per 300 Kb), whereas >48% G + C content, called H3 isochore, indicates the gene-rich segment (1 per 58 Kb) and that is equivalent to light bands of Giemsa banding of metaphase chromosomes (reviewed in Antonarakis et al., 2004).

2.1 Occurrence of trisomy 21

Human cells divide in two ways: by mitosis and meiosis. Mitosis is the somatic cell division where 2 cells are formed from 1 cell with the

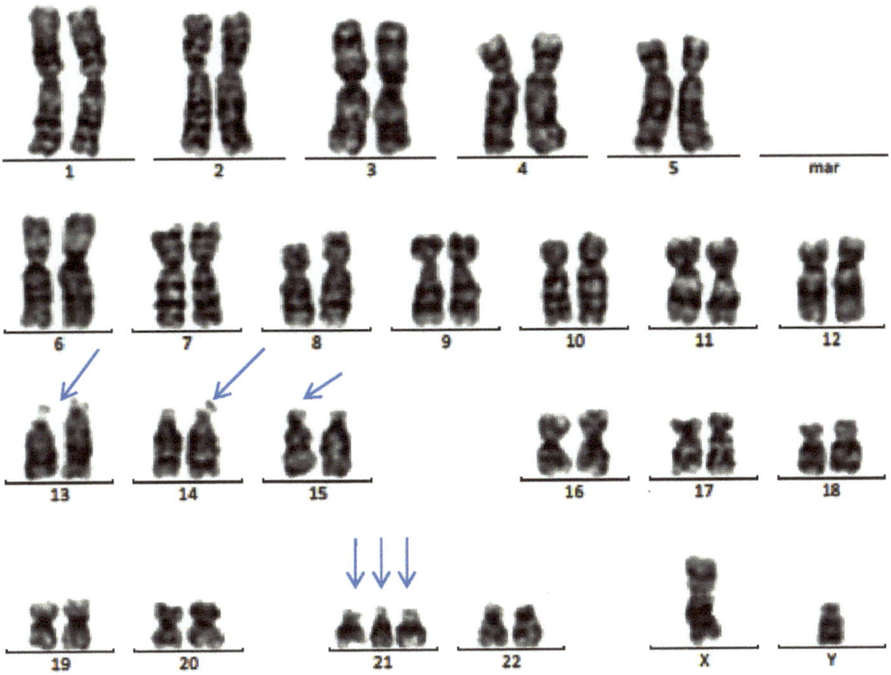

FIGURE 3.1 A human male karyotype with trisomy 21. Acrocentric chromosomes (*arrowed*) show satellites, satellite stalk, and the "q" arm. *Long arrows* indicate D-group (13, 14, and 15) chromosomes, and *short arrows* indicate "G"-group (21 and 22) chromosomes.

exactly same number and type of chromosomes as that of the parent cell. Meiosis takes place in the ovaries and testicles, where 4 cells are produced with 23 or haploid set of chromosomes from one diploid cell. Meiosis involves two phases with two different kinds of chromosomal division. In the first phase, haploid division takes place when 2 cells are formed from a single diploid cell with 50% of chromosomes in each. Each haploid cell of the first phase completes the second phase of meiosis with equal number of chromosomes; thus, a total of four haploid ($n = 23$,X/Y) cells produced from a single diploid cell ($2n = 46$,XX/XY). This process of cell division takes place for male and female gamete formation (male and female gametogenesis), and produce normal egg and sperm cells with 23 chromosomes in each cell instead of 46. Union of one each of egg and sperm cell produces a diploid zygote cell with 46 chromosomes carrying maternal and paternal genome in equal half.

Occurrence of trisomy 21 has been characterized with three genetic mechanisms: the first and most common mechanism is called meiotic nondisjunction, where there is an entire extra chromosome 21 is present in all cells; the second one is 'mosaicism', where aneuploid cells with trisomy 21 are mixed with a second cell line, usually "normal" (46,XX or 46,XY). Mosaic DS individuals have slightly milder form of syndromic features in their appearance; however, that depends on the proportion of normal cells present in their genome. The third mechanism is 'translocation', where part or all of chromosome 21 is translocated to another acrocentric chromosome, frequently with 14, 21, or 22. Translocation could be inherited from one of the parents carrying it or acquired de novo. A de novo mutation leads to 75% of DS population with translocation, and a balanced translocation in a parent leads to 25% of these cases.

In case of meiotic nondisjunction, both 21s are passed down to a single cell, most likely in the

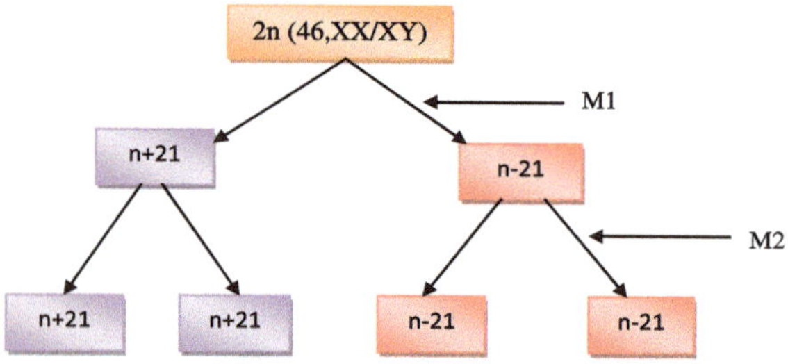

FIGURE 3.2 Schematic diagram of nondisjunction in meiosis phase I (M1) and origin of gametes with 24 chromosomes having one extra chromosome 21. After M1 division, two haploid gametes (cells) are produced and both are affected by nondisjunction of chromosome 21: one with n + 21 and the other one with n − 21. After M2, 4 gametes are produced from the two gametes of M1 following equal division of chromosomes. Thus, 4 haploid gametes are formed from one diploid cell, but unequal number of chromosomes. The cells with n + 21 have received extra 21 or both 21s and the cells with n − 21 are null for chromosome 21 (will die). Fertilization of cell with n + 21 with a normal gamete from one parent with an apparently normal gamete with n (with one copy of 21) will produce an aneuploid zygote (n+21 + n = 47,XX,+21/XY+21). 2n (46) = diploid cell; n(23) = haploid cell; M1 (Meiosis phase 1) = haploid division; M2 (Meiosis phase 2) = equal division.

first phase (M1; maternal: 75%; paternal: 40%) of meiosis (Fig. 3.2), and the other cell will not receive any 21st chromosome and die. Thus, two viable gametes produced at the end of gametogenesis will have 24 chromosomes, instead of 4 gametes with 23 chromosomes, and following fertilization with the other normal gamete having 23 chromosomes, will form an aneuploid cell with 47 chromosomes with an extra 21 (Hassold and Sherman, 2000). The embryo thus has 47 chromosomes, with three copies of chromosome 21 in all the cells of the body (constitutive abnormality). The trisomy 21 karyotype figure shows the chromosomal arrangement, with the prominent extra chromosome 21 (Karyotype of DS) (Fig. 3.3). The nondisjunction may also occur at the second phase (M2; maternal:25%; paternal:60%) of meiotic cell division (Fig. 3.4), which may produce gametes with zero or two 21s resulting in trisomy 21 (free) following fertilization with a normal gamete. The following diagram shows the two possible types of nondisjunction in meiosis.

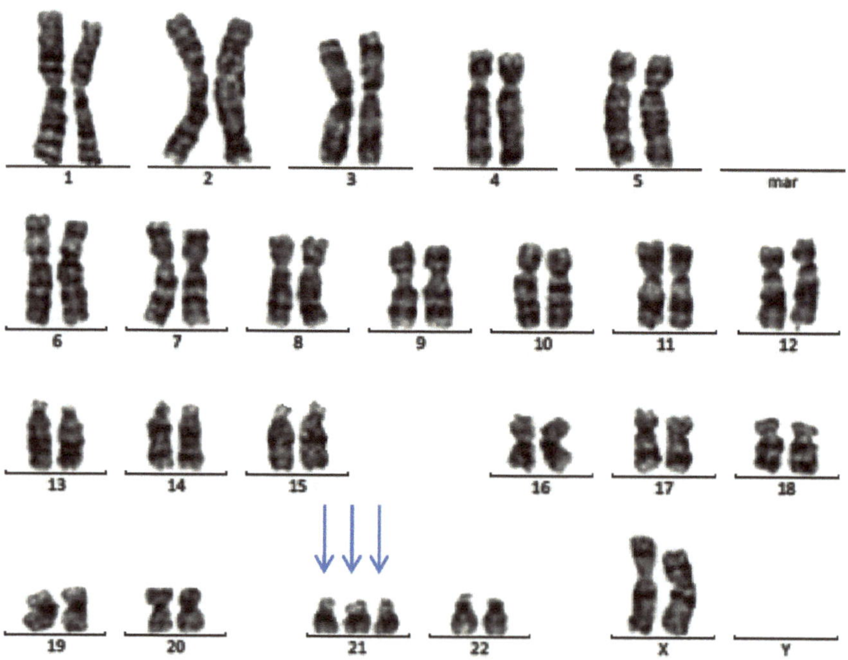

FIGURE 3.3 karyotype of a female with Down syndrome showing trisomy 21.

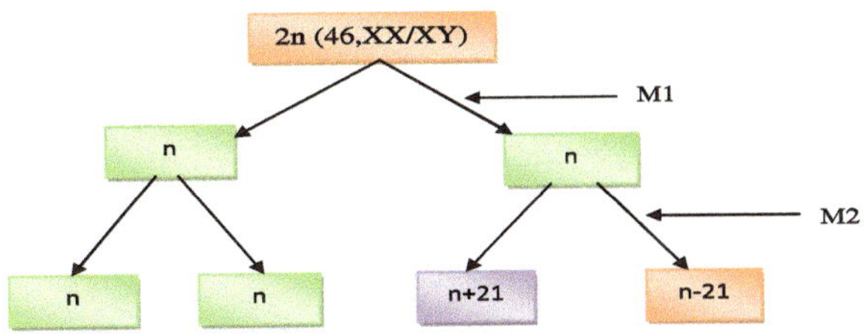

FIGURE 3.4 Schematic diagram of nondisjunction in Meiosis phase II and origin of gametes with 24 chromosomes having one extra chromosome 21. After M1 division, two haploid gametes (cells) are produced with equal half of 46 chromosomes of a diploid cell (2n = 46). After M2, 4 gametes are produced from the two gametes of M1 following equal division of chromosomes in one clone, while the other clone was affected with nondisjunction at M2 and produced 2 cells with n + 21 and n − 21 pattern. Thus, 4 haploid gametes are formed from one diploid cell, but unequal number of chromosomes. The cells with n + 21 have received extra 21 or both 21s and the cells with n − 21 are null for chromosome 21 (lethal). Fertilization of cell with n + 21 with a normal gamete from one parent with an apparently normal gamete with n (with one copy of 21) will produce an aneuploid zygote (n + 21 + n = 47,XX,+21/XY+21). 2n (46) = diploid cell; n (23) = haploid cell; M1 (Meiosis phase 1) = haploid division; M2 (Meiosis phase 2) = equal division.

Nondisjunction may occur in the first meiotic phase (M1) or the second meiotic phase (M2). Trisomy 21 is the cause of approximately 95% of observed DSs, with 80%–90% coming from nondisjunction during maternal gametogenesis and 10%–20% from paternal nondisjunction. Among the DS cases of maternal origin, approximately, 75% result from M1 nondisjunction and 25% from M2, whereas 40% of paternal nondisjunction occurs in M1 and 60% in M2 (Antonarakis, 1998; Antonarakis et al., 2020; Yoon et al., 1996). There is no maternal age difference between maternal M1 and M2 nondisjunction (Yoon et al., 1996; Sherman et al., 2007). The effect of paternal age on paternal nondisjunction is contradictory (Savage et al., 1998). The study explained that a higher proportion of males than females with DS occurred from paternal M2 nondisjunction.

2.2 Classical Down syndrome

The classical DS has been defined as the cases where each cell present in an individual body has 47 chromosomes with three copies of chromosome 21 or free trisomy 21. Thus, it is apparent that the body in the form of single cell zygotic stage received 47 chromosomes following fertilization of the two gametes—one carrying two copies of 21 and the whole body, including all organs, develops from this single cell. Thus, all cells present in a body of a DS individual will have 47 chromosomes, including the extra 21 in free form. Their karyotypes are represented as 47,XX,+21 or 47,XY,+21 for females and males, respectively (Fig. 3.5), and the genetic condition is called "classical." This situation is also called a constitutive abnormality, indicating the abnormality present in the genomic architecture of the individual having trisomic HSA21. Classical DS constitutes 95% of all births with DS.

2.3 Variant Down syndrome

DS could also have the extra 21 in various other combinations and configurations. Approximately, 5% of all DSs have +21 in a different form such as mosaicism in combination with normal cells, and 21 translocated with other accrocentric chromosomes of D (13, 14, and 15) or G-group (21 and 22).

2.3.1 Mosaic Down syndrome

A small percentage of the DS have an extra copy of chromosome 21 only in some of the cells in the body where the other cells in the body are normal; the proportion of normal and abnormal aneuploid cells with trisomy 21 may vary but all developed from the single zygote (Fig. 3.6). In these individuals, the condition is called mosaicism or chimerism, and mosaicism with trisomy 21 is called mosaic DS such as presence of 46,XX/47,XX,+21 cells lines in the same individual (Papavassiliou et al. 2014). Mosaicism is produced through nondisjunction or malsegregation of chromosomes, which occurs most likely at postfertilization mitosis. In general, mitotic nondisjunction after fertilization or conception during embryonic growth is believed to be the reason of mosaicism. This can happen in two possible ways: the mitotic nondisjunction occurs during an early cell division in a normal embryo resulting in a fraction of the cells with

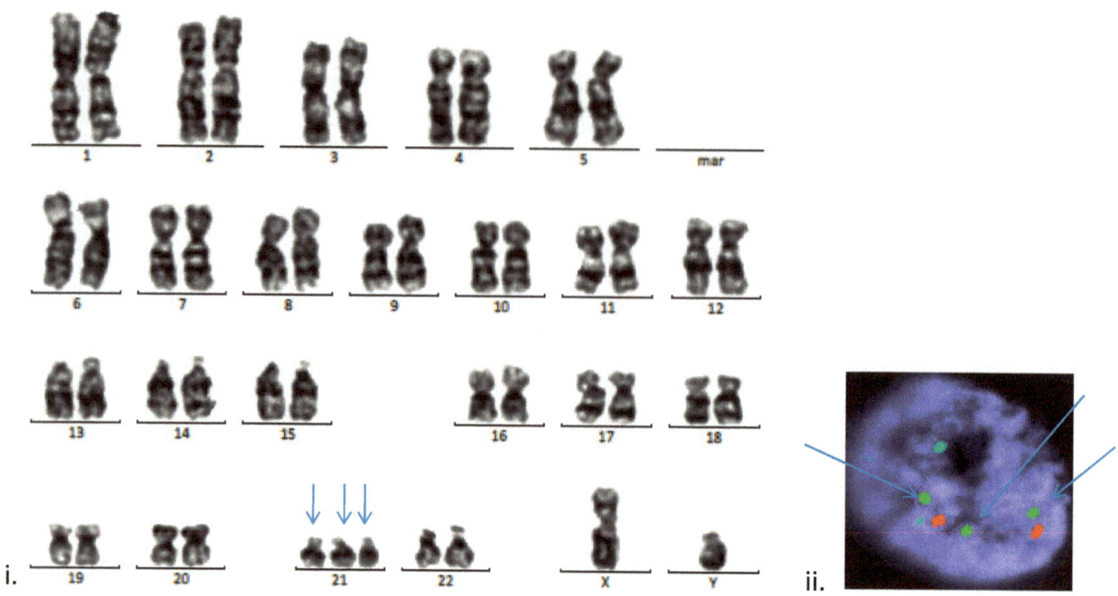

FIGURE 3.5 Down syndrome the three copies of chromosome 21. i. Conventional karyogram; ii. fluorescence in situ hybridization (FISH) image.

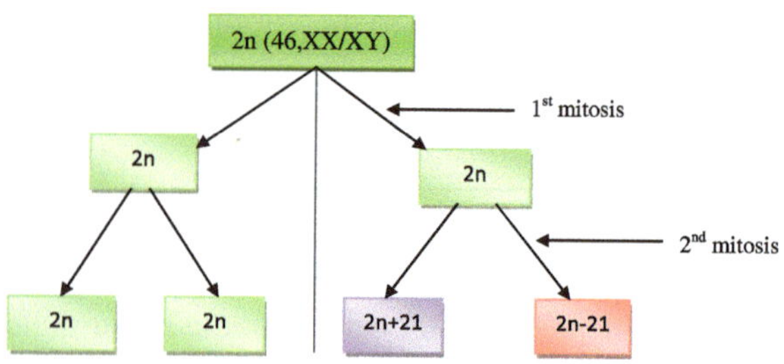

FIGURE 3.6 Schematic diagram of mitotic nondisjunction in a zygote cell showing origin of mosaic cell lines and mosaic Down syndrome. After first mitosis, two cells were developed from a single zygote cell with equal number of chromosomes or genetic material. After second mitosis, one cell followed normal division and maintained a normal diploid clone with 46 chromosomes, while the other cell developed two abnormal cells: one with 2n+21 (viable) and the other one with 2n−21 (lethal) due to nondisjunction. Thus, there will be two distinct clones: one normal with 46 chromosomes and one aneuploid with 47,+21 chromosomes. These individuals are called "mosaic Down syndrome." The severity of syndromic features will vary and depend on the size of the two clones in this Down syndrome population.

trisomy 21 and a fraction with normal karyotype; or nondisjunction occurs in a DS embryo where some of the cells revert to the normal chromosomal arrangement in the embryo due to nondisjunction, and the other cells of this clone carrying four copies of 21 (3 of its own and fourth one from the other cells reverted back to normal with two 21s) might be lethal and lost. There is considerable variability in the fraction of cells having trisomy 21 in mosaic DS, both as a whole and among tissues. It is believed that the degree of disabilities and prominence of DS features will have milder effects than constitutive DS, which varies depending on the frequency of abnormal cells carrying trisomy 21 and the size of clone having normal genetic makeup. Parental mosaicism may also cause mosaic DS.

The frequency of mosaic DS appeared to be 1 in 16,670 births and 1.3%–5% of all DS (Jackson-Cook, 2011). Mosaicism can be present in certain tissues, whereas other tissues may have normal cell lines. Low-degree mosaicism may remain undetected clinically or prenatally. Therefore, conventional blood karyotyping may present a normal karyotype, whereas skin fibroblasts can have trisomy 21. Hence, diagnosis of mosaic DS can be missed in cases with smaller size of a trisomy 21 clone. In such context, based on the physical features, other tissues (skin/bone marrow) are recommended for testing or confirmation of DS (Papavassiliou et al., 2009). It is noteworthy that testing skin cells reflect the karyotypic status of the brain cells because skin and brain cells arise from the ectoderm at the beginning of fetal development. Low-degree mosaicism can be confirmed using special DNA markers following FISH, microarray-based comparative genomic hybridization (CGH), or next generation sequencing technologies (Biesecker and Spinner, 2013). Mosaic DS children are equally prone to transient myeloproliferative disorder (TMPD) and subsequent transformation to overt acute megakaryocytic leukemia (AMKL) (Hu et al., 2005).

2.3.2 Trisomy 21 with Robertsonian translocation

Robertsonian translocations (RTs) represent the most common structural alteration between acrocentric chromosomes of human, with an incidence estimated at 1 in 1000 in the general population, who do not have specific symptoms or syndromic features. Such chromosomal exchanges may occur at the centromere region, in the proximal short-arm satellite DNA, or within the nucleolar-organizing region (NOR), resulting in monocentric or dicentric chromosomes, depending on the site of the recombination event, and the configuration of the two acrocentric chromosomes changes to metacentric (D;D or G;G) or sub-metacentric chromosomes (D;G). This kind of fusion of two acrocentric chromosomes was first described by William Rees Brebner Robertson in grasshoppers in 1916, and thus, such rearrangements are named after him. However, such chimeric aberrations are also called as whole-arm translocations or centric-fusion translocations or unbalanced translocations. In human, RTs were first described by Hecht and coworkers for clinical management. DS also occurs when part of chromosome 21 is attached (translocated) to another acrocentric chromosome of D (13, 14, and 15) or G-group (21 and 22) during the formation of reproductive cells or egg and sperm cells, or in very early stages of embryonic development, resulting in loss of the centromere and satellite of one acrocentric chromosome (Antonarakis et al., 2004). DS people affected with RT have two free copies of chromosome 21, plus extra chromosome 21 attached to another chromosome in head-on-head configuration. In this case, the long arm of chromosome 21 is attached to another chromosome, often chromosome 14 (46,XX,+21, t(14p;21q)) (Fig. 3.7), or itself (called an isochromosome, 46,XX,+21,t(21q;21q) (Fig. 3.8). These DS individuals with this genetic change have total 46 chromosomes instead of 47 because one 21 is translocated/fused with other chromosome as a whole and present in all cells of the body. However, they have DS since three copies of 21 are present in their genome. RT involving 21 with trisomy 21 causes DS in ~2%–3% of all cases with trisomic HSA21. This mechanism is not specifically influenced by maternal age and randomly comes from the carrier father or mother indicating that the paternal and maternal factors may equally be responsible. The incidence of DS with RT has been estimated as 1 in about 14,000 live births (Hsu, 1998).

RT in variant DS can occur de novo (not inherited but occurs at the time of an individual's conception) or be inherited from a parent carrying a RT (Antonarakis et al., 2004). In ~25% DS individuals with RT, the translocation is inherited from one parent, and that is called "familial" DS with two different abnormalities, including trisomy 21 and an unbalanced RT (Nussbaum et al., 2001). Parents carrying such

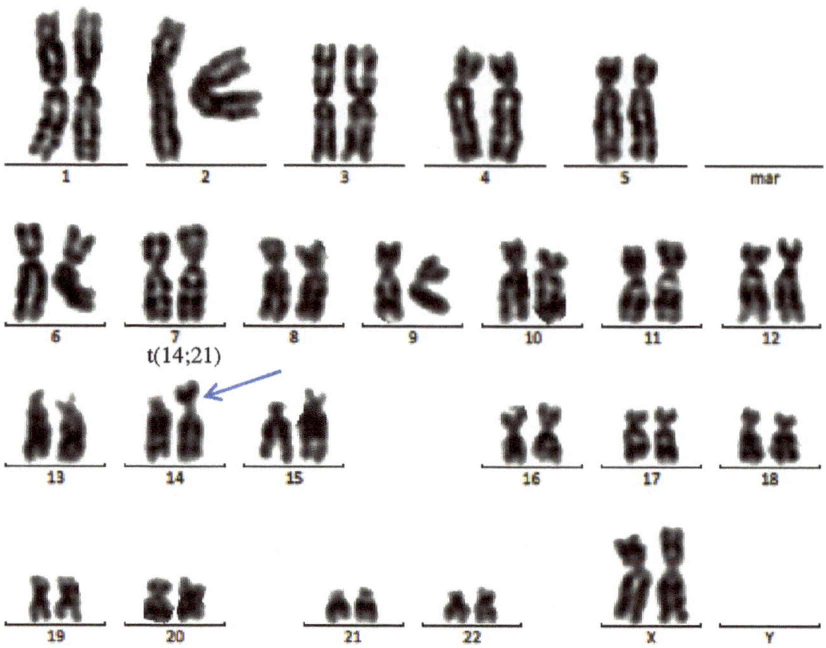

FIGURE 3.7 Karyotype of Down syndrome with t(14;21) Robertsonian translocation. Note the three copies of 21q (the long arm of chromosome 21). *Reproduced from: Ganguly and Kadam (2016). https://doi.org/10.1007/s13237-016-0187-y.*

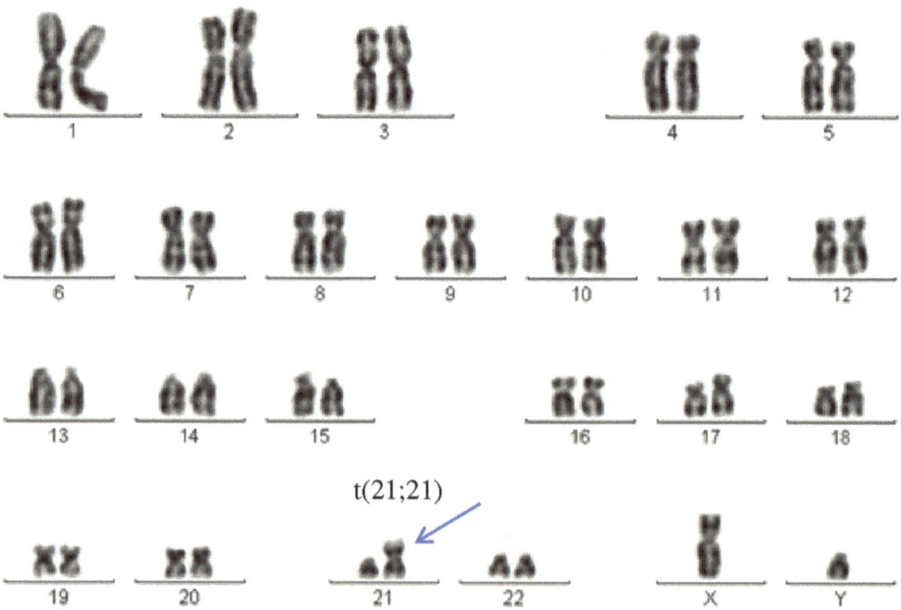

FIGURE 3.8 Robertsonian translocation between two 21s (called an isochromosome of 21) with 46,XX,+21,t(21q;21q) karyotypic configuration.

translocation will have total 45 chromosomes in their karyotype due to fusion of two acrocentric chromosomes with (monocentric) or without (dicentric) loss of one centromere and the short arm (p) including satellites of one of the two acrocentric chromosomes. Carriers of such translocations may not have any discernible clinical or phenotypic effect; however, they are at a greater risk of experiencing meiotic nondisjunction and thus, reproducing DS children with RT. If the carrier mother or father has a RT involving two 21s (Fig. 3.8), all of their children (100%) will have trisomy 21 or DS.

2.3.3 Balanced translocation

Partial or segmental trisomy 21 (approximately >5 Mb in size), though occurs as 1 in 1800 births, may appear as an unbalanced rearrangement through abnormal meiotic segregation in carriers of balanced translocations (Antonarakis et al., 2004). Such segmental trisomies may involve a segment of the genome equivalent to $\geq$ one chromosomal band, and these trisomies are less frequent than full trisomies. In Fig. 3.9, presence of a balanced translocation between two acrocentric chromosomes shows one normal copy of 14 and 21 and one

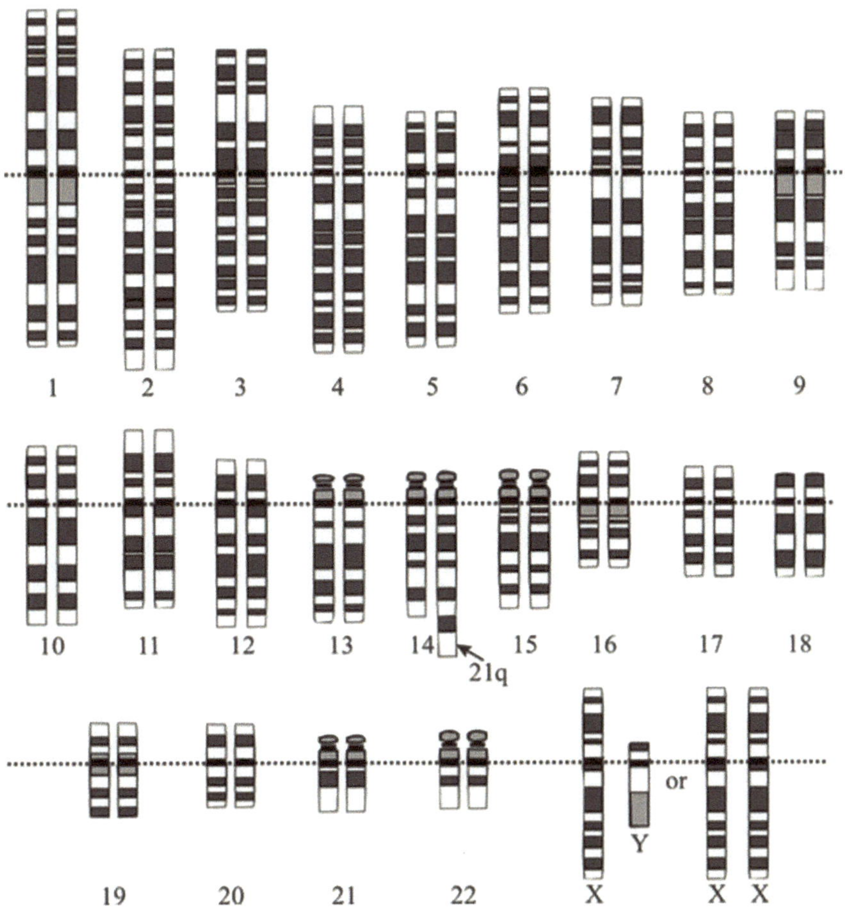

FIGURE 3.9 Schematic presentation of a karyotype with an unbalanced t(14;21). From the National Institutes of Health, part of the United States Department of Health and Human Services.

recombinant 14, which contains the long arm (q) of a 21 and a tiny recombinant 21. During meiosis, the chromosomal arrangement interferes with normal separation of chromosomes. Possible gametic arrangements are: normal 14 and normal 21; translocated 14/21 and normal 21; translocated 14/21 and normal 14; and both translocated 14/21. When combined with a normal gamete from the other parent, the last is lethal, leading to spontaneous abortion. The first, combined with a normal gamete from the other parent, gives rise to a normal child. The second and third will lead to translocation carrier, like the parent. These parents are phenotypically normal. However, this translocation can lead to nondisjunction where the child will be a DS with a balanced translocation between 14 and 21 (Fig. 3.9).

2.3.4 *Partial duplication of chromosome 21*

Segmental trisomy may occur in small sizes (≤3–5 Mb) due to duplication that results through unequal crossovers in meiosis and mediated by the presence of interchromosomal "duplicons" or "low copy repeats" (10–100 Kb) (Antonarakis et al., 2004). Duplicons constitute ~5% of the human genome, promote unequal recombination, and produce microtrisomies and micromonosomies (Bailey et al., 2002). Taken together, segmental partial trisomy also results in imbalance in gene dosage and may contribute to variable clinical/phenotypic expressions. A region of chromosome 21 may undergo a duplication event. This will lead to making extra copies of some, but not all, of the genes encoded on chromosome 21 (46,XX,dup(21q)) (Fig. 3.10). If the duplicated region has genes that are responsible for physical and mental characteristics of DS, the individuals will have manifestation of the syndrome. Conventional G-banding analysis may not detect such cryptic rearrangements, and hence, such incidence remains mostly unrecognized. Such events are very rare, and thus, no epidemiological rate has been established. Such variant DS

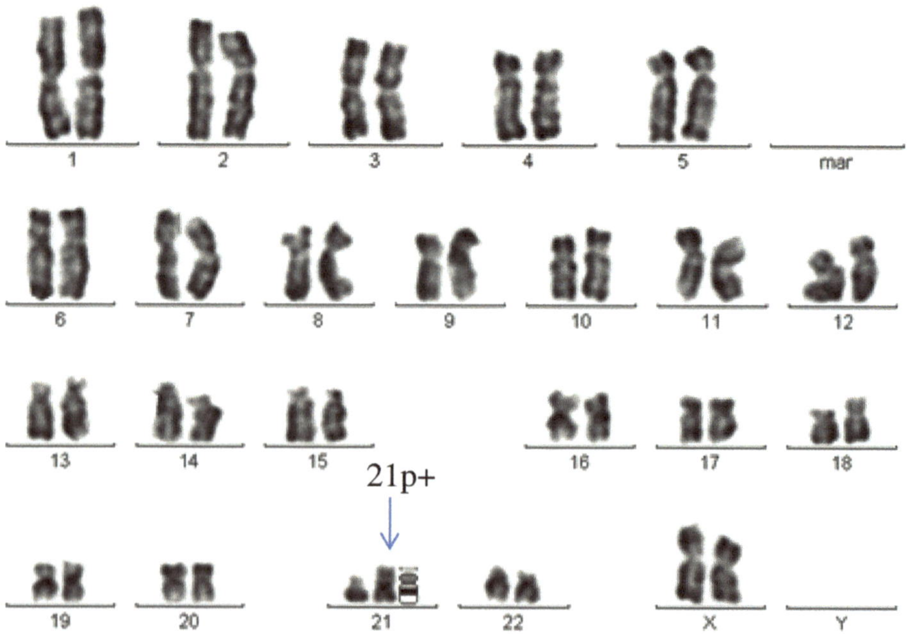

FIGURE 3.10 Shows partial duplication of 21 (21p+).

could be identified using aCGH or sequencing techniques. Five cryptic duplications of 21q were detected in patients with DS that were inherited from parental balanced translocations by Garcia-Heras and Rao (1999). All cases were identified by FISH and/or DNA diagnosis because the phenotype was inconsistent with the conventional cytogenetic studies. These rearrangements seem to escape detection without expanded testing and are probably more frequent than observed.

Duplications of single genes or single functional genomic elements may produce triplication of genes, which may be pathogenic, as duplicated PLP1 and PMP22 cause Pelizaeus-Mertzbacher disease and Charcot Marie Tooth disease, type 1A (CMT1A), respectively (Ellis and Malcolm, 1994). Transgenic mice developed with Plp1 or Pmp22 expressed phenotypic abnormalities due to duplication of the two genes (Readhead et al., 1994; Magyar et al., 1996). However, the genomic duplications with these two genes may also encompass additional coding sequences. Additionally, duplication of nongenic but functional, genomic elements may contribute to pathogenesis as observed in preaxial polydactyly in sasquatch mouse variant caused by duplication of ~30 Kb conserved regulatory element that maps ~1 Mb from the Sonic hedgehog (Shh) locus (Lettice et al., 2003). Such segmental duplication in HSA21 is also likely to cause DS in individuals, but cannot be easily detected phenotypically or genetically, since many of the diagnoses of DS are managed by conventional karyotyping and/or FISH.

3. Genes of HSA21

Initial concept described the 21q with 33.5 Mb, which constitutes approximately 225 genes (127 known genes and 98 putative novel genes predicted in silico) and 59 pseudogenes (Hattori et al., 2000). Over 30 pathogenic chromosomal aberrations have been assigned to 21q21-21q22, which has been established as the DS critical region (DSCR) in HSA21 (Korenberg et al., 1994). The 21st chromosome in the human genome was declared to harbor ~400 genes (Hattori et al., 2000), 1/3rd of which is overexpressed and 2/3rd compensated. Understanding of the genotype–phenotype correlation and molecular mechanisms for every single phenotype has been a major concern with a view to developing effective therapeutics for DS population. Defects in neurogenesis, synaptogenesis, and cell signaling pathways were thought to affect the learning, memory, and brain development (Belichenko et al., 2009). Some of the genotype–phenotype associations are listed below (adapted from Ganguly and Kadam, 2016):

Later, sequencing of HSA21 had measured ~364 genes (Gardiner et al., 2003), though a precise catalog of HSA21 encoded genes has not yet been conclusively established. Its protein coding genes (~3%) encode transcription factors, regulators, and modulators (18 genes); interferons and immune response (9 genes); kinases (8 genes); channels (7 genes); proteases and protease inhibitors (6 genes); RNA processing (5 genes); receptors (5 genes); adhesion molecules (4 genes); energy metabolism (4 genes); and ubiquitin pathway (4 genes) (Fig. 3.11). And ~1% corresponds to conserved nongenic sequences, which are not transcribed and do not code for proteins (Dermitzakis et al., 2002, 2004). The most recent manual annotation of the HSA21q, which spans ~35 Mb, identified ~500 genes, including 164 classical protein-coding genes, 49 members of the gene family of keratin-associated proteins (KRTAPs), and ~300 additional transcripts that encode ambiguous open reading frames (ORFs; short ORFs, multiple ORFs, or no ORFs), that may or may not be functional pseudogenes (Gupta et al., 2016; Moyer et al., 2021). Expression-wise, KRTAP-family genes are restricted to skin and may include many pseudogenes. Among the non-KRTAP protein-coding genes, including the

FIGURE 3.11 Functional attributes of genes of chromosome 21 (i) and their association with systemic phenotypes (ii) of Down syndrome. *Image created from different literatures available in public domain.*

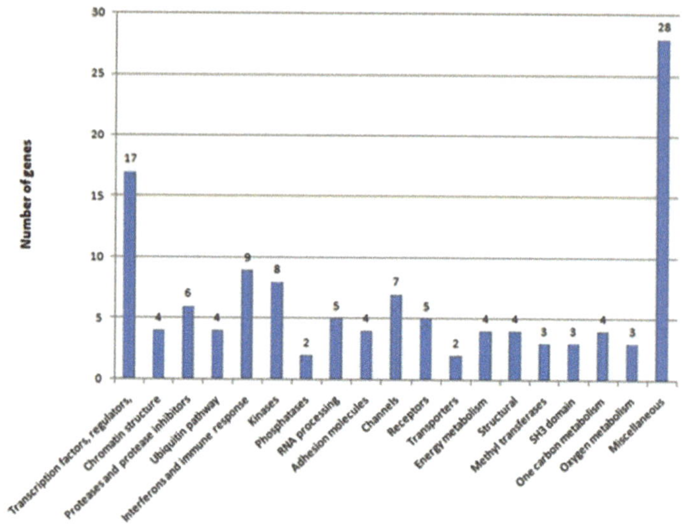

i.

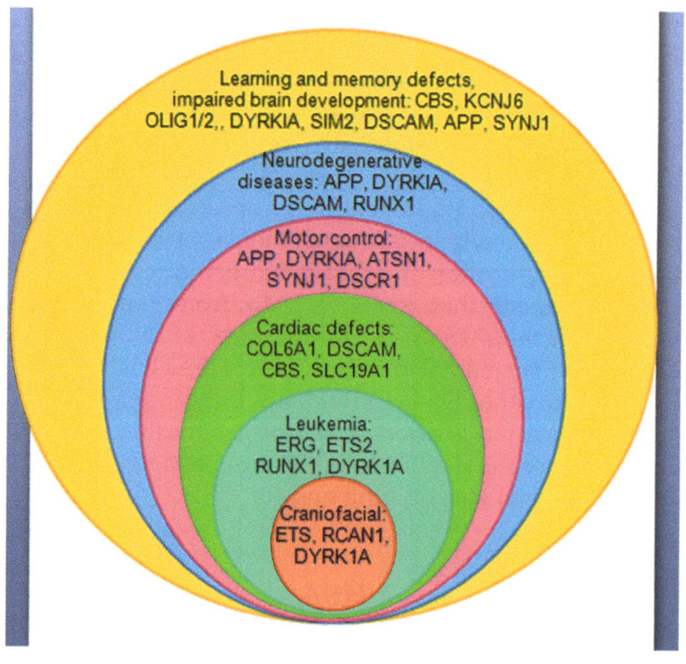

ii.

transcription factors/modulators, components of the ubiquitin/proteasome, and RNA processing pathways, there are several genes attributed to sex differences in functional properties, and many are involved in Alzheimer disease (AD) with the amyloid precursor protein (APP) and proteins that modulate APP processing or trafficking (reviewed in Moyer et al., 2021). The functional properties of most of these are merely predicted from sequence similarities and protein production matching with genes throughout the genome. Few of these genes have been individually studied for understanding the consequences of overexpression in the context of trisomy 21 in DS patients.

Each prominent characteristic of individuals with DS varies widely, and thus, the identification and characterization of HSA21 genes appeared critical for understanding the molecular basis of the expression of syndromic features. Delineation of the DSCR was successful to pinpoint DSCR gene and its sub-types (DSCR1, DSCR2, and so on) responsible for specific features of DS. This DSCR measured as ~5.4 Mb was extensively studied for deciphering genotype-phenotype association precisely. The overexpression of DSCR1 protein, renamed as RCAN1 ("regulator of calcineurin 1"), in the human brain and heart suggested that triplication of RCAN1 may be involved in the phenotypes of DS, especially for intellectual disability, cardiac defects, and neurodegeneration. Its overexpression was evidenced in the brain of DS fetuses leading to physical and functional interaction of calcineurin A, the catalytic subunit of the Ca(2+)/calmodulin-dependent protein phosphatase PP2B (Fuentes et al., 2000; Harris et al., 2005). Its sub-cellular location led to hypothesize that functionally RCAN1 could act as a regulator of transcription (Pfister et al., 2002; Silveira et al., 2004). It has been demonstrated that the synergistic action of the genes RCAN1 and DYRK1A, both mapped in the DSCR, leads to prevention of the nuclear occupancy of NFATc transcription factors (Arron et al., 2006). The co-operating effect of the two genes results in destabilization of a regulatory circuit and that reduces the NFATc activity and manifests many of the features of DS.

Regulation of the phenotypes of DS was thought to be associated with the dysregulation of transcription factors, particularly the nuclear factor of activated T-cells (NFAT), which is partly controlled by DSCR1 and DYRK1A proteins (Epstein, 2001). People with trisomy 21 have 1.5 times of these proteins higher than normal euploids (Arron et al., 2006). DYRK1A (dual-specificity tyrosine phosphorylation-regulated kinase 1A) inhibits cell proliferation and promotes premature neuronal differentiation by regulating NOTCH signaling (Hammerle et al., 2011). DYRK1A has been proposed to be an important candidate gene responsible for various DS phenotypic traits because of its involvement in multiple cellular pathways, including nuclear factor activation and basic cellular metabolism. This gene also inhibits nuclear translocation of the NFAT family in association with DSCR1 (or RCAN1) protein, resulting in neuronal and cardiac defects in individuals with DS (Arron et al., 2006). Small segmental duplications in DYRK1A did not establish specific effect of 21 that leads to severe reductions in IQ in DS. Increase in oxidative stress and generation of free radicals such as reactive oxygen species (ROS) are associated in DS, wherein several genes of 21, including APP, BACH1, DYRK1A, DSCR1, ETS2, SOD1, and S100B have been implicated with ROS generation (Campos and Casado, 2015). However, variation in phenotype of DS was noticed in different individuals, as observed with overexpression of APP gene that encodes the precursor of the amyloid protein.

APP gene has been established as the candidate gene for neuropathological features of AD and DS. Individuals with DS and AD have exhibited neurofibrillary tangles and amyloid plaques in common, wherein the layer II of the entorhinal cortex and the subiculum appeared critical for memory consolidation and damage first, followed by gradual reduction of the nerve cells of the cortex. Similar observation was noticed in the Ts65Dn mouse model of human DS whose chromosome 16 is syntenic to human chromosome 21. Mutations of APP contribute to early onset of AD in DS because proteolysis of APP produces the amyloid-β plaques in the brains of AD (Reynolds et al., 2010). The cooperating effect of other HSA21 genes with APP-overexpression has been thought of for degeneration of basal forebrain cholinergic neurons (BFCN) in DS. The cooperating effect of DYRK1A has been described in DS with higher levels of phosphorylated APP produced by DYRK1A protein in transgenic mice (Margueron et al., 2008).

The SOD1 encodes superoxide dismutase (SOD) enzyme and catalyzes the conversion of oxygen radicals into hydrogen peroxide and water. Free oxygen radicals produced in cells are harmful for the cellular structures. It was thought that disproportionate increase in SOD activity could be responsible for removal of hydrogen peroxide (e.g., glutathione peroxidase), which may result in peroxide damage in the cells. Elevation of superoxide dismutase activity was hypothesized in DS due to triplication of SOD1 gene mapped onto 21q22. Therefore, scientists believed that the treatment of DS with free radical scavengers can substantially prevent oxidative damage in neurons and neuronal degeneration and thus, early onset of brain aging and AD. The alterations, including reduction in terminal axon area to postsynaptic membrane, and hyperplastic secondary folds in neuromuscular junctions were reported in muscles of transgenic aging mice and humans with DS (Avraham et al., 1988).

The overexpression of DSCR2 gene in proliferating tissues and cells, including fetal tissues, adult testis, and cancer cell lines (Vidal-Taboada et al., 2000), and its intracellular localization and proteolytic cleavage of the protein, have been studied extensively (Vesa et al., 2005). DSCR2 has been designated as "proteosome assembling chaperone-1" (PAC1), wherein PAC1 and PAC2 chaperones function as heterodimers in the maturation of mammalian 20S proteosomes (Hirano et al., 2005). Involvement of DSCR2 protein has been ascribed with the correct assembly of 20S proteosomes from the mechanism of accelerated formation of precursor proteosomes (due to overexpression of PAC1 or PAC2), and its impairment via knockdown by siRNA, that finally results in poor maturation of 20S proteosomes (Hirano et al., 2005).

The extensive variation in the characteristics of DS indicates variability in expression of genes of chromosome 21, owing to varying degrees of expression of overlapping and/or nonoverlapping genes contributing to features of DS, and many of them are compensated (Aït Yahya-Graison et al., 2007; Prandini et al., 2007). Thus, it was apparent that overexpressed genes lead to minimal expression overlap and are associated with the constant features of DS (Fig. 3.12). Also, the triplication of genes on HSA21 and its expression might alter the global gene expression pattern in DS resulting in dysregulation of the disomic sets of non-HSA21 genes (Tang et al., 2004). Besides the fact that the clinical features of DS are not attributable to any single gene, the cooperating effects of the dysregulated genes with the primary gene-dosage effect on phenotype has been reported in fetal organs of DS. Fetal heart defects in DS have been attributed to downregulation of genes encoding mitochondrial enzymes and upregulation of genes encoding extracellular matrix proteins through functional clustering of dysregulated genes (Conti et al., 2007). Moreover, characterization of tissue specific proteome with respect to dysregulated expression of different sets of disomic genes in presence of triplicated HSA21 would precisely detect the

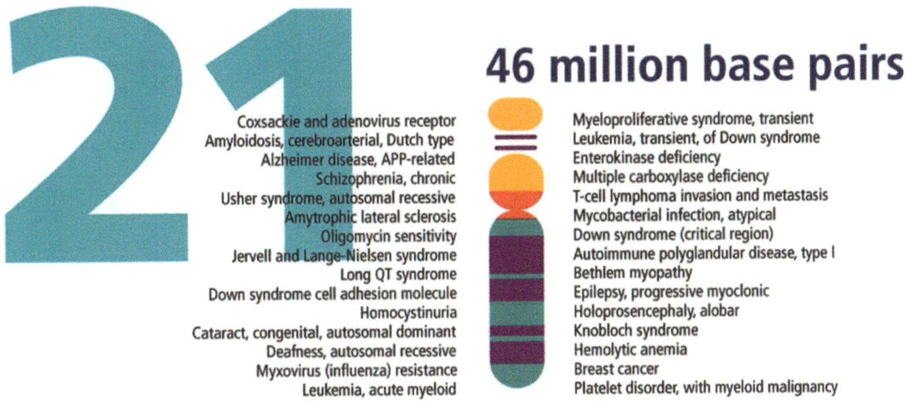

FIGURE 3.12 Chromosome 21 and human diseases. *From U.S. Department of Energy, URL: http://www.ornl.gov/hgmis accessed June 1 2021, available in public domain.*

1. *CAF1A* (OMIM 601245)
2. *CBS* (OMIM 236200)
3. *DYRK1A* (OMIM 600855)
4. *GART* (OMIM 138440)

(1–4: DNA synthesis and repair — Alzheimer disease, intellectual disability)

5. *COL6A1* (OMIM 120220): Heart defects
6. *CRYA1* (OMIM 123580): Mental retardation, cataracts
7. *ETS2* (OMIM 164740): Leukemia & skeletal abnormalities
8. *IFNAR* (OMIM 107450): Interferon/immune dysfunction
9. *SOD1* (OMIM 147450): Premature aging/immune system
10. addfsdfNRIP1/GABPA/DYRK1A/SUMO3: Down syndrome phenotype
11. *APP* (OMIM 104760): Alzheimer disease
12. *GLUR5* (OMIM 138245)
13. *KCNJ6* (OMIM 600877)
14. *PFKL* (OMIM 171860)
15. *S100B* (176990)
16. *Synaptojanin* (OMIM 604297)
17. *TIAM1* (OMIM 600687)

(12–17: Down syndrome phenotype)

candidate genes and/or sets of genes participating in extending cooperating effects for organ-specific features of DS. Serial analysis of gene expression (SAGE) in lymphocytes with trisomy 21 identified genes involved in transcription, RNA processing, signaling, immune response, and lipid metabolism suggesting a modest dysregulation of disomic genes in association with trisomy 21 perturbs the immune system in DS (Sommer et al., 2008).

References

Aït Yahya-Graison, E., Aubert, J., Dauphinot, L., Rivals, I., Prieur, M., Golfier, G., Rossier, J., Personnaz, L., Créau, N., et al., 2007. Classification of human chromosome 21 gene-expression variations in Down syndrome: impact on disease phenotypes. Am. J. Hum. Genet. 81 (3), 475–491.

Antonarakis, S.E., 1998. Down syndrome. In: Jameson, J.L. (Ed.), Principles of Molecular Medicine. Humana Press, Totowa, NJ. https://doi.org/10.1007/978-1-59259-726-0_119.

Antonarakis, S.E., Lyle, R., Dermitzakis, E.T., Reymond, A., Deutsch, S., 2004. Chromosome 21 and Down syndrome: from genomics to pathophysiology. Nat. Rev. Genet. 5, 725–738.

Antonarakis, S.E., Skotko, B.G., Rafii, M.S., Strydom, A., Pape, S.E., Bianchi, D.W., Sherman, S.L., Reeves, R.H., February 6, 2020. Down syndrome. Nat. Rev. Dis. Prim. 6 (1), 9. https://doi.org/10.1038/s41572-019-0143-7.

Arron, J.R., Winslow, M.M., Polleri, A., Chang, C.P., Wu, H., Gao, X., et al., 2006. NFAT dysregulation by increased dosage of DSCR1 and DYRK1A on chromosome 21. Nature 441, 595–600.

Avraham, K.B., Schickler, M., Sapoznikov, D., Yarom, R., Groner, Y., September 9, 1988. Down syndrome: abnormal neuromuscular junction in tongue of transgenic mice with elevated levels of human Cu/Zn-superoxide dismutase. Cell 54, 823–829.

Bailey, J., Gu, Z., Clark, R.A., Reinert, K., Samonte, R.V., Schwartz, S., et al., 2002. Recent segmental duplications in the human genome. Science 297, 1003–1007.

Belichenko, N.P., Belichenko, P.V., Kleschevnikov, A.M., Salehi, A., Reeves, R.H., Mobley, W.C., 2009. The "Down syndrome critical region" is sufficient in the mouse model to confer behavioral, neurophysiological, and synaptic phenotypes characteristic of Down syndrome. J. Neurosci. 29, 5938–5948.

Biesecker, L.G., Spinner, N.B., 2013. A genomic view of mosaicism and human disease. Nat. Rev. Genet. 14, 307–320.

Campos, C., Casado, Á., 2015. Oxidative stress, thyroid dysfunction & Down syndrome. Indian J. Med. Res. 142, 113–119.

Conti, A., Fabbrini, F., D'agostino, P., Negri, R., Greco, D., Genesio, R., D'armiento, M., Olla, C., Paladini, D., Zannini, M., Nitsch, L., 2007. Altered expression of mitochondrial and extracellular matrix genes in the heart of human fetuses with chromosome 21 trisomy. BMC Genom. 8, 268.

Dermitzakis, E.T., Reymond, A., Lyle, R., Samuffa, N., Ucla, C., Deutsch, S., Stevenson, B.J., Flegel, V., Bucher, P., Jongeneel, C.V., Antonarakis, S.E., 2002. Numerous potentially functional but non-genic conserved sequences on human chromosome 21. Nature 420 (6915), 578–582.

Dermitzakis, E.T., Kirkness, E., Schwarz, S., Birney, E., Reymond, A., Antonarakis, S.E., 2004. Comparison of human chromosome 21 conserved nongenic sequences (CNGs) with the mouse and dog genomes shows that their selective constraint is independent of their genic environment. Genome Res. 14 (5), 852–859.

Ellis, D., Malcolm, S., 1994. Proteolopid protein gene dosage effect in Pelizaeus-Mertzbacher disease. Nat. Genet. 6, 333–334.

Epstein, C.J., 2001. Down syndrome (trisomy 21). In: Scriver, C.R., Beaudet, A.L., Sly, W.S., Valle, D. (Eds.), The Metabolic and Molecular Bases of Inherited Diseases. McGraw-Hill, Inc., NewYork, pp. 1223–1256.

Fuentes, J.J., Genesca, L., Kingsbury, T.J., Cunningham, K.W., Perez-riba, M., Estivill, X., de la Luna, S., 2000. DSCR1, Overexpressed in Down syndrome, is an inhibitor of calcineurin-mediated signaling pathways. Hum. Mol. Genet. 9 (11), 1681–1690.

Ganguly, B.B., Kadam, N.N., 2016. Down syndrome: from the age of characterization to the era of curative approach. Nucleus. https://doi.org/10.1007/s13237-016-0187-y.

Ganguly, B.B., Dolai, T.K., De, R., Kadam, N.N., 2016. Spectrum of complex chromosomal aberrations in a myelodysplastic syndrome and a brief review. J. Cancer Res. Therapeut. 12, 1203–1206.

Garcia-Heras, J., Rao, N.P., 1999. A brief review of cryptic duplication of 21q as an emerging cause of Down syndrome: practical considerations for accurate detection. Clin. Genet. 55 (3), 207–211. https://doi.org/10.1034/j.1399-0004.1999.550310.x.

Gardiner, K., Fortna, A., Bechtel, L., Davisson, M.T., 2003. Mouse models of Down syndrome: how useful can they be? Comparison of the gene content of human chromosome 21 with orthologous mouse genomic regions. Gene 318, 137–147.

Gupta, M., et al., 2016. Mouse models of Down syndrome: gene content and consequences. Mamm. Genome 27, 538–555.

Hammerle, B., Ulin, E., Guimera, J., Becker, W., Guillemot, F., Tejedor, F.J., 2011. Transient expression of Mnb/Dyrk1a couples cell cycle exit and differentiation of neuronal precursors by inducing p27KIP1 expression and suppressing NOTCH signaling. Development 138, 2543–2554.

Harris, C.D., Ermak, G., Davies, K.J., 2005. Multiple roles of the *DSCR1* (*Adapt78* or *RCAN1*) gene and its protein product Calcipressin 1 (or RCAN1) in disease. Cell. Mol. Life Sci. 62 (21), 2477–2486.

Harrow, J.L., et al., 2014. The vertebrate genome annotation browser 10 years on. Nucleic Acids Res. 42, D771–D779.

Hassold, T., Sherman, S., 2000. Down syndrome: genetic recombination and the origin of the extra chromosome 21. Clin. Genet. 57 (2), 95–100.

Hattori, M., Fujiyama, A., Taylor, T.D., et al., 2000. Chromosome 21 mapping and sequencing consortium the DNA sequence of human chromosome 21. Nature 405, 311–319.

Hattori, M., et al., 2001. The DNA sequence of human chromosome 21. Tanpakushitsu Kakusan Koso 46, 2254–2261.

Hirano, Y., Hendil, K.B., Yashiroda, H., Iemura, S., Nagane, R., Hioki, Y., Natsume, T., Tanaka, K., Murata, S., 2005. A heterodimeric complex that promotes the assembly of mammalian 20S proteasomes. Nature 437 (7063), 1381–1385.

Hsu, L.Y., 1998. In: Milunsky, A. (Ed.), In Genetic Disorders and the Fetus. Jhons Hopkins University Press, Baltimore, pp. 179–248.

Hu, J., Shekhter-Levin, S., Shaw, P.H., Bay, C., Kochmar, S., Surti, U., 2005. A case of myelodysplastic syndrome with acquired monosomy 7 in a child with a constitutional t(1;19) and a mosaicism for trisomy 21. Cancer Genet. Cytogenet. 156A, 62–67.

Jackson-Cook, C., 2011. Constitutional and acquired autosomal aneuploidy. Clin. Lab. Med. 31, 481–511.

Korenberg, J.R., Chen, X.-N., Schipper, R., Sun, Z., Gonsky, R., Gerwehr, S., et al., 1994. Down syndrome phenotypes: the consequences of chromosomal imbalance. Proc. Natl. Acad. Sci. U.S.A. 91, 4997–5001.

Lejeune, J., Gautier, M., Turpin, R., 1959. Etudes des chromosomes somatiques de neuf enfants mongoliens. Comptes Renus de l' Academic les Sciences 248, 1721–1722.

Lettice, L.A., et al., 2003. A long-range Shh enhancer regulates expression in the developing limb and fin and is associated with preaxial polydactyly. Hum. Mol. Genet. 12, 1725–1735.

Magyar, J.P., et al., 1996. Impaired differentiation of Schwann cells in transgenic mice with increased Pmp22 gene dosage. J. Neurosci. 16, 5351–5360.

Margueron, R., Li, G., Sarma, K., Blais, A., Zavadil, J., Woodcock, C.L., et al., 2008. Ezh1 and Ezh2 maintain repressive chromatin through different mechanisms. Mol. Cell 32, 503–518.

Mégarbané, A., Ravel, A., Mircher, C., Sturtz, F., Grattau, Y., et al., 2009. The 50th anniversary of the discovery of trisomy 21: the past, present and future of research and treatment of Down syndrome. Genet. Med. 11 (9), 611–616.

Moyer, A.J., Gardiner, K., Reeves, R.H., 2021. All creatures great and small: New approaches for understanding down syndrome genetics. Trends Genet. 37 (5), 444–459. https://doi.org/10.1016/j.tig.2020.09.017.

Nussbaum, R.L., McInnes, R.R., Willard, H.F., 2001. Thompson & Thompson: Genetics in Medicine, sixth ed. WB Saunders Company, Philadelphia, pp. 157–162.

Papavassiliou, P., York, T.P., Gursoy, N., Hill, G., Nicely, L.V., Sundaram, U., et al., 2009. The phenotype of persons having mosaicism for trisomy 21/Down syndrome reflects the percentage of trisomic cells present in different tissues. Am. J. Med. Genet. 149A, 573–583.

Papavassiliou, P., Charalsawadi, C., Rafferty, K., Jackson-Cook, C., 2014. Mosaicism for trisomy 21: a review. Am. J. Med. Genet. 167A, 26–39, 2015.

Parker, S.E., Mai, C.T., Canfield, M.A., et al., 2010. Updated national birth prevalence estimates for selected birth defects in the United Sates, 2004-2006. Birth Defects Res. A Clin. Mol. Teratol. 88, 1008–1016.

Pfister, S.C., Machado-Santelli, G.M., Han, S.W., Henrique-Silva, F., 2002. Mutational analyses of the signals involved in the subcellular location of DSCR1. BMC Cell Biol. 3, 24.

Prandini, P., Deutsch, S., Lyle, R., Gagnebin, M., Vivier, C.D., Delorenzi, M., et al., 2007. Natural gene-expression variation in Down syndrome modulates the outcome of gene-dosage imbalance. Am. J. Hum. Genet. 81 (2), 252–263.

Readhead, C., Schneider, A., Griffiths, I., Nave, K.A., 1994. Premature arrest of myelin formation in transgenic mice with increased proteolipid protein gene dosage. Neuron 12, 583–595.

Reynolds, L.E., Watson, A.R., Baker, M., Jones, T.A., D'Amico, G., Robinson, S.D., et al., 2010. Tumor angiogenesis is reduced in the Tc1mouse model of Down's syndrome. Nature 465, 813–817.

Savage, A., Petersen, M., Pettay, D., et al., 1998. Elucidating the mechanisms of paternal non-disjunction of chromosome 21 in humans. Hum. Mol. Genet. 7, 1221–1227.

Sherman, S.L., Allen, E.G., Bean, L.H., Freeman, S.B., 2007. Epidemiology of down syndrome. Ment. Retard. Dev. Disabl. Res. 13, 221–227.

Silveira, H.C.S., Sommer, C.A., Soares-Costa, A., Henrique-Silva, F., 2004. A calcineurin inhibitory protein overexpressed in Down's Syndrome interacts with the product of a ubiquitously expressed transcript. Braz. J. Med. Biol. Res. 37 (6), 785–789.

Sommer, C.A., Pavarino-Bertelli, E.C., Goloni-Bertollo, E.M., Henrique-Silva, F., 2008. Identification of dysregulated genes in lymphocytes from children with Down syndrome. Genome 51 (1), 19–29.

Tang, Y., Schapiro, M.B., Franz, D.N., Patterson, B.J., Hickey, F.J., Schorry, E.K., Hopkin, R.J., Wylie, M., et al., 2004. Blood expression profiles for tuberous sclerosis complex 2, neurofibromatosis type 1, and Down's syndrome. Ann. Neurol. 56 (6), 808–814.

Vesa, J., Brown, Y., Greenfield, D., Korenberg, J.R., 2005. Molecular and cellular characterization of the Down syndrome critical region protein 2. Biochem. Biophys. Res. Commun. 328 (1), 235–242.

Vidal-Taboada, J.M., Lu, A., Pique, M., Pons, G., Gil, J., Oliva, R., 2000. Down Syndrome Critical Region Gene 2: expression during mouse development and in human cell lines indicates a function related to cell proliferation. Biochem. Biophys. Res. Commun. 272, 156–163.

Yoon, P.W., Freeman, S.B., Sherman, S.L., Taft, L.F., Gu, Y., Pettay, D., et al., 1996. Advanced maternal age and the risk of Down syndrome characterized by the meiotic stage of the chromosomal error: a population based study. Am. J. Hum. Genet. 58, 628–633.

CHAPTER

4

Possible factors contributing to chromosomal nondisjunction and formation of trisomic HSA21

1. Introduction

Mis-segregation of chromosome 21 during gametogenesis results in gamete with aneuploidy, which, following fertilization with a normal haploid gamete, produces an aneuploid zygote with an extra chromosome 21 or trisomy 21. Such mis-segregation often occurs at meiosis I but may also occur in meiosis II. Either case will lead to production of a zygote with trisomy 21 at the genomic level. However, mitotic nondisjunction during embryonic development may generate cells of two clones, such as normal diploid cell and aneuploid cell with trisomy 21 resulting in mosaicism (Ganguly and Kadam, 2017). The etiology of such misdivision of chromosomes could be many folds, of which an advanced maternal age factor has been established as the most prominent one, though younger mothers also give birth to children with Down syndrome (DS), and in that case, the source of aneuploid gamete has not been ascertained. Interestingly, the research on this topic could not draw much of attention as revealed through the literature, which could mainly be due to transition of focus toward betterment of life of individuals born with DS. Indeed, the major attention has been concentrated on understanding the genetic consequences of the extra 21 at the backdrop of 46 disomic chromosomes, which has heralded the knowledge of gene−gene interaction and its implications on health and life of DS.

Besides advanced maternal age and altered recombination risk factors for DS, the impact of low maternal socioeconomic status (SES) and other environmental and genetic factors on nondisjunction of the chromosome 21 is unclear. A significant association between low maternal SES, especially low household income, and nondisjunction of chromosome 21 in meiosis II has been demonstrated (Hunter et al., 2013). However, the finding was independent of the maternal age effect with a range of younger and older than 35 years of maternal age. Nevertheless, the etiology of maternal nondisjunction is clearly multifactorial, which can be discerned from further studies on the biological mechanisms involved in the association between maternal SES and nondisjunction (Hunter et al., 2013).

This chapter will discuss about the major factors contributing to nondisjunction of chromosomes during meiosis for gamete production

https://doi.org/10.1016/B978-0-323-90456-8.00003-X

or mitosis of early embryonic growth. The factors, including maternal age and antenatal complexity, exposure to environmental, occupational and lifestyle factors, and meiotic recombination have been considered. However, there could be many more intrinsic and/or extrinsic factors that result in chromosomal misdivision and birth of DS. Moreover, the impact of paternal and maternal severe acute respiratory syndrome coronavirus 2 (SARS-CoV-2) status on this mechanism of nondisjunction stands totally unknown. *Per se* SARS-CoV-2 infection in adults with DS have experienced 4 times higher fatality than the general population.

2. Factors that can cause nondisjunction of chromosomes

2.1 Age factor

The gene–environment interaction, parental infection with microorganisms, advanced maternal age, exposure to radiation and chemicals, and many other extrinsic and intrinsic factors may contribute to alteration in the structure and function of chromosomes and genes and finally establish genetic instability resulting in an increased risk of disease development, including DS. Among all the known risk factors, advanced maternal age has been established as the most common one (Hecht and Hook, 1996; Yoon et al., 1996), though 80% of the DS children are born to women aged less than 35 years (Fig. 4.1). Interestingly, the influence of paternal age, especially beyond 42 years, has also been demonstrated to increase the risk of DS in pregnancies of older mothers (Mclntosh et al., 1995; Yang et al., 2007; García-Ferreyra et al., 2018); however, the risk of advanced paternal age has failed to establish such an association when adjusted with maternal age (Zhu et al., 2005; Stoll et al., 1990; Steiner et al., 2015). In addition, an association of age of the maternal grandmother at the time of mother's birth has been reported to increase the risk of DS (Aagesen et al., 1984; Malini and Ramachandra, 2006). Since female meiosis begins during fetal life, it is hypothesized that nondisjunction in the first meiotic (M1) division might be induced during fetal period, especially if the mother was older. This might lead to development of genetic abnormality in the gonad, which eventually be transmitted to the offspring. Notably, DS is a random event that occurs during the formation of gametes or embryonic development, irrespective of the parental behavior or environmental factors. DS does not discriminate the ethnic groups or economic classes. Some of the known factors favoring trisomy 21 and birth of DS have been discussed.

The physiological status of the ovaries or the quality of eggs could be the maternal age factor to contribute in increasing the risk of DS and other chromosome aberrations (Freeman et al., 2000; Allen et al., 2009; Kurahashi et al., 2012). Additionally, delayed fertilization, change in hormone levels, and relaxed selection at older age could potentially supplement the age effect (Hassold et al., 1980). Incidence of reduced ovarian complement (congenital absence or removal of an ovary) and hormonal dysfunction due to polycystic ovarian disease or other factor may also influence nondisjunction, resulting in trisomy 21 and pregnancy with DS.

The hypothesis of aneuploid oocytes arise from a concatenation of events, including hormonal imbalance that causes a less-than-optimal microvasculature to develop around the maturing and mature follicles resulting in decreased size of the perifollicular capillary bed and reduction in the volume of blood flow through the area. The situation leads to oxygen deficiency and a concomitant increase in carbon dioxide and anaerobic products (such as lactic acid) inside the follicle. The acidic intracellular pH of the oocyte diminishes the size of the spindle, with subsequent displacement and nondisjunction of a chromosome. Such compromised microcirculation hypothesis explains the

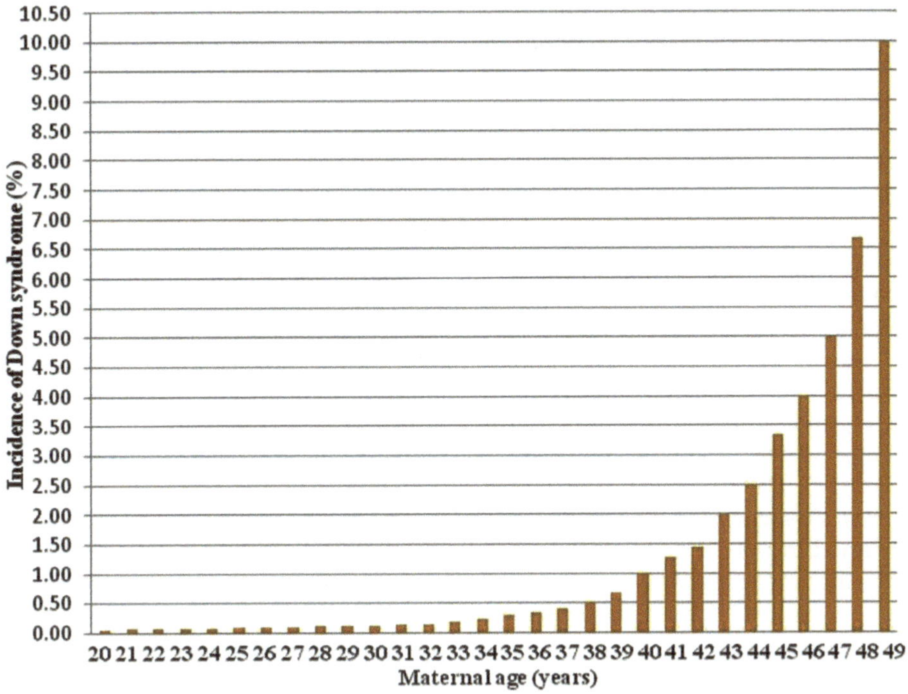

FIGURE 4.1 Association of maternal age and incidence of DS. *Data source: https://www.ndss.org/about-down-syndrome/down-syndrome/. (Accessed 3 July 2021).*

occasion of aneuploidy in primary and secondary oocytes and sperm precursor cells, indicating the possibility of why women of all reproductive ages may have a DS child (Gaulden, 1992). The hypothesis also suggests that nondisjunction occurs in M1 in ≥80% cases with subsequent completion in ovaries.

2.2 Antenatal complexity

Pregnancies with DS have been reflected with lower birth weight, premature delivery, and intrauterine growth retardation of the fetus (Lapunzina et al., 2002; Rasmussen et al., 2001). Although effect of multiparity on DS birth exists, that disappears when maternal age is considered (Chan et al., 1998). Also the order of pregnancy, particularly the first one, has a higher risk of DS birth, which is independent of maternal age (Hay and Barbano, 1972); however, contradictory reports exist in this regard (Stoll et al., 1990).

A short gap between pregnancies has been described to increase the risk of DS (Brender and Suarez, 1990; Neuhäuser and Krackow, 2007 https://doi.org/10.1007/s00114-006-0165-3). It is reported that conceptions occurring during the transition from anovulation to the onset of regular ovulation after childbirth could be responsible for maternal meiotic nondisjunction (Jongbloet and Vrieze, 1985). Anovulatory activity followed by conception might also increase the risk of DS. In addition, increased incidence of pregnancies with DS has been observed among the infants conceived by intracytoplasmic sperm injection or other mode of infertility treatment (Aboulghar et al., 2001).

It was assumed that seasonal variation might increase the delivery or conception of DS based on the fact that higher frequency of diagnostic tests is required for newborns during monsoon in India or June–September (Stoll et al., 1990; Castilla et al., 1990; Ganguly, unpublished data). However, the association could not be established due to multiple variable factors, such as season or month of conception, duration of pregnancy, parental age, use of contraception before conception of DS, etc.

Chromosomal abnormalities such as trisomy in maternal gonad, Robertsonian or unbalanced translocation involving 21 in any parent, parental mosaicism or balanced translocation with trisomy 21, and previous affected child in the same sibship might increase the risk of conception with DS (Panalos et al., 1992; Sherman et al., 1991). Association between parental consanguinity and birth of DS does not exist (Rittler et al., 2001), though autosomal recessive mutations might predispose nondisjunction (Stoll et al., 1990). Maternal health, including thyroid disorders, common cold with/without fever during first trimester gestation, diabetes, or epilepsy does not increase pregnancy with DS (Patterson, 1987; Stoll et al., 1990); however, conflicting reports on maternal hypo- or hyperthyroidism, viral infections such as measles and chicken pox, and TORCH (toxoplasma, rubella, cytomegalovirus, herpes simplex virus) status may influence the risk of DS (Chen et al., 1999; Hook, 1984; Ognjanovic et al., 2009). Family history of Alzheimer disease is also likely to increase the risk of DS birth (Schupf et al., 1994). Maternal smoking did not exhibit a clear association with incidence of DS pregnancy after adjusting maternal age (Chen et al., 1999; Torfs & Christianson, 2000).

Nevertheless, mitochondrial function may also contribute to meiotic mis-segregation of chromosomes. Abnormalities such as deletion, mutation, or replication abnormalities in mitochondrial DNA (mtDNA) could cause hypoxia, oxidative stress, and mitochondrial dysfunction leading to abnormalities in the inherent genomic constitution and advancement of age. Indeed, deletions in mtDNA have been demonstrated frequently in the oocytes of women of advanced reproductive age (Keefe et al., 1995). Mitochondrial dysfunction in the oocytes and early embryonic cells can perturb the ATP generation, which consequently can increase the risk of abnormal chromosomal segregation or developmental arrest (Van Blerkom et al., 1998).

SARS-CoV-2 pandemic has created a global warning on health consequences for all ages. However, the new science of infection control and health management for this virus was not sufficient to control the spike of mortality and morbidity associated with 'long coronavirus disease 2019' (COVID-19). There are reports on newborn infected with the virus, including one with DS (Piersigilli et al., 2020; Zeng et al., 2020; Zhang et al., 2020; Xia et al., 2020; Pontesa et al., 2020), and a possible horizontal transmission from parents; however, that happened after birth-related hospitalization (Espinosa, 2020; Zeng et al., 2020). The incidence of transplacental transmission of COVID-19 has not been reflected in newborns with or without DS due mainly to the short course of the new SARS-COV-2 outbreak observed so far (since December 2019). Nevertheless, DS has intrinsic overactivation of interferon, increased levels of inflammatory cytokines, and chronic immune dysregulation and that could lead to a prolonged and exacerbated development of more intense cytokine storm associated with COVID-19 (Espinosa, 2020). Notably, all COVID-19-infected newborns developed leukopenia, lymphopenia, and elevated C-reactive protein like others of adult age. Although there is paucity of reports on parental transmission of COVID-19, the infection may contribute to meiotic nondisjunction in co-operation with other intrinsic or extrinsic factors. The susceptibility of DS to COVID-19 has been discussed in Chapter 2.

A significant association of an abnormal folate metabolism and mutations in the *methylenetetrahydrofolate reductase* and *methionine synthase reductase* genes were found in women having pregnancies with DS (O'Leary et al., 2002; Hassold et al., 2001; Hobbs et al., 2000; Bosco et al., 2003). The result suggested periconceptional folic acid supplementation or fortification that may reduce the risk of occurrence of cells with trisomy 21 during pre- or postfertilization division; however, that gained controversial outcome (Czeizel and Rácz, 1990; Hollis et al., 2013).

2.3 Effects of environment, occupation, and lifestyle

The advent of industrial development and augmentation of environmental pollution are going hand in hand. Conflicting reports have been published on epidemiologic evidence on environmental exposure to radiation and birth of DS. Reports of increased occurrence of chromosome abnormalities in Hiroshima and Nagasaki population (Awa, 1991; Neel et al., 1990) are inconsistent. A similar report on population of Kolam in kerala, India, met with severe criticism (Kochupillai et al., 1976; Uchida and Freeman, 1977; Koya et al., 2012). The accident in Chernobyl nuclear reactor was presented as an explanation for a cluster of DS in Berlin (Sperling et al., 1994), though significant clustering was not evidenced from other birth defect registries (Harjulehto-Mervaala et al., 1992; de Wals et al., 1988). However, low-dose ionizing radiation from atomic weapon testing was correlated with increased birth of DS in a time-series study in England (Bound et al., 1995). In experimental animals, ionizing radiation had induced nondisjunction (Hook, 1984; Hook and Porter, 1976). An association between diagnostic exposure to X-rays and maternal age to influence nondisjunction has been found due to accumulation of radiation over a period of time (Alberman et al., 1972); however, that was contradicted by Francis and Snee (1991). Doyle et al. (2000) reported no association between paternal occupational exposures to ionizing radiation in nuclear industry and risk of DS. Nevertheless, more data are required to conclude these risk and incidences. Karyotyping of the third abortus of the three recurrent first trimester miscarriages revealed trisomy 11 (Fig. 4.2), where male partner was a cleaner of the reactor in an atomic energy institute (Ganguly, unpublished data).

A number of studies were conducted on parental occupational exposure and risk of DS birth (Coppedè, 2016). The reports were inconsistent and conflicting due mainly to many other significant factors, which were not taken into consideration such as exposure to insecticides and cleaning compounds, sanitary condition, drinking habit, small number of cases tested, etc. Significant association was reported on fathers working in different field at the time of conception (Brender and Suarez, 1990; Keen et al., 2020; Olshan et al., 1989). Parental exposure to metal industries, sales, teaching, agriculture, etc. was not associated with increase in risk of DS (Irgens et al., 2000). Other studies have reported significant contribution to high rates of DS (Olshan et al., 1989). Use of census data for such analysis could be the limiting factor for such variations (Morris et al., 2002). Similar conflicting reports have been published on proximity to various industries and hazardous landfill sites and risk of chromosome anomalies, particularly DS (Castilla and Orioli, 2004; Vrijheid et al., 2003).

Maternal use of spermicidal contraceptives was not found to affect DS birth (Louik et al., 1987). High coffee consumption has been hypothesized to reduce the viability of pregnancy with DS leading to loss of fetuses in early stage of pregnancy. With regards to occupational or environmental exposure, there could be many other confounding variables, which must be considered in association with intrinsic biological factors for analysis of such association with the increase of DS birth since the global environment is daily exacerbated with the heavy load of complex chemical compounds, particulate matters, and gaseous composition. Actually, the incidence of DS birth is on the rise, particularly in

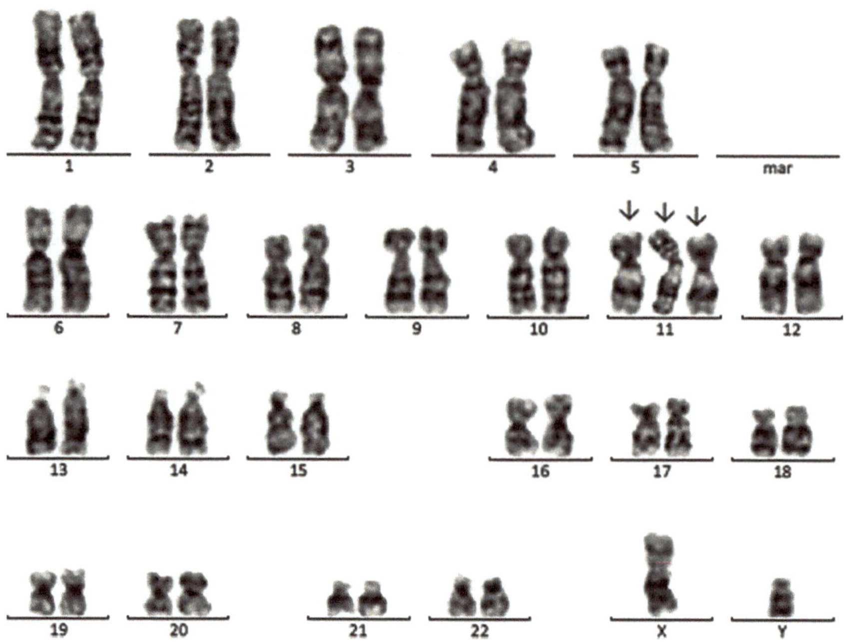

FIGURE 4.2 Karyotype of an abortus with 47,XX,+11 [25]. The father of this pregnancy was engaged in cleaning of an atomic energy reactor (Ganguly, unpublished data).

India, and if the aborted products of early pregnancies are also taken into account, the ratio of pregnancies with the syndrome would be much higher. The existing conflicting findings underlie the need for national surveillance programs and international comparisons to identify the environmental risk factors causing DS. Therefore, contribution of substance abuse and environmental factors during preconception period needs to be addressed to reduce the major health risks and ascertaining a healthy pregnancy. Studies have demonstrated the aversion and prevention of exposure effects in improving the health of reproductive women and their offspring. Notably, there is insufficient knowledge on such effects, and that need to be revisited with redesigned studies considering both paternal and maternal factors. Unfortunately, most of the available reports are concentrated around maternal factors, especially advanced maternal age at conception and recombination errors as certain risk factors for the birth of a child with DS.

2.4 Meiotic recombination

Although advanced maternal age influences occurrence of trisomy 21, the molecular mechanism of meiotic nondisjunction is not clear even after 50 years of identification of trisomy 21 as the underlying facor for DS. Since M1 anaphasic separation occurs after M1-metaphase, it has been speculated that junctions of "crossing over" could lead to unequal separation of chromosomes (Hassold and Sherman, 2000; Vranekovic' et al., 2012). The concept was further supported by the presence of variable phenotypes observed among the DS patients, such as presence or absence of congenital heart defects. Meiotic recombination and maternal age produce trisomy 21 approximately in 90% eggs. The risk of nondisjunction was ~50% for chromosomes lacking recombination, resulting in nondisjunction of "achiasmate bivalents," and that increased the risk of aneuploidy, frequently with malsegregation of chromosome 21 (Fig. 4.3) (Hultén, 1990). Further, the association between recombination site and nondisjunction has been established in humans (Hultén et al., 2008) and lower organisms. Additionally, epidemiological studies have identified many extrinsic risk factors such as alcoholism, exposure to radiation, toxic chemicals, use of fertility drugs, oral contraceptives and spermicides, viral infection, etc., to cause errors in cell division. However, unequivocal proof is still lacking for understanding the contribution of extrinsic and intrinsic risk factors on meiotic or mitotic nondisjunction, and therefore, further research is essential on large database for risk assessment of meiotic nondisjunction and/or recombination.

The basis of age-dependent increase in meiosis I errors has been an enigma for decades. The biological phenomenon was studied through genomic analyses of polymorphic markers of the chromosome 21 for determination of the frequency and/or location of the crossovers that ensure faithful chromosome segregation. Notably,

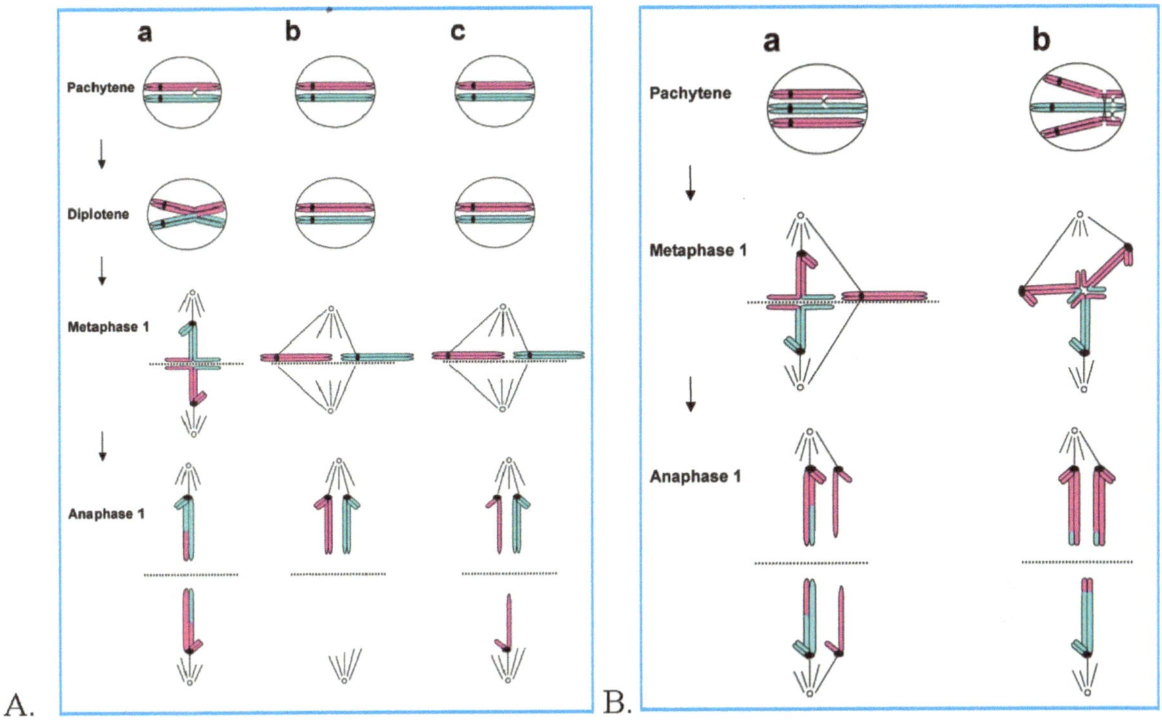

FIGURE 4.3 (A) **Schematic presentation of different types of meiosis I segregation that may take place in a normal disomy 21 oocyte:** (a). Normal chromosome pairing and crossing-over, attachment of the movement centers (kinetochores) at metaphase 1 and separation at anaphase 1; (b). lack of crossing-over and chiasma formation may lead to primary nondisjunction at anaphase 1; (c). lack of a chiasma can also lead to the same type of segregation at anaphase 1 as during mitosis (precocious meiotic disjunction). (B). **Schematic representation of different types of secondary nondisjunction that may take place in a trisomic 21 oocyte:** (a). Formation of a bivalent plus a univalent, where the univalent is undergoing precocious disjunction leading to a chromosome 21 plus a chromatid in each of the daughter cells; (b). formation of a trivalent with chiasmata in an aberrant position, that is, in this example distally within the long arm. Secondary nondisjunction at anaphase 1 will lead to two chromosomes 21 traveling into one of the daughter cells and one chromosome 21 into the other. *Reproduced from: (A) and (B) Hultén et al. (2008).*

cytogenetic analyses of the unfertilized oocytes in vitro could not draw any definitive conclusions regarding mis-segregation in meiosis I (Kurahashi et al., 2012). Nevertheless, studies of meiosis-specific cohesin have been described in conditional knockout mice. Furthermore, shortening of the telomere length has also been hypothesized to result in molecular aging of oocytes that might jeopardize the proper segregation of chromosomes in the MI phase of oogenesis (Treff et al., 2011).

References

Aagesen, L., Grinsted, J., Mikkelsen, M., 1984. Advanced grandmaternal age on the mother's side – a risk of giving rise to trisomy 21. Ann. Hum. Genet. 48 (4), 297–301. https://doi.org/10.1111/j.1469-1809.1984.tb00843.x.

Aboulghar, H., Aboulghar, M., Mansour, R., Serour, G., Amin, Y., Al-Inany, H., August 2001. A prospective controlled study of karyotyping for 430 consecutive babies conceived through intracytoplasmic sperm injection. Fertil. & Steril. 76 (2), 249–253.

Alberman, E., Polani, P.E., Fraser Roberts, J.A., Spicer, C.C., Elliott, M., Armstrong, E., 1972. Parental exposure to x-irradiation and Down's syndrome. Ann. Hum. Genet. 36, 195–207.

Allen, E.G., Freeman, S.B., Druschel, C., Hobbs, C.A., O'Leary, L.A., Romitti, P.A., Royle, M.H., Torfs, C.P., Sherman, S.L., February 2009. Maternal age and risk for trisomy 21 assessed by the origin of chromosome nondisjunction: a report from the Atlanta and National Down Syndrome Projects. Hum. Genet. 125 (1), 41–52. https://doi.org/10.1007/s00439-008-0603-8.

Awa, A.A., 1991. Persistent chromosome aberrations in the somatic cells of A-bomb survivors, Hiroshima and Nagasaki. J. Radiat. Res. 265–274. Supplement.

Bosco, P., Guéant-Rodriguez, R.M., Anello, G., Barone, C., Namour, F., Caraci, F., Romano, A., Romano, C., Guéant, J.L., 2003. Methionine synthase (MTR) 2756 (A! G) polymorphism, double heterozygosity methionine synthase 2756 AG/methionine synthase reductase (MTRR) 66 AG, and elevated homocysteinemia are three risk factors for having a child with down syndrome. Am. J. Med. Genet. 121A, 219–224.

Bound, J.P., Francis, B.J., Harvey, P.W., 1995. Down's syndrome: prevalence and ionising radiation in an area of north west England 1957-91. LOF Epidemiol. & Commun. Health 49, 164–170.

Brender, J.D., Suarez, L., 1990. Paternal occupation and anencephaly. Am. J. Epidemiol. 131, 517–521. https://doi.org/10.1093/oxfordjournals.aje.a115526.

Castilla, E.E., Orioli, I.M., 2004. ECLAMC: the Latin-American Collaborative study of congenital malformations. Commun. Genet. 7, 76–94. https://doi.org/10.1159/000080776.

Castilla, E.E., Orioli, I.M., Lugarinho, R., Graca, P.D., Lopez-Camelo, J.S., Campana, H.E., Spagnolo, A., Mastroiacovo, P., June 1990. Monthly and seasonal variations in the frequency of congenital anomalies. Int. J. Epidemiol. 19 (2), 399–404. https://doi.org/10.1093/ije/19.2.399.

Chan, A., McCaul, K.A., Keane, R.J., Haan, E.A., October 3, 1998. Effect of parity, gravidity, previous miscarriage, and age on risk of Down's syndrome: population based study. BMJ 317 (7163), 923–924. https://doi.org/10.1136/bmj.317.7163.923.

Chen, C.L., Gilbert, T.J., Daling, J.R., 1999. Maternal smoking and down syndrome: the confounding effect of maternal age. Am. J. Epidemiol. 149, 442–446.

Coppedè, F., December 2016. Risk factors for Down syndrome. Arch. Toxicol. 90 (12), 2917–2929. https://doi.org/10.1007/s00204-016-1843-3. Epub 2016 Sep 7. PMID: 27600794.

Czeizel, A., Rácz, J., November 1990. Evaluation of drug intake during pregnancy in the Hungarian case-control surveillance of congenital anomalies. Teratology 42 (5), 505–512. https://doi.org/10.1002/tera.1420420507.

de Wals, P., Bertrand, F., de is Mata, I., Lechat, M.F., 1988. Chromosomal anomalies and chemobyl. Inijilpidemniol 17, 230–231.

Doyle, P., Maconochie, N., Roman, E., Davies, G., Smith, P.G., Beral, V., 2000. Fetal death and congenital malformation in babies born to nuclear industry employees: report from the nuclear industry family study. Lancet 356, 1293–1299.

Espinosa, J.M., 2020. Down syndrome and COVID-19: a perfect storm? Cell Rep. Med. 1 (2), 100019.

Francis, J., Snee, M., May 1991. A case control study of trisomy 21 and maternal pre-conceptual radiography. Clin. Radiol. 43 (5), 343–346.

Freeman, S.B., Yang, Q., Katherine Allran, K., Taft, L.F., Sherman, S.L., 2000. Women with a reduced ovarian complement may have an increased risk for a child with down syndrome. Am. J. Hum. Genet. 66, 1680–1683.

Ganguly, B.B., unpublished. Factors contributing to Down syndrome. Down Syndrome: An Experience From Real Time Diagnostic Practice. MGM Institute of Health Sciences, India.

Ganguly, B.B., Kadam, N.N., 2017. Down syndrome from the age of characterization to the era of curative approach. Nucleus 60, 197–208. https://doi.org/10.1007/s13237-016-0187-y.

García-Ferreyra, J., Hilario, R., Dueñas, J., 2018. High percentages of embryos with 21, 18 or 13 trisomy are related to advanced paternal age in donor egg cycles. JBRA Assisted Reprod. 22 (1), 26–34. https://doi.org/10.5935/1518-0557.20180004.

Gaulden, M.E., 1992. Maternal age effect: the enigma of Down syndrome and other trisomic conditions. Mutat. Res. Rev. Genet. Toxicol. 296 (1–2), 69–88.

Harjulehto-Mervaala, T., Salonen, R., Aro, T., Saxén, L., March 1992. The accident at Chernobyl and trisomy 21 in Finland. Mutat. Res./DNAg. 275 (2), 81–86. https://doi.org/10.1016/0921-8734(92)90011-D.

Hassold, T.J., Sherman, S., 2000. Down syndrome: genetic recombination and the origin of the extra chromosome 21. Clin. Genet. 57 (2), 95–100. https://doi.org/10.1034/j.1399-0004.2000.570201.x.

Hassold, T., Jacobs, P., kline, J., Stein, Z., Warburton, D., July 1980. Effect of maternal age on autosomal trisomies. Ann. Hum. Genet. 44 (1), 29–36.

Hassold, T.J., Burrage, L.C., Chan, E.R., Judis, L.M., Schwartz, S., James, S.J., Jacobs, P.A., Thomas, N.S., 2001. Maternal folate polymorphisms and the etiology of human nondisjunction. Am. J. Hum. Genet. 69, 434–439.

Hay, S., Barbano, H., December 1972. Independent effects of maternal age and birth order on the incidence of selected congenital malformations. Teratology 6 (3), 271–279. https://doi.org/10.1002/tera.1420060304. PMID: 4345335.

Hecht, C.A., Hook, E.B., April 24, 1996. Rates of Down syndrome at livebirth by one-year maternal age intervals in studies with apparent close to complete ascertainment in populations of European origin: a proposed revised rate schedule for use in genetic and prenatal screening. Am. J. Med. Genet. 62 (4), 376–385. https://doi.org/10.1002/(SICI)1096-8628(19960424)62:4<376::AID-AJMG10>3.0.CO;2-L.

Hobbs, C.A., Sherman, S.L., Yi, P., Hopkins, S.E., Torfs, C.P., Hine, R.J., Pogribna, M., Rozen, R., James, S.J., 2000. Polymorphisms in genes involved in folate metabolism as maternal risk factors for Down syndrome. Am. J. Hum. Genet. 67, 623–630.

Hollis, N.D., Allen, E.G., Oliver, T.R., Tinker, S.W., Druschel, C., Hobbs, C.A., O'Leary, L.A., Romitti, P.A., Royle, M.H., Torfs, C.P., Freeman, S.B., Sherman, S.L., Bean, L.J., March 2013. Preconception folic acid supplementation and risk for chromosome 21 nondisjunction: a report from the National Down Syndrome Project. Am. J. Med. Genet. 161A (3), 438–444. https://doi.org/10.1002/ajmg.a.35796.

Hook, E.B., October 1984. Parental age and unbalanced Robertsonian translocations associated with Down syndrome and Patau syndrome: comparison with maternal and paternal age effects for 47, +21 and 47, +13. Ann. Hum. Genet. 48 (4), 313–325. https://doi.org/10.1111/j.1469-1809.1984.tb00845.x.

Hook, E.B., Porter, I.H., 1976. Human population cytogenetics: comments on racial differences in frequency of chromosome abnormalities, putative clustering of Down's syndrome, and radiation studies. In: Population Cytogenetics. Academic Press, New York, pp. 353–365.

Hultén, M.A., 1990. The origin of aneuploidy: bivalent instability and the maternal age effect in trisomy 21 Down syndrome. Am. J. Med. Genet. Suppl. 7, 160–161. https://doi.org/10.1002/ajmg.1320370732.

Hultén, M.A., Patel, S.D., Tankimanova, M., Westgren, M., Papadogiannakis, N., Jonsson, A.M., Iwarsson, E., 2008. On the origin of trisomy 21 Down syndrome. Mol. Cytogenet. 1, 21. https://doi.org/10.1186/1755-8166-1-21.

Hunter, J., Allen, E., Shin, M., et al., 2013. The association of low socioeconomic status and the risk of having a child with Down syndrome: a report from the National Down Syndrome Project. Genet. Med. 15, 698–705. https://doi.org/10.1038/gim.2013.34.

Irgens, A., Krüger, K., Skorve, A.H., Irgens, L.M., 2000. Birth defects and paternal occupational exposure. Hypotheses tested in a record linkage based dataset. Acta Obstet. Gynecol. Scand. 79, 465–470.

Jongbloet, P.H., Vrieze, O.J., 1985. Down syndrome: increased frequency of maternal meiosis I nondisjunction during the transitional stages of the ovulatory seasons. Hum. Genet. 71 (3), 241–248. https://doi.org/10.1007/BF00284583. PMID: 2933319.

Keefe, D.L., Niven-Fairchild, T., Powell, S., Buradagunta, S., 1995. Mitochondrial deoxyribonucleic acid deletions in oocytes and reproductive aging in women. Fertil. Steril. 64, 577–583.

Keen, C., Hunter, J.E., Allen, E.G., Rocheleau, C., Waters, M., Sherman, S.L., January 2020. The association between maternal occupation and down syndrome: a report from the national Down syndrome project. Int. J. Hyg Environ. Health 223 (1), 207–213. https://doi.org/10.1016/j.ijheh.2019.09.001.

Kochupillai, N., Verma, I.C., Grewal, M.S., Ramalingaswamy, V., 1976. Down syndrome and related abnormalities in an area of high background radiation in coastal Kerala. Nature 262, 60–61. https://doi.org/10.1038/262060a0.

Koya, P., Chougaonkar, M., Predeep, P., Jojo, P., Cheriyan, V., Mayya, Y., Seshadri, M., 2012. Effect of low and chronic radiation exposure: a case-control study of mental retardation and cleft lip/palate in the monazite-bearing coastal areas of Southern Kerala. Radiat. Res. 177 (1), 109–116.

Kurahashi, H., Tsutsumi, M., Nishiyama, S., Kogo, H., Inagaki, H., Ohye, T., 2012. Molecular basis of maternal age-related increase in oocyte aneuploidy. Congenital. Anom. 52, 8–15. https://doi.org/10.1111/j.1741-4520.2011.00350.x.

Lapunzina, P., Delicado, A., de Torres, M.L., Mor, M.A., Perez-Pacheco, R.F., Lopes, P.I., 2002. Autosomal recessive hydrocephalus due to aqueduct stenosis: report of a further family and implications for genetic counselling. J. Matern. Fetal Neonatal Med. 12, 64–66.

Louik, C., Mitchell, A.A., Werler, M.M., Hanson, J.W., Shapiro, S., August 20, 1987. Maternal exposure to spermicides in relation to certain birth defects. N. Engl. J. Med. 317 (8), 474–478. https://doi.org/10.1056/NEJM198708203170803.

Malini, S.S., Ramachandra, N.B., 2006. Influence of advanced age of maternal grandmothers on Down syndrome. BMC Med. Genet. 7, 4. https://doi.org/10.1186/1471-2350-7-4.

McIntosh, G.C., Olshan, A.F., Baird, P.A., 1995. Paternal age and the risk of birth defects in offspring. Epidemiology 6 (3), 282–288.

Morris, J.K., Mutton, D.E., Alberman, E., 2002. Revised estimates of maternal age specific live birth prevalence of Down syndrome. J. Med. Screen 9, 2–6.

Neel, J.V., Schullt, W.J., Awa, A.A., Satoh, C., Katot, H., Otaket, M., Yoshimoto, Y., 1990. The children of parents exposed to atomic bombs: estimates of the genetic doubling dose of radiation for humans. Am. J. Hum. Genet. 46, 1053–1072.

O'Leary, V.B., Parle-McDermott, A., Molloy, A.M., Kirke, P.N., Johnson, Z., Conley, M., Scott, J.M., Mills, J.L., January 15, 2002. MTRR and MTHFR polymorphism: link to Down syndrome? Am. J. Med. Genet. 107 (2), 151–155. https://doi.org/10.1002/ajmg.10121. First published: 28 November 2001.

Neuhäuser, M., Krackow, S., 2007. Adaptive-filtering of trisomy 21: risk of Down syndrome depends on family size and age of previous child. Naturwissenschaften 94 (2), 117–121. https://doi.org/10.1007/s00114-006-0165-3.

Ognjanovic, S., Puumala, S., Spector, L.G., Smith, F.O., Robison, L.L., Olshan, A.F., Ross, J.A., May 2009. Maternal health conditions during pregnancy and acute leukemia in children with Down syndrome: a Children's Oncology Group study. Pediatr. Blood Cancer 52 (5), 602–608. https://doi.org/10.1002/pbc.21914.

Olshan, A.F., Baird, P.A., Teschke, K., May 1989. Paternal occupational exposures and the risk of Down syndrome. Am. J. Hum. Genet. 44 (5), 646–651.

Pangalos, C., Théophile, D., Sinet, P.M., Marks, A., Stamboulieh-Abazis, D., Chettouh, Z., Prieur, M., Verellen, C., Rethoré, M.O., Lejeune, J., et al., December 1992. No significant effect of monosomy for distal 21q22.3 on the Down syndrome phenotype in "mirror" duplications of chromosome 21. Am. J. Hum. Genet. 51 (6), 1240–1250. PMID: 1463008; PMCID: PMC1682936.

Patterson, D., August 1987. The causes of Down syndrome. Sci. Am. 257 (2), 52–61. https://www.jstor.org/stable/24979443.

Piersigilli, F., Carkeek, K., Hocq, C., van Grambezen, B., Hubinont, C., Chatzis, O., Van der Linden, D., et al., 2020. COVID-19 in a 26-week preterm neonate. Lancet Child Adolesc. Health 4 (6), 476–478.

Pontesa, G.C.L., Lobatob, A.C.R., da Silvac, S.M., da Silvad, D.F.A., da Silvad, A.F., Rochae, R.S.B., de Oliveiraa, B.D., Martinsf, T.O.C., Chermontg, A.G., da Costa Cunha, K., 2020. COVID-19 in a down syndrome newborn. Int. J. Clin. Padiatr. 9 (4), 116–119. https://doi.org/10.14740/ijcp396.

Rasmussen, P., Börjesson, O., Wentz, E., Gillberg, C., 2001. Autistic disorders in Down syndrome: background factors and clinical correlates. Dev. Med. Child Neurol. 43 (11), 750–754. https://doi.org/10.1017/S0012162201001372.

Rittler, M., Liascovich, R., López-Camelo, J., Castilla, E.E., July 22, 2001. Parental consanguinity in specific types of congenital anomalies. Am. J. Med. Genet. 102 (1), 36–43. https://doi.org/10.1002/1096-8628(20010722)102:1<36::aid-ajmg1394>3.0.co;2-m. PMID: 11471170.

Schupf, N., Lee, J.H., Schupf, N., Ottman, R., Mayeux, R., et al., 1994. Increased risk of Alzheimer's disease in mothers of adults with Down's syndrome. The Lancet 344 (8919), 353–356. Own's syndrome.

Sherman, S.L., Takaesu, N., Freeman, S.B., et al., 1991. Trisomy 21: association between reduced recombination and nondisjunction. Am. J. Hum. Genet. 49 (3), 608–620.

Sperling, K., Pelz, J., Wegner, R.D., et al., 1994. Significant increase in trisomy 21 in Berlin nine months after the Chernobyl reactor accident: temporal correlation or causal relation? BMJ 309, 158–162.

Steiner, B., Masood, R., Rufibach, K., Niedrist, D., Kundert, O., Riegel, M., Schinzel, A., 2015. An unexpected finding: younger fathers have a higher risk for offspring with chromosomal aneuploidies. Eur. J. Hum. Genet. 23, 466–472. https://doi.org/10.1038/ejhg.2014.122.

Stoll, C., Alembik, Y., Dott, B., Roth, M.P., 1990. Epidemiology of Down syndrome in 118,265 consecutive births. Am. J. Med. Genet. Suppl. 7, 79–83. https://doi.org/10.1002/ajmg.1320370715.

Torfs, C.P., Christianson, R.E., December 15 , 2000. Effect of maternal smoking and coffee consumption on the risk of having a recognized down syndrome pregnancy. Am. J. Epidemiol. 152 (12), 1185–1191. https://doi.org/10.1093/aje/152.12.1185.

Treff, N.R., Su, J., Taylor, D., Scott Jr., R.T., 2011. Telomere DNA deficiency is associated with development of human embryonic aneuploidy. PLoS Genet. 7, e1002161.

Uchida, I., Freeman, C., 1977. Radiation-induced nondisjunction in oocytes of aged mice. Nature 265, 186–187. https://doi.org/10.1038/265186a0.

Van Blerkom, J., Sinclair, J., Davis, P., 1998. Mitochondrial transfer between oocytes: potential applications of mitochondrial donation and the issue of heteroplasmy. Hum. Reprod. 3, 2857–2868.

Vraneković, J., Bozović, I.B., Grubić, Z., Wagner, J., Pavlinić, D., Dahoun, S., et al., 2012. Down syndrome: parental origin, recombination, and maternal age. Genet. Test. Mol. Biomark. 16 (1), 70–73. https://doi.org/10.1089/gtmb.2011.0066. Mary Ann Liebert Inc.

Vrijheid, M., Armstrong, B., Dolk, H., van Tongeren, M., Botting, B., 2003. Risk of hypospadias in relation to maternal occupational exposure to potential endocrine disrupting chemicals. Occup. Environ. Med. 60, 543–550.

Xia, W., Shao, J., Guo, Y., Peng, X., Li, Z., Hu, D., 2020. Clinical and CT features in pediatric patients with COVID-19 infection: different points from adults. Pediatr. Pulmonol. 55 (5), 1169–1174.

Yang, Q., Wen, S.W., Leader, A., Chen, X.K., Lipson, J., Walker, M., 2007. Paternal age and birth defects: how strong is the association? Hum. Reprod. 22 (3), 696–701. https://doi.org/10.1093/humrep/del453.

Yoon, P.W., Freeman, S.B., Sherman, S.L., Taft, L.F., Gu, Y., Pettay, D., Flanders, W.D., Khoury, M.J., Hassold, T.J., 1996. Advanced maternal age and the risk of Down syndrome characterized by the meiotic stage of chromosomal error: a population-based study. Am. J. Hum. Genet. 58 (3), 628–633.

Zeng, L.K., Tao, X.W., Yuan, W.H., Wang, J., Liu, X., Liu, Z.S., 2020. [First case of neonate infected with novel coronavirus pneumonia in China]. Zhonghua Er Ke Za Zhi 58 (0), E009.

Zeng, L., Xia, S., Yuan, W., Yan, K., Xiao, F., Shao, J., Zhou, W., 2020. Neonatal early-onset infection with SARS-CoV-2 in 33 neonates born to mothers with COVID-19 in Wuhan, China. JAMA Pediatr. 174 (7), 722–725. https://doi.org/10.1001/jamapediatrics.2020.0878.

Zhang, Z.J., Yu, X.J., Fu, T., Liu, Y., Jiang, Y., Yang, B.X., Bi, Y., 2020. Novel coronavirus infection in newborn babies aged <28 days in China. Eur. Respir. J. 55 (6).

Zhu, J.L., Madsen, K.M., Vestergaard, M., Olesen, A.V., Basso, O., Olsen, J., 2005. Paternal age and congenital malformations. Hum. Reprod. 20 (11), 3173–3177. https://doi.org/10.1093/humrep/dei186.

CHAPTER

5

Prevention of Down syndrome: genetic counseling, prenatal screening, testing, awareness, attitude, and socio-cultural and public health aspects

1. Introduction

The proverb "Prevention is better than cure" is true for the genetic defects, since many of the genetic disorders have no cure till date. While research on understanding the origin of trisomy 21 (HSA21) and deciphering the gene responsible for specific characteristics of Down syndrome is going on, preventive approach has gained importance alongside for controlling the birth of Down syndrome (Ganguly and Kadam, 2016). Indeed, prenatal screening and testing and genetic counseling can reduce and control recurrence of Down syndrome in future pregnancies. Prenatal testing will be applicable to pregnancies with indicative screening results, pregnancies at advanced maternal age, and parents having past pregnancy with Down syndrome or with family history of the syndrome. Since the sampling procedure is invasive, many couples may not want to participate in prenatal testing and know the status or act on the information to be extracted.

Noninvasive biochemical screening in maternal serum including β-human chorionic gonadotropin (β-HCG), pregnancy-associated plasma protein (PAPPA), α−fetoprotein (AFP), Estriol, and Inhibin during first and second trimester of pregnancy, respectively (Table 5.1) (Ganguly and Kadam, 2019), has become a standard practice of screening of trisomy 21 and neural tube defect. If the AFP screening value is low, it suggests that the risk of Down syndrome in the fetus is equal to the risk of a woman of age 35 years. In cases with positive indication, prenatal testing is considered involving chorionic villus sampling (CVS) or amniocentesis for confirmation of the screening status and type of trisomy 21. However, these invasive sampling procedures have some empirical risk of miscarriage, though facilitate confirmation of Down syndrome with over 99% accuracy. In the recent era, combined screening of biochemical markers and high-resolution ultrasound imaging has proved wider acceptance for detection of pregnancies with Down syndrome. Furthermore, detection of trisomy 21 has been revolutionized considering cell-free-fetal-DNA (cffDNA) from noninvasive collection of maternal blood (NIPT) and that has been implemented into clinical practice in many developed countries

https://doi.org/10.1016/B978-0-323-90456-8.00010-7

TABLE 5.1 Chronological development of preventive approach.

Chronology	Developments
1950–59	Amniocentesis procedure was developed
1960	First amniocentesis was performed
1968	First Down syndrome was detected using amniocentesis
1976	Amnios became more common. Amniocentesis came into common use in the United States. The first abortion after mid-trimester amnios was reported.
1981	"Down syndrome preventative medical checklist," was edited by William I. Cohen.
1983	Influence of maternal age on Down syndrome was understood. Women over 35 years were offered prenatal testing.
1984	Biochemical markers in maternal blood was linked with pregnancies with Down syndrome.
1999	The Down syndrome preventive checklist was first revised after 1981.
2007	American college of Obstetrics and Gynecology (ACOG) issued new recommendation that all pregnant women, regardless of their age, should participate in prenatal screening for Down syndrome and other genetic disorders. This changed ACOGs long-standing recommendation of prenatal testing only to pregnant women over 35 years of age.
2011	NIH formed "Down syndrome consortium." Official guidelines on genetic counseling for catering better assistance and guidance to parents on present and future risk or having a newborn with Down syndrome were released.
Close 2012	Prenatal testing pamphlet was published for pregnant women.
2012–13	Combined screening of biochemical markers and ultrasound imaging were proved useful.
2014	NIPT was established for screening of Down syndrome in cffDNA obtained from the maternal blood.

depending on its regulatory issues of prenatal diagnostic practices. The chronological development and approach of preventive practice is listed below (Ganguly and Kadam, 2016):

Currently, sequencing of trisomy 21 in cffDNA is turning fate of prenatal genetic tests into choice. Noninvasive sampling and advanced technology have made prenatal genetic testing of trisomy 21 more acceptable by many more pregnant women to choose whether or not to have a Down syndrome child. Confirmation of positive screen result is accomplished with CVS or amniocentesis (Table 5.2).

2. Genetic counseling

Screening and testing of suspected genetic illness play crucial roles in the field of perinatology for reduction and prevention of genetic defects and their recurrent transmission. However, lack of awareness among the lay public appears to be a major concern observed in many surveys and screening programs of genetic diseases (Morgan et al., 2004; Qureshi et al., 2006). In addition, associated expenses and/or lack of reimbursement facility create further barriers (Morgan et al., 2004; Stark et al., 2013). Here comes the role of genetic counseling, which plays a pivotal role for making an understandable background of screening and/or testing of the genetic condition of a growing fetus. Genetic counseling will explain the genetic information in an understandable language and thus, help couples or parents taking a judicious decision on adoption of preconception and prenatal screening with a clear understanding of the significance. Notably, genetic counseling helps in

TABLE 5.2 Modalities of screening and testing for prevention of Down syndrome.

Name of the test	Sample	Technique	Time of screening	Detection	Diseases screened	Noninvasive/ Invasive	Risk: False + ve/− ve	Recommendation	Insurance coverage
NIPT	Cell-free fetal DNA (cff-DNA) in maternal plasma	Molecular techniques (a-CGH/ SNP-array/ sequencing)	Anytime after 8 weeks	Validated for trisomy 13/ 18/21 (detection of gene mutations possible)	Down/Patau/ Edward syndromes (does not detect NTD)	Noninvasive	Low, 99% accurate for Down syndrome	1. Integration with first trimester NT screen, and CVS/ amniocentesis for confirmation 2. Recommended only for ≥35 years women and/or having h/o high risk, and not for low-risk women (ACOG, 2015)	Most likely not covered by insurance companies. In absence of national coverage determination, varies on gestational age/ resource/ countries
Double screen	Maternal blood serum	Biochemical screening	10–13 weeks	HCG, PAPP-A	Down/Edward syndrome, cardiac defects	Noninvasive	High	Integration with first trimester NT screen	May be covered as a part of routine prenatal care
Nuchal Translucency (NT) screening	Expecting mother's womb	Ultrasound imaging	10–13 weeks	Collection of fluid beneath the fetal skin in the region of the fetal neck	Trisomy 21/ other chromosomal anomalies	Noninvasive	High	Integration with first trimester double screening	Covered as a part of routine prenatal care
Triple screen: Normally considered for ≥35 years old women or other history and indicative NT/double screen	Maternal blood serum	Biochemical screening	16–18 weeks	AFP, HCG, EU	↑AFP: Spina bifida or anencephaly; ↓AFP and abnormal HCG/EU: Down/Edward syndrome, other chromosomal aberrations	Noninvasive	High	CVS/ amniocentesis for confirmation; ultrasound imaging	Most likely covered by insurance companies

Continued

TABLE 5.2 Modalities of screening and testing for prevention of Down syndrome.—cont'd

Name of the test	Sample	Technique	Time of screening	Detection	Diseases screened	Noninvasive/ Invasive	Risk: False + ve/− ve	Recommendation	Insurance coverage
Quad screen	Maternal blood serum	Biochemical screening	14–20 weeks	AFP, HCG, EU, Inhibin A	Chromosomal aberrations and NTD	Noninvasive	High	CVS/ amniocentesis (less accurate than integrated screen; however, time goes for this screen)	Most likely covered by insurance companies
Chorionic villi sampling (CVS): Diagnostic testing: In cases with indicative NT/NIPT/ double screen, and women of ≥35 years age	CVS	Genetic (tissue culture)	11–13 weeks	Chromosomal aberrations and gene mutations (does not detect NTD and anatomical defects)	Numerical/ structural aberrations, and recessive/ dominant disorders	Invasive (trans-vaginal or transabdominal); carries risk (>1%) of miscarriage	Nil (~99–100% accurate)	NT screen for NTD, Genetic counseling for whether it is right for the parents and/or the best next steps if the results are positive	Generally covers CVS but it may not if women is below 35, "low-risk" for certain problems, and/ or normal results from first trimester screens
Amniocentesis: Diagnostic testing: In cases with indicative NT/NIPT/ double/triple/ quad screen, and women of ≥35 years age	Amniotic fluid	Genetic (tissue culture)	16–20 weeks	Chromosomal aberrations, gene mutations, and NTDs (does not detect anatomical defects)	Numerical/ structural aberrations, and recessive/ dominant disorders; NTD	Invasive (transabdominal); carries less risk (<1%) of miscarriage	Nil (~99–100% accurate)	NT screen for NTD Genetic counseling for whether it is right for the parents and/or the best next steps if the results are positive	Generally covers but it may not if women is below 35, "low risk" for certain problems, and/ or normal results from first- or second-trimester screens

AFP, alfa-fetoprotein; *CVS*, chorionic villus sampling; *EU*, unconjugated estriol; *HCG*, human chorionic gonadotropin; *NIPT*, noninvasive prenatal testing; *NT*, nuchal translucency; *NTDs*, neural tube defects (spina bifida); *PAPPA*, pregnancy-associated plasma protein A.

Reused with permission Ganguly and Kadam (2019).

breaking the news of diagnosis of Down syndrome, which otherwise is frightful to the parents and families carrying some heritable genetic defects and risk of having a pregnancy with anomalies. Moreover, while disclosing such news, the following points should be considered important:

a. dictate the diagnostic result as soon as possible in a straightforward approach;
b. discuss the pros and cons and genetic consequences of the syndrome, include both parents and some relatives having adequate knowledge to understand the discussion, if parents cannot do that;
c. maintain adequate privacy and confidentiality during counseling;
d. use of clear and understandable language; discussion should not use negative sentences but should be balanced with the point of view of the condition;
e. allow parents to ask questions and clear their doubts; provide educational materials, sources of further information, and phone numbers of the genetic counselor or a contact person for clarification of further queries;
f. discuss the necessity of parents' examinations and future pregnancies, and so on.

Genetic counseling with the parents on discussion of the facts and future of trisomy 21 in their child with Down syndrome gives an overview of the overall health of the child, clinical implications, risk of disease consequences, guidance for follow-up evaluations, and early intervention of therapeutics for supporting the development of growth and milestones. Genetic counseling can also facilitate access to resources including books, articles, or websites, educate the parents about the need of special schooling, and guide them for joining in the parent's group of Down syndrome children. Genetic counselor also explains about the consequences of screening of future pregnancies.

Physicians' familiarity with the available array of genetic tests, sensitivity and limitations of technological intervention, interpretation of the results, risk and associated trauma, societal stigmatization and professional discrimination, disclosure of the genetic result to the biological relatives and their reaction, and so on would be helpful to clear the doubts of the patients. Pre- and posttest genetic counseling and increased genetic knowledge might reduce the psychological effects and anxiety. It was experienced through several study groups that counseling should be provided by a clinician for communicating the genetic information, as lay people undergoing genetic screening are not familiar with the diseases screened for and do not understand a positive result (abnormality) (Cho et al., 2013). In cases of known genetic illness, prenatal diagnosis (PND) or preimplantation genetic diagnosis (PGD) will control the birth with the prevailing genetic diseases, provided the reduction of the disease burden or prevention may be regarded as an explicit goal of the family or the affected community.

The American College of Obstetrics and Gynecology (ACOG) and American College of Medical Genetics (ACMG) recommend every individual undergoing a genetic test or screening shall be explained about the purpose, importance, and appropriateness of the tests or screening, when and which test to order, what information is expected, limitations, sensitivity, and specificity of the tests, risk of false-positive or negative results in light of patient's medical and family history, the options of medical management or other possible remedial actions available, facts and future consequences of genetic defects, possible implications of the test results, potential risks, and benefits of the procedure, including psychosocial trauma of discrimination and stigmatization (WHO, 1998). The possibility of psychological harms can be minimized through genetic counseling about the screening and consequences of being a carrier of some genetic defect.

Moreover, public shall be educated with genetic information for creating a positive attitude.

A detail discussion of the screening is the prerequisite for making informed decisions, which shall be expressed on "informed consent." Patients should be informed about the present health status and future risk, carrier, marriage, or reproductive choices and also the uncertainties regarding test-reliability, penetrance of genes, unavailability of efficacious interventions to reduce the consequences of the genetic diseases, and its impact on blood-linked kins. The consent may also protect the inadvertent and involuntary disclosure of genetic information to kindred and unauthorized people. The physicians may articulate the circumstances under which they would consider disclosure obligatory. The ownership of the genetic information, confidentiality, security, and future access by parents and child may be clarified in the consent form and that may be left with the parents and the child. Counseling should also address the anticipated use of genetic samples, when will they be destroyed, whether used for research and diagnostic test development, state and federal regulations governing genetic information, confidentiality of information, etc (ACMG Statement, 1995). Couple's informed consent for prenatal screening or testing is mandatory.

3. Preconception screening and prenatal diagnosis

Preconception screening of the parents can rule out the possibility of carrying translocation of 21 or any other chromosomes, and that can prevent birth of children with unforeseen deleterious inheritance from the parents and needless suffering from the disorder (Committee Opinion No. 442, 2009; Grody et al., 2013). However, screening is generally considered for a known family history and/or opportunistic assessment of risk of genetic illness, more commonly for Down syndrome due to its high prevalence. Moreover, pregnancies suspected to be carrying an affected child, in which the consequences are severe and detrimental, shall be considered for PND.

PND involves a group of procedures to look for trisomy 21 in a developing fetus (Table 5.2; Fig. 5.1) (Bianchi et al., 2012; Norton et al., 2012; Canick et al., 2012). The significance of prenatal screening (indicates likelihood) and testing (confirms a diagnosis) must be understood with a view to rendering appropriate antenatal management. In addition, PND relieves the parents' from anxiety and unfulfilled expectations from a child born with a genetic defect. ACOG recommendations include PND and PGD for couples who already have a child with Down syndrome, pregnant women older than 35 years, and couples with ≥ 2 miscarriages (Committee Opinion No. 545, 2012; Committee Opinion No. 640, 2015). The PND/PGD also facilitates preparation of birth and care of a potentially affected pregnancy, if the couple opts for delivering such pregnancy or may move toward termination. PGD on gametes collects information on genetic abnormalities prior to fertilization and thus, directs implantation of genetically normal embryos. PGD is also employed on embryos at preimplantation stage. Notably, the approach of PND/PGD may minimize the consequences of bringing a helpless baby into this world, who may suffer a serious and fatal destiny for which the parent as well as the child will have no control, especially until intervention with effective treatments is established. ACOG and the Society for Maternal-Fetal Medicine (SMFM) recommended confirmation of Down syndrome following CVS or amniocentesis and chromosome analysis (Committee Opinion No. 545, 2012; Committee Opinion No. 640, 2015; Soothill and Lo, 2014). For couples having an affected child with the syndrome, prior knowledge about the structure of the extra 21 (free or translocation) is important for considering parental karyotyping by selecting an appropriate technology, with a view to studying the structure of the extra 21. Such awareness, ideally prior to the second conception, would be more

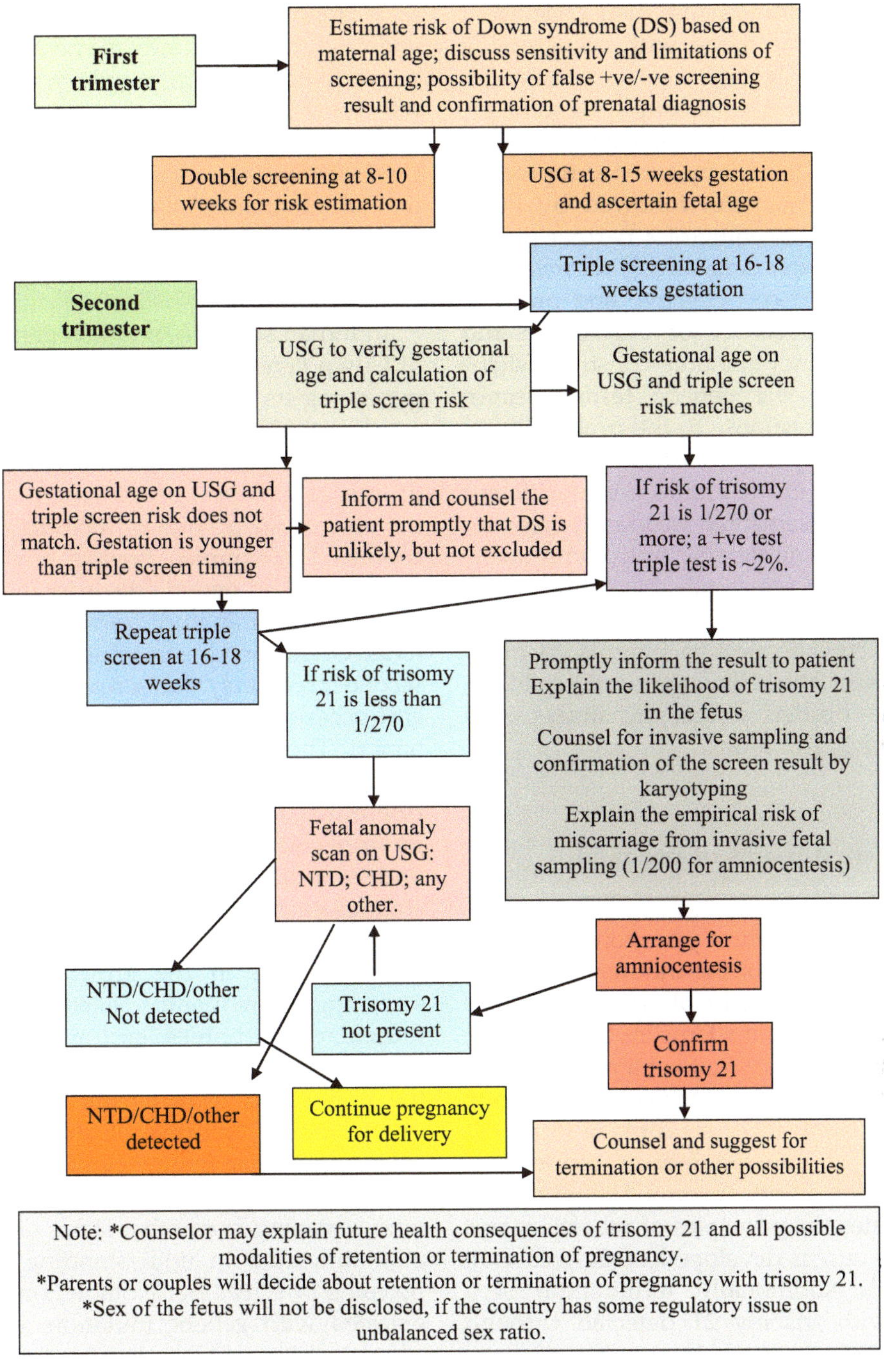

FIGURE 5.1 Algorithm of prenatal risk assessment and diagnosis of Down syndrome.

cost-effective and time-saving than concurrent screening and testing along with the growing fetus of suspect.

Further, carrier status of a couple for some genetic defects may guide to achieve a parenthood status biologically and/or by assisted reproduction with donors' gametes followed by PGD. Ideally, a positive screening of a partner guides for the other partner's screening. Preconception screening would be cost-effective and practical than concurrent screening of the couple and PND of the pregnancy at a time-constraint situation. Collectively, the genetic testing leaves certain choices in response to the probability of having a genetically healthy or unhealthy child. Therefore, the healthcare authorities of different countries have implemented various approaches of antenatal screening ranging from community-based population screening to families with known history of genetic illness, with a view to empowering the prospective parents to make informed choices about having normal children, lowering the burden of genetic illness, and improving the health index.

3.1 Anticipated issues of prenatal screening

The practice of "in vitro" abortion lowers the emotional attachment of the mother. Nevertheless, such practice of in vitro abortion of embryo or in vivo termination of a fetus involves human individual "presentient" (not yet a person), ethically and in principle (Medical Policy No. 4.01.21, 2015). Conflicts of concepts of terminating the lives of inflicted fetuses report that eliminating a fetus with trisomy 21 that cannot be cured dilutes the principles of medicine, and thus, until its cure is developed, such lives shall be preserved. Additionally, termination of a pregnancy with trisomy 21 detected through PND raises debates on selective breeding or eugenics, which might one day eliminate population with Down syndrome entirely because they cannot meet the criteria of an ideal society, and thus, the community may emerge into world of supremacy where people with the extra 21 will be scorned. However, a small section of the pediatricians somewhat thought that Down syndrome is not a legitimate reason for termination of pregnancy, though 15%–21% of them thought that current prenatal screenings in general are somewhat based on eugenic thinking (Williams, 2011).

In Indian scenario, prenatal genetic screening of Down syndrome is primarily advised by the gynecologists to all couples, irrespective of their age, but obviously to women of $\geq$35 years of age and having an affected child with Down syndrome, at the appropriate gestations to exclude the risk of aneuploidies, including Down syndrome. The couples having a child affected with identified genetic disorder are advised for carrier screening and testing of the growing fetus. Often carrier screening is opted while the second pregnancy has already been conceived, and occasionally before planning further pregnancy. The primary issue is affordability for three tests (parents and the affected child to rule out structural alteration in one 21) at one go since concept of medical insurance is beyond understanding of many families. Hence, many couples even directly decide for termination of pregnancy, without confirmation of the presence of trisomy 21 in the growing fetus. Carrier screening is frequently asked for thalassemia, muscular dystrophies, and autism by the families having a prevailing history. Screening at the population level is still not an approach due to limited information available on prevalence and carrier frequency of genetic disorders in the country. Nevertheless, outsourcing of genetic screening from other states or overseas creates difficulties in understanding the result or outcome of screening. Notably, once a carrier is detected with genetic mutation, undue fear of stigmatization and psychological trauma, blame by the spouse, and affected offspring and in-laws affect the family environment to a large extent. A positive result at pre-employment or premarital stage also poses serious risk, which can be minimized by genetic counseling. Notably, the overall uptake rates for individual and/or family screening or couples of blood lineage planning a pregnancy are much lower, even though such screening has tremendous significance for prevention of Down syndrome, and even if such screening is offered at zero cost (Ganguly et al., 2017).

There is undue fear of feeling guilty or blame and/or fear of discrimination, particularly in reserved communities, if a woman screens positive or delivers a child with Down syndrome (Clarke, 2014). Parents of Down syndrome should be informed about the present health status and future risk and also choices of marriage or reproduction and its impact on blood-linked kins. The public shall also be educated with genetic information for creating a positive attitude toward families and children with Down syndrome.

Nevertheless, confidentiality of test results and genetic discrimination by insurers may cause detrimental effects, including pressure on individual choice, particularly in socially tight communities, high schools, and workplace. Also, detection of a positive carrier may devalue the lives of affected individuals or impedes the search for a cure (Godard et al., 2003). However, a negative test result may relieve the patients and all blood-linked members from anxiety. As there is a possibility of a small residual risk in carrier screening or diagnosis of translocation involving 21 at the molecular level, mostly due to incomplete sensitivity of the test, or incomplete penetrance of the genes of partial trisomy after a favorable test result, the expecting couples may thus be explained with the unexpected birth of a child with this genetic disease. Similarly, in cases of normal karyotype in children born with significant congenital malformations, or their parents, there could be possibility of genetic imbalance or partial trisomy of chromosome 21. And without having any knowledge of a suspected genetic mutation for a clinical condition, molecular screening and/or testing, though feasible with today's technologies, would be a long-term research for an affected family. Healthcare professionals, thus, need to explain the potential misperceptions that screening can guarantee a genetically healthy child.

Taken together, careful weighing of certain advantages and disadvantages of genetic screening and testing makes a straightforward decision-making process of having a genetically healthy or impaired child. Societal discrimination, including health insurance, employment, and adoption, already exists for the genetically affected children. The practice of elective abortion of pregnancies with Down syndrome is widely accepted in the United States (91%–93% of pregnancies with Down syndrome) and many other countries. Otherwise, birth of a child with such genetic impairment may develop distress, psychological, and emotional harm and sufferings, early death of the child, loss of opportunities and freedom, stigmatization or isolation, etc., and that too at the cost of added financial expenses. There could be potential threat on the relationship between parents and the family due to parents' unawareness or lack of preparation for dealing with the special medical interventions that the child may be subjected to. Therefore, PND shall be considered to rule out Down syndrome at the unborn age.

4. The attitude and awareness of the parents

In all circumstances, pre- and posttest counseling on the clinical significance of test or screening shall be advised to the couples for explaining the details of present and future health consequences of the child born with Down syndrome. However, the couple's frankness on sharing of the information on family and their reproductive history, and also their willingness in prenatal screening of Down syndrome has a crucial role on reproductive outcome and birth of Down syndrome. This is very true for the countries having unbalanced sex ratio due to couple's participation more on fetal sex determination than determination of the genetic condition, which is necessary for the mothers older than 35 years. The couple's sole interest on knowing the fetal sex is to terminate the female pregnancies. Therefore, there are countries that have restrictions on fetal sex determination and implemented punishable acts toward it, such as India.

One couple was seeking for genetic counseling after the LSCS delivery of a male child at 3 weeks pre-term. The treating Gynecologist guided them for a mandatory karyotyping of the fetus in CVS or amniotic fluid because the mother was 39-year old. The clinician had explained the gestational age for sampling and procedure for the purpose and the associated empirical risk of miscarriage. The couple had neither revealed the number of previous children or pregnancies nor did they participate in prenatal screening or diagnosis. They delivered the male child who was diagnosed with Down syndrome after birth. Meanwhile, the couple had also preserved the umbilical cord blood stem cells of the child for future use and signed an agreement for 21 years of banking with a private facility. Now, the couple has learnt about the health consequences of trisomy 21 and presence of extra 21 in the stem cells.

While drawing the pedigree of the family and collecting their family history, the couple mentioned to the genetic counselor that they were afraid of sampling-related empirical risk of miscarriage of their only male pregnancy after delivering three daughters (Fig. 5.2). Hence, they did ot opt for fetal karyotyping. However, knowing the fact that their only male child will have typical facial features and impaired cognition besides the risk of other chronic illnesses, the couple now is actually looking for a rehabilitation center where they can drop their only

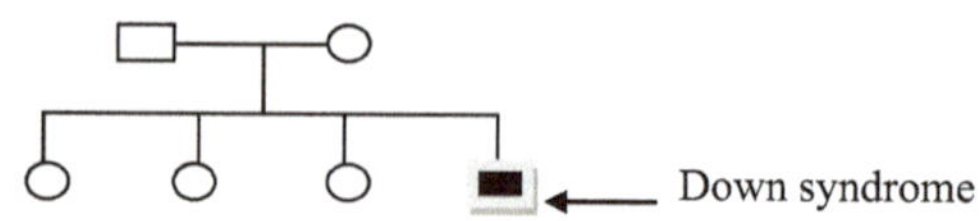

FIGURE 5.2 The couple had three daughters, did not participate in screening of Down syndrome, though the mother was 39-year old. After delivery, the confirmation of Down syndrome led the couple to decide to drop the child at rehabilitation center because the child will have mongoloid features, chronic illnesses, and most importantly, cognitive deficiency.

male child (straight from the hospital) forever and at any cost.

The couple might have done sex determination secretly without the knowledge of their treating Gynecologist, and since that was a male pregnancy, they bypassed fetal sampling and karyotyping, though the mother was 39-year old. And, now, knowing the Down syndrome child would be different from other normal children, the couple was unwilling to take the child home from the hospital and nurture the newborn who will require more attention and care.

Given the situation, antenatal counseling may be made mandatory by the country's administrative system, which can reduce the burden of births with irreparable genetic defects, and can give the prospective parents a reproductive choice to deliver a child with Down syndrome or not. Ideally, the regulation can mandate prenatal testing in cases of mothers older than $\geq$35 years, by passing biochemical screening of aneuploidies. PND can also facilitate in utero medical intervention, if required. Also, banking of umbilical cord blood stem cells shall include fetal/neonatal karyotyping in prebanking work up; otherwise, the stem cells with genetic impairment can be of no use in future for the affected child or any sibs. Moreover, use of such stem cells can implant genetic defect in otherwise normal karyotypic context. If the stem cell banking achieves a success like banking of other blood components, use of HLA-matched cells having aneuploidy may create additional complication.

4.1 Unawareness: a blessing in disguise

Sometimes unawareness works better than being aware, particularly if there is some abnormality, which cannot be cured. Such concept of unawareness could be considered as a myth. There was a family of two brothers (Fig. 5.3A): the elder brother had one elder son and one younger daughter, who appeared to be Down syndrome, and the younger brother had three elder daughters and one younger son, who had some kind of cognitive deficiency. The suspected Down syndrome girl had typical facial characteristics, drooling saliva till teenage or perhaps more, and cognitive impairment to severe extent. She did not attend any school and got occupied in household activities. She grew up, attained menarche, and survived till ~25 years. Her brother's children were normal. Her cousin did not have typical features of Down syndrome, used to attend a normal school and studied till fourth class, and got occupied in family farming. However, he did not look normal at 40 years age (Fig. 5.3B), and he had one son with Down syndrome (Fig. 5.3C and D). The family lived in rural India, where no awareness about Down syndrome–related concept exists. No genetic diagnosis was considered for all three members of the family.

Hence, there was no societal discrimination for the family. The Down syndrome child is studying at eighth standard along with other normal children at a normal school and never received any special attention for managing hypotonia or other health problems. The only fact about these three members of the two families known to the society was their dysmorphic look and impaired cognitive ability. The Down

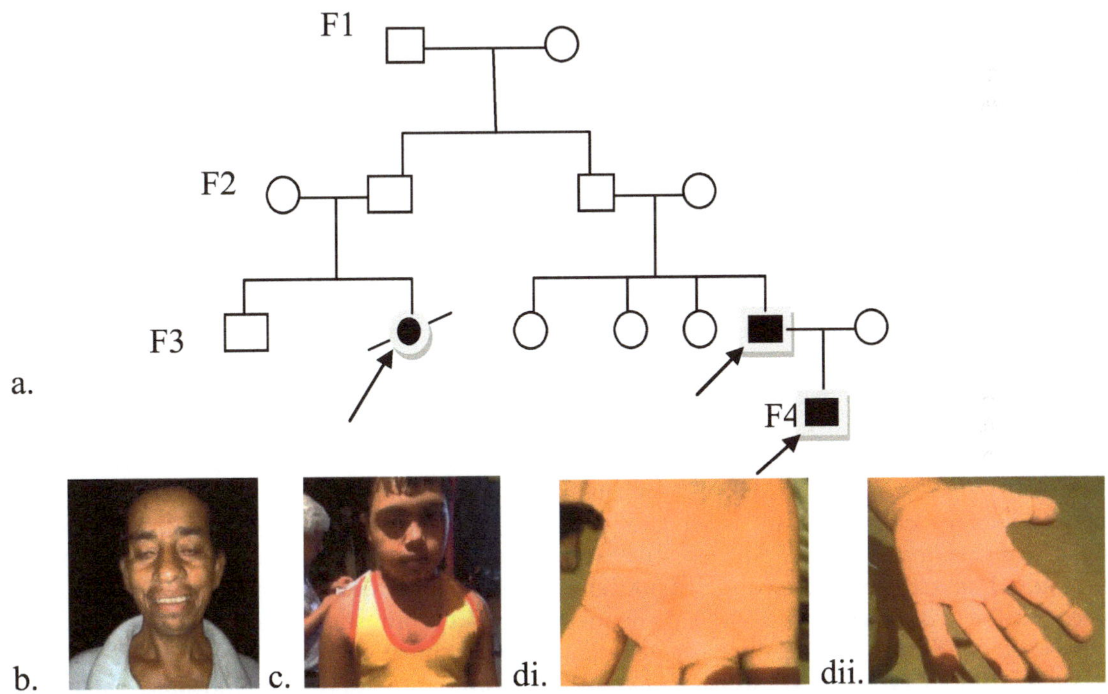

FIGURE 5.3 The man with unidentified genetic defects fathered the child with Down syndrome: (A) the pedigree of the family; (B) the father; (C) the child with Down syndrome, and (D) (i–ii). the palmer creases of the child with Down syndrome. The child is attending a normal school in the absence of the special school facility in the locality.

syndrome child could manage to score 70% −75% marks in aggregate in final school examinations. He also takes part in dance competitions, reciting poems, and in other cultural acts organized in school or their housing society. However, neither the family nor the society is aware about their genetic defects and future health risks.

4.2 What would be the impact of their unawareness?

There is science of heredity at the backdrop of this family history. The F2 generation boy could be having a chromosomal abnormality, more so an unbalanced translocation inherited from his father who could be a carrier of a balanced translocation with or without involving chromosome 21 and has fathered the Down syndrome child. He is completely unaware about his present health status except his compromised eyesight (Fig. 5.3B). His son could be having trisomy 21 and also the unbalanced rearrangement, if present in the father. The F2 generation girl with features of Down syndrome has expired at early age (Fig. 5.3A), but the family did not have any knowledge about her illness. However, her brother and his kids are apparently normal. There is no facility of medical investigation of their condition nor the facility of primary diagnosis of Down syndrome. Now the question is about the life of the Down syndrome child. If he gets married and fathers a child, what will be the impact? Only consolation is if his trisomy 21 is free from rearrangement, then 50% of his gametes will carry trisomy 21, and more dangerously, if his trisomy 21 status induces further aneuploidy. I have diagnosed two male children with co-existence of trisomy 21 and XXY, meaning presence of 48 chromosomes in their genetic makeup with 48,XXY,+21 karyotype in Down syndrome coexisting with Klinefelter syndrome (Fig. 5.4).

Trisomy 21 renders stunted growth in Down syndrome, while XXY confers long stature in Klinefelter syndrome. Hence, the pattern of growth and expression of syndrome-specific features in such case would be an important observation. However, at the time of diagnosis, the characteristics of Down syndrome was prominent, and thus, diagnosis of Down syndrome was asked for. Nevertheless, Barr body of the X-heterochromatin in Klinefelter syndromes indicates gene-dosage compensation. Therefore, overexpression of triplicated 21 would likely to dominate over XXY condition. However, a close observation of phenotype and expression of the genes in such complex scenario can only draw the conclusion of compound chromosomal syndromes.

Consanguineous marriage stands as a religious custom in many states in India. The system is known to multiply the risk of hereditary condition of recessive mutations in future generations and affects births with genetic impairments. Thalassemia is one such prominent example for India, Jews Genetic Diseases (JDG) for Ashkenazi Jews population, and many such racial recessive genetic disorders have been established through consanguineous marriages (Ganguly and Kadam, 2019).

Fig. 5.5 of a partial karyotype depicts a unique example of hereditary transmission and/or preservation of a Robertsoian translocation through consanguineous marriage and increased risk of Down syndrome in the family. The mother was married to her maternal cousin and had a male child with Down syndrome, and two other apparently normal children, including a girl. The girl was the eldest one and appeared with primary amenorrhoea at 20 years of age and was detected with two copies of Rob(14;21) having 44 chromosomes along with two normal X chromosomes in 44,XX,2xRob(14;21) karyotypic genome. The karyotype indicated parental consanguinity and presence of one copy of the translocation in each parent. Further counseling and

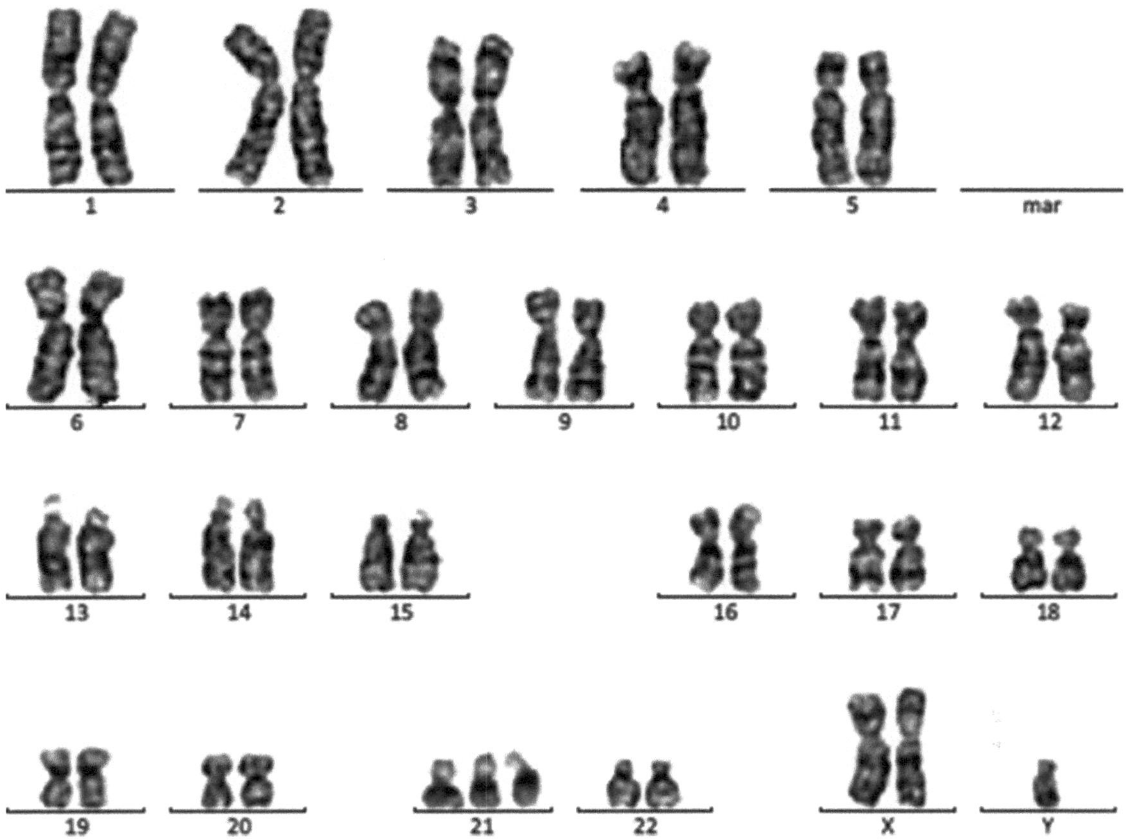

FIGURE 5.4 A karyogram of a male child with 48,XXY,+21, who has both Down syndrome and Klinefelter syndrome. His parent's karyotyping was not performed.

karyotyping of the mother, one maternal uncle, and the two brothers revealed: mother and the uncle with the translocation as carrier; the brother with Down syndrome with one copy of the translocation and trisomy 21; and another brother was carrier of one copy of the translocation. Hence, the girl was fortunate enough to inherit the translocation from both parents, and interestingly, she did neither possess a normal 14 nor a normal 21 (both 14 and 21s were fused in the form of Rob(14;21)). Father was deceased at the time of this study; however, the presence of the translocation in both mother and maternal uncle and presence of two copies in the girl retrospectively indicate its presence in the father also.

5. Ethical aspects

Ethical issues and guidelines of screening and testing of Down syndrome and genetic defects have been established for clinicians, patients, counselors, insurers, and policy makers for protecting the use of genetic information and rights of patients carrying genetic illness, especially for preconception and PND with a view to reproductive decision-making. Always, couple's

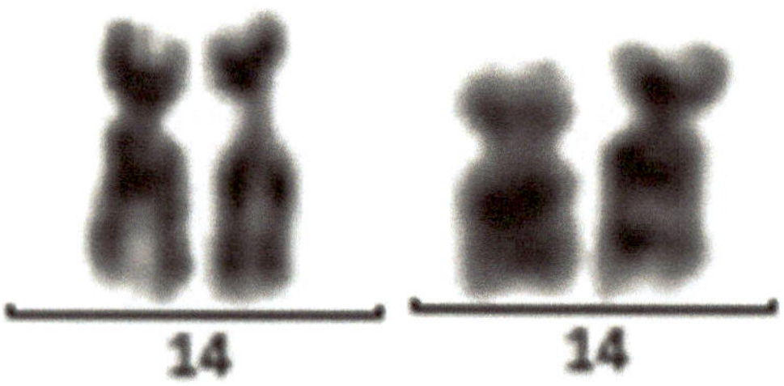

FIGURE 5.5 A family history of consanguineous marriage and a partial karyotype with two copies of Rob(14; 21): (A) pedigree of the family; (B) partial karyogram of the girl showing two copies of the Robertsonian translocation.

preference and decision hold the right to deliver or terminate a pregnancy with potential or dominant genetic disorder (Committee Opinion No. 410, 2008). However, invasive testing of pregnancies in absence of indicated predisposing factors and more prominently thorny issues of sex selection and sex determination have brought in regulatory and ethical issues pertaining to preconception and prenatal testing. Restriction on sex selection has been imposed in several countries with unbalanced sex ratio, except sex-linked genetic conditions (Braude, 2006; Committee Opinion No. 360, 2007). The physician's ethics shall pay unlimited obeisance to a patient's confidentiality (American Medical Association, 2006). The Health Insurance Portability and Accountability Act (HIPPA) prevented insurance companies from refusing health care based on predictive testing (Fluda and Lykens, 2006).

In the present era, the noninvasive prenatal testing (NIPT) using cff-DNA/RNA from maternal blood plasma, which is already established as a direct-to-consumer product, involves serious ethical concerns about pretest counseling and obtaining consent from the couple (Lo et al., 1997; ACOG, 2020). The regulations of states and/or countries have yet be framed and implemented into practice for controlling sex determination in cff-DNA (Benn and Chapman, 2010; Bianchi, 2006; Hall et al., 2009). Taken together, test regulation, professional guidelines, and policies of the advisory groups shall be defined to meet the challenges of preconception and PND in maternal blood (Committee Opinion No. 640, 2015; Medical Policy No. 4.01.21, 2015), so that genetic testing can be performed with particular caution and an explicit "informed consent" for rendering protection against discrimination by insurers, psychological distress, individual privacy, etc (Committee Opinion No. 410, 2008).

Disclosure of genetic test result of PND, PGD, or carrier screening is an important ethical concern. The result of PND and PGD shall be disclosed to the expecting couples and/or their authorized relatives. For carrier screening, the patient may voluntarily disclose the genetic condition as recommended by the American Society of Human Genetics (ASHG), and if the result is likely to be serious for the at-risk kindred, the healthcare providers may disclose (Offit et al., 2004). Hiding paternal identity, on one hand, might lead to transmission of autosomal recessive mutations or unnecessary amniocentesis in all future pregnancies in cases where a different person is the biological father and accordingly creating unnecessary concern for other family members.

6. Policies and guidelines

Policies and guidelines shall be framed on the issues of genetic screening and testing, especially for PND/PGD, to protect the couple's right to decide whether to abort or deliver a genetically affected child, irrespective of the sex of the fetus (Hill et al., 2016). Employment of the appropriate technologies for the purpose may be recommended by the policies of the states or countries, such as the ACCE-framework (analytic validity, clinical validity, clinical utility, ethical, legal and social implications) developed by the Centers for Disease Control and Prevention (CDC) and National Office of Public Health and the Foundation of Blood Research, USA, for genetic screening, with a view to lowering the genetic disease burden. Interestingly, the positive predictive screening value may differ across different disorders as a function of penetrance, expressivity, and may be mutation dependent and the triggering effects of epigenetic modifiers. Also a given mutation may be associated with variable clinical severity of disease manifestation within the same family. Commercial service providers develop panels of genetic screening for public as well as for the targeted screening for a limited number of mutations depending on their available infrastructure and hence, cannot address molecular patho-mechanism appropriately either with the spectrum of tested genes or the spectrum of mutations (Kingsmore, 2012).

In India, the load of genetic disorders and congenital malformations are steadily rising contributing as the third most common cause of mortality among the urban newborns (WHO, 2004). Interestingly, an estimated 80% of public healthcare funding comes from the states, whereas policies and plans are defined by the Ministry of Health and Family Welfare of the central Government. Furthermore, healthcare delivery is mostly managed by the private sectors with variable panels of genetic tests available in individual genetic testing laboratories (WHO, 2006). Also variable technology is applied for a particular genetic test by different laboratories, and in many cases, genetic testing is being carried out by diagnostic facilities isolated from patient-care centers and thus, mostly lacking in genetic counseling facility or understanding the interpretative comments (Ganguly and Kadam, 2018). Notably, the diagnostic expenses are generally met out-of-pocket by the individuals, and only 21.3% of healthcare expenditure of 6.1% of the gross domestic product (GDP) is provided by the government (WHO, 2005). Furthermore, education and practice of medical genetics and genetic counseling are at incipient stage in over 90% of the Indian medical institutions (Ghosh and Mohanty, 2002; Ganguly and Chakraborty, 2009).

Hence, the necessity of the policies and regulatory issues on consistent professional judgment and its practical implementation, quality control, and assurance system on genetic diagnostic and counseling services are applicable to both public and private services. Clear messages are essential from the regulators on what genetic testing and suitable technology are applicable or affordable for a given genetic condition. Nevertheless, the discovery of disease-causing genes and technological challenges in the field of medical genetics will create even more debatable issues. Genetic predisposition of a fetus may raise a wide array of questions and issues that must be confronted while developing policies to deal with genetic screening. To prepare citizens for informed personal decision-making, public education and counseling will become inevitable to understand the new concepts of genetic expressions. Although these tests are usually advised for high-risk families, nowadays, most parents are opting for prenatal genetic screening, especially for cff-DNA, to know whether their child has a normal genetic makeup or not. This has led to several debates between people who support prenatal genetic screening and those who are against it. Government intervention on standards and policies will be helpful for regulation of genetic screening and testing, which may truly eradicate and/or minimize the burden of genetic diseases.

7. Public health point of view toward Down syndrome

Ascertainment of genetic defects stands crucial for every country to implement a corrective measure for prevention of recurrence of birth defects and chromosomal syndromes. The population-based birth defects are generally measured through data-collection for census; however, surveillance systems including the pre-/postnatal data sources; administrative databases of hospitals such as discharge data and vital records; and Mediclaim databases; in-/outpatient medical records; and pre- and postnatal cytogenetic investigations were recommended by Mai et al. (2013). Birth defect programs shall ideally employ trained staff or paramedics who can review medical records for case finding (active case finding), in parallel with relying on hospital reporting and/or administrative databases (passive case finding). Data collected by the census staff generally based on verbal interaction with the family members, wherein these staff may not have appropriate understanding of genetic factors of birth defects. An accurate and sizable amount of data can be collected by the public health administration of every country by attaching due importance, which can accumulate data from Panchayat-to-District-to-State level of health offices depending on the country's administrative system.

The collected data on birth defects associated with genetic factors shall be analyzed with a view to serious understanding of the underlying factors, which ultimately can guide to prevent the hereditary transmission of translocations or de novo acquisition of aneuploidies and other genetic alterations. However, variability was observed in prevalence of trisomies across the United States (Mai et al., 2013), which occurred through differences in case ascertainment methodology, ability of a surveillance system to capture all affected cases, and state variations for risk factors such as the distribution of maternal age and awareness level in different populations. A significant increase of chromosomal abnormalities was demonstrated while prenatal records from perinatal and maternal–fetal medicine departments were included to the surveillance registry data (Cragan and Gilboa, 2009). Nevertheless, differences can exist in state legislation's system of data acquisition, access, and resources in different countries.

There is a great deal of variation reported in ascertainment of the prevalence of Down syndrome, which may include only live births, live and still births, elective termination of pregnancies with trisomy 21, or outcome of all pregnancy. The aim of such programs actually directs the type of data to be considered, such as follow-up programs will only include the cases among live births; however, there can be differences in program purpose, legal authority, and the availability of data. Collectively, it is important to consider all pregnancy outcomes for understanding the full impact of trisomy 21 on the population. An increase of 4.8 pregnancy with Down syndrome per 10,000 pregnancies (11.5/10,000 live births vs. 16.3/10,000 for all pregnancies) was reported when all pregnancy outcomes were considered for ascertainment of the prevalence of trisomy 21 (Jackson et al., 2014). Also, maternal age–related conceptions with trisomy 21 shall be ascertained through all pregnancy outcomes, since there is more chances of spontaneous abortions, still births, and neonatal deaths.

Data of antenatal markers screened in maternal serum and conventional and color-Doppler ultrasonography (USG) collected on every check up of pregnancies regardless of the maternal age would contribute significantly for estimating the prevalence of genetic disorders precisely (Ekelund et al., 2008; Nakata et al., 2010; Zhou et al., 2021). In the recent era, screening of double, triple, and/or quadruple markers and USG at every gynecological check up has become a standard screening practice for trisomies and neural tube defects in many countries, including India; however, structured programs on quantification of birth defects are yet to be implemented.

Additionally, data of direct prenatal testing from cytogenetic laboratories or from medical records of surveillance systems would allow a more accurate ascertainment of cases with Down syndrome and other genetic defects. Also, elective termination following PND has shown to affect the live birth prevalence of Down syndrome (Moorthie et al., 2018; Natoli et al., 2012) and other trisomies (Crider et al., 2008; Guon et al., 2014). Additionally, data of noninvasive prenatal screening using cffDNA collected from maternal blood plasma would increase the frequency of Down syndrome, Edward syndrome and Patau syndrome with trisomy 21, trisomy 18, and trisomy 13, respectively (Langlois et al., 2013); however, this screening tool is not commonly followed for all pregnancies. Nevertheless, the screening technique of Down syndrome as a decision-making factor for elective termination of pregnancy varies among the states in the United States and other countries, which also differs by age, race/ethnicity, religion, socioeconomic status, marital status, and type of health insurance (Jones and Jerman, 2014; Pazol et al., 2012).

Information on trends and increasing prevalence of Down syndrome has been demonstrated (Shin et al., 2009; Cocchi et al., 2010), and such trend of birth prevalence was consistently reported by the US National Birth Defects Prevention Network (NBDPN) among the states that included live births and still births and all pregnancy outcomes. An increase in the prevalence of live births (from 9.0 to 11.8 per 10,000 live births) of Down syndrome was reported during 1979–2003 (Shin et al., 2009). Additionally, the improved survival of the live births (94% with 1-year survival) with Down syndrome has been reported (Kucik et al., 2013; Zhu et al., 2013). The survival is largely affected by low birth weight or presence of a congenital heart defect, and that results in death at infant and early childhood (Shin et al., 2010; Zhu et al., 2013).

The countries may implement population-based surveillance programs to capture state-wise data on birth defects, including Down syndrome, which could be a continued source of data to understand the impact of versatile conditions on prevalence of Down syndrome and other birth defects. The practice of ascertainment and documentation of various condition shall highlight the factors contributing to prevalence and/or affecting the life and survival of Down syndrome. Such comprehensive state or country-level information would be important for clinicians, healthcare regulators, and other administrative officials on the growing public health importance of chromosomal aneuploidies. Additionally, such surveillance programs would indicate the need and direction for more research on the role of prenatal detection for improving postnatal health and planning of health service that addresses the lifetime needs of a growing population.

7.1 Public Health Department's support

The life of people with Down syndrome is critical mainly for the compromised immune system due to altered expression of all disomic as well as trisomy 21 genes. Besides suffering from chronic illnesses, discrimination in employment and societal stigmatization take a huge toll for families having children with Down syndrome. However, the administrative system of a country can alleviate the trauma of such families and can make the life of Down syndrome better. The developed nations, including the United States, have policies to support the living of people with Down syndrome and encouraging programs to facilitate such people to take part in sports and other extracurricular activities. As such, nurturing a child with Down syndrome is an expensive affair for many families, where health insurance companies of different countries have limitation to support the medical affairs for a congenital health condition. The policies are yet to come up to frame the rules and regulations of such conditions, so that the people who require more support do not get excluded from medical insurance. Reservation in employment shall also be included in the country's policy, especially for the developing nations. Thus, the Public Health Department of a country has to play crucial roles in making policies to support the living of people born with congenital disorders and protect them from deprivation due to discrimination.

Mr. Sylvester (an US native) and Mrs. Sylvester (an Indian) got their first child with Down syndrome, when Mrs. Sylvester was 38 years' old. Being unaware about the importance of pregnancy monitoring, neither had they consulted a Gynecologist for regular check up nor did they get the PND to rule out Down syndrome. The child was delivered uneventfully at full term in our hospital in Mumbai, India. Upon Neonatologist's examination, routine blood karyotyping was considered and that confirmed the diagnosis of Down syndrome with free trisomy 21. The family was not upset knowing the facts and future of life of a Down syndrome. They just requested for the best possible options of medical support for their child, particularly for growth and cognitive development. He requested for the soft copy of the report with physical signature with seal of the Geneticist and the hospital that he had to send to the United States for getting the financial support as per the administrative rule of the country. Within the next 2 years, the family had adopted a girl who was a "deaf and dumb" with an intention to give a better life to that girl. The family was comprised of father, mother, one son with Down syndrome, and one daughter (adopted) with hearing and speech impairment (Fig. 5.6).

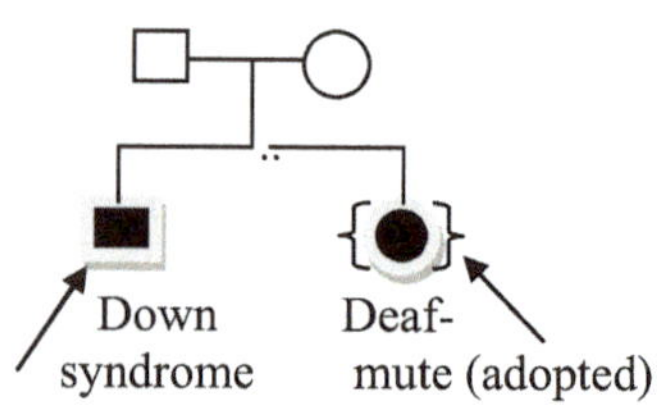

FIGURE 5.6 A happy family with a son having Down syndrome and a daughter (adopted) having speech and hearing impairment. The family is financially supported by the US Government, though living in India. However, Indian Government has yet to implement such law to support such families having children with Down syndrome or other irreparable genetic disorders.

8. Socio-cultural aspects of Down syndrome

The typical facial appearance of Down syndrome and cognitive impairment often discriminate such children or adults in the society. The families bear with the situation of social stigmatization, besides struggling with the medical condition of these patients for giving a better quality of life. Therefore, as genetic counseling is essential for the families having Down syndrome for understanding the health implications and recurrence of trisomy 21, it is also necessary for the public and the society at large to recognize these children or adults as other human beings, and their families with respect. Having trisomy 21 or giving birth of Down syndrome is not a crime, and it happens unknowingly and unintentionally. Public or the society shall be receptive and friendly with these individuals, as happens with other diseases such as diabetes or heart attack. The perspectives of living with Down syndrome and society's outlook toward it can improve the quality of their life in a progressive society of any nation.

Knowledge and skill for parenting children with Down syndrome can extend additional educational support and care for various factors. As such, there are strides for these people in education, employment, housing, and more importantly, social settings that create a supportive environment for their living. Although housing of Down syndrome and other people having any form of mental or physical disability in isolated institutions and excluded colonies of early 20th century has changed and higher-quality opportunities have increased, the parents and families having child with Down syndrome still pose a challenge and often encounter with lack of patronizing attitudes and discrimination in the wider community. Nevertheless, societies of developed nations have changed the mentality and advocates for the rights of equal opportunities for these people (Fig. 5.7). In India, the

a. b. c.

FIGURE 5.7 Professionalism of Down syndrome: (A) 23-year-old man with Down syndrome becomes award-winning travel photographer. (B) an actor with Down syndrome, to announce the Oscar 2020 in the best live action short category. (C) a man with Down syndrome is performing in Dance-India-Dance competition and he is narrating his dream of becoming a dancer and an actor. *Reproduced from: (A) The Logical Indian Crew, April 9, 2021. https://thelogicalindian.com/uplifting/23-yr-old-us-man-with-down-syndrome-27784 Accessed July 3, 2021; (B) https://newzhook.com/story/actor-down-syndrome-oscars-presenter-disabled-zaack-gottagen-inclusion/ Accessed July 3, 2021; (C) photography of live program on Indian television channel.*

Self Advocates Forum of India (SAFI) is meant for the people born with some kind of disabilities. These people voice their opinion through SAFI and attract the attention of the regulators. In fact, the "disability" has been changed to "specially able".

The syndrome is honored with celebration of the "World Down Syndrome Day," which was first held on March 21, 2006, indicating trisomy (the third month, March) 21 (date). The concept was first proclaimed by the European Down Syndrome Association during their European congress in 2005 held in Palma de Mallorca. Every country commemorates the day in different month of the year such as the National Down Syndrome Society of the United States observes "Down Syndrome Month" every October. The target of the forum is to dispel the stereotypes and provide accurate information by raising awareness of the potential of individuals with Down syndrome (Tamblyn et al., 2016). South Africa celebrates the "Down Syndrome Awareness Day" on every October 20 (Gamis and Smith, 2012). Nowadays, there are several kinds of organizations to provide training in sports, dance, music, computer work, and other vocational training and education round the year for individuals with intellectual disabilities such as Down syndrome.

References

ACMG statement, 1995. Statement of storage and use of genetic materials. American College of Medical Genetics, Storage of Genetics Materials Committee. Am. J. Hum. Genet. 57, 1499–1500.

ACOG, Number 226, 2020. Screening for fetal chromosomal abnormalities: ACOG practice bulletin summary, number 226. Obstet. Gynecol. 136 (4), 859–867, 32976375.

American Medical Association, 2006. Confidentiality. In: Code of Medical Ethics of the American Medical Association: Current Opinions with Annotations. 2006–2007. AMA, Chicago (IL), pp. 136–150.

Benn, P.A., Chapman, A.R., 2010. Ethical challenges in providing noninvasive prenatal diagnosis. Curr. Opin. Obstet. Gynecol. 22, 128–134.

Bianchi, D., 2006. At-home fetal DNA gender testing. Obstet. Gynecol. 107, 216–219.

Bianchi, D.W., Platt, L.D., Goldberg, J.D., Abuhamad, A.Z., Sehnert, A.J., Rava, R.P., 2012. Genome- wide fetal aneuploidy detection by maternal plasma DNA sequencing. MatErnal Blood Is Source to Accurately Diagnose Fetal Aneoploidy (MELISSA) Study Group. Obstet. Gynecol. 119, 890–901.

Braude, P., 2006. Preimplantation diagnosis forgenetic susuceptibility. N. Engl. J. Med. 355, 541–543.

Canick, J.A., Kloza, E.M., Lambert-Messerlian, G.M., et al., 2012. DNA sequencing of maternal plasma to identify Down syndrome and other trisomies in multiple gestations. Prenat. Diagn. https://doi.org/10.1002/pd.3892.

Cho, D., McGowan, M.L., Metcalfe, J., Sharp, R.R., 2013. Expanded carrier screening in reproductive healthcare: perspectives from genetics professionals. Hum. Reprod. 28, 1725–1730.

Clarke, A., 2014. Genetic testing and genomic screening. In: Kumar, D., Eng, C. (Eds.), The Principles and Practice of Genomic Medicine. Oxford University Press, p. 224. Ch. 14.

Cocchi, G., Gualdi, S., Bower, C., Halliday, J., Jonsson, B., Myrelid, A., Mastroiacovo, P., Amar, E., et al., 2010. International trends of down syndrome 1993–2004: births in relation to maternal age and terminations of pregnancies. Birth Defects Res. (Part A) Clin. & Mol. Teratol. 88, 474–479. https://doi.org/10.1002/bdra.20666.

Committee Opinion No 360, 2007. Sex selection. American College of Obstetricians and Gynecologists. Obstet. Gynecol. 109, 475–478.

Committee Opinion No 410, 2008. Ethical Issues in Genetic Testing. American College of Obstetricians and Gynecologists.

Committee Opinion No 442, 2009. Preconception and Prenatal Carrier Screening for Genetic Diseases in Individuals of Eastern European Jewish Descent. American College of Obstetricians and Gynecologists.

Committee Opinion No 545, 2012. Noninvasive Prenatal Testing for Fetal Aneuploidy. American College of Obstetricians and Gynecologists, and The Society for Maternal-Fetal Medicine.

Committee Opinion No 640, 2015. Cell-free DNA Screening for Fetal Aneuploidy. American College of Obstetricians and Gynecologists, and The Society for Maternal-Fetal Medicine.

Cragan, J.D., Gilboa, S.M., 2009. Including prenatal diagnoses in birth defects monitoring: experience of the metropolitan atlanta congenital defects program. Birth Defects Res. Part A 85 (1), 20–29. https://doi.org/10.1002/bdra.20508. A Special Issue: A Festschrift in Honor of Professor Lewis B. Holmes.

Crider, K.S., Olney, R.S., Cragan, J.D., 2008. Trisomies 13 and 18: population prevalences, characteristics, and prenatal diagnosis, metropolitan Atlanta, 1994–2003. Am. J. Med. Genet. 146A (7), 820–826. https://doi.org/10.1002/ajmg.a.32200.

Ekelund, C.K., Jørgensen, F.S., Petersen, O.B., Sundberg, K., Tabor, A., Danish Fetal Medicine Research Group, 2008. Impact of a new national screening policy for Down's syndrome in Denmark: population based cohort study. BMJ 337. https://doi.org/10.1136/bmj.a2547.

Fluda, K.G., Lykens, K., 2006. Ethical issues in predictive genetic testing: a public health perspective. J. Med. Ethics 32, 143–147.

Gamis, A.S., Smith, F.O., 2012. Transient myeloproliferative disorder in children with Down syndrome: clarity to this enigmatic disorder. Br. J. Haematol. 159 (3), 277–287. https://doi.org/10.1111/bjh.12041.

Ganguly, B.B., Chakrabarti, D., 2009. Urge of prenatal diagnosis in modernized civilization. Eur. J. Hum. Genet. 17 (Suppl. 2), 160.

Ganguly, B.B., Kadam, N.N., 2019. Genetic race: prevalence of diseases and guidelines for prevention. MGM J. Med. Sci. 6 (1), 32–41. https://doi.org/10.5005/jp-journals-10036-1228.

Ganguly, B.B., Kadam, N.N., 2018. Pros and cons of carrier screening for prevention of genetic disorders and other associated issues: review of current status. MGM J. Med. Sci. 5 (3), 135–142.

Ganguly, B.B., Daruwalla, D., Kadam, N.N., 2017. Attitude of parents to carrier screening for genetically transmitted diseases in India. MGM J. Med. Sci. 4 (4), 187–190.

Ganguly, B.B., Kadam, N.N., October 21, 2016. Down syndrome: from the age of characterization to the era of curative approach. Nucleus. https://doi.org/10.1007/s13237-016-0187-y.

Ghosh, K., Mohanty, D., 2002. Teaching of medical genetics in the medical colleges of India – way ahead. Ind. J. Med. Genet. 8, 43–44.

Godard, B., Ten Kate, L.P., Evers-Kiebooms, G., Aymé, S., 2003. Population genetic screening programmes:

principles, techniques, practices, and policies. Eur. J. Hum. Genet. 11 (Suppl. 2), S49–S87.

Grody, W.W., Thompson, B.H., Gregg, A.R., et al., 2013. ACMG position statement on prenatal/preconception expanded screening. Genet. Med. 15 (6), 482–483.

Guon, J., Wilfond, B.S., Farlow, B., Brazg, T., Annie Janvier, A., 2014. Our children are not a diagnosis: the experience of parents who continue their pregnancy after a prenatal diagnosis of trisomy 13 or 18. Am. J. Med. Genet. 164 (2), 308–318. https://doi.org/10.1002/ajmg.a.36298.

Hall, A., Bostanci, A., John, S., 2009. Ethical, legal and social issues arising from cell-free fetal DNA technologies. In: Appendix III to the Report: Cell-free Fetal Nucleic Acids for Noninvasive Prenatal Diagnosis. PHG Foundation, Cambridge. www.phgfoundation.org.

Hill, M., Johnson, J.A., Langlois, S., Lee, H., Winsor, S., Dineley, B., Horniachek, M., Lalatta, F., et al., 2016. Preferences for prenatal tests for Down syndrome: an international comparison of the views of pregnant women and health professionals. Eur. J. Hum. Genet. 24, 968–975. https://doi.org/10.1038/ejhg.2015.249.

Jackson, J.M., Crider, K.S., Cragan, J.D., Rasmussen, S.A., Olney, R.S., January 2014. Frequency of prenatal cytogenetic diagnosis and pregnancy outcomes by maternal race–ethnicity, and the effect on the prevalence of trisomy 21, Metropolitan Atlanta, 1996–2005. Am. J. Med. Genet. 0 (1), 70–76. https://doi.org/10.1002/ajmg.a.36247.

Jones, R.K., Jerman, J., 2014. Abortion incidence and service availability in the United States, 2011. Perspect. Sex. Reprod. Health 46 (1), 3–14.

Kingsmore, S., 2012. Comprehensive carrier screening and molecular diagnostic testing for recessive childhood diseases. PLoS Curr. e4f9877ab8ffa9.

Kucik, J.E., Shin, M., Siffel, C., Marengo, L., Correa, A., 2013. For the congenital anomaly multistate prevalence and survival collaborative. Trends in survival among children with down syndrome in 10 regions of the United States. Pediatrics 131 (1), e27–e36. https://doi.org/10.1542/peds.2012-1616.

Langlois, S., Brock, J.A., Genetics Committee, February 2013. Current status in non-invasive prenatal detection of Down syndrome, trisomy 18, and trisomy 13 using cell-free DNA in maternal plasma. J. Obstet. Gynaecol. Can. 35 (2), 177–181. https://doi.org/10.1016/S1701-2163(15)31025-2. English, French.

Lo, Y.M., Corbetta, N., Chamberlain, P.F., Rai, V., Sargent, I.L., Redman, C.W., et al., 1997. Presence of fetal DNA in maternal plasma and serum. Lancet 350, 485–487.

Mai, C.T., Kucik, J.E., Isenburg, J., Feldkamp, M.L., Marengo, L.K., Bugenske, E.M., et al., 2013. For the national birth defects prevention Network. Congenital malformation surveillance report: a report from the national birth defects prevention network. Selected birth defects data from population-based birth defects surveillance programs in the United States, 2006 to 2010: featuring trisomy conditions. Birth Defect. Res. (Part A) 97, 709–725.

Medical Policy No 4.01.21, 2015. Premera Noninvasive Prenatal Screening for Fetal Aneuploidies and Microdeletions Using Cell-free. DNA.

Moorthie, S., Blencowe, H., Darlison, M.W., et al., 2018. Chromosomal disorders: estimating baseline birth prevalence and pregnancy outcomes worldwide. J. Comm. Genet. 9, 377–386. https://doi.org/10.1007/s12687-017-0336-2.

Morgan, M.A., Driscoll, D.A., Mennuti, M.T., Schulkin, J., 2004. Practice patterns of obstetrician- gynecologists regarding preconception and prenatal screening for cystic fibrosis. Genet. Med. 6, 450–455.

Nakata, N., Wang, Y., Bhatt, S., 2010. Trends in prenatal screening and diagnostic testing among women referred for advanced maternal age. Prenat. Diagn. 30 (3), 198–206. https://doi.org/10.1002/pd.2434.

Natoli, J.L., Ackerman, D.L., McDermott, S., Edwards, J.G., 2012. Prenatal diagnosis of Down syndrome: a systematic review of termination rates (1995–2011). Prenat. Diagn. 32, 142–153.

Norton, M.E., Brar, H., Weiss, J., et al., 2012. Non-Invasive Chromosomal Evaluation (NICE) study: results of a multicenter, prospective, cohort study for detection of fetal trisomy 18. Am. J. Obstet. Gynecol. https://doi.org/10.1016/ajog.2012.05.021.

Offit, K., Groeger, E., Turner, S., Wadsworth, E.A., Weiser, M.A., 2004. The "duty to warn" a patient's family members about hereditary disease risks. JAMA 292, 1469–1473.

Pazol, K., Creanga, A.A., Zane, S.B., Burley, K.D., Jamieson, D.J., 2012. Abortion surveillance-United States, 2009. MMWR (Morb. Mortal. Wkly. Rep.) 61 (8), 1–44.

Qureshi, N., Armstrong, S., Modell, B., 2006. GPs' opinions of their role in prenatal genetic services: a cross-sectional survey. Fam. Pract. 23, 106–110.

Shin, M., Besser, L.M., Kucik, J.E., Lu, C., Siffel, C., Correa, A., 2009. The congenital anomaly multistate prevalence and survival (CAMPS) collaborative. Prevalence of down syndrome among children and adolescents in 10 regions of the United States. Pediatrics 124 (6), 1565–1571. https://doi.org/10.1542/peds.2009-0745.

Shin, M., Siffel, C., Correa, A., 2010. Survival of children with mosaic Down syndrome. Am. J. Med. Genet. 152A, 800–801.

Soothill, P.W., Lo, Y.M.D., 2014. Non-invasive Prenatal Testing for Chromosomal Abnormality Using Maternal Plasma DNA. Scientific Impact Paper No. 15. Royal College of Obstetricians & Gynecologists.

Stark, Z., Massie, J., McClaren, B., et al., 2013. Current practice and attitudes of Australian obstetricians toward population-based carrier screening for inherited conditions. Twin Res. Hum. Genet. 16, 601–607.

Tamblyn, J.A., Norton, A., Spurgeon, L., Donovan, V., Bedford Russell, A., Bonnici, J., Perkins, K., Vyas, P., Roberts, I., Kilby, M.D., 2016. Prenatal therapy in transient abnormal myelopoiesis: a systematic review. Arch. Dis. Child. Fetal Neonatal Ed. 101 (1), F67–F71.

Williams Textbook of Endocrinology Expert Consult, twelfth ed., 2011. Elsevier Health Sciences, London, ISBN 978-1-4377-3600-7. Archived from the original on 2017-01-23.

World Health Organization (WHO), 1998. Proposed International Guidelines on Ethical Issues in Medical Genetics and Genetic Services.

WHO, 2004. Identifying Regional Priorities in the Area of Human Genetics in SEAR: Report of an Intercountry Consultation, Bangkok, Thailand, 23–25 September 2003. World Health Organization Office for South East Asia, New Delhi (SEA-RES-121).

WHO, 2006. World Health Organization,. http://www.who.int/genomics/policy/india/en/index7.html#content: Case Study:India.

WHO, 2005. The World Health Report 2005. Make Every Mother and Child Count. World Health Organization, Geneva.

Zhou, X.J., Lin, Y.J., Chen, X.W., Zheng, J.H., Zhou, Y.J., 2021. Prenatal diagnosis of harlequin ichthyosis by ultrasonography: a case report. Ann. Transl. Med. 9 (2), 183. https://doi.org/10.21037/atm-20-8223.

Zhu, J.L., Hasle, H., Correa, A., Schendel, D., Friedman, J.M., Olsen, J., Rasmussen, S.A., 2013. Survival among people with down syndrome: a nationwide population-based study in Denmark. Genet. Med. 15 (1), 64–69. https://doi.org/10.1038/gim.2012.93.

CHAPTER

6

Gene-dosage imbalance due to trisomic HSA21 and genotype–phenotype association in Down syndrome

1. Introduction

The human chromosome 21 (HSA21) has a prominent long arm ("q") of 33.6 Mb size that represents ~1% of the total genome (3020.3 MB of NCBI Build 34, version 1); however, total number of genes present on HSA21 and their characteristics have not been fully uncovered. Initially, 225 genes (1/3rd is overexpressed and 2/3rd compensated) were estimated on HSA21 (Hattori et al., 2000), which were later enumerated as 364 protein coding transcripts (Gardiner et al., 2003). These genes act as transcription factors, regulators and modulators, interferons in immune response, kinases, channels, proteases and protease inhibitors, receptors, adhesion molecules, and in RNA processing, energy metabolism, and ubiquitin pathway. Some of the HSA21 genes have been presented in Supplementary Table S6.1, which has been cataloged from the data available in the public domain and through Google search. Indeed, the total number of genes comprising of protein-coding and noncoding RNAs on HSA21 (Fig. 6.1), meaning the catalog of genes, has not yet been conclusively established. However, that is very important for understanding the association of genes contributing to phenotypes in individuals with Down syndrome (DS), especially for cognitive and memory deficits, and onset of neurodegenerative diseases such as Alzheimer disease (AD) at early age. Practically, genotype–phenotype correlation and molecular mechanisms for every single phenotype have been targeted for elucidation of pathogenesis (Table 6.1), with a view to developing effective therapeutics for rendering a better quality of life to DS population. Besides protein coding genes, initial computational analysis has documented the presence of five microRNAs (miRNAs) in HSA21, but with no precise knowledge about their function and potential contribution to DS phenotypes. However, recent era has extracted enormous information on miRNAs and noncoding long RNAs (lncRNAs) on pathomechanism of DS. Indeed, extensive efforts have been made on understanding the contribution of the specific genes located on the HSA21 to express the characteristic phenotypes of DS and narrowing down the actual chromosomal region/genes that contain the specific codons responsible for specific phenotypes (Fig. 6.2). Over 30 pathogenic chromosomal aberrations have been delineated to pin-point

Genetics and Neurobiology of Down Syndrome
https://doi.org/10.1016/B978-0-323-90456-8.00005-3

the 21q21-21q22 segment that might contain the critical region for DS pathogenesis (Korenberg et al., 1994).

The molecular genetics study on genotype–phenotype association in DS has been expanded in the last 2 decades; however, real-time gene mapping on DS patients has been a constraint owing to limited number of DS patients with partial trisomy or other variant DS patients available to support drawing a derivation of genes and phenotypes one-on-one basis. Alongside gene mapping studies on human DS, mouse models have been created with segmental trisomies, transchromosomal and transgenic engineering, and also deletion/duplication models aiming toward tapering down genes responsible for highly derogatory phenotypes, such as learning and memory impairment, early onset of AD and age-related dementia, and leukemia (detailed in Chapter 14). Among all mouse models, Ts65Dn, Ts1Cje, Ts1Rhr, and deletion/duplication models are extensively employed for the gene-dosage expression analysis. These genetically engineered trisomic DS models were designed in combination with gene knockouts and YAC or BAC transgenics assimilated with individual dosage-sensitive genes for assigning with different phenotypes of DS. Additionally, studies of cell lines and induced pluripotent stem cells (iPSCs) derived from DS patients are highlighting insights into the gene-dosage imbalance in patients with DS. However, an immense depth of the problem remains unresolved due to complex interactions of HSA21 and non-HSA21 genes, and the resultant expression against a variable biological backdrop such as inherent genetic makeup, allelic variation of coding and noncoding genes, epigenetic and environmental influences and stochastic events (Antonarakis, 2004). Indeed, the research on DS is not only revolving around the smallest autosome, that is, HSA21, but also extracting the insights of the complex gene-expression mechanisms responsible for typical phenotypes of DS.

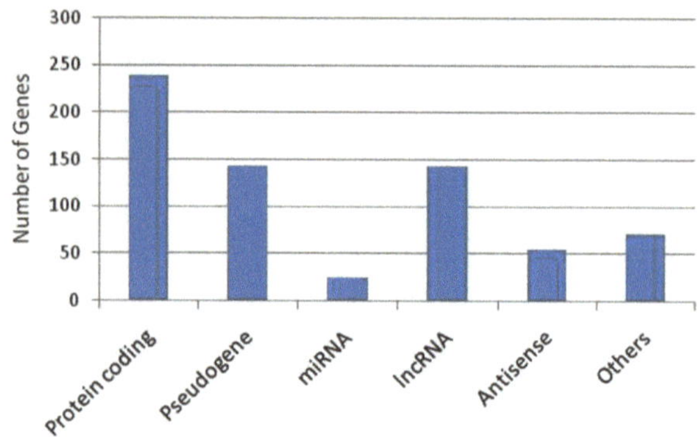

FIGURE 6.1 Frequency of coding and noncoding genes mapped on to HSA21. *The bar diagram is prepared from the information available in Antonarakis (2017). https://doi.org/10.1038/nrg.2016.154.*

Indeed, understanding the genetic basis of the large variation in DS phenotypes would identify more candidate genes and thus, dissect the effects of individual and/or combinations of alleles both on HSA21 and on other disomic chromosomes. Genetic variations at the population or ethnic level would further complicate the inferences of genotype-phenotype correlations. Therefore, collaborative studies must be planned at international level involving high level of cooperation between countries to study the clinical and psychometric condition of individuals with DS by maintaining unified and harmonized protocols. Notably, there is a huge challenge for dissecting the molecular genetics of DS with genetic engineering tools for recognition of the individual dosage-sensitive genes and for understanding the biological effects of the trisomy 21. Moreover, the highlights of the extraordinarily variable copy numbers present in all humans would challenge the importance of gene dosage to everyone, not just to DS population. Thus, the molecular examination of HSA21 genes of DS would be a daunting task for understanding the euploid and aneuploid biology. The basic question on the implication of dosage-sensitive genes of HSA21 and the synergistic or antagonistic interaction of the global disomic genes on

TABLE 6.1 Genes of HSA21 overexpressed in Down syndrome.[a]

Genes	Name	Chromosomal location	Functional categories[a]/description	Human diseases
APP	Amyloid precursor protein	21q21.2	Miscellaneous, especially amyloid precursor protein (β-APP)	AD; poor motor control; learning and memory deficits
DSCAM	DS cell adhesion molecule	21q22.2	Adhesion molecule	Defects in heart and brain development; learning and memory deficits
DSCR1	Down syndrome critical region 1	21q22.2	Miscellaneous	DS-specific phenotype, brain, heart; leukemia; poor motor control
BACH1	BTB domain and CNC homolog 1	21q21.3	Transcription factors	ROS production; leukemia
DYRK1A	Dual specificity tyrosine phosphorylation regulated kinase 1A	21q22.13	Kinases; regulation of cell proliferation in the brain	Causes of intellectual disability; neurodegeneration
S100β	S100 calcium binding protein B	21q22.3	Miscellaneous	ROS production; brain
ERG	ETS transcription factor ERG	21q22.2	Transcription factors	AMKL; TMPD
ETS2	ETS proto-oncogene 2, transcription factor	21q22.2	Transcription factors	AMKL; TMPD; craniofacial anomaly
SOD1	Superoxide dismutase 1	21q22.1	Oxygen metabolism	ALS1; ROS generation
IFNAR1/2	Interferon alpha and beta receptor subunit 1/2	21q22.11	Receptors	AD
BACE2	Beta-secretase 2	21q22.2-q22.3	Proteases	AD
SYNJ1	Synaptojanin 1	21q22.11	Phosphatases	Cognitive disability; poor motor control
PRKCBP2	Protein kinase C-binding protein 2	21q22.1	RACK17, OLIG2	T(14; 21); T-cell ALL
KCNE1/2	Potassium voltage-gated channel subfamily E regulatory subunit 1/2	21q22.1	KCNE1: IKs slow voltage gated P-channel protein; KCNE2: K-channel voltage gated, MiRP1	KCNE1:Larveil and lange-nelson syndrome; KCNE2: Long QT syndrome
RUNX1	RUNX family transcription factor 1	21q22.12	Acute myeloid leukemia 1 protein (AML1); CBFA2	AML, t(18; 21); FPDMM
RCAN1	Regulator of calcineurin 1	21q22.12	KCNE2 34364.024	
KCNJ6/15	Potassium voltage-gated channel subfamily J member 6/15	21q22.2	G-protein activated K-channel; GIRK2; promotes inward flow of potassium ions to cells of the heart and brain	Weaver mouse

(Continued)

TABLE 6.1 Genes of HSA21 overexpressed in Down syndrome.[a]—cont'd

Genes	Name	Chromosomal location	Functional categories[a]/description	Human diseases
CBS	Cystathionine β-synthase	21q22.3	Heart defect	Homocystinurea
CSTB	Cystatin B (Stein B) (CST6)	21q22.3	No data	EPM1
COL6A1/2	Collagen type VI alpha 1 chain/2	21q22.3	Collagen VI, alpha-1/2 chain	Bethlem myopathy; heart defects
CHAF1B	Chromatin assembly factor 1 subunit B	21q22.12-q22.13	DNA replication	No data
CLDN14	Claudin 14	21q22.13	Watertight seals around cell margins	Cause of deafness
PIGP	Phosphatidylinositol glycan anchor biosynthesis class P, formerly DSCR5	21q22.13	Synthesis of glycolipid anchor to blood cells; tongue development	No data
RIPPLY3	Ripply transcriptional repressor 3, formerly DSCR6		No data	No data
SIM2	Single-minded family bHLH transcription factor 2	21q22.13	Regulates the generation of nerve cells; discovered in fruit flies	AD
TTC3	Tetratricopeptide repeat domain 3		Expressed in embryonic neuron development	No data
KRTAP10-1/2-12	Keratin associated protein 10-1/2-12	21q22.3	KRTAP-family genes are restricted to skin, and may include many pseudogenes	AD

[a] *The table has been prepared from functional assignments extracted from several publications available in the public domain, and all such publications are credited with proper citations and included in reference section. Therefore, gene-specific referencing has been avoided to minimize space utilization.*

DS patients deserves an answer first for attending the clinico–pathologic phenotypes of dosage effects. Taken together, the disorder is highly complex from several aspects, and understanding is always underway since 1959, but new therapies are being targeted and under trials for giving a so-called normal life to individuals with DS. The first attempt is to enhance learning and memory and to combat complications of AD in DS individuals. Therefore, "someday" will come as "today" to achieve these benefits.

2. Gene-dosage imbalance

The discovery of trisomy 21 causing DS by Lejeune et al., in 1959, has established the fact that the genes present in HSA21 are all triplicated, and thus, an obvious overexpression was hypothesized to be 1.5-fold compared to normal diploid state. DS phenotypes were hypothesized to be linked to two possible theories such as "developmental instability" and "gene-dosage effect" (Ait Yahya-Graison et al., 2007). The first

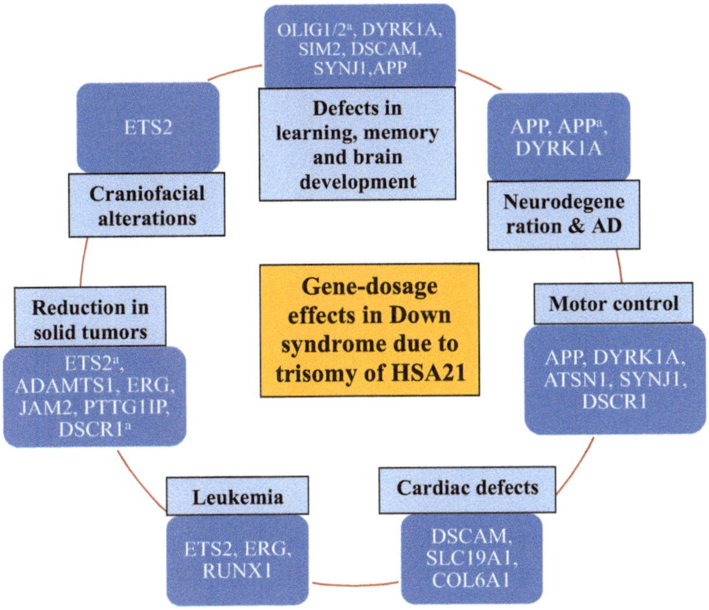

FIGURE 6.2 Dosage-sensitive genes in Down syndrome. Genes included only for which direct evidence is established that increased dose or allelic variation (SLC19A1, COL6A1) of the gene contributes to a phenotype. [a]indicated genes whose reduction from three to two copies reverses the phenotype in a mouse model. *The figure is prepared from the table presented in Lana-Elola et al. (2011).*

theory explained that the genomic trisomy of HSA21 globally perturbs the normal stability of gene expression in DS cells during development (Saran et al., 2003) and exerts a nonspecific alteration of regulation and function of the genome (Vilardell et al., 2011) leading to a disruption of homeostasis throughout the genome. The second presumption of the "gene-dosage effect" correlates the overexpression of triplicated genes on HSA21 directly to different aspects of DS phenotype (Korenberg, 1990). Indeed, a great deal of investigation has been conducted on gene-expression studies in human tissues or cell lines and/or mouse models using several methods, including microarrays, serial analysis of gene expression (SAGE), real-time reverse-transcriptase polymerase chain reaction (RT-PCR), RNA-sequencing, or proteomic approaches (Prandini et al., 2007; Volk et al., 2013; Kong et al., 2014; Zhao et al., 2016; Liu et al., 2017); however, the outcome could not conclusively select and/or justify the hypothesis applicable to the etiology of DS.

With respect to gene expression, approximately 29% transcripts were proportional to the gene-dosage effect or amplified, 15% were highly variable, and 56% were compensated as observed in lymphoblastoid cells derived from DS patients (Ait Graison-Yahya et al., 2007). Expression of HSA21 genes was categorized into four classes in lymphoblastoid cells (Fig. 6.3), which include class I genes (no. = 30) showed 1.5-fold expression level in DS versus controls, and were responsible for DS features directly, or indirectly as secondary effect of cis- and transacting genes; class II (no. = 9) showed >1.64 expression ratio between DS and controls relative to amplification of initial gene dosage where SAMSN1, SNF1LK, STCH, and BTG3 had an expression range of 1.64–2.27; class III (no. = 77) genes were abundant with an expression ratio of <1.4 (DS vs. control), which were thought to be compensated in DS (similar to Lyle et al., 2004 with 45% compensated), (compensation is likely due to modulation of transcriptional activity or stability of mRNA; expression of compensated genes could be regulated by mechanisms that are not affected in DS); and class IV (no. = 15) genes showed expression level of one or 1.5-times, and were highly variable in both cells of DS patients and controls

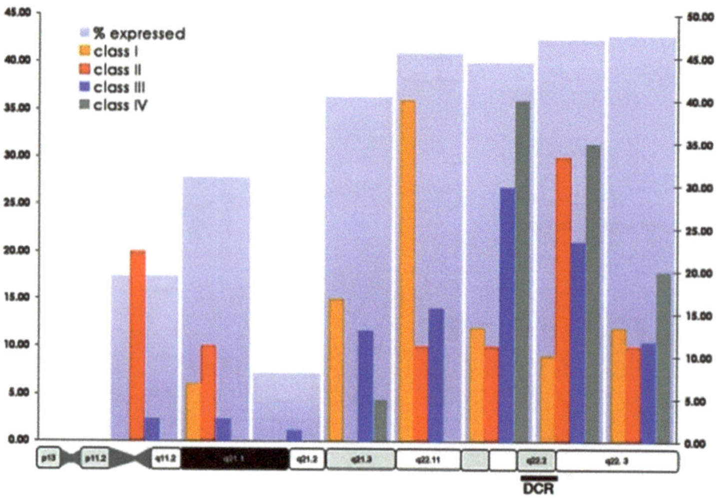

FIGURE 6.3 Expression of class I, II, III, and IV genes along the HSA21. The left Y-axis depicts the proportion of expressed genes in each 5-Mb interval, and the right Y-axis presents the proportion of class I, II, III, and IV genes in each 5-Mb interval. *Reprinted from: Ait Graison-Yahya et al. (2007) with permission from Elsevier and Copyright Clearance Center.*

(e.g., CBR1, PRDM15, ADARB1). These lymphoblastoid cells also showed copy number variations (CNVs) (Redon et al., 2006). Gene dosage amplification of class II genes was hypothesized to result from a cascading effect involving regulation of cis- or trans-acting genes as observed in various signaling pathways through involvement of mitogen-activated HSA21 kinases in silico (Pellegrini-Calace and Tramontano, 2006), and in lymphoblastoid cells derived from DS patients (e.g., expression of DYRK1A (gene dosage effect), SNF1LK (amplified), DSCR3 (compensated), but not RIPK4) (Ait Graison-Yahya et al., 2007). Overexpressed DYRK1A might regulate amplification of SNF1LK gene expression, and DSCR3 might act as a scaffolding protein. In this connection, overexpression of 16 genes was demonstrated in human trisomic fetal heart (Li et al., 2006). The presence of genes of class I, II, and III observed in all tissues suggested that gene-dosage, amplification, and compensation are general features observed in DS lymphoblastoid cells (Ait Graison-Yahya et al., 2007). Interestingly, a nonuniform gene density has been demonstrated along the long arm of chromosome 21, where most of the expressed genes were located to the distal part of the chromosome (Gardiner et al., 2003). The density of class III genes in the 21q terminal (telomeric) region indicates higher ratio of gene predictions that are located inside the introns.

The development of many mouse models for partial or full trisomy and/or iPSCs and other cell lines developed from individuals with trisomy of HSA21 have facilitated understanding of the molecular pathophysiology of DS and circumventing the difficulties associated with genotype–phenotype features. Testing of Ts65Dn mice having trisomy of 78 genes on MMU16 has described 1.5-fold upregulation of ~37% genes; 18% with >1.5-fold upregulation, ~45% with downregulation, and ~9% genes with no remarkable change (Lyle et al., 2004). Collectively, the knowledge gathered indicated complex regulation of gene expression of the trisomic HSA21 and its orthologs in human and mouse models of DS (Fig. 6.4). Notably, the cellular and mouse models are extremely useful for preclinical development of therapeutic targets.

Interestingly, gene expression variations were suggested to be controlled by mechanisms involving cis- and transregulators, directly or via the gene-regulation networks. To explain, it was hypothesized that the transinhibitors could directly act on expression of the

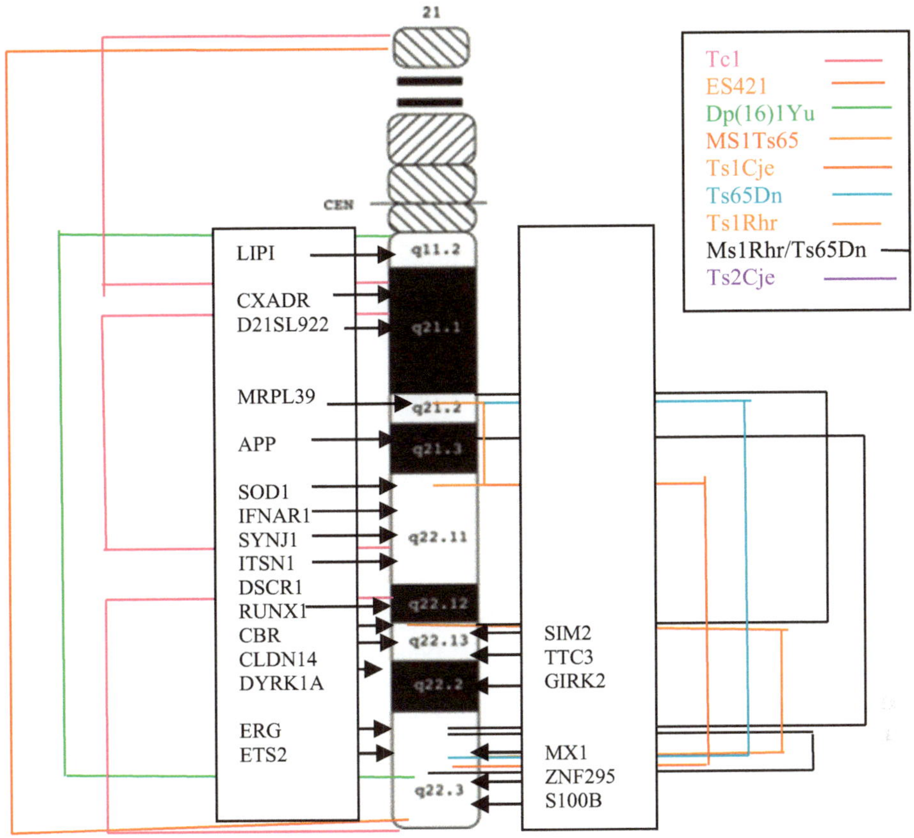

FIGURE 6.4 Location of genes on human chromosome 21 and mouse models showing orthologous regions on HSA21. *Modified from Rachidi and Lopes (2011).*

compensated genes, and transactivators would activate inhibitors that are triplicated in HSA21 and reduce the expression level of targeted genes of HSA21 (Brichler et al., 2005). In addition, similar gene expression was reflected in the alternative transcripts of the same gene, whereas that was not always observed in sense and antisense transcripts. Variability in gene expression was thought to be responsible for variation in severity of DS phenotypes.

Information collected based on several theories, including dosage compensation, epigenetic gene expression variation, or negative feedback, could apparently explain the discordance between the genomic dosage imbalance and the expression levels of HSA21 genes (Antonarakis, 2017; Prandini et al., 2007). Indeed, identification of the mechanisms that control the expression pattern of HSA21 genes and, also, whether nontrisomic gene deregulation is stochastic or it is the influence of HSA21 genes on specific non-HSA21 genes remains one of the crucial factors of DS research. Such studies would further characterize the molecular pathways and molecular targets for potential therapeutic drugs. Interestingly, observation such as a faction of HSA21 genes escapes dosage effects and overexpression of HSA21 genes in particular segments, in particular 21q22, revealing a high prevalence of overexpression over other genomic regions, in DS samples compared to genes from other chromosomes, suggests a decisive role in the pathogenesis of DS. Since 1970s, the concept of DS critical region (DSCR) responsible for DS phenotypes has changed (3.8—6.5 Mb, localized in the distal segment of the long arm of chromosome 21 at 21q22.11—21q22.2), and it has been described that genes located outside DSCR also contribute to DS-features (Korenberg et al., 1994). Therefore, there is no single gene or region of HSA21 responsible for expression of DS-specific features; however, still there is no consensus about the regions or genes as determinant for clinical manifestations in DS.

Summarily, the specific region of HSA21, that is, 21q22 might contain most sensitive overexpressed genes involved in DS pathogenesis, and a complex interaction between trisomic genes and other regions of the genome, which might be disomic and dysregulated in presence of trisomy 21, could establish a direct association between the HSA21 genes and DS phenotypes (Pelleri et al., 2018). Indeed, the dysregulation of gene expression in trisomy 21 is not just limited to genes on HSA21 but has been hypothesized as a genome-wide effect. This dysregulation may be linked to chromatin domains. Notably, DS characteristics in people with partial duplication of HSA21 are not contributed by single trisomic region; rather several susceptible regions (when presented in three copies) are responsible for DS-specific features. Also, complex rearrangements have been presented in several individuals, including contiguous or noncontiguous deletions or duplications, CNVs of other regions or the duplication of genes located in the p-arm of HSA21. These rearrangements can significantly impact the phenotypic expression of the HSA21 duplication and add noise to the genetic association of clinical manifestations related to human trisomy 21. Additionally, exploration of the functional genomic consequences of HSA21 in the postsequencing years has provided an unprecedented opportunity to decipher the molecular expression of genomic dosage imbalance. Altogether, the genic interaction and expression toward overproduction result in deregulation of biochemical pathways that are implicated in clinical aspects and posterior complications (Pritchard et al., 2008).

A study using DS mouse models observed some of the DS features, including aberrant behavioral and cognitive performance, brain morphology, and hippocampal gene expression in a standardized and controlled manner (Table 6.2) (Duchon et al., 2021). The partial duplication of the MMU16 region homologous to HSA21 appeared in consensus with human genetic analysis (Korbel et al., 2009; Lyle et al., 2009; Delabar et al., 1993; Korenberg et al., 1994). The report suggests possibilities of multiple genetic interactions between different regions of HSA21 contribute to changes in behavioral, morphological, and phenotypes of molecular pathways. Dissection of all these genetic interactions has highlighted the convergence in the biological cascades altered in overlapping DS models, and that centered around six main hubs, such as DYRK1A, GSK3β, NPY, SNARE, RHOA and NPAS4 (Fig. 6.5). These hubs were postulated to be the center of biological cascade in DS (Duchon et al., 2021). An altered gene—gene crosstalk and molecular mechanisms were demonstrated in these six highly specific interconnected DS hubs in the mouse models.

3. Genotype—phenotype correlation

It is an obvious understanding that the triplication of HSA21 genes of trisomy HSA21 is responsible for DS-specific characteristics. Importantly, analysis of the spatial or temporal expression of genes and/or the resultant functional proteins has indicated wide variability in gene expression, which might be responsible for wide variation in DS phenotypes. On the other hand, genes with no

TABLE 6.2 Differential expression analysis results from the seven mouse models analyzed.

Mouse lines	Dp1yey	Dp3Yah	Ts65Dn (Mmu16)	Dp5/Dp1	Dp5Yah	Dp1Rhr	Tg(*Dyrk1a*)	Ts65Dn (Mmu17)
Nb of probes detected by the array	35,556							
Nb of annotated probes detected (without control probes)	27,359							
Number of trisomic expressed genes (TEGs)	155	21	130	127	87	40	1	40
Number of differentially expressed genes (DEGs)	711	826	1074	922	736	1306	850	
Number of differential trisomic expressed genes TEGs	66	13	64	54	39	18	1	28
% of TEGs[a]	43%	62%	49%	43%	45%	45%	100%	70%
% of compensated trisomic genes detected	57%	38%	51%	57%	55%	55%	0%	
Number of GAGE KEGG, and GOs (CC, BP, MF) terms disregulated in the trisomic model FDR <0.1	244	67	12	111	318	225	231	
Number of GAGE KEGG and GOs (CC, BP, MF) terms upregulated in the trisomic model FDR <0.1	207	60	3	33	4	132	222	
Number of GAGE KEGG and GOs (CC, BP, MF) terms downregulated in the trisomic model FDR <0.1	37	7	9	78	314	93	9	
Number of GAGE KEGG and GOs (CC, BP, MF) terms in the trisomic model unique to each mice line FDR <0.1	80	1	2	30	195	107	64	

GO are Go functional terms involved in cellular compartment (CC), molecular function (MF), and biological processes (BP); *GAGE*, generally applicable gene set enrichment for pathway analysis; *KEGG*, Kyoto Encyclopedia of Genes and Genomes pathway; *TEGs*, trisomic expressed genes.
Reproduced from Duchon et al. (2021). https://doi.org/10.1093/hmg/ddab012 under the terms of the Creative Commons CC by license.

variation in expression might underlie the obligatory phenotypes that are present in all DS patients. Additionally, studies on individuals with partial trisomy of HSA21 have demonstrated DS-specific phenotypes delineating a DSCR that contains genes specific for cognitive deficits. Interestingly, DS patients having partial triplication outside this critical region display some of the DS-specific features. Also, the nontriplicated and non-HSA21 transcripts have been demonstrated to contribute to DS phenotypes in microarray analysis of cells from tissues and organs with trisomy HSA21. Besides involvement of protein coding genes, contribution of functional noncoding sequences to dosage related DS phenotypes is apparent, in particular, cis- or transregulators or that contains variability in functional consequences (reviewed in Antonarakis et al., 2004). Identification of the noncoding sequences having potential gene regulatory function may pinpoint the obligatory genomic elements contributing to dosage imbalance, while

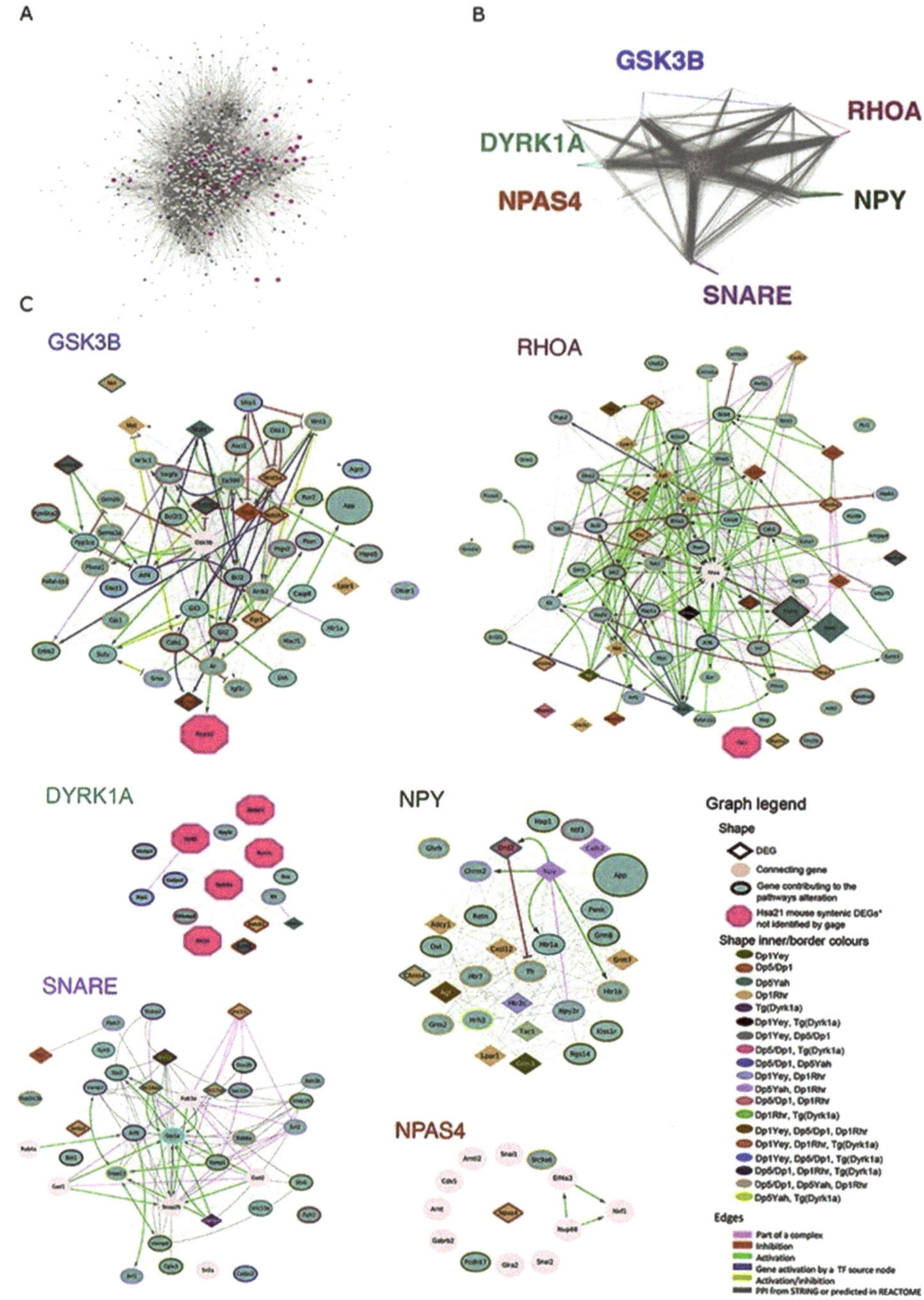

FIGURE 6.5 Protein–protein interaction networks involving differentially expressed genes (DEGs) linked to synaptic function in Down syndrome: (A) STRING04 MiniPPINet of genes involved in synaptic function has been highlighted by using edge-weighed spring embedded layout. The shapes of the nodes indicate the following information: Shapes: i. *Pallid pink ellipses*

trisomy of some of these elements and/or their genomic variation might be linked to certain DS features. Investigation of functional genes or noncoding sequences of transacting nonHSA21 elements may additionally indicate their predisposition to DS characteristics, and also, to DS-like phenotypes in individuals having nontrisomic HSA21 (e.g., presence of Simian crease in individuals without trisomic HSA21).

Analysis of dosage sensitivity and potential involvement of miRNAs and other lncRNAs related to DS-features may indicate the association and/or interaction with trisomic functional genes. Altogether, global gene expression in trisomic tissues and organs in association with environmental and epigenetic factors, including the allelic variations, could really define the alterations in biological processes and/or dysfunctional systems underlying DS phenotypes (Fig. 6.6). This could further be augmented by studies of monozygotic twins with DS, discrepant identical twins (one with trisomy HSA21 and other one normal HSA21), or mosaic DS. Therefore, more and more information on genic and nongenic contribution on DS-specific characteristics extracted through high-throughput investigations would appear with better clarity on genotype—phenotype association and be challenging for targeting potential therapeutic interventions. Moreover, the context of timing and cellular-specific expression puts an urge to understand the phenotypic changes and/or expression of functional genes relative to the developmental stages of DS individuals, which was possible in analogous models of the syndrome, such as mouse models (Ts65Dn, Ts1Cje, and so on).

3.1 Genomic variability for DS phenotypes

Phenotypic variability among individuals with DS is thought to be caused by imbalance in gene-dosage, primarily due to triplication of HSA21 and variability in gene expression. Furthermore, considerable variability in the presence of Hirschsprung disease, duodenal atresia/stenosis, AMKL, variation in the degree of learning and cognitive impairment, and craniofacial appearance add to the question of variability in expression of genes and/or proteins among DS individuals. To delineate the genotype—phenotype association, several

represent connecting proteins added for assuring the full connectivity of the network; ii. *pink octagons* represent HSA21 syntenic genes in mice not observed in GAGE contributing to meta-pathway dysregulation; and iii. *green inner colored ellipses*, contributing, even slightly, to any dysregulated pathway inside the meta-pathway. The identical sizes of the octagons are HSA21 syntenic genes in mice. In addition, the *border color* indicates that the genes are altered in the mouse model multigroup; iv. *diamonds* indicate DEGs identified by GAGE and FCROS. (B) Network structure decomposition of the STRING04 MiniPPINet highlighted interaction of GSK3β, NPY, SNARE proteins, DYRK1A, and RHOA in different colors, and the color of NPAS4 indicates the first level of interactions. (C) Six RegPPINets were extracted from the selection of each of the following proteins and their second interactors from STRING04MinPPINet: RHOA, DYRK1A, GSK3β, NPY, SNARE proteins, and NPAS4. Then, those were further annotated with regulatory information using REACTOME (Duchon et al., 2021). The shapes of the nodes represent the following information: Shapes: i. *Pallid pink ellipses* represent connecting proteins added for assuring the full connectivity of the network; ii. *pink octagons* represent HSA21 syntenic genes in mice not observed in GAGE contributing to meta-pathway dysregulation; and iii. *green inner colored ellipses*, contributing, even slightly, to any dysregulated pathway inside the meta-pathway. The identical sizes of the octagons are HSA21 syntenic genes in mice. In addition, the *border color* indicates that the genes are altered in the mouse model multi group; iv. *diamonds* indicate DEGs identified by GAGE and FCROS. The *edge-colors* represent the type of interaction annotated by following the PathPPI classification (Tang et al., 2015), and ReactomeFIViz annotations as follows (i) The GErel edges indicated as expression in blue and repression in yellow. (ii) PPrel edges indicated activation in green color, inhibition in red. (iii) Interactions between proteins known to be part of complexes in violet. (iv) Predicted interactions were represented in gray including the PPI interactions identified by STRING DB (Szklarczyk et al., 2017), after merging both networks. *Reprinted from: Duchon et al. (2021). https://doi.org/10.1093/hmg/ddab012.*

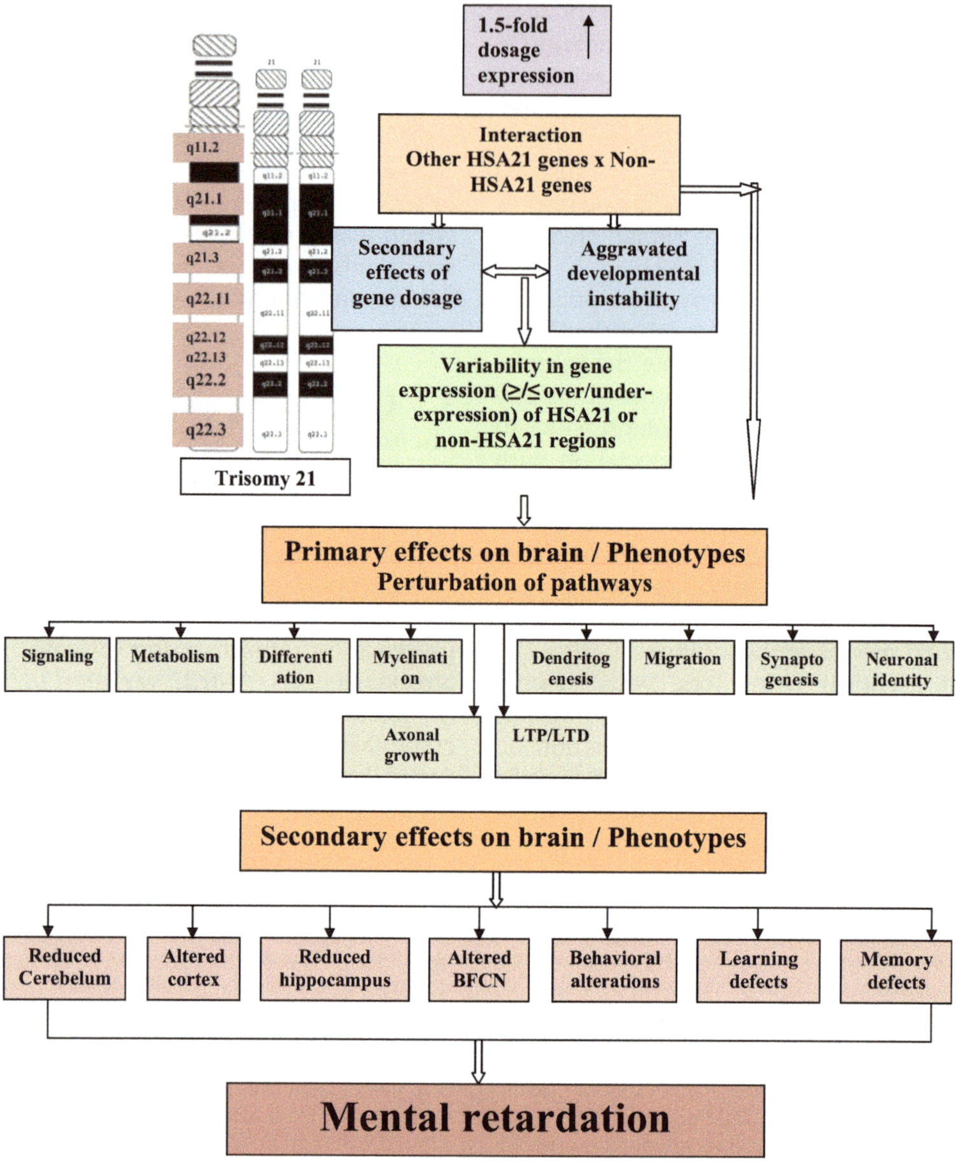

FIGURE 6.6 Gene-dosage effects on neurological phenotypes and mental retardation in Down syndrome. Interactive genetic mechanisms with the overdosage of HSA21 genes induce global alterations of gene expression in the brain. Firstly, the primary effect of the 1.5-fold overexpression of HSA21 genes and non-HSA21 genes alter neurological phenotypes at functional and cellular level producing primary phenotypes, and secondly, the overexpressed genes modify the expression level of other trisomic genes on HSA21 and of disomic genes on other chromosomes, by a secondary gene effect or by a more general alteration of the transcriptional homeostasis (amplified developmental instability) and result in DS-specific and variable phenotypes. These neuronal alterations result in neuromorphological, neurological, and behavioral alterations, leading to production of secondary phenotypes. Collectively, the neurological alterations caused by the presence of trisomy 21 leads to the mental retardation in individuals with DS.

gene-expression studies have demonstrated a global upregulation of HSA21 genes in trisomic HSA21. The so called 1.5-fold overexpression was documented in most organs of DS; however, a complete map of gene-dosage imbalance has not been established in all kinds of organs/tissues/cells and at all developmental stages. Animal models have displayed differential expression of the trisomic genes orthologous to HSA21 (Fig. 6.4). A list of the upregulated genes has been assigned with their functional contribution and involvement in systemic deregulation in DS individuals (Table 6.1). Molecular understanding of genotype–phenotype association in DS has speculated whether overexpression of restricted number of genes is responsible for variation in phenotypic abnormalities (Giannone et al., 2004; Yahya-Graison et al., 2007). Polymorphism in ~1.26% of HSA21 containing short repetitive sequences and 24,386 SNPs (identified through chromosome-wide linkage disequilibrium studies) were hypothesized to be responsible for variability in phenotypic manifestations (Antonarakis et al., 2004). Recent studies, including comparative sequence-analysis with orthologous animals have identified some of the novel genes and putative transcripts (Gardiner et al., 2003; Mural et al., 2002; Weijerman and de Winter, 2010).

It was thought from triplicated HSA21 that DS phenotypes are caused by single dosage-sensitive genes, which are independent of the influence of other genes or environmental and others factors to either cooperate or exacerbate the gene-dosage effect (Fig. 6.6). To decipher this effect, a number of transgenic mice were engineered with HSA21 genes or their mouse orthologues, and/or segmental trisomy for the same (plus other) genes, though the expression of the phenotypes was not compared quantitatively. The concept of single dosage-sensitive gene led to hypothesize the effect of "critical regions" on HSA21 for specific DS phenotypes. However, individuals with partial trisomy 21 carrying rearrangement involving 21 were identified with some phenotypes of DS and that led to identify a common region of overlap, the so-called DSCR of about 5.4 Mb on HSA21 (Delabar et al., 1993). The DSCR was associated with most of the major DS phenotypes, including protruding tongue and flat face (largely contributed by hypoplastic mandible and craniofacial skeleton respectively), short stature, mental retardation, joint hyperlaxity, muscle hypotonia, and a variety of dermatoglyphic abnormalities. Thus, the trisomic genes of DSCR appeared sufficient to manifest DS features. In this sense, the gene is Mendelian in its function. The dosage-sensitive expression of single gene for endostatin that inhibits angiogenesis in tumor may explain the possible mechanism of resistance of DS to cancer of the solid tumors (Zorick et al., 2001). However, variable DS phenotypes might not be contributed by single dosage-sensitive gene, which could rather be the cooperative expression of multiple genes.

Interestingly, several features of DS attributed to the DSCR have also been measured precisely in Ts65Dn mice of DS models. On the basis of the phenotypic expression caused by a critical region gene or genes, transgenic mice were created with re-engineered MMU16 corresponding to the deletion or duplication of DSCR of HSA21 (Olson et al., 2004a). Mice carrying the duplication of MMU16 segment and the segmental trisomy of the critical region genes did not display the features of stature, the midface hypoplasia, small mandible, or dysmorphology of the skull presented by the DSCR concept and that indicated that gene(s) from this DSCR region was not sufficient to produce these phenotypes. Interestingly, these mice carrying deletion of the critical region segment were crossed to Ts65Dn mice (which carry all of the DS phenotypes) to enable returning of the critical region gene dosage to normal in the mice that carried the majority of Ts65Dn segment genes in trisomic status. The phenotypes seen in Ts65Dn were somewhat attenuated in these mice, indicating that trisomy of DSCR genes can contribute for some of the DS phenotypes but not all. Thus, the hypothesis of involvement of multiple genes was established for expression of complex and variable characteristics in individuals with DS.

Variability in DS phenotypes might also be contributed by the existence of variability in molecular dissection of the HSA21. In relevance to the variation documented in short nucleotide polymorphisms (SNPs) and CNVs spanning the whole genome of humans, the HSA21 contributes ~1.26% short sequence polymorphic repeats and 105,334 probable SNPs (reviewed in Antonarakis et al., 2004). Studies involving linkage disequilibrium of polymorphic sites and sequencing using tiled oligonucleotide arrays of haploid human genome from different ethnic groups observed 24,386 SNPs with rare alleles, observed at least twice, in addition to 11,603 nucleotide variants (Patil et al., 2001). Altogether, 4135 inferred haplotype blocks were reported with an average length of 7.83 Kb per block. Such maps on chromosome-wide linkage disequilibrium would determine the SNPs involved in gene expression variations and their predisposing potential to different variable phenotypic manifestations and also, to common complex phenotypes in individuals with DS.

Regulation of the phenotypes of DS was thought to be associated with the dysregulation of transcription factors, particularly the nuclear factor of activated T-cells (NFAT), which is partly controlled by DSCR1 and DYRK1A proteins (Epstein, 2001). Notably, the people with trisomy 21 have 1.5 times of these proteins higher than normal euploid individuals (Arron et al., 2006). DYRK1A (dual-specificity tyrosine phosphorylation-regulated kinase 1A) has been proposed to be an important candidate gene responsible for various phenotypic traits of DS because of its involvement in multiple cellular pathways, including nuclear factor activation and basic cellular metabolism (Park et al., 2009). DYRK1A inhibits cell proliferation and promotes premature neuronal differentiation by regulating NOTCH signaling (Hammerle et al., 2011). This gene also inhibits nuclear translocation of the NFAT family in association with DSCR1 (or RCAN1) protein, leading to neuronal and cardiac defects in individuals with DS (Arron et al., 2006). Small segmental duplications in DYRK1A did not establish specific effect of 21 that leads to severe reductions in IQ in DS population (Korbel et al., 2009).

APP gene has been established as the candidate gene for neuro-pathological features of sporadic AD and also in AD of DS patients (DSAD). Individuals with DS and AD have exhibited neurofibrillary tangles and amyloid-β plaques in common, wherein the layer II of the entorhinal cortex and the subiculum appeared critical for memory consolidation and damaged first, followed by gradual reduction of the nerve cells of the cortex. Similar observation was noticed in the Ts65Dn mouse model of human DS whose chromosome 16 is syntenic to HSA21. Mutations of the APP contribute to early onset of AD through proteolysis of APP producing amyloid-β plaques in the brains of AD (McNaughton et al., 2012). The cooperating effect of other HSA21 genes with APP-overexpression has been thought of for degeneration of basal forebrain cholinergic neurons (BFCN) in DS (Salehi et al., 2006). The cooperating association of DYRK1A has been described in DS with higher levels of phosphorylated APP produced by DYRK1A-protein and transgenic mice (Korbel et al., 2009; Ryoo et al., 2008).

Elevation of superoxide dismutase (SOD) activity was reported in DS due to triplication of HSA21. The SOD1 gene, mapped onto 21q22, encodes SOD enzyme to catalyze the conversion of oxygen radicals into hydrogen peroxide and water. Free oxygen radicals affect the cellular integrity. It was hypothesized that a disproportionate increase in SOD activity could be responsible for removal of hydrogen peroxide (e.g., glutathione peroxidase), which may result in peroxide damage in the cells. Thus, treatment of DS with free radical scavengers was thought of for prevention of oxidative damage in neurons and neuronal degeneration substantially, and thus, early onset of brain aging and AD. A reduction in terminal axon area to postsynaptic membrane and hyperplastic secondary folds in neuromuscular junctions were reported in muscles of transgenic aging mice and humans with DS (Avraham et al., 1988).

Significant effects of RCAN1, mapped to HSA21, have been demonstrated on mitochondrial integrity in cardiomyocytes and other types of cells (Parra et al., 2018). RCAN1-encoded protein inhibits the Ca^{2+}-activated protein phosphatase calcineurin (CN) through a direct binding to its catabolic subunit and protein—protein interaction (Fuentes et al., 2000; Rothermel et al., 2000), which is important for cardiac remodeling. Thus, downregulation of RCAN1 increases CN-dependent dephosphorylation of the profission protein DRP1 (dynamin-1-like; a GTPase) and the rate of mitochondrial fission and pathological hypertrophy of the heart (Cereghetti et al., 2008). In individuals with DS, RCAN1 is constitutively elevated by triplication of HSA21 and attributes to metabolically more active mitochondrial network, including enhanced fusion, increased ROS-generation and oxidative stress, and interfere with the process of mitochondrial repair (Helguera et al., 2013). RCAN1 maintains mitochondrial integrity through a coordinated inhibition of fission and promotion of fusion mechanism through co-ordinated expression of OPA1 (optic atrophy 1), PINK1, and MFN2 (mitofusin 2) proteins (Corbalan and Kitsis, 2018; Parra et al., 2018). RCAN1 interacts with adenosine nucleotide translocase protein, both directly and indirectly, which can uncouple the inner mitochondrial membrane proton gradient (Azzu et al., 2010). Of the two isoforms of RCAN1, the exon one isoform (RCAN1.1) is abundant in brain and striated muscle with relatively constant expression, whereas, expression of the exon four isoform (RCAN1.4) is controlled by the CN, which thus acts as a feedback inhibitor of CN activity (Yang et al., 2000). An increase in mitochondrial fusion and protection from ischemia/reperfusion (I/R) was described in RCAN1.1 depleted and siRNA transfected ventricular myocytes of neonatal rat (NRVM) (Parra et al., 2018).

RCAN1 deficiency in cardiomyocytes exhibits a deficiency in mitochondrial Ca^{2+} uptake with concomitant increase in cytosolic Ca^{2+} concentrations and also decrease in the lowers the rate of O_2 consumption and ROS generation (Corbalan and Kitsis., 2018). Thus, depletion of RCAN1 in cardiomyocytes reduces mitochondrial membrane potential ($\Delta\Psi_m$) and metabolic activity, including decrease in mitochondrial Ca^{2+} uptake, O_2 consumption, and generation of ROS by mitochondrial fission. Thus, mitochondrial fission reduces the capacity to buffer cytosolic Ca^{2+}, thereby, increase the risk of heart attack or stroke (Parra et al., 2018). Although mitochondrial fusion reduces CAPN effects in cardiomyocytes after reperfusion, a continual increase in fusion may increase the capacity of ROS generation and cellular damage and reduce the potential for mitochondrial repair. Increase in oxidative stress and generation of free radicals are associated in DS, where several genes of 21, including APP, BACH1, DYRK1A, DSCR1, ETS2, SOD1, and S100B have been implicated with ROS generation (Campos and Casado, 2015).

3.2 Allelic variation on HSA21

The expressivity and sensitivity of trisomic HSA21 genes can be explained with allelic variants. A mutant allele may lower the levels of gene product; however, in cases of recessive inheritance, the wild-type allele would be sufficient to carry on normal function of the genes. A loss-of-function mutation in two copies with one wild-type allele in trisomic condition may not appreciably alter the phenotype; in contrast, a gain-of-function mutation in two copies may supersede the buffering effect of the normal allele in a trisomic individual resulting in prominent expression of the corresponding phenotype. The heterotrisomy is another possible manifestation of trisomy at the molecular level (when alleles from three grandparents are present in every cell due to nondysfunctional error in meiosis I) (Baptista et al., 2000). The collagen-like multimeric proteins are formed from the combinations of multimers. The distal part of HSA21 contains COL6A1, COL6A2, and COL18A1, and a heterotrisomic region between D21S167 and HMG14 on HSA21 was demonstrated frequently in individuals with DS and CHD (Baptista et al., 2000).

3.3 Modifier genes

Most of the DS features are variable in severity (expressivity) and, in occurrence (penetrance), except the characteristic phenotypes of DS such as facial appearance, cognitive disability, and early onset of AD. Some of the commonly described DS phenotypes are reported in euploid normal individuals such as CHD (Epstein, 2001). Such wide variation of phenotypes and severity of expression was explained by the effects of gene–environmental interaction and the genetic background of an individual. Although trisomy 21 appears to be the largest risk factor of CHD, variation of its occurrence (~50% of all DS) indicates involvement of other factors. The CHD in DS majorly involves septal defects. A complete atrioventricular canal occurs in 1/5th of the DS individuals and one in 10,000 of the euploid population (Ferencz et al., 1989). However, 80% of the DS do not have complete atrioventricular canal, and 50% of the DS do not have clinical presentation of heart defects. Thus, it was apparent that trisomy 21 alone is not sufficient to cause CHD in DS. In mouse models, mutations in disomic genes were evidenced to affect septation, which may establish a link with the occurrence of CHD in individuals with DS. The variable occurrence of CHD, leukemia, and Hirschsprung disease in DS suggests involvement of additional genetic factors in association with trisomy 21. Also, cooperating effects of predisposing modifier genes with the effects of trisomy 21 may produce a threshold effect and demonstrable phenotype. Notably, all predisposing conditions caused by trisomy 21 may not exert a negative impact as a lower incidence of solid tumors and atherosclerosis were reported in individuals with DS (Hasle, 2001; Yla-Herttuala et al., 1989). A challenge for genotype–phenotype characterization of these effects might indicate therapeutic approaches to reduce incidences of cancer or cardiovascular disease in all people.

3.4 Coincidental small effects

The DS phenotype is altered due to overexpression of one or a few trisomic genes of major impact. Also, small additive effects of the many genes overexpressed in every trisomic cell may contribute to DS phenotypes. Notably, the transcript levels of the majority of trisomic genes are increased about 1.5-fold in a few tissues of DS individuals (Mao et al., 2003), and similar observation was reported in a broad range of tissues in trisomic mouse models (Kahlem et al., 2004; Lyle et al., 2004). It was postulated that an individual triplicated gene might not put in on DS-specific phenotype alone; however, the collective effect of many such genes of multiple cellular processes may exert significant effect on DS phenotypes (Pritchard et al., 2008). In support of that hypothesis, several phenotypes were precisely correlated in Ts65Dn mouse having trisomy of about 130 HSA21 orthologs, and Ts1Cje mice having 91 triplicated genes orthologous to HSA21. The phenotypes, including the structure and function of the brain, dysmorphology of the skull, and gene expression in the cerebellum appeared similar in the two models (Ts65Dn and Ts1Cje), though attenuation of the phenotypes was observed in Ts1Cje due to trisomy of fewer genes, which might be augmented with the cooperating effects of the neighboring trisomic genes (Potier et al., 2006; Olson et al., 2004b; Richtsmeier et al., 2002; Siarey et al., 2005).

It is established that trisomic genes are overexpressed in tissues of individuals with trisomy 21. However, perturbation of gene expression levels is demonstrated in disomic genes also, which significantly affects the expression levels of transcripts in trisomic tissues in mice (Dauphinot et al., 2005; Saran et al., 2003). Alteration of one-third of the disomic gene transcripts was described in a study of trisomic mouse cerebellum (Saran et al., 2003). However, conflicting results have been reported on perturbation of disomic gene expression in human studies (Mao et al., 2003; FitzPatrick et al., 2002). The use of different analytical approaches in different studies for array analysis has augmented the controversy.

3.5 Interaction of genes for DS phenotypes

The advent of sequencing technologies and knowledge gathered on genotypic contribution of triplicated 21 has reported the spatial and temporal pattern of expression of genes and proteins. "Gene-rich" (>48% G + C) and "gene-poor" (<43% G + C) regions in the DSCR and interindividual variability in overall individual genomic makeup are expected to display the divergent and uniform features of DS (Antonarakis et al., 2004). Molecular characterization of segmental trisomies has further depicted the specific genomic regions for DS-features (Barlow et al., 2001). Co-operation and interaction of non-HSA21 transcripts with trisomy 21 could likely contribute to heterogeneity in DS phenotypes (Fig. 6.7) and thus, be investigated for potential therapeutic developments. Further, identification of conserved nonfunctional sequences, especially the *cis*- and *trans*regulators, or variation in functional consequences could also be considered for understanding of dosage-related phenotypic dysmorphism (Antonarakis et al., 2004). Segmental analysis of the triplicated gene dosage has explained the variability in dosage sensitivity among the expressed transcripts (29%: sensitive to the gene-dosage effect or amplified; 56%: compensated; 15%: highly variable among DS individuals), wherein upregulated genes contributed to DS features but not by the compensated genes (Yahya-Graison et al., 2007).

There are conflicting reports on contribution of the 21st chromosome for CHD observed in >40% of all DS (Gittenberger-de Groot et al., 2003; Korbel et al., 2009). The *COL6A1*, mapped onto HSA21, encodes the α1-chain of collagen VI and is overexpressed in the atrio-ventricular cushions of the trisomic fetal heart and that results in ASD in DS. However, cooperating effects of mutations in cell adhesion molecules such as *DSCAM* and *CRELD1* (HSA3) have been speculated to contribute to and augment ASD in DS (Korbel et al., 2009). Interestingly, similar cooperation has been observed with polymorphisms in folate-pathway genes such as *SLC19A1* (HSA21: 21q22. 3) and *MTHFR* (HSA1: 1p36. 3) that also causes ASD in DS (Locke et al., 2010).

Experimental approaches for understanding of the divergence from normal patterns of development in a cell that exists only for a defined period during embryogenesis may decipher how trisomy 21 affects the steady-state functions in a terminally differentiated cell, including signaling or metabolic pathways. The outcome will ultimately indicate potential therapeutic approaches such as neuronal response to stimulation and so on. The altered functions of a mature cell may not be the direct implication of overexpression of trisomy 21 but the consequences of downstream effects of the global genome in DS individuals that perturbs the cellular functions. Thus, a specific phenotype of DS may be a consequence of but not a direct effect of trisomic gene expression (Roper and Reeves, 2006). Therefore, trisomic effects on development may indicate alteration in functions of cells after end point of differentiation.

For elucidation of the impact of gene expression in the context of triplicated HSA21 genes, animal models of DS played critical roles for segregating the direct and cascading effects of trisomy 21 on development and divergent functions. A variety of DS characteristics have been explained for quantitative tracing of their origins in development in mouse models. The Ts65Dn mouse is trisomic for a segment that contains orthologs of about half of HSA21 genes, while Ts1Cje mice are trisomic for about two-thirds of the genes that are trisomic in Ts65Dn (Gardiner et al., 2003; Antonarakis et al., 2004). A variety of powerful mouse models were created by manipulating the set of genes that are trisomic in a mouse. The mouse chromosomes 10 (MMU10), 16 (MMU16), and 17 (MMU17) are syntenic with regions on the long "q" arm of human chromosome 21. With the advent of complete sequencing for HSA21 and its mouse orthologs and systematic study of the prenatal consequences of trisomy 21, the gene catalog has further supported to understand the genetic contributions to DS phenotypes in the mouse.

Trisomic gene content of HSA21 has been manipulated in mice to add or delete trisomic segments by chromosome engineering technique (Olson et al., 2004a) and that facilitated creation of a transchromosomal mouse model (Tc1) with a copy of a nearly intact HSA21 (O'Doherty et al., 2005). Mosaicism occurred in these mice with random loss of the human chromosome from subsets of mouse cells during development led to understand that the gene content of the trisomic cells provide a unique representation of the genetic condition in different tissues throughout development, maturation, and aging of DS individuals (Kahlem et al., 2004; Lyle et al., 2004). These mice demonstrated a number of developmental problems those are analogous to those observed in DS, including similarities in defective development of heart, which are not present in mice trisomic only for MMU16 orthologs of HSA21 genes.

It was thought that DS phenotypes might be caused by direct effect of trisomic HSA21 and/or altered expression of other disomic genes, while both might undergo interaction with several epigenetic and environmental factors. The contribution of single dosage-sensitive genes – alone or in combination, the possible effects of multiple recessive alleles and heterotrisomy, small cooperating effects of dozens of genes, and effects of disomic modifier genes are collectively attributed to DS phenotypes.

The increased risk of acute megakaryocytic leukemia (AMKL), in DS, is always associated with a loss-of-function mutation in the GATA1 gene. Somatic mutations in the *GATA1* transcription factor occur in most of the AMKL and almost every case of transient myeloproliferative disorder (TMPD) in individuals with DS but not in other leukemias of DS and AMKL of euploid individuals (Crispino, 2005). Interestingly, GATA1 is encoded on the X chromosome. Apparently, the megakaryoblasts produced by trisomy 21 appears highly sensitive to *GATA1* mutations; otherwise, the same mutations have no deleterious consequences in cells of euploid individuals. Thus, a cooperating effect of the *GATA1* mutation is essential for development of AMKL in DS (Hitzler and Zipurskey, 2005; Watanabe et al., 2004); however, the exact mechanism of such co-operative interaction with such non-HSA21 mutations with the triplicated HAS21 genes remains to be elucidated. A complex karyotype with multiple chromosomal rearrangements was reported in a child with DS (Ganguly et al., 2017). Notably, bone marrow cytogenetics of the child showed 38 chromosomes in 98% cells with involvement of two 21s in complex translocations.

Sequencing of TMPD and myeloid leukemia samples obtained from individuals with DS revealed presence of truncating mutations in

GATA1 in TMPD with trisomy 21, whereas clonal and subclonal variants were present in DS with myeloid leukemia (Labuhn et al., 2019). The authors have identified a recurrent gain-of-function mutation in myeloid cytokine receptor CSF2RB. Additionally, loss-of-function mutation in 18 different genes was demonstrated to develop myeloid leukemia in the murine model of TMPD, which were similar to myeloid leukemia of DS phenotypically, genetically, and transcriptionally (Labuhn et al., 2019) (Fig. 6.7).

The mutations in GATA1 uniformly results in loss of the full-length protein and synthesis of a truncated protein termed GATA1S (Wechsler et al., 2002); however, that allows production of a shorter isoform (GATA1s), which lacks the N-terminal activation domain (Roberts and Izraeli, 2014). Modeling of hematopoietic abnormalities with patient-derived iPSCs harboring different combinations of genes encoded on chromosome 21 and GATA1 isoforms has demonstrated a synergistic mechanism of these genetic alterations (Banno et al., 2016). It is apparent that combined effect of trisomy 21 and elevated expression of GATA1s accelerates

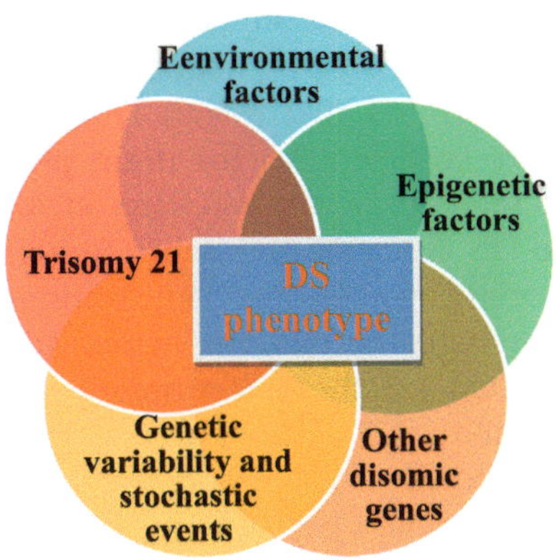

FIGURE 6.7 Phenotypic consequences of gene action in Down syndrome. The trisomic HSA21 gene or genes might directly contribute to phenotype of DS; trisomic genes may interact with other disomic genes or contribute to DS phenotype in presence of trisomy 21; and epigenetic and/or environmental factors might interact at multiple points to initiate, ameliorate, or exacerbate the phenotypes.

the production of immature cells, which are susceptible to leukemic transformation. Nevertheless, AMKL requires cooperation of trisomic HSA21, GATA1 mutation, and perhaps, additional mutations of non-HSA genes (Hitzler and Zipurskey, 2005). However, reports on co-occurrence of trisomy 21/GATA1 mutation in DS leukemia appeared conflicting in iPSC lines derived from individuals with DS (Byrska-Bishop et al., 2015; McNulty and Crispino, 2016). Notably, upregulation of both the long and short GATA1 isoforms was observed in human fetal liver samples with trisomy 21 (Tunstall-Pedoe et al., 2008).

In addition, triplication of the spanning region from RUNX1 to CBS, including ERG and ETS2 contributes to development of AMKL and aberrant megakaryopoiesis of TMPD in DS (McNulty and Crispino, 2016). The transcription factor RUNX1 was proposed to play a pivotal role for predisposition of AMKL in triploid condition; however, its contribution to myeloproliferative disorder (MPD) was not evidenced from its subtraction from three to two copies through cross breeding of trisomic Ts65Dn mice (Kirsammer et al., 2008). Interestingly, genes of interferon-pathway co-operate with genes of the DSCR or 21q22 region, including the ETS-family, RUNX1, BACH1, DYRK1A, SON, SLC37A1, BRWD1, HLCS, HMGN1 for leukemogenesis in DS (Mateos et al., 2015). Furthermore, overexpression of ETS2 induced apoptosis in thymus, spleen, and brain cells (Wolvetang et al., 2003) and also the mitochondrial death pathway in neurons of DS patients (Helguera et al., 2013). Additionally, overexpression of ETS-2 enhanced the cytoplasmic translocation of cytochrome *c*.

There can be interaction of multiple dosage-sensitive genes, which can lead to variable degrees of DS characteristics. The co-expression of multiple genes having major effect on development of cells might alter the biochemical pathways; however, the alterations in functions due to interactions of most HSA21 and other disomic genes are not delineated yet. Thus, the combinatorial possibilities for testing of single and/or multiple genes may present an obvious challenge to direct interrogation for expression of DS features. However, only few characteristics are defined with appreciable precision to consistently detect the incremental contribution of one or two genes to the trisomic phenotype in mice (detailed in CHAPTER 14). Nevertheless, understanding of the pathways and processes that are perturbed due to trisomy of HSA21 might be considered as logical approach for correction of the unbalanced pathways or introduction of therapies to reduce the complex consequences of health status of DS.

4. Transcriptomes of trisomy 21

The transcripts of trisomic genes are increased by about 50% in different cells and tissues and at different developmental stages and that indicates overexpression of those trisomic genes in all cells throughout the development. Thus, alteration of function of the overexpressed genes in fully differentiated cells of trisomic adults might indicate their contribution to DS phenotype. However, overexpression of a given gene may not perturb the development and function of every cell type and at every developmental time point. It was demonstrated that detrimental expression of overexpression of some genes exists in specific cell types and at specific timepoint of development. Also, a trisomy-induced alteration of function of 1 cell population is likely to affect the function of neighboring cells as a secondary consequence of trisomy, resulting in aberrant expression of DS phenotype (Roper and Reeves, 2006). Moreover, overexpression of regulatory genes may cause severe detrimental effects on several signaling pathways and establish aberrant patterns of development (Brichler et al., 2005).

Understanding the transcriptional consequences caused by trisomy 21 was expected to unfold the effects of overexpression of trisomy 21 (FitzPatrick, 2005). The overexpression of genes that encode transcription factors demonstrated induction of a misregulation of disomic genes, resulting in production of some phenotypic effects. Also, the contribution of the triplicated noncoding sequences could not be ruled out in mouse models of DS (Amano et al., 2004; Dauphinot et al., 2005; Lyle et al., 2004). Interestingly, studies conducted on human trisomic tissues indicated that only a subset of HSA21 genes is overexpressed (Tang et al., 2004; Mao et al., 2005), and the set of overexpressed HSA21 genes differs across the trisomic cell types (Li et al., 2006). These findings indicated that the presence of three copies of a gene alone does not necessarily result in overexpression; however, cooperating effects of other factors, such as developmental stage, tissue-specific differences, etc., also contribute to dysregulation the trisomic gene expression.

The genes that encode transcription factors and co-regulators/modulators of transcription and its dysregulation in trisomic status were paid prime importance for assessment of their contribution to the molecular pathogenesis of DS. Despite the efforts made for identification of the critical region on chromosome 21 for attributing to the clinical features of DS, it has been demonstrated that trisom DSCR is necessary but not sufficient for the brain phenotypes observed in trisomic mice (Olson et al., 2007). Thus, the DS abnormalities were thought to be multifactorial, besides the contributions of trisomy 21, where the genetic factors might also be augmented by the environmental and stochastic influences (Reeves et al., 2001).

Initial studies using microarrays or serial analysis of gene expression (SAGE) have revealed the differences in global transcriptome expression between trisomic HSA21 and euploid state (FitzPatrick et al., 2002; Mao et al., 2003). There were upregulation of 187 and downregulation of 85 transcripts reported in human amniocytes (FitzPatrick et al., 2002) and differential regulation of 679 and 725 transcripts in human astrocyte cell lines and human frozen fetal cerebrum, respectively (Mao et al., 2003, 2005). However, owing to paucity of human samples for such investigation, analysis of mouse models of DS was considered, and similar studies indicated >2-fold consistent difference in the cerebellum (Saran et al., 2003), and 330 differently expressed RNA tags in brains of Ts65Dn mice (Chrast et al., 2000). Another study demonstrated an elevated transcript level of all triplicated genes in most adult tissues, besides tissue-specific downregulation, compensation, or strong overexpression of few genes (Lyle et al., 2004; Kahlem et al., 2004).

A transcriptome analysis on four different cell types obtained from people with and without trisomic HSA21 demonstrated that trisomy 21 causes widespread alterations in gene expression across the genome, including prominent and consistent activation of the interferon transcriptional response (Sullivan et al., 2016; Gold et al., 2010). An activated interferon response associated with trisomy 21 in human cells of different origins was consistent with overexpression of the four interferon receptors encoded on chromosome 21; however, the impact of trisomy 21 on systemic signaling cascades in individuals with DS remains largely unelucidated.

The PKNOX1/PREP1, another HSA21 gene, encodes for a homeodomain transcription factor that regulates multiple aspects of embryonic development by regulating the homeobox protein PBX activity (Ferretti et al., 2006). The fibroblasts of human DS expressing more PREP1 appeared more sensitive to genotoxic stress. The observation indicated that PREP1 may be involved in the apoptosis of DS tissues (Micali et al., 2009).

5. Proteomes of trisomy 21

Trisomy 21 caused a consistent global change in the circulating proteomes. Of the 299 proteins differentially detected in individuals with DS compared to euploid controls, 178 proteins were downregulated and 121 upregulated. In this direction, of the 50 proteins, for which aptamers were available, nine were observed markedly upregulated in the SOMAscan® assay. The result reflects that most of the proteome changes correspond to the proteins encoded elsewhere in the genome in people with DS, and these proteins are associated with signaling pathways (Sullivan et al., 2017). Other reports were consistent with the fact that only a fraction of HSA21 genes are significantly upregulated in human and mouse models (Scarpato et al., 2014; Walus et al., 2015). However, individual chromosomes did not demonstrate any obvious spatial relationships among the deregulated proteins. Notably, HSA21 encoded proteins displayed a consistent trend toward upregulation. The β-2-microglobulin (B2M), a subunit of the major histocompatibility complex I (MHC I), is consistently elevated in the circulating proteome of people with trisomy 21, whereas immunoglobulin E (IgHE) is consistently downregulated, indicating an increase in interferon signaling (Wedgwood et al., 1988).

5.1 Proteomes detected in trisomic fetus

Proteomic analysis in amniotic fluid of 17th week pregnancies with DS revealed splicing factor arginine/serine-rich 4 (SFRS4; Q08170) and elevation of alpha-1-microglobulin (AMBP; P02760), collagen alpha 1 (I) chain (CO1A1; P02452), collagen alpha 1 (III) chain (CO3A1; P02461), collagen alpha 1 (V) chain (CO5A1; P20908), and basement membrane–specific heparin sulfate proteoglycan core protein (PGBM; P98160) (Tsangaris et al., 2006). However, 40% decrease in protein IBP-1 (P08833) was observed in DS fetuses compared with chromosomally disomic 21 fetuses. The four proteins, including CO1A1, CO3A1, CO5A1, and PGBM, appeared as fragments.

The proteomic analyses with isobaric tags for relative and absolute quantification (iTRAQ) technique, combined with matrix-assisted laser desorption/ionization (MALDI) time of flight (TOF)/TOF and gene ontology (GO), identified the proteins that are differentially expressed in DS and their predicted functions (Sui et al., 2015). The five proteins [apolipoprotein E (ApoE), complement factor B (CFB), amyloid P component, serum (APCS), matrin-3, and osteopontin (OPN)] were consistent with DS, of which first three (ApoE, CFB, and APCS) were significantly upregulated and OPN and matrin-3 were downregulated in DS. Other proteins, including leucine-rich repeat-containing protein 47, 26S proteasome non-ATPase regulatory subunit 13, nucleosome assembly protein 1-like 1, lethal (2) giant larvae protein homolog 2, endoplasmic oxidoreductin-1-like (ERO1-like) protein α, 2-oxoglutarate dehydrogenase-like, mitochondrial, thymosin β-10, ATPase ASNA1, large neutral amino acids transporter small subunit 2, ectonucleotide pyrophosphatase/phosphodiesterase family member 4, thioredoxin-related transmembrane protein 4, prefoldin subunit 2, AMP deaminase 2 and receptor expression-enhancing protein 5 displayed the highest ratios (60.27 and 0.01), compared to healthy controls (Sui et al., 2015).

CFB is a key component involved in AD and associated with the brain damage in DS (Head et al., 2001). CFB is significantly increased in the maternal serum and umbilical cord blood of fetuses with DS (Sui et al., 2015; Yu et al., 2012). APCS and the Igλ chain C region were also reported in the sera of DS patients (Kolla et al., 2010). The APCS glycoprotein is encoded by a gene mapped on the HSA1q23.2. APCS causes immunoreactivity in DS, AD, and other amyloid disorders (Perlmutter et al., 1995; Rowe et al., 1984). The apolipoprotein E gene was 3.75 times higher in the DS (Sui et al., 2015), of which the E4 allele is reported as a putative risk factor for AD in the general population.

OPN, a glycoprotein with an arginine-glycine-aspartic acid (RGD) sequence, binds to a number of receptors, including the integrins and certain variant forms of CD44. Perinatal liver fibrosis is caused by OPN in hepatocytes and macrophage-like cells in DS (Tokairin et al., 2008). Matrin-3 was significantly decreased in the fetal DS brain indicating its aberrant transcription (Bernert et al., 2002) and in the plasma of umbilical cord blood (Sui et al., 2015).

5.2 Proteomic changes of chronic autoinflammation in people with DS

Functional association among the 299 differential proteins was related to control of the immune system, including inflammatory and humoral immune response, complement activation, regulation of coagulation, growth factor signaling, regulation of proteolysis, regulation of neurogenesis, leukocyte mediated immunity, and so on, with some overlapping functions. The largest functional category, control of immune system, is consisted of 150 differential proteins, which revealed protein—protein interactions within this functional category. The physically interacting proteins appear as the key factors in antigen presentation, autoimmunity, and T-cell regulation (Arayaa et al., 2019), including B2M, LILRB1, LILRB2, FCRL3, ZAP70, and ERAP2; many of which have been linked to autoimmune disorders (Table 6.3) (Sullivan et al., 2017).

Differences in immune signaling have been identified in several independent reports of transcriptome analysis of cells and tissues from individuals with trisomy 21 and mouse models of DS (Alexander et al., 2016; Ling et al., 2014). B2M is a known interferon stimulated gene (ISG), and the plasma levels of B2M, which is required for peptide loading, increase during therapeutic administration of interferon (IFN-α) (Wedgwood et al., 1988; Remes et al., 1996). Downregulated LILRB1 and LILRB2 (leukocyte immunoglobulin-like receptors B1 and B2) are negative regulators of MHC class I signaling in

TABLE 6.3 Differential proteins and associated medical conditions.

Medical condition	Proteins expressed
Alzheimer disease	Complement (C1QA, C1S, C1R, C3, C6)
Type I interferonopathies	Complement (C1QA, C1S, C1R, C3, C6)
Systematic lupus erythematosus	Complement (C1QA, C1S, C1R, C3, C6); FCRL3; IL-10; IL-17; TNF-α; TNFRSF1A; TNFRSF1B
Autoimmune thyroid disease	FCRL3
Rheumatoid arthritis	FCRL3; IL-6; TNF-α
Neuromyelitis optica	FCRL3
Sjogren's syndrome	IL-6; TNF-α
Aicardi-Goutieres syndrome	IP-10; TNF-α; VEGF
Ankylosing spondylitis	M1 Amiopeptidases (ERAP2)
Inflammatory bowel disease	M1 Amiopeptidases (ERAP2)
Behçet's disease	M1 Amiopeptidases (ERAP2)
Autoimmune type 1 diabetes	M1 Amiopeptidases (ERAP2)
Psoriasis	M1 Amiopeptidases (ERAP2)
Down syndrome	TNF-α
Severe combined immunodeficiency disorder	ZAP70
Immunodeficiency	ZAP70
Infantile-onset of multisystem autoimmune disease 2	ZAP70
Autoimmune arthritis (mice)	ZAP70
Myeloproliferative disorders	FGFR1
Craniofacial abnormalities	FGFR1
Hirschsprung's disease	RET

Modified from Sullivan et al. (2017).

individuals with trisomy 21 (Sullivan et al., 2017; Hirayasu and Arase, 2015). Therefore, elevated B2M and reduced LILRBs could be indicative of a regular active immune system in DS, which may lead to enhanced antigen presentation, increased signaling, downstream of MHC class I engagement, and potential predisposition to autoimmunity, influencing onset of autoimmune disorders.

FCRL3, ERAP2, and ZAP70 appeared significantly downregulated in individuals with DS (Sullivan et al., 2017). FCRL3, a member of the immunoglobulin receptor superfamily and one of several Fc receptor-like glycoproteins, is involved in immune control (Davis et al., 2002). Several clinical expressions, including autoimmune thyroid disease, rheumatoid arthritis, and systemic lupus erythematosus (SLE) are caused by mutations and polymorphisms in this gene (Chistiakov and Chistiakov, 2007; Wang et al., 2016). ZAP70, Src-related kinase, SRK, and a tyrosine kinase, physically interacts with both FCRL3 and the zeta-chain (CD247) of the T-cell receptor (TCR) (Sjolin-Goodfellow et al., 2015). ZAP70 regulates motility, adhesion, and cytokine expression of mature T cells and thymocyte development (Au-Yeung et al., 2009; Cuvelier et al., 2016). Development of autoimmune and autoinflammatory diseases, including immunodeficiency-48 (IMD48) and infantile-onset of multisystem autoimmune disease have been associated with mutations in ZAP70 (Chan et al., 2016). Missense mutations in ZAP70 caused chronic autoimmune arthritis in mice (Sakaguchi et al., 2003). ERAP2, a member of the M1 aminopeptidase family, is associated with trimming antigenic epitopes for presentation by MHC class I molecules (Tsujimoto and Hattori, 2005; Tanioka et al., 2003), and these aminopeptidases cause several immune-mediated diseases, including psoriasis, ankylosing spondylitis, inflammatory bowel disease, Behçet's disease, and type I diabetes (Agrawal and Brown, 2014; Evnouchidou et al., 2012; Hinks et al., 2013). Therefore, functions of the FCRL3, ZAP70, and ERAP2 genes led to the hypothesis that deregulation of these genes could be linked to the high propensity of autoimmune conditions in the population with DS.

A proteomics study using the SOMAscan proteomics platform in 165 DS individuals identified dozens of proteins that are consistently deregulated by trisomy 21, of which numerous factors involved in immune control, the complement cascade, and growth factor signaling were most prominent (Sullivan et al., 2017). The SOMAscan proteomics platform showed obvious signs of increased TNF-α signaling in people with DS along with enhanced circulating levels of the two TNF-α receptors, TNFRSF1A and TNFRSF1B, and the protein kinase RFK (Sullivan et al., 2017), which is involved in ROS production and downstream of TNF-α receptor engagement (Yazdanpanah et al., 2009). Higher levels of TNF-α ligand were also displayed in DS. Overall, a significant increase in levels of potent inflammatory cytokines and chemokines, which are known to act downstream of IFN signaling, including IL-6, IL-22, MCP-1 (CCL2), TNF-α, and VEGF-A, have been reported in DS (Sullivan et al., 2017). Importantly, upregulated expression of many of these factors was noted in the circulating proteomes of individuals with chronic inflammatory diseases (Table 6.3), such as systemic lupus erythematosus (SLE) (a condition associated with hyperactive IFN signaling) is linked to increased levels of TNF-α, the two soluble TNF-α receptors, IL-10, and IL-17 (Hinks et al., 2013; Yazdanpanah et al., 2009; Mozo et al., 2014; Ferreira et al., 2016), and also in DS (Nateghi Rostami et al., 2012). Another inflammatory cytokine, CCL17 (TARC), was also significantly elevated in people with DS. Notably, many of the proinflammatory cytokines pronounced complement consumption, resembling changes seen in type I interferonopathies and other autoinflammatory conditions were higher in people with DS. The study has hypothesized that increased interferon signaling caused by trisomy 21 leads to chronic immune dysregulation in DS. In addition, the outcome of immunomodulatory therapeutics has proposed DS as an interferonopathy.

IL22, a cytokine of IL-10 family related to IFN-γ, is upregulated and known as a causal factor of psoriasis (Hao, 2014). Elevated levels of TNF-α and IL-6 are demonstrated in patients with rheumatoid arthritis and Sjogren's syndrome (Szodoray et al., 2004; Meyer et al., 2010). Upregulated TNF-α, IP-10, and VEGF were reported in the cerebrospinal fluid or plasma of patients with Aicardi-Goutieres syndrome (AGS), a canonical monogenic type I interferonopathy (Takanohashi et al., 2013).

Significant enrichment of interferon-related proteins was differentially detected in people with DS, wherein many other changes could be indirectly linked to enhance interferon signaling (Sullivan et al., 2017). IFN-α/β could strikingly lower the levels of IgE for suppressing allergic inflammatory processes through prevention of granulocyte activation and IL-4-mediated isotype switching to IgE (Gonzales-van Horn and Farrar, 2015). Also, elevated circulating levels of cystatin 3 (CST3), which is significantly increased in people with DS, result in reduced glomerular filtration (Sullivan et al., 2017). Enhanced interferon signaling is known to disturb renal function during lupus nephritis (Lech and Anders, 2013) and viral glomerulonephritis (Migliorini et al., 2013). Clearly, a proinflammatory signature is postulated to work in the circulating proteome with multiple ties with interferon signaling in people with DS.

A group of 21 proteins of the complement and coagulation cascade is deregulated in individuals with DS, of which 18, including C1R, C1QA, C3, C1S, C6, etc., are downregulated in the circulating proteome indicating a complement consumption phenotype. Chronic activation of interferon signaling could cause complement consumption, and such interferon-driven expression has been linked to DS phenotypes, such as risk of bacterial lung pneumonia and otitis media (Ram and Chinen, 2011). Interestingly, removal of *Streptococcus pneumoniae* by complement depletion was evidenced. Elevated levels of CST3, B2M, and TFF3 proteins in plasma may indicate perturbed kidney function in DS (Du et al., 2013). Since kidneys are prone to complement-mediated injury and that could further be influenced by the accumulation of immune complexes and complement activation, an apparent risk of kidney problems corroborated with the increase of their life expectancy could be speculated for individuals with DS (Sullivan et al., 2017; Thurman and Le Quintrec, 2016).

Notably, the complement system mediated by the activation of classic complement cascade involving β-amyloid deposits and soluble B2M contributes to AD in people with DS (Nybo et al., 1999). Also, microglia-mediated synapse destruction by a number of the complement subunits leads to progression of AD (Hong et al., 2016). In addition, downregulation of multiple proteins within the coagulation cascade was reported in samples having trisomy HSA21, such as factor II (F2), fibrinogen (FGA), and the inhibitor SERPINC1, whereas upregulation of an inhibitor, protein C (PROC), of numerous coagulation factors was noticed (Sullivan et al., 2017). Altogether, it is apparent that hypocomplementia could modulate progression of AD in DS individuals.

5.3 Proteins and growth factor receptors involved in neurogenesis

Highly deregulated growth factor signaling proteins, including proteins of FGF, VEGF, and EGF pathways, have been reported in people with trisomic HSA21. Downregulation of the receptors such as KDR and VEGFR of VEGF signaling was noted in DS people, while its ligand VEGFA was up-regulated. Noticeably, VEGFA induction has been reported as a hallmark of inflammatory processes leading to healing of wounds (Roth et al., 2006). In DS, FGFR1 was up-regulated, whereas its ligand FGF1 was down-regulated (Sullivan et al., 2017). Notably, DS people are prone to hematopoietic malignancies, which might be contributed by the gain-of-function mutations in FGFR1, since these mutations are associated with myeloproliferative disorders leading to leukemia (Ibrahimi et al., 2004; Jackson et al., 2010).

Downregulation of several components of the EGF signaling cascade is reported in DS, including the receptors EGFR, ERBB3, and ERBB4, and the ligand itself (HBEGF). Interestingly, 17 of the 75 proteins of the growth factor signaling, including EGFR, ERBB4, NTRK2, NTRK3, and RET were demonstrated linked to regulation of neurogenesis (Sullivan et al., 2017). Downregulated NTRK2 and NTRK3, the transmembrane receptors for the brain derived neurotrophic factor (BDNF) and neurotrophin 3, respectively, were reported in the plasma of individuals with DS. Downregulation of a third neurotrophin receptor required for normal GABAergic synapse development, SLITRK383, was also noticed. Notably, several studies have reported a significant link of the altered GABAergic signaling to the cognitive deficits in individuals with trisomic HSA21 (Contestabile et al., 2017). Additionally, a receptor tyrosine kinase of the cadherin superfamily having key roles in neural crest development, RET, displayed lowered circulating levels in people with DS. Importantly, Hirschsprung disease, which is significantly prevalent in people with DS, is known to be caused by the loss-of-function mutations in RET (Delalande et al., 2008; Friedmacher and Puri, 2013; Tomuschat and Puri, 2015). Therefore, the proteomics datasets have generated significant information on how trisomy of HSA21 causes profound immune dysregulation, with many of the changes being highly important for those observed in type I interferonopathies and other autoinflammatory conditions (Table 6.3).

The knowledge of the circulating proteome in people with DS also provides a new framework of understanding the etiology of DS-associated developmental characteristics and co-morbidities (Sullivan et al., 2017). Most importantly, these findings further enable mechanistic information for deciphering the molecular underpinnings of the developmental problems and cognitive deficits in people with DS and support the concept that DS people could benefit from therapeutic strategies for controlling the interferon signaling.

For decades, researchers thought that DS is caused by the genes present on the extra copy of chromosome 21; however, the scientists of the recent era have recognized that trisomy 21 causes genome-wide changes in gene expression (Vilardell et al., 2011). A distinct pattern of gene dysregulation was reported across the genome of the fetus with DS of a set of monozygotic twins, wherein the other twin was normal (Antonarakis, 2017). The genetic mechanism of the disrupted expression remained a mystery. The authors measured the levels of mRNA from four separate areas of the skin from each of the twins, with and without DS, and observed that an alternating up- and down-regulated pattern of expression throughout the genome of the twin with DS in contrast to the normal twin having some groups of genes being expressed and some were suppressed. The differential expression between the twins is organized in either up-regulated or down-regulated domains along all chromosomes (Letourneau et al., 2014). The gene expression dysregulation domains (GEDDs) were remarkably stable and persisted in iPSCs derived from skin fibroblasts of twins. Surprisingly, comparison of the transcriptome of the Ts65Dn mouse model of DS and normal littermate mouse fibroblasts showed GEDDs that affected the same genes in mouse chromosomes were syntenic in human. Normally, the GEDDs were overlapped with specific chromosomal regions called lamina-associated domains (LADs) (Guelen et al., 2008) and early replication domains (ERDs) (Hansen et al., 2010); whereas LADs have low gene expression of suppressive patterns in the local chromatin, ERDs have highly active regions in the genome. Authors noticed unusually active LADs in the trisomic twin as well as the Ts65Dn mouse strain, whereas their ERDs were subdued (Letourneau et al., 2014).

The overall position of LADs in trisomic cells of the twin was not altered; however, methylation of histone H3 (H3K4me3) was increased in the areas of up-regulated genes in trisomic fibroblasts, which along with widespread chromatin changes (due to disrupted transcriptional regulation) could explain some of the phenotypes in DS (Letourneau et al., 2014); though further investigations have been recommended in mosaic DS. GEDDs were noticed in DS cells and in fibroblasts and hippocampus of Dp1Tyb mouse (Ahlfors et al., 2019). Interestingly, GEDDs did not depend on the DS genotype but occurred in conjunction with gene expression changes, suggesting that GEDDs are not a DS-specific feature; however, it results from the clustering of co-regulated genes with respect to

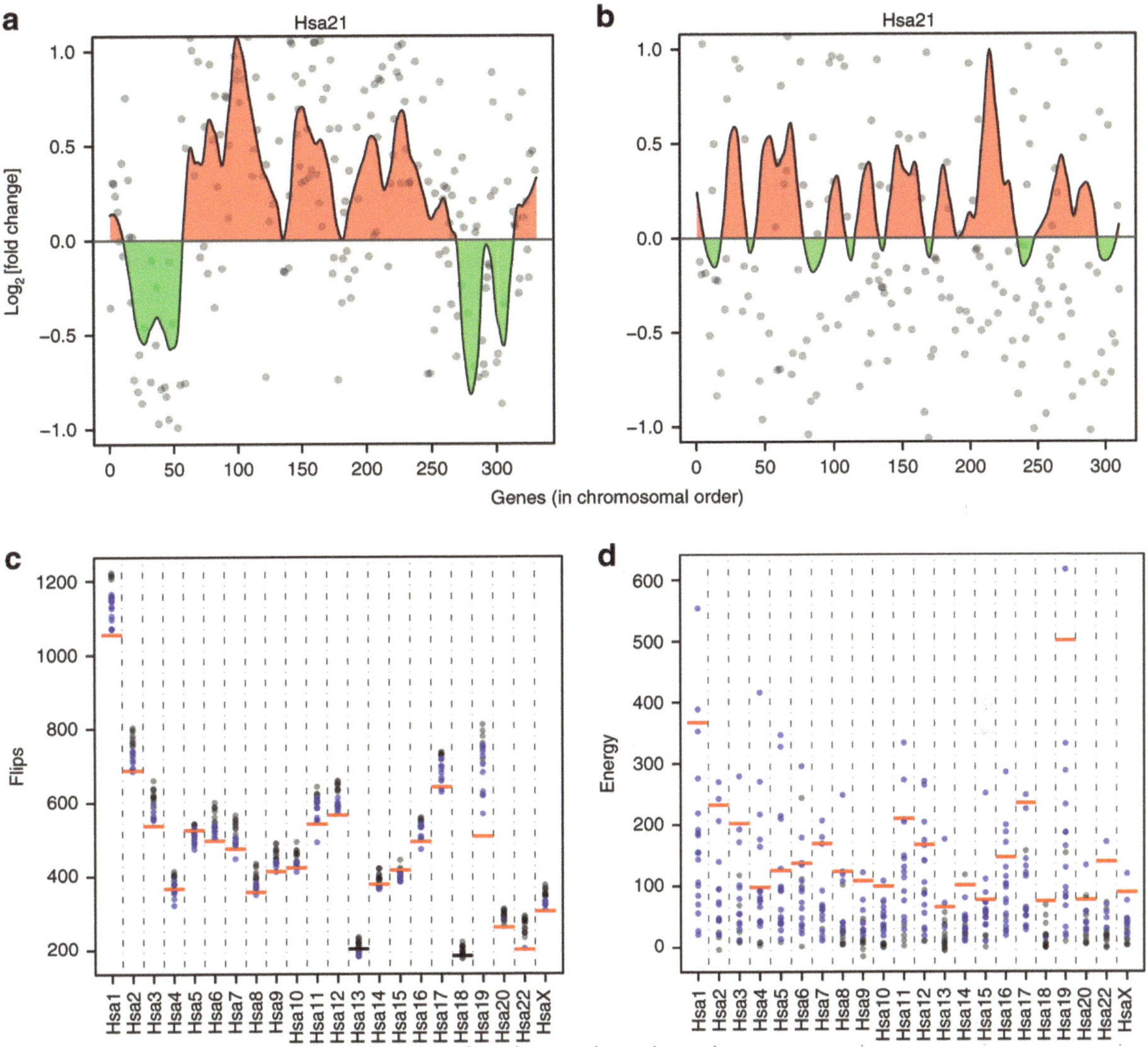

FIGURE 6.8 Gene expression dysregulation domains do not depend on the DS genotype. (A) Plot showing fold change of gene expression on Hsa21 in DS versus euploid human fibroblasts3. Expressed genes (*gray dots*) are plotted in chromosomal order. A Loess smoothing curve is superimposed with regions that are up-regulated or down-regulated indicated in red or green, respectively. (B) Same as (A) but the fold changes are computed between two DS and two euploid samples versus between two different DS and two different euploid samples. Note apparent regions of upregulation and downregulation in this no-genotype-difference comparison. (C, D) Plots of flips (C) and energy (D) for comparisons of different combinations of DS and euploid human fibroblast samples arranged such that there is no genotype difference in the comparison. Each *dot* is a comparison of different combinations of samples with blue indicating combinations that are significantly different (>2 SD) from the mean of bootstrapped chromosomes and black for not significant. *Horizontal lines* indicate the values obtained for the correct comparison of four DS versus four euploid samples (red, significant; black, not significant). *Reprinted from: Ahlfors et al. (2019) under a Creative Commons Attribution 4.0 International License.*

SUPPLEMENTARY TABLE S6.1 Functional genes mapped onto HSA21.

HAS21			
Ideogram	**Location**	**Genes**	**Name**
Cen	21p11.1	BAGE	B Melanoma antigen
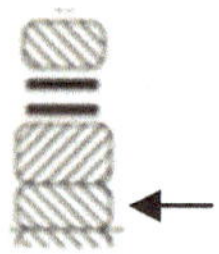	21p11.2	KCNE1B	Potassium voltage-gated channel subfamily E regulatory subunit 1B
	21p11.2	TPTE	Transmembrane phosphatase with tensin homology
Cen	**21p12**	**CBSL**	**Cystathionine beta-synthase like**
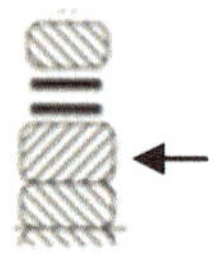	**21p12**	**U2AF1L5**	**U2 small nuclear RNA auxiliary factor 1 like 5**
	21p12	LOC106780825	523 FOXM1-regulated, gastric cancer-associated-like
	21p12	CRYAA2	Crystallin alpha A2
	21p12	SIK1B	Salt inducible kinase 1B (putative)
	21p12	LOC102723996	ICOS ligand
	21q11.2	POTED	POTE ankyrin domain family member D
	21q11.2	**LIPI**	**Lipase I**
	21q11.2	RBM11	RNA binding motif protein 11
	21q11.2	HSPA13	Heat shock protein family A (Hsp70) member 13
	21q11.2	SAMSN1/AS1	SAM domain, SH3 domain and nuclear localization signals 1/SAMSN1 antisense RNA 1
	21q11.2	CYP4F29P	Cytochrome P450 family 4 subfamily F member 29, pseudogene
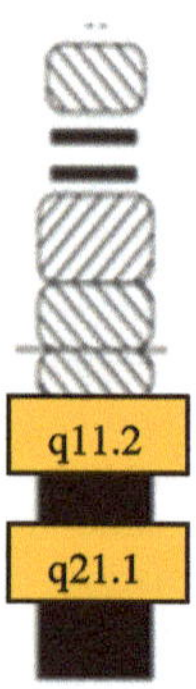	**21q11.2-q21.1**	**NRIP1**	**Nuclear receptor interacting protein 1**
	21q21.1	**USP25**	**Ubiquitin specific peptidase 25**
	21q21.1	**NCAM2**	**Neural cell adhesion molecule 2**
	21q21.1	**CXADR**	**CXADR Ig-like cell adhesion molecule**
	21q21.1	BTG3/AS1	BTG antiproliferation factor 3/BTG3 antisense RNA 1
	21q21.1	SNORD74B	Small nucleolar RNA, C/D box 74B
	21q21.1	C21orf91-OT1	C21orf91 overlapping transcript 1
	21q21.1	C21orf91	Chromosome 21 open reading frame 91
	21q21.1	CHODL-AS1	CHODL antisense RNA 1
	21q21.1	CHODL	Chondrolectin
	21q21.1	TMPRSS15	Transmembrane serine protease 15

SUPPLEMENTARY TABLE S6.1 Functional genes mapped onto HSA21.—cont'd

HAS21			
Ideogram	**Location**	**Genes**	**Name**
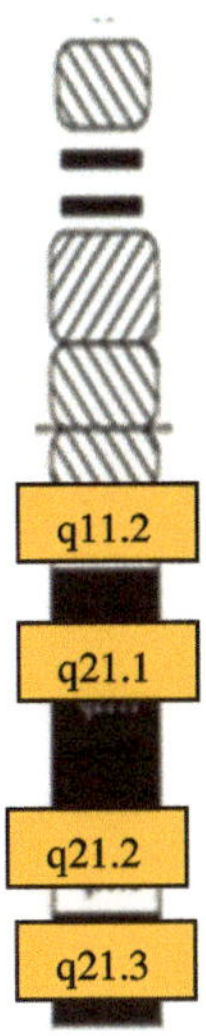	**21q21.3**	**MRPL39**	**Mitochondrial ribosomal protein L39**
	21q21.3	JAM2	Junctional adhesion molecule 2
	21q21.3	ATP5PF	ATP synthase peripheral stalk subunit F6
	21q21.3	GABPA	GA binding protein transcription factor subunit alpha
	21q21.3	**APP**	**Amyloid beta precursor protein**
	21q21.3	**BACH1**	**BTB domain and CNC homolog 1**
	21q21.3	**GRIK1/AS1/AS2**	**Glutamate ionotropic receptor kainate type subunit 1/GRIK1 antisense RNA 1/2**
	21q21.3	**CLDN17**	**Claudin 17**
	21q21.3	**ADAMTS1**	**ADAM metallopeptidase with thrombospondin type 1 motif 1**
	21q21.3	**ADAMTS5**	**ADAM metallopeptidase with thrombospondin type 1 motif 5**
	21q21.3	**USP16**	**Ubiquitin specific peptidase 16**
	21q21.3	RWDD2B	RWD domain containing 2B
	21q21.3	**LTN1**	**Listerin E3 ubiquitin protein ligase 1**
	21q21.3	N6AMT1	N-6 adenine-specific DNA methyltransferase 1
	21q21.3	CYYR1	Cysteine and tyrosine rich 1
	21q21.3	MAP3K7CL	MAP3K7 C-terminal like
	21q21.3	**BACH1-IT2/3**	**BACH1 intronic transcript 2/3**
	21q21.3	CCT8	Chaperonin containing TCP1 subunit 8
	21q22.11	DNAJC28	DnaJ heat shock protein family (Hsp40) member C28
	21q22.11	KRTAP24-1	Keratin-associated protein 24-1
	21q22.11	KRTAP25-1 (23-1 to 27-1)	Keratin-associated protein 25-1(23-1 to 27-1)
	21q22.11	KRTAP13 (1-4)	Keratin-associated protein 13(1-4)
	21q22.11	KRTAP15-1	Keratin-associated protein 15-1
	21q22.11	KRTAP19-(1-8)	Keratin-associated protein 19-(1-8)
	21q22.11	KRTAP22-(1-2)	Keratin-associated protein 22-(1-2)
	21q22.11	KRTAP6-(1-3)	Keratin-associated protein 6-(1-3)
	21q22.11	KRTAP20-(1-4)	Keratin-associated protein 20–(1-4)
	21q22.11	KRTAP21-(1-2)	Keratin-associated protein 21-(1-2)
	21q22.11	KRTAP8-1	Keratin-associated protein 8-1

(*Continued*)

SUPPLEMENTARY TABLE S6.1 Functional genes mapped onto HSA21.—cont'd

HAS21			
Ideogram	**Location**	**Genes**	**Name**
	21q22.11	KRTAP7-1	Keratin-associated protein 7-1 (gene/pseudogene)
	21q22.11	KRTAP11-1	Keratin-associated protein 11-1
	21q22.11	**TIAM1**	**TIAM Rac1-associated GEF 1**
	21q22.11	**SOD1**	**Superoxide dismutase 1**
	21q22.11	**HUNK**	**Hormonally up-regulated neu-associated kinase**
	21q22.11	CFAP298	Cilia-5z and flagella-associated protein 298
	21q22.11	**SYNJ1**	**Synaptojanin 1**
	21q22.11	**OLIG1/2**	**oligodendrocyte transcription factor 1/2**
	21q22.11	**IL10RB**	**Interleukin 10 receptor subunit beta**
	21q22.11	**IL10RB-DT**	**IL10RB divergent transcript**
	21q22.11	**IFNAR1/2**	**Interferon alpha and beta receptor subunit 1/2**
	21q22.11	IFNGR2	Interferon gamma receptor 2
	21q22.11	TMEM50B	Transmembrane protein 50B
	21q22.11	GART	Phosphoribosylglycinamide formyltransferase, phosphoribosylglycinamide synthetase, phosphoribosylaminoimidazole synthetase
	21q22.11	SON	SON DNA binding protein
	21q22.11	SCAF4	SR-related CTD associated factor 4
	21q22.11	MIS18A	MIS18 kinetochore protein A
	21q22.11	MRAP	Melanocortin 2 receptor accessory protein
	21q22.11	DONSON	DNA replication fork stabilization factor DONSON
	21q22.11	CRYZL1	Crystallin zeta like 1

SUPPLEMENTARY TABLE S6.1 Functional genes mapped onto HSA21.—cont'd

Ideogram	HAS21 Location	Genes	Name
	21q22.11	ITSN1	Intersectin 1
	21q22.11	ATP5PO	ATP synthase peripheral stalk subunit OSCP
	21q22.11	URB1	URB1 ribosome biogenesis homolog
	21q22.11	SNORA80A	Small nucleolar RNA, H/ACA box 80A
	21q22.11	EVA1C	Eva-1 homolog
	21q22.11	C TCP10L	t-complex 10 like
	21q22.11	PAXBP1/AS1	PAX3 and PAX7 binding protein 1/PAXBP1 antisense RNA 1
	21q22.11	C21orf62	Chromosome 21 open reading frame 62
	21q22.11	MRPS6	Mitochondrial ribosomal protein S6
	21q22.11	SLC5A3	Solute carrier family 5 member 3
	21q22.11	**KCNE1/2**	**Potassium voltage-gated channel subfamily E regulatory subunit 1/2**
	21q22.11-q22.12	**SMIM11A/B**	**Small integral membrane protein 11A/B**
	21q22.12	CLIC6	Chloride intracellular channel 6
	21q22.12	**RCAN1**	**Regulator of calcineurin 1**
	21q22.12	**RUNX1**	**RUNX family transcription factor 1**
	21q22.12	SETD4	SET domain containing 4
	21q22.12	CBR1/3/AS1	Carbonyl reductase 1/3/CBR1 antisense RNA1
	21q22.12	PPP1R2P2	Protein phosphatase 1 regulatory inhibitor subunit 2 pseudogene 2
	21q22.12	DOP1B	DOP1 leucine zipper like protein B
	21q22.12	MORC3	MORC family
	21q22.12	FAM243A/B	Family with sequence similarity 243 member A/B
	21q22.12-q22.13	**CHAF1B**	**Chromatin assembly factor 1**
	21q22.13	CLDN14	Holocarboxylase synthetase

(*Continued*)

SUPPLEMENTARY TABLE S6.1 Functional genes mapped onto HSA21.—cont'd

	HAS21		
Ideogram	**Location**	**Genes**	**Name**
	**21q22.13**	**PIGP**	**Phosphatidylinositol glycan anchor biosynthesis class P**
	21q22.13	VPS26C	VPS26 endosomal protein sorting factor C
	21q22.13	**DYRK1A**	**Dual specificity tyrosine phosphorylation regulated kinase 1A**
	21q22.13	**KCNJ6/15**	**Potassium inwardly rectifying channel subfamily J member 6/15**
	21q22.13	**SIM2**	**SIM bHLH transcription factor 2**
	21q22.13	**RIPPLY3**	**Ripply transcriptional repressor 3**
	21q22.13	**TTC3/AS1**	**Tetratricopeptide repeat domain 3/TTC3 antisense RNA 1**
	21q22.2	PSMG1	Proteasome assembly chaperone 1
	21q22.2	BRWD1/2/AS1-2	Bromodomain and WD repeat domain containing 1/2/BRWD1/2 antisense RNA 1/2
	21q22.2	HMGN1	High mobility group nucleosome binding domain 1
	21q22.2	LCA5L	Lebercilin LCA5 like
	21q22.2	SH3BGR	SH3 domain binding glutamate rich protein
	21q22.2	B3GALT5	Beta-1,3-galactosyltransferase 5
	21q22.2	PLAC4	Placenta enriched 4
	21q22.2	**ERG**	**ETS transcription factor ERG**
	21q22.2	**ETS2**	**ETS proto-oncogene 2, transcription factor**
	21q22.2	IGSF5	Immunoglobulin superfamily member 5
	21q22.2	PCP4	Purkinje cell protein 4
	21q22.2	**DSCAM/AS1**	**DS cell adhesion molecule/DSCAM antisense RNA 1**
	21q22.2	**DSCAM-IT1**	**DSCAM intronic transcript 1**
	21q22.2-q22.3	**BACE2**	**Beta-secretase 2**
	21q22.3	KRTAP12-(1-4)	Keratin-associated protein 12-(1-4)
	21q22.3	KRTAP10-12	Keratin-associated protein 10-12

SUPPLEMENTARY TABLE S6.1 Functional genes mapped onto HSA21.—cont'd

HAS21			
Ideogram	**Location**	**Genes**	**Name**
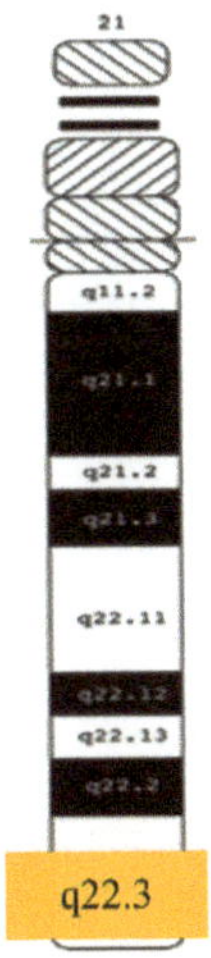	21q22.3	FAM3B	FAM3 metabolism regulating signaling molecule B
	21q22.3	MX2/1	MX dynamin like GTPase 2/1
	21q22.3	TMPRSS2/3	Transmembrane serine protease 2/3
	21q22.3	**RIPK4**	**Receptor interacting serine/threonine kinase 4**
	21q22.3	**ZBTB21**	**Zinc finger and BTB domain containing 21**
	21q22.3	**ABCG1**	**ATP binding cassette subfamily G member 1**
	21q22.3	**CBS**	**Cystathionine beta-synthase**
	21q22.3	**U2AF1**	**U2 small nuclear RNA auxiliary factor 1**
	21q22.3	FRGCA	FOXM1-regulated, gastric cancer associated
	21q22.3	PDXK	Pyridoxal kinase
	21q22.3	**CSTB**	**Cystatin B**
	21q22.3	RRP1	Ribosomal RNA processing 1
	21q22.3	TRAPPC10	Trafficking protein particle complex 10
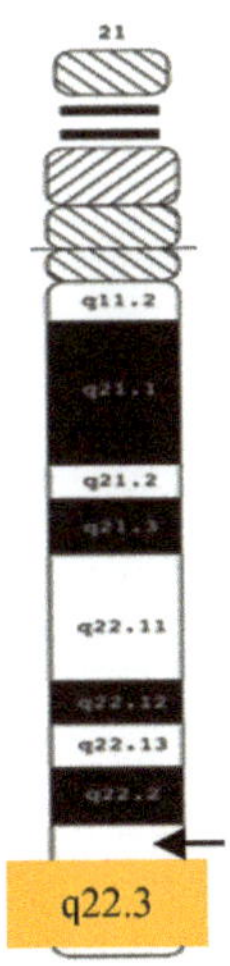	21q22.3	GATD3A/3B	Glutamine amidotransferase like class 1 domain containing 3A/3B
	21q22.3	**DNMT3L**	**DNA methyltransferase 3 like**
	21q22.3	TRPM2	Transient receptor potential cation channel subfamily M member 2
	21q22.3	KRTAP10-1/2-12	Keratin associated protein 10-1/2-12
	21q22.3	**SUMO3**	**Small ubiquitin like modifier 3**
	21q22.3	COL18A1	Collagen type XVIII alpha 1 chain
	21q22.3	**COL6A1**	**Collagen type VI alpha 1 chain**
	21q22.3	**COL6A2**	**Collagen type VI alpha 2 chain**
	21q22.3	PCNT	Centrosomal pericentrin
	21q22.3	**S100B**	**S100 calcium binding protein B**
	21q22.3	**PRMT2**	**Protein arginine methyltransferase 2**
	21q22.3	RRP1B	Ribosomal RNA processing 1B
	21q22.3	PRDM15	PR/SET domain 15
	21q22.3	C2CD2	C2 calcium dependent domain containing 2
	21q22.3	SNORA91	Small nucleolar RNA, H/ACA box 91
	21q22.3	UMODL1	Uromodulin like 1
	21q22.3	TFF1-3	Trefoil factor1- 3

(Continued)

SUPPLEMENTARY TABLE S6.1 Functional genes mapped onto HSA21.—cont'd

HAS21			
Ideogram	Location	Genes	Name
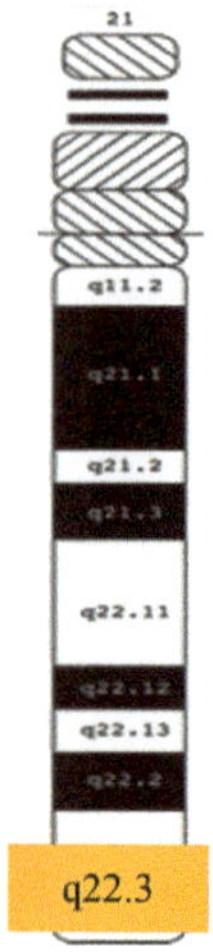	**21q22.3**	**UBASH3A**	**Ubiquitin-associated and SH3 domain containing A**
	21q22.3	RSPH1	Radial spoke head component 1
	21q22.3	SLC37A1	Solute carrier family 37 member 1
	21q22.3	PDE9A	Phosphodiesterase 9A
	21q22.3	WDR4	WD repeat domain 4
	21q22.3	NDUFV3	NADH:ubiquinone oxidoreductase subunit V3
	21q22.3	ERVH48-1	Endogenous retrovirus group 48 member 1
	21q22.3	PKNOX1	PBX/knotted 1 homeobox 1
	21q22.3	CRYAA	Crystallin alpha A
	21q22.3	SIK1	Salt inducible kinase 1
	21q22.3	AGPAT3	1-Acylglycerol-3-phosphate O-acyltransferase 3
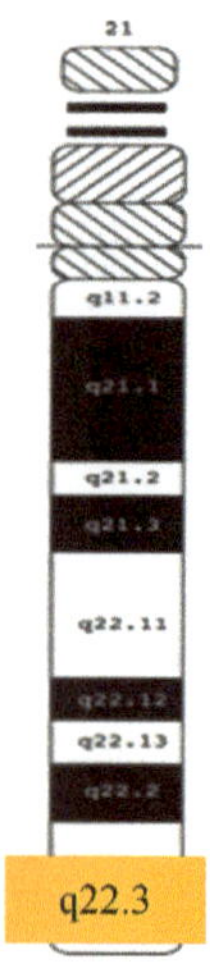	21q22.3	PWP2	PWP2 small subunit processome component
	21q22.3	ICOSLG	Inducible T-cell costimulator ligand
	21q22.3	AIRE	Autoimmune regulator
	21q22.3	PFKL	Phosphofructokinase, liver type
	21q22.3	CFAP410	Cilia and flagella associated protein 410
	21q22.3	TRPM2-AS	TRPM2 antisense RNA
	21q22.3	LRRC3-DT	LRRC3 divergent transcript
	21q22.3	LRRC3	Leucine-rich repeat containing 3
	21q22.3	TSPEAR/AS1	Thrombospondin type laminin G domain and EAR repeats/ TSPEAR antisense RNA 1
	21q22.3	UBE2G2	Ubiquitin conjugating enzyme E2 G2
	21q22.3	PTTG1IP	PTTG1 interacting protein
	21q22.3	ITGB2	Integrin subunit beta 2
	21q22.3	FAM207A	Family with sequence similarity 207 member A
	21q22.3	ADARB1	Adenosine deaminase RNA specific B1
	21q22.3	POFUT2	Protein O-fucosyltransferase 2
	21q22.3	COL18A1-AS2	COL18A1 antisense RNA 2

SUPPLEMENTARY TABLE S6.1 Functional genes mapped onto HSA21.—cont'd

HAS21			
Ideogram	**Location**	**Genes**	**Name**
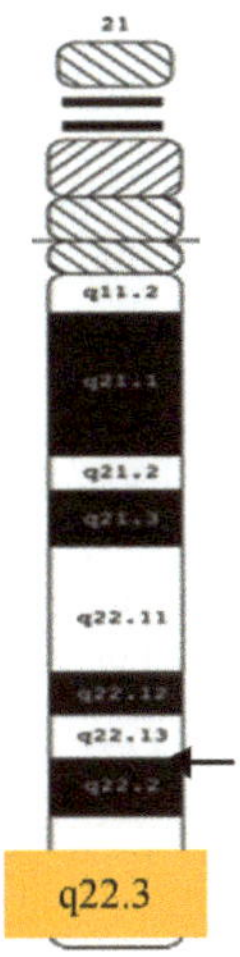	21q22.3	SLC19A1	Solute carrier family 19 member 1
	21q22.3	PCBP3	poly(rC) binding protein 3
	21q22.3	FTCD/AS1	Formimidoyltransferase cyclodeaminase/FTCD antisense RNA 1
	21q22.3	SPATC1L	Spermatogenesis and centriole associated 1 like
	21q22.3	LSS	Lanosterol synthase
	21q22.3	SNORD159	Small nucleolar RNA, C/D box 159
	21q22.3	MCM3AP/AS1	Minichromosome maintenance complex component 3 associated protein/MCM3AP antisense RNA 1
	21q22.3	YBEY	ybeY metalloendoribonuclease
	21q22.3	C21orf58	Chromosome 21 open reading frame 58
	21q22.3	DIP2A/IT1	Disco interacting protein 2 homolog A/DIP2A intronic transcript 1
	21q22.3	**S100B**	**S100 calcium binding protein B**
	21q22.3	PRMT2	Protein arginine methyltransferase 2
	21q22.3	HSF2BP	Heat shock transcription factor 2 binding protein
		LOC110091776	Ankyrin repeat domain 18B pseudogene

Genes described somewhere in the text of this document are presented in **bold.**
Source: The table is prepared from the information available in the public domain.

a function of organization of the mammalian genome (Fig. 6.8).

Moreover, alterations in epigenetic mechanism, pathomechanism of disease development, and identification of druggable candidates are discussed in different contexts (discussed in detail in Chapter 13) (Jin et al., 2013; Jones et al., 2013). A global hypermethylation in all autosomes and the global genome in the placenta of DS pregnancies demonstrated that epigenetic perturbation may be one of the important mechanisms that link the triplication of HSA21 with phenotypic expression of DS (Lyle et al., 2009; Malinge et al., 2013). It has been hypothesized that overexpression of the triplicated *DYRK1A* may induce global epigenetic changes through downregulation of the repressor element one silencing transcription factor (*REST*) (Jin et al., 2013). The Enhancer of Zeste Homolog 2 (*EZH2*), the main catalytic member of the Polycomb Repressive Complex 2 (PRC2) that encodes a histone methyltransferase and is responsible for mono-, di-, and trimethylation of lysine 27 of histone 3 (H3K27me1, me2, m3, respectively) (Margueron et al., 2008), was previously assigned on 21q22 and assumed to be the candidate gene for some of the phenotypes of DS. The PRC2-complex contains *EZH1* (in both dividing and nondividing cells) and *EZH2* (only in actively dividing cells) as its catalytic subunits. Exon sequencing mapped the *EZH2* to 7q35 and demonstrated that the sequence previously isolated from 21q22 corresponds to a "pseudogene" (Cardoso et al., 2000). The report has suggested about the possibility of considering the small-molecule inhibitors of *EZH2* as potential targeted drugs for treatment of DS leukemia. Missense mutations, possibly dominant negative or gain-of-function, have been described in the *U2AF1* (U2 small nuclear RNA auxillary factor 1), an mRNA-splicing factor located at 21q22.3, and hypothesized to cause myeloproliferative disorders in patients having del20q11 and *ASXL1* mutations (Ganguly and Kadam, 2016). Thus, the triplicated *U2AF1* may most likely increase the risk of leukemogenesis in DS population, and thus, identification of targeted small-molecule inhibitors may direct development of novel therapeutics for treatment of DS leukemia.

References

Agrawal, N., Brown, M.A., 2014. Genetic associations and functional characterization of M1 aminopeptidases and immune-mediated diseases. Gene Immun. 15, 521–527.

Ahlfors, H., Anyanwu, N., Pakanavicius, E., et al., 2019. Gene expression dysregulation domains are not a specific feature of Down syndrome. Nat. Commun. 10, 2489. https://doi.org/10.1038/s41467-019-10129-9.

Aït Yahya-Graison, E., Aubert, J., Dauphinot, L., Rivals, I., Prieur, M., Golfier, G., Rossier, J., Personnaz, L., Créau, N., et al., 2007. Classification of human chromosome 21 gene-expression variations in Down syndrome: impact on disease phenotypes. Am. J. Hum. Genet. 81 (3), 475–491.

Alexander, M., et al., 2016. Morbidity and medication in a large population of individuals with Down syndrome compared to the general population. Dev. Med. Child Neurol. 58, 246–254.

Amano, K., Sago, H., Uchikawa, C., Suzuki, T., Kotliarova, S.E., Nukina, N., Epstein, C.J., Yamakawa, K., 2004. Dosage-dependent over-expression of genes in the trisomic region of Ts1Cje mouse model for Down syndrome. Hum. Mol. Genet 13 (13), 1333–1340.

Antonarakis, S.E., 2004. Comparison of human chromosome 21 conserved nongenic sequences (CNGs) with the mouse and dog genomes shows that their selective constraint is independent of their genic environment. Genome Res. 14 (5), 852–859.

Antonarakis, S.E., Lyle, R., Dermitzakis, E.T., Reymond, A., Deutsch, S., 2004. Chromosome 21 and Down syndrome: from genomics to pathophysiology. Nat. Rev. Genet. 5, 725–738.

Antonarakis, S., 2017. Down syndrome and the complexity of genome dosage imbalance. Nat. Rev. Genet. 18, 147–163. https://doi.org/10.1038/nrg.2016.154.

Arayaa, P., Waugha, K.A., Sullivana, K.D., Núñezb, N.G., Rosellib, E., Smitha, K.P., Granratha, R.E., et al., 2019. Trisomy 21 dysregulates T cell lineages toward an autoimmunity-prone state associated with interferon hyperactivity. PNAS 116, 24231–24241, 48. www.pnas.org/cgi/doi/10.1073/pnas.1908129116.

Arron, J.R., Winslow, M.M., Polleri, A., Chang, C.P., Wu, H., Gao, X., et al., 2006. NFAT dysregulation by increased dosage of DSCR1 and DYRK1A on chromosome 21. Nature 441, 595–600.

Au-Yeung, B.B., et al., 2009. The structure, regulation, and function of ZAP-70. Immunol. Rev. 228, 41–57.

Avraham, K.B., Schickler, M., Sapoznikov, D., et al., 1988. Down syndrome: abnormal neuromuscular junction in tongue of transgenic mice with elevated levels of human Cu/Zn-superoxide dismutase. Cell 54, 823–829.

Azzu, V., Jastroch, M., Divakaruni, A.S., Brand, M.D., June-July 2010. The regulation and turnover of mitochondrial uncoupling proteins. Biochim. Biophys. Acta 1797 (6–7), 785–791. https://doi.org/10.1016/j.bbabio.2010.02.035.

Banno, K., Omori, S., Hirata, K., Nawa, N., Nakagawa, N., Nishimura, K., Ohtaka, M., Nakanishi, M., Sakuma, T., Yamamoto, T., et al., 2016. Systematic cellular disease models reveal synergistic interaction of trisomy 21 and GATA1 mutations in hematopoietic abnormalities. Cell Rep. 15 (6), 1228–1241. https://doi.org/10.1016/j.celrep.2016.04.031.

Baptista, M.J., Fairbrother, U.L., Howard, C.M., Farrer, M.J., Davies, G.E., et al., 2000. Heterotrisomy, a significant contributing factor to ventricular septal defect associated with Down syndrome? Hum. Genet. 107, 476–482.

Barlow, G.M., Chen, X.N., Shi, Z.Y., Lyons, G.E., Kurnit, D.M., Celle, L., et al., 2001. Down syndrome congenital heart disease: a narrowed region and a candidate gene. Genet. Med. 3, 91–101.

Bernert, G.1, Fountoulakis, M., Lubec, G., 2002. Manifold decreased protein levels of matrin 3, reduced motor protein HMP and hlark in fetal Down's syndrome brain. Proteomics 2, 1752–1757.

Brichler, J.A., Riddle, N.C., Auger, D.L., Veitia, R.A., 2005. Dosage balance in gene regulation: biological implications. Trends Genet. 21, 219–226.

Byrska-Bishop, M., VanDorn, D., Campbell, A.E., Betensky, M., Arca, P.R., Yao, Y., Gadue, P., Costa, F.F., Nemiroff, R.L., Blobel, G.A., et al., 2015. J. Clin. Invest. 125, 993–1005.

Campos, C., Casado, Á., 2015. Oxidative stress, thyroid dysfunction & Down syndrome. Indian J. Med. Res. 142, 113–119.

Cardoso, C., Mignon, C., Hetet, G., Grandchamps, B., Fontes, M., Colleaux, L., 2000. The human EZH2 gene: genomic organisation and revised mapping in 7q35 within the critical region for malignant myeloid disorders. Eur. J. Hum. Genet. 8, 174–180. https://doi.org/10.1038/sj.ejhg.5200439.

Cereghetti, G.M., Stangherlin, A., Martins de Brito, O., Chang, C.R., Blackstone, C., Bernardi, P., Scorrano, L., 2008. Dephosphorylation by calcineurin regulates translocation of Drp1 to mitochondria. Proc. Natl. Acad. Sci. U.S.A. 105, 15803–15808. https://doi.org/10.1073/pnas.0808249105.

Chan, A.Y., et al., 2016. A novel human autoimmune syndrome caused by combined hypomorphic and activating mutations in ZAP-70. J. Exp. Med. 213, 155–165.

Chrast, R., et al., 2000. The mouse brain treanscriptome by SAGE: difference in gene expression between P30 brain of the partial trisomy 16 mouse model of Down syndrome (Ts65Dn) and normals. Genome Res. 10, 2006–2021.

Chistiakov, D.A., Chistiakov, A.P., 2007. Is FCRL3 a new general autoimmunity gene? Hum. Immunol. 68, 375–383.

Contestabile, A., Magara, S., Cancedda, L., 2017. The GABAergic hypothesis for cognitive disabilities in down syndrome. Front. Cell. Neurosci. 11, 54.

Corbalan, J.J., Kitsis, R.N., 2018. RCAN1–calcineurin axis and the set-point for myocardial damage during ischemia-reperfusion. Circ. Res. 122, 796–798. https://doi.org/10.1161/CIRCRESAHA.118.312787.

Crispino, J.D., 2005. GATA1 mutations in Down syndrome: implications for biology and diagnosis of children with transient myeloproliferative disorder and acute megakaryoblastic leukemia. Pediatr. Blood Cancer 44, 40–44.

Cuvelier, G.D.E., Rubin, T.S., Wall, D.A., Schroeder, M.L., 2016. Long-term outcomes of hematopoietic stem cell transplantation for ZAP70 deficiency. J. Clin. Immunol. 36, 713–724.

Dauphinot, L., Lyle, R., Rivals, I., Dang, M.T., Moldrich, R.X., et al., 2005. The cerebellar transcriptome during postnatal development of the Ts1Cje mouse, a segmental trisomy model for Down syndrome. Hum. Mol. Genet. 14, 373–384.

Davis, R.S., et al., 2002. Fc receptor homologs: newest members of a remarkably diverse Fc receptor gene family. Immunol. Rev. 190, 123–136.

Delabar, J.M., Theophile, D., Rahmani, Z., Chettouh, Z., Blouin, J.L., Prieur, M., Noel, B., Sinet, P.M., 1993. Molecular mapping of twenty-four features of Down syndrome on chromosome 21. Eur. J. Hum. Genet. 1, 114–124.

Delalande, J.-M., et al., 2008. The receptor tyrosine kinase RET regulates hindgut colonization by sacral neural crest cells. Dev. Biol. 313, 279–292.

Duchon, A., del Mar Muniz Moreno, M., Lorenzo, S.M., de Souza, M.P.S., Chevalier, C., Nalesso, V., Meziane, H., de Sousa, P.L., Noblet, V., Armspach, J.-P., Brault, V., Herault, Y., 2021. Multi-influential genetic interactions alter behaviour and cognition through six main biological cascades in Down syndrome mouse models. Hum. Mol. Genet. 30 (9), 771–788. https://doi.org/10.1093/hmg/ddab012.

Du, T., et al., 2013. Circulating serum trefoil factor 3 (TFF3) is dramatically increased in chronic kidney disease. PLoS One 8, e80271.

Epstein, C.J., 2001. Down syndrome (trisomy 21). In: Scriver, C.R., Beaudet, A.L., Sly, W.S., Valle, D. (Eds.), The Metabolic and Molecular Bases of Inherited Diseases. McGraw-Hill, Inc., NewYork, pp. 1223–1256.

Evnouchidou, I., et al., 2012. A common single nucleotide polymorphism in endoplasmic reticulum aminopeptidase 2 induces a specificity switch that leads to altered antigen processing. J. Immunol. 189, 2383–2392.

Ferencz, C., Neill, C.A., Boughman, J.A., Rubin, J.D., Brenner, J.I., et al., 1989. Congenital cardiovascular malformations associated with chromosome abnormalities: an epidemiologic study. J. Pediatr. 114, 79–86.

Ferreira, G.A., Teixeira, A.L., Calderaro, D.C., Sato, E.I., 2016. Atorvastatin reduced soluble receptors of TNF-alpha in systemic lupus erythematosus. Clin. Exp. Rheumatol. 34, 42–48.

Ferretti, E., Villaescusa, J.C., Di Rosa, P., Fernandez-Diaz, L.C., Longobardi, E., Mazzieri, R., et al., 2006. Hypomorphic mutation of the TALE gene Prep1 (pKnox1) causes a major reduction of Pbx and Meis proteins and a pleiotropic embryonic phenotype. Mol. Cell Biol. 26, 5650–5662.

FitzPatrick, D.R., Ramsay, J., McGill, N.I., Shade, M., Carothers, A.D., et al., 2002. Transcriptome analysis of human autosomal trisomy. Hum. Mol. Genet. 11, 3249–3256.

FitzPatrick, D.R., 2005. Transcriptional consequences of autosomal trisomy: primary gene dosage with complex downstream effects. Trends Genet. 21 (5), 249–253.

Fuentes, J.J., Genesca, L., Kingsbury, T.J., Cunningham, K.W., Perez-riba, M., Estivill, X., de la Luna, S., 2000. DSCR1, overexpressed in Down syndrome, is an inhibitor of calcineurin-mediated signaling pathways. Hum. Mol. Genet. 9 (11), 1681–1690.

Friedmacher, F., Puri, P., 2013. Hirschsprung's disease associated with Down syndrome: a meta-analysis of incidence, functional outcomes and mortality. Pediatr. Surg. Int. 29, 937–946.

Ganguly, B.B., Kadam, N.N., 2016. Mutations of myelodysplastic syndromes (MDS): an update. Mutat. Res. 769, 47–62.

Ganguly, B.B., Kadam, N.N., Mandal, P.K., 2017. Complexity of chromosomal rearrangements in Down syndrome leukemia. J. Can. Res. Ther. 13, 381–383.

Gardiner, K., Fortna, A., Bechtel, L., Davisson, M.T., 2003. Mouse models of Down syndrome: how useful can they be? Comparison of the gene content of human chromosome 21 with orthologous mouse genomic regions. Gene 318, 137–147.

Giannone, S., Strippoli, P., Vitale, L., Casadei, R., Canaider, S., Lenzi, L., et al., 2004. Gene expression profile analysis in human T lymphocytes from patients with Down syndrome. Ann. Hum. Genet. 68, 546–554.

Gittenberger-de Groot, A.C., Bartram, U., Oosthoek, P.W., Bartelings, M.M., Hogers, B., Poelmann, R.E., et al., 2003. Collagen type VI expression during cardiac development and in human fetuses with trisomy 21. Anat. Rec. A Discov. Mol. Cell Evol. Biol. 275, 1109–1116.

Gold, L., et al., 2010. Aptamer-based multiplexed proteomic technology for biomarker discovery. PLoS One 5, e15004.

Gonzales-van Horn, S.R., Farrar, J.D., 2015. Interferon at the crossroads of allergy and viral infections. J. Leukoc. Biol. 98, 185–194.

Guelen, L., et al., 2008. Domain organization of human chromosomes revealed by mapping of nuclear lamina interactions. Nature 453, 948–951.

Hammerle, B., Ulin, E., Guimera, J., Becker, W., Guillemot, F., Tejedor, F.J., 2011. Transient expression of Mnb/Dyrk1a couples cell cycle exit and differentiation of neuronal precursors by inducing p27KIP1 expression and suppressing NOTCH signaling. Development 138, 2543–2554.

Hansen, R.S., et al., 2010. Sequencing newly replicated DNA reveals widespread plasticity in human replication timing. Proc. Natl. Acad. Sci. U.S.A. 107, 139–144.

Hao, J.-Q., 2014. Targeting interleukin-22 in psoriasis. Inflammation 37, 94–99.

Hasle, H., 2001. Pattern of malignant disorders in individuals with Down's syndrome. Lancet Oncol. 2, 429–436.

Hattori, M., et al., 2000. The DNA sequence of human chromosome 21. Nature 405, 311–319.

Head, E., Azizeh, B.Y., Lott, I.T., Tenner, A.J., Cotman, C.W., Cribbs, D.H., 2001. Complement association with neurons and beta-amyloid deposition in the brains of aged individuals with Down syndrome. Neurobiol. Dis. 8, 252–265.

Helguera, P., Seiglie, J., Rodriguez, J., Hanna, M., Helguera, G., Busciglio, J., 2013. Adaptive downregulation of mitochondrial function in down syndrome. Cell Metabol. 17, 132–140.

Hinks, A., et al., 2013. Dense genotyping of immune-related disease regions identifies 14 new susceptibility loci for juvenile idiopathic arthritis. Nat. Genet. 45, 664–669.

Hirayasu, K., Arase, H., 2015. Functional and genetic diversity of leukocyte immunoglobulin-like receptor and implication for disease associations. J. Hum. Genet. 60, 703–708.

Hitzler, J.K., Zipurskey, A., 2005. Origins of leukemia in children with Down syndrome. Nat. Rev. Cancer 5, 11–20.

Hong, S., et al., 2016. Complement and microglia mediate early synapse loss in Alzheimer mouse models. Science 352, 712–716.

Ibrahimi, O.A., Zhang, F., Eliseenkova, A.V., Linhardt, R.J., Mohammadi, M., 2004. Proline to arginine mutations in FGF receptors 1 and 3 result in Pfeiffer and Muenke craniosynostosis syndromes through enhancement of FGF binding affinity. Hum. Mol. Genet. 13, 69–78.

Jackson, C.C., Medeiros, L.J., Miranda, R.N., 2010. 8p11 myeloproliferative syndrome: a review. Hum. Pathol. 41, 461–476.

Jin, S., Lee, J.K., Lim, Y.C., Zheng, Z., Lin, X.M., Ng, D.P., et al., 2013. Global DNA hypermethylation in Down syndrome placenta. PLoS Genet. 9, e1003515.

Jones, M.J., Farré, P., McEwen, L.M., MacIssac, J.L., Watt, K., Neumann, S.M., et al., 2013. Distinct DNA methylation patterns of cognitive impairment and trisomy 21 in Down syndrome. BMC Med. Genom. 6, 58.

Kahlem, P., Sultan, M., Herwig, R., Steinfath, M., Balzereit, D., et al., 2004. Transcript level alterations reflect gene dosage effects across multiple tissues in a mouse model of down syndrome. Genome Res. 14, 1258–1267.

Kirsammer, G., Jilani, S., Liu, H., Davis, E., Gurbuxani, S., Le Beau, M.M., Crispino, J.D., January 15, 2008. Highly penetrant myeloproliferative disease in the Ts65Dn mouse model of Down syndrome. Blood 111 (2), 767–775. https://doi.org/10.1182/blood-2007-04-085670.

Kolla, V., Jenö, P., Moes, S., Tercanli, S., Lapaire, O., Choolani, M., Hahn, S., 2010. Quantitative proteomics analysis of maternal plasma in Down syndrome pregnancies using isobaric tagging reagent (iTRAQ). J. Biomed. Biotechnol. 2010, 952047.

Kong, X.D., Liu, N., Xu, X.J., 2014. Bioinformatics analysis of biomarkers and transcriptional factor motifs in Down syndrome. Braz. J. Med. Biol. Res. 47, 834–841. https://doi.org/10.1590/1414-431X20143792.

Korbel, J.O., Tirosh-Wagner, T., Urban, A.E., Chen, X.N., Kasowski, M., Dai, L., Grubert, F., Erdman, C., Gao, M.C., Lange, K., et al., 2009. The genetic architecture of Down syndrome phenotypes revealed by high-resolution analysis of human segmental trisomies. Proc. Natl. Acad. Sci. U.S.A. 106, 12031–12036.

Korenberg, J.R., 1990. Molecular mapping of the Down syndrome phenotype. Prog. Clin. Biol. Res. 360, 105–115.

Korenberg, J.R., Chen, X.-N., Schipper, R., Sun, Z., Gonsky, R., Gerwehr, S., et al., 1994. Down syndrome phenotypes: the consequences of chromosomal imbalance. Proc. Natl. Acad. Sci. U.S.A. 91, 4997–5001.

Labuhn, M., Perkins, K., Matzk, S., Varghese, L., Garnett, C., Papaemmanuil, E., Metzner, M., Kennedy, A., Amstislavskiy, V., Risch, T., Bhayadia, R., Samulowski, D., Hernandez, D.C., Stoilova, B., Iotchkova, V., Oppermann, U., Scheer, C., Yoshida, K., Schwarzer, A., Taub, J.W., Crispino, J.D., Weiss, M.J., Hayashi, Y., Taga, T., Ito, E., Ogawa, S., Reinhardt, D.,

Yaspo, M.-L., Campbell, P.J., Roberts, I., Constantinescu, S.N., Vyas, P., Heckl, D., Klusmann, J.-H., 2019. Mechanisms of progression of myeloid preleukemia to transformed myeloid leukemia in children with down syndrome. Cancer Cell 36 (2), 123–138. https://doi.org/10.1016/j.ccell.2019.06.007. E10.

Lana-Elola, E., Watson-Scales, S.D., Fisher, E.M.C., Tybulewicz, V.L.J., 2011. Down syndrome: searching for the genetic culprits. Dis. Models & Mech. 4, 586–595. https://doi.org/10.1242/dmm.008078.

Lech, M., Anders, H.-J., 2013. The pathogenesis of lupus nephritis. J. Am. Soc. Nephrol. 24, 1357–1366.

Letourneau, A., Santoni, F.A., Bonilla, X., Sailani, M.R., Gonzalez, D., Kind, J., Chevalier, C., et al., 2014. Domains of genome-wide gene expression dysregulation in Down's syndrome. Nature 508, 345–350. https://doi.org/10.1038/nature13200.

Li, C.M., Guo, M., Salas, M., et al., 2006. Cell-typespecific overexpression of chromosome 21 genes in fibroblasts and fetal hearts with trisomy 21. BMC Med. Genet. 7, 24.

Ling, K.-H., et al., 2014. Functional transcriptome analysis of the postnatal brain of the Ts1Cje mouse model for Down syndrome reveals global disruption of interferon-related molecular networks. BMC Genom. 15, 624.

Liu, Y., Borel, C., Li, L., Muller, T., Williams, E.G., Germain, P.L., et al., 2017. Systematic proteome and proteostasis profiling in human Trisomy 21 fibroblast cells. Nat. Commun. 8, 1212. https://doi.org/10.1038/s41467-017-01422-6.

Locke, A.E., Dooley, K.J., Tinker, S.W., Cheong, S.Y., Feingold, E., Allen, E.G., et al., 2010. Variation in folate pathway genes contributes to risk of congenital heart defects among individuals with Down syndrome. Genet. Epidemiol. 34, 613–623.

Lyle, R., Gehring, C., Neergaard-Henrichsen, C., Deutsch, S., Antonarakis, S.E., 2004. Gene expression from the aneuploid chromosome in a trisomy mouse model of Down syndrome. Genome Res. 14, 1268–1274.

Lyle, R., Béna, F., Gagos, S., Gehrig, C., Lopez, G., Schinzel, A., Lespinasse, J., Bottani, A., Dahoun, S., Taine, L., et al., 2009. Genotype-phenotype correlations in Down syndrome identified by array CGH in 30 cases of partial trisomy and partial monosomy chromosome 21. Eur. J. Hum. Genet. 17, 454–466.

Mao, R., Zielke, C.L., Ronald Zielke, H., Pevsner, J., 2003. Global up-regulation of chromosome 21 gene expression in the developing Down syndrome brain. Genomics 81, 457–467.

Mao, R., Wang, X., Spitznagel-Junior, E.L., Frelin, L.P., Ting, J.C., Ding, H., Kim, J.W., Ruczinski, I., Downey, T.J., Pevsner, J., 2005. Primary and secondary transcriptional effects in the developing human Down syndrome brain and heart. Genome Biol. 6 (13), R107.

Malinge, S., Chlon, T., Doré, L.C., Ketterling, R.P., Tallman, M.S., Paietta, E., et al., 2013. Development of acute megakaryoblastic leukemia in Down syndrome is associated with sequential epigenetic changes. Blood 122, e33–e43.

Mateos, M.K., Barbaric, D., Byatt, S.-A., Sutton, R., Marshall, G.M., 2015. Down syndrome and leukemia: insights into leukemogenesis and translational targets. Transt Pediatr. 4, 76–92.

McNulty, M., Crispino, J.D., 2016. iPSCs offer a new look at GATA1-trisomy 21 cooperation. Cell Stem Cell 18, 563–564. https://doi.org/10.1016/j.stem.2016.04.007.

McNaughton, D., Knight, W., Guerreiro, R., Ryan, N., Lowe, J., Poulter, M., et al., 2012. Duplication of amyloid precursor protein (APP), but not prion protein (PRNP) gene is a significant cause of early onset dementia in a large UK series. Neurobiol. Aging 33, 426.e13-21.

Margueron, R., Li, G., Sarma, K., Blais, A., Zavadil, J., Woodcock, C.L., et al., 2008. Ezh1 and Ezh2 maintain repressive chromatin through different mechanisms. Mol. Cell 32, 503–518.

Meyer, P.W.A., et al., 2010. Circulating cytokine profiles and their relationships with autoantibodies, acute phase reactants, and disease activity in patients with rheumatoid arthritis. Mediat. Inflamm. 2010, 158514.

Micali, N., Ferrai, C., Fernandez-Diaz, L.C., Blasi, F., Crippa, M.P., 2009. Prep1 directly regulates the intrinsic apoptotic pathway by controlling Bcl-XL levels. Mol. Cell Biol. 29, 1143–1151.

Migliorini, A., et al., 2013. The antiviral cytokines IFN-α and IFN-β modulate parietal epithelial cells and promote podocyte loss: implications for IFN toxicity, viral glomerulonephritis, and glomerular regeneration. Am. J. Pathol. 183, 431–440.

Mozo, L., López, P., Caminal-Montero, L., Rodríguez-Carrio, J., Suárez, A., 2014. Anti-ribosomal P antibodies are associated with elevated circulating IFNα and IL-10 levels in systemic lupus erythematosus patients. Lupus 23, 1477–1485.

Mural, R.J., Adams, M.D., Myers, E.W., Smith, H.O., Miklos, G.L., Wides, R., et al., 2002. A comparison of whole-genome shortgun-derived mouse chromosome 16 and the human genome. Science 296, 1661–1671.

Nateghi Rostami, M., Douraghi, M., Miramin Mohammadi, A., Nikmanesh, B., 2012. Altered serum pro-inflammatory cytokines in children with Down's syndrome. Eur. Cytokine Netw. 23, 64–67.

Nybo, M., Nielsen, E.H., Svehag, S.E., 1999. Complement activation by the amyloid proteins A beta peptide and beta 2-microglobulin. Amyloid 6, 265–272.

Olson, L.E., Richtsmeier, J.T., Leszl, J., Reeves, R.H., 2004a. A chromosome 21 critical region does not cause specific down syndrome phenotypes. Science 306, 687–690.

Olson, L.E., Roper, R.J., Baxter, L.L., Carlson, E.J., Epstein, C.J., et al., 2004b. Down syndrome mouse models Ts65Dn, Ts1Cje, and Ms1Cje/Ts65Dn exhibit variable severity of cerebellar phenotypes. Dev. Dynam. 230, 581–589.

Olson, L.E., Roper, R.J., Sengstaken, C.L., Peterson, E.A., Aquino, V., Galdzicki, Z., Siarey, R., Pletnikov, M., Moran, T.H., Reeves, R.H., April 1, 2007. Trisomy for the Down syndrome 'critical region' is necessary but not sufficient for brain phenotypes of trisomic mice. Hum. Mol. Genet. 16 (7), 774–782. https://doi.org/10.1093/hmg/ddm022.

O'Doherty, A., Ruf, S., Mulligan, C., Hildreth, V., Errington, M.L., et al., 2005. An aneuploid mouse strain carrying human chromosome 21 with down syndrome phenotypes. Science 309, 2033–2037.

Park, J., Song, W.J., Chung, K.C., October 2009. Function and regulation of Dyrk1A: towards understanding Down syndrome. Cell. Mol. Life Sci. 66 (20), 3235–3240. https://doi.org/10.1007/s00018-009-0123-2.

Parra, V., Altamirano, F., Hernández-Fuentes, C.P., et al., 2018. Down syndrome critical region 1 gene, *Rcan1*, helps maintain a more fused mitochondrial network. Circ. Res. 122 (6), e20–e33. https://doi.org/10.1161/CIRCRESAHA.117.311522.

Patil, N., et al., 2001. Blocks of limited haplotype diversity revealed by high-resolution scanning of human chromosome 21. Science 294, 1719–1723.

Pellegrini-Calace, M., Tramontano, A., 2006. Identification of a novel putative mitogen-activated kinase cascade on human chromosome 21 by computational approaches. Bioinformatiocs 22, 775–778.

Pelleri, M.C., Cattani, C., Vitale, L., Antonaros, F., Strippoli, P., Locatelli, C., Cocchi, G., Piovesan, A., Caracausi, M., 2018. Integrated quantitative transcriptome maps of human trisomy 21 tissues and cells. Front. Genet. 9, 125. https://doi.org/10.3389/fgene.2018.00125.

Perlmutter, L.S., Barrón, E., Myers, M., Saperia, D., Chui, H.C., 1995. Localization of amyloid P component in human brain: vascular staining patterns and association with Alzheimer's disease. J. Comp. Neurol. 352, 92–105.

Pritchard, M., Reeves, R.H., Dierssen, M., Patterson, D., Gardiner, K.J., 2008. Down syndrome and the genes of human chromosome 21: current knowledge and future potentials. Report on the Expert workshop on the biology of chromosome 21 genes: towards gene-phenotype correlations in Down syndrome. Washington, DC, September 28–October 1, 2007. Cytogenet. Genome Res. 121, 67–77.

Potier, M.C., Rivals, I., Mercier, G., Ettwiller, L., Moldrich, R.X., et al., 2006. Transcriptional disruptions in Down syndrome: a case study in the Ts1Cje mouse cerebellum during post-natal development. J. Neurochem. 97 (Suppl. 1), 104–109.

Prandini, P., Deutsch, S., Lyle, R., Gagnebin, M., Vivier, C.D., Delorenzi, M., et al., 2007. Natural gene-expression variation in Down syndrome modulates the outcome of gene-dosage imbalance. Am. J. Hum. Genet. 81 (2), 252–263.

Rachidi, M., Lopes, C., 2011. Mental retardation and human chromosome 21 gene overdosage: from functional genomics and molecular mechanisms towards prevention and treatment of the neuropathogenesis of down syndrome. In: Clelland, J. (Ed.), Genomics, Proteomics, and the Nervous System, Advances in Neurobiology, vol. 2. Springer, New York, NY. https://doi.org/10.1007/978-1-4419-7197-5_2.

Ram, G., Chinen, J., 2011. Infections and immunodeficiency in Down syndrome. Clin. Exp. Immunol. 164, 9–16.

Redon, R., Ishikawa, S., Fitch, K.R., Feuk, L., Perry, G.H., et al., 2006. Global variation in copy number in the human genome. Nature 444, 444–454.

Reeves, R.H., Baxter, L.L., Richtsmeier, J.T., February 2001. Too much of a good thing: mechanisms of gene action in Down syndrome. Trends Genet. 17 (2), 83–88. https://doi.org/10.1016/s0168-9525(00)02172-7.

Remes, K., Tienhaara, A., Pelliniemi, T.-T., 1996. Effects of alpha-interferon on serum beta-2-microglobulin. Leuk. Lymphoma 21, 233–238.

Richtsmeier, J.T., Zumwalt, A., Carlson, E.J., Epstein, C.J., Reeves, R.H., 2002. Craniofacial phenotypes in segmentally trisomic mouse models for Down syndrome. Am. J. Med. Genet. 107, 317–324.

Rothermel, B., Vega, R.B., Yang, J., Wu, H., Bassel-Duby, R., Williams, R.S., 2000. A protein encoded within the down syndrome critical region is enriched in striated muscles and inhibits calcineurin signaling. J. Biol. Chem. 275 (12), 8719–8725. https://doi.org/10.1074/jbc.275.12.8719.

Ryoo, S.R., Cho, H.J., Lee, H.W., Jeong, H.K., Radnaabazar, C., Kim, Y.S., et al., 2008. Dual-specificity tyrosine(Y)-phosphorylation regulated kinase 1A-mediated phosphorylation of amyloid precursor protein: evidence for a functional link between Down syndrome and Alzheimer's disease. J. Neurochem. 104, 1333–1344.

Roberts, I., Izraeli, S., 2014. Br. J. Haematol. 167, 587–599.

Roper, R.J., Reeves, R.H., 2006. Understanding the basis for down syndrome phenotypes. PLoS Genet 2 (3), e50. https://doi.org/10.1371/journal.pgen.0020050.

Roth, D., et al., 2006. Plasmin modulates vascular endothelial growth factor-A-mediated angiogenesis during wound repair. Am. J. Pathol. 168, 670–684.

Rowe, I.F., Jennson, O., Lewis, P., Candy, J., Tennent, G.A., Pepys, M.B., 1984. Immunohistochemical demonstration of amyloid P component in cerebro-vascular amyloidosis. Neuropathol. Appl. Neurobiol. 10, 53–61.

Sakaguchi, N., et al., 2003. Altered thymic T-cell selection due to a mutation of the ZAP-70 gene causes autoimmune arthritis in mice. Nature 426, 454–460.

Salehi, A., Delcroix, J.D., Belichenko, P.V., Zhan, K., Wu, C., Valletta, J.S., et al., 2006. Increased App expression in a mouse model of Down's syndrome disrupts NGF transport and causes cholinergic neuron degeneration. Neuron 51, 29–42.

Saran, N.G., Pletcher, M.T., Natale, J.E., Cheng, Y., Reeves, R.H., 2003. Global disruption of the cerebellar transcriptome in a Down syndrome mouse model. Hum. Mol. Genet. 12, 2013–2019. https://doi.org/10.1093/hmg/ddg217.

Scarpato, M., et al., 2014. AnaLysis of Expression on human chromosome 21, ALE-HSA21: a pilot integrated web resource. Database 2014, bau009.

Siarey, R.J., Villar, A.J., Epstein, C.J., Galdzicki, Z., 2005. Abnormal synaptic plasticity in the Ts1Cje segmental trisomy 16 mouse model of Down syndrome. Neuropharmacology 49, 122–128.

Sjolin-Goodfellow, H., et al., 2015. The catalytic activity of the kinase ZAP-70 mediates basal signaling and negative feedback of the T cell receptor pathway. Sci. Signal. 8 ra49-ra49.

Sui, W., Zhang, R., Chen, J., He, H., Cui, Z., Ou, M., Li, W., et al., 2015. Quantitative proteomic analysis of Down syndrome in the umbilical cord blood using itraq. Mol. Med. Rep. 11, 1391–1399.

Sullivan, K.D., et al., 2016. Trisomy 21 consistently activates the interferon response. Elife 5, e16220.

Sullivan, K.D., Evans, D., Pandey, A., Hraha Thomas, H., Smith Keith, P., Neil, M., Rachubinski Angela, L., Kristine, W.-W., Francis, H., Espinosa Joaquin, M., Thomas, B., 2017. Trisomy 21 causes changes in the circulating proteome indicative of chronic autoinflammation. Sci. Rep. 7, 14818. https://doi.org/10.1038/s41598-017-13858-3.

Szklarczyk, D., Morris, J.H., Cook, H., Kuhn, M., Wyder, S., Simonovic, M., Santos, A., Doncheva, N.T., Roth, A., Bork, P., et al., 2017. The STRING database in 2017: qualitycontrolled protein-protein association networks, made broadly accessible. Nucleic Acids Res. 45, D362–D368.

Szodoray, P., Alex, P., Brun, J.G., Centola, M., Jonsson, R., 2004. Circulating cytokines in primary Sjogren's syndrome determined by a multiplex cytokine array system. Scand. J. Immunol. 59, 592–599.

Takanohashi, A., et al., 2013. Elevation of proinflammatory cytokines in patients with Aicardi-Goutieres syndrome. Neurology 80, 997–1002.

Tang, H., Zhong, F., Liu, W., He, F., Xie, H., 2015. PathPPI: an integrated dataset of human pathways and proteinprotein interactions. Sci. China Life Sci. 58, 579–589.

Tang, Y., Schapiro, M.B., Franz, D.N., Patterson, B.J., Hickey, F.J., Schorry, E.K., Hopkin, R.J., Wylie, M., et al., 2004. Blood expression profiles for tuberous sclerosis complex 2, neurofibromatosis type 1, and Down's syndrome. Ann. Neurol. 56 (6), 808–814.

Tanioka, T., et al., 2003. Human leukocyte-derived arginine aminopeptidase: the third member of the oxytocinase subfamily of aminopeptidases. J. Biol. Chem. 278, 32275–32283.

Thurman, J.M., Le Quintrec, M., 2016. Targeting the complement cascade: novel treatments coming down the pike. Kidney Int. 90, 746–752.

Tomuschat, C., Puri, P., 2015. RET gene is a major risk factor for Hirschsprung's disease: a meta-analysis. Pediatr. Surg. Int. 31, 701–710.

Tokairin, T., Nishikawa, Y., Watanabe, H., et al., 2008. Osteopontin expression in the liver with severe perisinusoidal fibrosis: autopsy case of Down syndrome with transient myeloproliferative disorder. Pathol. Int. 58, 64–68.

Tunstall-Pedoe, O., Roy, A., Karadimitris, A., de la Fuente, J., Fisk, N.M., Bennett, P., Norton, A., Vyas, P., Roberts, I., 2008. Blood 112, 4507–4511.

Tsangaris, G.T., Karamessinis, P., Kolialexi, A., Garbis, S.D., Antsaklis, A., Mavrou, A., Fountoulakis, M., 2006. Proteomic analysis of amniotic fluid in pregnancies with Down syndrome. Proteomics 6 (15), 4410–4419. https://doi.org/10.1002/pmic.200600085.

Tsujimoto, M., Hattori, A., 2005. The oxytocinase subfamily of M1 aminopeptidases. Biochim. Biophys. Acta Protein Proteonom. 1751, 9–18.

Vilardell, M., Rasche, A., Thormann, A., Maschke-Dutz, E., Perez-Jurado, L.A., Lehrach, H., et al., 2011. Meta-analysis of heterogeneous Down Syndrome data reveals consistent genome-wide dosage effects related to neurological processes. BMC Genom. 12, 229. https://doi.org/10.1186/1471-2164-12-229.

Volk, M., Maver, A., Lovrecic, L., Juvan, P., Peterlin, B., 2013. Expression signature as a biomarker for prenatal diagnosis of Trisomy 21. PLoS One 8, e74184. https://doi.org/10.1371/journal.pone.0074184.

Walus, M., Kida, E., Rabe, A., Albertini, G., Golabek, A.A., 2015. Widespread cerebellar transcriptome changes in Ts65Dn down syndrome mouse model after lifelong running. Behav. Brain Res. 296, 35–46.

Wang, X., et al., 2016. Significant association between Fc receptor-like 3 polymorphisms (−1901A>G and −658C>T) and neuromyelitis optica (NMO) susceptibility in the Chinese population. Mol. Neurobiol. 53, 686–694.

Watanabe, H., Fujiyama, A., Hattori, M., Taylor, T.D., Toyoda, A., Kuroki, Y., et al., 2004. DNA sequence and comparative analysis of chimpanzee chromosome 22. Nature 429, 382–388.

Wechsler, J., Greene, M., McDevitt, M.A., Anastasi, J., Karp, J.E., Le Beau, M.M., Crispino, J.D., 2002. Acquired mutations in GATA1 in the megakaryoblastic leukemia of Down syndrome. Nat. Genet. 32 (1), 148−152.

Wedgwood, J.F., Hatam, L., Bonagura, V.R., 1988. Effect of interferon-gamma and tumor necrosis factor on the expression of class I and class II major histocompatibility molecules by cultured human umbilical vein endothelial cells. Cell. Immunol. 111, 1−9.

Weijerman, M.E., de Winter, J.P., 2010. Clinical practice. The care of children with Down syndrome. Eur. J. Pediatr. 169, 1445−1452.

Wolvetang, E.J., Wilson, T.J., Sanij, E., Busciglio, J., Hatzistavrou, T., Seth, A., et al., 2003. ETS2 overexpression in transgenic models and in down syndrome predisposes to apoptosis via the p53 pathway. Hum. Mol. Genet. 12, 247−255.

Yahya-Graison, E.A., Aubert, J., Dauphinot, L., Rivals, I., Prieur, M., Golfier, G., et al., 2007. Classification of human chromosome 21 gene-expression variations in Down syndrome: impact on disease phenotypes. Am. J. Hum. Genet. 81, 475−491.

Yang, J., Rothermel, B., Vega, R.B., Frey, N., McKinsey, T.A., Olson, E.N., Bassel-Duby, R., Williams, R.S., 2000. Independent signals control expression of the calcineurin inhibitory proteins MCIP1 and MCIP2 in striated muscles. Circ. Res. 87, E61−E68.

Yazdanpanah, B., et al., 2009. Riboflavin kinase couples TNF receptor 1 to NADPH oxidase. Nature 460, 1159−1163.

Yla-Herttuala, S., Luoma, J., Nikkari, T., Kivimaki, T., 1989. Down's syndrome and atherosclerosis. Atherosclerosis 76, 269−272.

Yu, B., Zhang, B., Wang, J., Wang, Q.W., Huang, R.P., Yang, Y.Q., Shao, S.H., 2012. Preliminary proteomic-based identification of a novel protein for Down's syndrome in maternal serum. Exp. Biol. Med. 237, 530−539.

Zhao, J., Zhang, Z., Ren, S., Zong, Y., andKong, X., 2016. Co-expressionnetwork analysis of Down's syndrome based on microarray data. Exp. Ther. Med. 12, 1503−1508. https://doi.org/10.3892/etm.2016.3462.

Zorick, T.S., Mustacchi, Z., Bando, S.Y., Zatz, M., Moreira-Filho, C.A., et al., 2001. High serum endostatin levels in Down syndrome: implications for improved treatment and prevention of solid tumours. Eur. J. Hum. Genet. 9, 811−814.

CHAPTER

7

Development of specific phenotypes and genetic consequences in Down syndrome: malformations in craniofacial and cardiac anatomy and development of leukemia and cognitive and learning deficits

1. Introduction

All individuals with Down syndrome (DS) have typical stigmatic phenotypes, including craniofacial malformations, memory and learning difficulties, and muscle hypotonia, but all do not have congenital heart defects, leukemia, or gut abnormalities. Moreover, the severity of the symptoms is variable as evidenced with the degree of cognitive impairment between individuals with DS, though all DS individuals possess a complete or partially extra copy of the chromosome 21 (HSA21) (Pennington et al., 2003). These symptoms are obviously the result of the dosage imbalance of genes on HSA21, regardless of HSA21 haplotypes, genetic background, or stochastic events. The extra HSA21 carrying ~310 genes (Ensembl release 62, including known and newly identified protein coding and RNA genes but excluding pseudogenes; http://21:1=www.ensembl.org/Homo_sapiens/Location/Chromosome?r 48129895) has been labeled as the culprit of DS pathology with variability in penetrance of the overexpressed genes (Epstein et al., 1991). One of the completely penetrant features is the typical facial characteristic of DS, which is largely a product of the underlying craniofacial skeleton.

The prevailing hypothesis of dosage sensitivity has been guiding to identify the specific dosage-sensitive genes contributing to each of the DS phenotypes. The hope is that identification of such genes and cross-talk with other disomic genes will facilitate better understanding of the molecular mechanisms underlying the DS pathologies and hence, development of better effective therapeutics (Antonarakis et al., 2004). The search for the DS-phenotype-specific genes involved both human and experimental models, especially mouse models created for studying the genetic expression of the dosage sensitivity in DS individuals. Early studies in humans considering partial trisomies of HSA21 to narrow down the chromosomal regions for recognition of the phenotype-specific and dosage-sensitive genes hypothesized

Genetics and Neurobiology of Down Syndrome
https://doi.org/10.1016/B978-0-323-90456-8.00002-8

that the DS critical region (DSCR) (DSCR1 on chromosome 22q11.2 and DSCR2 on chromosome 21q22) might contain one or more dosage-sensitive genes responsible for many of the DS phenotypes (Belichenko et al., 2009). However, the hypothesis of a single DSCR was argued by the later studies that included more detailed genetic mapping on more number of patients having partial trisomy to delineate different regions of HSA21 contributing to different DS characteristics (Korbel et al., 2009; Lyle et al., 2009). Altogether, such studies are limited with rarity of partial trisomies to support identification of dosage-sensitive genes and also heterogeneity of the specific phenotype and genetic variation between DS individuals. Interestingly, gene–dosage expression in individuals with triplicated HSA21 is significantly modified by several other non-HSA21 genes, epigenetic modifiers, inherent genetic makeup, environmental factors, and stochastic events (Fig. 7.1) (Antonarakis et al., 2004).

DS, per se, has been demonstrated as an intractable and complex disorder having underlying genetic complexity and a great deal of variability in phenotypic expression. This chapter includes recent reports on breakthroughs in understanding of the impact of the gene dosage of trisomic HSA21 on development of tissues of different organs, in particular for cranio-facial

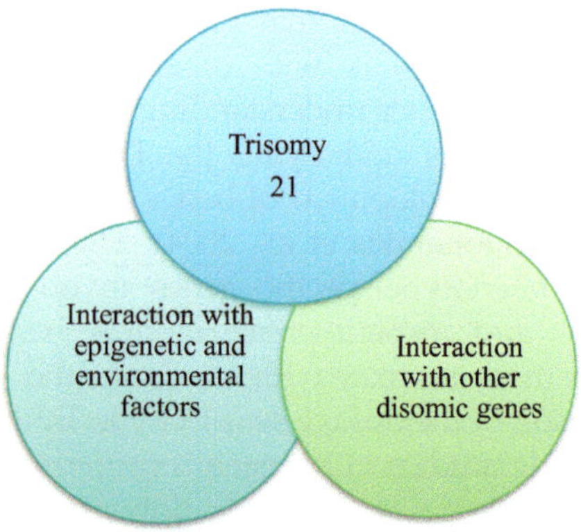

FIGURE 7.1 Interaction of trisomic HSA21 genes with other global disomic genes and environmental factors.

malformation, heart defects, cognitive impairment and learning and memory deficits, and leukemia, in detail. The information illustrating these genotype–phenotype associations would certainly make significant strides to understand the expression and interaction of genes and to develop therapeutics to improve the debilitating features in individuals with DS.

2. Genes of developmental stages

To define the consequences of expression of trisomic genes in different developmental stages, understanding of its direct and indirect interaction with other disomic genes in divergent environmental context is important. In particular, the development of many organs, most importantly, the heart and the brain have been considered for elucidation of the genotype–phenotype association and amelioration of the critical phenotypes of DS. The transcripts of trisomic genes are increased by about 50% in different cells and tissues at different developmental stages and that indicates that overexpression of those trisomic genes will occur in all cells throughout the development. Thus, alteration of function of the overexpressed genes in fully differentiated cells of trisomic adults might indicate their contribution to DS phenotype. However, overexpression of a given gene may not perturb the development and function of every cell type and at every developmental time point. It was demonstrated that detrimental overexpression of some genes exists in specific cell types and at specific time point of development. Also, a trisomy-induced alteration of function of some cell population is likely to affect the function of neighboring cells as a secondary consequence of trisomy, resulting in aberrant expression of DS phenotype (Roper et al., 2009). Moreover, overexpression of regulatory genes may cause severe detrimental effects on several signaling pathways and establish aberrant patterns of development (Birchler et al., 2005).

The interplay between DYRK1A and RCAN1 was thought to have an impact on the development of tissues of multiple organs (Arron et al., 2006). DYRK1A, a priming kinase, activates further phosphorylation of numerous proteins in association with other kinases, while RCAN1 regulates the protein phosphatase calcineurin (Fig. 7.2): both are upregulated in a number of tissues in people with DS in trisomic state. Coexistence of trisomic and upregulated DYRK1A and RCAN1 was postulated to act synergistically and modulate signaling using the NFAT family of transcription factors (Arron et al., 2006). Overexpression of DYRK1A was demonstrated to repress the expression of RE1-silencing transcription factor (REST) (Canzonetta et al., 2008), resulting in perturbation of development of many cells types. Nevertheless, REST is essential to sustain pluripotency and differentiation of neuronal cells (Wiseman et al., 2009). Importantly, heart defects and learning and memory deficits have been attributed to overexpression of DYRK1A in some animal models of DS but not in all (Arron et al., 2006; Ryoo et al., 2008).

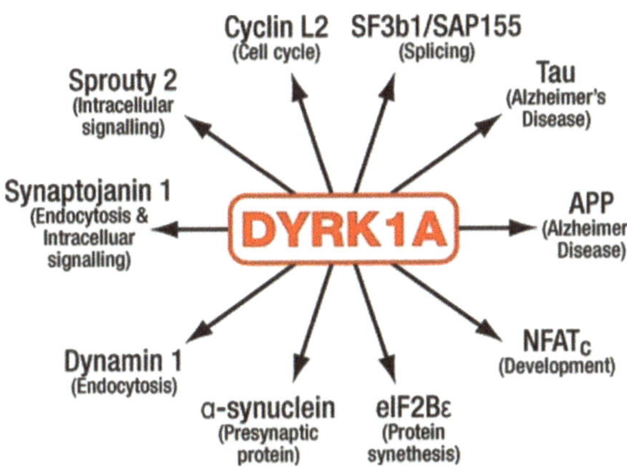

FIGURE 7.2 Phosphorylation targets of DYRK1A. The HSA21-encoded kinase DYRK1A has been shown to phosphorylate a multitude of targets, which have been implicated in a number of biological processes and Down syndrome–associated phenotypes, including endocytosis and Alzheimer disease. *Reprinted from Wiseman et al. (2009), under the terms of the Creative Commons Attribution License.*

Therefore, influence of polymorphismic variations in these genes or interaction with other trisomic or disomic genes was explained for the variable expression and/or outcome of upregulated DYRK1A in DS.

Complications of congenital gut disorders, which frequently include Hirschprung disease (~12%), and 260- and 33-times higher incidence of duodenal stenosis (DST) and imperforate anus (IA) respectively (Korbel et al., 2009), have been reported in individuals with DS. Hirschprung disease is caused by defective enteric nervous system in colon through a concerted effect of trisomic genes and nontrisomic and non-HSA21 (~10) genes (Amiel et al., 2008). Involvement of a 13 Mb region in HSA21 has been identified responsible for Hirschprung disease through analysis of human segmental trisomies. This region harbors the DSCAM gene, which is expressed in the neural crest cells (NCCs) and involved in the enteric nervous system (Korbel et al., 2009). DST and IA are caused by the altered expression of genes present in an overlapping region (Korbel et al., 2009).

Individuals with trisomy of HSA21 have been characterized with smaller size of the brain, hippocampus, and cerebellum being particularly affected (Pearlson et al., 1998). The Ts65Dn mouse model of DS showed a similar phenotype, in particular, disproportionately small cerebellum due to reduced density of neuronal cells in the Purkinje and granule cell layers (Baxter et al., 2000). A close monitoring in mice through developmental stages to identify the earliest stage of divergence of trisomic and euploid cerebellar development demonstrated that the number of mitotic cells was significantly reduced in trisomic mice, though the number of granule cell precursors remains the same at the day of birth. The diminished response to the mitogen, Sonic hedgehog, of the granule cell precursors in both genetic marker crosses, and primary culture assays suggested a likeliness of an agonistic activity of the hedgehog pathway and that corrected the granule cell deficiency through (at least) the 1/3rd of cerebellar development

(Roper et al., 2009). Although the responsible genes of mouse chromosome 16 (MMU16; syntenic to HSA21) for blocking the mitogenic response are yet to be identified, amelioration of the structural deficits in the cerebellum, and perhaps other brain regions, would be possible for understanding the interface of developmental processes and divergent genetic mechanisms of trisomy. Hypocellularity of the hippocampus has also been implicated to a developmental origin, and additionally, alterations in cell cycle, neocortical neurogenesis, and apoptosis have been associated with this phenotype (Contestabile et al., 2007). In Ts65Dn mice, deficiency of neurogenesis in hippocampus was rescued by therapeutic application of the antidepressant fluoxetine (Clark et al., 2006). Fluoxetine, a serotonin reuptake inhibitor, was demonstrated to promote neurogenesis through a series of potential mechanisms, including its direct effect on maintaining the serotonin levels or its indirect effect on behavior. Nevertheless, similar effect of fluoxetine during embryonic development has yet to be ascertained. The Ts65Dn littermates also exhibited delayed development of several milestones similar to that observed in children with DS, such as grip by forelimbs and the righting reflex. Use of neuroprotective peptides for treatment of Ts65Dn embryos lowered the risk of delay in the development of sensory and motor milestones during early postnatal age (Toso et al., 2008).

The craniofacial dysmorphology in Ts65Dn and Tc1 models, including a reduced mandible as observed in human DS, was manifested at early postnatal development and that can be corroborated with specific alterations in bone development (Parsons et al., 2007). The reduced mandible in human DS was demonstrated to be caused by migration and proliferation defects in NCCs (precursor cells of mandible) during embryonic development, which might be affected by an altered response to the mitogenic action of Sonic hedgehog (Roper et al., 2009). It may be noteworthy that in animal models of DS, the implication of segmental trisomies has been described on development as expression of individual trisomic gene or in cooperation with other factors. However, the precise extrapolation to humans for clinical application necessitates defined studies to uncover the myriad sources of divergence from normal development and function initiated by the trisomic status or the interface of genotype and phenotype in DS.

3. Trisomic segments in mouse models: genotyping of DS phenotypes

Creation of different mouse models of DS has enabled further search of dosage-sensitive genes underlying DS characteristics. The approach required detection and demonstration of compelling parallels between phenotypes in mouse and human. HSA21 shares conserved synteny with orthologous regions on three mouse chromosomes: MMU10, MMU16, and MMU17 (Fig. 7.3; Lana-Elola et al., 2011). The first two mouse models, Ts65Dn and Ts1Cje, were generated with duplications of parts of MMU16 that are orthologous to HSA21. The Ts65Dn mice carry an extra chromosome that has a region of MMU16 translocated onto a small segment of MMU17, leading to a trisomic status for ~104 genes on MMU16 that are orthologous to HSA21 genes (Gardiner et al., 2003; Olson et al., 2004). Notably, the Ts65Dn strain also has three copies of 19 genes mapped on MMU17 that are not orthologous to HSA21 and that might implicate some phenotypes in this strain, which might not be related to human DS (Li et al., 2007). Alterations in cardiac, neurological, and craniofacial precursors of Ts65Dn mice have been linked to postnatal development of DS-associated phenotypes (Chakrabarti et al., 2007; Roper et al., 2009; Williams et al., 2008).

Conversely, the Ts1Cje mice have a duplication of a shorter region of MMU16, encompassing ~81 genes with orthologs of genes on HSA21 (Fig. 7.3) (Gardiner et al., 2003; Olson et al., 2004). Moreover, both strains exhibited some features of DS, including learning difficulties, and

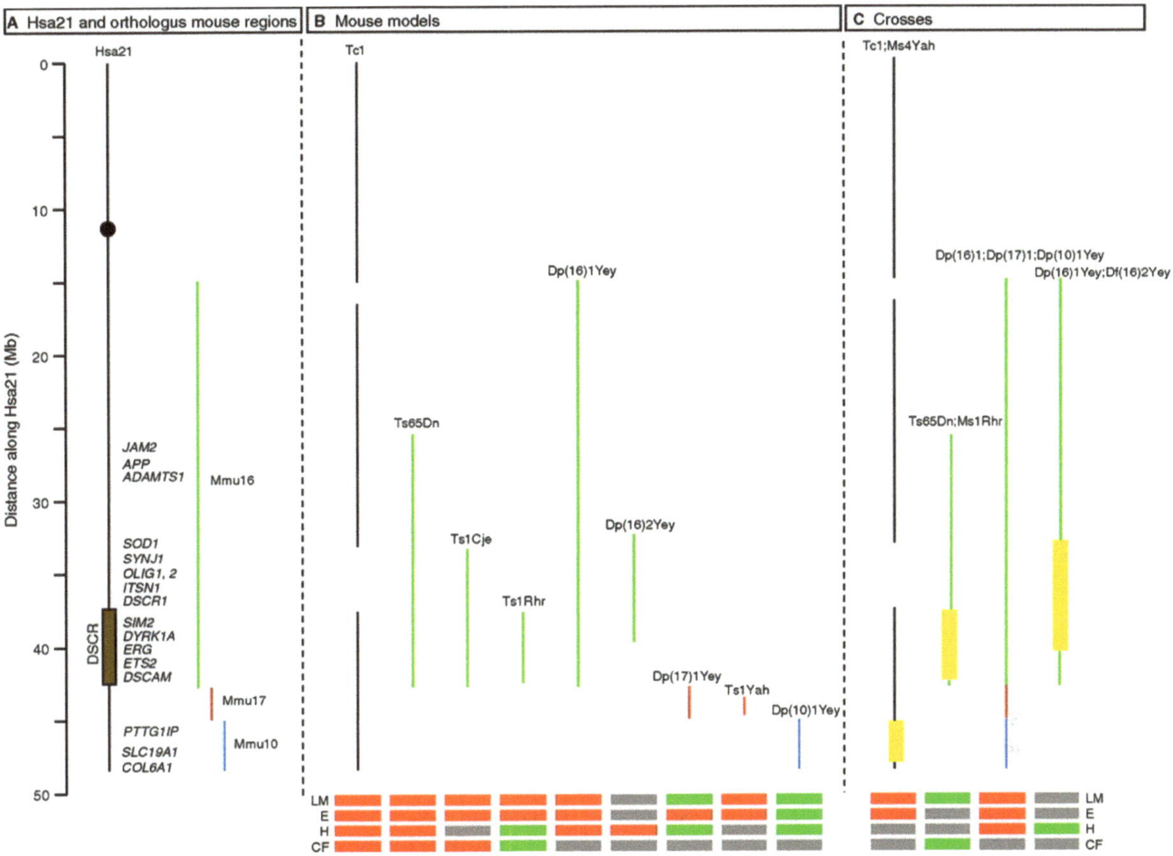

FIGURE 7.3 Mouse models of Down syndrome and orthologous mouse regions syntenic to HSA21. (A) The schematic diagram of HSA21 indicating the approximate positions of candidate dosage-sensitive genes and of orthologous regions on mouse chromosomes 10 (blue), 16 (green), and 17 (red). The *black circle* and the *brown rectangle* represent the centromere and the approximate location of the DSCR, respectively. (B) The extent of trisomy in the mouse models of DS: The Tc1 mouse carries a copy of HSA21 (with some deletions), whereas the other models all contain duplications of mouse regions that are orthologous to HSA21. (C) Crosses of mouse strains where a strain with extra copies of genes has been crossed to a deficiency [Ms1Rhr, Ms4Yah or Df(16)2Yey], the regions where gene dosage has been reduced from three to two copies are indicated by a *yellow box*. Beneath (B) and (C), *red and green boxes* indicate the presence or absence, respectively, of phenotypes in the following areas: learning and memory (LM), electrophysiology (E), heart (H), and craniofacial (CF). Gray boxes indicate that no data were available. Notably, use of different assays in different models and crosses made direct comparisons difficult, and in some cases, the phenotypes result in an improvement rather than a defect, e.g., Ts1Yah shows improved learning and LTP responses compared with euploid controls. *Reprinted from Lana-Elola et al. (2011), under the terms of the Creative Commons Attribution License (CC BY) (http://creativecommons.org/licenses/by-nc-sa/3.0).*

that led to more use of these mouse models to narrow down the search for dosage-sensitive genes. The recent use of Cre-loxP technology has enabled precise engineering of duplications of the HSA21-orthologous regions in Ts1Rhr mouse. The Ts1Rhr strain carries a duplication of the region of MMU16 that is orthologous to the human DSCR and is trisomic for ~33 genes with orthologs on HSA21 (Olson et al., 2004). The "triple trisomic" mouse carrying Dp(10)1Yey; Dp(16)

1Yey; Dp(17)1Yey has duplications of MMU10, MMU16, and MMU17 and thus, trisomy of all ~250 mouse genes that are orthologous to HSA21 genes (Fig. 7.3) (Yu et al., 2010). These compound strains of mouse model are currently available as the most complete model for DS research.

The Ts1Cje mouse model of DS having segmental trisomy 16 arose as a fortuitous translocation of MMU16 in a mouse line produced by homologous recombination at the Sod1 locus, which has a duplication of a shorter region of MMU16, encompassing ~81 genes with orthologs of genes of HSA21 (Gardiner et al., 2003; Olson et al., 2004). The Tc1 mouse model carries a freely segregating copy of HSA21 (Fig. 7.3) (O'Doherty et al., 2005; Lana-Elola et al., 2011). This strain has several deletions, leaving ~83% of the genes of HSA21, and thus, the Tc1 strain has the advantage of allowing the study of the HSA21 genes, including those which are not found in the orthologous regions of MMU10, MMU16, and MMU17 of mice strains (Gardiner et al., 2003). The Tc1 mice expressed a number of DS-like phenotypes, including learning and memory impairment, cardiac defects, craniofacial abnormalities, and decreased tumor angiogenesis (O'Doherty et al., 2005; Galante et al., 2009; Reynolds et al., 2010). Cross breeding of these mice with mouse strains carrying small deletions resulted in decrease in copy number of trisomic genes indicating precise association of genes with DS-characteristics from the deleted intervals. The cross breeding of Ms4Yah strain, which carries a deletion of the region of MMU10 that is orthologous to HSA21, with the Tc1 mice demonstrated that trisomic genes in this interval are not responsible for learning defects (Duchon et al., 2010). Such cross mapping has indicated the necessity of such crossing with even smaller deletions with a "knockout" for the relevant gene, which might eventually identify the individual dosage-sensitive genes in the mouse models of DS.

Taken together, imbalance in gene dosage in DS is underlying factor for the DS-specific phenotypes, but which genes of HSA21 contributing to which specific characteristic in DS has yet to be precisely assigned. Hence, the objective of this chapter is to pin-point the genes responsible for prominent DS-specific phenotypes and state of art on research and therapeutic developments with a view to lowering the clinical impact of the expression of genes in triplicated condition.

4. Craniofacial alterations in Down syndrome

The craniofacial characteristics of DS have been described from prenatal to postnatal age (Guihard-Costa et al., 2006), through questionnaire-based examination, clinical examination of computerized tomography (CT), anthropometrics, dental cast, photographs, radiographs, magnetic resonance imaging (MRI), and cephalometric features (reviewed in Jesuino and Valladares-Neto, 2013). The craniofacial features of DS broadly include: reduced head and facial dimensions representing microcephaly; reduced skull size; brachycephalic cranium (relatively wide neurocranium, reduction in most head dimensions but not in mediolateral dimensions, producing a broad, short brachycephalic skull); reduced orbital region mediolaterally; reduced bizygomatic breadth; shorter and flatter cranial base; reduced or absent frontal sinus and nasal bone; small maxilla; hypertelorism; epicanthic fold; seal-like nose; small ear; insufficient lip seal; high palate with reduction in its length; narrowed oropharynx; presence of mouth breathing causing obstructive sleep apnea; normal or reduced goniac angle; hypotonic and protrusive tongue, relative rather than absolute macroglossia; delayed dental eruption and hypodontia, dimensions, shape, and position of the teeth; and reduced, normal, or prognathic mandible (reduced distances among all mandibular landmark subsets); increased individual variability in mandible and mid-face deficiency (spanning orbits and maxillary alveolus); reduced facial height with reduced nasal protrusion; and so on (reviewed in Jesuino and Valladares-Neto, 2013; Richtsmeier et al., 2002; Vicente et al., 2020). A higher prevalence of malocclusion, with prominence of anterior and posterior open bite, posterior crossbite, dental protrusion, crowding, and pseudo and skeletal class III malocclusion predispose lesser facial convexity as a result of maxillary deficiency, mandibular protrusion, or both. On the whole, maxillary deficiency, also called "retrognathic maxilla," is a common finding in these individuals (Jesuino and Valladares-Neto, 2013). Craniofacial alterations may contribute to complications in breathing, eating, and communication (Faulks et al., 2008). Indeed, obstructive sleep apnea affects ~30%–~75% of individuals with DS.

The delay in intrasphenoidal synchondrosis fusion in postnatal life is suggested to be essential for the cranial base's flattening mechanism. This growth is gradually reduced with cessation of the spheno-occipital synchondrosis around 15 years of age (Alió et al., 2008). The facial features are associated with compressing of the skull base or platybasia with an increase of the skull base angle (Rodrigues et al., 2019). Maxillary deficiency accompanying with a flatter and shorter cranial base on its anterior site, general reduction in linear maxillary and mandibular length, and anterior facial heights was described in a cephalometric analysis of prepubertal DS children compared the age, gender, and ethnicity matched nontrisomic children (Jesuino and Valladares-Neto, 2013). Besides the presence of highly compromised maxillary length, the differential maxilla-mandibular-sagittal relationship was less deteriorated, and the maxillary deficiency was not so prominent owing to overall reduction of craniofacial dimensions in the face of DS children. Thus, an opportunity of orthopedic and orthodontic treatment may create an esthetic correction of the face in a carefully selected group of DS children to enhance their social acceptance. The dental agenesis is frequently reported in subjects with DS (~54.6%–58.5%) (Palaska and Antonarakis, 2016; Abeleira et al., 2014).

4.1 Hypotheses on craniofacial malformations

Little is known about the mechanism of action of trisomic HSA21 on alteration of craniofacial morphogenesis to create DS-specific phenotypes. The generalized genetic imbalance disturbs the evolutionarily conserved developmental pathways by changing the developmental homeostasis and precision throughout the developmental phases. The dosage imbalance of HAS21 genes contributes to clinical presentation of DS (Birchler et al., 2007; Veitia et al., 2008). Trisomic HSA21 is hypothesized to cause an "amplified developmental instability" (ADI) that results in asymmetrical phenotypic characteristics (Shapiro, 1992). Developmental instability (DI) leads to produce morphological alteration in response to developmental perturbations due to genetic or environmental stress (Richtsmeier et al., 2005). Consequently, bilateral features, including facial features, of an organism are developed as replicates or mirror images of the opposite side. Notably, genetic or environmental factors contribute to developmental noise causing minor/major departures from an ideal developmental program for a bilateral symmetry (Hallgrímsson et al., 2002). According to ADI hypothesis, DS-phenotypic traits, such as reduced facial dimensions, mid-facial hypoplasia, irregular sinus development, abnormal tooth morphogenesis, malocclusion, smaller palates, abnormal dermatoglyphics, etc. are developmentally labile and more likely to exhibit increased phenotypic variation when perturbed by genetic or environmental insults during development (Shapiro, 1992). Fluctuating asymmetry (FA) was demonstrated in DS faces as 140%–164% more than the faces of their biological siblings and unrelated age-matched euploid individuals and that was 2.7-to 6.9-fold between

DS siblings (Starbuck et al., 2013). Besides genotype–phenotype analysis, algorithm of machine learning has demonstrated the patterns of recognition of facial dysmorphism in individuals with DS (Fig. 7.4) (Agbolade et al., 2020; Kruszka et al., 2017).

Facial development occurs through series of highly coordinated spatio-temporal morphogenetic events directing the correct formation, expansion, and fusion of facial prominences (Brugmann et al., 2007). The pluripotent and transiently migratory cranial NCCs of the dorsal aspect of the neural tube form the tissues of the craniofacial complex during embryogenesis under unique molecular programs (Cordero et al., 2010). Individual genes such as Shh, Fgf's, Bmp's, Wnt's, Dlx, Msx, Otx, and signaling pathways and signaling centers (such as frontonasal ectodermal zone—FEZ, nasal pit) play pivotal roles in construction of craniofacial phenotypes (Liu et al., 2010; Young et al., 2010). Gene–dosage imbalance caused by trisomy 21 in DS perturbs the subtle changes of the biological processes of craniofacial morphogenesis that potentially affect the developmental stability, prominence formation, and facial development producing DS-specific craniofacial blueprint. However, the exact mechanism of action of gene–dosage imbalance to alter developmental stability and facial morphogenesis is not fully uncovered. Some of the hypothesis indicates that:

1. Triplication of HSA21 affects the spatiotemporal deployment of important genes involved in regulatory mechanisms of developing facial primordia. Notably, the stoichiometric interactions among the growth factors of the Wnt, Bmp, and Fgf families may play critical roles in shaping the maxillary and mandibular prominences (Foppiana et al., 2007). Also, alteration of facial gene expression is likely through cis- or transmisregulation caused by gene–dosage imbalance (Starbuck et al., 2013);
2. Trisomy 21 may facilitate inhibition of delamination and migration of cranial NCCs from the rhombomeres into pharyngeal arches 1 and 2 or interfere with cell type specification

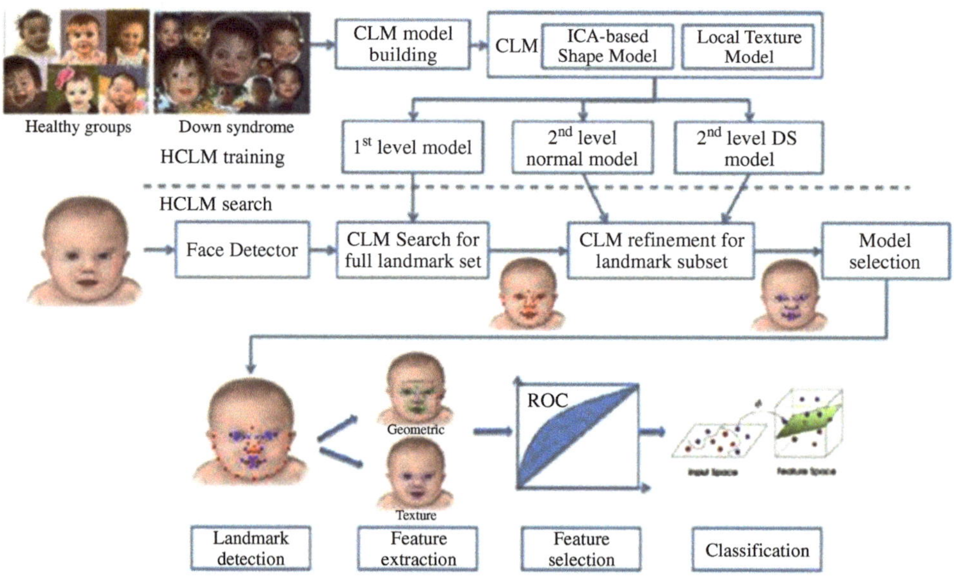

FIGURE 7.4 Algorithm for landmark detection in facial geography of Down syndrome. *Reprinted with permission from Agbolade et al. (2020), under the terms of the Creative Commons Attribution License (CC BY) (http://creativecommons.org/licenses/by-nc-sa/3.0).*

and apoptosis. Trisomy 21 acting as a neurocristopathogen is evidenced in Ts65Dn mouse that resulted in reduction of mitosis and cell proliferation and migration of cranial NCCs into the face during craniofacial morphogenesis (Roper et al., 2009). DS mouse models have displayed deficiencies in bone formation, low bone mineral density in trabecular and cortical bone, decreased bone volume and reduction in rod-like trabeculae (Blazek et al., 2010; Parsons et al., 2007), suggesting an impact of gene–dosage imbalance on decreased osteoblast and odontoblast activity, which culminates into facial morphology of DS. There is also possibility that gene–dosage imbalance impairs the morphogenesis of Meckel's cartilage, that forms the NCCs, through impairment of the spatiotemporal migration of NCCs and subsequent disruption of conversion of cartilage into bone (Wyganowska-Świątkowska and Przystańska, 2011);

3. In addition, gene–dosage imbalance may alter the growth rates of particular prominences (such as mandibular prominence) affecting the boundaries of developing facial regions or may lower the duration of contact of prominences and expedite their fusion (Rueda et al., 2012; Starbuck et al., 2013).

Nevertheless, a better understanding of the effects of gene–dosage imbalance on craniofacial morphogenesis may alleviate health disparities associated with the face and skull in DS. Moreover, alteration of the labile facial morphogenetic features indicates a higher developmental plasticity of facial primordia giving rise to the mandibular and maxillary region than the other regions of the face.

4.2 Development of craniofacial expressions

Majority of the tissues such as bone, cartilage, connective tissue, and peripheral nervous tissue in the head are formed from the NCCs (Knight and Schilling, 2006; Santagati and Rijli, 2003). Subsets of cranial NCC drift to the first pharyngeal arch (PA1) or mandibular precursor, and the PA1 contributes to the formation of the lower jaw and the maxilla using Meckel's cartilage (Knight and Schilling, 2006). Trisomic HSA21 is thought to disrupt the genetic pathways and cellular constituents, including NCC, in early developmental stages, and thus, affect the structure and function of craniofacial and other tissues associated with DS features (Roper et al., 2009). The craniofacial phenotypes in Ts65Dn mice include a diminished craniofacial skeleton and brachycephaly; a flattened occiput and small midface; and a reduced interorbital breadth as well as maxilla and mandible (Richtsmeier et al., 2000). The newborn Ts65Dn litters exhibited differences in the face, neurocranium, palate, and mandible relative to the euploid mice (Hill et al., 2007). Crania were significantly smaller in the Ts65Dn mice than euploid strain but varied in the magnitude of the difference. The growth curve of the mandible from neonatal to adult mice was significantly altered in this strain. The PA1 was reduced and appeared with fewer NCC in Ts65Dn than the euploid offspring at embryonic day 9.5 (E9.5) (Roper et al., 2009).

The linear distances of the mandibular landmarks appeared smaller in these mice with varying distances measured at coronoid process or the angular process localizing the extremes of mandibular dysmorphology to these sites of muscle attachment. Reduction of the coronoid process and consequent anterior displacement of the apex of the process increased the distance between it and the condyle but decreased the distances between it and all other mandibular landmarks. Analysis of the smaller subsets of landmarks consistently presented markedly smaller mandible in the Ts65Dn consistently than normal mice (Richtsmeier et al., 2000). Interestingly, mandible appeared "shape conservative" between euploid and Ts65Dn mice and may correspond with mandible of DS that is underdeveloped but less affected than the rest of the face (Allanson et al., 1993).

The neurocranium of Ts65Dn was generally reduced in size but varied in the magnitude of the difference; especially, the linear distances that span the rostro-caudal dimension were profoundly smaller compared to the euploid mice. The situation results in a relatively broad or brachycephalic neurocranium in Ts65Dn mice, which is similar to that of DS individuals (Allanson et al., 1993).This mouse exhibited flattened occiput like characteristic of DS, which overlie the position of the cerebellum, and is reduced in volume in both Ts65Dn and DS (Baxter et al., 2000). The strong relationship between developing brain and skull is shown by these corresponding findings in humans and mice.

Summarily, the craniofacial skeleton of Ts65Dn mice showed DS-like pattern of craniofacial dysmorphism, including an overall reduction in size; a smaller flattened face, a disproportionately reduced mid-face, maxilla, and mandible; reduced interorbital breadth; brachycephaly, a reduced flattened occiput and a smaller mandible when compared statistically to the skulls of normal littermates (Richtsmeier et al., 2000). Despite obvious craniofacial effects are invariant aspect of DS, a great amount of interindividual variation has been reported in the craniofacial features and in other metric phenotypes among DS individuals (Cronk and Reed, 1981).

The cranial variation was evidenced between two mouse strains (Ts65Dn and Ts1Cje) carrying segmental trisomy of MMU16 (Richtsmeier et al., 2002). Notably, both strains possess some features of DS, including learning difficulties, and that led to frequent use of these models for DS research to recognize the dosage-sensitive genes. The Ts1Cje mice are at a dosage imbalance for a subset of genes triplicated in Ts65Dn mice. This smaller segment corresponds to a region of HSA21 spanning 9.8 Mb that contains 79 genes or 73% of the number triplicated in Ts65Dn mice. A comparative analysis revealed similarities in craniofacial alterations between the Ts1Cje and Ts65Dn mice (Richtsmeier et al., 2002), suggesting that the trisomic region in Ts1Cje mice (Sod1-Znf295) is adequate to produce the craniofacial phenotype.

4.3 Implication of dosage sensitivity on craniofacial expression

Imbalance in gene dosage underlies the craniofacial dysmorphology in individuals with DS with almost complete penetrance, but genes of specific craniofacial features are yet to be correlated (Lyle et al., 2009). The complex skeletal structure of mammals is formed by cranium and mandible. In general, the overall shape of the skull depends on the coordinated development of different bony, dental, and cartilaginous elements and active soft-tissue components where a cascade of appropriate developmental processes dictates the normal development of a functioning skull. The developmental ontogeny is regulated by intricate harmony of the various genetic programs and individual genetic makeup and that provides the raw material for the expression of new phenotypes (Hall and Miyake, 1995). Thus, understanding the contribution of triplicated HSA21 on craniofacial development was experimented in different mouse models of DS such as Ts65Dn and Ts1Cje. The morphology of the entire skull and dentary was affected by the dosage imbalance in both mouse models, though the specific craniofacial components were disproportionately affected. The two segmentally trisomic mice showed decrease in face size and brachycephaly, which might have produced by differently perturbed developmental mechanisms because the Ts65Dn showed medio-laterally increased cranial vault, while Ts1Cje was unaffected (Richtsmeier et al., 2000, 2002). Such significant broadening of the medio-lateral dimensions of the cranial vault in Ts65Dn mice might be influenced by the presence of three copies of specific genes that are present in two copies in Ts1Cje mice. Interestingly, over 80% of the euploid-to-aneuploid differences and similarities measured in the skull and dentary were conserved in the two DS models, wherein Ts65Dn mouse has more triplicated genes (108 genes in Ts65Dn vs. 79 in Ts1Cje strain) and that result in further dosage imbalance in the two models. Thus, the larger genetic insult in Ts65Dn might destabilize a number of developmental pathways, including disruption of canalization of skull development, whereas a mechanistic homeostasis is maintained by the Ts1Cje genome. Interestingly, testing of the skulls of Ms1Cje mice that are segmentally trisomic for only that segment not shared by Ts1Cje and Ts65Dn might explain the hypothesis of genetic imbalance (Lana-Elola et al., 2011). Alternatively, construction of transgenic mice for each of the 28 genes present in this region would decipher whether one of these triplicated genes causes a specific quantitative broadening of the posterior cranium. The combinatorial approach with molecular and morphometric techniques could precisely measure the genotype–phenotype expression, craniofacial modulation, in particular.

The mandible appeared smaller as a whole in both models, as seen in DS patients. The molar alveolar segment was significantly smaller than normal in Ts1Cje, while Ts65Dn was unaffected, though the length of the first mandibular molar (M1) was reduced to the similar extent in the two mice strains when recorded directly from the lingual surface of left hemi-mandibles. Thus, the tooth development in the two models was affected to the similar extent, and the differences recorded on the molar alveolar could be the result of differences in the bone development due to differences in triplicated gene dosage (Richtsmeier et al., 2002).

While understanding the null effect of the aneuploidy, similar effects were observed in the two models local to the angular process. An appreciably amplified effect was demonstrated in local to the coronoid process in Ts65Dn mice. Ontogenically, the angular and coronoid processes develop an intimate contact with the sub-mental and temporalis muscles, respectively, which might occur from the same progeny of NCCs that form the muscles of mastication as demonstrated in experiments with quail-chick chimeras that a single population of NCCs produced specific craniofacial muscles, and those sections of the bone where the muscle attaches (Köntges and Lumsden, 1996). In Ts1Cje and Ts65Dn models, the angular and coronoid processes might have been influenced mechanically by craniofacial muscles and dynamics of neural crest migration for formation of the skull (Moore, 1973). The skulls of Ts65Dn mice were affected to a slightly greater degree than Ts1Cje; however, a marked difference was described in some locations such as the coronoid process of the mandible. This variation could be explained as the effects of specific morphogenetic skeletal units or gene expression that affects the local environment, including the influences of muscle size and effects on the developing bone, the interaction of developing tooth buds and surrounding skeletal components, and the influence of articulating skeletal units (Richtsmeier et al., 2002). Except for a few characters that were significantly affected only in Ts1Cje strain, a great deal of similarities was evident in the two models. Notably, the similarities and differences in the two mouse models of DS are suggested to be influenced substantially by the imbalanced gene dosage and not by the translocations that generated the segmental trisomies for disrupting the development of bones of the skull in a specific manner as affected in individuals with DS. Furthermore, perturbation in these divergent mammalian species suggests an evolutionarily conserved genetic pathway in analogous fashion.

Trisomy of HSA21 has implicated in an imbalance in gene dosage; however, which genes are responsible for which craniofacial characteristic are yet to be precisely assigned. Interestingly, mouse models recapitulate traits and the genotype–phenotype association with DS (Lana-Elola et al., 2011; Wiseman et al., 2009). To understand the conserved craniofacial expression in humans with DS, Ts65Dn mouse model of DS was created with a segmental trisomy chromosome 16 (MMU16) with a small translocation (Davisson et al., 1990), for a region spanning 15.6 Mb of HSA21 that contains about half of the gene (108 of the 225) expressed sequences demonstrated orthologs in the HSA21 gene catalog (Hattori et al., 2000; Reeves et al., 1995; Sturgeon and Gardiner, 2011). Notably, the distal segment of the MMU16 demonstrates conserved linkage with most of HSA21 genes from STCH, the most proximal known gene on HSA21q, to TMPRSS2, located in the proximal half of 21q22.3 (Reeves et al., 1998). The Ts65Dn mouse is at dosage imbalance for most of this segment (Reeves et al., 1995) and demonstrates several phenotypic characteristics similar to those of DS, including the analogous craniofacial abnormalities, reduced birth weight, perinatal lethality, cognitive and behavioral impairments, cardiovascular malformations, and neurological structural deficiencies resulting from a similar complex genetic insult present in human DS (Richtsmeier et al., 2000; Roper et al., 2006; Williams et al., 2008).

The Ts1Rhr strain of DS mouse model is trisomic for the Dscr and exhibited rostrocaudally elongated skulls and larger mandibles than its euploid littermates, a contrast phenotype from that seen in individuals with DS and in Ts65Dn mice. This expression indicated that trisomy of the genes of the DSCR are not the only factors contributing to craniofacial phenotype of the DS (Olson et al., 2004). Interestingly, crossing of Ts65Dn mice to Ms1Rhr with a view to reducing the Dscr from three to two copies, the brachycephaly phenotype was rescued; however, the overall pattern of craniofacial dysmorphology remained similar to that seen in the Ts65Dn mice. Therefore, it was hypothesized that a complex interplay of genetic effect of the DSCR and other trisomic, possibly also the disomic, genes results in DS-like craniofacial phenotypes.

Overexpression of Ets2 in mice was postulated to cause skeletal malformation that is very similar to individuals with DS (Sumarsono et al., 1996). However, removal of one copy of the Ets2 gene from three to two did not result in rescue of the craniofacial phenotype of Ts65Dn mice, indicating that trisomy of the Ets2 gene is not responsible for this skeletal phenotype (Hill et al., 2009).

Expression of Rcan1 has been described in developing brain and craniofacial structures in mice. Trisomy of Rcan1 in mouse models, including Ts65Dn, Ts1Cje, and Ts16, exhibited high incidence of abnormalities in development of brain, skull, and sensory motors and also cardiac valvuloseptal malformations, which were predominant in Ts16 mice. Trisomic mouse for an MMU region syntenic to an HSA21 segment (Ts1Rhr) (Olson et al., 2004), but not including the Rcan1 locus did not present cranial dysmorphologies, and that indicated partial or full contribution of Rcan1 for development of these features.

Interestingly, gene expression analyses in tissues from humans with DS and DS mouse models have demonstrated dysregulation of both trisomic and nontrisomic (euploid/disomic) genes (Laffaire et al., 2009; Rozovski et al., 2007; Sommer et al., 2008). Trisomic craniofacial development was hypothesized to be regulated by two trisomic genes such as DYRK1A and RCAN1 through the NFAT pathway (Arron et al., 2006). Yet, the cascade of the intermediate development through the involvement of trisomic genetic mechanisms that result in DS-associated craniofacial structures has to be uncovered. A spatial and temporal specific analysis of developmental and genetic changes has highlighted the importance of nontrisomic genes in mandibular development, suggesting that triplicated HSA21 alters genetic pathways to cause DS phenotypes in early developmental stages. Understanding the kinship between trisomic and nontrisomic gene expression and subsequent disruption of the developmental processes may unlock the genotype–phenotype association of craniofacial abnormalities of DS.

4.4 Developmental algorithm of craniofacial structure in trisomic mice

E9.5 embryos of Ts65Dn strain display deficit in generation, migration, and proliferation of NCCs (Roper et al., 2009). The alterations affect the developing mandible and associated tissues including Meckel's and hyoid cartilages in

E13.5 embryos resulting in craniofacial abnormalities in neonatal and adult mice (Hill et al., 2007; Richtsmeier et al., 2000). Interestingly, only about half of the genes of Ts65Dn mice are trisomic for HSA21 homologs; however, similarities in structural developments in humans and mice indicate that these trisomic genes can cause similar craniofacial abnormalities at pre- and postnatal ages in humans with DS (Faulks et al., 2008; Guihard-Costa et al., 2006). The small mandibular and normal tongue precursors in trisomic E13.5 embryos having a large NCC component suggested a complex mechanism for development of trisomic phenotypes, including tongue hyperprotrusion as a result of "relative macroglossia" (Guimaraes et al., 2008). The origin of the relative macroglossia was hypothesized to develop with a normal-sized tongue precursor in a small developing oral cavity in individuals with DS, and tissue subsets of NCC for tongue and the mandible contribute through differential regulation of specific gene pathways in trisomic state (Billingsley et al., 2013). Spatial and temporal regulation of Hoxa2 gene has been associated with NCC-derived skeletal elements. A generalized delay in development and size of all tissues, though results in smaller embryonic volume, does not lead to reduction in all developing tissues; however, that could be altered in temporal and tissue-specific manners.

Over 1.5-fold expression of seven of the 15 trisomic genes on MMU16 suggested involvement of these genes early in mandibular development in Ts65Dn mice that resulted in dysregulation of gene expression in homeobox containing and other nontrisomic genes in the trisomic E13.5 mandibular precursor (Billingsley et al., 2013). The PathGen transitive gene pathway generator (http://psoda4.cs.byu.edu/pathgen/pathgen.cgi) (Clement et al., 2010) has exhibited potential parsimonious linkages between trisomic genes of MMU16 that are homologous to those on HSA21 and 15 of the 20 homeobox genes (as well as Sox9). Of the 99 MMU16 genes, 45 were linked to one or more of the nontrisomic homeobox genes leading to an average linkage to six of the nontrisomic genes tested. Trisomic App, Atp5O, Cct8, Dyrk1a, Ets2, Gabpa, Hmgn1, Ifnar1, Ifnar2 Ifngr2, Itsn1, Mx1, Olig2, Runx1, and Tiam1 genes were linked to more than 10 nontrisomic homeobox genes, which may collectively initiate cellular and genetic changes in the trisomic mandibular precursor (Billingsley et al., 2013).

The differential expression of the 20 homeobox genes demonstrated dysregulation in the postnatal day 0 (P0), P15, and P30 trisomic cerebellum (Dauphinot et al., 2005). Variability in gene expression has been reported in trisomic genes, which may confer high phenotypic variability (Ait Yahya-Graison et al., 2007; Saran et al., 2003). Downregulation of six homeobox genes, including Hoxa5 a regulator of cell growth and apoptosis (Chen et al., 2007), b5, and d4 was observed in the trisomic E13.5 mandibular precursor and at P0. Hoxa5 was also downregulated in the cerebellum of adult Ts65Dn mice (Saran et al., 2003). Interaction of Hoxb2 with Sod1 and Hoxa5, b2, and d4 with 16, 28, and 26 trisomic genes respectively was predicted (Billingsley et al., 2013; Zhai et al., 2005). Ectopic expression of Hoxb6 was linked to craniofacial abnormalities in mice (Kaur et al., 1992). In general, Hox genes are well characterized for patterning roles; thus, dysregulation of Hox genes indicated an essential role for multiple DS phenotypes, including mandibular development. Additionally, the combined effects of upregulated genes of HSA21 with a homeobox DNA-binding domain might have a pivotal role in craniofacial development and mandibular precursor in Ts65Dn mice (Billingsley et al., 2013). The En2 and Otx2 genes are expressed at E8.25 and E9.5, respectively, in the mandibular arch, while a smaller jaw and other craniofacial malformations were linked to Otx2 heterozygous mutation (Degenhardt et al., 2002; Kimura et al., 1997; Matsuo et al., 1995). Interestingly, interaction of En1 with the Fgf signaling pathway regulated the differentiation of osteoblasts and signaling in skull development (Deckelbaum et al., 2006) and is expressed in mandibular precursor cells at E13.5 affecting the mandibular development (Zhong et al., 2010).

An elevated expression of Sox9 in E13.5 Ts65Dn embryos may be linked with the delayed proliferation and hypertrophy of the chondrocytes producing smaller Meckel's and hyoid cartilages (formed from chondrocytes) (Billingsley et al., 2013). Further endochondral ossification of Meckel's cartilage could alter hypoplastic mandibular development, indicating a significant role of Sox9 overexpression in Ts65Dn mouse model (Billingsley et al., 2013). Sox9, a transcription factor, regulates chondrogenesis and is necessary for endochondral bone formation (Akiyama et al., 2004; Mori-Akiyama et al., 2003). Its overexpression (~1.2 fold) showed reduced chondrocyte proliferation, interrupted transition of proliferating chondrocytes to hypertrophy, and dwarfism in nontrisomic mice (Akiyama et al., 2004).

Altered expression of trisomic and nontrisomic genes was reported in tissues from pharyngeal arch and heart regions of trisomic E10.5 embryos for MMU17 of HSA21 homologs (Liu et al., 2011) and that suggested the importance of some of the dysregulated genes in the mandibular precursor in trisomic mice. Indeed, trisomic genes may serve as an important initiator of differential gene expression of nontrisomic genes for dysregulated development of DS phenotypes. Lastly, small changes in trisomic gene expression that are not found to be significant might be very important in the resultant trisomic phenotypes. Interestingly, each of the individual DS phenotypes is also seen in individuals without DS to some extent. Thus, the disruption of nontrisomic genes and their genetic pathways influenced by trisomic HSA21 indicate the importance of these pathways in overall phenotypic development.

5. Cardiac defects in Down syndrome

Congenital heart defect (CHD) is one of the several variable phenotypes present in ~40%–61% of individuals with DS but imposes high morbidity and/or infant mortality (Yang et al., 2002; Vis et al., 2009). The atrioventricular septal defect (AVSD) appears as the most common CHD with an incidence of 1000-fold increase in DS individuals compared with non-DS population (Torfs and Christianson, 1998). However, conflicting reports exist on the contribution of HSA21 for CHD observed in DS (Gittenberger-de Groot et al., 2003; Korbel et al., 2009). Ventricular septal defect (VSD) affects ~35% of the patients with DS (Maslen et al., 2006). The hallmark of morphological alteration of an AVSD is described with the presence of a common atrioventricular junction as compared to the separate right and left atrioventricular junction seen in a normal heart. Additional defects of the muscular and membranous atrioventricular septum and an ovoid shape of the common atrioventricular junction have been demonstrated in DS. Inconsistent dimensions of the outlet and the inlet of the left ventricle, with the former being greater than the latter as compared to that of normal population with similar dimensions, are also reported (Craig, 2006). The defect of VSD lies in ventricular septum of the heart and that allows leakage of blood from the left ventricle to the right one causing pulmonary hypertension.

5.1 Morphogenesis of heart defects and contribution of Creld1

Notably, Creld1 has been reported to control the formation of the atrioventricular cushion (AVC), and the progenitor structure of the developing septa that might result in embryonic death at E11.5 due to a malformation in heart development. During AVC morphogenesis between E9.5 and E10.5 in mice, endocardial cells start delaminating within the heart tube and subsequently occupy the extracellular matrix (ECM), acquire mesenchymal properties, and proliferate. This process of endocardial–mesenchymal transformation (EMT) depends on intricate interactions between the endocardium and the underlying myocardium (reviewed in Beckert et al., 2021). Creld1 promotes calcineurin-dependent nuclear translocation of nuclear factor of activated T cells 1 (NFATc1), and, thereby, the Nfatc1 genes play a role in the proliferation of endocardial cells at E10.5 that depends on the vascular endothelial growth factor (VEGF) (reviewed in Beckert et al., 2021). Following completion of AVC formation, the heart is differentiated into the cardiac septa and valves. Despite Creld1 is expressed in both endocardium and myocardium, Nfatc1 translocation happens only in endocardial cells, raising an issue on functions of Creld1 in the endocardium. Beyond cushion formation, the endocardium and myocardium interaction also regulates the development and maturation of the ventricular wall and finally, development of the mature heart. The ventricular wall is made up of the outer wall and the trabeculae, during development, where an endocardial cell layer covers the myocardial cells. Actually, proliferation of the outer cell layer forms the compact ventricular wall, and thus, disturbance in compaction results in severe cardiac defects and ultimately heart failure (reviewed in Beckert et al., 2021). The Notch signaling pathway plays a crucial role in regulation of the trabeculation; however, involvement of Creld1 in regulating the later stages of myocardial development, especially after E11.5, is not uncovered. Studies of Creld1-knockout mice demonstrated that loss of Creld1 function in the endocardium was not essential for heart development, but Creld1 function was indispensable in the myocardium for heart function and survival (Fig. 7.5) (Beckert et al., 2021). Since Creld1 controls modeling of the ECM and trabeculation in the myocardium; its loss-of-function mutation resulted in myocardial hypoplasia. Notably, knockout mice with Creld1-deficient myocardium died shortly after birth. Therefore, it was envisaged that Creld1 controls the development of septa and valves of the heart at an early stage and controls maturation and function of the myocardium at a later stage during development, suggesting the importance of Creld1 in development and function of mammalian heart.

5.2 Genotyping of heart defects in Down syndrome

A partial trisomy of HSA21, a 2.82 Mb region ranging from DSCAM to CBS, has been identified sufficient to cause CHD in population with DS (Korbel et al., 2009). DSCAM acts as a cell adhesion molecule and is articulated during cardiac development. Collagen VI is reported to promote the stability of a heart and skeletal muscle development and is expected to be imbalanced with higher gene dosage in DS, similar to COL6A2 and DSCR1 (Dey et al., 2013). During cardiac development, collagen VI is manifested in the basement membranes of endothelia and in the endocardial cushions. Genetic variation of COL6A1, a telomeric gene lies outside this region, has also been documented to be associated with CHD in DS with trisomy 21 (Davies et al., 1995). COL6A1, encodes the α-1 chain of collagen VI, is overexpressed in the AVCs in the hearts of trisomic fetuses compared with the euploid controls (Gittenberger-de Groot et al., 2003). However, cooperating effects of other genes of HSA21 and genes of disomic chromosomes have been hypothesized to augment the susceptibility to CHD in DS population. In this direction, cooperating effect has been demonstrated to cause ASD in DS with polymorphisms in two folate-pathway genes, such as SLC19A1 (mapped onto HSA21) and MTHFR (mapped onto HSA1) (Locke et al., 2010). Interestingly, significant downregulation of endothelin1 (EDN1) and ryanodine receptor 1 (RYR1), which was attributed to congenital myopathies

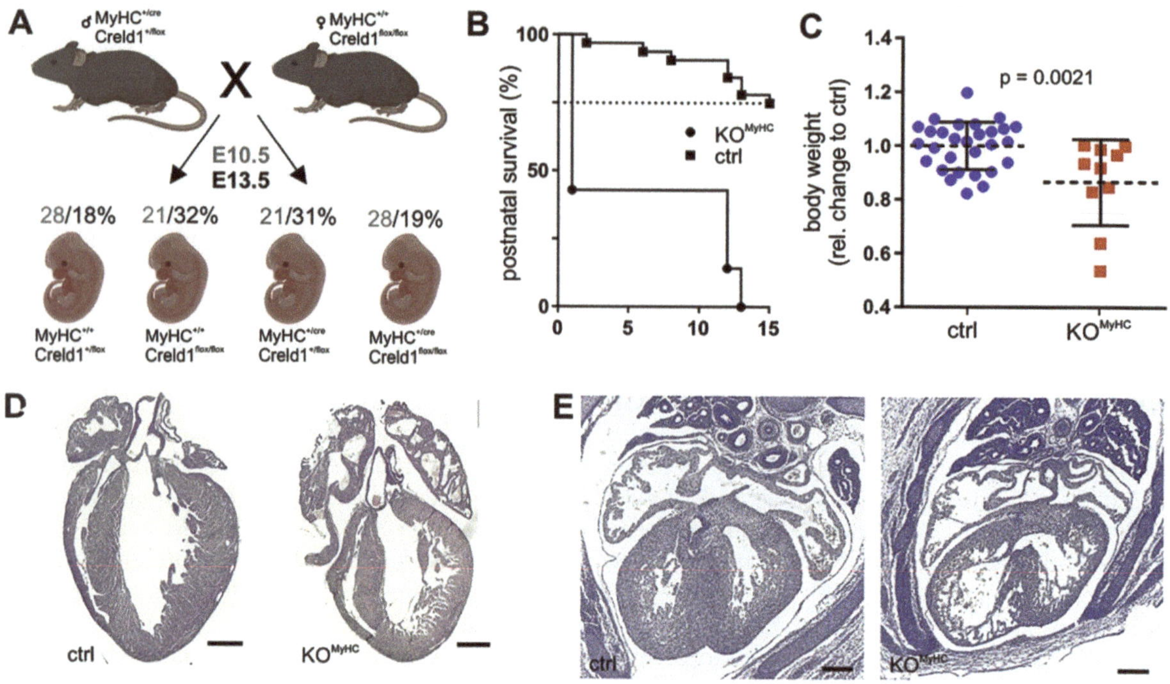

FIGURE 7.5 Creld1 expression in the myocardium is essential for cardiac function and survival. (A) Mating scheme for the generation of control and KO^{MyHC} knockout mice. Percentages of genotypes at E10.5 (gray, n = 53) and E13.5 (black, n = 138) are indicated. (B) Survival rate of wild-type (ctrl) and KO^{MyHC} mice after birth. Days after birth are indicated. (C) Relative body weight of control (ctrl) and KO^{MyHC} mice. Weights have been determined from P0–P13 and normalized to the mean weight of the respective control littermates. Statistical comparison has been performed using Student's t-test (two tailed, unpaired). Data are shown as mean ± SD and individual data points. (D) Hematoxylin-eosin staining of cardiac sections from wild-type (ctrl) and KO^{MyHC} mice at P7. Similar results were obtained from other mice at P0 and P5. Scale bar: 0.4 mm. (E) Hematoxylin-eosin staining of cardiac sections from wild-type (ctrl) and KO^{MyHC} mice at E13.5. Scale bar: 0.2 mm. All images are representatives of at least n = 3 per genotype. *Reprinted from Beckert et al. (2021), with permission from RightsLink Copyright Clearance Center, Inc.*

(Laforgia et al., 2018; Shepherd et al., 2004), was documented in trisomic mesodermal progenitor cells (Galat et al., 2020). Downregulated EDN1 in mesodermal progenitor cells and endothelial cells in DS was speculated to contribute to cardiac dysfunction.

Mutations in nontrisomic and non-HSA21 cell adhesion molecules such as CRELD1 (cysteine-rich EGF like domain1; mapped onto HSA3 at 3p25) gene have been postulated to cause protein misfolding and contribute to and augment the risk of AVSD in DS and also in non-DS family members (Maslen, 2004, 2006). The finding suggested that CRELD1 is not sufficient to cause AVSD but may augment the risk of developing AVSD making it a reasonable candidate or a susceptibility locus (modifier) Other mutant alleles as genetic modifiers such as HEY2, Tbx5, or Nkx2.5, have been described to perturb cardiac development as genetic modifiers in both DS individuals and DS model organisms. Indeed, the role of Creld1 on the myocardium, but not the endocardium, was demonstrated to be essential for development and function of heart and survival through embryonic till early postnatal age (Beckert et al., 2021).

Somatic mutations in HEY2 have been reported in cardiopathic DS individuals but not in cardiopathic euploid populations (Reamon-Buettner and Borlak, 2006), which was further supported by early postnatal fatality of Hey2$^{-/-}$ mice having cardiac abnormalities of septal defects (Gessler et al., 2002). Reduced expression of Hey2 generated through crossing of Ts65Dn mice with a null allele of Hey2 in a trisomic background expressed a deleterious additive effect on septal deformity in Hey2$^{+/-}$ pups (Li et al., 2012). Interestingly, crossing of a double heterozygous Creld1$^{+/-}$ and Hey2$^{+/-}$ mouse to Hey2$^{+/-}$ exhibited membranous VSDs in 9 of 10 pups appeared with homozygous for the Hey2 null allele (which is expected to produce membranous VSD) and heterozygous at Creld1. In addition, one had an ostium secundum ASD and one had a muscular VSD. Thus, reduced Creld1 expression affected the pattern of defects in homozygous Hey2 null mice, indicating an interaction between Creld1 and Hey2 (Li et al., 2012).

CRELD1 encodes a cell surface protein that acts as a cell adhesion molecule for cardiac cushion development, while Hey2 is a nuclear transcription factor; both are commonly expressed in heart during development and also throughout life. CRELD1 has 11 exons at ~12 kb span (Rupp et al., 2002). Two (p.R329C and p.E414K) heterozygous missense mutations in CRELD1 were observed in two and one of 39 DS subjects respectively having complete AVSD, while these mutations were absent in 30 DS individuals without CHD (Maslen et al., 2006; Rupp et al., 2002). Contribution of R329C variant to damage CRELD1 protein structure was demonstrated in an individual with sporadic partial AVSD and in DS individual with AVSD, suggesting involvement of CRELD1 in increasing likelihood of AVSD in DS. CRELD1 with deletion breakpoints on HSA3 has also been associated with AVSDs in people with 3p-syndrome. Additionally, Notch1 was reported to be downregulated in Creld1$^{+/-}$ mice with downregulated Hey2 expression and that suggested a possibility of CRELD1-induced perturbation of Hey2 expression through the Notch pathway (Li et al., 2012). Indeed, Notch1 is a major transcription factor for Hey2, while CRELD1 and Hey2, the two disomic modifiers, demonstrated co-regulation at the transcript level.

Additionally, involvement of Jam2 gene has been identified as a potentiator of the genetic modifier Creld1 in Ts65Dn mouse (Li et al., 2016). JAM2, a cell membrane protein with immunoglobulin-like domains, localized in both large and small vessels at cell-to-cell junctions of the endothelial cells of heart, caused defective angiogenesis in Tc1 mice (Reynolds et al., 2010). Upregulated Jam2 was identified in mouse embryonic stem (ES) cells; however, its downregulation started quickly with the differentiation of ES cells (Cunningham et al., 2000). Overexpression of Jam2 was observed to cause malformation of heart in screening of HSA21 gene effects on early embryonic zebrafish development (Li et al., 2016). JAM2 has been hypothesized to affect endocardial cushion formation, possibly by an indirect mechanism (not by direct physical interaction) through signaling pathways related to a crosstalk between Ts65Dn and Creld1 (Li et al., 2016). Altogether, CRELD1 mutations were demonstrated in ~2.5% (4/159) of complete AVSD cases studied, including heterotaxy-associated AVSD and DS with AVSD (Beckert et al., 2021). It is anticipated that introducing Creld1 or Hey2 mutations into other trisomic DS models having smaller segmental trisomies, such as Ts65Dn/Ms1Rhr, Ts1Cje, and Ts1Rhr, may localize the dosage-sensitive (trisomic) genetic modifiers present on the T65Dn marker chromosome (Olson et al., 2004). Overexpression of subsets of the genes upregulated in Ts65Dn mice have been highlighted in these mice. Importantly, the Ts65Dn mice without AVSD may indicate the fact that their trisomic segment is orthologous only for about half of HSA21 genes. Thus, the trisomic mouse models for HSA21-orthologous segments of MMU17 and MMU10 and for all of the HSA21 orthologs on MMU16 might be able to typify the involvement of dosage sensitive genes to CHD

(Yu et al., 2010). Mouse models with segmental duplications are characterized with seven coding genes, such as Setd4, Mx2, Tmprss2, Ripk4, Prdm15, C2cd2, and Zbtb21 (Lana-Elola et al., 2016).

"CHD critical region" or "CHD loci" has been mapped onto HSA21, one near the RIPK4 and another one includes ZBTB21, following genome-wide association studies (GWAS) (Sailani et al., 2013). The segment (6 kb) residing upstream of RIPK4 has been postulated to contain the regulatory elements for this gene. Overexpression of RIPK4 has been described in the heart tissue of the Ts65Dn mouse model (Lyle et al., 2004). Interestingly, analysis of copy number variants (CNVs) showed no single pathogenetic microrearrangement on HSA21 of DS patients with AVSD (Rambo-Martin et al., 2018). Furthermore, the Sonic hedgehog signaling has been documented to be linked with AVSD in DS and in other genetic syndromes with AVSD (Digilio et al., 2019). Particularly, studies on trisomic cells of cerebral, skin, liver, and intestine origin have exhibited a dysregulated mitogenic hedgehog activity on impairment of cell proliferation caused by overexpression of Ptch1—a receptor that normally represses the hedgehog pathway (Fuchs et al., 2012). Therefore, deciphering the minor or major contribution of multiple risk alleles is suggested to unravel the cumulative effect on the multifactorial development of CHD in DS (Calcagni et al., 2021).

5.3 Development of heart in trisomic DS mouse models

A number of DS models exhibit DS-like common heart defects, which indicated influence of trisomy of one or more of the approximately 100 genes to perturb heart development in the DS models (Williams et al., 2008; O'Doherty et al., 2005; Li et al., 2007). The embryos of transchromosomal model, Tc1 mouse, exhibited a range of cardiac defects, including ASD similar to that seen in humans with DS in high-resolution episcopic microscopy (HREM) and three-dimensional (3D) modeling studies (Dunlevy et al., 2010). Appearance of cardiac defects, not ASD, in the Ts65Dn and Dp(16)1Yey mouse models of DS but not in Dp(10)1Yey or Dp(17)1Yey mice, suggested that dosage-sensitive genes for CHDs lie on MMU16 (Li et al., 2007; Williams et al., 2008; Yu et al., 2010). Contribution of different sets of triplicated genes and/or genetic background differences might explain the absence of ASD in these mice compared with Tc1 mice. Alternatively, the chances of the defects being missed in Ts65Dn or Dp(16)1Yey mice could be apparent due to difficulty in detecting ASD by routine methods. Thus, application of the HREM technique to embryos of Ts65Dn and Dp(16)1Yey strains could trace down the ASD in these models.

Combined analysis of results from partial or segmental trisomy in human and data from mouse models led to hypothesize a "heart critical region" of 1.77 Mb spanning from DSCAM to PKNOX1 responsible for CHD in DS (Korbel et al., 2009; Li et al., 2007; O'Doherty et al., 2005). DSCAM being the only gene of this region expressed in the heart is suggested to be the candidate dosage-sensitive gene for CHDs in DS. However, the Ts1Rhr strain did not exhibit any significant cardiac malformations, though it is trisomic for 33 orthologous genes of the DSCR region, including Dscam. This indicated that three copies of Dscam are not sufficient to cause phenotype of CHD in these mice (Dunlevy et al., 2010; Liu et al., 2011). Additionally, duplication of a 5.43 Mb region of MMU16 from Tiam1 to Kcnj6 in the Dp(16)2Yey strain showed that segment was sufficient to cause cardiac defects (Fig. 7.3), which again excluded Dscam and further narrowed down the location of the causative dosage-sensitive genes (Liu et al., 2011). Nevertheless, further studies are necessary to identity other genes involved in development of heart in people with DS.

HREM and 3D modeling in the Dp1Tyb mouse strain of DS model revealed a marked increase in CHD in E14.5 embryos compared with the wild-type (WT) pups (Fig. 7.6A). The heart morphology in the mutant embryos revealed a range of defects, including abnormal arterial trunk arrangements (overriding aorta (OA) and double outlet right ventricle (DORV) with subaortic communication) and VSD either singly or in combination with other defects in ~18% and 62% of the mutant embryos (Lana-Elola et al., 2016). VSDs observed were either perimembranous (pVSD) (located in the membranous portion of the ventricular septum) or muscular or trabecular (mVSD) (opens to the inlet of the right ventricle) (Fig.7.6C). AVSD appeared with two bridging leaflets across the single AV junction and an "unwedged" morphology of the left outflow tract in ~25% of the Dp1Tyb embryos (Fig. 7.6C). Notably, the display of AVSD was always with a ventricular shunt, never with an atrial shunting, in the Dp1Tyb mice and that indicated the Dp1Tyb strain as a model of CHD seen in DS (Lana-Elola et al., 2016). The Dp1Tyb mouse strain was developed with a duplication region of 23 Mb spanning from Lipi to Zbtb21 on MMU16 and 148 coding genes by engineering the Cre/loxP mediated recombination. Further strains (no.3) were created with contiguous segmental duplications covering the region from Lipi to Hunk (Dp9Tyb), Mis18a to Runx1 (Dp2Tyb), and Mir802 to Zbtb21 (Dp3Tyb). The Dp3Tyb strain was further segmented as Dp4Tyb: Mir802 to Dscr3, Dp5Tyb: Dyrk1a to B3galt5, and Dp6Tyb: Igsf5 to Zbtb21 (Lana-Elola et al., 2016). Interestingly, embryos of both Dp9Tyb and Dp2Tyb showed no significant increase in CHD, whereas the Dp3Tyb strain displayed a significant increase in CHD (pVSD, mVSD, and AVSD) (Fig. 7.7A).

Interestingly, the most common subtype of AVSD in people with DS was reported to be a complete AVSD with both atrial and ventricular shunting, followed by AVSD with exclusive ventricular shunting as the next common subtype and that is similar to that appeared in the Dp1Tyb mouse (Freeman et al., 2008). Indeed, some DS individuals have complete AVSD, whereas others have subtypes with only atrial or ventricular shunting, and thus, AVSD is another variable feature of DS within cardiopathic DS population. Importantly, distinct mechanisms might contribute to defects in the atrial and ventricular septa. In this context, defects in the dorsal mesenchymal protrusion (DMP) have been associated with development of AVSD (Briggs et al., 2012). Therefore, genetic manipulations behind perturbation of DMP would result in AVSD with exclusive atrial shunting, suggesting the critical role of the DMP for the accurate development of the atrial side of the atrioventricular junction (Briggs et al., 2012). On the contrary, AVSD in Dp1Tyb mice was exclusively with ventricular shunting and normal DMP development (Lana-Elola et al., 2016). The plausible explanation behind such differential expression of AVSD was proposed as: the malformation of the DMP is responsible for the defects in the atrial septum causing the ostium primum type of AVSD, while perturbations in other tissues such as endocardial cushions or myocardium might cause ventricular defects. Furthermore, defects in myocardial signaling could lead to AVSD was hypothesized depending on the fact that Gata4-deficient mice displayed AVSD with a normal DMP, however, that showed perturbed myocardial signaling in the adjacent atrioventricular cushion mesenchyme (Misra et al., 2014). Notably, the Dp1Tyb and Dp3Tyb mouse strains were demonstrated with AVSD with exclusive ventricular shunting, suggesting further study to unravel the underlying pathological mechanisms of this subtype of AVSD.

Interestingly, CHD was not manifested in the embryonic hearts from the Ts1Rhr strain, which has a shorter duplication compared with Dp3Tyb strain by eight genes (7 coding genes: Setd4, Mx2, Tmprss2, Ripk4, Prdm15, C2cd2

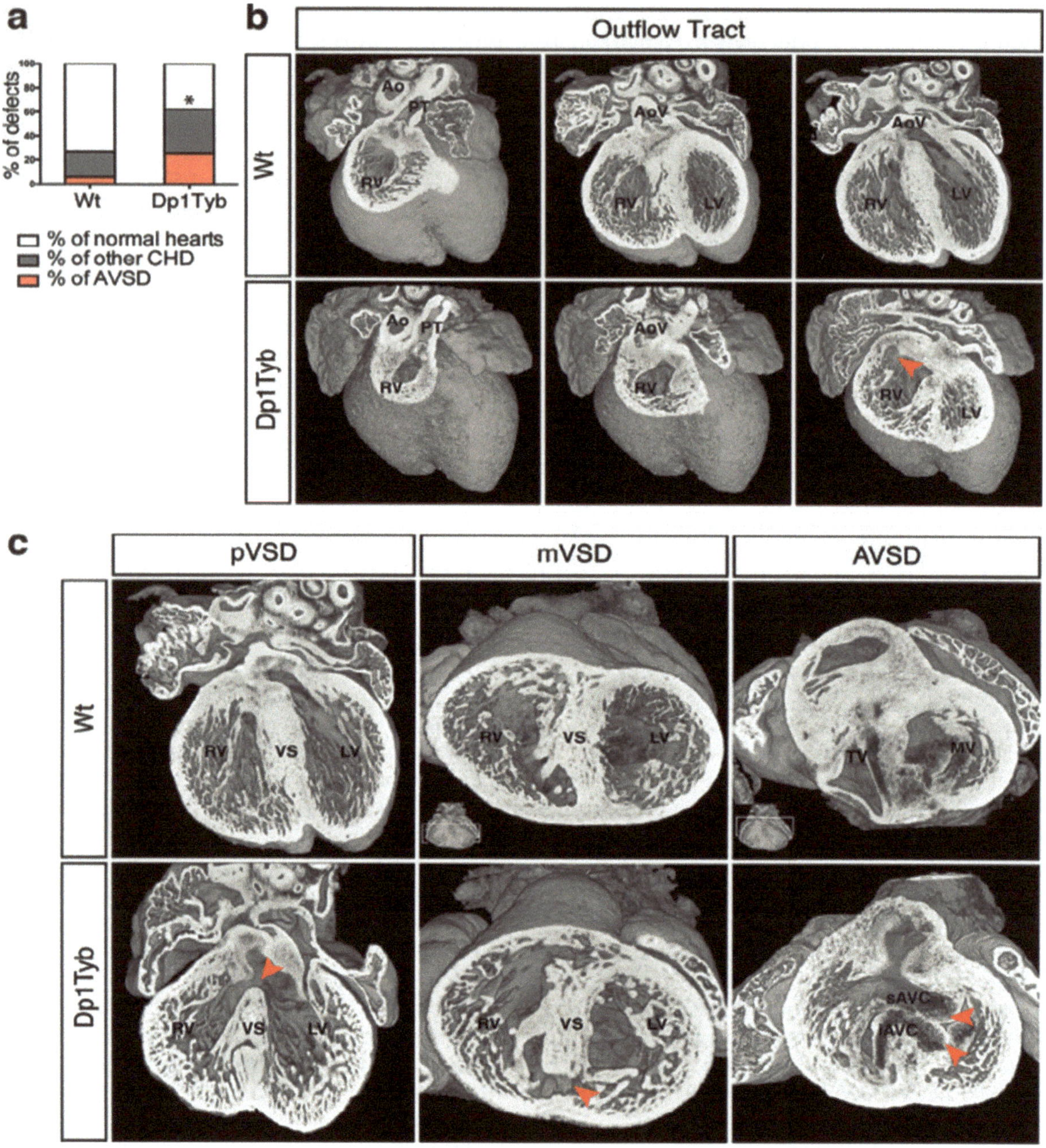

FIGURE 7.6 Dp1Tyb mice recapitulate the congenital heart defect (CHD) seen in Down syndrome (DS). (A) Graph shows the percentage of CHD in Dp1Tyb embryonic hearts at E14.5 compared to wild-type (WT) littermates; n = 26 WT and 39 Dp1Tyb embryos. Significant difference to corresponding wild-type incidence (Fisher's exact test) $*P < .05$. (B) 3D high-resolution episcopic microscopy (HREM) rendering of WT and Dp1Tyb hearts, eroded from anterior to posterior to show the four-chamber view in three successive planes. Outflow tract defects such as double outlet right ventricle are observed in

and Zbtb21, and one micro-RNA gene: Mir802) (Dunlevy et al., 2010; Olson et al., 2004). Also, backcrossing of Ts1Rhr strain to the C57BL/6JNimr genetic background did not reflect enhancement of CHD in the Ts1Rhr strain (Fig. 7.7C), suggesting that the extra copy of the eight genes duplicated in Dp3Tyb is required to cause CHD and that was shorter in Ts1Rhr. Further investigation of Dp4Tyb, Dp5Tyb, and Dp6Tyb mice, which carry duplication covering the whole of the duplicated region in Dp3Tyb mice, revealed no increased incidence of CHD in any of the three strains and that indicated involvement of at least two dosage-sensitive loci in the Dp3Tyb mouse to cause CHD (Fig. 7.7C) (Lana-Elola et al., 2016). Summarily, the three copies of the 4.9 Mb region of MMU16 spanning from Mir802 to Zbtb21 have been demonstrated sufficient to produce DS-like heart defects in mice. Importantly, mapping has indicated that trisomy of two or more loci within this region are required to generate CHD, and at least one of these loci to be located within the eight genes duplicated in Dp1Tyb but not Ts1Rhr mice (Lana-Elola et al., 2016). The duplication at the centromeric end (from Mir802 to Setd4) or at the telomeric end (from Mx2 to Zbtb21) was sufficient to cause CHD in the trisomic Dp3Tyb strain. Notably, a 3.7 Mb genomic region (Ifnar1 toKcnj6) of MMU16 in Dp(16)4Yey, which overlaps by 1.9 Mb from Mir802 to Kcnj6 at the centromeric end of the duplicated region of Dp3Tyb, appeared sufficient to cause CHD (Liu et al., 2013). In one individual, partial trisomy of a 7.7 Mb region from PSMG1 to PRMT2 caused VSD, and a 10.0 Mb region from HLCS to PRMT2 caused AVSD (Korbel et al., 2009; Lana-Elola et al., 2016). Overexpression of DSCAM was involved in causing CHD (Korbel et al., 2009). Conversely, duplication of *Dscam*gene, not being encoded within the minimal Ifnar1-Kcnj6 region, appeared neither necessary nor sufficient to cause CHDs in the Dp6Tyb strain (Lana-Elola et al., 2016). Taken together, the mouse and human data showed that no single gene being absolutely sufficient to cause CHD, but several dosage-sensitive genes might be responsible for this phenotype.

Additional, assessment of the DMP, the mesenchymal structure at the venous pole of the developing heart derived from the second heart field (SHF), revealed abnormalities in the DMP in trisomic human fetal hearts (Briggs et al., 2012), linking malformation of this tissue with the AVSD in DS. For analyzing DMP development in the Dp1Tyb mouse model of DS, two different mouse strains were created: crossing of the Isl1Cre strain to the Rosa26RLacZ reporter mice for identification of cells with Isl1 expression and the Isl1nLacZ strain for elucidation of the prevailing expression of Isl1. Analysis at E11.5 of both strains showed expansion of Isl1-expressing mesenchymal tissue formed DMP in the cavity of the right atrium (Fig. 7.8), and at E12.5, the DMP protruded into the inferior atrioventricular cushion. Notably, both reporters exhibited active expression of Isl1; however, expression of Isl1 reduced at E13.5 and disappears by E14.5, though development of the DMP was still detectable using the Isl1Cre/Rosa26RLacZ fate reporter strain. The E14.5 revealed formation of ventro-caudal buttress by DMP at the core of the atrioventricular junction compacted between the atrial septum and the endocardial cushions and subsequent

Dp1Tyb hearts (*red arrow head* indicates the aortic valve connecting to the right ventricle). (C) 3D reconstructions of WT and Dp1Tyb E14.5 hearts with different types of CHD (*red arrowheads*): perimembranous ventricular septal defects (pVSD), muscular ventricular septal defect (mVSD), atrioventricular septal defect (AVSD). *Ao*, aorta; *AoV*, aortic valve; *iAVC*, inferior atrioventricular cushion; *LV*, left ventricle; *MV*, mitral valve; *PT*, pulmonary trunk; *RV*, right ventricle; *sAVC*, superior atrioventricular cushion; *TV*, tricuspid valve; *VS*, ventricular septum. *Reprinted from: Lana-Elola et al. (2016) under the terms of the Creative Commons Attribution License.*

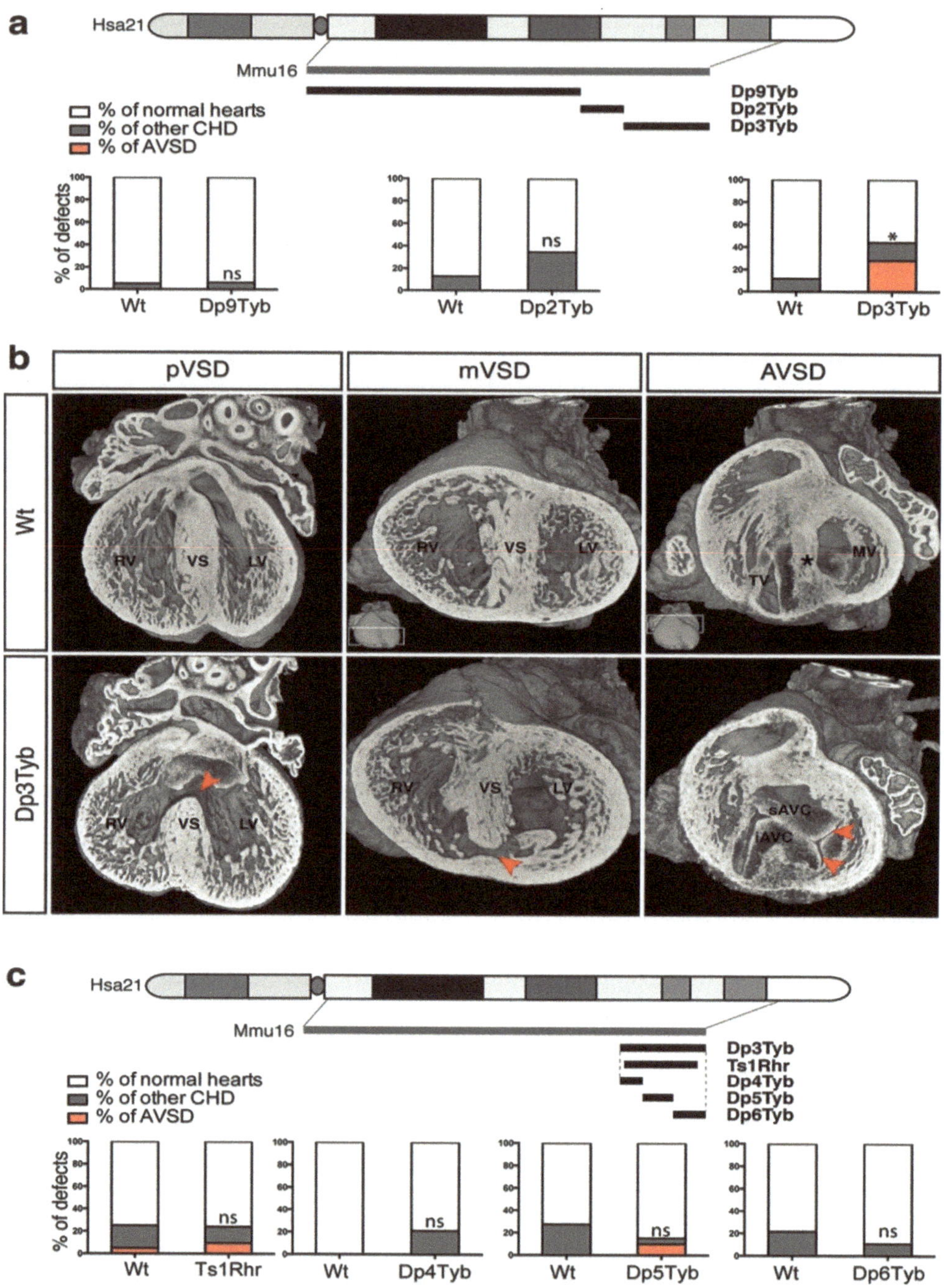

FIGURE 7.7 Genetic dissection of Down syndrome–associated congenital heart defect. (A) Representation of Hsa21 showing region of conserved synteny with Mmu16 in gray and the extent of the duplications in Dp9Tyb, Dp2Tyb, and Dp3Tyb in black. Graphs show the incidence of CHD in each strain; n = 22 WT, 17 Dp9Tyb; n = 16 WT, 26 Dp2Tyb and n = 26 WT, 25

development of the tricuspid and mitral valves (Fig. 7.8) (Lana-Elola et al., 2016). In the Isl1n-LacZ reporter strain, the location, size, shape, and volume of DMP appeared similar at E11.5 and E12.5 in embryos of both WT and Dp1Tyb, though mild reduction in size was noticed in the mutant mice at E12.5 (Lana-Elola et al., 2016).

A comparative analysis of hearts of WT and Dp1Tyb at E11.5 and E12.5 revealed AVSD in Dp1Tyb, despite presence of DMP in the correct position at atrioventricular junction. The result indicated that the present subtype of AVSD might have been caused by defects in cardiac tissues other than DMP. Interestingly, the Dp(16)1Yey mouse strain carrying similar genetic background to that of Dp1Tyb also exhibited heart defects during embryogenesis (Li et al., 2007). Also, similar cardiac phenotypes were demonstrated in Dp1Tyb and the transchromosomic Tc1 strain, even though the gene content of HSA21 ortholog and the origin of the duplicated genes (mouse in Dp1Tyb mice, human in Tc1 mice) were different in the two strains (Dunlevy et al., 2010). At weaning, 50% of the E14.5 Dp1Tyb embryos recovered, while 25% decrease in recovery was reported with the Dp3Tyb embryos. The severity of loss could be associated with severity of CHD and AVSD (Lana-Elola et al., 2016).

Summarily, heart development, per se, is a complex and multifactorial mechanism, where a complete genetic explanation, including genetic background (modifiers) has remarkable influence on both the nature and incidence of heart defects for a given mutation. Also, the changes in relevance to timing or location of the malformation in the heart of DS population may depend on incomplete penetrance and variations in genetic background and possible variations at non-HSA21 disomic chromosomes and may collectively act as genetic modifiers in conjunction with trisomic HSA21 genes. Variants of genetic modifiers, including CRELD1, CRELD2, GATA4, GATA5, FBLN2, FRZB, and BMP4 have been documented as susceptibility factors of AVSD (Maslen et al., 2006). In this relevance, genetic background of different mouse models has exhibited versatility of CHD as an ascertainment bias, as highest mortality rate was observed with C57BL/6 or 129 Sv/J inbred backgrounds (Li et al., 2012). Additionally, overexpression of several microRNAs (miRNAs) of HSA21 or other disomic human chromosomes have been reported in DS, including miR-155, miR-802, miR-125b-2, let-7c, miR-99a, and the miR-99a/let-7c cluster (Bras et al., 2018). Altogether, a large heterogeneity of genetic predisposing factors, not only within the extra HSA21 but at global genomic level, has been documented in molecular and experimental characterization of anatomic cardiac tissues in DS. Thus, it is imperative that co-operating effects of multiple genes are deeply implicated in cardiogenesis and complexity of CHD in the setting of DS and other genetic syndromes and also in nonsyndromic patients.

6. Cognitive development and learning and memory deficits

Every DS child is invariably affected with cognitive impairment but to a variable extent. Speech and expressive language skills are also delayed in most DS children, which are often encountered with their difficulty in expression.

Dp3Tyb embryos. Significant difference to corresponding wild-type incidence (Fisher's exact test) *$P < .05$. *ns*, not significant. (B) 3D reconstructions of WT and Dp3Tyb E14.5 hearts with different types of CHD (*red arrowheads*) as described in Fig. 7.6C. (C) Representation of Hsa21 showing region of conserved synteny with Mmu16 in gray and the extent of the duplications in the indicated strains in black. Graphs show incidence of CHD. n = 20 WT, 21 Ts1Rhr; n = 15 WT, 19 Dp4Tyb; n = 11 WT, 20 Dp5Tyb, and n = 28 WT, 19 Dp6Tyb embryos analyzed. *ns*, not significant (Fisher's exact test). *iAVC*, inferior atrioventricular cushion; *LV*, left ventricle; *MV*, mitral valve; *RV*, right ventricle; *sAVC*, superior atrioventricular cushion; *TV*, tricuspid valve; *VS*, ventricular septum. *Reprinted from Lana-Elola et al. (2016) under the terms of the Creative Commons Attribution License.*

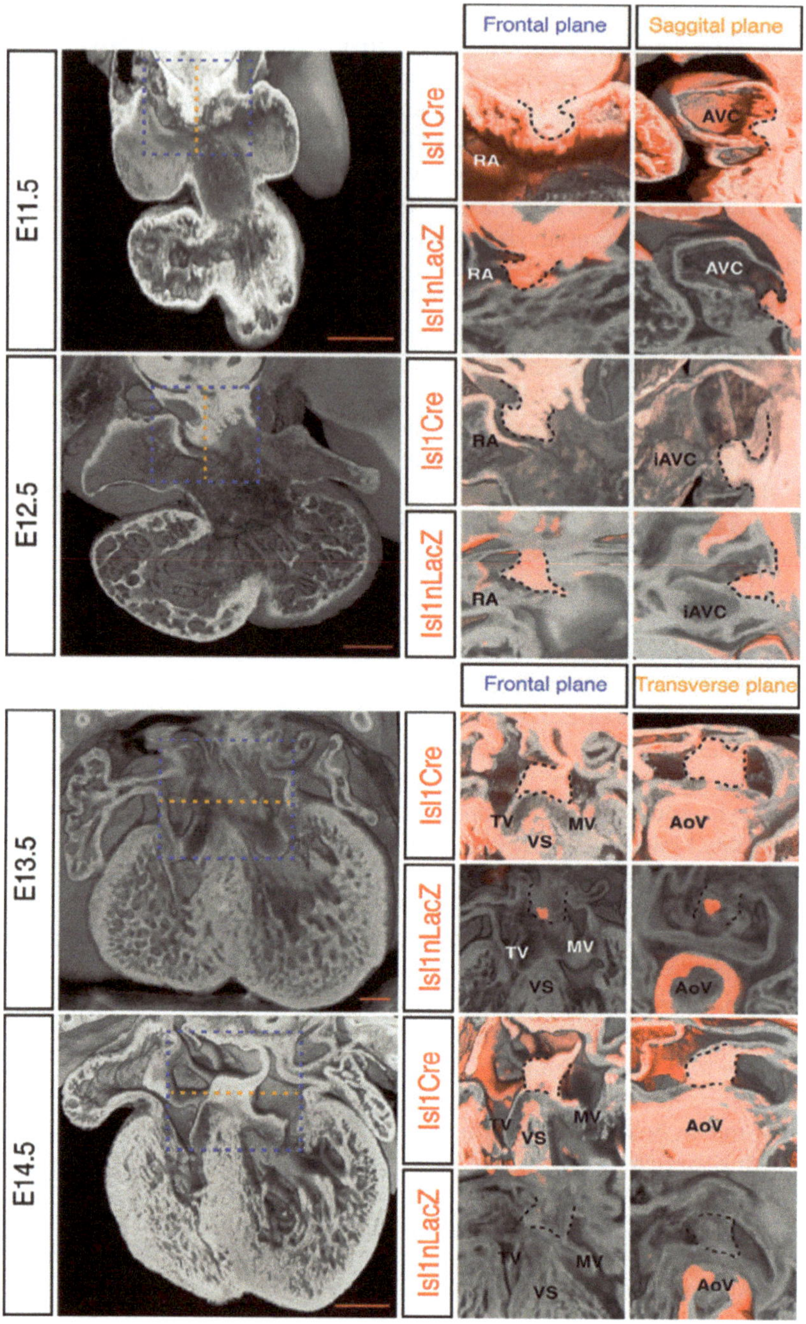

FIGURE 7.8 Development of the dorsal mesenchymal protrusion. Left panels show a series of 3D four-chamber views of hearts at embryonic stages E11.5, 12.5, 13.5, and 14.5. Middle panels show a close up of the AV junction (frontal plane view corresponds to *blue dashed box* in the left panels) in the Isl1Cre/Rosa26RLacZ (Isl1Cre) and the Isl1nLacZ strains; β-galactosidase expression is pseudocolored in red. Right panels show a sagittal view of the AV junction in E11.5 and E12.5 hearts and a transverse view in E13.5 and E14.5 hearts; sagittal and transverse planes shown as orange dashed line in left panels; β-galactosidase expression is pseudocolored in red. In the middle and right panels, the DMP is marked with a black dashed contour. The number of biological replicates used for each of the developmental stages, E11.5, E12.5, E13.5, and E14.5 respectively, were: n = 2, 2, six and four for the Isl1Cre/Rosa26RLacZ and n = 10, 9, seven and five for the Isl1nLacZ. *AoV*, aortic valve; *AVC*, atrioventricular canal; *iAVC*, inferior atrioventricular cushion; *MV*, mitral valve; *RA*, right atrium; *TV*, tricuspid valve; *VS*, ventricular septum. Scale bar, 200 μm. *Reprinted from Lana-Elola et al. (2016) under the terms of the Creative Commons Attribution License.*

An in-born delay in fine and gross motor skills may contribute to cognitive development (Patterson et al., 2013). Taken together, the individuals with DS vary considerably in their language and communication skills and that directs the necessity of mainstreaming education (special education) of DS children. These children may not age emotionally/socially and intellectually at the same pace compared with non-DS children and their chronological peers. The cognitive deficits in DS children differ from that seen in other syndromes characterized with mental retardation, such as Fragile X or Williams Syndrome (Dierssen and Ramakers, 2006). Moreover, despite having high levels of social intelligence with DS people, their impaired expressive communication and explicit memory affect the quality of life, which is often exacerbated by sleep apnea.

6.1 Morphogenesis of cognitive impairment

Hypothesis derived from the mouse models of DS indicated that defects in pathways, including neurogenesis, synaptogenesis, synaptic transmission, and cell signaling, might all potentially contribute to this phenotype, by producing disproportionate level of inhibitory neurotransmission (Korenberg et al., 1994; Siarey et al., 2005; Chakrabarti et al., 2007). Notably, antagonists of γ-aminobutyric acid (GABA) reduced the inhibitory action of neurons and reversed the cognitive deficit in DS mouse models (Fernandez et al., 2007).

Interestingly, the cognitive performances are sub-served by structures of the prefrontal cortex, hippocampus, and cerebellum that mature relatively late in development. Of these, hippocampal functions have been demonstrated to be profoundly impaired, including exploration, context learning, and spatial cognition. By 18 months of postnatal age, abnormalities in the development of dendritic spines, including decreased spine density, have been reported along with either larger or thinner than normal spine underlying alterations in neuronal and synaptic plasticity and finally affecting the cognitive ability (Torres et al., 2018; Valbuena et al., 2019). Dendritic abnormalities are also seen in mouse models of DS (Benavides-Piccione et al., 2004; Haas et al., 2013), with significantly fewer thin spines at 3 weeks and 3 months of age in Ts1Rhr and Tc1mice, respectively. Remarkably fewer mushroom spines, the morphology associated with synaptic inputs and learning and memory, was accompanied by a significant increase in the number of stubby spines, and that suggests important contribution of dendritic spine abnormalities to cognitive deficits in DS models, rather than defects in the overall neuronal architecture. Increase in spine size throughout the brain and decrease in spine density were observed in the hippocampus. Such abnormalities in synaptic plasticity (i.e., long-term potentiation: LTP) seen in mouse models are consistent with the memory deficits present in human DS. Trisomic gene dosage of HSA21 may lead to up- or down-regulation of a number of synaptic proteins, those are critical to normal spine development and synapse formation, in DS patients.

A key question on contribution of the defects in synaptic development as initiators of the development of the DS cognitive phenotype, or as a secondary consequence of other, is unanswered. Global changes in circuit activity, prior to synaptogenesis, may lead to synaptic abnormalities, including reduced neuron numbers (of cerebellar granule cells and cortical interneurons) have been observed during fetal development in humans. In mouse models, abnormalities in the timing of neuronal cell birth in the cortex and in the development of thalamo-cortical axons indicate the precise sequence of developmental abnormalities in DS. Creation of better mouse models and/or developing simple in vitro assays may elucidate the nature of the synaptic dysfunctions in DS population. Also, knowledge on synaptic defects outside the hippocampus in the Ts65Dn mouse require further studies on the behavioral phenotypes in mouse model of DS. Systematic assay of other major brain areas may capture an overall global picture of synaptic function in these DS animals. Defects in transmission at individual synapses may likely perturb the neurophysiological and neuroanatomical systems, resulting in cognitive deficits in DS. The Ts65Dn mice have presented a defect in LTP at hippocampal synapses. LTP has been plausibly hypothesized to underlie memory formation, which is significantly impaired in DS patients.

Involvement of synaptic vesicle trafficking is largely controlled by the molecules of HSA21, which is triplicated in DS, and thus, gene dosage imbalance might cause deficits in trafficking, spine morphology, and synaptic transmission. Additionally, enlarged endosomes, appear at 28 weeks of gestation, constitute one of the earliest pathological hallmarks in human DS and in the Ts65Dn mice. A number of synaptic proteins may be up- or down-regulated in DS that are critical to normal spine development and synapse formation. However, information is limited concerning the level of expression and function of the trafficking proteins in tissues of human DS or mouse models. Thus, the synaptic vesicle trafficking system could be focused for analyzing the functional impact of dysregulated proteins, including the expression of other genes involved in the same molecular pathways.

6.2 Genotype of learning and memory deficits

Learning and memory impairment is an obligatory phenotype noticed in all DS individuals; however, its severity varies greatly among the DS patients (Lott and Dierssen, 2010). Besides understanding the underlying genetic consequences of trisomy 21, environmental and epigenetic factors and nontrisomic genes might also interact and contribute to variability in cognitive function and development in individuals with DS (Fig. 7.1). Indeed, the cellular and molecular patho-biology in conjunction with functions of trisomic HSA21 genes may unravel the most mechanistic context of the obligatory and variable features of DS. In addition, the defects in cognitive development become more pronounced with early-onset dementia of the AD by early middle age. Overlapping of cognitive disability with dementia further adds to variability of this phenotype. As such, the age of onset of dementia varies from 40 years to 60–70 years. Indeed, the variable penetrance of the dementia, and dementia plus cognitive phenotype has yet to be discerned with the genetic consequences and biology of AD in DS.

Studies of partial trisomy of HSA21 demonstrated that genes of multiple regions in HSA21, especially the proximal portion of 21q, involved in several pathways, might be responsible for cognitive deficits in DS (Korbel et al., 2009; Lyle et al., 2009). Arguably, a trisomic region from 37.94 to 38.64 Mb in 21q (Lyle et al., 2009) or trisomy of the most telomeric segment (4.6 Mb) of the HSA21 (Korbel et al., 2009) caused the maximum IQ deficit in DS; however, contribution of this region for defects in learning and memory was contradicted by the analysis of mouse cross-breeds having orthologous genes on the MMU10 and MMU17 (Duchon et al., 2010; Yu et al., 2010). In reality, the Ts1Yah mouse strain having trisomy of 12 genes located on HSA21 orthologous chromosome MMU17 indicated improved performance in the Morris water maze (a test used to assess spatial learning) compared with the euploid mice, and both Ts1Yah and Dp(17)1Yey mouse strains showed increased hippocampal LTP. LTP is a form of synaptic plasticity that is suggested to be the physiological basis of learning and memory (Pereira et al., 2009; Yu et al., 2010).

The contribution of triplicated DSCR on neurological defects was not linked to intellectual disability in humans with partial trisomy of HSA21, though this study did not exclude its connection with this phenotype in DS (Korbel et al., 2009; Lyle et al., 2009). Studies involving mouse models with or without trisomy of DSCR showed that trisomy of this region was necessary but insufficient to reflect this defect in Morris water maze test (Olson et al., 2007). However, the novel-object-recognition (NOR) test, a different assay of learning and memory, demonstrated the potential of the triplicated DSCR to cause cognitive defects (Belichenko et al., 2009). Thus, more studies are necessary to determine the contribution of the trisomic DSCR for learning disability. Nevertheless, genes including DYRK1A, SIM2, DSCAM, and KCNJ6 located in the DSCR have been projected as aspirant dosage-sensitive genes associated with brain phenotypes in DS population.

DYRK1A, expressed in the developing and adult nervous system (Hammerle et al., 2011), potentially inhibits the cell proliferation and promotes premature neuronal differentiation by regulating the NOTCH signaling (Hammerle et al., 2011). DYRK1A inhibits the nuclear translocation of the NFAT family of transcription factors in association with the DSCR1 protein, also known as RCAN1, encoded on HSA21. This result led to hypothesize that excess inhibition of the NFAT pathway might contribute to both neuronal and cardiac defects in DS (Arron et al., 2006).

Severe learning difficulties and spatial memory deficits were evidenced in transgenic mice holding a 180 kb yeast artificial chromosome (YAC) that includes DYRK1A, indicating the potential of overexpressed Dyrk1a alone in mice (Smith and Rubin, 1997; Altafaj et al., 2001; Ahn et al., 2006). Extracellular hippocampal recordings exhibited altered induction of LTP and long-term depression (LTD; indicates the selective weakening of synapses during the learning and memory process) in transgenic mice with overexpression of Dyrk1a; however, the alterations of LTP and LTD appeared in opposite direction relative to that observed in other more complete trisomic mouse models, such as Ts65Dn and Ts1Cje (Kleschevnikov et al., 2004; Ahn et al., 2006). Nonetheless, these results indicated that DYRK1A might alter the balance between excitatory (glutamatergic) and inhibitory (GABAergic) transmission. Interestingly, small segmental duplications in HSA21 revealed that trisomy of DYRK1A do not cause severe impairment in IQ levels (Korbel et al., 2009).

The Drosophila single-minded transcription factor (Sim2), a master regulator of development, and orthologous of human single minded (SIM2), is expressed in the developing human brain (Rachidi and Lopes, 2011). Overexpression of Sim2 in transgenic mice caused mild learning and memory impairments (Ema et al., 1999); however, no defect was observed in transgenic mouse developed using a bacterial artificial chromosome (BAC) in which a physiologically more relevant level of overexpression of Sim2 occurred from its endogenous promoter (Chrast et al., 2000). Surprisingly, SIM2 may induce repression of Drebrin expression through binding to its promoter directly (Ooe et al., 2004). Drebrin is diminished in the cortex of sporadic and DS-AD patients and affects the dendritic spine structure and neuritogenesis (Shim and Lubec, 2002; Geraldo et al., 2008).

Overexpression of the adhesion molecule, DSCAM, is reported in the hippocampal neurons in vitro and hampers branching of dendrite (Alves-Sampaio et al., 2010). Within the neuronal dendrites, translation of DSCAM protein is regulated by NMDA-receptor (N-methyl-D-aspartate receptor, a glutamate, and ion channel protein receptor) signaling, an important component of learning and memory, and is likely to facilitate synaptic plasticity. Interestingly, triplicated Dscam has affected its local translation in neurons of the Ts1Cje mouse (Alves-Sampaio et al., 2010).

In Ts65Dn mouse model, overexpression of Kcnj6 (GIRK2; G protein-activated inward rectifying potassium channel subunit 2) in the hippocampus caused elevation of GIRK channel density in the hippocampal neurons along with increase in GIRK current in response to inhibitory GABA-B (γ-aminobutyric acid-B receptor) signaling (Best et al., 2007; Harashima et al., 2006). Indeed, an upregulated gene dosage of both Kcnj6 and Dyrk1a together might enlighten the synaptic phenotype in Ts1Rhr mice (Belichenko et al., 2009).

Genes located outside the DSCR have also been connected with DS-specific neurological phenotypes. An association of the transcription factors, Olig1 and Olig2 encoded on HSA21, has been demonstrated in neurogenesis and oligodendrogenesis (Takebayashi et al., 2000; Lu et al., 2002; Zhou and Anderson, 2002). Subtraction from three to two copies of both Olig genes has rectified the overproduction of inhibitory interneurons along with the associated elevation of the inhibitory neurotransmission in the forebrain of Ts65Dn mouse model; however, its subsequent effect on improvement of behavioral performance could not be established (Chakrabarti et al., 2010). Such protocol of reduction of copy number of the specific dosage-sensitive genes involved in causing DS-specific phenotypes was considered for understanding genotype-phenotype association.

SYNJ1 (synaptojanin 1 mapped onto HSA21) gene is involved in dephosphorylation of the lipid phosphatidylinositol (4,5)-bisphosphate [PtdIns(4,5)P2] and plays a key role in normal neurotransmission, where an altered PtdIns(4,5) P2 metabolism affected the brains in Ts65Dn mice. Interestingly, reduction of Synj1 from three to two copies could normalize the defective process (Voronov et al., 2008). Notably, overexpression of Synj1 resulted in defects in learning and memory in transgenic mice (Voronov et al., 2008).

Trisomic amyloid precursor protein (APP) leads to elevation of amyloid and Aβ proteins contributing to neurological phenotypes of DS and facilitates early onset of Alzheimer disease (EOAD) and dementia in human. Proteolysis of the dosage-sensitive gene APP generates the main constituent of amyloid plaques (amyloid-β or Aβ) in brains and that underlie Alzheimer disease (AD). Mutations, duplications, or triplication of APP have been associated with EOAD, both in sporadic and DS cases (McNaughton et al., 2010). Degeneration of basal forebrain cholinergic neurons (BFCNs) has also been reported in both humans with AD (Lehericy et al., 1993) and in the Ts65Dn mouse of DS model (Holtzman et al., 1996). Neurons in Ts65Dn mice exhibited a defect in traditional transport of the neurotrophic nerve growth factor (NGF) to the cell soma, and this might cause their degeneration (Cooper et al., 2001). Partial rescue of the affected NGF retrograde transport and BFCN degeneration was possible in Ts65Dn mice by subtracting from three to two copies through cross breeding (Salehi et al., 2006). Interestingly, overexpression of APP alone was sufficient to cause defect in NGF transport but insufficient to cause degeneration of BFCN (Fig. 7.9). Taken together, these events indicated that upregulation of APP may contribute to degeneration of BFCN in DS, but cooperating effect of other HSA21 genes results in manifestation of AD phenotype.

Phosphorylation of APP, mediated by DYRK1A protein, may identify DYRK1A gene as one of the contributing factors (reviewed in Lana-Elola et al., 2011). In this line, higher levels of phosphorylated APP and Aβ, activated by the overexpressed Dyrk1a, were noticed in transgenic mice (Ryoo et al., 2008). Additionally, it was hypothesized that DYRK1A might phosphorylate Tau and contribute to the AD-like pathology, directing further phosphorylation by

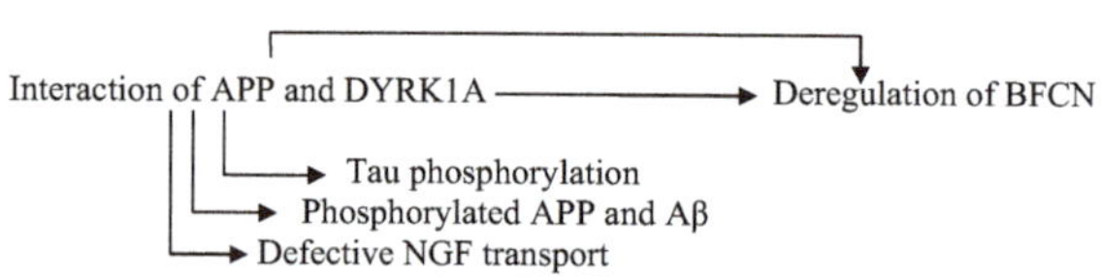

FIGURE 7.9 Schematic explanation of the interaction of DYRK1A and amyloid precursor protein.

glycogen synthase kinase 3β, and following aggregation of Tau into neuro-fibrillary tangles in brain (Liu et al., 2008). Studies on partial trisomies of HSA21 showed consistent result suggesting a role of increased dosage of APP but not of DYRK1A in the EOAD described in DS (Korbel et al., 2009).

Notably, a decreased level of APP metabolites was noticed in Ts65Dn mice against use of selective inhibitors, and application of inhibitors improved the learning and memory in these mice. Indeed, that confirms involvement of APP for this DS-specific phenotype (Netzer et al., 2010). However, segmental trisomies could not establish a direct association of APP on learning and memory observed in humans (Rovelet-Lecrux et al., 2006; Korbel et al., 2009).

6.3 Proteins linked to cognitive performance

Involvement of highly deregulated growth factor signaling proteins, including components of the FGF, VEGF, and EGF pathways, has been demonstrated in people with trisomic HSA21 with both downregulation (VEGF; FGF1 (ligand of FGFR1); EGF-signaling cascade: EGFR, ERBB3, ERBB4, and its ligand HBEGF, and upregulation (VEGFA-ligand of VEGF); FGFR1) of the factors and their ligands (Sullivan et al., 2017). Indeed, 17 of the 75 growth factor signaling proteins were linked to regulation of neurogenesis, such as downregulation of EGFR, ERBB4, NTRK2, NTRK3, BDNF, SLITRK383, and RET. Interestingly, some of these factors are essential for normal GABA-ergic synapse development, where downregulation and altered GABA-ergic signaling have been demonstrated to cause cognitive deficits in individuals with DS (Contestabile et al., 2017; Sullivan et al., 2017). Mutations in RET is linked to neural crest development and Hirschsprung disease, and these are variable phenotypes but reported in DS. Nevertheless, the proteomic datasets have collected significant information on dysregulation of several crucial signaling cascades.

In addition, deposition of amyloid plaques in the brain of DS patients owing to defects in processing and transport of APP causes AD in DS patients, which is identical to that of sporadic AD. Notably, both APP and one of its processing enzymes (BACE 2) are encoded on HSA21; thus, triplication of these components may worsen the patho-mechanism of AD in DS (DSAD). Indeed, the possible impact of misprocessing and mistrafficking of APP and the consequences of neuronal function of overexpression on DSAD may lead to targeted therapeutics for DS. Additionally, mitochondrial dysfunction has been associated with amyloid deposition in AD of DS patients and mouse models (detailed in Chapter 8). Thus, antagonism of mitochondrial dysfunction could also be aimed as key therapeutic goal.

Notably, the complement system, mediated by the activation of classic complement cascade involving β-amyloid deposits and soluble B2M, contributes to AD in people with DS (Nybo et al., 1999). Also, microglia-mediated synapse destruction by a number of the complement subunits was described in progression of AD (Hong et al., 2016). Thus, it is apparent that hypocomplementia could modulate AD progression in DS individuals.

7. Genetics of leukemia in Down syndrome

The genes of HSA21 increase the risk of leukemia by 10-50-fold in individuals with DS compared with the non-DS population (Hasle et al., 2000; Ganguly and Kadam, 2016). Notably, DS is associated with a 500-fold increased risk of acute megakaryocytic leukemia (AMKL) (Al-Kasim et al., 2002) and 20-fold increased risk of acute lymphoid leukemia (ALL) of B-lineage (Hasle et al., 2000). A transient myeloproliferative disorder (TMPD) affects ~10% of the newborns with DS, which is characterized by a clonal population of megakaryoblasts in the blood. TMPD generally resolves spontaneously, still it is transformed into AMKL in 10%–20% of the DS children by 4 years of age. Interestingly, AMKL in DS is always associated with a loss-of-function mutation in the X-linked transcription factor GATA1 gene with the triplicated set of HSA21 genes. This indicates the cooperating effects of GATA1 and triplicated HAS21 for the development of AMKL in DS (Hitzler and Zipurskey, 2005; Saida, 2016). The GATA1 mutations always have the same effect in TMPD and AMKL, leading to translation at the second ATG of the coding region and production of a shorter protein, called GATA1s. Seemingly, trisomy of HSA21 perturbs hematopoiesis, making megakaryocyte-erythroid progenitors (MEPs) vulnerable to the effects of GATA1s, and thereby, promoting development of TMPD. Trisomy of HSA21 alone has indicated an expansion of the MEP population in fetal livers from human DS abortuses, even in the absence of GATA1s (Khan et al., 2011). Thus, progression of this disease occurs through multistep leukemogenesis mechanism, including cooperating genetic events and complex relationship between trisomy of HSA21 and non-HSA21 genes (such as GATA1); however, the exact contribution of mutations remains obscure. Leukemia in a DS child appeared with complex karyotype with multiple chromosomal rearrangements (Ganguly et al., 2017).

7.1 Megakaryopoiesis and transient myeloproliferative disorders in Down syndrome

Somatic mutations in the GATA1 occur in most of the AMKL and almost all TMPDs in individuals with DS; however, GATA1 mutations are not reported in other DS leukemias and AMKL of euploids (Crispino, 2005). Apparently megakaryoblasts are subjected to the increased sensitivity produced by trisomy 21 to GATA1 mutations (Fig. 7.10A). Interestingly, these same mutations do not affect the cells of euploid individuals, unless HSA21 becomes triplicated in those cells. Thus, the frequency of GATA1 mutations in megakaryoblasts having trisomic HSA21 and in euploid blasts remains to be elucidated further. To delineate the mechanism of leukemic transformation, exome and targeted resequencing of 111 TMPD and 141 myeloid leukemia samples with DS revealed presence of truncating mutations in GATA1 in TMPD with trisomy 21, whereas clonal and subclonal variants were detected in DS-myeloid leukemia (Labuhn et al., 2019).

Conflicting reports exist on the expression of trisomic HSA21 genes in DS and co-occurrence of trisomy 21/GATA1 mutant hematopoietic cells. Several studies using induced pluripotent stem cell (iPSC) lines derived from individuals with DS have demonstrated that trisomy 21 alone is enough to alter hematopoietic differentiation in myeloid, erythroid, and megakaryocyte lineages (Banno et al., 2016; Chou et al., 2012; Maclean et al., 2012). Another report indicated an increase in hematopoietic stem cells (HSCs), megakaryocyte-erythroid progenitors (MEPs), megakaryocytes, and megakaryocyte-erythroid potential in human fetal liver hematopoiesis having trisomy 21 (Roy et al., 2012). Conversely, a GATA1 mutation rendered a more pronounced megakaryopoiesis independent of the trisomy 21 in iPSCs along with a marked reduction in erythropoiesis indicating requirement of a full-length GATA1 for development of this lineage (Byrska-Bishop et al., 2015; McNulty and Crispino, 2016). Banno et al. (2016) noticed a much larger proportion of CD34+, CD41+, and immature megakaryocytes in co-occurrence of GATA1s and trisomy 21 in the iPSCs than that lacking a GATA1 mutation. Upregulation of both the long and short GATA1 isoforms was described by Tunstall-Pedoe et al. (2008) in human fetal liver samples with trisomy 21.

A recurrent gain-of-function mutation in the myeloid cytokine receptor CSF2RB was demonstrated, in addition to the loss-of-function mutations in 18 different genes in TMPD of murine models, which were similar to DS-myeloid leukemia. Mutation in Janus kinase 3 (JAK3) was observed in a small proportion of TMPD/AMKL patients with DS (Khan et al., 2011; Kiyoi et al., 2007). However, both loss- and gain-of-function mutations of JAK3 in this group of patients debarred its consideration for tailoring targeted therapy for DS leukemia. The genetic affairs involved in ALL with DS are more obscure than that of AMKL of DS, which may include chromosomal abnormalities, gain-of-function mutations in JAK2, and sub-microscopic deletions of genes such as ETV6, CDKN2A, and PAX5 (Kearney et al., 2009).

Mapping of an 8.35 Mb segment in partial trisomy indicated involvement of RUNX1, ETS-gene family such as ERG, ETS2, and FLI1, and CBS (cystathionine beta synthase) in AMKL (Hitzler and Zipurskey, 2005; Korbel et al., 2009; Margueron et al., 2008). RUNX1, encodes as a transcription factor, plays a crucial role for megakaryopoiesis and maintenance of hematopoietic stem cells and thus, predisposes AMKL in DS (Roy et al., 2012). Two- to three-fold increased expression of genes, including the transcription factors RUNX1, ETS2, and ERG, that are clustered within a 4 Mb region on chromosome 21, were reported in trisomiciPSCs, where deletion of that 4 Mb region reversed most of the hematopoietic phenotypes of the trisomic cells and also suppressed the increased expression of GATA1s (Banno et al., 2016). Their study has linked the trisomy of RUNX1, ETS2, and ERG in expanded hematopoiesis and aberrant megakaryopoiesis of TMPD and further demonstrated that the concomitant upregulation of GATA1s stimulates the disease. However, the mechanism that the extra 21 contributes to increase the production of GATA1s, and the contribution of increased GATA1s, as a major contributor or a marker, remain uncovered for aberrant megakaryopoiesis (McNulty and Crispino, 2016). Further understanding on the co-operating association of genes of HSA21, elevated expression of GATA1s and other tertiary mutations such as mutations of cohesin complex, EZH2, and RAS-pathway on leukemogenesis in DS would be important for targeted drug development.

7.2 ALL in Down syndrome (DS-ALL)

Somatic mutations or rearrangements of CRLF2 (cytokine receptor-like factor 2) or genes involved in other lymphocyte differentiation factors, detected in a large number of DS-ALL, initiate tumorigenesis for ALL in DS (Mullighan et al., 2009; Hertzberg et al., 2010). The driver mutations in JAK2 or RAS (NRAS and KRAS) genes, which co-occur with CRLF2 and other primary driver mutations, but which are often characterized by smaller variant allele frequency, act as secondary factor for progression of the disease (Fig. 7.10B). JAK2 and RAS mutations are mutually exclusive and present in ~1/3rd of DS-ALL (Nikolaev et al., 2014). The secondary driver mutations switch between primary tumors and relapse, such as JAK2 or PTPN11 to RAS, but the primary driver mutations remain unaltered (Nikolaev et al., 2014). DS-ALL children have poorer survival and higher incidence of relapse. Interestingly, these driver mutations occur in non-HSA21 genes and thus, raise question on the role of trisomic HSA21. The transgenic mouse Ts1Rhr (DS model) showed triplication of a genomic interval that contains 31 genes orthologous to HSA21 and confers self-renewal of progenitor B-cells. Of these, overexpression of HMGN1 gene, that encodes a nucleosome remodeling protein, is linked to altered chromatin marks, such as suppression of the repressive mark H3K27me3, and proliferation of B-lymphocytes. Notably, upregulated HMGN1 is linked to gene expression dysregulation domain (GEDD) in DS-ALL, and thus, necessitates further investigation for elucidation of DS-ALL.

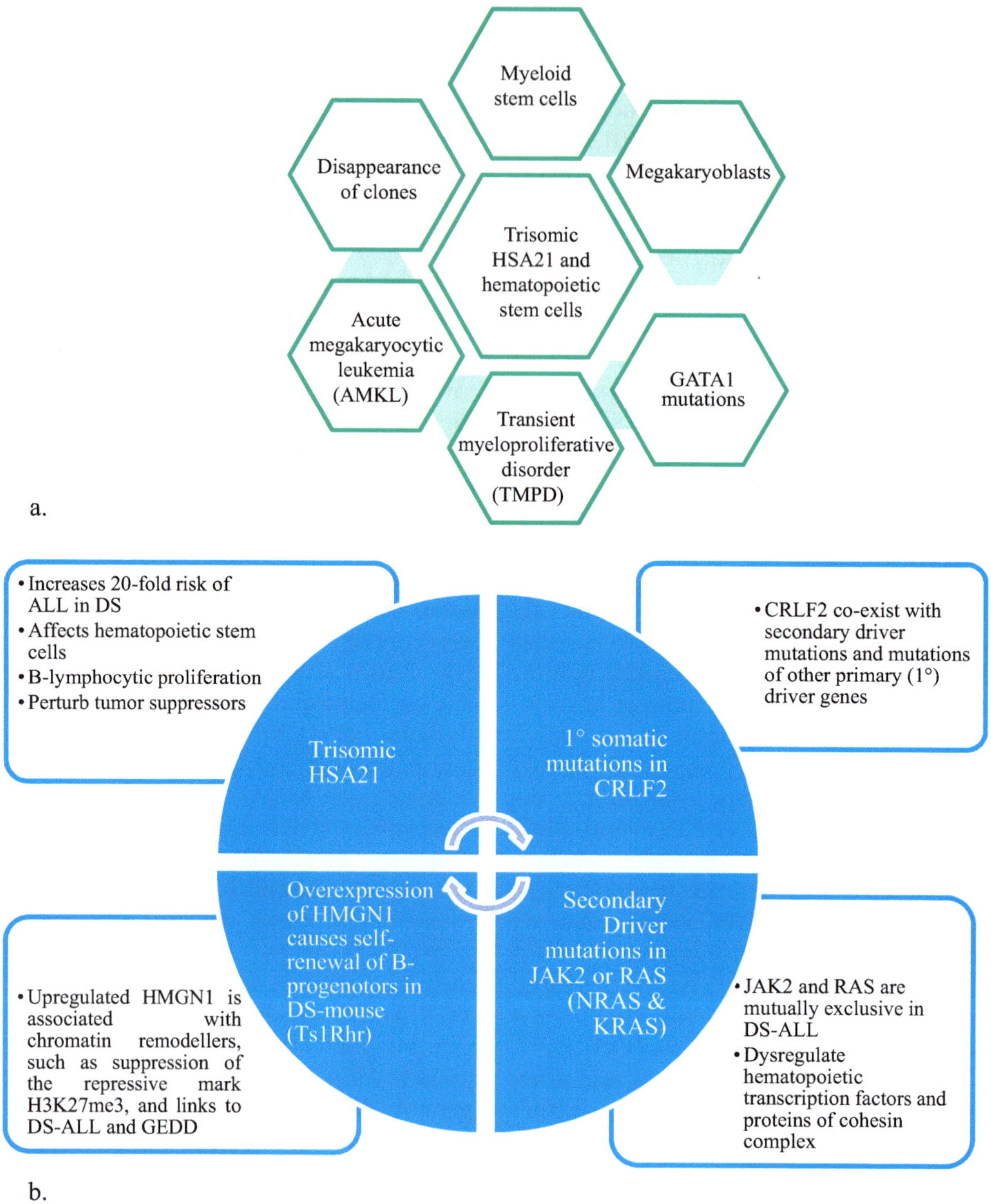

FIGURE 7.10 Leukemogenesis in individuals with Down syndrome: (A) Acute megakaryocytic leukemia (AMKL); (B) acute lymphoid leukemia (ALL).

7.3 Gene expression in mouse models for DS leukemia

For discerning of leukemogenesis in DS leukemia, mouse models of DS have been considered with a view to establishing which of the trisomic HSA21 and nontrisomic genes are needed for initiation of the disease. Studies in mouse models of DS have demonstrated alterations in megakaryopoiesis in Tc1, Ts65Dn, and Ts1Cje; however, leukemia was not revealed in anyof the strains (Kirsammer et al., 2008; Carmichael et al., 2009; Alford et al., 2010). The Tc1 strain showed macrocytic anemia, where splenomegaly with increased megakaryopoiesis and erythropoiesis was developed in older animals. Interestingly, acute leukemia was not apparent even through crossing with a strain expressing GATA1S (Alford et al., 2010). The Ts65Dn and Ts1Cje strains also showed macrocytic anemia; however, myeloproliferative disorder (MPD) was reflected along with an increase in megakaryopoiesis and erythropoiesis only in Ts65Dn mice (Kirsammer et al., 2008; Carmichael et al., 2009). The result suggested involvement of one or more of the 23 genes present in three copies in Ts65Dn mice for the MPD because these genes were included in the Ts1Cje mice. Nevertheless, contribution of the genes of the shared trisomic region between the two mouse strains was hypothesized to cause macrocytic anemia. It may be apparent that the Ts65Dn model does not have trisomy of the relevant dosage-sensitive genes required for the development of AMKL and/or expression of a mutant form of GATA1 is required to increase the sensitivity to leukemogenesis in these mice. The RUNX1, a transcription factor mapped onto HSA21, plays a pivotal role in megakaryopoiesis and hematopoietic stem cell maintenance (Okuda et al., 1996) and was proposed to be responsible for the predisposition to AMKL in triploid condition; however, its reduction from three to two copies through crossing of trisomic and disomic mice demonstrated that Runx1 did not affect MPD in Ts65Dn mice (Kirsammer et al., 2008). Ts65Dn mouse showed late-onset of MPD but that did not progress to leukemia.

Taken together, genes located in the DSCR or 21q22, including the ETS family, RUNX1, BACH1, DYRK1A, SON, SLC37A1, BRWD1, HLCS, HMGN1, and genes of interferon-pathway, contribute to leukemogenesis in DS in co-existence with many other HSA21 genes and genes of disomic chromosomes (Mateos et al., 2015). Notably, DS phenotypes and clinical expressions led to hypothesize that DS could be considered as an "immune disorder." Treatment with interferon consistently activated the cells obtained from individuals with DS, indicating insights of the patho-mechanism of autoimmune disorders, AD, and leukemia in DS. It was thus postulated that an increase in interferon signaling triggers the protective defenses of the immune system in DS (Sullivan et al., 2017). Nevertheless, overexpression of ETS2 in transgenic mice resulted in development of abnormalities in thymus and lymphocyte, which was similar to the features observed in DS patients (Seewald et al., 2012). Induction of apoptosis was also attributed to ETS-2 overexpression in thymus, spleen, and brain cells (Wolvetang et al., 2003). Furthermore, ETS-2 influenced the mitochondrial death pathway in neurons of DS patients and also manifested apoptotic features in normal human cortical neurons (Helguera et al., 2013). Upregulation of FGFR1 has been reported in DS (Sullivan et al., 2017), and its gain-of-function mutations were associated with MPDs and consequent development of leukemia (Ibrahimi et al., 2004; Jackson et al., 2010). Similar incidence could be likely in DS suggesting its contribution to leukemogenesis in patients with DS.

8. Less of solid tumors in Down syndrome

Contrary to the increased risk of acute leukemia in DS, a lower risk of developing almost all kinds of malignant solid tumors are apparent in individuals with DS (Yang et al., 2002), indicating a protection rendered by the trisomic HSA21 and tumor suppressor genes encoded on HSA21 for growth of the solid tumors. A cross breeding of the ApcMin mouse model for colorectal cancer with either Ts65Dn or Ts1Rhr trisomic mouse model of DS resulted in reduced frequency of tumors in the trisomic mice relative to the euploid controls. Conversely, crossing of ApcMin mice with Ms1Rhr mouse model of DS (which has segmental monosomy for the same 33-gene region that is triplicated in the Ts1Rhr strain), increased the number of tumors significantly compared with controls, indicating the influence of gene dosage on tumor growth (Sussan et al., 2008). This effect was assigned to the Ets2 proto-oncogene, which in three copies inhibits tumor formation and might act as a tumor suppressor, and conversely, its reduction to one copy results in increased tumor frequency (Sussan et al., 2008). Consistent with this, studies using the NPcis mouse model (which develops lymphomas, sarcomas, and carcinomas at high frequency and known as mouse model of cancer) indicated that Ts65Dn genotype might delay tumor development (Yang and Reeves, 2011). However, dosage effect of Ets2 was not reflected in this case.

Notably, reduced angiogenesis was hypothesized to lower the incidence of solid tumors in individuals with DS (Baek et al., 2009; Reynolds et al., 2010). Reduction in tumor growth in the ectopically implanted tumor lines was observed in Ts65Dn mice compared with euploid controls due to the effects in the supporting stroma in the trisomic mice (Baek et al., 2009). Decreased micro-vessel density observed in the histological examination of the tumors suggested a defect in tumor angiogenesis in Ts65Dn mice. The HSA21-encoded DSCR1 protein had inhibited angiogenesis induced by VEGF by attenuating NFAT signaling, which was thus a candidate gene for this effect. This result was further supported by the fact that reduction of Dscr1 from three to two copies resulted in increased tumor volume and tumor angiogenesis. Studies using the Tc1 mouse also suggested that reduced angiogenesis in DS leads to reduction in tumor growth due to the presence of elevated dosage of Adamts1, Erg, Jam2, and Pttg1ip (Reynolds et al., 2010). Interestingly, the Tc1 mice are not trisomic for Dscr1, and thus, Dscr1 does not contribute to reduce angiogenesis in Tc1 mice. On the contrary, the crossing of Ts65Dn and NPcis mice revealed no effect on tumor angiogenesis, suggesting its inappropriateness as a universal mechanism to describe the low rate of solid tumors in DS (Yang and Reeves, 2011). Summarily, it has been suggested that the reduced tumor frequency in DS might be due to increased dosage of several genes in the tumor cells as well as in the supporting stroma.

Differentiated endothelial cells (build blood vessels and the vascular system) and mesodermal cells (responsible for connective tissues and muscle development) in the iPSCs generated from skin samples of DS patients expressed downregulation of genes that interferes with the processes required for development of solid tumors. Interestingly, these genes create a microenvironment that is not conducive for solid tumors, by producing impeded cell movement, slow proliferation, and reduced inflammatory response (Galat et al., 2020). However, the result should be verified in other DS models.

9. Genes of motor control and hypotonia

Hypotonia is an invariable feature of DS. Incidence of muscle hypotonia in neonates with DS and some extent of motor impairment often refer to clumsiness or deficits in fine motor control were reported in many DS patients (Lana-Elola et al., 2011; Moldrich et al., 2007). However, these motor and muscle phenotypes have not been cataloged to the level of penetrance or severity like cognitive and cardiac defects in DS. Two regions of a span of 37.4–38.4 Mb and from 46.5 Mb mapped onto HSA21qter are required in trisomy to manifest hypotonia in DS patients (Lyle et al., 2009). DS-associated motor deficits were expressed as fewer number of granule neurons in the cerebellum, the section of the brain important for fine motor control in DS patients and trisomic mouse models

(O'Doherty et al., 2005; Galante et al., 2009; Baxter et al., 2000). Such diminished cerebellar neurogenesis was thought to be caused by defective Sonic hedgehog (SHH) signaling in neuronal precursors, which might be enhanced from overexpression of APP, resulting in enhanced expression of the SHH receptor patched 1, and that acts as an inhibitor of the SHH signaling pathway (Trazzi et al., 2011). Overexpression of Dyrk1a in transgenic mice (TgDyrk1a) caused delay in neuromotor development, indicating impaired performance in motor tasks (Altafaj et al., 2001). Notably, this defect was rescued by intrastriatal administration of viral vectors expressing shRNA against Dyrk1a into TgDyrk1a mice (Ortiz-Abalia et al., 2008). Conversely, overexpression of Dyrk1a in BAC transgenic mice revealed no significant motor deficits, which might explain that a much higher level of Dyrk1a overexpression might impair motor phenotypes in TgDyrk1a mice than those observed in DS (Ahn et al., 2006). Such variation was explained by the allelic variation between mouse Dyrk1a and human DYRK1A and/or the site of transgene insertion in mouse strains. Additionally, defects in the synapse morphology and vesicle recycling at the neuromuscular junction (NMJ) could cause motor delay in DS, and this hypothesis was supported by the overexpression of the Drosophila homologs of Itsn1, Synj1, and Dscr1 in transgenic flies that caused locomotor defects and impaired vesicle recycling at the NMJ. This result suggested that these Itsn1, Synj1, and Dscr1 genes, in cooperation with upregulation of APP and DYRK1A, are the dosage-sensitive signature genes for causing motor defects in DS individuals (Chang and Min, 2009).

10. Trisomic HSA21 genes and hypocomplementia

Hypocomplementia is a hallmark of diseases of immune complex, varieties of inflammatory conditions characterized by antigen–antibody deposition and consequent complement activation, including interferon-driven expression such as systemic lupus erythematosus (SLE) (Nydegger, 2007; Ramos-Casals et al., 2004). Hypocomplementia is reported as a common feature of type I interferono-pathies (Crow and Manel, 2015), indicating chronic activation of interferon signaling that could cause complement consumption. It could increase the risk of bacterial lung pneumonia and otitis media in the people with DS (Ram and Chinen, 2011), as evidenced from the prevention of removal of *Streptococcus pneumoniae* by complement depletion (Gross et al., 1978). Elevated CST3, B2M, and TFF3 proteins in DS may indicate kidney function, since these were increased in plasma levels upon kidney injury (Du et al., 2013). Also, kidney is reported to be prone to complement-mediated injury (Thurman and Le Quintrec, 2016). Therefore, kidney problems are apparent in DS along with the increase in their life expectancy (de Winter et al., 2014), which could be influenced by the accumulation of immune complexes and complement activation in the kidney (Sullivan et al., 2017). A distinct link between the complement system and AD in people with DS is mediated by activation of the classic complement cascade by both β-amyloid deposits and soluble B2M (Nybo et al., 1999). Additionally, microglia-mediated synapse destruction is attenuated by a number of the complement subunits leading to progression of AD (Hong et al., 2016; Jiang et al., 1994; Veerhuis et al., 1996). Collectively, downregulation of multiple proteins within the coagulation cascade were evidenced in samples with trisomic HSA21, including factor II (F2), fibrinogen (FGA), and the inhibitor SERPINC1, while upregulation of an inhibitor of numerous coagulation factors, protein C (PROC), was noticed (Sullivan et al., 2017). Summarily, it is apparent that hypocomplementia could modulate AD progression in DS individuals.

11. Concluding remarks

Genotype–phenotype association in DS involved mostly molecular genetic analysis of trisomic HSA21; however, gene mapping was mostly concerned around the limited number of human DS with partial trisomy. The shortcomings led to develop mouse strains, initially with segmental trisomy orthologous to HSA21 and later plenty of deletion and duplications models for precise recapitulation of dosage sensitivity responsible for specific phenotypes of human DS. Further, cross-breeding of these strains with mice bearing deletions of chromosomal segments syntenic to HSA21 (such as Ms1Yah and Ms1Rhr) allowed subtraction of the anticipated responsible gene that eventually helped in identification of the sensitive genes associated with DS phenotype. Crucially, exploitation of trisomic mouse strains, in combination with YAC or BAC transgenics and gene knockouts, identified responsible dosage-sensitive genes for different DS-specific phenotypes and helped in therapeutic development, mostly for improvement of cognition and AD-related dementia. Notably, assay of the entire smallest autosome, including sequencing could merely contribute to change the scale of the problem of dosage sensitivity on the disorder known for over 6 decades. Nevertheless, the advent of chromosome engineering to create trisomic animals with the entire stretches of HSA21 or the orthologous mouse regions has become realistic tools and resource models to unravel the dosage-sensitive genes involved in human DS. However, besides the effects of individual dosage-sensitive gene, it is likely that complex interaction of multiple HSA21 genes and global disomic genes contribute to the DS phenotypes. In addition, the matter of the large variation in DS phenotypes appeared more challenging and largely, allelic variation in HSA21 and non-HSA21 genes, individual genetic background and genetic modifiers are taken into consideration for understanding the puzzle of variability of obligatory phenotypes in fully trisomic DS. Thus, dissection of the individual alleles responsible for phenotypic expressions is the need of the present situation, which can only be mated out with intense level of cooperation between different countries for exchange of DS samples of different race and geography, and technological collaborations. However, the catch is to comply with the ethical and regulatory issues on cross-border exchange of DNA samples, though the system is already in place for diagnostic purposes.

In addition, the significant variability in copy number observed in human genomic studies needs a synergistic corroboration for molecular dissection of implications of the dosage sensitive genes and causative pathways involved in gene expression at the backdrop of trisomic HSA21. Therefore, deciphering the genotype–phenotype association remains a challenging endeavor for understanding the insights into euploid and aneuploid biology of DS. Nevertheless, development of therapeutics based on the knowledge of the dosage-sensitive genes is another challenge for this complex disorder. Importantly, a number of novel drugs have already been formulated for enhancement of learning and memory in people with DS. Indeed, a range of therapeutics is in clinical trials for ameliorating the DS-associated AD.

Therapeutically, a number of potential compounds have shown improvement in learning in theTs65Dn mouse of DS model. These mice presented improvement in hippocampal-based learning and LTP deficits from chronic treatment with picrotoxinor pentylenetetrazole and also after discontinuation of treatment (Fernandez et al., 2007), via reduction in GABA-mediated inhibition in the hippocampus development. The treatment with noncompetitive N-methyl-D-aspartic acid receptor (NMDAR) antagonist, memantine, has also improved learning in this strain. Memantine has been demonstrated to partially inhibit the opening of the NMDAR and is proposed to mitigate the effect of trisomic RCAN1 on the function of the receptor (Costa et al., 2008). Further studies and clinical trials would be important for additional investigation of the potential of many more druggable candidates with a view to improving the cognition in people with DS. Indeed, the development of new targeted therapeutics, consequences of gene dosage, and expression of DS phenotypes are remarkably necessary. In addition, a precisely standard definition of phenotype is essential. Also, formulation of standardized protocols would be important, which can be universally applied by multiple centers on sufficiently large number of samples for a meaningful genotype–phenotype analysis through availability of carefully designed and curated bio-bank of detailed phenotypic description along with DNA and tissue samples of participating DS individuals. Pooling of such large-size data sets derived from individual candidate gene and genome-wide analyses will change much to the present understanding of DS. Undoubtedly, the current progress in DS research has facilitated unparalleled comparison of data collected from DS population and myriads of mouse models for testing genetic associations, exploration of biological mechanisms and therapeutic trials.

References

Abeleira, M.T., et al., 2014. Dimensions of central incisors, canines, and first molars in subjects with Down syndrome measured on cone-beam computed tomographs. Am. J. Orthod. Dentofacial Orthop. 146, 765–775. https://doi.org/10.1016/j.ajodo .2014.08.016.

Agbolade, O., Nazri, A., Yaakob, R., Ghani, A.A., Cheah, Y.K., 2020. Down syndrome face recognition: a review. Symmetry 12, 1182. https://doi.org/10.3390/sym12071182.

Ahn, K.J., Jeong, H.K., Choi, H.S., Ryoo, S.R., Kim, Y.J., Goo, J.S., Choi, S.Y., Han, J.S., Ha, I., Song, W.J., 2006. DYRK1A BAC transgenic mice show altered synaptic plasticity with learning and memory defects. Neurobiol. Dis. 22, 463–472.

Ait Yahya-Graison, E., Aubert, J., Dauphinot, L., Rivals, I., Prieur, M., Golfier, G., Rossier, J., Personnaz, L., Creau, N., Blehaut, H., Robin, S., Delabar, J.M., Potier, M.C., 2007. Classification of human chromosome 21 gene-expression variations in Down syndrome: impact on disease phenotypes. Am. J. Hum. Genet. 81, 475–491.

Akiyama, H., Lyons, J.P., Mori-Akiyama, Y., Yang, X., Zhang, R., Zhang, Z., Deng, J.M., Taketo, M.M., Nakamura, T., Behringer, R.R., McCrea, P.D., de Crombrugghe, B., 2004. Interactions between Sox9 and betacatenin control chondrocyte differentiation. Genes Dev. 18, 1072–1087.

Alford, K.A., Slender, A., Vanes, L., Li, Z., Fisher, E.M., Nizetic, D., Orkin, S.H., Roberts, I., Tybulewicz, V.L., 2010. Perturbed Hematopoiesis in the Tc1 Mouse Model of Down Syndrome. Blood115, pp. 2928–2937.

Alió, J.J., Lorenzo, J., Iglesias, C., 2008. Cranial base growth in patients with Down syndrome: a longitudinal study. Am. J. Orthod. Dentofacial Orthop. 133, 729–737. https://doi.org/10.1016/j.ajodo .2006.03.036.

Al-Kasim, F., et al., 2002. Incidence and treatment of potentially lethal diseases in transient leukemia of Down syndrome. Pediatric Oncology Group Study. J. Pediatr. Hematol. Oncol. 24, 9–13.

Allanson, J., O'Hara, P., Farkas, L., Nair, R., 1993. Anthropometric craniofacial pattern profiles in Down syndrome. Am. J. Med. Genet. 47, 748–752.

Altafaj, X., et al., 2001. Neurodevelopmental delay, motor abnormalities and cognitive deficits in transgenic mice overexpressing Dyrk1A (minibrain), a murine model of Down's syndrome. Hum. Mol. Genet. 10, 1915–1923.

Alves-Sampaio, A., Troca-Marin, J.A., Montesinos, M.L., 2010. NMDA-mediated regulation of DSCAM dendritic local translation is lost in a mouse model of Down's syndrome. J. Neurosci. 30, 13537–13548.

Amiel, J., Sproat-Emison, E., Garcia-Barcelo, M., Lantieri, F., Burzynski, G., Borrego, S., Pelet, A., Arnold, S., Miao, X., Griseri, P., Brooks, A.S., Antinolo, G., de Pontual, L., Clement-Ziza, M., Munnich, A., Kashuk, C., West, K., Wong, K.K., Lyonnet, S., Chakravarti, A., Tam, P.K., Ceccherini, I., Hofstra, R.M., Fernandez, R., January 2008. Hirschsprung Disease Consortium. Hirschsprung disease, associated syndromes and genetics: a review. J. Med. Genet. 45 (1), 1–14.

Antonarakis, S.E., Lyle, R., Dermitzakis, E.T., Reymond, A., Deutsch, S., 2004. Chromosome 21 and Down syndrome: from genomics to pathophysiology. Nat. Rev. Genet. 5, 725–738.

Arron, J.R., Winslow, M.M., Polleri, A., Chang, C.P., Wu, H., Gao, X., Neilson, J.R., Chen, L., Heit, J.J., Kim, S.K., Yamasaki, N., Miyakawa, T., Francke, U., Graef, I.A., Crabtree, G.R., 2006. NFAT dysregulation by increased dosage of DSCR1 and DYRK1A on chromosome 21. Nature 441, 595–600.

Baek, K.H., Zaslavsky, A., Lynch, R.C., Britt, C., Okada, Y., Siarey, R.J., Lensch, M.W., Park, I.H., Yoon, S.S.,

Minami, T., et al., 2009. Down's syndrome suppression of tumour growth and the role of the calcineurin inhibitor DSCR1. Nature 11261130.

Banno, K., Omori, S., Hirata, K., Nawa, N., Nakagawa, N., Nishimura, K., Ohtaka, M., Nakanishi, M., Sakuma, T., Yamamoto, T., et al., 2016. Systematic cellular disease models reveal synergistic interaction of trisomy 21 and GATA1 mutations in hematopoietic abnormalities. Cell Rep. 15 (6), 1228–1241. https://doi.org/10.1016/j.celrep.2016.04.031.

Baxter, L.L., Moran, T.H., Richtsmeier, J.T., Troncoso, J., Reeves, R.H., January 22, 2000. Discovery and genetic localization of Down Syndrome cerebellar phenotypes using the Ts65Dn mouse. Hum. Mol. Genet. 9 (2), 195–202. https://doi.org/10.1093/hmg/9.2.195.

Beckert, V., Rassmann, S., Kayvanjoo, A.H., Klausen, C., Bonaguro, L., et al., 2021. Creld1 regulates myocardial development and function. J. Mol. Cell. Cardiol. 156, 45–56. https://doi.org/10.1016/j.yjmcc.2021.03.008.

Belichenko, N.P., Belichenko, P.V., Kleschevnikov, A.M., Salehi, A., Reeves, R.H., Mobley, W.C., May 6, 2009. The "down syndrome critical region" is sufficient in the mouse model to confer behavioral, neurophysiological, and synaptic phenotypes characteristic of down syndrome. J. Neurosci. 29 (18), 5938–5948. https://doi.org/10.1523/JNEUROSCI.1547-09.2009.

Benavides-Piccione, R., Ballesteros-Yáñez, I., de Lagrán, M.M., Elston, G., Estivill, X., Fillat, C., Defelipe, J., Dierssen, M., October 2004. On dendrites in Down syndrome and DS murine models: a spiny way to learn. Prog. Neurobiol. 74 (2), 111–126. https://doi.org/10.1016/j.pneurobio.2004.08.001.

Best, T.K., Siarey, R.J., Galdzicki, Z., 2007. Ts65Dn, a mouse model of Down syndrome, exhibits increased GABAB-induced potassium current. J. Neurophysiol. 97, 892–900.

Billingsley, C.N., Allen, J.R., Baumann, D.D., Deitz, S.L., Blazek, J.D., Newbauer, A., Darrah, A., Long, B.C., Young, B., Clement, M., Doerge, R.W., Roper, R.J., August 2013. Non-trisomic homeobox gene expression during craniofacial development in the Ts65Dn mouse model of Down syndrome. Am. J. Med. Genet. 161A (8), 1866–1874. https://doi.org/10.1002/ajmg.a.36006.

Birchler, J.A., Riddle, N.C., Auger, D.L., Veitia, R.A., 2005. Dosage balance in gene regulation: biological implications. Trends Genet. 21, 219–226.

Birchler, J.A., Yao, H., Chudalayandi, S., 2007. Biological consequences of dosage dependent gene regulatory systems. Biochim. Biophys. Acta 1769, 422–428.

Blazek, J.D., Billingsley, C.N., Newbauer, A., Roper, R.J., 2010. Embryonic and not maternal trisomy causes developmental attenuation in the Ts65Dn mouse model for Down syndrome. Dev. Dynam. 239, 1645–1653.

Bras, A., Rodrigues, A.S., Gomes, B., Rueff, J., 2018. Down syndrome and microRNAs. Biomed. Rep. 8, 11–16.

Briggs, L.E., Kakarla, J., Wessels, A., 2012. The pathogenesis of atrial and atrioventricular septal defects with special emphasis on the role of the dorsal mesenchymal protrusion. Differentiation 84, 117–130. https://doi.org/10.1016/j.diff.2012.05.006.

Brugmann, S.A., Goodnough, L.H., Gregorieff, A., Leucht, P., ten Berge, D., Fuerer, C., Clevers, H., Nusse, R., Helms, J., 2007. Wnt signaling mediates regional specification in the vertebrate face. Development 134, 3283–3295.

Byrska-Bishop, M., VanDorn, D., Campbell, A.E., Betensky, M., Arca, P.R., Yao, Y., Gadue, P., Costa, F.F., Nemiroff, R.L., Blobel, G.A., et al., 2015. J. Clin. Invest. 125, 993–1005.

Calcagni, G., Pugnaloni, F., Digilio, M.C., Unolt, M., Putotto, C., Niceta, M., Baban, A., Piceci Sparascio, F., Drago, F., De Luca, A., et al., 2021. Cardiac defects and genetic syndromes: old uncertainties and new insights. Genes 12, 1047. https://doi.org/10.3390/genes12071047.

Canzonetta, C., Mulligan, C., Deutsch, S., Ruf, S., O'Doherty, A., Lyle, R., Borel, C., Lin-Marq, N., Delom, F., Groet, J., et al., 2008. DYRK1A-dosage imbalance perturbs NRSF/REST levels, deregulating pluripotency and embryonic stem cell fate in Down syndrome. Am. J. Hum. Genet. 83, 388–400.

Carmichael, C.L., Majewski, I.J., Alexander, W.S., Metcalf, D., Hilton, D.J., Hewitt, C.A., Scott, H.S., 2009. Hematopoietic defects in the Ts1Cje mouse model of Down syndrome. Blood 1929–1937.

Chakrabarti, L., Galdzicki, Z., Haydar, T.F., 2007. Defects in embryonic neurogenesis and initial synapse formation in the forebrain of the Ts65Dn mouse model of Down syndrome. J. Neurosci. 27, 11483–11495.

Chakrabarti, L., Best, T.K., Cramer, N.P., Carney, R.S.E., Isaac, J.T.R., Galdzicki, Z., Haydar, T.F., 2010. Olig1 and Olig2 triplication causes developmental brain defects in down syndrome. Nat. Neurosci. 13, 927–934. https://doi.org/10.1038/nn.2600.

Chang, K.T., Min, K.T., 2009. Upregulation of three Drosophila homologs of human chromosome 21 genes alters synaptic function: implications for Down syndrome. Proc. Natl. Acad. Sci. USA 106, 17117–17122.

Chen, H., Zhang, H., Lee, J., Liang, X., Wu, X., Zhu, T., Lo, P.K., Zhang, X., Sukumar, S., 2007. HOXA5 acts directly downstream of retinoic acid receptor beta and contributes to retinoic acid-induced apoptosis and growth inhibition. Cancer Res. 67, 8007–8013.

Chou, S.T., Byrska-Bishop, M., Tober, J.M., Yao, Y., Vandorn, D., Opalinska, J.B., Mills, J.A., Choi, J.K., Speck, N.A., Gadue, P., et al., 2012. Proc. Natl. Acad. Sci. USA 109, 17573–17578.

Chrast, R., et al., 2000. The mouse brain treanscriptome by SAGE: difference in gene expression between P30 brain of the partial trisomy 16 mouse model of Down syndrome (Ts65Dn) and normals. Genome Res. 10, 2006–2021.

Clark, S., Schwalbe, J., Stasko, M.R., Yarowsky, P.J., Costa, A.C., 2006. Fluoxetine rescues deficient neurogenesis in hippocampus of the Ts65Dn mouse model for Down syndrome. Exp. Neurol. 200, 256–261.

Clement, K., Gustafson, N., Berbert, A., Carroll, H., Merris, C., et al., February 1, 2010. PathGen: a transitive gene pathway generator. Bioinformatics 26 (Issue 3), 423–425. https://doi.org/10.1093/bioinformatics/btp661.

Contestabile, A., Fila, T., Ceccarelli, C., Bonasoni, P., Bonapace, L., Santini, D., Bartesaghi, R., Ciani, E., 2007. Cell cycle alteration and decreased cell proliferation in the hippocampal dentate gyrus and in the neocortical germinal matrix of fetuses with Down syndrome and in Ts65Dn mice. Hippocampus 17, 665–678.

Contestabile, A., Magara, S., Cancedda, L., 2017. The GABAergic hypothesis for cognitive disabilities in Down syndrome. Front. Cell. Neurosci. 11, 54.

Cooper, J.D., Salehi, A., Delcroix, J.D., Howe, C.L., Belichenko, P.V., ChuaCouzens, J., Kilbridge, J.F., Carlson, E.J., Epstein, C.J., Mobley, W.C., 2001. Failed retrograde transport of NGF in a mouse model of Down's syndrome: reversal of cholinergic neurodegenerative phenotypes following NGF infusion. Proc. Natl. Acad. Sci. USA 98, 10439–10444.

Cordero, D.R., Brugmann, S., Chu, Y., Bajpai, R., Jame, M., Helms, J.A., 2010. Cranial neural crest cells on the move: their roles in craniofacial development. Am. J. Med. Genet. 155, 270–279.

Costa, A.C., Scott-McKean, J.J., Stasko, M.R., 2008. Acute injections of the NMDA receptor antagonist memantine rescue performance deficits of the Ts65Dn mouse model of Down syndrome on a fear conditioning test. Neuropsychopharmacology 33, 1624–1632.

Craig, B., 2006. Atrioventricular septal defect: from fetus to adult. Heart 92 (12), 1879–1885.

Crispino, J.D., 2005. GATA1 mutations in Down syndrome: implications for biology and diagnosis of children with transient myeloproliferative disorder and acute megakaryoblastic leukemia. Pediatr. Blood Cancer 44, 40–44.

Cronk, C., Reed, R., 1981. Canalization of growth in Down syndrome children three months to six years. Hum. Biol. 53, 383–398.

Crow, Y.J., Manel, N., 2015. Aicardi-Goutières syndrome and the type I interferonopathies. Nat. Rev. Immunol. 15, 429–440.

Cunningham, S.A., Arrate, M.P., Rodriguez, J.M., Bjercke, R.J., Vanderslice, P., et al., 2000. A novel protein with homology to the junctional adhesion molecule. Characterization of leukocyte interactions. J. Biol. Chem. 275, 34750–34756.

Dauphinot, L., Lyle, R., Rivals, I., Dang, M.T., Moldrich, R.X., Golfier, G., Ettwiller, L., Toyama, K., Rossier, J., Personnaz, L., Antonarakis, S.E., Epstein, C.J., Sinet, P.M., Potier, M.C., 2005. The cerebellar transcriptome during postnatal development of the Ts1Cje mouse, a segmental trisomy model for Down syndrome. Hum. Mol. Genet. 14, 373–384.

Davies, G.E., Howard, C.M., Farrer, M.J., Coleman, M.M., Bennett, L.B., Cullen, L.M., Wyse, R.K., Burn, J., Williamson, R., Kessling, A.M., 1995. Genetic variation in the COL6A1 region is associated with congenital heart defects in trisomy 21 (Down's syndrome). Ann. Hum. Genet. 59, 253–269.

Davisson, M.T., Schmidt, C., Akeson, E., 1990. Segmental trisomy of murine chromosome 16: a newmodel system for studying Down syndrome. Prog. Clin. Biol. Res. 360, 263±280.

de Winter, C.F., Echteld, M.A., Evenhuis, H.M., 2014. Chronic kidney disease in older people with intellectual disability: results of the HA-ID study. Res. Dev. Disabil. 35, 726–732.

Dierssen, M., Ramakers, G.J.A., 2006. Dendritic pathology in mental retardation: from molecular genetics to neurobiology. Gene Brain Behav. 5 (Suppl. 2), 48–60.

Deckelbaum, R.A., Majithia, A., Booker, T., Henderson, J.E., Loomis, C.A., 2006. The homeoprotein engrailed 1 has pleiotropic functions in calvarial intramembranous bone formation and remodeling. Development 133, 63–74.

Degenhardt, K., Rentschler, S., Fishman, G., Sassoon, D.A., 2002. Cellular and cis-regulation of En-2 expression in the mandibular arch. Mech. Dev. 111, 125–136.

Dey, A., et al., 2013. Down syndrome related muscle hypotonia: association with COL6A3 functional SNP rs 2270669. Front. Genet. 4, 57.

Digilio, M.C., Pugnaloni, F., De Luca, A., Calcagni, G., Baban, A., Dentici, M.L., Versacci, P., Dallapiccola, B., Tartaglia, M., Marino, B., 2019. Atrioventricular canal defect and genetic syndromes: the unifying role of sonic hedgehog. Clin. Genet. 95, 268–276.

Du, T., et al., 2013. Circulating serum trefoil factor 3 (TFF3) is dramatically increased in chronic kidney disease. PLoS One 8, e80271.

Duchon, A., Pothion, S., Brault, V., Sharp, A.J., Tybulewicz, V.L., Fisher, E.M., Herault, Y., 2010. The telomeric part of the human chromosome 21 from Cstb to Prmt2 is not necessary for the locomotor and short-term memory deficits observed in the Tc1 mouse model of Down syndrome. Behav. Brain Res. 217, 271–281.

Dunlevy, L., Bennett, M., Slender, A., Lana-Elola, E., Tybulewicz, V.L., Fisher, E.M., Mohun, T., 2010. Down's syndrome-like cardiac developmental defects in embryos of the transchromosomic Tc1 mouse. Cardiovasc. Res. 88, 287–295.

Ema, M., Ikegami, S., Hosoya, T., Mimura, J., Ohtani, H., Nakao, K., Inokuchi, K., Katsuki, M., Fujii-Kuriyama, Y., 1999. Mild impairment of learning and memory in mice overexpressing the mSim2 gene located on chromosome 16, an animal model of Down's syndrome. Hum. Mol. Genet. 8, 1409–1415.

Epstein, C.J., Korenberg, J.R., Anneren, G., Antonarakis, S.E., Ayme, S., Courchesne, E., Epstein, L.B., Fowler, A., Groner, Y., Huret, J.L., et al., 1991. Protocols to establish genotype-phenotype correlations in Down syndrome. Am. J. Hum. Genet. 49, 207–235.

Faulks, D., Collado, V., Mazille, M.N., Veyrune, J.L., Hennequin, M., 2008. Masticatory dysfunction in persons with Down's syndrome. Part 1: aetiology and incidence. J. Oral Rehabil. 35, 854–862.

Fernandez, F., Morishita, W., Zuniga, E., Nguyen, J., Blank, M., Malenka, R.C., Garner, C.C., 2007. Pharmacotherapy for cognitive impairment in a mouse model of Down syndrome. Nat. Neurosci. 10, 411–413.

Foppiana, S., Hu, D., Marcucio, R.S., 2007. Signaling by bone morphogenetic proteins directs formation of an ectodermal signaling center that regulated craniofacial development. Dev. Biol. 312, 103–114.

Freeman, S.B., Bean, L.H., Allen, E.G., Tinker, S.W., Locke, A.E., Druschel, C., Hobbs, C.A., Romitti, P.A., Royle, M.H., Torfs, C.P., Dooley, K.J., Sherman, S.L., 2008. Ethnicity, sex, and the incidence of congenital heart defects: a report from the national down syndrome project. Genet. Med. 10, 173–180. https://doi.org/10.1097/GIM.0b013e3181634867.

Fuchs, C., Ciani, E., Guidi, S., Trazzi, S., Bartesaghi, R., 2012. Early-occurring proliferation defects in peripheral tissues of the Ts65Dn mouse model of Down syndrome are associated with patched1 over expression. Lab. Invest. 92, 1648–1660.

Galante, M., Jani, H., Vanes, L., Daniel, H., Fisher, E.M., Tybulewicz, V.L., Bliss, T.V., Morice, E., 2009. Impairments in motor coordination without major changes in cerebellar plasticity in the Tc1 mouse model of Down syndrome. Hum. Mol. Genet. 18, 1449–1463.

Galat, Y., Perepitchka, M., Elcheva, I., Iannaccone, S., Iannaccone, P.M., Galat, V., 2020. iPSC-derived progenitor stromal cells provide new insights into aberrant musculoskeletal development and resistance to cancer in Down syndrome. Sci. Rep. 10 (1), 13252. https://doi.org/10.1038/s41598-020-69418-9.

Ganguly, B.B., Kadam, N.N., October 21, 2016. Down syndrome: from the age of characterization to the era of curative approach. Nucleus. https://doi.org/10.1007/s13237-016-0187-y.

Ganguly, B.B., Kadam, N.N., Mondal, P.K., 2017. Complexity of chromosomal rearrangements in Down syndrome leukemia. J. Cancer Res. Therapeut. 13, 381–383.

Gardiner, K., Fortna, A., Bechtel, L., Davisson, M.T., 2003. Mouse models of Down syndrome: how useful can they be? Comparison of the gene content of human chromosome 21 with orthologous mouse genomic regions. Gene 318, 137–147.

Geraldo, S., Khanzada, U.K., Parsons, M., Chilton, J.K., Gordon-Weeks, P.R., 2008. Targeting of the F-actin-binding protein drebrin by the microtubule plus-tip protein EB3 is required for neuritogenesis. Nat. Cell Biol. 10, 1181–1189.

Gessler, M., Knobeloch, K.P., Helisch, A., Amann, K., Schumacher, N., Rohde, E., et al., 2002. Mouse gridlock: No aortic coarctation or deficiency, but fatal cardiac defects in hey2 −/− mice. Curr. Biol. 12, 1601–1604.

Gittenberger-de Groot, A.C., Bartram, U., Oosthoek, P.W., Bartelings, M.M., Hogers, B., Poelmann, R.E., et al., 2003. Collagen type VI expression during cardiac development and in human fetuses with trisomy 21. Anat. Rec. A Discov. Mol. Cell Evol. Biol. 275, 1109–1116.

Gross, G.N., Rehm, S.R., Pierce, A.K., 1978. The effect of complement depletion on lung clearance of bacteria. J. Clin. Invest. 62, 373–378.

Guihard-Costa, A.M., Khung, S., Delbecque, K., Menez, F., Delezoide, A.L., 2006. Biometry of face and brain in fetuses with trisomy 21. Pediatr. Res. 59, 33–38.

Guimaraes, C.V., Donnelly, L.F., Shott, S.R., Amin, R.S., Kalra, M., 2008. Relative rather than absolute macroglossia in patients with Down syndrome: implications for treatment of obstructive sleep apnea. Pediatr. Radiol. 38, 1062–1067.

Hall, B.K., Miyake, T., 1995. Divide, accumulate, differentiate: cell condensation in skeletal development revisited. Int. J. Dev. Biol. 39 (6), 881–893.

Hattori, M., Fujiyama, A., Taylor, T.D., Watanabe, H., Yada, T., Park, H.S., Toyoda, A., Ishii, K., Totoki, Y., Choi, D.K., et al., 2000. The DNA sequence of human chromosome 21. Nature 405, 311–319.

Haas, M.A., Bell, D., Slender, A., Lana-Elola, E., Watson-Scales, S., Fisher, E.M.C., et al., 2013. Alterations to dendritic spine morphology, but not dendrite patterning, of cortical projection neurons in Tc1 and Ts1Rhr mouse models of down syndrome. PLoS One 8 (10), e78561. https://doi.org/10.1371/journal.pone.0078561.

Hallgrímsson, B., Willmore, K., Hall, B.K., 2002. Canalization, developmental stability, and morphological integration in primate limbs. Am. J. Phys. Anthropol. Suppl. 35, 131–158.

Hammerle, B., Ulin, E., Guimera, J., Becker, W., Guillemot, F., Tejedor, F.J., 2011. Transient expression of Mnb/Dyrk1a couples cell cycle exit and differentiation of neuronal precursors by inducing p27KIP1 expression and suppressing NOTCH signaling. Development 138, 2543–2554.

Harashima, C., Jacobowitz, D.M., Witta, J., Borke, R.C., Best, T.K., Siarey, R.J., Galdzicki, Z., 2006. Abnormal expression of the G-protein-activated inwardly rectifying potassium channel 2 (GIRK2) in hippocampus, frontal cortex, and substantia nigra of Ts65Dn mouse: a model of Down syndrome. J. Comp. Neurol. 494, 815–833.

Hasle, H., Clemmensen, I.H., Mikkelsen, M., 2000. Risks of leukemia and solid tumors in individuals with Down's syndrome. Lancet 355, 165–169.

Helguera, P., Seiglie, J., Rodriguez, J., Hanna, M., Helguera, G., Busciglio, J., 2013. Adaptive downregulation of mitochondrial function in down syndrome. Cell Metabol. 17, 132–140.

Hertzberg, L., et al., 2010. Down syndrome acute lymphpoblastic leukemia, a highly heterogeneous disease in which aberrant expression of CRLF2 is associated with mutated JAK2: a report from the International BFM Study Group. Blood 115, 1006–1017.

Hill, C.A., Reeves, R.H., Richtsmeier, J.T., 2007. Effects of aneuploidy on skull growth in a mouse model of Down syndrome. J. Anat. 210, 394–405.

Hill, C.A., Sussan, T.E., Reeves, R.H., Richtsmeier, J.T., 2009. Complex contributions of Ets2 to craniofacial and thymus phenotypes of trisomic "Down syndrome" mice. Am. J. Med. Genet. 149A (10), 2158–2165. https://doi.org/10.1002/ajmg.a.33012.

Hitzler, J.K., Zipurskey, A., 2005. Origins of leukemia in children with Down syndrome. Nat. Rev. Cancer 5, 11–20.

Holtzman, D.M., Santucci, D., Kilbridge, J., Chua-Couzens, J., Fontana, D.J., Daniels, S.E., Johnson, R.M., Chen, K., Sun, Y., Carlson, E., et al., 1996. Developmental abnormalities and age-related neurodegeneration in a mouse model of Down syndrome. Proc. Natl. Acad. Sci. USA 93, 13333–13338.

Hong, S., et al., 2016. Complement and microglia mediate early synapse loss in Alzheimer mouse models. Science 352, 712–716.

Ibrahimi, O.A., Zhang, F., Eliseenkova, A.V., Linhardt, R.J., Mohammadi, M., 2004. Proline to arginine mutations in FGF receptors 1 and 3 result in Pfeiffer and *Muenke craniosynostosis* syndromes through enhancement of FGF binding affinity. Hum. Mol. Genet. 13, 69–78.

Jackson, C.C., Medeiros, L.J., Miranda, R.N., 2010. 8p11 myeloproliferative syndrome: a review. Hum. Pathol. 41, 461–476.

Jesuino, F.A.S., Valladares-Neto, J., 2013. Craniofacial morphological differences between Down syndrome and maxillary deficiency children. Eur. J. Orthod. 35 (1), 124–130. https://doi.org/10.1093/ejo/cjr105.

Jiang, H., Burdick, D., Glabe, C.G., Cotman, C.W., Tenner, A., 1994. J. beta-amyloid activates complement by binding to a specific region of the collagen-like domain of the C1q A chain. J. Immunol. 152, 5050–5059.

Kaur, S., Singh, G., Stock, J.L., Schreiner, C.M., Kier, A.B., Yager, K.L., Mucenski, M.L., Scott Jr., W.J., Potter, S.S., 1992. Dominant mutation of the murine Hox-2.2 gene results in developmental abnormalities. J. Exp. Zool. 264, 323–336.

Kearney, L., Gonzalez De Castro, D., Yeung, J., Procter, J., Horsley, S.W., Eguchi-Ishimae, M., Bateman, C.M., Anderson, K., Chaplin, T., Young, B.D., Harrison, C.J., Kempski, H., So, C.W., Ford, A.M., Greaves, M., January 15, 2009. Specific JAK2 mutation (JAK2R683) and multiple gene deletions in Down syndrome acute lymphoblastic leukemia. Blood 113 (3), 646–648. https://doi.org/10.1182/blood-2008-08-170928.

Khan, I., Malinge, S., Crispino, J., 2011. Myeloid leukemia in Down syndrome. Crit. Rev. Oncog. 16 (1–2), 25–36. https://doi.org/10.1615/critrevoncog.v16.i1-2.40.

Kimura, C., Takeda, N., Suzuki, M., Oshimura, M., Aizawa, S., Matsuo, I., 1997. Cis-acting elements conserved between mouse and pufferfish Otx2 genes govern the expression in mesencephalic neural crest cells. Development 124, 3929–3941.

Kirsammer, G., Jilani, S., Liu, H., Davis, E., Gurbuxani, S., Le Beau, M.M., Crispino, J.D., 2008. Highly penetrant myeloproliferative disease in the Ts65Dn mouse model of Down syndrome. Blood 111, 767–775.

Kiyoi, H., Yamaji, S., Kojima, S., et al., 2007. JAK3 mutations occur in acute megakaryoblastic leukemia both in Down syndrome children and non-Down syndrome adults. Leukemia 21, 574–576. https://doi.org/10.1038/sj.leu.2404527.

Kleschevnikov, A.M., Belichenko, P.V., Villar, A.J., Epstein, C.J., Malenka, R.C., Mobley, W.C., 2004. Hippocampal long-term potentiation suppressed by increased inhibition in the Ts65Dn mouse, a genetic model of Down syndrome. J. Neurosci. 24, 8153–8160.

Knight, R.D., Schilling, T.F., 2006. Cranial neural crest and development of the head skeleton. Adv. Exp. Med. Biol. 589, 120–133.

Köntges, G., Lumsden, A., 1996. Rhombencephalic neural crest segmentation is preserved throughout craniofacial ontogeny. Development 122, 3229–3242.

Korbel, J.O., Tirosh-Wagner, T., Urban, A.E., Chen, X.N., Kasowski, M., Dai, L., Grubert, F., Erdman, C., Gao, M.C., Lange, K., et al., 2009. The genetic architecture of Down syndrome phenotypes revealed by high-resolution analysis of human segmental trisomies. Proc. Natl. Acad. Sci. USA 106, 12031–12036.

Korenberg, J.R., Chen, X.N., Schipper, R., Sun, Z., Gonsky, R., Gerwehr, S., Carpenter, N., Daumer, C., Dignan, P.,

Disteche, C., 1994. Down syndrome phenotypes: the consequences of chromosomal imbalance. Proc. Natl. Acad. Sci. USA 91, 4997–5001.

Kruszka, P., Porras, A.R., Sobering, A.K., Ikolo, F.A., La Qua, S., Shotelersuk, V., Chung, B.H., Mok, G.T., Uwineza, A., Mutesa, L., 2017. Down syndrome in diverse populations. Am. J. Med. Genet. 173, 42–53.

Labuhn, M., Perkins, K., Matzk, S., Varghese, L., Garnett, C., Papaemmanuil, E., Metzner, M., Kennedy, A., Amstislavskiy, V., Risch, T., Bhayadia, R., Samulowski, D., Hernandez, D.C., Stoilova, B., Iotchkova, V., Oppermann, U., Scheer, C., Yoshida, K., Schwarzer, A., Taub, J.W., Crispino, J.D., Weiss, M.J., Hayashi, Y., Taga, T., Ito, E., Ogawa, S., Reinhardt, D., Yaspo, M.-L., Campbell, P.J., Roberts, I., Constantinescu, S.N., Vyas, P., Heckl, D., Klusmann, J.-H., 2019. Mechanisms of progression of myeloid preleukemia to transformed myeloid leukemia in children with down syndrome. Cancer Cell 36 (2), 123–138. https://doi.org/10.1016/j.ccell.2019.06.007. E10.

Laffaire, J., Rivals, I., Dauphinot, L., Pasteau, F., Wehrle, R., Larrat, B., Vitalis, T., Moldrich, R.X., Rossier, J., Sinkus, R., Herault, Y., Dusart, I., Potier, M.C., 2009. Gene expression signature of cerebellar hypoplasia in a mouse model of Down syndrome during postnatal development. BMC Genom. 10, 138.

Laforgia, N., et al., 2018. A rare case of severe congenital RYR1-associated myopathy. Case Rep. Genet. 2018, 6184185.

Lana-Elola, E., Watson-Scales, S.D., Fisher, E.M.C., Tybulewicz, V.L.J., 2011. Down syndrome: searching for the genetic culprits. Dis. Models & Mech. 4, 586–595. https://doi.org/10.1242/dmm.008078.

Lana-Elola, E., Watson-Scales, S., Slender, A., et al., January 14, 2016. Genetic dissection of Down syndrome-associated congenital heart defects using a new mouse mapping panel [Published correction appears in Elife. 2020 Jul 14;9:] Elife 5, e11614. https://doi.org/10.7554/eLife.11614.

Lehericy, S., Hirsch, E.C., Cervera-Pierot, P., Hersh, L.B., Bakchine, S., Piette, F., Duyckaerts, C., Hauw, J.J., Javoy-Agid, F., Agid, Y., 1993. Heterogeneity and selectivity of the degeneration of cholinergic neurons in the basal forebrain of patients with Alzheimer's disease. J. Comp. Neurol. 330, 15–31.

Li, H., Cherry, S., Klinedinst, D., et al., 2012. Genetic modifiers predisposing to congenital heart disease in the sensitized Down syndrome population. Circ. Cardiovasc. Genet. 5 (3), 301–308. https://doi.org/10.1161/CIRCGENETICS.111.960872.

Li, H., Edie, S., Klinedinst, D., Jeong, J.S., Blackshaw, S., Maslen, C.L., Reeves, R.H., 2016. Penetrance of Congenital heart disease in a mouse model of Down syndrome depends on a trisomic potentiator of a disomic modifier. Genetics 203, 763–770.

Li, Z., Yu, T., Morishima, M., Pao, A., LaDuca, J., Conroy, J., Nowak, N., Matsui, S., Shiraishi, I., Yu, Y.E., 2007. Duplication of the entire 22.9 Mb human chromosome 21 syntenic region on mouse chromosome 16 causes cardiovascular and gastrointestinal abnormalities. Hum. Mol. Genet. 16, 1359–1366.

Liu, B., Rooker, S.M., Helms, J.A., 2010. Molecular control of facial morphology. Semin. Cell Dev. Biol. 21, 309–313.

Liu, C., Morishima, M., Jiang, X., Yu, T., Meng, K., Ray, D., Pao, A., Ye, P., Parmacek, M.S., Yu, Y.E., 2013. Engineered chromosome-based genetic mapping establishes a 3.7 mb critical genomic region for down syndrome-associated heart defects in mice. Hum. Genet. 133, 743–753. https://doi.org/10.1007/s00439-013-1407-z.

Liu, C., Belichenko, P.V., Zhang, L., Fu, D., Kleschevnikov, A.M., Baldini, A., Antonarakis, S.E., Mobley, W.C., Yu, Y.E., 2011. Mouse models for down syndrome-associated developmental cognitive disabilities. Dev. Neurosci. 33, 404–413. https://doi.org/10.1159/000329422.

Liu, F., Liang, Z., Wegiel, J., Hwang, Y.W., Iqbal, K., Grundke-Iqbal, I., Ramakrishna, N., Gong, C.X., 2008. Overexpression of Dyrk1A contributes to neurofibrillary degeneration in Down syndrome. Faseb. J. 22, 3224–3233.

Locke, A.E., Dooley, K.J., Tinker, S.W., Cheong, S.Y., Feingold, E., Allen, E.G., et al., 2010. Variation in folate pathway genes contributes to risk of congenital heart defects among individuals with Down syndrome. Genet. Epidemiol. 34, 613–623.

Lott, I.T., Dierssen, M., June 2010. Cognitive deficits and associated neurological complications in individuals with Down's syndrome. Lancet Neurol. 9 (6), 623–633.

Lu, Q.R., Sun, T., Zhu, Z., Ma, N., Garcia, M., Stiles, C.D., Rowitch, D.H., 2002. Common developmental requirement for Olig function indicates a motor neuron/oligodendrocyte connection. Cell 109, 75–86.

Lyle, R., Gehrig, C., Neergaard-Henrichsen, C., Deutsch, S., Antonarakis, S.E., 2004. Gene expression from the aneuploid chromosome in a trisomy mouse model of Down syndrome. Genome Res. 14, 1268–1274.

Lyle, R., Bena, F., Gagos, S., Gehrig, C., Lopez, G., Schinzel, A., Lespinasse, J., Bottani, A., Dahoun, S., Taine, L., et al., 2009. Genotype-phenotype correlations in Down syndrome identified by array CGH in 30 cases of partial trisomy and partial monosomy chromosome 21. Eur. J. Hum. Genet. 17, 454–466.

Maslen, C.L., 2004. Molecular genetics of atrioventricular septal defects. Curr. Opin. Cardiol. 19, 205–210.

Maslen, C.L., Babcock, D., Robinson, S.W., Bean, L.J., Dooley, K.J., Willour, V.L., et al., 2006. CRELD1 mutations contribute to the occurrence of cardiac atrioventricular septal defects in Down syndrome. Am. J. Med. Genet. 140, 2501–2505.

Matsuo, I., Kuratani, S., Kimura, C., Takeda, N., Aizawa, S., 1995. Mouse Otx2 functions in the formation and patterning of rostral head. Genes Dev 9, 2646–2658.

Maclean, G.A., Menne, T.F., Guo, G., Sanchez, D.J., Park, I.H., Daley, G.Q., Orkin, S.H., 2012. Proc. Natl. Acad. Sci. USA 109, 17567–17572.

McNaughton, D., Knight, W., Guerreiro, R., Ryan, N., Lowe, J., Poulter, M., Nicholl, D.J., Hardy, J., Revesz, T., Rossor, M., et al., 2010. Duplication of amyloid precursor protein (APP), but not prion protein (PRNP) gene is a significant cause of early onset dementia in a large UK series. Neurobiol. Aging. https://doi.org/10.1016/j.neurobiolaging.2010.1010.1010 [Epub ahead of print].

McNulty, M., Crispino, J.D., 2016. iPSCs offer a new look at GATA1-trisomy 21 cooperation. Cell Stem Cell 18, 563–564. https://doi.org/10.1016/j.stem.2016.04.007.

Margueron, R., Li, G., Sarma, K., Blais, A., Zavadil, J., Woodcock, C.L., et al., 2008. Ezh1 and Ezh2 maintain repressive chromatin through different mechanisms. Mol. Cell 32, 503–518.

Mateos, M.K., Barbaric, D., Byatt, S.-A., Sutton, R., Marshall, G.M., 2015. Down syndrome and leukemia: insights into leukemogenesis and translational targets. Transt. Pediatr. 4, 76–92.

Misra, C., Chang, S.-W., Basu, M., Huang, N., Garg, V., 2014. Disruption of myocardial Gata4 and Tbx5 results in defects in cardiomyocyte proliferation and atrioventricular septation. Hum. Mol. Genet. 23, 5025–5035. https://doi.org/10.1093/hmg/ddu215.

Moldrich, R.X., Dauphinot, L., Laffaire, J., Rossier, J., Potier, M.C., 2007. Down syndrome gene dosage imbalance on cerebellum development. Prog. Neurobiol. 82, 87–94.

Moore, W.J., 1973. An experimental study of the functional components of growth in the rat mandible. Acta Anat. 85, 378–385. https://doi.org/10.1159/000144005.

Mori-Akiyama, Y., Akiyama, H., Rowitch, D.H., de Crombrugghe, B., 2003. Sox9 is required for determination of the chondrogenic cell lineage in the cranial neural crest. Proc. Natl. Acad. Sci. USA 100, 9360–9365.

Mullighan, C.G., et al., 2009. Rearangement of CRLF2 in B-progenitor- and Down syndrome-associated acute lymphblastic leukemia. Nat. Genet. 41, 1243–1246.

Netzer, W.J., Powell, C., Nong, Y., Blundell, J., Wong, L., Duff, K., Flajolet, M., Greengard, P., 2010. Lowering beta-amyloid levels rescues learning and memory in a Down syndrome mouse model. PLoS One 5, e10943.

Nikolaev, S.I., et al., 2014. Frequent cases of RAS-mutated Down syndrome acute lymphoblastic leukemia lack JAK2 mutations. Nat. Commun. 5, 4654.

Nybo, M., Nielsen, E.H., Svehag, S.E., 1999. Complement activation by the amyloid proteins A beta peptide and beta 2-microglobulin. Amyloid 6, 265–272.

Nydegger, U.E., 2007. Immune complex pathophysiology. Ann. N. Y. Acad. Sci. 1109, 66–83.

O'Doherty, A., Ruf, S., Mulligan, C., Hildreth, V., Errington, M.L., Cooke, S., Sesay, A., Modino, S., Vanes, L., Hernandez, D., et al., 2005. An aneuploid mouse strain carrying human chromosome 21 with Down syndrome phenotypes. Science 309, 2033–2037.

Okuda, T., van Deursen, J., Hiebert, S.W., Grosveld, G., Downing, J.R., 1996. AML1, the target of multiple chromosomal translocations in human leukemia, is essential for normal fetal liver hematopoiesis. Cell 84, 321–330.

Olson, L.E., Richtsmeier, J.T., Leszl, J., Reeves, R.H., 2004. A chromosome 21 critical region does not cause specific Down syndrome phenotypes. Science 306, 687–690.

Olson, L.E., Roper, R.J., Sengstaken, C.L., Peterson, E.A., Aquino, V., Galdzicki, Z., Siarey, R., Pletnikov, M., Moran, T.H., Reeves, R.H., April 1, 2007. Trisomy for the Down syndrome 'critical region' is necessary but not sufficient for brain phenotypes of trisomic mice. Hum. Mol. Genet. 16 (7), 774–782. https://doi.org/10.1093/hmg/ddm022.

Ooe, N., Saito, K., Mikami, N., Nakatuka, I., Kaneko, H., 2004. Identification of a novel basic helix-loop-helix-PAS factor, NXF, reveals a Sim2 competitive, positive regulatory role in dendritic-cytoskeleton modulator drebrin gene expression. Mol. Cell Biol. 24, 608–616.

Ortiz-Abalia, J., Sahun, I., Altafaj, X., Andreu, N., Estivill, X., Dierssen, M., Fillat, C., 2008. Targeting Dyrk1A with AAVshRNA attenuates motor alterations in TgDyrk1A, a mouse model of Down syndrome. Am. J. Hum. Genet. 83, 479–488.

Palaska, P.K., Antonarakis, G.S., 2016. Prevalence and patterns of permanent tooth agenesis in individuals with Down syndrome: a meta-analysis. Eur. J. Oral Sci. 124, 317–328. https://doi.org/10.1111/eos.12282.

Parsons, T., Ryan, T.M., Reeves, R.H., Richtsmeier, J.T., 2007. Microstructure of trabecular bone in a mouse model for Down syndrome. Anat. Rec. 290, 414–421.

Patterson, T., Rapsey, C.M., Glue, P.(A., 2013. Systematic review of cognitive development across childhood in Down syndrome: implications for treatment interventions. J. Intellect. Disabil. Res. 57 (4), 306–318. https://doi.org/10.1111/j.1365-2788.2012.01536.x.

Pearlson, G.D., Breiter, S.N., Aylward, E.H., Warren, A.C., Grygorcewicz, M., Frangou, S., Barta, P.E., Pulsifer, M.B., 1998. MRI brain changes in subjects with Down syndrome with and without dementia. Dev. Med. Child Neurol. 40, 326–334.

Pennington, B.F., Moon, J., Edgin, J., Stedron, J., Nadel, L., 2003. The neuropsychology of Down syndrome: evidence for hippocampal dysfunction. Child Dev. 74, 75–93.

Pereira, P.L., Magnol, L., Sahun, I., Brault, V., Duchon, A., Prandini, P., Gruart, A., Bizot, J.-C., Chadefaux-Vekemans, B., Deutsch, S., Trovero, F., Delgado-Garcia, J.M., Antonarakis, S.E., Dierssen, M., Herault, Y., Duchon, A., Prandini, P., Gruart, A., Bizot, J.C., Chadefaux-Vekemans, B., et al., 2009. A new mouse model for the trisomy of the Abcg1-U2af1 region reveals the complexity of the combinatorial genetic code of down syndrome. Hum. Mol. Genet. 18, 4756–4769. https://doi.org/10.1093/hmg/ddp438.

Rachidi, M., Lopes, C., 2011. Mental retardation and human chromosome 21 gene overdosage: from functional genomics and molecular mechanisms towards prevention and treatment of the neuropathogenesis of Down syndrome. In: Clelland, J. (Ed.), Genomics, Proteomics, and the Nervous System, Advances in Neurobiology, vol. 2. Springer, New York, NY. https://doi.org/10.1007/978-1-4419-7197-5_2.

Ram, G., Chinen, J., 2011. Infections and immunodeficiency in Down syndrome. Clin. Exp. Immunol. 164, 9–16.

Rambo-Martin, B.L., Mulle, J.G., Cutler, D.J., Bean, L.J.H., Rosser, T.C., Dooley, K.J., Cua, C., Capone, G., Maslen, C.L., Reeves, R.H., et al., 2018. Analysis of copy number variants on chromosome 21 in Down syndrome-associated congenital heart defects. G3 8, 105–111.

Ramos-Casals, M., et al., 2004. Hypocomplementemia in systemic lupus erythematosus and primary antiphospholipid syndrome: prevalence and clinical significance in 667 patients. Lupus 13, 777–783.

Reamon-Buettner, S.M., Borlak, J., 2006. Hey2 mutations in malformed hearts. Hum. Mutat. 27, 118.

Reeves, R.H., Rue, E., Yu, J., Kao, F.-T., 1998. Stch maps to mouse chromosome 16 extending the conserved synteny with human chromosome 21. Genomics 49, 156–157.

Reeves, R.H., Irving, N., Moran, T., Wohn, A., Kitt, C., Sisodia, S., Schmidt, C., Bronson, R., Davisson, M., 1995. A mouse model for Down syndrome exhibits learning and behavior deficits. Nat. Genet. 11, 177–184.

Reynolds, L.E., Watson, A.R., Baker, M., Jones, T.A., D'Amico, G., Robinson, S.D., Joffre, C., Garrido-Urbani, S., Rodriguez-Manzaneque, J.C., Martino-Echarri, E., et al., 2010. Tumour angiogenesis is reduced in the Tc1 mouse model of Down's syndrome. Nature 465, 813–817.

Richtsmeier, J.T., Zumwalt, A., Carlson, E.J., Epstein, C.J., Reeves, R.H., 2002. Craniofacial phenotypes in segmentally trisomic mouse models for Down syndrome. Am. J. Med. Genet. 107, 317–324.

Richtsmeier, J.T., Baxter, L.L., Reeves, R.H., 2000. Parallels of craniofacial Maldevelopment in down syndrome and Ts65Dn mice. Dev. Dynam. 217, 137–145.

Richtsmeier, J.T., Lele, S., Cole III., T.M., 2005. An invariant approach to the study of fluctuating asymmetry: developmental instability in a mouse model for Down syndrome. In: Slice, D. (Ed.), Modern Morphometrics in Physical Anthropology. Kluwer Academic/Plenum Publishers Series, New York, pp. 187–212.

Rodrigues, M., Nunes, J., Figueiredo, S., Martins de Campos, A., Geraldo, A.F., May 20, 2019. Neuroimaging assessment in Down syndrome: a pictorial review. Insights Imag. 10 (1), 52. https://doi.org/10.1186/s13244-019-0729-3.

Roper, R.J., VanHorn, J.F., Cain, C.C., Reeves, R.H., 2009. A neural crest deficit in Down syndrome mice is associated with deficient mitotic response to Sonic hedgehog. Mech. Dev. 126, 212–219. https://doi.org/10.1016/j.mod.2008.11.002.

Roper, R.J., St John, H.K., Philip, J., Lawler, A., Reeves, R.H., 2006. Perinatal loss of Ts65Dn Down syndrome mice. Genetics 172, 437–443.

Rovelet-Lecrux, A., Hannequin, D., Raux, G., Le Meur, N., Laquerriere, A., Vital, A., Dumanchin, C., Feuillette, S., Brice, A., Vercelletto, M., et al., 2006. APP locus duplication causes autosomal dominant early-onset Alzheimer disease with cerebral amyloid angiopathy. Nat. Genet. 38, 24–26.

Roy, A., Cowan, G., Mead, A.J., Filippi, S., Bohn, G., Chaidos, A., Tunstall, O., Chan, J.K., Choolani, M., Bennett, P., et al., 2012. Perturbation of fetal liver hematopoietic stem and progenitor cell development by trisomy 21. Proc. Natl. Acad. Sci. USA 109, 17579–17584.

Rozovski, U., Jonish-Grossman, A., Bar-Shira, A., Ochshorn, Y., Goldstein, M., Yaron, Y., 2007. Genome-wide expression analysis of cultured trophoblast with trisomy 21 karyotype. Hum. Reprod. 22, 2538–2545.

Rueda, N., Flórez, J., MartInez-Cue, C., 2012. Mouse models of Down syndrome as a tool to unravel the causes of mental disabilities. Neural Plast. 2012, 1–26.

Rupp, P.A., Fouad, G.T., Egelston, C.A., Reifsteck, C.A., Olson, S.B., Knosp, W.M., et al., 2002. Identification, genomic organization and mRNA expression of creld1, the founding member of a unique family of matricellular proteins. Gene 293, 47–57.

Ryoo, S.R., Cho, H.J., Lee, H.W., Jeong, H.K., Radnaabazar, C., Kim, Y.S., Kim, M.J., Son, M.Y., Seo, H., Chung, S.H., et al., 2008. Dual-specificity tyrosine(Y)phosphorylationregulated kinase 1A-mediated phosphorylation of amyloid precursor protein: evidence for a functional link between Down syndrome and Alzheimer's disease. J. Neurochem. 104, 1333–1344.

Saida, S., April 2016. Evolution of myeloid leukemia in children with Down syndrome. Int. J. Hematol. 103 (4), 365–372. https://doi.org/10.1007/s12185-016-1959-5.

Sailani, M.R., Makrythanasis, P., Valsesia, A., Santoni, F.A., Deutsch, S., Popadin, K., Borel, C., Migliavacca, E., Sharp, A.J., Duriaux Sail, G., et al., 2013. The complex SNP and CNV genetic architecture of the increased risk of congenital heart defects in Down syndrome. Genome Res. 23, 1410–1421.

Salehi, A., Delcroix, J.D., Belichenko, P.V., Zhan, K., Wu, C., Valletta, J.S., Takimoto-Kimura, R., Kleschevnikov, A.M., Sambamurti, K., Chung, P.P., et al., 2006. Increased App expression in a mouse model of Down's syndrome disrupts NGF transport and causes cholinergic neuron degeneration. Neuron 51, 29–42.

Santagati, F., Rijli, F.M., 2003. Cranial neural crest and the building of the vertebrate head. Nat. Rev. Neurosci. 4, 806–818.

Saran, N.G., Pletcher, M.T., Natale, J.E., Cheng, Y., Reeves, R.H., 2003. Global disruption of the cerebellar transcriptome in a Down syndrome mouse model. Hum. Mol. Genet. 12, 2013–2019.

Seewald, L., Taub, J.W., Maloney, K.W., McCabe, E.R., 2012. Acute leukemias in children with Down syndrome. Mol. Genet. Metabol. 107 (1–2), 25–30.

Shapiro, B.L., 1992. Development of human autosomal aneuploid phenotypes (with an emphasis on Down syndrome). Acta Zool. Fennica 191, 97–105.

Shepherd, S., Ellis, F., Halsall, J., Hopkins, P., Robinson, R., 2004. RYR1-mutations in UK central core disease patients: more than just the C-terminal transmembrane region of the RYR1 gene. J. Med. Genet. 41, e33.

Shim, K.S., Lubec, G., 2002. Drebrin, a dendritic spine protein, is manifold decreased in brains of patients with Alzheimer's disease and Down syndrome. Neurosci. Lett. 324, 209–212.

Siarey, R.J., Villar, A.J., Epstein, C.J., Galdzicki, Z., 2005. Abnormal synaptic plasticity in the Ts1Cje segmental trisomy 16 mouse model of Down syndrome. Neuropharmacology 49, 122–128.

Smith, D.J., Rubin, E.M., 1997. Functional screening and complex traits: human 21q22.2 sequences affecting learning in mice. Hum. Mol. Genet. 6, 1729–1733.

Sommer, C.A., Pavarino-Bertelli, E.C., Goloni-Bertollo, E.M., Henrique-Silva, F., 2008. Identification of dysregulated genes in lymphocytes from children with Down syndrome. Genome 51, 19–29.

Sumarsono, S.H., Wilson, T.J., Tymms, M.J., et al., March 1996. Down's syndrome-like skeletal abnormalities in Ets2 transgenic mice. Nature 379 (6565), 534–537. https://doi.org/10.1038/379534a0.

Starbuck, J.M., Cole III, T.M., Reeves, R.H., Richtsmeier, J.T., 2013. Trisomy 21 and facial developmental instability. Am. J. Phys. Anthropol. 151 (1). https://doi.org/10.1002/ajpa.22255.

Sturgeon, X., Gardiner, K.J., 2011. Transcript catalogs of human chromosome 21 and orthologous chimpanzee and mouse regions. Mamm. Genome 22, 261–271.

Sullivan, K.D., Evans, D., Pandey, A., Hraha Thomas, H., Smith Keith, P., Neil, M., Rachubinski Angela, L., Wolter-Warmerdam, K., Francis, H., Espinosa Joaquin, M., Thomas, B., 2017. Trisomy 21 causes changes in the circulating proteome indicative of chronic autoinflammation. Sci. Rep. 7, 14818. https://doi.org/10.1038/s41598-017-13858-3.

Sussan, T.E., Yang, A., Li, F., Ostrowski, M.C., Reeves, R.H., 2008. Trisomy represses Apc(Min)-mediated tumours in mouse models of Down's syndrome. Nature 451, 73–75.

Takebayashi, H., Yoshida, S., Sugimori, M., Kosako, H., Kominami, R., Nakafuku, M., Nabeshima, Y., 2000. Dynamic expression of basic helix-loop-helix Olig family members: implication of Olig2 in neuron and oligodendrocyte differentiation and identification of a new member, Olig3. Mech. Dev. 99, 143–148.

Thurman, J.M., Le Quintrec, M., 2016. Targeting the complement cascade: novel treatments coming down the pike. Kidney Int. 90, 746–752.

Torfs, C.P., Christianson, R.E., 1998. Anomalies in Down syndrome individuals in a large population-based registry. Am. J. Med. Genet. 77, 431–438.

Torres, M.D., Garcia, O., Tang, C., Busciglio, J., January 2018. Dendritic spine pathology and thrombospondin-1 deficits in Down syndrome. Free Radic. Biol. Med. 114, 10–14. https://doi.org/10.1016/j.freeradbiomed.2017.09.025.

Toso, L., Cameroni, I., Roberson, R., Abebe, D., Bissell, S., Spong, C.Y., 2008. Prevention of developmental delays in a Down syndrome mouse model. Obstet. Gynecol. 112, 1242–1251.

Trazzi, S., Mitrugno, V.M., Valli, E., Fuchs, C., Rizzi, S., Guidi, S., Perini, G., Bartesaghi, R., Ciani, E., 2011. APP-dependent up regulation of Ptch1 underlies proliferation impairment of neural precursors in Down syndrome. Hum. Mol. Genet. 20, 1560–1573.

Tunstall-Pedoe, O., Roy, A., Karadimitris, A., de la Fuente, J., Fisk, N.M., Bennett, P., Norton, A., Vyas, P., Roberts, I., 2008. Blood 112, 4507–4511.

Valbuena, S., García, Á., Mazier, W., et al., 2019. Unbalanced dendritic inhibition of CA1 neurons drives spatial-memory deficits in the Ts2Cje Down syndrome model. Nat. Commun. 10, 4991. https://doi.org/10.1038/s41467-019-13004-9.

Veerhuis, R., Janssen, I., Hack, C.E., Eikelenboom, P., 1996. Early complement components in Alzheimer's disease brains. Acta Neuropathol. 91, 53–60.

Veitia, R.A., Bottani, S., Birchler, J.A., 2008. Cellular reactions to gene dosage imbalance: genomic, transcriptomic and proteomic effects. Trends Genet. 24, 390–397.

Vicente, A., Bravo-González, L.A., López-Romero, A., et al., 2020. Craniofacial morphology in down syndrome: a systematic review and meta-analysis. Sci. Rep. 10, 19895. https://doi.org/10.1038/s41598-020-76984-5.

Vis, J.C., Duffels, M.G., Winter, M.M., Weijerman, M.E., Cobben, J.M., Huisman, S.A., Mulder, B.J., 2009. Down syndrome: a cardiovascular perspective. J. Intellect. Disabil. Res. 53, 419–425.

Voronov, S.V., Frere, S.G., Giovedi, S., Pollina, E.A., Borel, C., Zhang, H., Schmidt, C., Akeson, E.C., Wenk, M.R., Cimasoni, L., et al., 2008. Synaptojanin 1-linked phosphoinositide dyshomeostasis and cognitive deficits in mouse models of Down's syndrome. Proc. Natl. Acad. Sci. USA 105, 9415–9420.

Williams, A.D., Mjaatvedt, C.H., Moore, C.S., 2008. Characterization of the cardiac phenotype in neonatal Ts65Dn mice. Dev. Dynam. 237, 426–435.

Wiseman, F.K., Alford, K.A., Tybulewicz, V.L., Fisher, E.M., 2009. Down syndrome–recent progress and future prospects. Hum. Mol. Genet. 18, R75–R83. https://doi.org/10.1093/hmg/ddp010.

Wolvetang, E.J., Wilson, T.J., Sanij, E., Busciglio, J., Hatzistavrou, T., Seth, A., et al., 2003. ETS2 overexpression in transgenic models and in down syndrome predisposes to apoptosis via the p53 pathway. Hum. Mol. Genet. 12, 247–255.

Wyganowska-Świątkowska, M., Przystańska, A., 2011. The Meckel's cartilage in human embryonic and early fetal periods. Anat. Sci. Int. 86, 98–107.

Yang, A., Reeves, R.H., 2011. Increased survival following tumorigenesis in Ts65Dn mice that model Down Syndrome. Cancer Res. 71, 3573–3581.

Yang, Q., Rasmussen, S.A., Friedman, J.M., 2002. Mortality associated with Down's syndrome in the USA from 1983 to 1997, a population-based study. Lancet 359, 1019–1025.

Young, N.M., Chong, J.H., Hu, D., Hallgrímsson, B., Marcucio, R.S., 2010. Quantitative analyses link modulation of sonic hedgehog signaling to continuous variation in facial growth and shape. Development 137, 3405–3409.

Yu, T., Li, Z., Jia, Z., Clapcote, S.J., Liu, C., Li, S., Asrar, S., Pao, A., Chen, R., Fan, N., et al., 2010. A mouse model of Down syndrome trisomic for all human chromosome 21 syntenic regions. Hum. Mol. Genet. 19, 2780–2791.

Zhai, J., Lin, H., Canete-Soler, R., Schlaepfer, W.W., 2005. HoxB2 binds mutant SOD1 and is altered in transgenic model of ALS. Hum. Mol. Genet. 14, 2629–2640.

Zhou, Q., Anderson, D.J., 2002. The bHLH transcription factors OLIG2 and OLIG1 couple neuronal and glial subtype specification. Cell 109, 61–73.

Zhong, S.C., Chen, X.S., Cai, Q.Y., Luo, X., Chen, X.H., Liu, J., Yao, Z.X., 2010. Dynamic expression and heterogeneous intracellular location of En-1 during late mouse embryonic development. Cells Tissues Organs. 191, 289–300.

CHAPTER

8

Triplicated HSA21 genes and mtDNA on mitochondrial dysfunction: consequential changes in mitochondrial dynamics, biogenesis, bioenergetics, redox potential and Ca-homeostasis

1. Introduction

Hundreds to thousands of mitochondria are present in each cell and produce the key energy currency required for metabolism, and thus, mitochondria are called the "engines" of cells. Mitochondria are crucial for various cellular functions, including synthesis of the adenosine triphosphate (ATP) for all energy-requiring cellular activities through the oxidative phosphorylation (OXPHOS) mechanism, maintenance of cellular calcium homeostasis, and detoxification of reactive oxygen species (ROS) generated from the cellular respiration. Defective mitochondria cause damage in the most energy consuming cells of the body, such as muscles, nerves, and the brain. Abnormalities in structure and activity have been documented in human subjects with Down syndrome (DS), animal models of DS and in cultured cells having trisomic chromosome 21 (HSA21), such as neurons, astrocytes, pancreatic β cells, endothelial cells, smooth muscle cells, lymphocytes, and fibroblasts (Bambrick and Fiskum, 2008; Pallardo et al., 2010; Piccoli et al., 2013; Izzo et al., 2017a). Indeed, significant reduction in respiratory capacity, ATP production, mitochondrial membrane potential, oxido-reductase activity, and impairment in mitochondrial dynamics were demonstrated in individuals with DS. Nevertheless, understanding the contribution of the extra set of HSA21 genes on phenotype and pathogenesis of DS remains yet undeciphered fully. Anticipation of the synergistic and antagonistic interaction among the triplicated HSA21 genes and cooperation of the genome-wide disomic genes might influence pathogenetic mechanisms, is revolving around DS models of in vivo and in vitro studies.

In human, mitochondrial DNA (mtDNA) consists of a self-replicating 16.5-kb double-stranded circular molecule. Each mitochondrial genome

Genetics and Neurobiology of Down Syndrome
https://doi.org/10.1016/B978-0-323-90456-8.00012-0

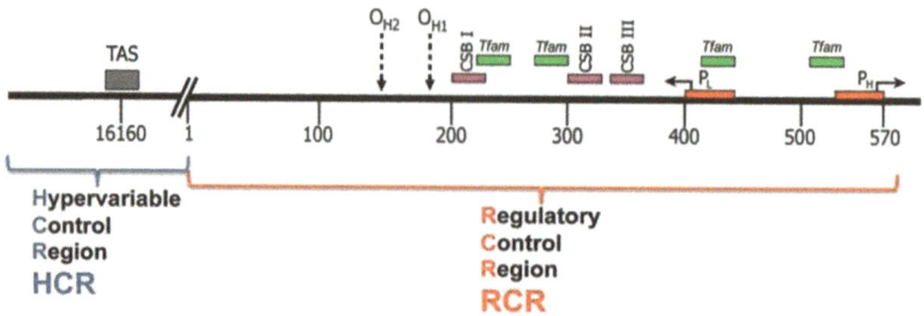

FIGURE 8.1 Schematic representation of human mitochondrial DNA control region. Complete mtDNA control region is divided into the regulatory control region (RCR, located between nt 1 and 576) and the hypervariable control region (HCR, located between nt 16,024 and 16,569). *CSBs*, conserved sequence blocks I, II, and III; *HSP*, the promoter for the H-strand; *LSP*, the promoter for the light (L)-strand; O_H, the origin of the heavy (H)-strand DNA replication; *TAS*, termination- associated sequence; *Tfam*, transcription factor A mitochondrial binding sites. *Reprinted from Coskun et al. (2010) with permission from IOS press.*

contains a noncoding control region (CR), 13 genes encoding essential polypeptides for the OXPHOS-respiratory chain system, and 24 genes specifying two rRNA genes and 22 tRNA for mitochondrial translation and synthesis of 13 polypeptides. The CR further consists of regulatory control region (RCR) and hypervariable control region (HCR). (Fig. 8.1) (Coskun et al., 2010). The mtDNA encodes a small but essential number of proteins required for energy production in cells.

The mtDNA is inherited solely from the mother because the paternal mtDNA present in the sperm are destroyed following fertilization with the egg. Hence, inheritance of mitochondrial diseases does not follow the rules of the Mendelian genetics. The eggs are formed at a very early stage of development. A precursor egg cell divides into a number of egg cells, and the mitochondria from the precursor cell get distributed randomly in all egg cells resulting in variability in mtDNA amount present in different egg cells. Such hypothesis indicated that there could be difference in the amount of mutant genetic material in the mtDNA that is passed on to the next generation with risk of variable disease spectrum among siblings (Cree et al., 2008). In cases with mutation in mtDNA, a mixture of mutant and normal mtDNA can be present, where the proportion of mutant mtDNA determines the severity of the disease. Deletion and replication abnormalities of mtDNA could be caused by nuclear genetic abnormalities, hypoxia, oxidative stress, or old age leading to mitochondrial dysfunction. Indeed, frequent deletions in mtDNA are reported in the oocytes of women of advanced reproductive age (Keefe et al., 1995). Mitochondrial dysfunction in turn can alter ATP generation in the oocytes and early embryo, which can lead to aberrant chromosomal segregation or developmental arrest (Van Blerkom et al., 1998).

Besides the mechanism of dual genetic control (mitochondrial and nuclear genetic control) for mitochondrial operation, there are four other unique features of this organelle, which are important for understanding the mitochondrial function causing neurodegeneration and many other disease consequences (Pagano and Castello, 2012; Schon and Manfredi, 2003). These include: 1. there are hundreds to thousands of mtDNAs in each cell, with approximately five mtDNAs per organelle unlike two sets of chromosomes in the nucleus; 2. division of mitochondrial organelles and replication of mtDNA occur independently of the cell cycle, both in the dividing cells (such as glia) and in postmitotic nondividing cells (such as neurons); 3. the mitochondria and their mtDNAs are randomly partitioned between the daughter cells during mitotic segregation following cell division; and 4. the number of mitochondrion varies among cells, which largely depend on the metabolic requirements of the specific cells, such as skin fibroblasts contain a few hundred mitochondria, whereas neurons may contain thousands and cardiomyocytes tens of thousands of mitochondrial organelles. Altogether, hundreds of mitochondria are present in every cell, and each mitochondrion contains multiple copies of mtDNA (Ye et al., 2014).

The motile mitochondrial organelles are highly complex and metabolically active and are crucial for maintenance of cellular energy-production processes and redox homeostasis. Mitochondrial function contributes to several metabolic pathways related to the signaling networks of neurogenesis and neuroplasticity. The cellular OXPHOS and many other fundamental cellular functions occur in the mitochondrial organelles, and thus, the mitochondrial metabolic networks are considered as master regulators of life-and-death of cells (Ryan and Hoogenraad, 2007; Benard et al., 2010). Therefore, alterations in mitochondrial structure and function might cause irreversible damage to mitochondria-dependent cellular functions, which might generate some of the critical stress-initiating activities and finally contribute to neurodegenerative and neurodevelopmental diseases with many others, including aging in human (Valenti et al., 2014). Besides the key role in oxygen metabolism and cellular energy production by the OXPHOS mechanism, mitochondria play a pivotal role to produce ROS, regulate cytosolic Ca^{2+} homeostasis, and control cell death or apoptosis through activation of the mitochondrial permeability transition pore (mtPTP) (Wallace, 2005). Therefore, damage to mtDNA may result in mitochondrial dysfunction, and the dysfunctional mitochondria in turn lead to generation of stress in cells. Collectively, mitochondrial dysfunction contributes to a number of metabolic and degenerative diseases, which is pronounced in the metabolically active tissues such as heart and brain. Mitochondrial dysfunction also contributes to cancer and aging (Wallace, 2005).

Such dysfunctional mitochondrial mechanism in DS, due to gene-dosage effects of triplicated HSA21, might represent the major causal factors of neurodevelopmental delay, intellectual disability, early neurodegeneration, and onset of Alzheimer disease (AD), dementia, and early aging in DS population (Perluigi and Butterfield, 2012). Recent reports on the molecular mechanisms evidenced in several DS-derived human cells and animal models of DS, on mitochondrial damage, and the resultant mitochondrial dysfunction highlighted the pathomechanism of clinico-phenotypic manifestations in DS. Mitochondrial energy metabolism, redox homeostasis, and mitochondrial dynamics are impaired in all types of cell analyzed in DS ranging from peripheral to central nervous system (CNS), which are characterized by lowered ATP production through OXPHOS, reduced respiratory capacity and efficacy to generate mitochondrial membrane potential (Piccoli et al., 2013; Izzo et al., 2018; Valenti et al., 2013). Furthermore, mitochondria-derived insights are leading to identification of potential naturally occurring and chemically synthesized druggable candidates and suggesting likeliness of new therapeutic tools for modulation of mitochondrial function, attenuation of oxidative stress, and mitigation of energy deficits in DS (Vacca and Valenti, 2015; Valenti et al., 2018). Therefore, mitochondrial dysfunction has been established as an obvious phenomenon in DS (Valenti et al., 2014;

Perluigi and Butterfield, 2012). Nevertheless, mitochondrial dysfunction has been proposed as an adaptive response to maintain optimal cellular functions and protect from injuries (Helguera et al., 2013).

This chapter has included discussion on the mitochondrion: its dynamics and genome; altered mitochondrial dynamics in DS; integrity of mitochondrial network; HSA21 genes involved in mitochondrial cascade; mitochondrial dysfunction, and the consequences of oxidative stress. The mitochondrial involvement in pathobiology of DS has been described in Chapter 9, and mitochondrial involvement in neurodegeneration and AD has been explained in Chapters 10 and 11. Moreover, the gene expression elaborated in this chapter primarily highlighted the contribution of mitochondrial dysfunction, altered OXPHOS, oxidative stress, and implication of ROS in individuals with DS would be an important context of therapeutic development for DS population.

2. Mitochondrial dynamics and genome

Mitochondria are highly dynamic organelles that undergoes continuous change in number and size by mitochondrial fission and mitochondrial fusion processes within a cell (Chan, 2012; Knott et al., 2008). The outer membrane of the mitochondria is permeable for ions and small molecules, whereas the impermeable inner membrane forms an electron gradient that is crucial for mitochondrial function. The division of mitochondria is enabled by an incision in the inner and outer membrane, division and distribution of the mtDNA and other critical mitochondrial components between the daughter mitochondria, which are finally transformed into isolated, small, round organelles. Importantly, the precise mechanisms of mitochondrial division remain to be fully elucidated. Mitochondrial dynamics in cell organelles are essential; however, disarray in its structure and function leads to abnormalities in energy production and increase in generation of free radicals. Therefore, de novo mitochondrial biogenesis and clearance of abnormal mitochondrion are highly crucial for mitochondrial homeostasis. The surveillance and quality control mechanisms are mediated through proteolytic activities of the mitochondrial proteases for scavenging the damaged, misfolded and surplus proteins, and regulating the maturation cleavage and assembly of polypeptides (Ni et al., 2014; Levytskyy et al., 2016). In addition, dynamics of mitochondrial fusion/fission mechanism maintains the balance of the damaged components via the fission process and distribution of mitochondrial content via fusion process in this organelle (Fig. 8.2) (Misrani et al., 2021). The event of asymmetrical mitochondrial fission is controlled by sequestration and elimination of the damaged mitochondrial components by autophagy (Shirihai et al., 2015). Studies conducted by photo-labeling experiments showed that mitochondrial fission/fusion is a continuous cycle leading to production of two distinguishable sets of mitochondria having divergent transmembrane potential (Shirihai et al., 2015). The daughter mitochondrion with the higher transmembrane potential of an asymmetrical division proceeds to fusion, whereas the component having lower transmembrane potential will experience degradation of the depolarized fusion proteins of the inner and outer membranes (Zhang et al., 2014). The unbalanced transmembrane potential condition will unlikely allow the daughter mitochondrion of lower transmembrane potential for a fusion event, while, the solitary mitochondrion undergoes degradation through autophagy (Twig et al., 2008). However, the partially damaged mitochondria having a moderate drop in transmembrane potential might still get a possibility of reincorporation into the mitochondrial network. Therefore, mitochondrial fusion represents a selective mechanism to rescue the partly impaired mitochondria.

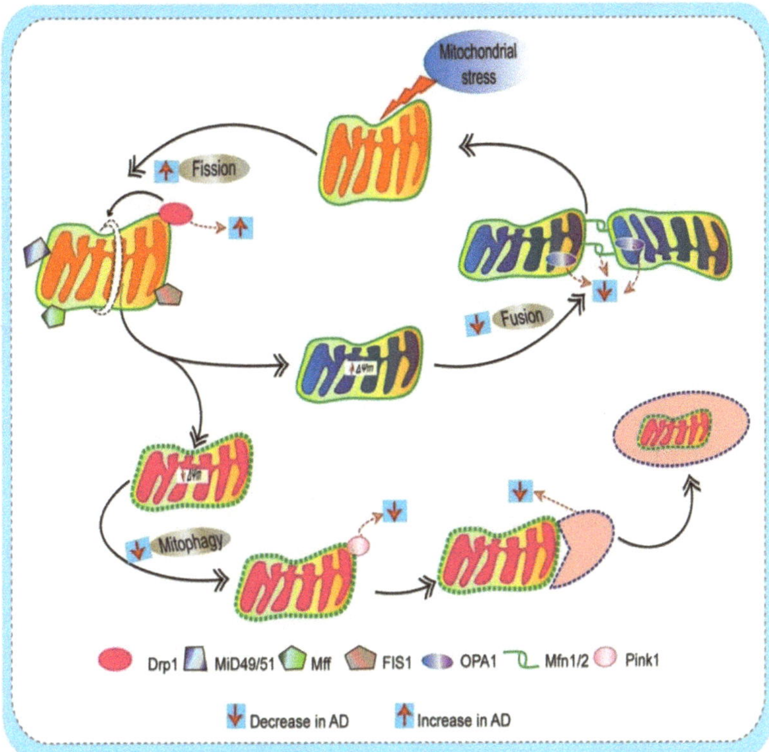

FIGURE 8.2 Schematic diagram depicting defects of mitochondrial dynamics (fission and fusion) and mitophagy mechanism in Alzheimer disease (AD). Mitochondria are dynamic, and they undergo frequent changes in shape, size, number, and location to maintain mitochondrial biology and quality control. Actions of outer membrane Drp1 control mitochondrial fission. Drp1 is recruited by mitochondrial fission factor (Mff), mitochondrial fission 1 protein (Fis1), mitochondrial dynamics protein 49/51 (MiD49/51) to promote the mitochondrial fission process. On the other hand, mitochondrial fusion is regulated by mitofusin (Mfn) 1 and 2 and optical atrophy protein 1 (OPA1). This allows for the exchange of material (matrix components and damaged mtDNA), as well as promoting a balance in bioenergetic properties (e.g., mitochondrial membrane potential 19m). These fission/fusion processes also involve mitophagy mechanisms (removal of damaged mitochondria) to maintain quality control. When mitochondria become damaged due to cellular stress, sustained depolarization of their inner membrane occurs, resulting in the loss of 19m, which stabilizes PINK1 at the outer membrane to initiate mitophagy. In AD, due to excessive ROS burden on neurons, impaired fission/fusion balance occurs, resulting in defective mitophagy. *Reprinted from Misrani et al. (2021), under the terms of the Creative Commons Attribution License (CC BY).*

The fact is that the mitochondrion organelles require function of more than 37 genes producing about 850 polypeptides contributing to ~75 structural components of the respiratory complexes (Schon and Manfredi, 2003), and at least another 20 are involved in assembling and maintaining the structural components in working order. Interestingly, all of these are encoded by nuclear DNA (nDNA), for production and maintenance of the functioning organelle. Moreover, these proteins are obviously produced in the cytoplasm as happened with other nuclear encoded proteins and are transported to the mitochondrion, where they are segregated into the four major compartments of this organelle, viz. the outer mitochondrial membrane (OMM), the inner mitochondrial membrane (IMM), the intermembrane space (IMS), and the matrix—all are located in the organelle's cytoplasm. Hence, a dual genetic mechanism control and operate mitochondrial dynamics. The five complexes of the respiratory chain/OXPHOS system are located in the IMM, which includes complexes I (NADH ubiquinone oxidoreductase), II (succinate ubiquinone oxidoreductase), III (ubiquinone—cytochrome c reductase), IV (cytochrome *c* oxidase), and V (ATP synthase). In addition, two electron carriers, ubiquinone, also called coenzyme Q, and cytochrome *c* are located in the IMM and IMS, respectively.

The functioning of the mitochondrial respiratory chain (MRC) complex is regulated by ~13 proteins encoded by the mitochondrial genome or mtDNA. The mtDNA has lower repair capacity and thus, has a greater risk of mutations compared to nDNA. However, fusion/fission process acts as effective quality control mechanisms for maintaining homogeneity in mitochondrial population (Chan, 2012). The fusion of the mitochondrial outer membrane is controlled by dynamin-related GTPases, Mfn1 and Mfn2, whereas, that of the inner membrane is modulated by the optic atrophy 1 (OPA1) in the mitochondrial organelles (Shirihai et al., 2015; Schrepfer and Scorrano, 2016; MacVicar and Langer, 2016). Both Mfn1 and Mfn2 are mitochondrial transmembrane proteins localized to the OMM and apparently play similar roles in mitochondrial fusion, even though they have independent function and different rates of GTP hydrolysis (Ishihara et al., 2004). During fusion, Mfn11 and Mfn2 could form homo-oligomeric and hetero-oligomeric complexes through interactions of their coiled-coil domains and thus, bind outer membranes of the neighboring mitochondria (Ishihara et al., 2004). OPA1, the inner membrane protein, is involved in maintaining the normal cristae morphology and assembling the electron transport chain complexes; thus, the function of OPA1 helps in maintaining the normal mitochondrial function (Cogliati et al., 2016). Mfn2 proteins mediate mitochondrial fusion under some experimental conditions. These two genes (Mfn2 and OPA1) play crucial role in maintenance and remodeling of cristae morphology in addition to its involvement in mitochondrial fusion mechanism (Frezza et al., 2006; Zick et al., 2009). For inner membrane fusion, OPA1 faces intermembrane space and primarily requires Mfn1, but not Mfn2, to mediate this process (Angajala et al., 2018) (Fig. 8.3).

The fission of mitochondria involves a large GTPase, dynamin-like protein 1 (DLP1, also referred to as Drp1, a functionally inert cytoplasmic protein) and a small molecule, Fis1 in mammals through translocation of Drp1 to the outer membrane for commissioning of the fission process (Chan, 2012; Knott et al., 2008; Smirnova et al., 2001). Drp1, a member of the conserved dynamin large GTPase superfamily, is primarily a cytosolic protein and is recruited to control membrane tabulation and fission. Drp1 can also oligomerize and form large complexes. It binds to the OMM through interaction of four mitochondrial receptors such as Fis1, Mff, MID49, and MID51 and polymerizes into spirals (Fig. 8.3) (Angajala et al., 2018) and punctuates spots on the mitochondrial surface in a GTP-dependent manner resulting in fission of both inner and outer membranes (Otera et al., 2010; Palmer et al., 2011; Losón et al., 2013). It is thought that once a ring-like complex structure is formed along the mitochondrial surface, Drp1 uses hydrolysis of GTP to constrict and twist tubule to initiate fission (Smirnova et al., 2001). Fis1, an outer membrane protein of mitochondria, is evenly resides on the surface of mitochondria and acts as a receptor to recruit Drp1 to mitochondria during fission. The mitochondrial fission creates two spherical daughter mitochondria along with cristae and other inner structure remodeling (Chan, 2012; Knott et al., 2008). Thus, it has been demonstrated that mitochondrial integrity is maintained by a coordinated inhibition of fission and promotion of fusion mechanism (Corbalan and Kitsis, 2018). RCAN1 has been reported to be involved in

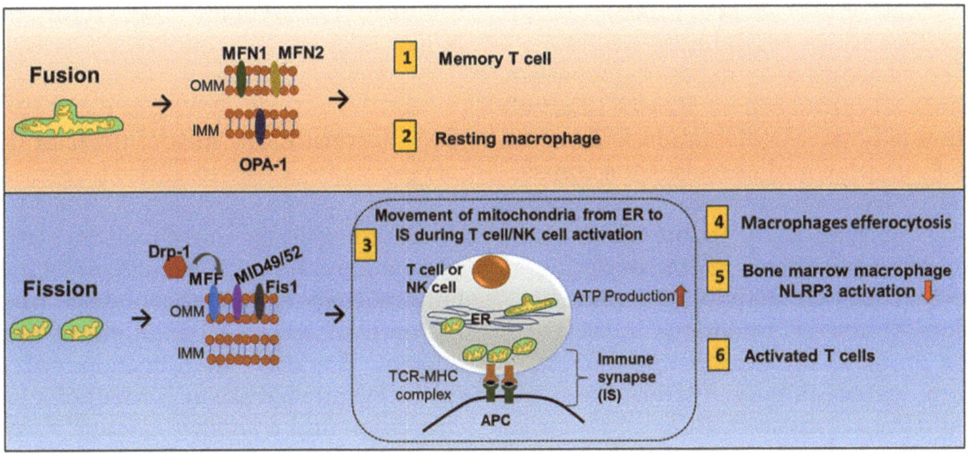

FIGURE 8.3 Mitochondrial fission and fusion contribute to immune responses. Mitochondrial fission and fusion controls mitochondrial mass. Nutrient deprivation induces an increase in mitochondrial fusion and suppression of mitophagy. On the other hand, prolonged DNA damage leads to mitochondrial fission. Fusion (tubular mitochondria) is controlled by two dynamin-like GTPases for the outer membrane proteins (mitofusin MFN1 and MFN2) and inner membrane proteins (OPA1). Fission (fragmented mitochondria) is coordinated by four outer membrane proteins (Drp1, Mff, Mid49, and mid52). Fusion and fission of mitochondria contribute to immune response. (1) Memory T cells have increased fusion generating more oxidative phosphorylation (OXPHOS) and FAO. (2) Clearance of apoptotic cells by phagocytes is known as efferocytosis. Resting macrophages have displayed tubular mitochondria. (3) Fission contributes to mitochondria mobility. During antigen-specific T-cell and NK-cell activation, mitochondria fission increased, and mitochondria move toward the immune synapse (IS) and facilitate ATP production at the IS. (4) Uptake of apoptotic cells by macrophages during efferocytosis requires fission and increased Drp1. (5) In mouse-derived bone marrow macrophages, knocking down Drp1 initiates NLRP3 inflammasomal activation and inducing mitochondrial fission attenuated NLRP3 inflammasomal assembly and activation. (6) Activated effector T cells have increased mitochondrial fission and exhibit higher rates of glycolysis. *Reprinted from Angajala et al. (2018) under the terms of the Creative Commons Attribution License (CC BY).*

mitochondrial fission/fusion dynamics via phosphorylation of Drp-1 and altered expression of mitochondrial factors through inhibition of nuclear translocation of transcription factors NFKB/NFAT (Cereghetti et al., 2010).

Notably, translocation of Drp1 from cytosol to mitochondria is influenced by specific Drp1-dependent posttranslational modifications, including phosphorylation/dephosphorylation, S-nitrosylation, ubiquitination, and SUMOylation/deSUMOylation during cell division, apoptosis, mitophagy, and mitochondrial remodeling (reviewed in Elgass et al., 2013). The posttranslational modification for functioning of the Drp1 is regulated by several enzymes, including calcium/calmodulin-dependent protein kinase 1a, protein kinase A or Ca^{2+} calcineurin (CN) (Corbalan and Kitsis, 2018). Drp1 can be phosphorylated at least at Ser616 and Ser637, wherein phosphorylation at Ser616 is activated by mitosis promoting factor (MPF, also referred to as cyclin dependent kinase 1) and that facilitates mitochondrial fission (Reviewed in Su et al., 2010). Conversely, phosphorylation of Drp1 at Ser637 by protein kinase A (PKA) inhibits mitochondrial fission, whereas its calcium-dependent dephosphorylation by CN influences mitochondrial fission (Cereghetti et al., 2010). Also, translocation of Drp1 to mitochondria and enhancement of mitochondrial fission was reported controlled by calcium-dependent phosphorylation of this site by CaM kinase 1α (Han et al., 2008). Ubiquitination of Drp1 was thought to facilitate its degradation (Nakamura et al., 2006), whereas, its degradation is protected by SUMO-1 though sumoylation of Drp1. SUMOylation widens the stabilized Drp1 pool and enhances the translocation of Drp1 from cytosol to the mitochondria (Reviewed in Su et al., 2010). Additionally, post-translational modification by S-nitrosylation of Drp1 activates GTPase activity and mitochondrial fission. Moreover, mutations in the Drp1 and fission protein 1 are demonstrated to cause nonuniform distribution of mtDNA, reduced ATP production, increased ROS production, and increased vulnerability of cells to apoptotic cell death (Knott et al., 2008; Oettinghaus et al., 2016).

3. Alterations in mitochondrial dynamics in DS

The integrity of mitochondrial structure and number, and function is maintained by the equilibrium between mitochondrial fusion and fission mechanism (Cogliati et al., 2016). A significant alteration in mitochondrial structure, including an increase in number of shorter mitochondrion and a smaller average mitochondrial volume, has been evidenced in primary cultures of astrocytes, neurons, and fibroblast cells

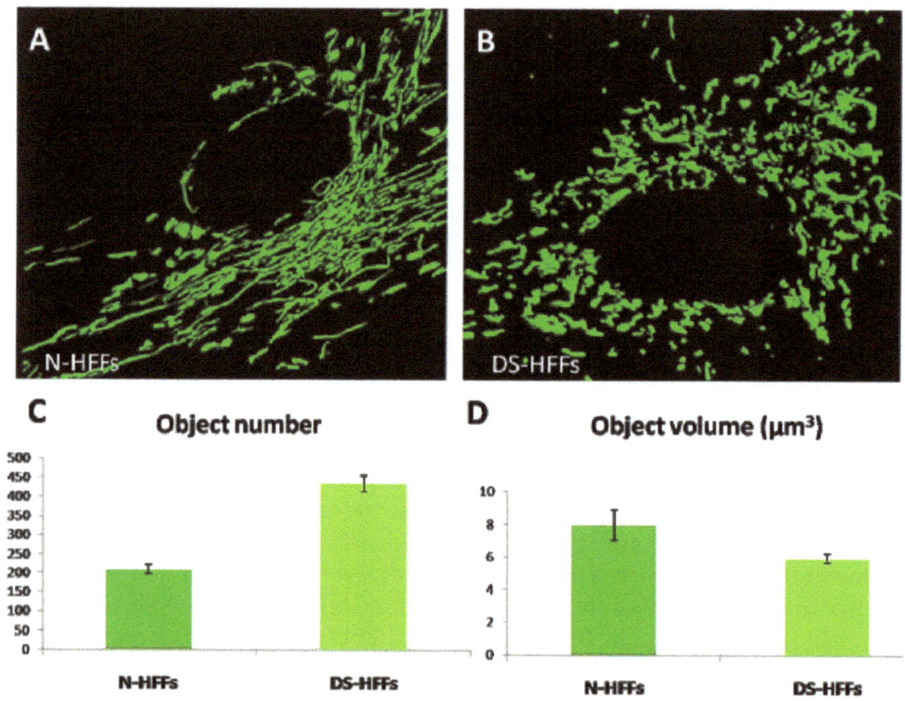

FIGURE 8.4 Human fetal fibroblasts with Down syndrome (DS-HFFs) mitochondrial network appears fragmented. Representative images showing that the mitochondrial network is less fragmented in euploid (N)-HFFs (A) than in DS-HFFs (B). The number of mitochondria is significantly higher in trisomic cells compared with nontrisomic cells (C). The mitochondrial volume is significantly lower in DS cells compared with nontrisomic cells (D). The bars show mean values ± SEM of two nontrisomic and five trisomic cell cultures. *Reprinted from Mollo et al. (2017), with permission from Anna Conti.*

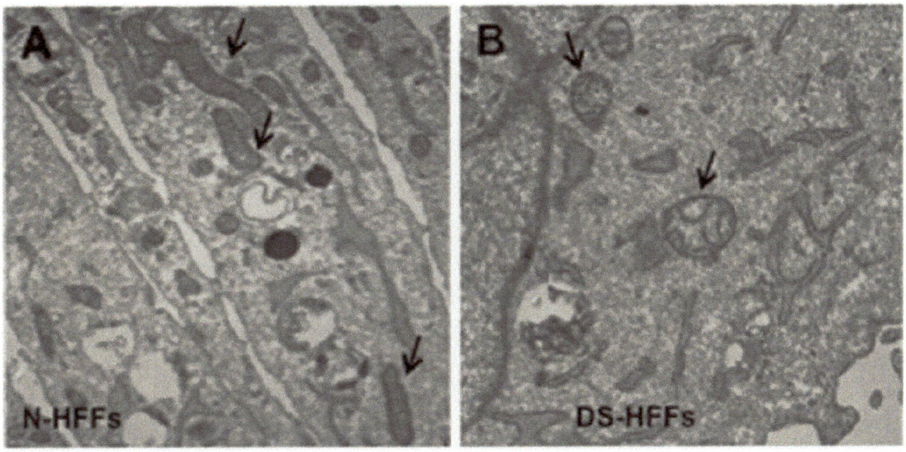

FIGURE 8.5 Mitochondrial ultrastructure is altered in human fetal fibroblasts with Down syndrome (DS-HFFs). Representative ultrastructural images of N-HFFs showing mitochondria with an intact morphology (arrowheads), without any particular changes in the structure (A), and of DS-HFFs showing damaged mitochondria characterized by a range of mitochondrial alterations (arrowheads). Scale bar 1 μm. *Reprinted from Mollo et al. (2017), with permission from Anna Conti.*

carrying trisomic HSA21 compared to normal human cells (Fig. 8.4) (Helguera et al., 2013; Mollo et al., 2017). The ultrastructure of the human fetal fibroblasts with DS (DS-HFF) revealed a substantial degree of mitochondrial damage, which appeared broken and shorter with concentric and/or highly swollen cristae (Fig. 8.5) (Mollo et al., 2017). Presence of destructured cristae observed in DS-HFF mitochondria and its correlation with reduced respiratory activity might account for dysfunctioning of the OXPHOS system in trisomic samples (Piccoli et al., 2013; Darshi et al., 2011; Zick et al., 2009). Notably, an impaired OXPHOS competence, reduced respiratory activity, and morphological alterations, particularly destructured cristae in mitochondria were demonstrated in trisomic fibroblasts of cardiopathic foeti (Piccoli et al., 2013; Darshi et al., 2011; Zick et al., 2009).

Altered mitochondrial morphology was evidenced in brain tissues of human DS and mouse models (Piccoli et al., 2013). The two genes of the mitochondrial fusion machinery, OPA1 and Mfn2, are downregulated in HSA21 trisomic cells, wherein decreased function of the OPA1 has been linked to the cristae formation in mitochondria (Izzo et al., 2017a). Conversely, RCAN1 is constitutively elevated by triplication of HAS21 and attributes to metabolically more active mitochondrial network, including enhanced fusion, increased ROS generation and oxidative stress, and interfered with the process of mitochondrial repair in DS individuals (Helguera et al., 2013). RCAN1-depleted cells exhibited reduction in OPA1 protein levels, which was associated with lower mitochondrial membrane potential influencing OPA1 cleavage and degradation (Parra et al., 2018). Rcan1-depleted neonatal rat ventricular myocytes (NRVM) showed decrease in OPA1 protein levels and was associated with lower mitochondrial membrane potential ($\Delta\Psi m$), which influences OPA1 cleavage and degradation. Thus, in NRVM, deficiency in OPA1 expression triggered the mitochondrial fission in Rcan1-deficient cells. Conversely, increase in PINK1 protein was observed with reduced $\Delta\Psi m$ and mitochondrial degradation (Parra et al., 2018). Thus, it has been established that RCAN1 maintains mitochondrial integrity through a coordinated inhibition of fission and promotion of fusion mechanism (Corbalan and Kitsis, 2018). Of the two isoforms encoded by RCAN1, the exon 1 isoform (RCAN1.1) is abundant in brain and striated muscle with relatively constant expression, whereas, expression of the exon 4 (RCAN1.4) isoform is controlled by the CN, which thus acts as a feedback inhibitor of CN activity (Yang et al., 2000). An increase in mitochondrial fusion and protection from ischemia/reperfusion (I/R) was described in RCAN1.1-depleted and siRNA-transfected NRVM (Parra et al., 2018). The iPSCs derived from DS individuals exhibited more fused mitochondria with higher O_2 consumption; however, coupling efficiency and metabolic flexibility were compromised compared to iPSCs of disomic state (Parra et al., 2018). A sustained generation and accumulation of mitochondrial damage may explain the mechanism of increase in uncoupling in trisomic iPSC and NRVM infected with ad-hRCAN1.1. RCAN1 interacts with adenosine nucleotide translocase protein, both directly and indirectly, which can uncouple the IMM proton gradient (Azzu and Brand, 2010). The knockdown of RCAN1 has highlighted that this characteristic of RCAN1 in trisomic HSA21 cells was not associated with triplication of other genes of HSA21. Therefore, overexpression of RCAN1 was identified as a crucial factor for the elevated levels of oxidative stress in patients with DS.

Also, the down-regulation of OXPHOS genes reduces the efficiency of mitochondrial energy production apparatus, which results in overproduction of ROS in mitochondria of DS, and that too in the context of significant alteration in mitochondrial structure and dynamics. The incidence of oxidative stress has been documented early in embryonic life as well as in fetal fibroblasts of DS (Piccoli et al., 2013). Such a complex situation in mitochondria of DS individuals is further multiplexed by the contemporary dysregulation of ATP translocators, ATP synthase, and adenylate kinase (Piccoli et al., 2013; Valenti et al., 2010). Therefore, it is imperative that the impairment in mitochondrial function and dynamics and the unbalanced production of ROS as a consequence of mitochondrial dysfunction might collectively contribute to some phenotypic anomalies in DS.

4. Regulation of mitochondrial bioenergetics in DS

Mitochondria act as power houses in most of the catabolic tissues, which provide energy to cells for growth and important biochemical pathways that consume approximately 90% of mammalian oxygen to synthesize ATP (Rolfe and Brown, 1997). The five complexes of the mitochondrial OXPHOS system are organized across the IMMs in a supermolecular organization to organize the MRC complex. The mitochondrial synthesis of ATP from cytosolic ADP and adenylate kinase (AK) and the interconversion of adenylate nucleotides are metabolically interconnected with the MRC, wherein export of ATP from mitochondria to the cytosol is activated by the mitochondrial transporters such as adenine nucleotide translocators (ANTs: ANT1, ANT2, ANT3) (Valenti et al., 2010). Also, mitochondria are involved in catabolism of major macromolecules into smaller metabolites such as pyruvate, fatty acids and amino acids, and their subsequent conversion into NADH and/or $FADH_2$ via β-oxidation and Krebs cycle. The oxidation reaction activates donation of electrons to the mitochondrial electron transport chain in the mitochondrial inner membrane, and the exchange of electrons in the MRC between complex I (and/or II) and complex IV influences the translocation of protons from the matrix to the IMS of the mitochondria (Mazat et al., 2013). An electrochemical proton gradient produced through this electron transport mechanism is required for F1-F0 ATP synthase (i.e., complex V) activity and ATP synthesis. The transport of NADH facilitated by specific mitochondrial "shuttle" systems has also been demonstrated (Palmieri et al., 1996).

The trisomic HSA21 leads to downregulation of genes encoding for complexes I, III, and V of the MRC subunits and affects the OXPHOS apparatus in the heart of DS fetuses and brain regions of subjects with DS. Analysis of mitochondrial function in fibroblasts and lymphoblastoid cells isolated from DS patients revealed multi-level defects in the OXPHOS machinery, which selectively include the MRC complex I, the ATP synthase, the ADP/ATP translocator, and the adenylate kinase enzyme that results in cell energy deficits and increase in generation of free radicals from mitochondria (Valenti et al., 2010, 2011, 2013; Valenti and Atlante, 2021). The imbalance of expression of the triplicated gene dosage of some HSA21 genes, including the activity of the ATP5O and ATP5J subunits of ATP synthase and NDUFV3 subunit of MRC complex I (Granese et al., 2013), regulatory activity of the RCAN1 on protein CN/NFAT pathway (Kipanyula et al., 2016), and the activity of DYRK1A on CREB/PKA signaling pathways (Yang et al., 2001) could also affect mitochondrial activity and alter the OXPHOS pathways in DS. The gene-dosage effect has been described also to increase certain enzymes of the Krebs cycle such as aconitase and NADP-linked isocitrate dehydrogenase (Kim et al., 2001; Conti et al., 2007).

The murine model of DS or the Ts65Dn mice has exhibited severe bioenergetic deficiency in the neural progenitor cells (NPCs) and a significant deficit in proliferation of cells in the hippocampal tissue, which could not be substantiated by increasing the glycolytic compensation as measured by the increase in the basal levels of L-lactate, indicating a marked reduction of the overall ATP content culminating into reduced proliferation of NPCs (Contestabile et al., 2013; Valenti et al., 2016). The impairment of bioenergetic processes of the MRC complex (in particular complex I and ATP synthase) was hypothesized to perturb both mitochondrial and respiration-mediated ATP generation, and hence, its deficiency significantly affects the bioenergetic function of mitochondria (Telford et al., 2009). The bioenergetic deficiency is linked to decreased basal levels of cAMP and reduced PKA activity leading to alteration of cAMP/PKA signaling pathway. The hippocampus of Ts65Dn mice expressed a downregulation of PKA activity and cAMP signaling (Dierssen et al., 1997). The cAMP-dependent phosphorylation of the NDUFS4 subunit of complex I was reduced due to depression of cAMP-dependent PKA activity, which led to consequent reduction in the complex I activity in skin fibroblasts of DS subjects (Valenti et al., 2011, 2013). In human DS, elevation of oxidative stress and ROS has been associated with triplication of several HSA21 genes, which include DYRK1A, DSCR1, SOD1, ETS2, S100B, APP, and BACH1 (Vilardell et al., 2011; Lott, 2012).

5. Mitochrondrial dysfunction in DS: increased oxidative stress and inflammation

Oxidation is a normal and important physiological system for a living organism. However, an imbalance between free radicals and antioxidants activity induces oxidative stress. The uneven number of electrons of the free radicals containing oxygen molecules stimulates free reaction with other molecules in oxidative chemical reactions in large chain. Antioxidants donate an electron to a free radical for stabilizing the free radicals and making it less reactive; however, antioxidants remain stable. Free radicals can be beneficial or harmful. In general, free radicals fight off the pathogens; however, they can potentially damage the fatty tissue, DNA, proteins, etc., leading to initiation of various diseases over time, such as diabetes, atherosclerosis, hypertension, cancer, and neurodegenerative diseases, including AD and Parkinson disease.

Oxidative stress is caused either by overproduction of radical species and/or reduced antioxidant response of the cells and inappropriate clearance of the oxidized proteins, which collectively creates an imbalance in cellular redox status (Tramutola et al., 2017). As the main site of ROS generator, the mitochondrial organelles house the mechanism of cellular free radical generation and also as targets of ROS and thus, play the pivotal role in causing oxidative stress and mitochondrial dysfunction. The event of oxidative stress and its accumulation in various cells has been established as the essential etiology of pathomechanism of majority of the phenotypes and their early onset in DS (Butterfield et al., 2014). Indeed, the markers of oxidative stress were evidenced in amniotic fluid of the trisomic fetuses (Perluigi et al., 2011). The complex double-membrane organization of the mitochondria, its involvement in critical pathways and tightly controlled fusion/fission dynamics make the organelle particularly prone to its functional alterations and/or dysfunctions (Zamponi and Helguera, 2019). Thus, upregulation of ROS production and mitochondrial dysfunction has been anticipated to commence during embryonic life and early developmental stages in DS (Busciglio and Yankner, 1995). In addition, primary cultured astrocytes and neurons from Ts1CJe mouse model of DS showed increased ROS levels and that was associated with mitochondrial dysfunction (Ishihara et al., 2015).

An intricate association between altered mitochondrial function and oxidative stress has been explained in DS (Helguera et al., 2013), though the impact of morphological changes on mitochondrial functions has not been fully elucidated in trisomic tissues. An imbalance between increased oxidative stress and decreased mitochondrial metabolism has been demonstrated in different DS-derived primary cells, such as fibroblasts of fetal and adult origin (Valenti et al., 2010, 2011, 2013), cortical neurons, astrocytes, and pancreatic β cells (Helguera et al., 2013). These imbalances are responsible for the increase in basal levels of ROS, deficits in electron transport chain components (Valenti et al., 2011, 2013), and a downregulated performance of the mitochondrial activities (Helguera et al., 2013). The reduced catalytic activity of the mitochondrial complex I was postulated to be the primary cause of increased mitochondrial ROS generation in peripheral cells of human DS, which further supported the mitochondrial origin of oxidative stress in DS (Valenti et al., 2011, 2013). Additionally, the complex I of the Ts16 neurons showed the alterations in ROS homeostasis as the key site of oxidative stress (Capone et al., 2002).

The overexpression of HSA21-derived APP was reported as another important player involved in the oxidative stress and pathobiology of brain damage in DS patients (Perluigi and Butterfield, 2012). Upregulated APP enhances mitochondrial oxidative stress and influences the intrinsic apoptotic cascade (Bartley et al., 2012; Picca et al., 2020). Alteration of redox homeostasis has been influenced also by deregulation of iron metabolism in DS (Barone et al., 2017). Progressive accumulation of APP protein, specifically fragments of Aβ42 protein, within the mitochondrial matrix, has been reported to increase free radicals and impairment of mitochondrial metabolism (Manczak et al., 2010). Free radicals, in turn, directly induce toxicity within the cells, leading to Ca^{2+} dysregulation, mitochondrial dysfunction, and induction of oxidative stress (Demuro et al., 2010). Additionally, overexpressed APP has been demonstrated to translocate to the mitochondria in human cortical neuronal cell line.

The increasing evidences on the association of oxidative stress and overexpression of some of the HSA21 genes, which affects lipid metabolism and results in ROS production in DS, involves an unelucidated molecular mechanism (Convertini et al., 2016). A significant increase in ROS generation was demonstrated with increased mitochondrial fusion in trisomic iPSCs (Parra et al., 2018). The transgenic overexpression of

RCAN1 increased mitochondrial ROS in neurons and pancreatic islets under normal glucose conditions (Peiris et al., 2012). Involvement of nuclear sirtuin 1 (Sirt1), particularly Sirt1 downstream such as peroxisome proliferator–activated receptor-γ coactivator 1 alpha (PGC-1α), and the transcription factors, forkhead box O1 (FOXO1) and O3 (FOXO3) have also been demonstrated in mitochondrial energy sensing network (Cantó and Auwerx, 2009). Release of ROS has been correlated with an overload of Ca^{2+} in the mitochondria, and blockade of the major mitochondrial Ca-transporting system resulted in substantial depression of ROS overproduction (Feissner et al., 2009). In DS fibroblasts, the impact of reduced OXPHOS caused severe bioenergetic failure compromising cell growth (Piccoli et al., 2013), which could be explained by an adaptive compensatory increase in the glycolytic flux (Valenti et al., 2010) and by the gene-dosage effect of the HSA21-encoded regulatory glycolytic enzyme phosphofructokinase, PFKL.

Assessment of intracellular ROS level by confocal microscopy revealed enhanced ROS production in every trisomic human fetal fibroblasts (DS-HFF) samples when compared with normal HFF (N–HFF), with a higher redox imbalance in cardiopathic DS-HFF (CDS-HFF) (Fig. 8.6A). The intensity of the fluorescent signal was significantly brighter in CDS-HFF when compared with noncardiopathic DS-HFF (NCDS-HFF) (Fig. 8.6B). For each individual HFF sample, plot of the pixel intensity of the dichlorofluorescein (DCF) fluorescence versus the normalized complex I activity indicated the presence of a threshold value of complex I activity, below which extra-ROS production was generated (Fig. 8.6C) (Piccoli et al., 2013). Interestingly, treatment of HFF cells in vitro with a pan inhibitor of flavin-containing oxidases (including complex I; DPI) resulted in significant reduction of overproduction of ROS in representative cells of both NCDS- and CDS-HFF; however, the treatment was ineffective in N–HFF (Piccoli et al., 2013). Notably, respiratory chain complex I regulated production of ROS is promoted by the presence of a $\Delta\Psi m$. Additionally, short-term treatment with the uncoupler carbonilcyanide p-triflouromethoxyphenylhydrazone (FCCP) markedly lowered the release of ROS in cells of both NCDS- and CDS-HFF with a larger effect in CDS-HFF (Piccoli et al., 2013).

Citrate pathway and mitochondrial energy production are regulated by several kinase effectors and transcription factors. Activation of AMP-activated protein kinase (AMPK), the key cell energy sensor, and central regulator of cellular metabolism in eukaryotes is facilitated by the changes in cellular AMP/ATP ratio. The activated AMPK contributes to phosphorylation reactions that further activate catabolism of ATP synthesis and block the anabolic pathways that utilize ATP (Hardie, 2007). In DS, activation of AMPK was significantly affected resulting in impairment of bioenergetic processes (Valenti et al., 2016). A significant reduction in the activity of complex I in fetal fibroblasts of DS was attributed to defective cAMP/PKA-dependent phosphorylation and labeled as a major ROS generator (Valenti et al., 2011). ROS production was substantially suppressed by db-cAMP treatment, suggesting that the deregulation of post-translational modification of complex I is involved in the redox imbalance in DS-HFF.

Modulation of PKA-dependent signaling has been reported to affect functioning of the OXPHOS system (Valenti et al., 2011; Papa et al., 2012). Furthermore, incubation of cells with the cAMP analog, db-cAMP significantly decreased the production of ROS in both NCDS- and CDS-HFF; however, the redox state was not fully renormalized in CDS-HFF to the level of N-HFF (Piccoli et al., 2013). An increased production of ROS led to a remarkable alteration

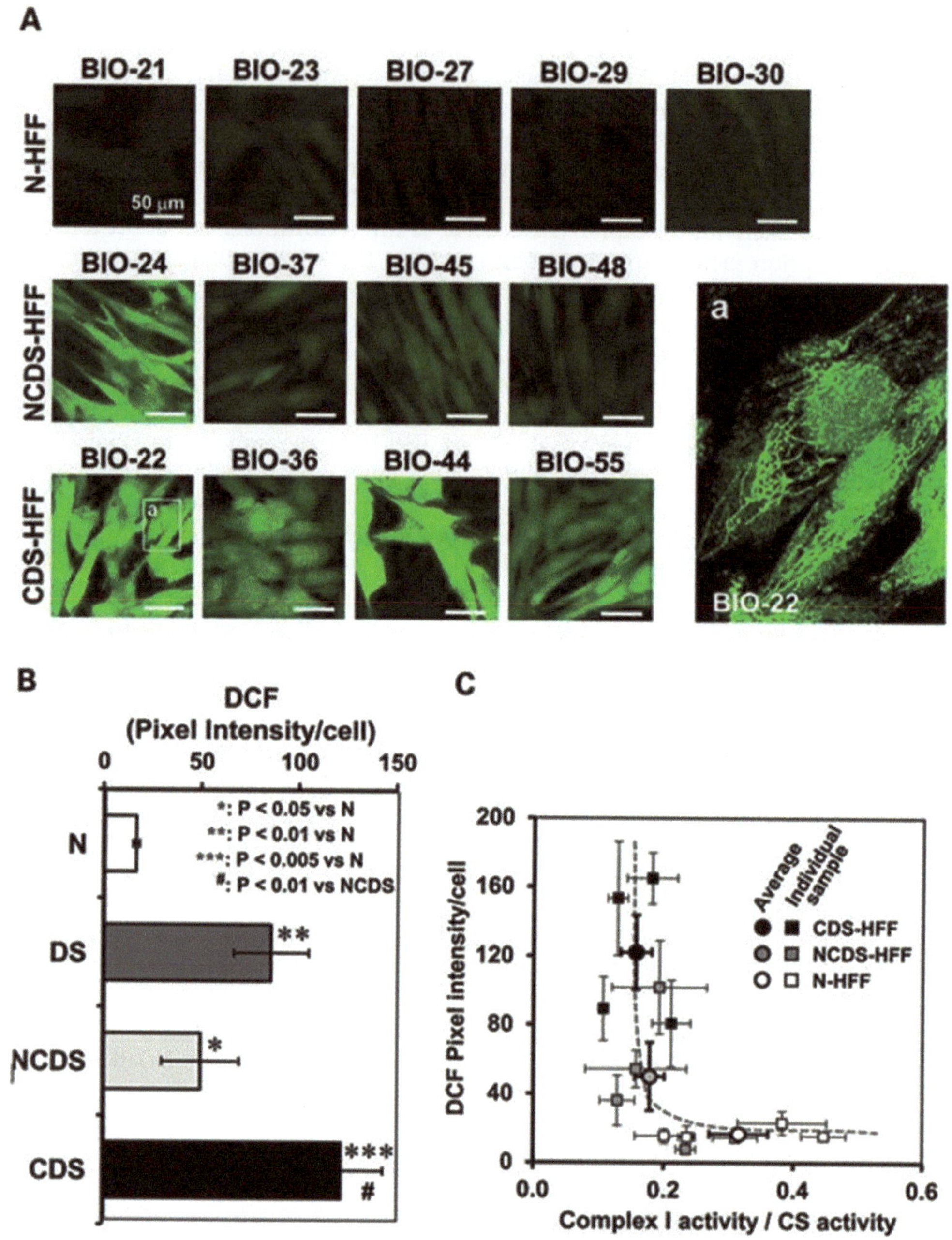

FIGURE 8.6 Confocal microscopy analysis of reactive oxygen species production in Down syndrome live fibroblasts. (A) Representative laser scanning confocal microscopy imaging of the dichlorofluorescein (DCF)-related fluorescence of euploid (N-human fetal fibroblast (HFF)) and DS (noncardiopathic DS (NCDS)-HFF and CDS-HFF) fibroblasts. A representative magnification of an intracellular selected detail (*white rectangle*) of the indicated CDS sample is shown displaying compartmentalization of the brighter DCF fluorescence signal. (B) Statistical analysis of the fluorescence intensity per cell as resulting from the averaged values ± SEM of about 100 randomly selected different cells for each sample from at least in-duplicate experiments; statistical analysis of the differences is also shown. (C) Correlation plot between the Complex I activity normalized to the CS activity and the DCF-related fluorescence signal/cell for individual and averaged fibroblast samples (see symbol legend, means ± SEM). *CDS-HFF*, cardiopathic DS-HFF; *DS-HFF*, Down syndrome-HFF; *HFF*, human fetal fibroblasts; *NCDS-HFF*, noncardiopathic DS-HFF; *N-HFF*, normal euploid foeti. *Reprinted from Piccoli et al. (2013), with permission from Oxford University Press and Copyright Clearance Center.*

in the redox homeostasis in DS-HFF demonstrating its localization to an intracellular compartment resembling the mitochondrial network, which was sensitive to the uncoupler FCCP and to the complex I inhibitor DPI. These features indicated that complex I is a major ROS generator in DS-HFF and that is maintained by a "forward electron transfer" mechanism (Selivanov et al., 2011).

An increase in mitochondrial Ca^{2+} levels has been reported in heart and fibroblasts of trisomic fetuses (Piccoli et al., 2013). An overload of Ca^{2+} in the mitochondria has been reported to be responsible for release of ROS; however, the mechanism of this association remains to be satisfactorily uncovered (Adam-Vizi and Starkov, 2010). In this line, a steady level of intramitochondrial Ca^{2+} ($mtCa^{2+}$) appeared higher in DS-HFF and the highest in CDS-HFF when compared with N−HFF (Piccoli et al., 2013). The study showed a linear positive correlation between $mtCa^{2+}$ and ROS generation in the three HFF samples: N-HFF, DS-HFF, and CDS-HFF. Most importantly, substantial depression of ROS overproduction was noticed in DS-HFF by blocking the major mitochondrial Ca^{2+} transporting system, that was ineffective in N-HFF. Collectively, interplay among the chronic intramitochondrial Ca^{2+} levels, inhibition of complex I and mitochondrial ROS production is consistent with mitochondrial dysfunction. Nevertheless, a cross-talk between PPARγ and Ca^{2+} mobilization/signaling, and depression of PGC-1α activity might consequently affect the transcriptional efficiency of PPARs-transcription factor family, mainly PPARγ in trisomic cells (Oyekan, 2011).

Mitochondria act as "Ca^{2+} buffer" by taking up Ca^{2+} mainly via a specific ruthenium red (RR)-inhibitable uniporter (DeStephani et al., 2011). The cells derived from DS patients or in murine models of DS displayed deregulation of Ca^{2+} homeostasis and Ca^{2+}-mediated signaling (Corbalan and Kitsis, 2018). Evaluation of the intramitochondrial level of $mtCa^{2+}$ using the specific probe Rhod-1 demonstrated a significant increase in intense Rhod-1-related fluorescence in cardiopathic DS-HFF compared to that of N-HFF in confocal microscopic imaging (Fig. 8.7A and B) (Piccoli et al., 2013). The brighter signal largely displayed the steady-state of mitochondrial Ca^{2+} level, and the increased $mtCa^{2+}$ load in DS-HFF correlated positively with the enhancement of ROS production (Fig. 8.7C). This incidence was consistent with the notion that calcium entry in mitochondria induces alterations of redox state (Piccoli et al., 2013). Inhibition of the mitochondrial Ca^{2+} porter in representative RR-treated trisomic samples caused a substantial inhibition of ROS production, indicating that the entry of Ca^{2+} in the mitochondrial compartment was at least partially responsible for higher redox imbalance in trisomic cells (Piccoli et al., 2013). Notably, RR treatment resulted in full recovery of the respiratory activity in DS-HFF. Redox imbalance was significantly higher in cardiopathic foeti irrespective of the similar degree of inhibition of complex I in NCDS- and CDS-HFF. In addition, RR treatment resulted in enhancement of the respiratory activity in DS-HFF to the level of N−HFF. Interestingly, treatment with a calcium-sensible photoprotein, the aequorin, exhibited analogous results in evaluating mitochondrial calcium levels (Piccoli et al., 2013).

6. Triplicated HSA21 genes: implication on mitochondrial biogenesis and function

6.1 Trisomic HSA21 genes involved in mitochondrial dysfunction

Trisomy 21 contributes to upregulation of approximately 77 genes, which are involved in

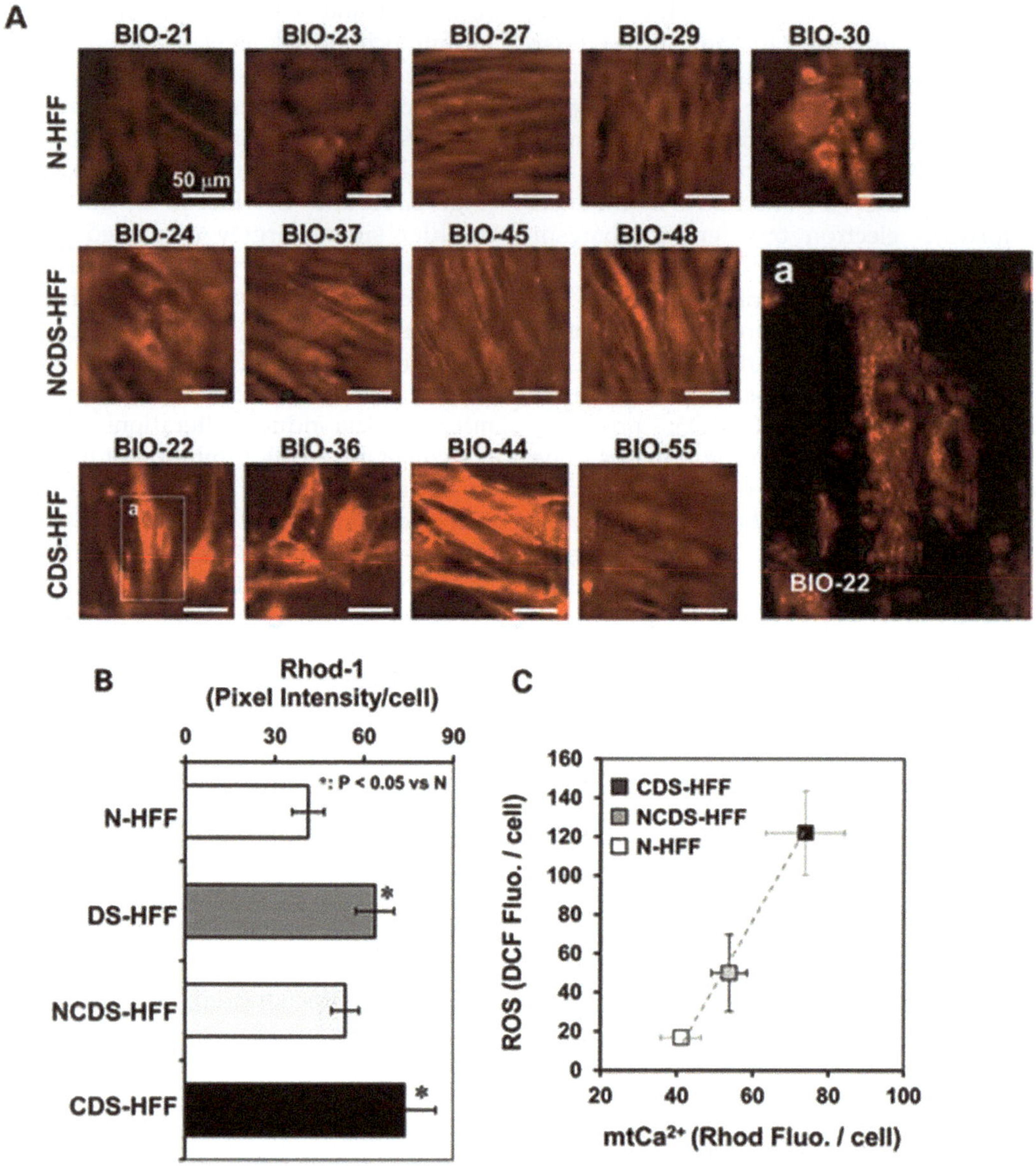

FIGURE 8.7 Confocal microscopy analysis of mitochondrial Ca^{2+} in Down syndrome live fibroblasts. (A) Representative laser scanning confocal microscopy imaging of the Rhod-1-related fluorescence of euploid (N-human fetal fibroblast (HFF)) and DS (noncardiopathic DS (NCDS)-HFF and CDS-HFF) fibroblasts. A representative magnification (*white rectangle*) of the indicated CDS sample is shown displaying the punctuate compartmentalization of the Rhod-1-fluorescence signal. (B) Statistical analysis of the fluorescence intensity per cell as resulting from the averaged values ± SEM of about 100 randomly selected different cells for each sample from at least in-duplicate experiments; when statistically significant, the difference when compared with the euploid samples is shown. (C) Correlation plot between the Rhod-1- and dichlorofluorescein (DCF)-related fluorescence signal/cell for averaged fibroblast samples (see symbol legend, means ± SEM). *CDS-HFF*, cardiopathic DS-HFF; *DS-HFF*, Down syndrome-HFF; *HFF*, human fetal fibroblasts; *NCDS-HFF*, noncardiopathic DS-HFF; *N–HFF*, normal euploid foeti. *Reprinted from Piccoli et al. (2013), with permission from Oxford University Press and Copyright Clearance Center.*

TABLE 8.1 List of genes of HSA21 functionally associated to the mitochondrial abnormalities in DS (Izzo et al., 2018; Panagaki et al., 2019 with partial modifications).

HSA21 genes and transcripts involved in mitochondrial function	Expression of mitochondrial phenotype
NRIP1/RIP140—nuclear receptor interacting protein 1	Decreases respiratory efficiency and alters morphology of mitochondria.
SUMO3—small ubiquitin-like modifier 3	Modulates NRIP1 repressive activity and attenuates the transcriptional activity of PGC-1α.
DYRK1A—dual-specificity tyrosine phosphorylation-regulated kinase 1A	Controls PGC-1α via the calcineurin/NFAT pathway.
DSCR1/RCAN1—Down syndrome critical region gene 1	Controls PGC-1α via the calcineurin/NFAT pathway and is associated with calcium overloading.
SOD1—superoxide dismutase 1	Associated with oxidative stress.
APP—amyloid beta precursor protein	Induces mitochondrial oxidative stress and mitochondrial dysfunction.
CBS—Cystathionine-β-synthase	Enhances H_2S production and mitochondrial dysfunction; influences the mitochondrial redox activity negatively regulating methylation processes.
ETS2—V-ETS avian erythroblastosis virus E26 oncogene homolog 2	Promotes the activation of a mitochondrial death pathway.
ITSN1—intersectin 1	Regulates the mitochondrial apoptotic pathway.
PREP1—PBX-regulating protein 1	Inhibits the OXPHOS by negatively regulating PGC-1α and mitochondrial fusion genes OPA1 and Mfn2.
S100B—S100 calcium-binding protein, beta	Overexpression induces ROS formation, activation of stress response kinases and increased programmed cell death
BACH1—BTB domain and CNC homolog 1	Contributes to the early increase of oxidative stress in DS through the inhibition of the HO-1/BVR-A axis
Mir-155 (microRNA)	Affects mitochondrial biogenesis by targeting TFAM.
Let-7c (microRNA)	May affect mitochondrial function by targeting ANT1.

OXPHOS, oxidative phosphorylation.

characterization of DS phenotype (Vilardell et al., 2011). Some of these genes such as APP, CBS, DYRK1A, NRIP1, RCAN1, SUMO3, and SOD1 are directly or indirectly associated with mitochondrial function (Table 8.1), and thus, dysregulation of these genes in trisomic status might cause mitochondrial dysfunction in DS. Despite overexpression of several genes in DS, downregulation of over 80 genes, encoding mitochondrial enzymes and respiratory chain subunits, was demonstrated in cardiomyocytes of trisomic fetuses (Conti et al., 2007), of which 40% have consensus DNA binding sites in their 5′ flanking regions for the nuclear respiratory factor 1 (NRF1), and 20% showed a high resemblance for the estrogen-related receptor α (ERRα) (Piccoli et al., 2013). The copy number of mtDNA reduced to 400 in CDS-HFF and 500 in NCDS-HFF from ~600 copies in N–HFF per nuclear genome (i.e., per cell) (Piccoli et al., 2013). Assay of HSA21 genes by quantitative real-time PCR (qRT-PCR) showed significant upregulation of nuclear-embedded mitochondrial genes (BTG3, SOD1, ITSN1, DYRK1A, NRF2 and RIP140) in

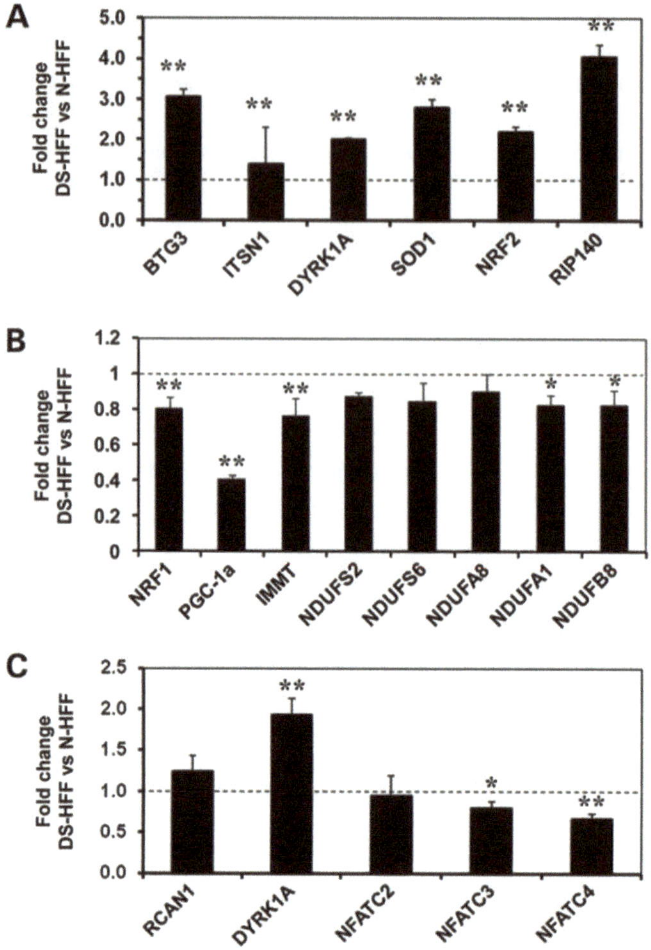

FIGURE 8.8 Gene expression is dysregulated in Down syndrome fibroblasts. Gene expression fold change in DS-human fetal fibroblast (HFF) samples versus N-HFF samples for HSA21 genes (A), nuclear-encoded mitochondrial genes (B) and calcineurin/NFAT-related genes (C) as obtained by qRT-PCR. Values represent the mean of three replicates ±SEM. $*P < .05$, $**P < .01$, HFF, human fetal fibroblasts; *DS-HFF*, Down syndrome-HFF; *N−HFF*, normal euploid foeti. *Reprinted from Piccoli et al. (2013), with permission from Oxford University Press and Copyright Clearance Center.*

DS-HFF compared to euploid (N−HFF) indicating imbalance of gene dosage (Fig. 8.8A) (Piccoli et al., 2013). In contrast, significant downregulation of genes involved in various mitochondrial functions, including the respiratory chain, mitochondrial biogenesis and maintenance of morphology, and also genes associated with related pathways, such as the CN/NFAT (Nuclear factor of activated T-cells) were observed in trisomic fibroblasts (Fig. 8.8B). Such downregulation of genes involved in mitochondrial functions demonstrated perturbation of the expression of genes linked to mitochondrial pathways by the trisomic state of the HSA21 genes. Interestingly, downregulation of NFATc3 and NFATc4 was co-occurred with the upregulation of DYRK1A and RCAN1 of HSA21 genes in trisomic fibroblasts, which are involved in regulation of the levels of NFATc phosphorylation (Fig. 8.8C) (Piccoli et al., 2013).

Network prediction analysis has indicated many HSA21 proteins interacting with components of the JAK/STAT pathway (Sturgeon et al., 2012). Mitochondrial abnormalities and reduction in COX activity were suggested to be induced by overproduction of beta APP (Askanas et al., 1996), a gene mapped to HSA21; however, the Ts1Cje mouse model with disomic APP showed decreased mitochondrial function and ATP production. Several HSA21 genes have been demonstrated to enhance APP-dependent toxic effects on the mitochondria. Furthermore, ETS-2 mapped onto HSA21 influenced the mitochondrial death pathway in neurons of patients with DS (Helguera et al., 2013). Another gene of HSA21, the intersectin-1s (ITSN1), has been demonstrated to regulate the mitochondrial apoptotic pathway in endothelial cells (Predescu et al., 2007). SLC25A4/ANT1, a gene among genes of several targets that are downregulated in trisomic fetal hearts, plays a major role in mitochondrial function. Across the mitochondrial membrane, this gene codes for the main translocator of ADP/ATP (Izzo et al., 2017a).

6.1.1 The co-activator PGC-1α

Notably, alterations in mitochondrial expression in DS cells have been attributed to frequency and activity of the transcriptional co-activator PGC-1α (peroxisome proliferator-activated receptor gamma co-activator 1-alpha or PPARGC1A). PGC-1α is the master modulator of mitochondrial biogenesis and respiratory function (Piccoli et al., 2013; Valenti et al., 2010, 2011; Papa et al., 2012), which is downregulated at both the mRNA and protein level in DS-HFF (Piccoli et al., 2013). It was demonstrated that mitochondrial dysfunction in DS is caused by downregulation of PGC-1α, and several nuclear-encoded mitochondrial genes (NEMGs) (Izzo et al., 2014). PGC-1α acts as a coactivator of the PPARs transcription factor family, and thus, depression of PGC-1α activity affects the transcriptional ability of PPAR-controlled genes (Fig. 8.9) (Mollo et al. 2020). In relevance, the

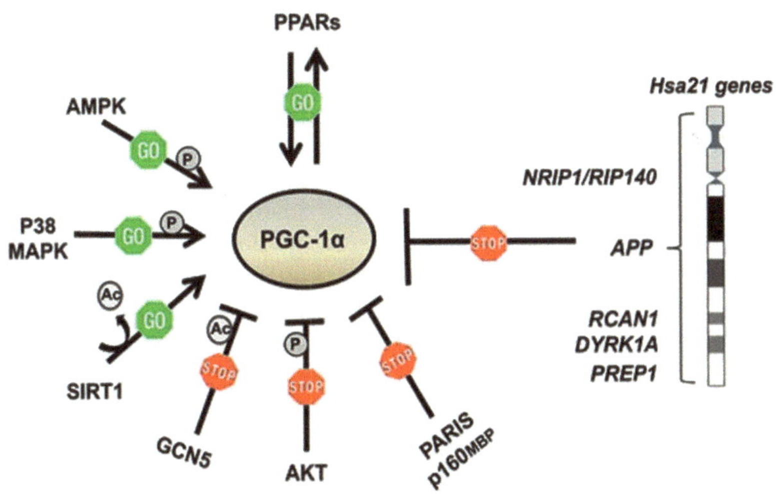

FIGURE 8.9 Factors regulating PGC-1α expression and/or activity. The peroxisome-proliferator-activated receptors (PPARs) regulate, and are regulated by, PGC-1α. AMPK, P38 MAPK and SIRT1 positively regulate PGC-1α at post-transcriptional levels. GCN5, AKT, PARIS and p160MBP negatively regulate PGC-1α. PGC-1α expression and/or activity is also regulated by genes mapping to chromosome 21. *Reprinted from Mollo et al., 2020 under the terms and conditions of the Creative Commons Attribution (CC BY) license, http://creativecommons.org/licenses/by/4.0/.*

function of PGC-1α is regulated by the nuclear receptor interacting protein 1 (NRIP1), and thus, repression of PGC-1α activity is caused by the overexpression of the NRIP1/RIP140 of HSA21 (Conti et al., 2007; Vilardell et al., 2011; Izzo et al., 2014), which is further supported by attenuation of NRIP1 in DS-HFF that increased the levels of transcription of PGC-1α and NEMGs and restored mitochondrial function (Izzo et al., 2014). This protein is known to further interact with transcriptional partners such as ERRα, NRF1, and NRF2 that regulate the expression of NEMGs. Down-regulation of its transcriptional partners and many other NEMGs has been reported in fetal hearts of DS (Conti et al., 2007).

Expression of PGC-1α is also controlled by two other HSA21 genes, the kinase DYRK1A and the regulator of CN 1 (DSCR1/RCAN1) through the binding of NFATc to the PGC-1α promoter in the CN/NFAT pathway (Arron et al., 2006). The concurrent overexpression of the trisomic HSA21 genes, including RIP140, SUMO3, RCAN1 and DYRK1A, and downregulation of NFATc genes, is expected to result in the depression of PGC-1α expression in DS samples (Arron et al., 2006).

Downregulation of PGC-1α protein in trisomic fetal fibroblasts was correlated well with the downregulation of complex I activity and with the decrease in copy number of mtDNA (more pronounced in CDS-HFF). The transgenic PGC-1α null mice showed reduced expression of mitochondrial genes in multiple tissues (Mitra et al., 2012). PGC-1α controls the expression of NEMGs in association with its transcriptional partners NRF1 and ERRα, which are also downregulated in DS. Significant downregulation of NRF1, PGC-1α, and IMMT is reported in DS-HFF (Piccoli et al., 2013). A noticeable change was reported, following silencing experiments, in the organization of the inner membrane that formed the concentric layers instead of organized tubular cristae, suggesting that downregulation of IMMT caused alterations in cristae patterns similar to those observed in DS samples (John et al., 2005; Piccoli et al., 2013).

6.1.2 The nuclear receptor interacting protein 1

In general, the expression of the NRIP1 (also known as RIP140), a transcriptional corepressor gene mapped onto HSA21, increases during neural differentiation of human embryonic stem cells and is negatively correlated with stem cell markers such as Oct4 and Sox2 during early stages of neuronal differentiation. Upregulation (1.5- to 4-fold) of this highly conserved gene was described in the heart, hippocampus and fibroblasts of cardiopathic fetus with DS (CDS-HFF) (Conti et al., 2007; Piccoli et al., 2013; Vilardell et al., 2011). The overexpression of NRIP1 plays a crucial role through reduction of respiratory efficiency and alteration in mitochondrial morphology in DS (Izzo et al., 2014; Powelka et al., 2006). NRIP1 has been suggested as a good candidate gene of HSA21 to confer downregulation of NEMGs. This gene binds to and repress many of the nuclear receptor super family and has been reported regulating expression of NEMGs in adipose tissue, heart, muscle, and liver (Fritah et al., 2010; Nautiyal et al., 2013) and also upregulated NRIP1 protein was described in the DS hippocampus (Gardiner, 2006). Co-upregulation of SUMO3 (another HSA21gene) and RIP140 is suggested to exert a synergistic effect in trisomic cells; however, sumoylation of RIP140 was hypothesized to modulate its repressive activity.

Thus, overexpressed NRIP1 or RIP140 affects the oxidative metabolism and mitochondrial biogenesis by negatively controlling the mitochondrial pathways through downregulation of PGC-1α and NEMGs (Izzo et al., 2014). Experiments on silencing and re-expression demonstrated significant effects on oxidative metabolism and mitochondrial biogenesis by

RIP140 expression (Powelka et al., 2006); even mild overexpression of RIP140 repressed expression of nuclear mitochondrial genes associated with all the respiratory chain complexes (Seth et al., 2007). The transcription factors such as NRF1, ERRα, and PPARs are repressed by trisomic NRIP1 and induced by PGC-1α in a dose-dependent manner in neonatal rat cardiomyocytes (Chen et al., 2005; Liu, 2012) and also in human fetal fibroblasts (Piccoli et al., 2013).

6.1.3 Interaction of PGC-1α and NRIP1

In cells of DS patients and mouse models, the activity of PGC-1α was repressed (Piccoli et al., 2013; Valenti et al., 2016; Izzo et al., 2017a). The amount of mtDNA was correlated with ~40%–50% reduction in the amount of PGC-1α gene transcripts and PGC-1α encoded protein level in trisomic NCDS-HFF and an even more marked decrease in CDS-HFF samples compared to N–HFF, suggesting a direct correlation between PGC-1α expression and amount of mtDNA (Fig. 8.8) (Piccoli et al., 2013). Overexpression of RCAN1, NRIP1, and miR-155 in trisomic HSA21 caused downregulation of PGC-1α and the mitochondrial transcription factor TFAM leading to lowering of the respective protein levels and mitochondrial biogenesis in different cells, including brain (Quiñones-Lombraña and Blanco, 2015). Reduced expression of PGC-1α activity leads to downregulation of Mnf-2 expression and those results in decrease in mitochondrial fusion in DS fibroblasts (Izzo et al., 2017a). Altogether, alterations in signaling pathways with concurrent impairment of PGC-1α expression collectively correlate with impairment in mitochondrial biogenesis. It is now apparent that the NRIP1-PGC-1α axis can be targeted for mitigation of mitochondrial dysfunction in DS.

Use of siRNAs repressed expression of NRIP1 in trisomic fetal fibroblasts, and that resulted in increased levels of PGC-1α and NEMGs, reversal of/increase in the mitochondrial function, reduction in ROS, and restoration of ATP production (Izzo et al., 2014). Thus, the interaction of NRIP1-PGC-1α indicated a potential therapeutic target for restoring the mitochondrial dysfunction in DS. Induction of PGC-1α has been demonstrated to upregulate hundreds of NEMGs in human osteoblast-like cells (Schreiber et al., 2004), of which at least 1/3 genes are NRIP1 targets. In contrast, PGC-1α null (Leone et al., 2005; Mitra et al., 2012), and NRIP1 knock-in (Seth et al., 2007) mice showed reduced expression of the same mitochondrial genes in multiple tissues. Thus, it is apparent that co-expression of NRIP1 and PGC-1α activates mutual antagonization in regulation of the mitochondrial metabolism. Accordingly, it was described that NRIP1 is significantly upregulated, whereas PGC-1α is significantly downregulated in DS fetal hearts and fibroblasts at both mRNA and protein level (Conti et al., 2007; Izzo et al., 2018; Piccoli et al., 2013). Thus, PGC-1α function is both antagonized and regulated by the nuclear receptor interacting protein NRIP1 or RIP140 encoded on HSA21.

6.1.4 SUMO3

Upregulation of SUMO3 is also demonstrated in fetal heart and DS-HFF, which affects mitochondrial function by modulating the NRIP1 repressive activity (Rytinki and Palvimo, 2009). The two conserved lysines, Lys756 and Lys1154, are located in the distinct repression domains of NRIP1 and thus, are subject to reversible SUMOylation. The SUMO3-acceptor lysines have a significant effect as a corepressor on the nuclear distribution of NRIP1 (Rytinki and Palvimo, 2009). Synergistic effects of simultaneous upregulation of NRIP1 or RIP140 and SUMO3 significantly affect oxidative metabolism and mitochondrial biogenesis through repression of mitochondrial genes of the respiratory chain complexes (Powelka et al., 2006; Seth et al., 2007). SUMOylation also affects the transcriptional activity of PGC-1α, by facilitating its transfer outside the nucleus (Rytinki and Palvimo, 2009). The SUMOylating effects on NRIP1 and PGC-1α have been ascribed for SUMO-1 and SUMO-2; however, an identical action of SUMO-3 has been presumed since that share a nearly identical structure with SUMO-2 (Mukhopadhyay and Riezman, 2007). Altogether, overexpression of SUMO-3 might therefore contribute to a concurrent improvement of NRIP1 function and decrease the activity of PGC-1α in DS (Izzo et al., 2018).

6.1.5 *DYRK1A*

In addition to NRIP1, the kinase DYRK1A and the regulator of CN 1 (DSCR1/RCAN1), control PGC-1α through the binding of NFATc to the PGC-1α promoter in the CN/NFAT pathway (Arron et al., 2006), and thus, have been proposed to contribute to mitochondrial dysfunction. Even the modest overexpression of DYRK1A demonstrated reduction in levels and activity of NFATc protein. The NFAT proteins may interact with MEF2C or MEF2D on the PGC-1α promoter (Handschin et al., 2003). The downregulation of NFATc3 and NFATc4 was observed in DS fetal hearts and fibroblasts associated with upregulation of DYRK1A and RCAN1 (Piccoli et al., 2013). In DS model, knockout of Nfatc3 and Nfatc4 demonstrated a decrease in enzymatic activity of complex II and IV of the respiratory chain and mitochondrial oxidative activity in cardiomyocytes (Bushdid et al., 2003). Interestingly, the concurrent overexpression of RIP140, SUMO3, RCAN1, and DYRK1A and the downregulation of NFATc genes result in the depression of PGC-1α expression in DS (Handschin et al., 2003; Arron et al., 2006).

6.1.6 *RCAN1*

RCAN1, also known as calcipressin, has been reported significantly overexpressed in the brain of both DS and sporadic AD patients (Wu and Song, 2013), and its overexpression has been linked to susceptibility to oxidative stress and mitochondrial dysfunction through altered Ca^{2+} buffering (Ermak et al., 2012). Neurons without RCAN1 expression displayed an increased resistance to cellular damage by ROS (Porta et al., 2007). Alteration in mitochondrial size and number, increased ROS production, and altered membrane potential were described in response to ROS in cells with overexpressed RCAN1 (Peiris et al., 2014). Overexpressed RCAN1 causes calcium overloading (Sun et al., 2014) and affects the function of mtPTP, which consequently results in incapability of Ca^{2+} retention, massive swelling of mitochondria and rupture of its outer membrane (Izzo et al., 2017b; Sun et al., 2014). RCAN1 induced autophagy of mitochondria and improved cell survival in cardiomyocytes (Duan et al., 2015). In mice, the overexpression of the brain-specific RCAN1.1S isoform induced deficiency of early age-dependent memory and synaptic plasticity, tau pathology, and dysregulation of dynamin-related protein 1 (DRP1) action linked to altered mitochondrial functions and oxidative stress (Wong et al., 2015). RCAN1 has also been reported to be responsible for changes in mitochondrial fission/fusion dynamics via phosphorylation of DRP-1 and altered expression of mitochondrial factors through inhibition of nuclear translocation of transcription factors NFKB/NFAT (Cereghetti et al., 2010; Tan et al., 2011). Additionally, RCAN1 acts as a stress-inducible factor that is modulated by oxidative stress (Ermak et al., 2012). Notably, an intermittent overexpression of RCAN1 appears to be protective, but its chronic upregulation is deleterious to cells through interaction with upregulated SOD1, another HSA21 gene.

6.1.7 *PREP1*

The PREP1 gene of HSA21, encoding a homeodomain transcription factor, has been demonstrated to regulate the mitochondrial OXPHOS components directly and/or indirectly (Kanzleiter et al., 2014). The indirect mechanism of PREP1 action controls the stability of p160 Myb binding protein, a powerful negative regulator of PGC-1α activity in mitochondria (Fan et al., 2004). Increase in Pgc-1α expression led to concomitant increase in mitochondrial capacity in the muscle of Prep1 ablated mice, which further showed an increase in abundance of mitochondrial OXPHOS proteins, an increased citrate synthase activity and also an improved maximal oxidative capacity (Kanzleiter et al., 2014). ChiP-seq analysis revealed the binding sites of Prep1 in the promoter regions of genes encoding 16 mitochondrial proteins, which were also upregulated in the skeletal muscle in response to Prep1 ablation. The report identified PREP1 as a direct negative transcriptional regulator of mitochondrial proteins in addition to its indirect effects via p160-PGC-1α (Kanzleiter et al., 2014). Furthermore, Opa1 and Mfn2, two genes of the mitochondrial fusion process, were significantly increased in the Prep1 ablated mice (Kanzleiter et al., 2014). These findings suggested that the PREP1 negatively regulates mitochondrial proteins and thus, affects both mitochondrial function and dynamics.

6.1.8 Humanin

Humanin (HN), a mitochondrial short 24-residue polypeptide, is expressed in response to mitochondrial stress (Ijiri et al., 2005). An increased expression of HN protein and HN-mRNA (mean mRNA levels in DS was 7.22 vs. 5.00 in normal subjects) levels has been reported in cultured fibroblasts from DS patients compared to normal controls, irrespective of the gender and age effects (Fig. 8.10) (Cantó et al., 2020; Salemi et al., 2020). Elevation of serum levels of HN was also reported in DS subjects compared to normal controls and that was positively correlated with aging in humans indicating contribution of HN in physical and biochemical parameters such as handgrip strength, insulin sensitivity, elevated triglycerides, and so on (Conte et al., 2018). The results had postulated that HN might have potential contribution to various aspects of the phenotypic expression in subjects with DS, including premature aging and several autoimmune factors related to DS. Moreover, the presence of mitochondrial dysfunction in DS could elicit the production of effective stress responses, including up-regulation of HN-expression and that has led HN to be considered as a potential therapeutic target for aging-related diseases. Indeed, exogenous HN treatment attenuated myocardial fibrosis and apoptosis in elderly mice, suggesting its role in cardioprotection (Qin et al., 2018). Also, HN's protective role in neurodegeneration against cognitive aging in humans (Yen et al., 2018) and age-related macular degeneration (AMD) (Solanki et al., 2018) represent a promising therapeutic opportunity for treatment of cognitive impairment, AD, and early aging in DS. Over 50% of patients with DS develop AD and dementia in their fifth or sixth decade of life. However, age effect on HN-expression in DS fibroblasts was not consistent in vitro (Salemi et al., 2020).

6.1.9 SOD1

Redox imbalance is caused by overexpression of Cu/Zn-superoxide dismutase, encoded by SOD-1 mapped to HSA21, and also by the cooperative effects and transcriptional regulation of genes located on other human chromosomes in vitro, ex vivo, and animal studies (Pagano and Castello, 2012; Epstein et al., 1987). In DS, an imbalance between the activities of SOD1 and glutathione peroxidase (GPX) leads to pathological expression of the oxidative stress. Overexpression of SOD1 (50% greater than normal) activity has been demonstrated (Pagano and Castello, 2012), with concomitant reduction in expression of catalase (CAT) and GPX in DS (Uttara et al., 2009). SOD1, the major SOD in mammalian cells, catalyzes the dismutation of

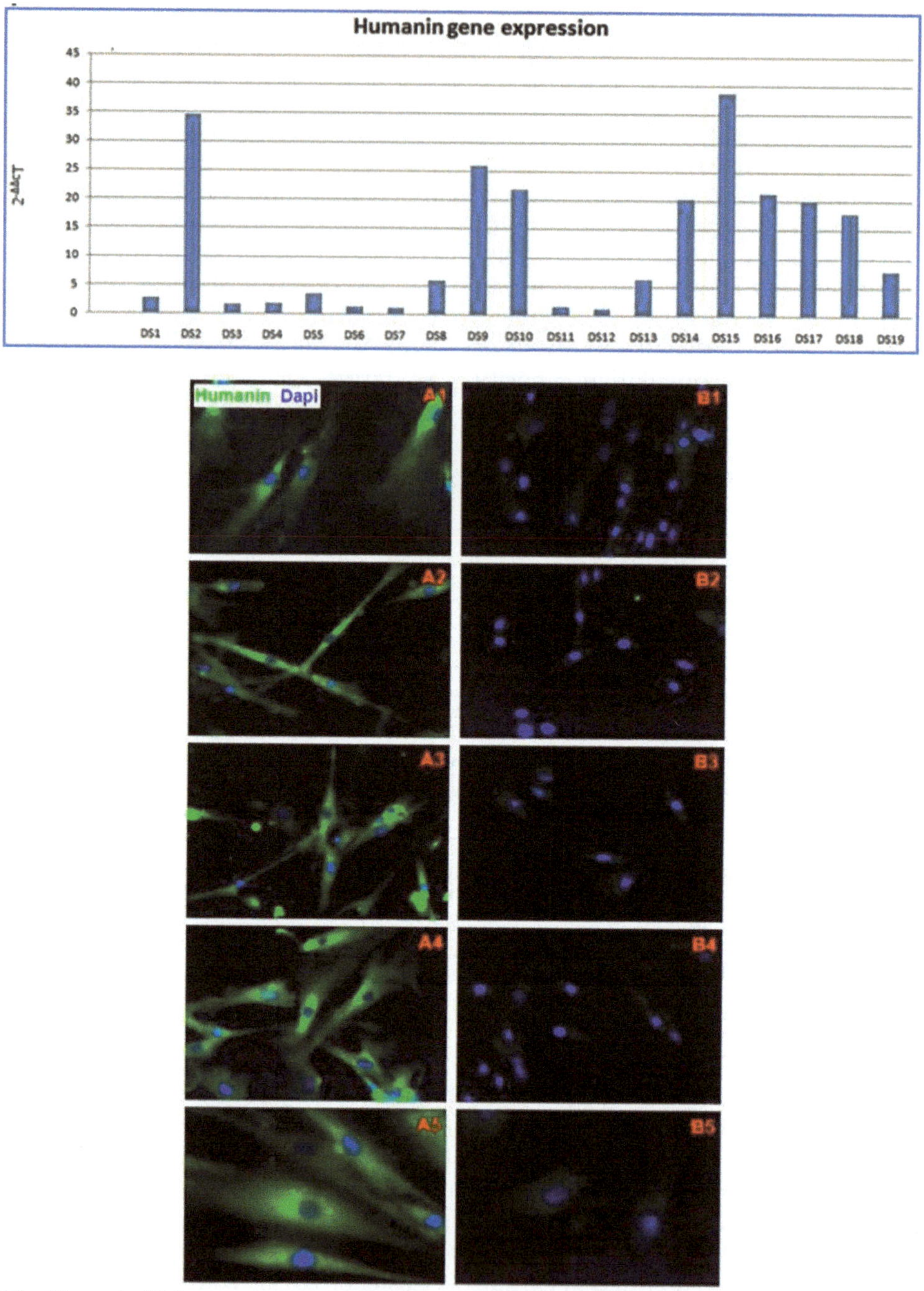

FIGURE 8.10 Humanin (HN) gene expression in Down syndrome (DS): (A) HN mRNA levels in cultured fibroblasts from DS patients and age and sex matched normal controls. Expression detected by quantitative real-time polymerase chain reaction (qRT-PCR); (B) immunofluorescence of fibroblasts obtained from oral biopsy of periodontal gingival tissue in DS subject

superoxide radicals to H_2O_2 and O_2, and this in turn results in a significant decrease in antioxidant defense mechanism in DS patients. An elevation in the oxidized/reduced glutathione ratio accompanying with increased endogenous peroxide production has been reported in the fibroblast cultures of human DS and that was consistent with reduced antioxidant activity in fetuses (Gimeno et al., 2014). In vivo alterations in mitochondrial function appeared consistent with a prooxidant state in Ts16 mice.

6.1.10 S100B

The S100B gene encodes the B-subunit of the calcium-binding S100 proteins and is present in the mammalian brain in the form of a homodimer. This protein is involved in glial cell activation, neuroinflammation and in the regulation and maintenance of the serotonergic nervous system, particularly on the hippocampus (reviewed in Shapiro et al., 2010). The neural precursors extracted from DS fetus demonstrated a constitutive overexpression of S100B in vitro, which led to increase in ROS formation and activation of stress response kinases and increased apoptosis (Shapiro et al., 2010).

6.1.11 Cystathionine-β-synthase

Perturbation in mitochondrial function, including the Complex IV, has been attributed to the profound repression of mitochondrial electron transport, oxygen consumption, and ATP generation in DS cells facilitated by overexpression of cystathionine-β-synthase (CBS) (mapped onto 21q22.3 and to the proximal half of mouse chromosome 17) and ensuing overproduction of the gaseous transmitter hydrogen sulfide (H_2S) and polysulfide (Kamoun et al., 2003; Panagaki et al., 2019). Higher expression of CBS was estimated in both cytosolic and mitochondrial compartments in fibroblasts derived from DS individuals compared with normal counterparts with concomitant slower proliferation of cells (Fig. 8.11) (Panagaki et al., 2019). Complex IV, the essential electron transport component, is best-characterized by its reversible (H_2S) and irreversible (cyanide) inhibitors (Nicholls et al., 2013). H_2S, an endogenous mammalian gaseous mediator, regulates mitochondrial function besides many cellular processes (Szabo, 2020). At lower concentration, H_2S influences cellular bioenergetics; conversely, its toxicological concentration suppresses mitochondrial function via inhibition of complex IV. Interestingly, treatment of DS cells with aminooxyacetate (AOAA) or siRNA-mediated silencing of CBS normalized cellular H_2S levels, improved mitochondrial electron transport and ATP synthesis, and restored complex IV activity, and cell proliferation. Thus, antagonism of CBS indicated a potential pharmacological approach for the correction of DS-associated mitochondrial dysfunction (Panagaki et al., 2019).

H_2S may directly affect mitochondrial ATP production and is likely to suppress cellular bioenergetics, O_2 consumption, tissue proliferation, etc., resulting in impairment of metabolic fitness in DS. Notably, overproduction of CBS has plausible effects on neurological features, including phenocopy of the DS-like neurocognitive deficits, mediated through H_2S overproduction and inhibition of mitochondrial complex IV and consequent perturbance of OXPHOS and production of ATP (Marechal et al., 2019). Moreover, pharmacological inhibition of CBS appears capable to reverse H_2S-mediated

(A1-A5), Humanin protein green fluorescence, nuclei were counter-stained in blue with 100 ng/mL 4,6-diamidino-2-phenylindole (DAPI), and Immunofluorescence of fibroblasts obtained from oral biopsy of periodontal gingival tissue in normal subject (B1–B5), Humanin protein green fluorescence, nuclei were counter-stained in blue with 100 ng/mL 4,6-diamidino-2-phenylindole (DAPI). *Reprinted from Salemi et al. (2020), under the terms of the Creative Commons Attribution License (CC BY).*

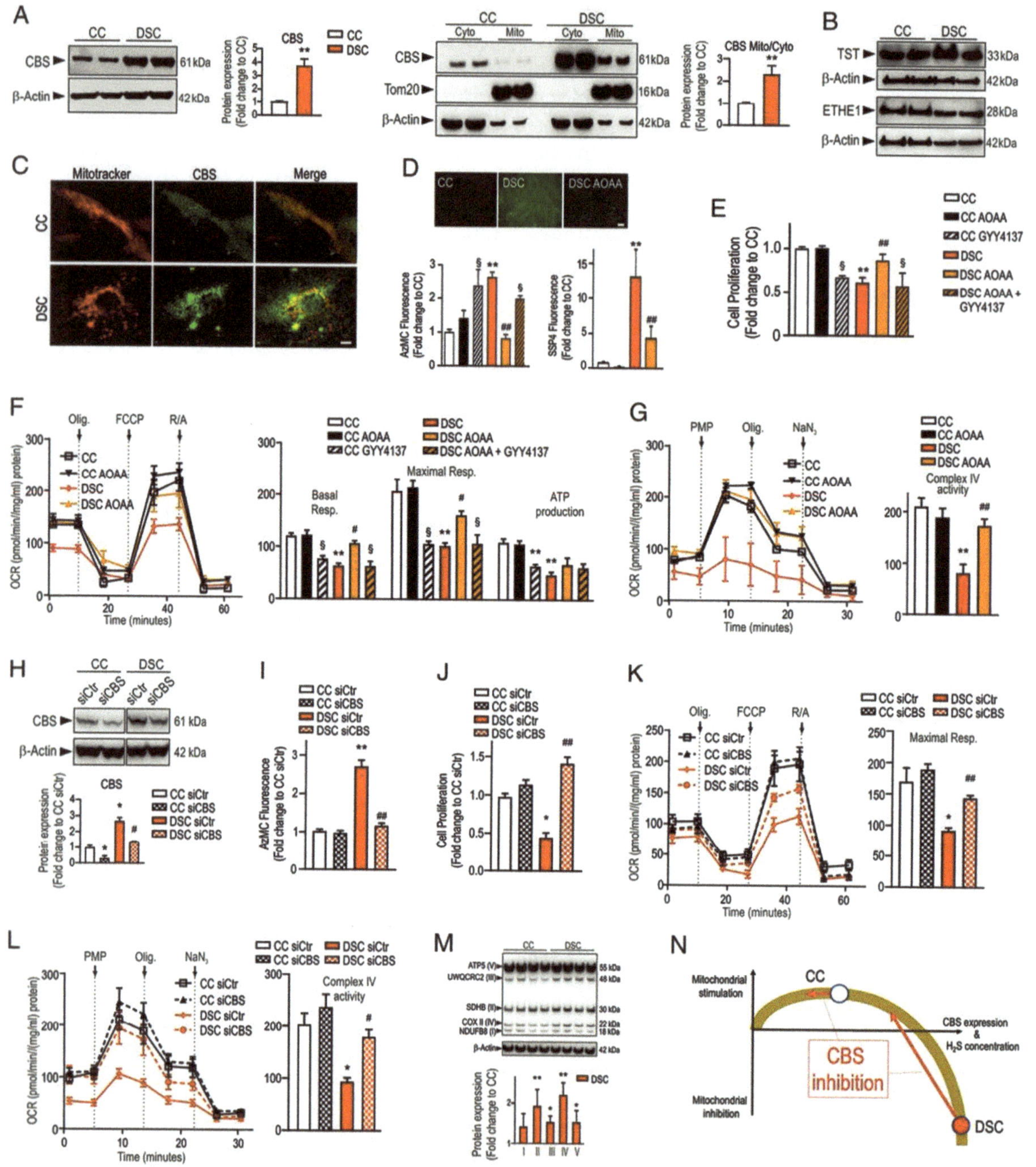

FIGURE 8.11 **Fribroblasts obtained from DS (DSC) show** markedly higher CBS expression compared with healthy controls (CC) ((A) Western blotting and (C) confocal microscopy). CBS was both cytosolically and mitochondrially localized. Expression of H_2S metabolizing enzymes TST and ETHE1 was unaltered in DSCs (B). DSCs exhibit increased amounts of intracellular H_2S and reactive polysulfides (D). DSCs show reduced cell proliferation rate, which is normalized by aminooxyacetate

inhibition of complex IV, and it is likely that some of the functional defects associated with DS could be reversed; however, stringent follow-up studies would be necessary to determine the reversal of the biochemical processes in vivo. Over the pastdecades, clinical trials have been considered with CBS inhibitors, including AOAA and various prodrugs of AOAA and natural products as CBS inhibitors such as apigenin. Notably, benserazide, a clinically approved drug, has been considered for its secondary pharmacological effect as a CBS inhibitor (Szabo and Papapetropoulos, 2017).

6.1.12 MicroRNA

Among the HSA21 encoded noncoding RNAs (ncRNAs), the long ncRNAs (lncRNA) have been expressed, while microRNAs (miRNAs) are less represented (Letourneau et al., 2014). Mitochondrial biogenesis is affected by mir-155-5p through the mitochondrial transcription factor A (TFAM) gene, which is downregulated in trisomic hearts (Quinones-Lombrana and Blanco, 2015). TFAM, a nuclear encoded protein, regulates transcription and maintenance of mtDNA, and thus, mitochondrial biogenesis. Another potential miRNA of HSA21, let-7c, is also responsible for mitochondrial anomalies. Thus, it has been hypothesized that dysregulation of both miR155 and let-7c might have significant effect on the mitochondrial phenotype in individuals with DS.

6.2 Mitochondrial biogenesis in DS

The mitochondrial activity depends on the integrity of mitochondrial biogenesis, including maintenance and survival of pre-existing mitochondrial function. Mitochondrial biogenesis includes membrane biosynthesis, replication of mtDNA, synthesis of nDNA and mtDNA-encoded proteins, and transport of nuclear-encoded proteins into the mitochondrial organelles (reviewed in Dominy and Puigserver, 2013). Mitochondrial biogenesis is a spontaneous response to the cells during division and differentiation, consequences of aging, and to combat with the environmental stressors such as exposure to heat and cold, exercise, calorie restriction, and oxidative stress. The reduction in mitochondrial biogenesis results in mitochondrial dysfunction, which in turn affects mitochondrial biogenesis.

Several ultrastructural studies have reported alterations in the network of the mitochondrial structure and dynamics both in DS mouse models

(AOAA) (3 μM); treatment of CCs with the H_2S donor GYY4137 (3 mM) phenocopies the inhibition of proliferation seen in DSCs (E). DSCs exhibit low basal and maximal mitochondrial oxygen consumption rates (OCR); these parameters are improved by AOAA; exposure of CCs to GYY4137 phenocopies the metabolic suppression seen in DSCs (F). Complex IV is blocked in DSCs, but it is restored by AOAA (G). The H_2S-releasing molecule P-(4-methoxyphenyl)-P-4-morpholinyl-phosphinodithioic acid (GYY4137) (Szabo and Papapetropoulos, 2017) phenocopied the effect of DS in CCs in terms of bioenergetics and proliferation and reversed the beneficial effects of AOAA in DSCs (E–G). SiRNA-mediated silencing of CBS in DSCs (H) suppresses H_2S production (I), normalizes DSC proliferation (J), and enhances OCR (K) and Complex IV activity (L). DSCs show increased expression of various mitochondrial electron transport chain proteins belonging to complexes II, III, and IV (M), likely representing a compensatory response. Comparison shows bell-shaped cellular bioenergetic role of H_2S in DSCs versus CCs and effect of CBS inhibition (N). Once CBS expression is normalized, the H_2S-mediated tonic suppression of Complex IV is lifted, and the cells regain their capacity to perform mitochondrial OXPHOS (N). Notably, CBS inhibition does not significantly affect proliferation or bioenergetics in healthy CCs, which are not under the tonic suppressive effect of high H_2S, (N). $*P < .05$ and $**P < .01$, significant difference between CCs and DSCs; $\#P < .05$ and $\#\#P < .01$, significant effect of AOAA in DSCs; $\S P < .05$, significant effect of GYY4137 in CCs or significant difference between AOAA and GYY4137+AOAA in DSCs. Data are shown as mean ± SEM of at least three experiments, and level of significance is derived from one- and 2-way ANOVAs, followed by a post-hoc Bonferroni test. *Reprinted from Panagaki et al. (2019), under the terms of the Creative Commons Attribution License (CC BY) as guided by PNAS.*

in vivo and cells of human DS in vitro. These include irregular-shaped mitochondria in cerebellar neurons of the Ts16 mice (Mollo et al., 2017), and other abnormalities, including irregular shape, predominant breaks within the inner membranes, alterations in the pattern of cristae along with an increased number of shorter mitochondria indicating fission and a smaller mitochondrial volume (Izzo et al., 2017a), increased mitochondrial fragmentation and breaking of mitochondrial network in the primary cultures of DS astrocytes and neurons (Helguera et al., 2013). The control of the maintenance of neuronal function, differentiation, and survival through the proliferation of adult neurons has been established with mitochondrial dynamics (Flippo and Strack, 2017), which led to postulate that alterations in mitochondrial network could be underlying for the neuronal dysfunctions in DS.

The biogenesis process is achieved through the coordinated expression of regulatory circuit, including signaling pathways and activation of nuclear-encoded transcriptional complexes for cellular energy production. The regulatory effects of several transcriptional factors of the nuclear genome, including the nuclear respiratory factor 1 and 2 (NRF-1/2) that regulates the expression of several proteins of the five respiratory complexes and controls the mitochondrial import and heme biosynthesis, are implicated in modulating the genomic expression of mitochondrial proteins in cells (Dominy and Puigserver, 2013). The NRF-1/2 are also involved in activation of the mitochondrial transcription factors such as TFAM, TFB1M, and TFB2M (Gleyzer et al., 2005), which are essential for basal transcription of mammalian mtDNA (Falkenberg et al., 2002). The regulation of the mitochondrial transcription factors by the transcriptional activity of NRFs is involved in nucleo-mitochondrial interactions. In Ts65Dn cells of the DS mouse model, mitochondrial biogenic system was overtly compromised, leading to reduction in the content of mtDNA and protein levels, particularly PGC-1α, NRF-1, and TFAM (Valenti et al., 2016). The OXPHOS activity was significantly declined in cells of the CNS, including the NPCs of the hippocampus of Ts65Dn mice (Valenti et al., 2016), and in skin fibroblasts of the cardiopathic DS fetuses (Piccoli et al., 2013), due to severe impairment of the mitochondrial biogenesis.

Mitochondrial biogenesis is regulated by additional transcriptional factors of the nuclear hormone receptor superfamily, including the peroxisome proliferator activated receptors (PPARs) that are activated by several fatty acid ligands. The PPAR family of nuclear receptor genes is involved as a master regulator of mitochondrial metabolism in different tissues and induces expression of the mitochondrial β-oxidation genes through interaction with specific transcription factors (reviewed in Mello et al., 2016). Of the three isoforms (PPAR-α, PPARβ/δ, and PPAR-γ) of PPAR family, PPAR-α and PPAR-δ act as regulators of lipid oxidation, and PPAR-γ promotes lipid synthesis and storage. PPAR-α cooperates with the PGC-1α cofactor in transcriptional control of nuclear genes, which encode the mitochondrial fatty acid oxidation enzymes (Vega et al., 2000), and is affected by the repressed activity of PGC-1α in trisomic status. The PPAR-γ and PPAR-δ also promote mitochondrial biogenesis in a cell type-specific manner, in addition to their involvement in lipid metabolism (Hock and Kralli, 2009). There are several potential PPAR agonists identified for their neuroprotective effects against mitochondrial alterations in neurodegenerative diseases (Agarwal et al., 2017; Skerrett et al., 2015). The PPAR-γ agonists have significantly induced the expression of endogenous PGC-1α and indicated that PPAR-γ affects mitochondrial biogenesis indirectly via activation of the transcription of PGC-1α. Since PGC-1α acts as a coactivator of PPAR-γ, PPAR-γ also enables PGC-1α to enhance its own expression in an autoregulatory fashion (Hock and Kralli, 2009). Notably, a cross-talk between PPARγ and Ca-mobilization/signaling has been suggested as the cause of the Ca^{2+}-homeostasis deregulation in trisomic cells; however, the association of Ca transport and ROS production has yet to be satisfactorily understood (Anastasiadis et al., 2007).

The ERRs, including ERR-α, ERR-β, and ERR-γ, are also involved in mitochondrial biogenesis through binding to hundreds of genes involved in mitochondrial functions evidenced in functional genomic analysis, such as OXPHOS, fatty acid oxidation, Krebs cycle, and fission/fusion dynamics of mitochondria (Eichner and Giguère, 2011). Mitochondrial biogenesis has been demonstrated to be controlled by several other nuclear transcription factors, such as CREB (Gopalakrishnan and Scarpulla, 1994), myocyte enhancer factor-2 (Wan and Moreadith, 1995), the GLI-Kruppel class of zinc finger protein YY1 (Cunningham et al., 2007), and c-Myc (Morrish and Hockenbery, 2014), through their binding to mitochondrial gene promoters in different signaling pathways.

The apparatus of mitochondrial energy production was less efficient in fetal fibroblasts with DS, which was caused by the dysregulation of adenine nucleotide translocator, ATP synthase and adenylate kinase, and the deficient complex I, and that contributes to overproduction of ROS in mitochondria of DS. These expressions were attributed to changes in the cAMP/PKA signaling pathway (Valenti et al., 2010, 2011), which affects the abundance of the transcriptional coactivator PGC-1α. This protein plays a pivotal role in regulation of mitochondrial biogenesis and respiratory function in association with other transcriptional partners, such as ERRα, NRF1, PPARs, and YY1, and is negatively controlled by the corepressor gene RIP140 (Scarpulla, 2011). However, further molecular studies would decipher the mitochondrial dysfunction at the transcriptional level and establish the contribution of trisomy of HSA21 genes on the mechanisms of mitochondrial dysfunction in DS.

The potentiation of the transcriptional activity of most of the nuclear transcription factors involved in regulating mitochondrial gene expression is controlled by the PGC-1 family, including PGC-1α and PGC-1β, and the polycomb repressor complexes (PRCs). The PGC-1α and PGC-1β enhance the transcriptional activity of NRFs, ERRs, PPARs, and YY1 (Dominy and Puigserver, 2013; Finck and Kelly, 2006) and regulate function and biogenesis of mitochondria (Handschin and Spiegelman, 2006). PGC-1 family of coactivators acts as the key factor for the mitochondrial functioning and overall energy metabolism, including response to meet the demand of different environmental conditions such as starvation, exercise, or cold. By considering the association between the transcriptional activity of PGC-1 family and environmental conditions, this has been established that PGC-1 activity is controlled by the environmental stress–regulating pathways such as gluconeogenesis, thermogenesis, muscle differentiation, and cell growth (Handschin and Spiegelman, 2006; Finck and Kelly, 2006). The coactivator PGC-1α plays the key role in oxidative metabolism in association with several signaling pathways (Ventura-Clapier et al., 2008). The control of the transcriptional activity of PGC-1α by calcium-dependent phosphatase CN in muscle provides a coordinated link between increased contractile activity and increase in mitochondrial function (Handschin et al., 2003). Another example explains the activation of the AMPK/Sirt1/PGC-1α pathway to meet energy metabolism in response to exercise and calorie restriction (Cantó and Auwerx, 2009).

Genes of mitochondrial biogenesis were considered as compelling therapeutic targets. However, therapeutic approaches have not been routinely undertaken with the drugs available for modulating the activity of PGC-1α and/or the downstream PPAR proteins to control the overall mitochondrial dysfunction in DS patients. It has been examined that PPARs, AMP-activated kinases (AMPKs), and the NAD-dependent deacetylase SIRT1 can control the PGC-1α activity (Cantó and Auwerx, 2009).

The posttranslational modifications of direct phosphorylation by AMPK to promote the PGC-1α-dependent induction at the PGC-1α promoter level (Jäger et al., 2007) and stimulation of PGC-1α activity through deacetylation by SIRT1 (Wenz, 2009), might induce mitochondrial biogenesis. The effective therapeutic activators for these proteins, such as use of metformin and resveratrol for induction of AMPK and SIRT1, respectively, have been tested for neurodegenerative diseases, including AD in mouse models in which mitochondrial alterations play the central role (Jäger et al., 2007). Interestingly, studies involving metformin treatment exhibited multiple positive outcome on the mitochondrial activity in human DS cells (Izzo et al., 2017b), and treatment with resveratrol and epigallocatechin3-gallate (EGCG) overturned the severe impairment of mitochondrial bioenergetics and biogenesis in hippocampal progenitor cells of Ts65Dn mice (Valenti et al., 2016). Additionally, thiazolidinediones, pioglitazones, and bezafibrates, which selectively stimulate PPARs of PGC-1α pathway, were also recommended for testing as additional pharmacologicals to reinstate mitochondrial function in DS. Such targets on new therapeutic candidates can improve the neurological perspectives in DS and thus, prevent associated pathologic manifestations, including AD, diabetes, and hypertrophic cardiopathy.

References

Adam-Vizi, V., Starkov, A.A., 2010. Calcium and mitochondrial reactive oxygen species generation: how to read the facts. J. Alzheimer Dis. 20 (Suppl. 2), S413–S426.

Agarwal, S., Yadav, A., Chaturvedi, R.K., 2017. Peroxisome proliferator-activated receptors (PPARs) as therapeutic target in neurodegenerative disorders. Biochem. Biophys. Res. Commun. 483, 1166–1177.

Anastasiadis, P.Z., Guerra, R., Luxon, B.A., Thompson, E.A., 2007. Functional genomic analysis reveals cross-talk between peroxisome proliferator-activated receptor gamma and calcium signaling in human colorectal cancer cells. J. Biol. Chem. 282, 23387–23401. https://doi.org/10.1074/jbc.M702708200.

Angajala, A., Sangbin Lim, S., Phillips, J.B., Kim, J.-H., Yates, C., You, Z., Tan, M., July 12, 2018. Diverse roles of mitochondria in immune responses: novel insights into immuno-metabolism. Front. Immunol. https://doi.org/10.3389/fimmu.2018.01605.

Arron, J.R., Winslow, M.M., Polleri, A., Chang, C.P., Wu, H., Gao, X., Neilson, J.R., Chen, L., Heit, J.J., Kim, S.K., et al., 2006. NFAT dysregulation by increased dosage of DSCR1 and DYRK1A on chromosome 21. Nature 441, 595–600. https://doi.org/10.1038/nature04678.

Askanas, V., McFerrin, J., Baque, S., Alvarez, R.B., Sarkozi, E., et al., 1996. Transfer of beta-amyloid precursor protein gene using adenovirus vector causes mitochondrial abnormalities in cultured normal human muscle. Proc. Natl. Acad. Sci. USA 93, 1314–1319.

Azzu, V., Brand, M.D., 2010. The on-off switches of the mitochondrial uncoupling proteins. Trends Biochem. Sci. 35, 298–307.

Bambrick, L.L., Fiskum, G., January 10, 2008. Mitochondrial dysfunction in mouse trisomy 16 brain. Brain Res. 1188, 9–16 (Research Support, U.S. Gov't, Non-P.H.S.).

Barone, E., Arena, A., Head, E., Butterfield, D.A., Perluigi, M., 2017. Disturbance of redox homeostasisin Down syndrome: role of iron dysmetabolism. Free Radic. Biol. Med. https://doi.org/10.1016/j.freeradbiomed.2017.07.009.

Bartley, M.G., Marquardt, K., Kirchhof, D., WilkinsHM, P.D., Linseman, D.A., 2012. Overexpression of amyloid-beta protein precursor induces mitochondrial oxidative stress and activates the intrinsic apoptotic cascade. J. Alzheimers Dis. 28, 855–868.

Benard, G., Bellance, N., Jose, C., Melser, S., Nouette-Gaulain, K., Rossignol, R., 2010. Multisite control and regulation of mitochondrial energy production. Biochim. Biophys. Acta 1797, 698–709. https://doi.org/10.1016/j.bbabio.2010.02.030.

Busciglio, J., Yankner, B.A., 1995. Apoptosis and increased generation of reactive oxygen species in Down's syndrome neurons in vitro. Nature 378, 776–779.

Bushdid, P.B., Osinska, H., Waclaw, R.R., Molkentin, J.D., Yutzey, K.E., 2003. NFATc3 and NFATc4 are required for cardiac development and mitochondrial function. Circ. Res. 92, 1305–1313.

Butterfield, D.A., Di Domenico, F., Swomley, A.M., Head, E., Perluigi, M., 2014. Redox proteomics analysis to decipher the neurobiology of Alzheimer-like neurodegeneration: overlaps in Down syndrome and Alzheimer disease brain. Biochem. J. 463, 177–189. https://doi.org/10.1042/BJ20140772.

Cantó, C., Auwerx, J., 2009. PGC-1alpha, SIRT1 and AMPK, an energy sensing network that controls energy expenditure. Curr. Opin. Lipidol. 20, 98–105. https://doi.org/10.1097/MOL.0b013e328328d0a4.

Cantó, C., Gerhart-Hines, Z., Feige, J.N., Lagouge, M., Noriega, L., Milne, J.C., Elliott, P.J., Salemi M, P., Ridolfo, F., Salluzzo, M.G., Cannarrella, R., Giambirtone, M., Caniglia, S., Tirolo, C., Ferri, R., Romano, C., 2020. Humanin gene expression in fibroblast of Down syndrome subjects. Int. J. Med. Sci. 17 (3), 320–324. https://doi.org/10.7150/ijms.39145. Available from: http://www.medsci.org/v17p0320.htm.

Capone, G., Kim, P., Jovanovich, S., Payne, L., Freund, L., Welch, K., Miller, E., Trush, M., 2002. Evidence for increased mitochondrial superoxide production in Down syndrome. Life Sci. 70, 2885–2895. https://doi.org/10.1016/S00243205(02)01538-2.

Cereghetti, G.M., Costa, V., Scorrano, L., 2010. Inhibition of Drp1-dependent mitochondrial fragmentation and apoptosis by a polypeptide antagonist of calcineurin. Cell Death & Diff. 17, 1785–1794.

Chan, D.C., 2012. Fusion and fission: interlinked processes critical for mitochondrial health. Annu. Rev. Genet. 46, 265–287. https://doi.org/10.1146/annurev-genet-110410-132529.

Chen, H., Chomyn, A., Chan, D.C., 2005. Disruption of fusion results in mitochondrial heterogeneity and dysfunction. J. Biol. Chem. 280, 26185–26192.

Cogliati, S., Enriquez, J.A., Scorrano, L., 2016. Mitochondrial cristae: where beauty meets functionality. Trends Biochem. Sci. 41, 261–273.

Conte, M., Ostan, R., Fabbri, C., Santoro, A., Guidarelli, G., Vitale, G., et al., June 27, 2018. Human aging and longevity are characterized by high levels of mitokines. J. Gerontol. A Biol. Sci. Med. Sci. https://doi.org/10.1093/gerona/gly153.

Contestabile, A., Greco, B., Ghezzi, D., Tucci, V., Benfenati, F., Gasparini, L., 2013. Lithium rescues synaptic plasticity and memory in Down syndrome mice. J. Clin. Invest. 123, 348–361. https://doi.org/10.1172/JCI64650.

Conti, A., Fabbrini, F., D'Agostino, P., Negri, R., Greco, D., Genesio, R., D'Armiento, M., Olla, C., Paladini, D., Zannini, M., et al., 2007. Altered expression of mitochondrial and extracellular matrix genes in the heart of human fetuses with chromosome 21 trisomy. BMC Genom. 8, 268. https://doi.org/10.1186/1471-2164-8-2683.

Convertini, P., Menga, A., Andria, G., Scala, I., Santarsiero, A., Castiglione Morelli, M.A., et al., 2016. The contribution of the citrate pathway to oxidative stress in down syndrome. Immunology 149, 423–431.

Corbalan, J.J., Kitsis, R.N., 2018. RCAN1–calcineurin axis and the set-point for myocardial damage during ischemia-reperfusion. Circ. Res. 122, 796–798. https://doi.org/10.1161/CIRCRESAHA.118.312787.

Coskun, P.E., Wyrembak, J., Derbereva, O., Melkonian, G., Doran, E., Lott, I.T., Head, E., Cotman, C.W., Wallace, D.C., 2010. Systemic mitochondrial dysfunction and the etiology of Alzheimer's disease and down syndrome dementia. J. Alzheimers Dis. 20 (Suppl. 2), S293–S310.

Cree, L.M., Samuels, D.C., Chuva de Sousa Lopes, S., Karur Rajasimha, H., Wonnapinij, P., Mann, J.R., M Dahl, H.-H., Chinnery, P.F., 2008. A reduction of mitochondrial DNA molecules during embryogenesis explains the rapid segregation of genotypes. Nat. Genet. https://doi.org/10.1038/ng.2007.63.

Cunningham, J.T., Rodgers, J.T., Arlow, D.H., Vazquez, F., Mootha, V.K., Puigserver, P., 2007. mTOR controls mitochondrial oxidative function through a YY1PGC-1alpha transcriptional complex. Nature 450, 736–740. https://doi.org/10.1038/nature06322.

Darshi, M., Mendiola, V.L., Mackey, M.R., Murphy, A.N., Koller, A., Perkins, G.A., Ellisman, M.H., Taylor, S.S., 2011. ChChd3, an inner mitochondrial membrane protein, is essential for maintaining crista integrity and mitochondrial function. J. Biol. Chem. 286, 2918–2932. https://doi.org/10.1074/jbc.M110.171975.

De Stefani, D., Raffaello, A., Teardo, E., Szabo, I., Rizzuto, R., 2011. A forty-kilodalton protein of the inner membrane is the mitochondrial calcium uniporter. Nature 476, 336–340. https://doi.org/10.1038/nature10230.

Demuro, A., Parker, I., Stutzmann, G.E., 2010. Calcium signaling and amyloid toxicity in Alzheimer disease. J. Biol. Chem. 285, 12463–12468.

Dierssen, M., Vallina, I.F., Baamonde, C., García-Calatayud, S., Lumbreras, M.A., Flórez, J., 1997. Alterations of central noradrenergic transmission in Ts65Dn mouse, a model for Down syndrome. Brain Res. 749, 238–244.

Dominy, J.E., Puigserver, P., 2013. Mitochondrial biogenesis through activation of nuclear signaling proteins. Cold Spring Harbor Perspect. Biol. 5. https://doi.org/10.1101/cshperspect.a015008.

Duan, H., Li, Y., Yan, L., Yang, H., Wu, J., Qian, P., et al., 2015. Rcan1-1L overexpression induces mitochondrial autophagy and improves cell survival in angiotensin IIexposed cardiomyocytes. Exp. Cell Res. 335, 99–106.

Eichner, L.J., Giguère, V., 2011. Estrogen related receptors (ERRs): a new dawn in transcriptional control of mitochondrial gene networks. Mitochondrion 11, 544–552. https://doi.org/10.1016/j.mito.2011.03.121.

Elgass, K., Pakay, J., Ryan, M.T., Palmer, C.S., 2013. Recent advances into the understanding of mitochondrial fission. Biochim. Biophys. Acta 1833, 150–161. https://doi.org/10.1016/j.bbamcr.2012.05.002.

Epstein, C.J., Avraham, K.B., Lovett, M., Smith, S., Elroy-Stein, O., Rotman, G., et al., 1987. Transgenic mice with increased cu/Zn-superoxide dismutase activity: animal model of dosage effects in down syndrome. Proc. Natl. Acad. Sci. USA 84, 8044–8048.

Ermak, G., Sojitra, S., Yin, F., Cadenas, E., Cuervo, A.M., Davies, K.J., 2012. Chronic expression of RCAN1-1L protein induces mitochondrial autophagy and metabolic shift from oxidative phosphorylation to glycolysis in neuronal cells. J. Biol. Chem. 287, 14088–14098.

Falkenberg, M., Gaspari, M., Rantanen, A., Trifunovic, A., Larsson, N.-G., Gustafsson, C.M., 2002. Mitochondrial transcription factors B1 and B2 activate transcription of human mtDNA. Nat. Genet. 31, 289–294. https://doi.org/10.1038/ng909.

Fan, M., Rhee, J., St-Pierre, J., Handschin, C., Puigserver, P., Lin, J., et al., 2004. Suppression of mitochondrial respiration through recruitment of p160 myb binding protein to PGC-1alpha: modulation by p38 MAPK. Genes Dev. 18, 278–289.

Feissner, R.F., Skalska, J., Gaum, W.E., Sheu, S.S., 2009. Crosstalk signalling between mitochondrial Ca^{2+} and ROS. Front. Biosci. 14, 1197–1218. https://doi.org/10.2741/3303.

Finck, B.N., Kelly, D.P., 2006. PGC-1 coactivators: inducible regulators of energy metabolism in health and disease. J. Clin. Invest. 116, 615–622. https://doi.org/10.1172/JCI27794.

Flippo, K.H., Strack, S., 2017. Mitochondrial dynamics in neuronal injury, development and plasticity. J. Cell Sci. 130, 671–681. https://doi.org/10.1242/jcs. 171017.

Frezza, C., Cipolat, S., Martins de Brito, O., Micaroni, M., Beznoussenko, G.V., et al., 2006. OPA1 controls apoptotic cristae remodeling independently from mitochondrial fusion. Cell 126, 177–189.

Fritah, A., Christian, M., Parker, M.G., 2010. The metabolic coregulator RIP140: an update. Am. J. Physiol. Endocrinol. Metab. 299, E335–E340.

Gardiner, K., 2006. Transcriptional dysregulation in down syndrome: predictions for altered protein complex stoichiometries and post-translational modifications, and consequences for learning/behavior genes ELK, CREB, and the estrogen and glucocorticoid receptors. Behav. Genet. 36, 439–453. https://doi.org/10.1007/s10519-006-9051-1.

Gimeno, A., García-Giménez, J.L., Audí, L., Toran, N., Andaluz, P., Dasí, F., Viña, J., Pallardó, F.V., 2014. Decreased cell proliferation and higher oxidative stress in fibroblasts from Down syndrome fetuses. Preliminary study. Biochim. Biophys. Acta 1842, 116–125. https://doi.org/10.1016/j.bbadis.2013.10.014.

Gleyzer, N., Vercauteren, K., Scarpulla, R.C., 2005. Control of mitochondrial transcription specificity factors (TFB1M and TFB2M) by nuclear respiratory factors (NRF-1 and NRF-2) and PGC-1 family coactivators. Mol. Cell Biol. 25, 1354–1366. https://doi.org/10.1128/MCB.25.4.1354-1366.2005.

Gopalakrishnan, L., Scarpulla, R.C., 1994. Differential regulation of respiratory chain subunits by a CREB-dependent signal transduction pathway. role of cyclic amp in cytochrome c and COXIV gene expression. J. Biol. Chem. 269, 105–113.

Granese, B., Scala, I., Spatuzza, C., Valentino, A., Coletta, M., Vacca, R.A., De Luca, P., Andria, G., 2013. Validation of microarray data in human lymphoblasts shows a role of the ubiquitin-proteasome system and NF-kB in the pathogenesis of Down syndrome. BMC Med. Genom. 6, 24. https://doi.org/10.1186/1755-87946-24.

Han, X.J., Lu, Y.F., Li, S.A., Kaitsuka, T., Sato, Y., Tomizawa, K., Nairn, A.C., Takei, K., Matsui, H., Matsushita, M., 2008. CaM kinase I alpha-induced phosphorylation of Drp1 regulates mitochondrial morphology. J. Cell Biol. 182, 573–585.

Handschin, C., Spiegelman, B.M., 2006. Peroxisome proliferator-activated receptor gamma coactivator 1 coactivators, energy homeostasis, and metabolism. Endocr. Rev. 27, 728–735. https://doi.org/10.1210/er.2006-0037.

Handschin, C., Rhee, J., Lin, J., Tarr, P.T., Spiegelman, B.M., 2003. An autoregulatory loop controls peroxisome proliferator-activated receptor gamma coactivator 1alpha expression in muscle. Proc. Natl. Acad. Sci. USA 100, 7111–7116. https://doi.org/10.1073/pnas.1232352100.

Hardie, D.G., 2007. AMP-activated/SNF1 protein kinases: conserved guardians of cellular energy. Nat. Rev. Mol. Cell Biol. 8, 774–785. https://doi.org/10.1038/nrm2249.

Helguera, P., Seiglie, J., Rodriguez, J., Hanna, M., Helguera, G., Busciglio, J., 2013. Adaptive downregulation of mitochondrial function in Down syndrome. Cell Metabol. 17, 132–140. https://doi.org/10.1016/j.cmet.2012.12.005.

Hock, M.B., Kralli, A., 2009. Transcriptional control of mitochondrial biogenesis and function. Annu. Rev. Physiol. 71, 177–203.

Ijiri, K., Tsuruga, H., Sakakima, H., Tomita, K., Taniguchi, N., Shimoonoda, K., et al., 2005. Increased expression of humanin peptide in diffuse type pigmented villonodular synovitis: implication of its mitochondrial abnormality. Ann. Rheum. Dis. 64, 816–823.

Ishihara, T., Ban-Ishihara, R., Maeda, M., et al., 2015. Dynamics of mitochondrial DNA nucleoids regulated by mitochondrial fission is essential for maintenance of homogeneously active mitochondria during neonatal heart development. Mol. Cell. Biol. 35 (1), 211–223. https://doi.org/10.1128/MCB.01054-14.

Ishihara, N., Eura, Y., Mihara, K., 2004. Mitofusin 1 and 2 play distinct roles in mitochondrial fusion reactions via GTPase activity. J. Cell Sci. 117, 6535–6546.

Izzo, A., Manco, R., Bonfiglio, F., Cali, G., De Cristofaro, T., et al., 2014. NRIP1/RIP140 siRNA-mediated attenuation counteracts mitochondrial dysfunction in Down syndrome. Hum. Mol. Genet. 23, 4406–4419.

Izzo, A., Manco, R., De Cristofaro, T., Bonfiglio, F., Cicatiello, R., Mollo, N., et al., 2017a. Overexpression of chromosome 21 miRNAs may affect mitochondrial function in the hearts of Down syndrome fetuses. Int. J. Genom. 2017, 8737649. https://doi.org/10.1155/2017/8737649.

Izzo, A., Nitti, M., Mollo, N., Paladino, S., Procaccini, C., et al., 2017b. Metformin restores the mitochondrial network and reverses mitochondrial dysfunction in Down syndrome cells. Hum. Mol. Genet. 26, 1056–1069. https://doi.org/10.1093/hmg/ddx016.

Izzo, A., Mollo, N., Nitti, M., Paladino, S., Calì, G., Genesio, R., Bonfiglio, F., Cicatiello, R., Barbato, M., Sarnataro, V., Conti, A., Nitsch, L., March 15, 2018. Mitochondrial dysfunction in down syndrome: molecular mechanisms and therapeutic targets. Mol. Med. 24 (1), 2. https://doi.org/10.1186/s10020-018-0004-y.

Jäger, S., Handschin, C., St-Pierre, J., Spiegelman, B.M., 2007. AMP-activated protein kinase (AMPK) action in skeletal muscle via direct phosphorylation of PGC-1alpha. Proc. Natl. Acad. Sci. USA 104, 12017–12022. https://doi.org/10.1073/pnas.0705070104.

John, G.B., Shang, Y., Li, L., Renken, C., et al., 2005. The mitochondrial inner membrane protein mitofilin controls cristae morphology. Mol. Biol. Cell 16, 1543–1554.

Kamoun, P., Belardinelli, M.C., Chabli, A., Lallouchi, K., Chadefaux-Vekemans, B., 2003. Endogenous hydrogen sulfide overproduction in Down syndrome. Am. J. Med. Genet. A. 116A, 310–311.

Kanzleiter, T., Rath, M., Penkov, D., Puchkov, D., Schulz, N., Blasi, F., et al., 2014. Pknox1/Prep1 regulates mitochondrial oxidative phosphorylation components in skeletal muscle. Mol. Cell Biol. 34, 290–298.

Keefe, D.L., Niven-Fairchild, T., Powell, S., Buradagunta, S., 1995. Mitochondrial deoxyribonucleic acid deletions in oocytes and reproductive aging in women. Fertil. Steril. 64, 577–583.

Kim, S.H., Vlkolinsky, R., Cairns, N., Fountoulakis, M., Lubec, G., 2001. The reduction of NADH ubiquinone oxidoreductase 24- and 75-kDa subunits in brains of patients with Down syndrome and Alzheimer's disease. Life Sci. 68, 2741–2750.

Kipanyula, M.J., Kimaro, W.H., Seke Etet, P.F., 2016. The emerging roles of the calcineurin-nuclear factor of activated T-lymphocytes pathway in nervous system functions and diseases. J. Aging Res. 2016, 5081021. https://doi.org/10.1155/2016/5081021.

Knott, A.B., Perkins, G., Schwarzenbacher, R., Bossy-Wetzel, E., 2008. Mitochondrial fragmentation in neurodegeneration. Nat. Rev. Neurosci. 9, 505–518. https://doi.org/10.1038/nrn2417.

Leone, T.C., Lehman, J.J., Finck, B.N., et al., 2005. PGC-1α deficiency causes multi-system energy metabolic derangements: muscle dysfunction, abnormal weight control and hepatic steatosis. PLoS Biol. 3 (4), e101. https://doi.org/10.1371/journal.pbio.0030101.

Letourneau, A., Santoni, F.A., Bonilla, X., Sailani, M.R., Gonzalez, D., Kind, J., Chevalier, C., et al., April 17, 2014. Domains of genome-wide gene expression dysregulation in Down's syndrome. Nature 508 (7496), 345–350. https://doi.org/10.1038/nature13200. Erratum in: Nature. 2016 Mar 17;531(7594):400.

Levytskyy, R.M., Germany, E.M., Khalimonchuk, O., 2016. Mitochondrial quality control proteases in neuronal welfare. J. NeuroImmune Pharmacol. Off. J. Soc. NeuroImmune Pharmacol. 11, 629–644. https://doi.org/10.1007/s11481-016-9683-8.

Liu, P., 2012. Roles of transcriptional corepressor RIP140 and coactivator PGC-1alphain energy state of chronically infarcted rat hearts and mitochondrial function of cardiomyocytes. Mol. Cell. Endocrinol. 362, 11–18. https://doi.org/10.1016/j.mce.2012.03.023.

Losón, O.C., Song, Z., Chen, H., Chan, D.C., Fis1, M., 2013. MiD49, and MiD51 mediate Drp1 recruitment in mitochondrial fission. Mol. Biol. Cell 24, 659–667. https://doi.org/10.1091/mbc.E12-10-0721.

Lott, I.T., 2012. Neurological phenotypes for Down syndrome across the life span. Prog. Brain Res. 197, 101–121.

MacVicar, T., Langer, T., 2016. OPA1 processing in cell death and disease – the long and short of it. J. Cell Sci. 129, 2297–2306. https://doi.org/10.1242/jcs.159186.

Manczak, M., Mao, P., Calkins, M.J., Cornea, A., Reddy, A.P., Murphy, M.P., et al., 2010. Mitochondria-targeted antioxidants protect against amyloid-beta toxicity in Alzheimer's disease neurons. J. Alzheimers Dis. 20 (Suppl. 2), S609–S631.

Marechal, D., et al., 2019. CBS overdosage is necessary and sufficient to induce cognitive phenotypes in mouse models of Down syndrome and interacts genetically with Dyrk1a. Hum. Mol. Genet. 28, 1561–1577.

Mazat, J.-P., Ransac, S., Heiske, M., Devin, A., Rigoulet, M., 2013. Mitochondrial energetic metabolism-some general principles. IUBMB Life 65, 171–179. https://doi.org/10.1002/iub.1138.

Mello, T., Materozzi, M., Galli, A., 2016. PPARs and mitochondrial metabolism: from NAFLD to HCC. PPAR Res. 2016, 7403230. https://doi.org/10.1155/2016/7403230.

Misrani, A., Tabassum, S., Yang, L., 2021. Mitochondrial dysfunction and oxidative stress in Alzheimer's disease. Front. Aging Neurosci. 13, 617588. https://doi.org/10.3389/fnagi.2021.617588.

Mitra, R., Nogee, D.P., Zechner, J.F., Yea, K., Gierasch, C.M., Kovacs, A., Medeiros, D.M., Kelly, D.P., Duncan, J.G., 2012. The transcriptional coactivators, PGC-1 alpha and beta, cooperate to maintain cardiac mitochondrial function during the early stages of insulin resistance. J. Mol. Cell. Cardiol. 52, 701–710. https://doi.org/10.1016/j.yjmcc.2011.10.010.

Mollo, N., Cicatiello, R., Aurilia, M., Scognamiglio, R., Genesio, R., et al., 2020. Targeting mitochondrial network architecture in down syndrome and aging. Int. J. Mol. Sci. 21 (9), 3134. https://doi.org/10.3390/ijms21093134.

Mollo, N., Izzo, A., Nitti, M., Paladino, S., Calì, G., Genesio, R., Bonfiglio, F., Cicatiello, R., Conti, A., Nitsch, L., 2017. Mitochondria in down syndrome: from organelle abnormalities to clinical phenotype. In: Down Syndrome, p. 15. www.smgebooks.com.

Morrish, F., Hockenbery, D., 2014. MYC and mitochondrial biogenesis. Cold Spring Harb. Perspect. Med. 4. https://doi.org/10.1101/cshperspect.a014225.

Mukhopadhyay, D., Riezman, H., 2007. Proteasome-independent functions of ubiquitin in endocytosis and signaling. Science 315, 201–205.

Nakamura, N., Kimura, Y., Tokuda, M., Honda, S., Hirose, S., 2006. MARCH-V is a novel mitofusin 2- and Drp1-binding protein able to change mitochondrial morphology. EMBO Rep. 7, 1019–1022.

Nautiyal, J., Christian, M., Parker, M.G., 2013. Distinct functions for RIP140 in development, inflammation, and metabolism. Trends Endocrinol. Metabol. 24, 451–459.

Ni, H.-M., Williams, J.A., Ding, W.-X., 2014. Mitochondrial dynamics and mitochondrial quality control. Redox Biol. 4, 6–13. https://doi.org/10.1016/j.redox.2014.11.006.

Nicholls, P., Marshall, D.C., Cooper, C.E., Wilson, M.T., 2013. Sulfide inhibition of and metabolism by cytochrome c oxidase. Biochem. Soc. Trans. 41, 1312–1316.

Oettinghaus, B., Schulz, J.M., Restelli, L.M., Licci, M., Savoia, C., Schmidt, A., Schmitt, K., Grimm, A., Morè, L., Hench, J., Tolnay, M., Eckert, A., D'Adamo, P., Franken, P., Ishihara, N., Mihara, K., Bischofberger, J., Scorrano, L., Frank, S., 2016. Synaptic dysfunction, memory deficits and hippocampal atrophy due to ablation of mitochondrial fission in adult forebrain neurons. Cell Death Differ. 23, 18–28. https://doi.org/10.1038/cdd.2015.39.

Otera, H., Wang, C., Cleland, M.M., Setoguchi, K., Yokota, S., Youle, R.J., Mihara, K., 2010. Mff is an essential factor for mitochondrial recruitment of Drp1 during mitochondrial fission in mammalian cells. J. Cell Biol. 191, 1141–1158. https://doi.org/10.1083/jcb.201007152.

Oyekan, A., 2011. PPARs and their effects on the cardiovascular system. Clin. Exp. Hypertens 33, 287–293. https://doi.org/10.3109/10641963.2010.531845.

Pagano, G., Castello, G., 2012. Oxidative stress and mitochondrial dysfunction in Down syndrome. In: Ahmad, S.I. (Ed.), Neurodegenerative Diseases. Springer, New York, NY, pp. 291–299. https://doi.org/10.1007/978-1-4614-0653-2_22. Adv. Exp. Med. Biol. 724.

Pallardo, F.V., Lloret, A., Lebel, M., d'Ischia, M., Cogger, V.C., et al., 2010. Mitochondrial dysfunction in some oxidative stress-related genetic diseases: Ataxia-Telangiectasia, Down syndrome, Fanconi Anaemia and Werner Syndrome. Biogerontology 11, 401–419.

Palmer, C.S., Osellame, L.D., Laine, D., Koutsopoulos, O.S., Frazier, A.E., Ryan, M.T., 2011. MiD49 and MiD51, new components of the mitochondrial fission machinery. EMBO Rep. 12, 565–573. https://doi.org/10.1038/embor.2011.54.

Palmieri, F., Bisaccia, F., Capobianco, L., Dolce, V., Fiermonte, G., Iacobazzi, V., Indiveri, C., Palmieri, L., 1996. Mitochondrial metabolite transporters. Biochim. Biophys. Acta 1275, 127–132.

Panagaki, T., Randia, E.B., Augsburgera, F., Szabo, C., 2019. Overproduction of H_2S, generated by CBS, inhibits mitochondrial Complex IV and suppresses oxidative phosphorylation in Down syndrome. Proc. Natl. Acad. Sci. USA 116 (38), 18769–18771. www.pnas.org/cgi/doi/10.1073/pnas.1911895116.

Papa, S., Rasmo, D.D., Technikova-Dobrova, Z., Panelli, D., Signorile, A., et al., 2012. Respiratory chain complex I, a main regulatory target of the cAMP/PKA pathway is defective in different human diseases. FEBS Lett. 586, 568–577.

Parra, V., Altamirano, F., Hernández-Fuentes, C.P., et al., 2018. Down syndrome critical region 1 gene, Rcan1, helps maintain a more fused mitochondrial network. Circ. Res. 122 (6), e20–e33. https://doi.org/10.1161/CIRCRESAHA.117.311522.

Peiris, H., Raghupathi, R., Jessup, C.F., Zanin, M.P., Mohanasundaram, D., Mackenzie, K.D., Chataway, T., Clarke, J.N., Brealey, J., Coates, P.T., Pritchard, M.A., Keating, D.J., 2012. Increased expression of the glucose-responsive gene, RCAN1, causes hypoinsulinemia, β-cell dysfunction, and diabetes. Endocrinology 153, 5212–5221.

Peiris, H., Dubach, D., Jessup, C.F., Unterweger, P., Raghupathi, R., et al., 2014. RCAN1 regulates mitochondrial function and increases susceptibility to oxidative stress in mammalian cells. Oxid. Med. Cell. Longev. 520316.

Perluigi, M., Butterfield, D.A., 2012. Oxidative stress and Down syndrome: a route toward Alzheimer-like dementia. Curr. Gerontol. Geriatr. Res. 2012, 724904. https://doi.org/10.1155/2012/724904.

Perluigi, M., di Domenico, F., Fiorini, A., Cocciolo, A., Giorgi, A., Foppoli, C., Butterfield, D.A., Giorlandino, M., Giorlandino, C., Schininà, M.E., Coccia, R., 2011. Oxidative stress occurs early in Down syndrome pregnancy: a redox proteomics analysis of amniotic fluid. Proteonomics Clin. Appl. 5, 167–178. https://doi.org/10.1002/prca.201000121.

Picca, A., Calvani, R., Coelho-Junior, H.J., Landi, F., Bernabei, R., Marzetti, E., 2020. Mitochondrial dysfunction, oxidative stress, and neuroinflammation: intertwined roads to neurodegeneration. Antioxidants 9 (8), 647. https://doi.org/10.3390/antiox9080647.

Piccoli, C., Izzo, A., Scrima, R., Bonfiglio, F., Manco, R., Negri, R., Quarato, G., Cela, O., Ripoli, M., Prisco, M., Gentile, F., Calì, G., Pinton, P., Conti, A., Nitsch, L., Capitanio, N., March 15, 2013. Chronic pro-oxidative state and mitochondrial dysfunctions are more pronounced in fibroblasts from Down syndrome foeti with congenital heart defects. Hum. Mol. Genet. 22 (6), 1218–1232. https://doi.org/10.1093/hmg/dds529.

Porta, S., Serra, S.A., Huch, M., Valverde, M.A., Llorens, F., et al., 2007. RCAN1 (DSCR1) increases neuronal susceptibility to oxidative stress: a potential pathogenic process in neurodegeneration. Hum. Mol. Genet. 16, 1039–1050.

Powelka, A.M., Seth, A., Virbasius, J.V., Kiskinis, E., Nicoloro, S.M., Guilherme, A., Tang, X., Straubhaar, J., Cherniack, A.D., Parker, M.G., et al., 2006. Suppression of oxidative metabolism and mitochondrial biogenesis by the transcriptional corepressor RIP140 in mouse adipocytes. J. Clin. Invest. 116, 125–136. https://doi.org/10.1172/JCI26040.

Predescu, S.A., Predescu, D.N., Knezevic, I., Klein, I.K., Malik, A.B., 2007. Intersectin-1s regulates the mitochondrial apoptotic pathway in endothelial cells. J. Biol. Chem. 282, 17166–17178.

Qin, Q., Mehta, H., Yen, K., Navarrete, G., Brandhorst, S., Wan, J., et al., 2018. Chronic treatment with the mitochondrial peptide humanin prevents age-related myocardial fibrosis in mice. Am. J. Physiol. Heart Circ. Physiol. 315, H1127–H1136.

Quinones-Lombrana, A., Blanco, J.G., 2015. Chromosome 21-derived hsa-miR-155-5p regulates mitochondrial biogenesis by targeting Mitochondrial Transcription Factor A (TFAM). Biochim. Biophys. Acta 1852, 1420–1427. https://doi.org/10.1016/j.bbadis.2015.04.004.

Rolfe, D.F., Brown, G.C., 1997. Cellular energy utilization and molecular origin of standard metabolic rate in mammals. Physiol. Rev. 77, 731–758.

Ryan, M.T., Hoogenraad, N.J., 2007. Mitochondrial-nuclear communications. Annu. Rev. Biochem. 76, 701–722. https://doi.org/10.1146/annurev.biochem.76.052305.091720.

Rytinki, M.M., Palvimo, J.J., 2009. SUMOylation attenuates the function of PGC-1alpha. J. Biol. Chem. 284, 26184–26193. https://doi.org/10.1074/jbc.M109.038943.

Salemi, M., Ridolfo, F., Salluzzo, M.G., Cannarrella, R., Giambirtone, M., Caniglia, S., Tirolo, C., Ferri, R., Romano, C., 2020. Humanin gene expression in fibroblast of Down syndrome subjects. Int. J. Med. Sci. 17 (3), 320–324. https://doi.org/10.7150/ijms.39145.

Scarpulla, R.C., 2011. Metabolic control of mitochondrial biogenesis through the PGC-1 family regulatory network. Biochim. Biophys. Acta 1813, 1269–1278. https://doi.org/10.1016/j.bbamcr.2010.09.019.

Schon, E.A., Manfredi, G., 2003. Neuronal degeneration and mitochondrial dysfunction. J. Clin. Invest. 111 (3), 303–312. https://doi.org/10.1172/JCI17741.

Schreiber, S.N., Emter, R., Hock, M.B., Knutti, D., Cardenas, J., Podvinec, M., et al., 2004. The estrogen-related receptor alpha (ERRalpha) functions in PPARgamma coactivator 1alpha (PGC-1alpha)-induced mitochondrial biogenesis. Proc. Natl. Acad. Sci. USA 101, 6472–6477.

Schrepfer, E., Scorrano, L., 2016. Mitofusins, from mitochondria to metabolism. Mol. Cell. 61, 683–694. https://doi.org/10.1016/j.molcel.2016.02.022.

Selivanov, V.A., Votyakova, T.V., Pivtoraiko, V.N., Zeak, J., Sukhomlin, T., Trucco, M., Roca, J., Cascante, M., 2011. Reactive oxygen species production by forward and reverse electron fluxes in the mitochondrial respiratory chain. PLoS Comput. Biol. 7, e1001115. https://doi.org/10.1371/journal.pcbi.1001115.

Seth, A., Steel, J.H., Nichol, D., Pocock, V., Kumaran, M.K., Fritah, A., Mobberley, M., Ryder, T.A., Rowlerson, A., Scott, J., et al., 2007. The transcriptional corepressor RIP140 regulates oxidative metabolism in skeletal muscle. Cell Metabol. 6, 236–245. https://doi.org/10.1016/j.cmet.2007.08.004.

Shapiro, L.A., Bialowas-McGoey, L.A., Whitaker-Azmitia, P.M., 2010. Effects of S100B on serotonergic plasticity and neuroinflammation in the Hippocampus in down syndrome and Alzheimer's disease: studies in an S100B overexpressing mouse model. Cardiov. Psychiat. & Neurol. 13. https://doi.org/10.1155/2010/153657. Article ID 153657.

Shirihai, O., Song, M., Dorn, G.W., 2015. How mitochondrial dynamism orchestrates mitophagy. Circ. Res. 116, 1835–1849. https://doi.org/10.1161/CIRCRESAHA.116.306374.

Skerrett, R., Pellegrino, M.P., Casali, B.T., Taraboanta, L., Landreth, G.E., 2015. Combined liver X receptor/peroxisome proliferator-activated receptor gamma agonist treatment reduces amyloid beta levels and improves behavior in amyloid precursor protein/Presenilin 1 mice. J. Biol. Chem. 290, 21591–21602.

Smirnova, E., Griparic, L., Shurland, D.L., van der Bliek, A.M., 2001. Dynamin-related protein Drp1 is required for mitochondrial division in mammalian cells. Mol. Biol. Cell 12, 2245–2256.

Solanki, A., Smalling, R., Parola, A.H., Nathan, I., Kasher, R., Pathak, Y., et al., 2018. Humanin nanoparticles for reducing pathological factors characteristic of age-related macular degeneration. Curr. Drug Deliv. 31. https://doi.org/10.2174/156720181566618103116 3111.

Sturgeon, X., Le, T., Ahmed, M.M., Gardiner, K.J., 2012. Pathways to cognitive deficits in Down syndrome. Prog. Brain Res. 197, 73–100 (Research support, N.I.H., Extramural Research Support, Non-U.S. Gov't).

Su, B., Wang, X., Ling, Z., George, P., Smith, M.A., Zhu, X., January 2010. Abnormal mitochondrial dynamics and neurodegenerative diseases. Biochim. Biophys. Acta 1802 (1), 135–142. https://doi.org/10.1016/j.bbadis.2009.09.013.

Sun, X., Wu, Y., Herculano, B., Song, W., 2014. RCAN1 overexpression exacerbates calcium overloading-induced neuronal apoptosis. PLoS One 9, e95471.

Szabo, C., 2020. The re-emerging pathophysiological role of the Cystathionine-β-synthase - hydrogen sulfide system in Down syndrome. FEBS J. 287, 3150–3160. https://doi.org/10.1111/febs.15214.

Szabo, C., Papapetropoulos, A., 2017. International union of basic and clinical pharmacology. CII: pharmacological modulation of H_2S levels: H_2S donors and H_2S biosynthesis inhibitors. Pharmacol. Rev. 69, 497–564.

Tan, A.R., Cai, A.Y., Deheshi, S., Rintoul, G.L., 2011. Elevated intracellular calcium causes distinct mitochondrial remodelling and calcineur independent fission in astrocytes. Cell Calcium 49, 108–114.

Telford, J.E., Kilbride, S.M., Davey, G.P., 2009. Complex I is rate-limiting for oxygen consumption in the nerve terminal. J. Biol. Chem. 284, 9109–9114. https://doi.org/10.1074/jbc.M809101200.

Tramutola, A., Di Domenico, F., Barone, E., Arena, A., Giorgi, A., di Francesco, L., Schininà, M.E., Coccia, R., Head, E., Butterfield, D.A., Perluigi, M., 2017. Polyubiquitinylation profile in Down syndrome brain before and after the development of Alzheimer neuropathology. Antioxid. Redox Signal. 26, 280–298. https://doi.org/10.1089/ars.2016.6686.

Twig, G., Elorza, A., Molina, A.J.A., Mohamed, H., Wikstrom, J.D., Walzer, G., Stiles, L., Haigh, S.E., Katz, S., Las, G., Alroy, J., Wu, M., Py, B.F., Yuan, J., Deeney, J.T., Corkey, B.E., Shirihai, O.S., 2008. Fission and selective fusion govern mitochondrial segregation and elimination by autophagy. EMBO J. 27, 433–446. https://doi.org/10.1038/sj.emboj.7601963.

Uttara, B., Singh, A.V., Zamboni, P., Mahajan, R.T., 2009. Oxidative stress and neuro degenerative diseases: are view of upstream and downstream antioxidant therapeutic options. Curr. Neuropharmacol. 7, 65–74. https://doi.org/10.2174/157015909787602823.

Vacca, R.A., Valenti, D., 2015. Green tea EGCG plus fish oil omega-3 dietary supplements rescue mitochondrial dysfunctions and are safe in a Down's syndrome child. Clin. Nutr. Edinb. Scotl. 34, 783–784. https://doi.org/10.1016/j.clnu.2015.04. 012.

Valenti, D., Tullo, A., Caratozzolo, M.F., Merafina, R.S., Scartezzini, P., Marra, E., Vacca, R.A., 2010. Impairment of F1F0-ATPase, adenine nucleotide translocator and adenylate kinase causes mitochondrial energy deficit in human skin fibroblasts with chromosome 21 trisomy. Biochem. J. 431, 299–310. https://doi.org/10.1042/BJ20100581.

Valenti, D., Manente, G.A., Moro, L., Marra, E., Vacca, R.A., 2011. Deficit of complex I activity in human skin fibroblasts with chromosome 21 trisomy and overproduction of reactive oxygen species by mitochondria: involvement of the cAMP/PKA signalling pathway. Biochem. J. 435, 679–688. https://doi.org/10.1042/BJ20101908.

Valenti, D., De Rasmo, D., Signorile, A., Rossi, L., de Bari, L., Scala, I., Granese, B., Papa, S., Vacca, R.A., 2013. Epigallocatechin-3-gallate prevents oxidative phosphorylation deficit and promotes mitochondrial biogenesis in human cells from subjects with Down's syndrome. Biochim. Biophys. Acta 1832, 542–552. https://doi.org/10.1016/j.bbadis.2012.12.011.

Valenti, D., Atlante, A., 2021. Mitochondrial bioenergetics in different pathophysiological conditions. Int. J. Mol. Sci. 22 (14), 7562. https://doi.org/10.3390/ijms22147562.

Valenti, D., Braidy, N., De Rasmo, D., Signorile, A., Rossi, L., et al., 2018. Mitochondria as pharmacological targets in Down syndrome. Free Radic. Biol. Med. 114, 69–83. https://doi.org/10.1016/j.freeradbiomed.2017.08.014.

Valenti, D., de Bari, L., De Filippis, B., Henrion-Caude, A., Vacca, R.A., 2014. Mitochondrial dysfunction as a central actor in intellectual disability-related diseases: an overview of Down syndrome, autism, Fragile X and Rett syndrome. Neurosci. Biobehav. Rev. 46 (Pt 2), 202–217. https://doi.org/10.1016/j.neubiorev. 2014.01.012.

Valenti, D., de Bari, L., de Rasmo, D., Signorile, A., Henrion-Caude, A., Contestabile, A., Vacca, R.A., 2016. The polyphenols resveratrol and epigallocatechin-3gallate restore the severe impairment of mitochondria in hippocampal progenitor cells from a Down syndrome mouse model. Biochim. Biophys. Acta 1862, 1093–1104. https://doi.org/10.1016/j.bbadis.2016.03.003.

Van Blerkom, J., Sinclair, J., Davis, P., 1998. Mitochondrial transfer between oocytes: potential applications of mitochondrial donation and the issue of heteroplasmy. Hum. Reprod. 3, 2857–2868.

Vega, R.B., Huss, J.M., Kelly, D.P., 2000. The coactivator PGC-1 cooperates with peroxisome proliferator-activated receptor alpha in transcriptional control of nuclear genes encoding mitochondrial fatty acid oxidation enzymes. Mol. Cell Biol. 20, 1868–1876.

Ventura-Clapier, R., Garnier, A., Veksler, V., 2008. Transcriptional control of mitochondrial biogenesis: the central role of PGC-1alpha. Cardiovasc. Res. 79, 208–217. https://doi.org/10.1093/cvr/cvn098.

Vilardell, M., Rasche, A., Thormann, A., Maschke-Dutz, E., Perez-Jurado, L.A., et al., 2011. Meta-analysis of heterogeneous Down Syndrome data reveals consistent genome-wide dosage effects related to neurological processes. BMC Genom. 12, 229.

Wallace, D.C., 2005. A mitochondrial paradigm of metabolic and degenerative diseases, aging, and cancer: a dawn for evolutionary medicine. Annu. Rev. Genet. 39, 359–407.

Wan, B., Moreadith, R.W., 1995. Structural characterization and regulatory element analysis of the heart isoform of cytochrome c oxidase VIa. J. Biol. Chem. 270, 26433–26440.

Wenz, T., 2009. PGC-1alpha activation as a therapeutic approach in mitochondrial disease. IUBMB Life 61, 1051–1062. https://doi.org/10.1002/iub.261.

Wong, H., Levenga, J., Cain, P., Rothermel, B., Klann, E., Hoeffer, C., 2015. RCAN1 overexpression promotes age-dependent mitochondrial dysregulation related to neurodegeneration in Alzheimer's disease. Acta Neuropathol. 130, 829–843.

Wu, Y., Song, W., 2013. Regulation of RCAN1 translation and its role in oxidative stress induced apoptosis. FASEB. J. 27, 208–221.

Yang, J., Rothermel, B., Vega, R.B., Frey, N., McKinsey, T.A., Olson, E.N., Bassel-Duby, R., Williams, R.S., 2000. Independent signals control expression of the calcineurin inhibitory proteins MCIP1 and MCIP2 in striated muscles. Circ. Res. 87, E61–E68.

Yang, E.J., Ahn, Y.S., Chung, K.C., 2001. Protein kinase Dyrk1 activates cAMP response element-binding protein during neuronal differentiation in hippocampal progenitor cells. J. Biol. Chem. 276, 39819–39824. https://doi.org/10.1074/jbc.M104091200.

Ye, K., Lu, J., Ma, F., Keinan, A., Gu, Z., 2014. Extensive pathogenicity of mitochondrial heteroplasmy in healthy human individuals. Proc. Natl. Acad. Sci. USA 111 (29), 10654–10659.

Yen, K., Wan, J., Mehta, H.H., Miller, B., Christensen, A., Levine, M.E., et al., 2018. Humanin prevents age-related cognitive decline in mice and is associated with improved cognitive age in humans. Sci. Rep. 21, 14212.

Zamponi, E., Helguera, P.R., 2019. The shape of mitochondrial dysfunction in Down syndrome. Develop. Neurobiol. 79 (7). https://doi.org/10.1002/dneu.22673.

Zhang, K., Li, H., Song, Z., 2014. Membrane depolarization activates the mitochondrial protease OMA1 by stimulating self-cleavage. EMBO Rep. 15, 576–585. https://doi.org/10.1002/embr.201338240.

Zick, M., Rabl, R., Reichert, A.S., 2009. Cristae formation-linking ultrastructure and function of mitochondria. Biochim. Biophys. Acta 1793, 5–19.

CHAPTER

9

Disarrayed mitochondrial function on pathobiology in Down syndrome and targeted therapeutics

1. Introduction

Aneuploidy of the chromosome 21 (HAS21) in Down syndrome (DS) results in an imbalance in the whole genome and that leads to a variability in severity of expression of a wide spectrum of syndrome-specific features. The expression of the extranuclear mitochondrial genome (mtDNA) is not spared by the dosage imbalance of nuclear genome. Individuals with DS exhibit mitochondrial dysfunction having triplication of the HSA21 genes (Arbuzova et al., 2002; Coskun and Busciglio, 2012). Perturbation in oxidative phosphorylation (OXPHOS) and fusion/fission dynamism of the mitochondria contributes to chronic oxidative stress, mutations in mtDNA, production of reactive oxygen species (ROS), and mitochondrial dysfunction in total, which emerge as important modulators of phenotypes in individuals with DS. As such, the mitochondria in DS primary cells are more fragmented causing impairment of its functions and energy production. However, the underlying mechanisms are not fully understood from the investigations of the genotype–phenotype relationship between abnormal energy metabolism and oxidative stress, based on transcriptional and functional alterations in DS cells. The association of mitochondrial morphology with its function is highly complex, which is further complicated by multiple factors. To summarize, the mitochondria play an important role in oxygen metabolism, and mitochondrial dysfunction occurs in a situation of redox imbalances. Indeed, interaction of nuclear genes with mtDNA perturbs several signaling pathways contributing to several pathomechanisms in individuals with extra HSA21. In DS mitochondria, a correlation of the dysfunctional mitochondrial activity and altered mitochondrial morphology has demonstrated that an increase in fusion capacity can potentially prevent morphological fragmentation but not the functional alterations (Helguera et al., 2013; Knott et al., 2008).

Phenotypic variability of DS is further complicated by an adaptive response of the cells toward stress. Seemingly, reduced mitochondrial functions prevent cellular injury and restore basic activities, and that expression has been explained as an adaptive response or hormesis to external stressors. A comprehensive analysis of signature pathways might recognize the key elements of hormesis in DS and thus, may predict the disease onset at early stage and facilitate initiation of

Genetics and Neurobiology of Down Syndrome
https://doi.org/10.1016/B978-0-323-90456-8.00001-6

valuable treatments for these individuals. A network analysis of the overexpressed genes displayed functional integration in pathways of energy metabolism and oxidative stress observed in DS (Helguera et al., 2013; Coskun and Busciglio, 2012).

An approach toward prevention of oxidative damage may contribute to mitochondrial downregulation and that may increase the risk of clinical conditions in DS individuals. Additionally, alteration in mitochondrial morphology and function appears context-specific and broadly depends on the cellular state and factor that affects the cells. With this notion, induction of mitochondrial fragmentation with a dominant negative form of mfn1 did not affect the ATP-dependent hormone secretion in insulinemia cells, however, that was significantly reduced by Fis1 overexpression along with generation of a fragmented phenotype (Park et al., 2008). Notably, overexpression of mitofusin can recover the fragmented morphology but cannot improve the mitochondrial membrane potential in trisomic cells. Stimulation with creatine enhanced mitochondrial activity and mitochondrial anterograde movement and also doubled ROS production and increased lipid peroxidation in trisomic cells with decrease in cell viability (Helguera et al., 2013). The sustained stimulation increased damage and death of DS cells, indicating an adaptive response catered by downregulation of the mitochondria to circumvent the cellular injury due to excess ROS generation. Indeed, upon stimulation, mitochondria in DS can potentially increase the energy metabolism through consistent increase in oxidoreductase activity, ATP levels, and matrix metalloproteinase (MMP) in DS cells.

Higher rate of mutations in the mtDNA, specifically of the control region, and reduction in copy numbers of the mtDNA have been reported in brains of DS patients and that has been associated with the etiology of early onset of dementia and premature aging in DS patients. Abnormalities in mitochondrial function and variations in sequences of the mtDNA are associated with a variety of classic degenerative diseases in specific body systems, including selective neuronal degeneration, cardiovascular disease, kidney failure, cancer, and endocrine, musculoskeletal, immunological, and neurological systems in DS (Coskun et al., 2010); however, the exact molecular mechanisms linking to mtDNA variation to diseases of DS are not yet characterized fully (Dai et al., 2012). Altogether, mitochondrial metabolism plays a crucial role in human health and aging, especially in highly oxidative organs such as the heart and brain (Corbalan and Kitsis, 2018; Parra et al., 2018; Valenti et al., 2018). Alteration in energy metabolism has been demonstrated to cause diseases such as premature aging, Alzheimer disease (AD), heart defects, type 2 diabetes, and several other conditions, which collectively distort the quality of life of DS individuals.

This chapter has focused on the diseases of mitochondrial dysfunction in DS, especially heart defects, hypotonia, immune disorders, type 2 diabetes and obesity, aging, intellectual deficiency and neurodegeneration, and AD. In addition, the concept of mitochondrial adaptive response and mechanism of mitophagy have also been demonstrated. The relevant context of genetics of mitochondrial dysfunction and redox imbalances has been demonstrated in the previous chapter (Chapter 8).

2. Diseases caused by mitochondrial dysfunction

2.1 Heart defects and mitochondria

Congenital heart defect (CHD) is one of the common (~50%) clinical expressions in DS, majorly as endocardial cushion defects and atrioventricular septal defects (AVSDs) followed by ventricular septal defects (VSDs) and tetralogy of Fallot. Trisomic heart tissue of DS fetus exhibited a global downregulation of nuclear-encoded mitochondrial genes (NEMGs) in all five respiratory chain complexes, more specifically, the enzymes involved in the OXPHOS detected in transcriptome analysis (Pilar Bayona-Bafaluy et al., 2021; Hefti et al., 2017), along with global upregulation of ~400 HSA21 genes and dysregulation of non-HSA21 genes (Conti et al., 2007). Notably, the cardiovascular system has been reported to be less affected by aging in DS, which included the low incidence of hypertension and ischemic cardiac disease in DS adults (Zigman, 2013). Indeed, the mitochondrial functions are key mediators for determination of reperfusion of ischemic tissue (ischemia/reperfusion: I/R) and risk of the heart to I/R damage (Parra et al., 2018).

Pronounced differences reported in mitochondrial respiration, complex I activity, and ROS production in fibroblasts of DS fetuses and suggested an association between mitochondrial function and cardiac phenotype (Piccoli et al., 2013). The complex I dysfunction resulted in alterations in mitochondrial function in the right ventricular cardiac hypertrophy (Wust et al., 2016). The CHD in DS patients is connected to pulmonary hypertension (PH) and patent ductus arteriosus and pulmonary stenosis (Park et al., 1977). PH in turn has been associated with both functional and structural alterations of mitochondria. The smooth muscle cells isolated from the pulmonary vessels of rats with PH demonstrated deficiencies in the function of complexes I-III, increased mitochondrial ROS production, and alteration in mitochondrial membrane potential. The muscle cells of patients with PH showed disruption of the normal mitochondrial filamentous reticulum (Bonnet et al., 2006).

It is noted that mitochondrial alterations are more pronounced in cardiopathic DS-HFF. Interestingly, all individuals with DS do not develop CHD; however, a developmental delay of the fetal heart was reported in all human embryos with DS at 8–10 weeks of gestational age. Thus, the variability in cardiac phenotype of DS is demonstrated to be the expression of impaired oxidative metabolism due to higher mitochondrial Ca-load and related ROS generation at a background of interindividual variation in genetic makeup, which might also be associated with different bioenergetic phenotypes (Piccoli et al., 2013). Consistent with the hypothesis that the Down syndrome critical region (DSCR) on HSA21 spanning between Tiam1 and Kcnj6 is responsible for variable cardiovascular development in DS (Liu et al., 2011). In addition, different bioenergetic phenotypes at the cellular level, a larger mitochondrial Ca^{2+} load, and related ROS production might be associated with more severe form of cardiac phenotype in DS as reported in CDS-HFF (Piccoli et al., 2013). To this effect, induction of oxidative stress in pregnant mice on day 7.5 displayed disruptions of cardiac neural crest migration and that caused outflow tract defects as observed in human DS. Interestingly, administration of antioxidant before the induction of oxidative stress prevented heart defects (Valenti et al., 2021).

The DYRK1A and RCAN1 genes of HSA21 affect the mitochondrial morphology and activity during heart development via the calcineurin (CN)-NFAT pathway. Gene dosage effect of DYRK1A and RCAN1 in fibroblasts and hearts of DS fetuses with CHD demonstrated marked prooxidative state in mitochondria than fetuses with cardiopathy, with concomitant downregulation of NFATc3 and NFATc4 expression (Piccoli et al., 2013). The dosage effect was responsible for regulating the levels of NFATc phosphorylation (Conti et al., 2007; Piccoli et al., 2013). NFATc-knockout mice showed human DS-like phenotypic anomalies, where 65% of NFATc1-4-null mice have exhibited endocardial cushion defects (Arron et al., 2006). The inhibition of the mitochondrial function in NFATc3−/− and Nfatc4−/− cardiomyocytes indicated perturbation in mitochondrial activity during heart development and the involvement of CN/NFAT pathway in cardiopathy (Bushdid et al., 2003). Additionally, the NRIP1-directed downregulation of genes involved in mitochondrial function may also be linked to the ventricular hypertrophy that often occurs in DS after birth and that indicates a possible link between reduced mitochondrial electron-transport chain and oxygen consumption. Overexpression of NRIP1 in a transgenic mouse exhibited cardiac hypertrophy (Fritah et al., 2010). The HSA21 encoded miR-155, which is known to be a repressor of TFAM gene, has been demonstrated as an inducer of cardiac hypertrophy, which led to hypothesize that its inhibition might counteract this pathophysiology in DS (Seok et al., 2014).

The patients with DS are reported to be more susceptible to the cardiotoxic effects of anticancer anthracycline drugs (Hefti and Blanco, 2016). Indeed, variability was reported in certain anthracycline metabolizing enzymes, such as AKR7A2, and mtDNA content in cardiac synthesis of drug metabolites. Enhancement of mtDNA variants (also called heteroplasmy) in cardiac tissue may confer cardiac dysfunction in DS (Hefti et al., 2017). A nonsignificant decrease (30%) in cardiac mtDNA and increase (30%) in "common" mtDNA4977 deletion reported in trisomic heart tissue appeared independent of age of DS individuals (Hefti et al., 2017). A higher frequency of mtDNA variants was apparent in the cardiac tissue of DS due to mitochondrial dysfunction and high frequency of multiorgan disorders; however, that can vary in other tissues (Coskun and Busciglio, 2012).

2.1.1 *Association of RCAN1 and trisomic cardiomyocytes*

A great deal of significant effects of RCAN1 on mitochondrial integrity on cardiomyocytes and other types of cells has been established (Parra et al., 2018). RCAN1-encoded protein inhibits the Ca^{2+}-activated protein phosphatase CN through a direct binding to its catabolic subunit and protein—protein interaction, which is important for cardiac remodeling. Thus, RCAN1-depletion increases CN-dependent dephosphorylation of the profission protein—DRP1 (dynamin-1-like; a GTPase) leading to mitochondrial fission and pathological hypertrophy of the heart. CN-mediated dephosphorylation of serine 637 in human DRP1 promotes translocation of DRP1 from cytosol to mitochondria, where it causes constriction in the outer mitochondrial membrane resulting in mitochondrial fission (Cereghetti et al., 2008; Corbalan and Kitsis, 2018). Thus, downregulated RCAN1 in cardiomyocytes reduces mitochondrial membrane potential and metabolic activity; lower the efficiency of mitochondrial Ca^{2+} uptake, lower the rates of O_2 consumption and generation of ROS by fission or fragmentation of the mitochondria. Mitochondrial fission reduces the capacity to buffer cytosolic Ca^{2+} and thereby, increases the susceptibility of the tissue to calpain (CAPN; Ca^{2+}-activated protease)-mediated damage after ischemia/reperfusion (I/R) and finally, culminate into heart attack or stroke (Parra et al., 2018). Both CN and CAPN are inactive in low pH environment. Although mitochondrial fusion reduces CAPN effects in cardiomyocytes after reperfusion, a continual increase in fusion may increase the capacity of ROS generation and cellular damage and reduce the potential for mitochondrial repair. I/R causes significant increase in cytoplasmic Ca^{2+}.

Excess of RACN1 promotes mitochondrial fusion and increases its metabolism. Cardiac-specific overexpression of RCAN1 protects the heart from a number of pathological stresses, including I/R; thereby, RCAN1-deficient hearts and brains are vulnerable to I/R (Rotter et al., 2018). In Rcan1-knockout mice, Rcan1 deficiency caused significant damage in cardiac and neuronal cells and forced expression and conferred protection in both cardiomyocytes and neurons (Rotter et al., 2018). Collectively, RCAN1 has a direct effect on mitochondrial functioning, and RCAN1 deficiency leads to mitochondrial fragmentation in a variety of cell types, including mouse embryonic fibroblasts and primary neonatal and adult cardiomyocytes in rodents (Fig. 9.1) (Corbalan and Kitsis, 2018).

Mitochondrial fission can be reversed by pharmacological inhibition of CN or DRP1. However, the role of mitochondrial fission in apoptosis remains unclear, though complex interactions have

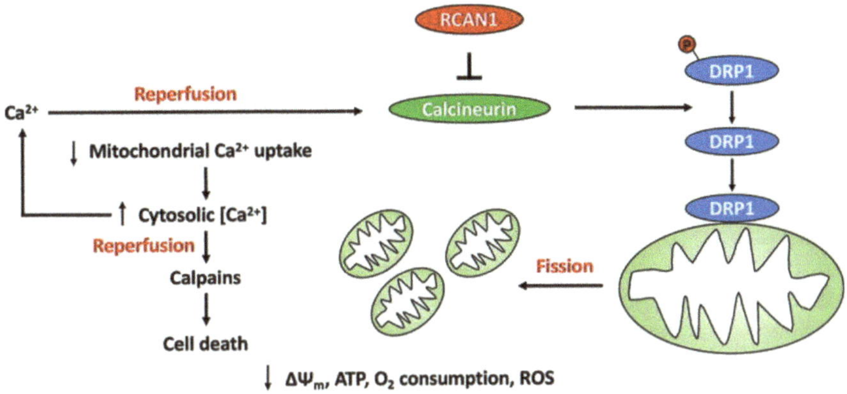

FIGURE 9.1 RCAN1 modulates tissue damage during myocardial ischemia-reperfusion. Depletion of RCAN1 disinhibits the Ca^{2+}-activated protein phosphatase CN resulting in CN-mediated dephosphorylation of serine 637 in human DRP1. This stimulates the translocation of DRP1 from the cytosol to the outer mitochondrial membrane where it promotes mitochondrial fission. Depletion of RCAN1 may also promote mitochondrial fragmentation through decreases in mitochondrial fusion proteins OPA1 and MFN2 (not shown). Fragmented mitochondria manifest decreased Ca^{2+} uptake, which may reflect their loss of electrical potential difference across the inner mitochondrial membrane ($\Delta\psi_m$) or perhaps a more direct action of CN on the mitochondrial Ca^{2+} import machinery. Impaired uptake of Ca^{2+} results in elevation of cytosolic Ca^{2+} concentrations, thereby predisposing to the activation of CAPNs, Ca^{2+}-activated proteases. Calpains cleave signaling and structural proteins to induce cell death. Activation of CN and calpains in this schema takes place during reperfusion because the activities of these enzymes are inhibited in the acidic environment of ischemia. *Reprinted from Corbalan and Kitsis (2018), with permission from Wolters Kluwer Health, Inc.*

been reported between DRP1 and BAX (BCL-2 (B-cell leukemia/lymphoma 2)–associated X) protein (Martinou and Youle, 2011).

2.1.2 The role of mitochondrial Ca^{2+} homeostasis inducing cell death

RCAN1 deficiency in cardiomyocytes exhibits a deficiency in mitochondrial Ca^{2+} uptake and concomitant increase in cytosolic Ca^{2+} concentrations. The mechanism of RCAN1 depletion effect on mitochondrial Ca^{2+} uptake, though not clear, may attribute to decrease in polarization across the inner mitochondrial membrane ($\Delta\psi$m), which is a driving force for the movement of Ca^{2+} into the mitochondrial matrix, through the mitochondrial calcium uniporter (Mammucari et al., 2017). Two possibilities have been hypothesized as to why $\Delta\psi$m is decreased in RCAN1-depleted cardiomyocytes: i. the decrease in $\Delta\psi$m merely reflects the poor health of the fragmented mitochondria, which also display decrease in ATP levels, and ii. RCAN1 deficiency primarily exerts a defect in mitochondrial Ca^{2+} uptake, which lowers the catabolism of nutrients (Corbalan and Kitsis, 2018). It was also postulated that RCAN1–CN axis directly regulates mitochondrial Ca^{2+} uptake. The reduction in mitochondrial Ca^{2+} uptake leads to a compromise in substrate oxidation in mitochondria, and thus, it seemingly appeared to be protective against Ca^{2+}-induced opening of the mtPTP leading to necrosis.

Beyond making ATP, mitochondria act as sites for Ca^{2+} uptake and control cell death or survival. For delineation of increased sensitivity in RCAN1-deficient cardiomyocytes to cell death, the mechanism of CAPNs-Ca^{2+}-stimulated proteases, including its signaling and structural molecules, were expected to be involved in apoptosis and cell death (Parra et al., 2018). Noticeably, pharmacological inhibition of these proteases or knockdown of CAPNs 1 or 2 reversed the increased sensitivity to simulated

I/R-induced cell death in RCAN1-knockdown experiments (Corbalan and Kitsis, 2018). The phenomenon indicated that high Ca^{2+} concentrations in the mitochondrial matrix have higher tolerance than in the cytosol. Overexpression of RCAN1 increased mitochondrial oxygen consumption and ROS levels in adenoviral-mediated transduction of cardiomyocytes indicating cardioprotection in vivo and in cardiomyocytes derived from trisomic iPSCs (which was prominent over the similarly derived cells of disomic counterparts). Also, its overexpression ameliorated the I/R-elicited death of cardiomyocytes. Nevertheless, similar mechanism of RCAN1-calceneurin network may exist in the brain to cause tissue damage. Importantly, brain-specific overexpression of RCAN1.1 has been associated with age-dependent cognitive impairments, which is similar to the early onset of dementia in DS (Wong et al., 2015). However, further research is necessary for precise elucidation of molecular mechanisms involved in calceneurin-mediated Ca^{2+} uptake and I/R-related myocardial damage.

2.2 Mitochondria and hypotonia

The basic mechanism of association of trisomic HSA21 and muscle hypotonia with altered motor coordination, often experienced by the DS patients, is still largely unknown. Since skeletal muscle is vulnerable to oxidative stress, a possible involvement of mitochondrial function on hypotonia was postulated. The fact is that the rapid and coordinated changes in energy supply and oxygen flux during contraction result in increased electron flux and leakage from the mitochondrial electron transport chain (ETC) (Mollo et al., 2019). Disarray in the mitochondrial dynamics, including mitochondrial dysfunction and fission, is suggested to be essential amplificatory loop of the muscular atrophy program in adult animals. Inhibition of the mitochondrial fission exhibited inhibition in muscle loss during fasting with overexpression of $FoxO_3$. Furthermore, apoptosis caused by the changes in mitochondrial morphology have been implicated with the regulation of muscle metabolism. The overexpression of PGC1-α in triplicated HSA21 has been indicated to inhibit muscle atrophy during fasting and denervation. Also, the overexpression of the HSA21 encoded miR-155 caused repressed expression of endogenous myogenic enhancer factor 2 (MEF2A) leading to inhibition of myoblast differentiation (Seok et al., 2014). The Ts65Dn mouse model exhibited motor dysfunctions similar to that of human DS patients (Cowley et al., 2012). The ultrastructure of the myofibrils of these trisomic mice demonstrated the evidence of changes in the mitochondrial structure, in addition to alteration in ATP biosynthesis, proteolysis, metabolism of glucose and fat, and neuromuscular transmission observed in a microarray analysis of the Ts65Dn muscle.

Furthermore, premature aging and associated changes in muscle strength and skin are significant and noteworthy in patients with DS compared to age matched euploid subjects. Such changes in muscle hypotonia and motor dysfunction in DS were supported by basic mechanisms of sarcopenia leading to the progressive age-related decrease in skeletal muscle mass, strength, and quality (Thompson, 2009). Additionally, significant decrease in the resynthesis kinetics of phosphocreatine (PCr) was reported in adults with DS compared to the non-DS counterpart, suggesting an impairment of muscle mitochondrial function in DS. The PCr depends on mitochondrial respiratory function and controls muscle mitochondrial function in vivo.

2.3 Mitochondria and type 2 diabetes and obesity

An increased risk of type 2 diabetes, childhood obesity, and various other endocrine disorders has been documented in children with DS.

Emerging evidences led to hypothesize that the mitochondrial dysfunction and impaired bioenergetics are potentially underlying for type 2 diabetes and childhood obesity in DS. A significant impairment of ETC activity of the subsarcolemmal mitochondria that is responsible for bioenergetics support of signal transduction, fat oxidation, and substrate transport have been attributed to the pathomechanism of insulin resistance in type 2 diabetes (Ritov et al., 2005). Mitochondria of the human skeletal muscle of both type 2 diabetes and obesity appeared smaller with reduced activity of the complex I. Diabetic complications are implicated with oxidative stress and consequent oxidative damage to DNA and proteins in both insulin deficiency and resistance states (Ruegsegger et al., 2018). Insulin regulates oxidative metabolism and ROS production; in return, ROS can affect insulin actions. Nevertheless, pathways involved in promoting insulin resistance may damage the mitochondrial function and increase ROS generation.

Moreover, interplay of the NRIP1 and PGC-1α in the transcriptional regulation of genes involved in energy homeostasis in triplicated HSA21 might play a crucial role in diabetes and obesity in DS. The mechanism of repressed expression of CIDEA, an important regulatory factor in adipose cell function and obesity, by NRIP1, which could be induced by PGC-1α, through ERRα and NRF-1 activity might contribute to obesity. The interaction of NRIP1 and PGC-1α are also thought to be involved in glucose uptake, and thus, responsible for pathophysiology of diabetes through the regulation of expression of the insulin-sensitive glucose transporter GLUT4 and its sub-cellular localization (Fritah et al., 2010). The report on the improvement of the metabolic parameters in both oxidative muscles and glycolytic processes through depletion of NRIP1 indicates the NRIP1 could be a potential drug target for the treatment of insulin resistance in type 2 diabetic and obese patients. Additionally, Nrip1-null mice are lean, resistant to high-fat diet-induced obesity, and showed higher oxygen consumption.

Mitochondrial fusion mechanism is pivotal in metabolism for maintenance of the mitochondrial dynamics. It was reported that reduced Mfn, particularly Mfn2, expression in both humans and mice is associated with obesity. Furthermore, the OPA1, an altered proteolytic processing of the GTPase, is associated with insulin resistance in humans and also loss of the mitochondrial protease OMA1, which alters processing of OPA1, resulted in obesity and defective thermogenesis in mice (Quiros et al., 2012).

2.4 Mitochondria and immune disorders

DS children are at a greater risk to infections, especially in the upper respiratory tract, autoimmune disorders, including hypothyroidism and celiac disease (health aspect is discussed in detail in Chapter 2). The impaired immune system in DS includes alteration of number of B and T cells, significant reduction in naive lymphocytes, abnormalities in functions and development of thymus, impairment of mitogen-induced T-cell proliferation, reduction in specific antibody responses to immunizations, and defects in neutrophil chemotaxis mechanism (Kusters et al., 2009). In the DS children, the slower lymphocyte respiration rate compared to euploid control group could envisage a relatively lower rate of mitochondrial energy conversion and alteration in the inner mitochondrial membrane potential, which might further explain the pathomechanism of this disorder. Additionally, mitochondrial fission was demonstrated during lymphocyte migration. Notably, signaling mechanism of mitochondrial remodeling was demonstrated to direct the programming of T-cell metabolism (Buck et al., 2016). The T-effector (TE) cells disrupt the mitochondria, whereas the T-memory (TM) cells maintain fused mitochondrial networks. Survival of TM lymphocytes decreased when the OPA1 was repressed, and the phenomenon was associated with altered cristae structure and reduced respiratory capacity. Thus, TE cells could be transformed to TM cells through the mitochondrial network. The report suggested that alteration in cristae morphology facilitated fusion in TM cells and that favored OXPHOS and fatty acid oxidation through the electron transport chain (ETC) organized into inner mitochondrial membrane, while fission in TE cells resulted in cristae expansion, reduction in ETC capacity, and induction of aerobic glycolysis (Buck et al., 2016).

2.5 Aging

Albeit DS is the most common aneuploidy that survives till birth, ~50% of the fetuses with trisomy HSA21 are lost during the first trimester of pregnancy (Leporrier et al., 2003). Indeed, ~50.3% of the total DS population survive till the first 18 years of life with a high preponderance (35.9%) of deaths during the first 5 years of life. The death rate gradually drops to 13.1% between 19 and 40 years of age, and DS individuals of 40+ years, especially noncardiopathic, can live beyond 60 years of age in developed countries (Bittles et al., 2007). In this context, it is noteworthy, that mtDNA is maternally inherited; thus, maternal inheritance of mtDNA mutations is expected to influence the frequency of DS in families and/or increase DS-pregnancies in females older than 35 years (Arbuzova et al., 2002). In addition, a normal mitochondrial cascade appeared necessary for chromosome segregation and spindle formation during meiosis and early embryogenesis (Eichenlaub-Ritter et al., 2011). Maternal age further contributes to mitochondrial function in follicular cells and oocytes. Notably, age-related increase in mtDNA mutations have been reported in oocytes (Eichenlaub-Ritter et al., 2011), and dysfunctional mitochondria have been encountered with meiotic nondisjunction of chromosomes in both first and second phase and mitotic anomalies during embryonic development. Additionally, mitochondrial errors affect the quality and development of aged oocytes and embryos in pregnancies of older age. Nevertheless, maternally inherited mtDNA mutations vary among the sibs and in different DS individuals. Therefore, individuals with higher mtDNA mutation rates would appear at a greater predisposition to age-related dementia and other geriatric diseases. Notably, DS with AD (DSAD) exhibited higher incidence of mtDNA mutations in the frontal cortex and lymphoblastoid cells compared to non-AD DS patients and age-matched controls (Coskun and Busciglio, 2012). Furthermore, aging is reflected with accumulation of mtDNA mutations. Higher rates of mtDNA mutations in specific nucleotides and in regulatory sequences of replication and transcription were observed in both sporadic AD and DSAD than in controls and that caused decrease in mtDNA levels and light strand gene expression in their brains (Coskun et al., 2004, 2010). Interestingly, defective repair mechanism of mtDNA damage was reported in DS.

Moreover, the DS population experience early aging and age-related ailments, including AD and dementia. Besides AD, other signs of geriatric illness appear prematurely in DS, which are collectively defined as "segmental progeroid syndrome" (Pallardó et al., 2010). Consistently, mitochondrial dysfunction and oxidative stress generate free radicals leading to structural damage in mitochondria and enhance signaling pathways connected with age-related diseases (Cenini et al., 2012). Albeit higher incidence of DSAD, an increased rate of mutations in mtDNA was demonstrated in the frontal cortex of DS and lymphoblastoid cells derived from DS, compared to age-matched controls (Coskun and Busciglio, 2012). The theory of telomere shortening and replicative senescence caused by cumulative effects of chronic superoxide production, oxidative damage, mitochondrial mutations, and mitochondrial dysfunction can explain the accelerated senescence of DS cells (Pallardo et al., 2010).

2.6 Mitochondrial dysfunction on neurodegeneration and Alzheimer disease

Mitochondrial function, fission–fusion dynamism, biogenesis, and degradation are crucial for neuronal function, including synaptogenesis, Ca^{2+} buffering, axonal transport, and mitochondrial bioenergetics (Fig. 9.2) (Oettinghaus et al., 2016; Valenti et al., 2018). Alteration in mitochondrial structure, increase in number of shorter mitochondria, and a smaller average mitochondrial volume have been documented in primary cultures of trisomic astrocytes and neurons indicating a potential link between mitochondrial dysfunction and neurodegeneration in DS individuals. In addition, generation of ROS in fetal neurons and higher levels of oxidative stress and apoptosis have been evidenced at all ages of DS compared to the general population (Cenini et al., 2012). Indeed, damage in mitochondrial function and structure incapacitate ATP production leading to onset of neurodegenerative process through production of proapoptotic factors and ROS. Taken together, mitochondrial anomalies, impaired ATP production, and oxidative stress increases cellular vulnerability and leads to death in DS brain and other tissues (Piccoli et al., 2013; Valenti et al., 2011).

People with DS are more vulnerable to AD due to trisomy of HSA21 and dysregulated gene expression. In DS, alterations in the Mfn1 and Drp1 proteins located in mitochondria perturb the regulation of mitochondrial morphology and activity in brains and cultured cells (Busciglio et al., 2002). Mitochondrial dysfunction was hypothesized to cause a persistent deficit in energy production and chronic oxidative stress, development of DS neuropathology, and AD in DS. The study of mitochondrial dysfunction in DS may include characterization of: the alterations in structure and function of mitochondria; mitochondrial alterations in lymphoblastoid cells; the molecular determinants of mitochondrial dysfunction; and determination of the mitochondrial dysfunction as a predictor of AD pathology.

DS brain is manifested with lower number of neurons in association with abnormal neuronal differentiation leading to intellectual disability and an increased risk for AD. People with DS ($\sim$75%) are afflicted with AD before the age 40 years compared to 60 or more years of the general population (Head et al., 2012). Oxidative stress and energy depletion induce amyloidogenic changes in amyloid β (Aβ)-precursor protein (AβPP) processing, indicating a link between mitochondrial dysfunction, oxidative stress, and Aβ production. The AD markers such as deposition of senile Aβ-plaques, hyperphosphorylated tau, and neurofibrilary tangles, collectively contributing to neurodegeneration, are associated with AD pathogenesis. Moreover, impaired mitochondrial function in DS astrocytes is associated with reduced mitochondrial redox activity and membrane potential resulting in breakdown of the cells and initiation of AD (Busciglio et al., 2002). Thus, impaired mitochondrial energy metabolism in the DS brain was suggested to be a significant factor of pathogenesis for DSAD.

A significant alteration in AβPP processing and Aβ trafficking was demonstrated in cortical DS astrocytes and neurons (Busciglio et al., 2002). The APP gene is upregulated and trims down the AβPP by a set of enzymes and causes accumulation of Aβ-42 and transfers less APP out of the brain cells observed in aborted fetuses with DS. The impaired energy metabolism was suggested to increase β-secretase cleavage of AβPP and altered Aβ trafficking, which results in intracellular accumulation of aggregated Aβ-42 in brain cells, which were higher in cells of DS brains compared to cells of normal individuals (Busciglio et al., 2002). AβPP has high affinity for β-secretase causing a drastic increase in Aβ production by the age of 55 years. This irony indicates a poor correlation between average ages of AD onset and the theoretical amount of Aβ production exacerbated by HSA21 genes.

Overexpression of DYRK1A caused downregulation of the neprilysin (NEP) activity, which is a major Aβ-degrading enzyme in the brain, and phosphorylation at the NEP cytoplasmic domain in transgenic mice (Asai et al., 2017). NEP activity is significantly reduced in fibroblasts derived from DS patients compared with that of healthy individuals. The CN-NFAT signaling pathway in trisomic HSA21 controlled by DYRK1A and RCAN1 plays fundamental role in development and degeneration of neurons. Overexpression of DYRK1A causes neurofibrillary degeneration and cell loss in the hippocampus area by affecting synaptic plasticity and memory consolidation (Guedj et al., 2014). DYRK1A inhibition rescued the impaired activity of NEP, which indicated that DYRK1A inhibitors could be effective against AD in DS adults, and also in sporadic AD patients. Multiple potential therapeutic approaches, including those targeting β-amyloid, tau, and the AβPP, await clinical trials for treatment and prevention of DSAD (Rafii, 2020).

Evaluation of mitochondrial structure and function and the molecular components involved in mitochondrial dysfunction in tissue samples and cultures of the brain, cortical neurons, astrocytes, and fibroblast, and lymphoblastoid cells from normal DS patients may provide novel information on mitochondrial structure and function in DS. Such studies may lead to understand the role of energy impairment in neurodegenerative disorders and guide for designing therapeutics toward prevention of neuronal dysfunction and the evolution of pathobiology of DSAD. Dysfunctional mitophagy has been implicated as the compelling genetic factor of neuropathomechanisms, including AD (Bordi et al., 2019).

3. Oxidative stress: an indication of adaptive response in DS

The mechanism of mitochondrial dysfunction on oxidative stress has been hypothesized to orchestrate an adaptive response rendering basic cellular functions and survival in DS. A dual role of oxidative stress and redox changes in DS demonstrated enhancement of cellular proliferation at low levels, while oxidative damage–induced apoptosis was expected at higher levels (Martin and Barrett, 2002). Based on the proposed theory, an adaptive response signaling, also known as hormesis, was expected to be triggered by a sublethal stress. A compensatory mechanism prepares the cells to fight off the higher levels of stress via stimulation of cellular functional changes toward protection from subsequent exposure to more severe stress (Martin and Barrett, 2002).

Indeed, dysfunctional mitochondria–induced ROS production plays the essential role in age-related oxidative damage. However, reports on reduced oxidative stress–associated increased life span (Schriner et al., 2005) are contradicted by interventions aimed to lower ROS level (Lott et al., 2011), or human studies failed to produce health beneficial effects (Ristow and Schmeisser, 2011). A randomized placebo-controlled daily trial on oral antioxidant supplementation over 2 years neither expressed improvement in cognitive functioning nor stabilized cognitive decline in DSAD (Lott, 2012). Thus, it is apparent that mitochondrial ROS production could potentially prompt cellular processes relevant to promotion of life span and health. It is noticeable that caloric restriction (CR) has established to exhibit mitochondrial hormesis (mitohormesis) by activating signaling events to increase mitochondrial respiration and elevation of mitochondrial ROS production but without alteration of ATP production (Sharma et al., 2011).

Prominently, although expression of the Nrf2/Keap1 genes and proteins in DS has not been linked to adaptive response, differential regulation of Nrf-2-associated oxidative stress response genes in astrocytes has indicated the presence of adaptive response in DS (Helguera et al., 2013). Additionally, elevated mitogen-activated protein kinases (MAPKs), including ERK1/2, SAPKs, and p38 have been demonstrated in brains of DS and AD (Swatton et al., 2004). Phosphorylation of Nrf2 facilitates its dissociation from the Nrf2/Keap1 complex by MAPKs (Taguchi et al., 2011). Noticeably, higher intellectual function and increased expression of glutathione peroxidase (GPx) in DS individuals indicated hormesis (Strydom et al., 2009). Under normal redox state, the transcription factor Nrf2 binds to Kelch-like ECH-associated protein 1 (Keap1) leading to its proteasomal degradation in the cytosol. Whereas under oxidative stress, conformational changes in Keap1 result in its dissociation from the Nrf2-Keap1 complex and translocation of free the Nrf2 into the nucleus where it can bind to antioxidant response element (ARE) regions in the genome, and influences the expression of stress response genes (reviewed in Coskun and Busciglio, 2012). Indeed, the Keap1-Nrf2-ARE signaling complex acts as a prominent sensor related to hormesis. Keap1 appears as a cysteine-rich protein and can sense the redox changes in cells. GPx and catalase are also Nrf2 target genes carrying ARE sequences in their promoters (reviewed in Coskun and Busciglio, 2012).

Moreover, Nrf2 interacts with PPARγ, PGC1α, and PI3K/Akt (Wang and Hai, 2011) and facilitates increase in mtDNA (Coskun et al., 2010) and mitochondrial biogenesis in DS cells (Valenti et al., 2011). Additionally, a generalized downregulation of mitochondrial activity has been linked to reduce ROS production and prevent cellular injury in different cell types of DS patients indicating a cellular adaptation (Helguera et al., 2013). However, concomitant exposure to additional stressors and/or challenges, including infections, seizures, age-related loss of function, etc., may cause exhaustion of strength of the functional adaptations to prevent cellular damage. In this perspective, oxidative stress and mitochondrial dysfunction may appear responsible for chronic respiratory infections, early senescence, and AD-like neurodegeneration. Therefore, the severity of the DS phenotype may depend on the background of mitochondrial mutations, the accumulation of oxidative damage, and the magnitude of the cellular adaptations triggered by the stressors. Thus, an early adaptive response triggered by sub-lethal levels of stress may confer a longer survival. However, the combinatorial expression of chronic stress and age-related health indices may indicate an early and accelerated development of age-related pathology such as dementia and AD.

Therapeutic interventions at the early stage of the stress or disease may confer cellular hormetic mechanism. In this context, long-term treatment with antioxidants may not reduce the level of oxidative stress or disease risk over the benefits that could be rendered by hormesis. Thus, it was hypothesized that antioxidants may not be helpful, rather may interfere with hormesis, under normal conditions. Thus, presence of multiple stressors may necessitate the use of the antioxidants (Ristow et al., 2009).

4. Clearance of dead cells: autophagy and mitophagy in DS

Mitochondrial dysfunction results in subsequent accumulation of damaged mitochondria, albeit cells have potential to clear the damaged cells by a mechanism called autophagy or macroautophagy. Autophagy majorly involves a lysosomal degradative pathway for recycling and removal of nonfunctional and damaged cytoplasmic components, including mitochondria (Galluzzi et al., 2017). The primary fibroblasts derived from DS patients were deficient in the similar system of "mitophagy" for removal of damaged mitochondria, which eventually leads to gathering of the damaged organelle and enhances oxidative stress.

Signaling events, including mitochondrial membrane depolarization for recruitment of specific adaptors, such as PINK1–PARKIN appears as the most well-characterized pathway for initiation of mitophagy. Activation of mitophagy is associated with accumulation of PINK1 on the mitochondrial outer membrane upon mitochondrial depolarization, followed by phosphorylating ubiquitin-promoting translocation of the E3 ubiquitin ligase PARKIN (PARK2) (Lazarou et al., 2015). Recruitment of specific autophagic cargo receptors, such as p62 (SQSTM1) and NDP52, facilitates sequestration into the autophagosomes (APs) following buildingup of ubiquitin chains on the mitochondrial external membrane proteins by PARKIN. Notably, similar sequestration and clearance steps work for mitophagy and bulk autophagy; however, the differential control of these processes is not fully resolved yet.

Downregulation of the ULK1 complex regulates autophagy negatively, mediated through the mTOR complex1 (mTORC1), a master regulatory protein kinase (Nazio et al., 2013). Additionally, mTORC1 prevents nuclear translocation of transcriptional factors (TFs) supporting activation of AP and lysosomal biogenesis. Importantly, hyperactivation of mTORC1 catalyzed the inactivating phosphorylation of ULK1 in DS fibroblasts, resulting in diminished autophagy induction and finally, a decline in mitophagy (Bordi et al., 2019). Interestingly, Ulk1-knockout mice exhibited defective mitophagy leading to accumulation of aberrant mitochondria (Egan et al., 2011). Positive regulation of the mTORC1 itself is facilitated by the small GTPase Rheb on lysosomes and by the PI3K/AKT cascade downstream of receptor tyrosine kinases (reviewed in Bordi et al., 2019). Activation of ULK1 complex renders inhibition of mTORC1 and stimulation of the downstream components of the autophagic pathway, such as PI3K—cIII complex and ATG-related proteins, leading to nucleation, elongation, and formation of APs. ATGs, including ATG7, ATG3, and ATG5, are represented as the essential proteins for the correct formation of APs, which are required in two ubiquitin-like systems and binding of LC3/GABARAPs with phosphatidylethanolamine (PE) (reviewed in Bordi et al., 2019). The components of two ubiquitin-like conjugation systems that are critical for initiation and maturation of AP were downregulated, including ATG7, ATG3, FOXO1 (as direct ATG7 interactor), ATG5, and ATG5-12 complex (Fig. 9.3) (Bordi et al., 2019), suggesting involvement of multiple levels, not just reduced ULK1 activity due to mTORC1-mediated S758 phosphorylation, for downregulation of autophagy induction. Notably, abnormal hyperactivation of the AKT/mTOR signaling pathway was observed by transcriptomic analysis in the frontal cortex and hippocampus of DS brain (Perluigi et al., 2014), indicating negative effects on mitochondrial turnover of the imbalances in autophagy flux regulation in DS. However, the molecular mechanism of the hyperactivated AKT-mTORC1 cascade in DS cells might be multifactorial and is not yet understood.

Additionally, elevated oxidative stress was hypothesized to further reduce the autophagy—mitophagy process, suggesting a negative regulatory feedback loop resulting from perturbed removal of dysfunctional mitochondria in DS cells. Notably, altered PARKIN function, disrupted E3 ligase activity, and compromised mitophagy were mediated by enhancement of ROS production. Indeed, oxidative stress inhibited autophagy directly by oxidizing ATG7 and ATG3, which are downregulated in DS fibroblasts (Frudd et al., 2018; Bordi et al., 2019). Impairment of PINK1/PARKIN and erroneous inhibition of macroautophagy have been identified as the underlying molecular mechanisms of mitophagy (Bordi et al., 2019).

Taken together, two hypotheses have been proposed: 1. Downregulated PARKIN and the mitophagic adaptor protein SQSTM1/p62 perturbs the induction of PINK1 to deactivate mitophagy following mitochondrial depolarization by CCCP or antimycin A plus oligomycin; 2. hyperactivation of mTOR, which leads to global suppression of macroautophagy activation and the transcriptional expression of proteins, such as ATG7, ATG3, and FOXO1, are critical for AP formation. Importantly, restoration of autophagy flux, initiation of mitophagy by PARKIN/PINK, and the clearance of damaged mitochondria by mitophagy could be achieved in DS cells through inhibition of mTOR complex 1 (mTORC1) and complex 2 (mTORC2) using AZD8055 (AZD) (Bordi et al., 2019), suggesting mTORC1–mTORC2 inhibitors as promising therapeutic targets for ameliorating deficits associated with DS. Interestingly, AZD restored expression of downregulated genes, including ATG7, FOXO1, and PARKIN, while significantly increased the PINK1 mRNA levels (underscoring its positive effect on mitophagy induction) in untreated DS (Bordi et al., 2019). AZD blocks mTORC2 that negatively regulates AKT and activates FOXO1 and FOXO3, which are essential for regulating the autophagic gene expression and transcription of PINK1 following deprivation of growth factors or diabetes-induced mitophagy (reviewed in Bordi et al., 2019). AZD8055 is a novel inhibitor of mTOR kinase activity.

5. Targeted therapies for restoration of mitochondrial network in DS

Induction of mitochondrial biogenesis with supplementation would be a beneficial modality for improving the expression and performance of mtDNA in the mitochondrial network. Therapeutic repression of the overexpressed HSA21 genes interacting with mitochondrial function would also be of value for controlling the DS phenotypes. Importantly, the time and duration of preconceptional, prenatal, and early postnatal treatment with bioenergetic compounds revealed the expected outcome in a number of investigations. Notably, activation of PPARGC1A and mitochondrial biogenesis protects against prenatal hypoxic–ischemic brain injury (Jia et al., 2020). Similar outcome is anticipated from therapeutic targets for DS phenotypes (Fig. 9.4) (Vacca et al., 2019; Valenti et al., 2018). With similar notion, the repression of the genes such as CBS of OXPHOS function, or DYRK1A, of mitochondrial biogenesis might be targeted as a potential early therapeutic agent. Thus, recent developments have been concentrated on the mitochondrial biogenesis and OXPHOS mechanisms, which are crucial for maintaining mitochondrial homeostasis that is required for improvement of DS phenotypes. In addition, since brain development, including neurogenesis and synapse formation

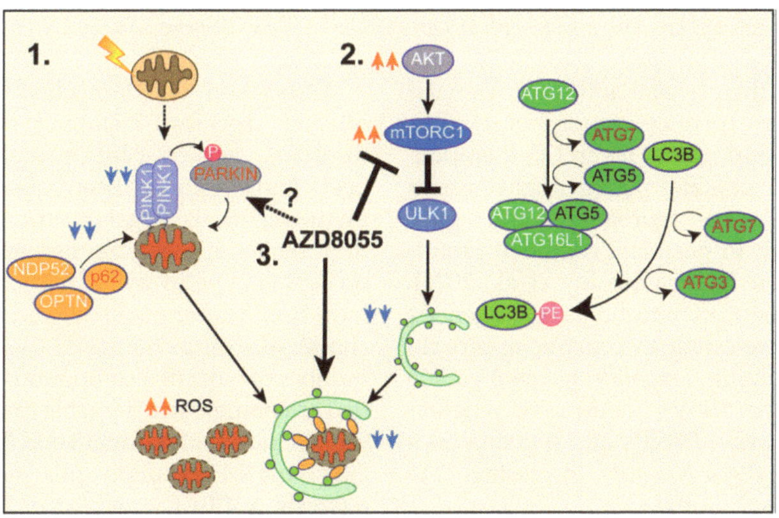

FIGURE 9.2 Implication of gene dosage effect of HSA21 on mitochondrial dynamics and oxidative phosphorylation in Down syndrome. Mitochondrial dysfunction and oxidative stress collectively contribute to pathologic manifestation, including early neurodegeneration leading to onset of Alzheimer disease in patients with Down syndrome. *Reproduced from: Valenti et al. (2018) with permission from RightsLink under the terms of Elsevier and Copyright Clearance Center.*

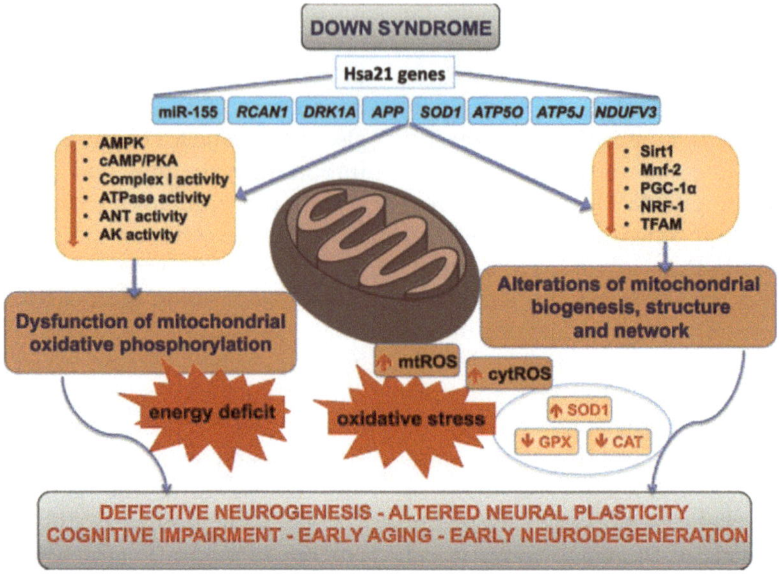

FIGURE 9.3 Proposed model illustrating the autophagy and mitophagy defects in DS fibroblasts. Our findings indicate that: 1. PARKIN/PINK1-mediated mitophagy pathway is altered in DS human fibroblasts leading to an accumulation of damaged mitochondria and thus to increase of ROS levels. We also find that macroautophagy is negatively regulated at different levels. 2. mTOR is hyperactivated leading to an inhibition of autophagy induction and crucial proteins involved in the AP formation, are downregulated, including ATG7 and FOXO1. 3. AZD8055 (AZD), an inhibitor of both of mTORC1 and mTORC2, rescues mitophagy and autophagy flux, thereby promoting the clearance of damaged mitochondria. The extensive autophagic deficits likely contribute in different ways to the severe symptoms described in DS individuals, leading to the supposition that mTORC1–mTORC2 inhibition can be a promising therapeutic option for DS. Factors downregulated in DS cells highlighted as follows: titles in red denote those downregulated at both the mRNA and protein level, black downregulated at the protein level, white are those that are unchanged. *Blue arrowheads* indicate those pathways downregulated while *orange arrowheads* those upregulated. *Reprinted from Bordi et al. (2019) under a Creative Commons Attribution 4.0 International License.*

mainly, occurs in the prenatal and early neonatal stages and also permanent neuromorphological alterations happen during fetal life in DS, therapeutic intervention at prenatal period might exert a major positive impact on neurogenesis and significant improvement in postnatal cognitive performance in DS (Pilar Bayona-Bafaluy et al., 2021). Altogether, promotion of mitochondrial biogenesis has become an interesting therapeutic approach, with drugs, including epigallocatechin-3-gallate (EGCG), resveratrol (RSV), metformin, pioglitazone, 2,4-dinitrophenol (DNP), and β-guanadinopropionic acid (β-GPA) for increasing the PPARGC1A levels.

Association of mitochondrial dysfunction causing diseases, including neurodegenerative disorders, cancer, cardiovascular diseases, diabetes, and obesity has been considered for targeted pharmaceuticals, largely to improve the mitochondrial biogenesis and bioenergetics in DS. Targets on signaling pathways of mitochondrial functions have been concentrated around to attenuate mitochondrial dysfunctions with new therapeutics. A number of bioactive compounds have been identified as compelling druggable candidates to control mitochondrial impairment and ROS production in DS (Table 9.1). Similarly, pharmacologically active agonists of PPARγ and/or of PGC-1α activators

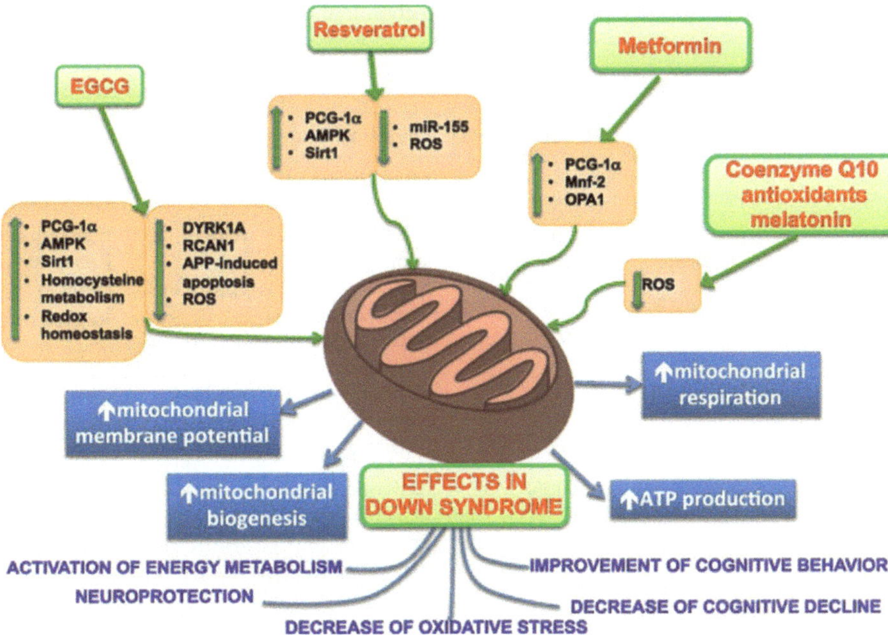

FIGURE 9.4 Therapeutic targets for amelioration of trplicated HSA21 genes in Down syndrome and druggable candidates for enhancement of mitochondrial membrane potential, mitochondrial ATP production, mitochondrial respiration and mitochondrial biogenesis. *Reproduced from: Valenti et al. (2018) with permission from RightsLink under the terms of Elsevier and Copyright Clearance Center.*

might correct mitochondrial dysfunctions in DS foeti/patients. The association of the microRNAs (miRNA), in particular miR-155, with respiratory chain function and ROS production, was considered to propose the potential roles of mitochondrial miRNAs (mitomiR) as novel therapeutic targets for DS patients. However, further studies on the transient or durable regulatory effects and delivery mechanisms of the mitomiRs to mitochondria may establish the therapeutic potential of miRNAs.

5.1 Polyphenols

Naturally occurring polyphenols have displayed multimodal action to modulate several pathways altered in DS, including gene expression (especially DYRK1A and RCAN1), and expression of several miRNAs; lipid oxidation; homocysteine metabolism; and signaling pathways that regulate mitochondrial functions; and redox homeostasis (reviewed in Valenti et al., 2018). Polyphenols are derived as secondary metabolites of plants largely found in fruits, vegetables, and cereals, which are essential components of normal human diet, and prevent humans from cancer, neurodegenerative, neurologic, cardiovascular, and metabolic diseases (Ajami et al., 2017). Epigallocatechin-3 gallate (EGCG) and RSV are the two polyphenols, of particular interest for DS.

5.1.1 Epigallocatechin-3-gallate

EGCG is a flavonoid class of polyphenol and is the most bioactive polyphenolic compound present in the green tea (*Camellia sinensis)* (Banerjee and Chatterjee, 2015). Its multipotent therapeutic features, including neuroprotective effects, free radical scavenger, crossing the blood–brain barrier, and so on, have been

TABLE 9.1 Therapeutic use of bioactive drugs targeting mitochondrial dysfunctions in Down syndrome (Valenti et al., 2018).

	Gene expression		Therapeutic actions	
Drugs	**Upregulated**	**Downregulated**	**On mitochondria**	**Others**
EGCG	PGC-1α; AMPK; Sirt1; Homocysteine metabolism; redox homeostasis	DYRK1A; RCAN1; APP-induced apoptosis; ROS	Mitochondrial ROS scavenging; Aβ-induced mitochondrial apoptosis; Rescue of MRC complex I and V activities; Activation of mitochondrial biogenesis; Promotion of ATP synthesis; Rescue of mitochondrial membrane potential	Modulation of homocysteine metabolism; Reduction of glutamate cytotoxicity
RSV	PGC-1α; AMPK; Sirt1	miR-155; ROS	Activation of mitochondrial biogenesis; Promotion of ATP synthesis; Promotion of respiratory chain activity	Reduction of Aβ peptides; Reduction of tau phosphorylation
Metformin	PGC-1α; MFN2; OPA1	NA	Promotion of mitochondrial network; Activation of mitochondrial bioenergetics; Activation of mitochondrial biogenesis	Control of blood sugar level
CoQ10	NA	NA	Activation of mitochondrial OXPHOS apparatus; Promotion of ATP synthesis; Mitochondrial ROS scavenging	Modulation of DNA repair mechanisms
Melatonin	NA	ROS	ROS scavenging; Reduction of oxidative damage	Regulation of sleep and wake cycles

Drugs tested in DS able to target mitochondria are indicated. Therapeutic actions relevant for DS not involving mitochondria are indicated with "Others". *NA.*, not available.

reported to be more effective than vitamin E and C. Vitamins rely on the presence of more steric structures than EGCG for stabilization of the phenoxy radical structure (Vacca et al., 2016). Indeed, polyphenol was detected in the brain tissue of mice following an oral administration of EGCG for 5 days (reviewed in Valenti et al., 2018).

EGCG has strong antioxidant properties and thus, is efficient to counteract the mitochondrial dysfunctions by targeting the signaling pathways, especially to control neural functions and minimize neurodegenerative diseases. Studies have demonstrated EGCG as a potent disease-modifying agent for maintaining mitochondrial homeostasis and scavenging of ROS; activating mitochondrial bioenergetics and biogenesis; and rendering protection from Aβ-amyloid-induced mitochondrial apoptosis and glutamate cytotoxicity, in cell and mouse models of DS (reviewed in Valenti et al., 2018).

EGCG's target on AMPK activation was reflected through enhanced mitochondrial ATP synthesis and proliferation of hippocampal neural progenitor cells in Ts65Dn mice treated with 20 μM EGCG, even after ischemic stroke in adult mice (Valenti et al., 2016; Zhang et al., 2017). EGCG has been reported to modulate expression of Sirt1, which acts as a master regulator for balancing genomic integrity and stability, and also as a key modulator of aging and neurodegeneration (Mazucanti et al., 2015). Interestingly, the synergism of Sirt1 and EGCG influences the phosphorylation/acetylation status of AMPK and the expression of PGC-1α. Indeed, 20 μM of EGCG-treatment influenced the respiratory chain function, mitochondrial biogenesis, and neural progenitor cell proliferation through upregulation of the AMPK/Sirt1/PGC-1α axis in Ts65Dn mice (Vacca et al., 2016). Importantly, similar treatment of DS-derived fibroblasts expressed stimulation of the complex I activity through modulation of the key signaling molecule, cAMP, and enhanced the cAMP-dependent PKA activity and phosphorylation of the NDUFS4 subunit of mitochondrial complex I (reviewed in Valenti et al., 2018). Thus, EGCG rescued the impaired complex I activity and decreased ROS generation. EGCG also acts as a DYRK1A inhibitor (Shi et al., 2018), which is overexpressed in DS causing impairment in brain development and cognitive deficits.

EGCG improved mitochondrial biogenesis and OXPHOS function in fibroblasts and lymphoblasts derived from DS individuals (Valenti et al., 2016). EGCG restored mitochondrial energy deficit and recovered the altered neurogenesis of NPCs in DS mice or trisomic human iPSCs (Valenti et al., 2016). Moreover, EGCG treatment at prenatal or neonatal age reversed neurogenesis and reduced cognitive defects in DS mouse models (Zhang et al., 2017). With relevance to sleep apnea of DS subjects, oral supplements of EGCG reduced the level of intermittent hypoxia in rodents, which is caused by oxidative stress and neuroinflammation (Burckhardt et al., 2008).

Interestingly, clinical studies conducted with EGCG demonstrated its safety and efficacy in DS, which includes: 1. the cognitive skill was potentiated in young adults with DS through supplementation of EGCG (9 mg/kg per day) for 12 months in a double-blind, randomized, placebo-controlled, phase two trial (de la Torre et al., 2016); 2. rescue of the mitochondrial dysfunction and improvement of some behavioral deficits in a DS child treated with dietary supplementation of EGCG (10 mg/kg per day) with fish oil omega-3 in a case study (Vacca and Valenti, 2015).

5.1.2 Resveratrol

Resveratrol is a polyphenol belonging to the class of stilbenoids, and found in grapes, peanuts, red fruits, and *Polygonum cuspidatum* plant. Deficits in mitochondrial bioenergetics and biogenesis were attenuated with 10 μM RSV in NPCs from Ts65Dn by modulating the PGC-1α/Sirt1/AMPK axis and that rescued the impaired neurogenesis in vitro (Valenti et al., 2016). Notably, RSV downregulates the HSA21-encoded miR-155 and improves mitochondrial biogenesis but does not interact with DYRK1A (Latruffe et al., 2015). The miR-155 controls TFAM. Thus, RSV can potentially normalize the impaired mitochondrial biogenesis and other metabolic alterations found in DS. Resveratrol reversed the mitochondrial energy deficit and rescued the altered neurogenesis of NPCs from a mouse model of DS. Similar to RSV, pioglitazone induced mitochondrial biogenesis and stimulated OXPHOS in DS fibroblasts (Mollo et al., 2019) and enhanced neurogenesis in SH-SY5Y cells of human neuroblastoma (Miglio et al., 2009).

The major drawback of polyphenol supplementation comprising of their rapid metabolism and extensive modifications occurring during absorption does not build a correlation between the doses and their bioavailability in the human body. Additionally, its concentration present in the natural food is low compared to their effective dose required for humans. Thus, in a broader perspective, different combinatorial strategies have been followed, such as co-administration with other polyphenols and/or other nutrients, encapsulation, prodrugs, polyphenol analogs, and alternative methods, other than oral administration, to elevate their delivery and bioavailability in humans (Vacca et al., 2016). In addition, dose–response studies would be essential before considering clinical translation, since increase of EGCG dose from 10 to 50 mg/kg per day appeared harmful to skeletal phenotypes of DS and did not improve the behavioral deficits in Ts65Dn mice (Stringer et al., 2017).

5.1.3 Omega-3 fatty acids

Omega-3 fatty acids have displayed enhancement of mitochondrial biogenesis in mouse models and cell lines (reviewed in Pilar Bayona-Bafaluy et al., 2021). Omega-3 alpha-linolenic acid was demonstrated essential in fetal brain development in mice, increased the PPARGC1A expression in mice fat and in human SH-SY5Y cells, and improved neuromorphological and cognitive alterations in Ts65Dn mice. Supplementation with omega-3 enriched fish oil at conception facilitated reduction in Rcan1 mRNA and protein levels in hippocampus of 1.5-month-old mice. Eicosapentaenoic (EPA) and docosahexaenoic (DHA) enhanced the expression of Ppargc1a in the mouse C2C12 myoblast cells (Lee et al., 2016; reviewed in Pilar Bayona-Bafaluy et al., 2021).

5.1.4 Metformin

Metformin, a therapeutic for type 2 diabetes, has increased expression of PGC-1α and fusion-inducing genes Mfn-2 and OPA1, and enhanced mitochondrial biogenesis and stimulated OXPHOS, indicating improvement of the dysfunctional mitochondrial network organization in fibroblasts of DS patients (Izzo et al., 2017). Thus, metformin was a potential therapeutic for treatment of DS. However, chronic use of metformin elevated the risk of AD by accumulating Aβ and tau proteins in the brain cortex and thus, influenced mitochondrial dysfunction (reviewed in Valenti et al., 2018). Importantly, the syndrome is per se tightly associated with high risk of early AD development, and thus, chronic therapy with metformin for DS patients should be considered with caution. Metformin enhanced mitochondrial biogenesis also in human aortic endothelial cells and stimulated neurogenesis in human bone marrow–mesenchymal stem cells (reviewed in Pilar Bayona-Bafaluy et al., 2021). Interestingly, metformin downregulated NRIP1 and augmented the Ppargc1a levels in mouse C2C12 cells (Fig. 9.5).

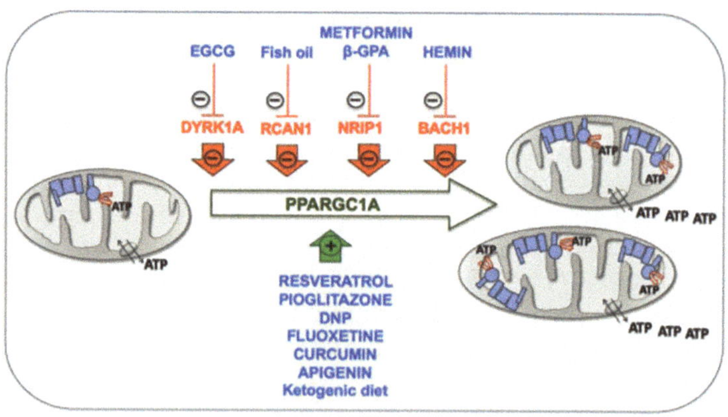

FIGURE 9.5 Inducers of mitochondrial biogenesis. Chromosome 21-encoded proteins that repress mitochondrial biogenesis are shown in red color. PPARGC1A, one of the main factors involved in mitochondrial biogenesis, is represented in green color. Drugs enhancing mitochondrial biogenesis are indicated in blue color. *Reprinted from Pilar Bayona-Bafaluy et al. (2021) under the Creative Commons Attribution License.*

5.2 Coenzyme Q10, antioxidants, and melatonin

Components, such as vitamins, coenzyme Q10 (CoQ10), and antioxidants, have been suggested for their possible therapeutic applications for DS diseases due to their potential target on mitochondrial dysfunctions (reviewed in Valenti et al., 2018). The bioenergetic enzyme, cofactor CoQ10, is involved in electron transfers from complexes I and II to the complex III in OXPHOS apparatus and scavenging ROS and confers protection as an antioxidant to the OXPHOS machinery (Acosta et al., 2016). It has displayed decrease in plasma levels of CoQ10 levels and a positive correlation with a low IQ score in DS individuals compared to controls. Additionally, a prolonged treatment protocol with CoQ10 supplementation (4 mg/kg per day) resulted in improvement in DNA oxidation in DS patients. CoQ10 supplementation potentially delayed disease progression in patients with Friedreich ataxia, Huntington, and Parkinson diseases. Antioxidants have gained potential therapeutic applications to specifically inhibit the mitochondrial alterations.

Antioxidants, such as α-tocopherol of vitamin E, ascorbic acid, and α-lipoic acid, were reported highly protective against oxidative imbalance and restored cognitive deficits in DS mouse models (Fig. 9.4) (reviewed in Valenti et al., 2018). However, its supplementation could not yield the replicable preclinical therapeutic effects in clinical trials on both children and young adults with DS (Lott, 2012). Melatonin, a sleep and wake cycle regulating hormone, appeared at low level in serum of the DS children compared to their normal counterparts. Owing to its strong antioxidant efficacies, testing of melatonin in DS mouse models revealed lowering of the oxidative damage and age-related degeneration of basal forebrain cholinergic neurons (BFCNs) and improvement of cognitive performance, suggesting melatonin as a therapeutic candidate for antioxidant supplementation in patients with DS (reviewed in Valenti et al., 2018). Flavonoids act as antioxidants to determine ROS synthesis suppression, inhibition of enzymes, or chelation of trace elements, those are responsible for free radical generation, scavenging ROS, and improvement of antioxidant defences (Pizzino et al., 2017). Genistein, a well-studied soy isoflavone, has been extensively employed as an antioxidant in myriads of studies, showing the potential to effectively scavenge ROS and broad pharmacological activities.

5.3 Other pharmacologic agents

Treatment with β-GPA feeding markedly reduced the expression of both Nrip1 protein and mRNA in the fast-twitch rat triceps muscles and increased mitochondrial proteins, such as COXI, COXIV, and CS and the mtDNA copy number in the cortex and ventral midbrain of mice, increased expression of several key metabolic enzymes of mitochondrial proliferation, and stimulated activation of Nfe2l2 signaling cascade (reviewed in Pilar Bayona-Bafaluy et al., 2021).

Hemin, the active ingredient of the FDA-approved drug Panhematin, binds to BACH1 and favors its release from DNA for nuclear export and subsequent degradation and thus, improves mitochondrial biogenesis. Hemin enhanced mitochondrial gene expression and stimulated oxygen consumption rate in triple-negative breast cancer cells (Fig. 9.5). Notably, it has been shown that dimethylfumarate (DMF) induced mitochondrial biogenesis and the amount of OXPHOS subunits and enhanced the transcription levels of mtDNA-encoded genes and increased the mtDNA copy number in mouse embryonic fibroblasts. DMF modulates the Kelch-like ECH associated protein 1 (KEAP1) to allow translocation of NFE2L2 to the nucleus and the dissociation of BACH1. KEAP1 otherwise sequesters NFE2L2 in the cytoplasm. Expression of NFE2L2 can be influenced by pharmacologic agents, including fluoxetine, curcumin, and apigenin, though none of them is related to HSA21 genes associated with suppression of mitochondrial biogenesis. Interestingly, prenatal treatment with these agents has shown to confer beneficial effects on neurogenesis and behavioral performance in DS mouse models and improvement of OXPHOS and mitochondrial biogenesis (reviewed in Pilar Bayona-Bafaluy et al., 2021).

DNP, a mild OXPHOS uncoupler that increased the synthesis of pyrimidine nucleotide–derived compounds to produce phospholipids, neuritogenesis, and neuronal differentiation, also activates mitochondrial biogenesis in mice, rats, and human rhabdomyosarcoma cells (Fig. 9.5) (reviewed in Pilar Bayona-Bafaluy et al., 2021). Interestingly, DNP showed significant reduction in phospholipid content in the central cortex and cerebellum of individuals with DS. It has been also reported that DNP decreased intracellular accumulation of App in trisomic cerebral cortex cells of mouse model.

The ketogenic diet enhanced the upregulation of the Ppargc1a mRNA and protein levels and other markers of mitochondrial biogenesis in mouse hippocampus (reviewed in Pilar Bayona-Bafaluy et al., 2021). The PPARGC1A levels and oxygen consumption could be supplemented by the β-hydroxybutyrate in cells from other species, including human fibroblasts. Prenatal and early postnatal ketosis improved brain development of pyruvate dehydrogenase complex–deficient mammalian progeny, and also, maternal ketosis increased fetal brain weight in pigs. Thus, ketogenic diet has been proposed to be a potential therapeutic opportunity for patients with DS.

Tryptolinamide, a PFK inhibitor, demonstrated activation of mitochondrial respiration and restored the neuronal differentiation of iPSCs carrying mutant mtDNAs. Indeed, PFK knockdown increased the neuronal differentiation of NSCs and influenced neurogenesis in the dentate gyrus of hippocampus in the mouse (reviewed in Pilar Bayona-Bafaluy et al., 2021).

Prenatal administration of choline has presented some beneficial effects on neurogenesis and behavioral performance in a DS mouse model (reviewed in Pilar Bayona-Bafaluy et al., 2021). Choline in association with cytidine triphosphate (CTP), derived from uridine triphosphate (UTP), is essential for phospholipid synthesis and brain development. Therefore, choline or uridine could be considered in therapeutic development for DS.

6. Concluding remarks

Albeit DS is a multigenic complex disorder with many altered cell signaling pathways, fetal onset of OXPHOS dysfunction might act as the key pathologic mechanism of intellectual disability and many other medical conditions in DS. Moreover, the impact of the ubiquitous OXPHOS and its impairment might affect the cellular functions in every organ of the body. Therefore, strategic development to improve the OXPHOS function could be considered as compelling therapeutic approach for individuals with DS. The present discussion is focussed on the impact of the mitochondrial dysfunctions and the resultant accumulation of ROS in cells and organs and associated pathobiology in DS. The disarrayed mitochondrial network is responsible for initiation of diseases of several systems in DS, such as neurodegenerative diseases, aging, heart defects, and other disorders associated with DS phenotype. Indeed, the mitochondrial alterations significantly perturb the brain and cardiac development because of their dependence on higher energy consumption and susceptibility to oxidative stress, starting from fetal age. Therefore, mitochondrial dysfunction was anticipated to confer the major etiological mechanisms for intellectual disability and cognitive decline and early onset of neurodegeneration and aging. Albeit the mitophagic efficacy of the cells for clearing the death and apoptotic cells, an adaptive response triggered by a longer survival has been demonstrated as the effect of the sub-lethal levels of stress on oxidative metabolism. Moreover, the combinatorial expression of chronic energy stress and lowered of ROS scavenging on age-related health indices may accelerate development and progression of DS pathobiology.

Therapeutic interventions aimed at reducing oxidative stress and improving mitochondrial function starting from the prenatal stage could be more beneficial to prevent neurobehavioral phenotypes. However, long-time treatment is recommended, since short-time prenatal or neonatal intervention appeared ineffective in adulthood. A number of natural bioactive compounds, such as polyphenol extracts from fruits or plants, are suggested for therapeutic applications in DS due to long-term retention of their safety and efficacy. Efficacy has also been tested from combination therapy such as recommendation of omega-3 fatty acids with polyphenols or other pharmacologic agents. Nevertheless, prior testing is essential before considering application on humans. Importantly, the challenge for improving mitochondrial function may ameliorate cognitive impairment and many other DS-clinical phenotypes, including AD. Several promising druggable candidates have already presented convincing amount of hope following testing on DS mouse models, trisomic cells lines derived from DS, and also on human. In principle, each of the triplicated genes that perturb the mitochondrial biogenesis, in particular neuronal cells, should be targeted for their contribution to OXPHOS dysfunction and neurogenesis. The prenatal testing of the proposed drugs on heart and brain development can be performed in DS animal models in vivo but that involves more time, effort, and resources for maintenance. Interestingly, few of the proposed drugs have been tested in pregnant women based on adequate preclinical evidence and appeared to be safe, which have been approved for the treatment of other pathologies in DS.

References

Acosta, M.J., Vazquez Fonseca, L., Desbats, M.A., Cerqua, C., Zordan, R., Trevisson, E., Salviati, L., 2016. Coenzyme Q biosynthesis in health and disease. Biochim. Biophys. Acta 1857, 1079–1085. https://doi.org/10.1016/j.bbabio.2016.03.036.

Ajami, M., Pazoki-Toroudi, H., Amani, H., Nabavi, S.F., Braidy, N., Vacca, R.A., Atanasov, A.G., Mocan, A., Nabavi, S.M., 2017. Therapeutic role of sirtuins in neurodegenerative disease and their modulation by polyphenols. Neurosci. Biobehav. Rev. 73, 39–47. https://doi.org/10.1016/j.neubiorev.2016.11.022.

Arbuzova, S., Hutchin, T., Cuckle, H., 2002. Mitochondrial dysfunction and Down's syndrome. Bioessays 24 (8), 681–684.

Arron, J.R., Winslow, M.M., Polleri, A., Chang, C.P., Wu, H., Gao, X., Neilson, J.R., Chen, L., Heit, J.J., Kim, S.K., et al., 2006. NFAT dysregulation by increased dosage of DSCR1 and DYRK1A on chromosome 21. Nature 441, 595–600. https://doi.org/10.1038/nature04678.

Asai, M., Kawakubo, T., Mori, R., Iwata, N., 2017. Elucidating pathogenic mechanisms of early-onset Alzheimer's disease in Down syndrome patients. Yakugaku Zasshi 137 (7), 801–805. https://doi.org/10.1248/yakushi.16-00236-2.

Banerjee, S., Chatterjee, J., June 2015. Efficient extraction strategies of tea (*Camellia sinensis*) biomolecules. J. Food Sci. Technol. 52 (6), 3158–3168. https://doi.org/10.1007/s13197-014-1487-3.

Bittles, A.H., Bower, C., Hussain, R., Glasson, E.J., 2007. The four ages of Down syndrome. Eur. J. Publ. Health 17 (2), 221–225.

Bonnet, S., Michelakis, E.D., Porter, C.J., Andrade-Navarro, M.A., Thebaud, B., et al., 2006. An abnormal mitochondrial-hypoxia inducible factor1alpha-Kv channel pathway disrupts oxygen sensing and triggers pulmonary arterial hypertension in fawn hooded rats: similarities to human pulmonary arterial hypertension. Circulation 113, 2630–2641.

Bordi, M., Darji, S., Sato, Y., Mellén, M., Berg, M.J., Kumar, A., Jiang, Y., Nixon, R.A., 2019. mTOR hyperactivation in Down Syndrome underlies deficits in autophagy induction, autophagosome formation, and mitophagy. Cell Death Dis. 10, 563. https://doi.org/10.1038/s41419-019-1752-5.

Buck, M.D., O'Sullivan, D., Klein Geltink, R.I., Curtis, J.D., Chang, C.H., et al., 2016. Mitochondrial dynamics controls T cell fate through metabolic programming. Cell 166, 63–76.

Burckhardt, I.C., Gozal, D., Dayyat, E., Cheng, Y., Li, R.C., Goldbart, A.D., Row, B.W., 2008. Green tea catechin polyphenols attenuate behavioral and oxidative responses to intermittent hypoxia. Am. J. Respir. Crit. Care Med. 177, 1135–1141. https://doi.org/10.1164/rccm.200701-110OC.

Busciglio, J., Pelsman, A., Wong, C., Pigino, G., Yuan, M., et al., 2002. Altered metabolism of the amyloid beta precursor protein is associated with mitochondrial dysfunction in Down's syndrome. Neuron 33, 677–688.

Bushdid, P.B., Osinska, H., Waclaw, R.R., Molkentin, J.D., Yutzey, K.E., 2003. NFATc3 and NFATc4 are required for cardiac development and mitochondrial function. Circ. Res. 92, 1305–1313.

Cenini, G., Dowling, A.L., Beckett, T.L., et al., 2012. Association between frontal cortex oxidative damage and beta-amyloid as a function of age in Down syndrome. Biochim. Biophys. Acta 1822 (2), 130–138.

Cereghetti, G.M., Stangherlin, A., Martins de Brito, O., Chang, C.R., Blackstone, C., Bernardi, P., Scorrano, L., 2008. Dephosphorylation by calcineurin regulates translocation of Drp1 to mitochondria. Proc. Natl. Acad. Sci. U.S.A. 105, 15803–15808. https://doi.org/10.1073/pnas.0808249105.

Conti, A., Fabbrini, F., D'Agostino, P., Negri, R., Greco, D., Genesio, R., D'Armiento, M., Olla, C., Paladini, D., Zannini, M., et al., 2007. Altered expression of mitochondrial and extracellular matrix genes in the heart of human fetuses with chromosome 21 trisomy. BMC Genom. 8, 268. https://doi.org/10.1186/1471-2164-8-2683.

Corbalan, J.J., Kitsis, R.N., March 16, 2018. RCAN1-Calcineurin Axis and the set-point for myocardial damage during ischemia-reperfusion. Circ. Res. 122 (6), 796–798. https://doi.org/10.1161/CIRCRESAHA.118.312787. PMID: 29700078.

Coskun, P.E., Busciglio, J., 2012. Oxidative stress and mitochondrial dysfunction in Down's syndrome: relevance to aging and dementia. Curr. Gerontol. Geriatr. Res. 2012, 383170. https://doi.org/10.1155/2012/383170.

Coskun, P.E., Beal, M.F., Wallace, D.C., 2004. Alzheimer's brains harbor somatic mtDNA control-region mutations that suppress mitochondrial transcription and replication. Proc. Natl. Acad. Sci. U.S.A. 101 (29), 10726–10731.

Coskun, P.E., Wyrembak, J., Derbereva, O., Melkonian, G., Doran, E., Lott, I.T., Head, E., Cotman, C.W., Wallace, D.C., 2010. Systemic mitochondrial dysfunction and the etiology of Alzheimer's disease and Down syndrome dementia. J. Alzheimers Dis. 20 (Suppl. 2), S293–S310.

Cowley, P.M., Keslacy, S., Middleton, F.A., DeRuisseau, L.R., Fernhall, B., et al., 2012. Functional and biochemical characterization of soleus muscle in Down syndrome mice: insight into the muscle dysfunction seen in the human condition. Am. J. Physiol. Regul. Integr. Comp. Physiol. 303, R1251–R1260.

Dai, D.F., Rabinovitch, P.S., Ungvari, Z., 2012. Mitochondria and cardiovascular aging. Circ. Res. 110 (8), 1109–1124.

de la Torre, R., de Sola, S., Hernandez, G., Farré, M., Pujol, J., Rodriguez, J., et al., 2016. TESDAD study group, safety and efficacy of cognitive training plus epigallocatechin-3-gallate in young adults with Down's syndrome (TESDAD): a double-blind, randomised, placebo-controlled, phase 2 trial. Lancet Neurol. 15, 801–810. https://doi.org/10.1016/S1474-4422(16)30034-5.

Egan, D.F., et al., 2011. Phosphorylation of ULK1 (hATG1) by AMP-activated protein kinase connects energy sensing to mitophagy. Science 331, 456–461.

Eichenlaub-Ritter, U., Wieczorek, M., Lüke, S., Seidel, T., 2011. Age related changes in mitochondrial function and new approaches to study redox regulation in mammalian oocytes in response to age or maturation conditions. Mitochondrion 11, 783–796.

Fritah, A., Steel, J.H., Nichol, D., Parker, N., Williams, S., et al., 2010. Elevated expression of the metabolic regulator receptor-interacting protein 140 results in cardiac hypertrophy and impaired cardiac function. Cardiovasc. Res. 86, 443–451.

Frudd, K., Burgoyne, T., Burgoyne, J.R., 2018. Oxidation of Atg3 and Atg7 mediates inhibition of autophagy. Nat. Commun. 9, 95.

Galluzzi, L., et al., 2017. Molecular definitions of autophagy and related processes. EMBO J. 36, 1811–1836.

Guedj, F., Bianchi, D.W., Delabar, J.M., 2014. Prenatal treatment of Down syndrome: a reality? Curr. Opin. Obstet. Gynecol. 26, 92–103.

Head, E., Silverman, W., Patterson, D., Lott, I.T., 2012. Aging and down syndrome. Curr. Gerontol. & Geriatr. Res. 412536.

Hefti, E., Blanco, J.G., 2016. Anthracycline-related cardiotoxicity in patients with acute myeloid leukemia and Down syndrome: a literature review. Cardiovasc. Toxicol. 16 (1), 5–13.

Hefti, E., Bard, J., Blanco, J.G., January 2017. Analysis of heteroplasmic variants in the cardiac mitochondrial genome of individuals with down syndrome. Hum. Mutat. 38 (1), 48–54. https://doi.org/10.1002/humu.23071.

Helguera, P., Seiglie, J., Rodriguez, J., Hanna, M., Helguera, G., Busciglio, J., 2013. Adaptive downregulation of mitochondrial function in Down syndrome. Cell Metabol. 17, 132–140. https://doi.org/10.1016/j.cmet.2012.12.005.

Izzo, A., Nitti, M., Mollo, N., Paladino, S., Procaccini, C., et al., 2017. Metformin restores the mitochondrial network and reverses mitochondrial dysfunction in Down syndrome cells. Hum. Mol. Genet. 26, 1056–1069. https://doi.org/10.1093/hmg/ddx016.

Jia, L., Wang, J., Cao, H., Zhang, X., Rong, W., Xu, Z., 2020. Activation of PGC-1α and mitochondrial biogenesis protects against prenatal hypoxic-ischemic brain injury. Neuroscience 432, 63–72. https://doi.org/10.1016/j.neuroscience.2020.02.035.

Knott, A.B., Perkins, G., Schwarzenbacher, R., Bossy-Wetzel, E., 2008. Mitochondrial fragmentation in neurodegeneration. Nat. Rev. Neurosci. 9, 505–518. https://doi.org/10.1038/nrn2417.

Kusters, M.A., Verstegen, R.H., Gemen, E.F., de Vries, E., 2009. Intrinsic defect of the immune system in children with Down syndrome: a review. Clin. Exp. Immunol. 156, 189–193.

Latruffe, N., Lançon, A., Frazzi, R., Aires, V., Delmas, D., Michaille, J.-J., Djouadi, F., Bastin, J., Cherkaoui-Malki, M., 2015. Exploring new ways of regulation by resveratrol involving miRNAs, with emphasis on inflammation. Ann. N. Y. Acad. Sci. 1348, 97–106. https://doi.org/10.1111/nyas.12819.

Lazarou, M., Sliter, D., Kane, L., et al., 2015. The ubiquitin kinase PINK1 recruits autophagy receptors to induce mitophagy. Nature 524, 309–314. https://doi.org/10.1038/nature14893.

Lee, M.-S., Shin, Y., Moon, S., Kim, S., Kim, Y., 2016. Effects of eicosapentaenoic acid and docosahexaenoic acid on mitochondrial DNA replication and PGC-1α gene expression in C2C12 muscle cells. Prev. Nutr. Food Sci. 21, 317–322. https://doi.org/10.3746/pnf.2016.21.4.317.

Leporrier, N., Herrou, M., Morello, R., Leymarie, P., 2003. Fetuses with Down's syndrome detected by prenatal screening are more likely to abort spontaneously than fetuses with Down's syndrome not detected by prenatal screening. J. Obstet. Gynaecol. 110 (1), 18–21.

Liu, C., Morishima, M., Yu, T., et al., 2011. Genetic analysis of Down syndrome-associated heart defects in mice. Hum. Genet. 130, 623. https://doi.org/10.1007/s00439-011-0980-2.

Lott, I.T., 2012. Antioxidants in down syndrome. Biochim. Biophys. Acta 1822 (5), 657–663.

Lott, I.T., Doran, E., Nguyen, V.Q., Tournay, A., Head, E., Gillen, D.L., 2011. Down syndrome and dementia: a randomized, controlled trial of antioxidant supplementation. Am. J. Med. Genet. 155 (8), 1939–1948.

Mammucari, C., Gherardi, G., Rizzuto, R., 2017. Structure, activity regulation, and role of the mitochondrial calcium uniporter in health and disease. Front. Oncol. 7, 139. https://doi.org/10.3389/fonc.2017.00139.

Martin, K.R., Barrett, J.C., February 2002. Reactive oxygen species as double-edged swords in cellular processes: low-dose cell signaling versus high-dose toxicity. Hum. Exp. Toxicol. 21 (2), 71–75.

Martinou, J.C., Youle, R.J., 2011. Mitochondria in apoptosis: Bcl-2 family members and mitochondrial dynamics. Dev. Cell 21, 92–101. https://doi.org/10.1016/j.devcel.2011.06.017.

Mazucanti, C.H., Cabral-Costa, J.V., Vasconcelos, A.R., Andreotti, D.Z., Scavone, C., Kawamoto, E.M., 2015. Longevity pathways (mTOR, SIRT, Insulin/IGF-1) as key modulatory targets on aging and neurodegeneration. Curr. Top. Med. Chem. 15, 2116–2138.

Miglio, G., Rattazzi, L., Rosa, A.C., Fantozzi, R., 2009. PPAR-gamma stimulation promotes neurite outgrowth in SH-SY5Y human neuroblastoma cells. Neurosci. Lett. 454, 134–138. https://doi.org/10.1016/j.neulet.2009.03.014.

Mollo, N., Nitti, M., Zerillo, L., Faicchia, D., Micillo, T., Accarino, R., Secondo, A., et al., 2019. Pioglitazone improves mitochondrial organization and bioenergetics in down syndrome cells. Front. Genet. 10, 606. https://doi.org/10.3389/fgene.2019.00606.

Nazio, F., et al., 2013. mTOR inhibits autophagy by controlling ULK1 ubiquitylation, self-association and function through AMBRA1 and TRAF6. Nat. Cell Biol. 15, 406–416.

Oettinghaus, B., Schulz, J.M., Restelli, L.M., Licci, M., Savoia, C., et al., 2016. Synaptic dysfunction, memory deficits and hippocampal atrophy due to ablation of mitochondrial fission in adult forebrain neurons. Cell Death Differ. 23, 18–28. https://doi.org/10.1038/cdd.2015.39.

Pallardo, F.V., Lloret, A., Lebel, M., d'Ischia, M., Cogger, V.C., et al., 2010. Mitochondrial dysfunction in some oxidative stress-related genetic diseases: Ataxia-Telangiectasia, Down Syndrome, Fanconi Anaemia and Werner Syndrome. Biogerontology 11, 401–419.

Park, S.C., Mathews, R.A., Zuberbuhler, J.R., Rowe, R.D., Neches, W.H., et al., 1977. Down syndrome with congenital heart malformation. Am. J. Dis. Child. 131, 29–33.

Park, K.S., Wiederkehr, A., Kirkpatrick, C., Mattenberger, Y., Martinou, J.C., Marchetti, P., Demaurex, N., Wollheim, C.B., 2008. Selective actions of mitochondrial fission/fusion genes on metabolism-secretion coupling in insulin-releasing cells. J. Biol. Chem. 283, 33347–33356.

Parra, V., Altamirano, F., Hernández-Fuentes, C.P., et al., 2018. Down syndrome critical region 1 gene, Rcan1, helps maintain a more fused mitochondrial network. Circ. Res. 122 (6), e20–e33. https://doi.org/10.1161/CIRCRESAHA.117.311522.

Perluigi, M., et al., 2014. Neuropathological role of PI3K/Akt/mTOR axis in Down syndrome brain. Biochim. Biophys. Acta (BBA) - Mol. Basis Dis. 1842, 1144–1153.

Piccoli, C., et al., 2013. Chronic pro-oxidative state and mitochondrial dysfunctions are more pronounced in fibroblasts from down syndrome foeti with congenital heart defects. Hum. Mol. Genet. 22, 1218–1232.

Pilar Bayona-Bafaluy, M., Garrido-Pérez, N., Meade, P., Iglesias, E., Jiménez-Salvador, I., et al., 2021. Down syndrome is an oxidative phosphorylation disorder. Redox Biol. 41, 101871. https://doi.org/10.1016/j.redox.2021.101871.

Pizzino, G., Irrera, N., Cucinotta, M., Pallio, G., Mannino, F., et al., 2017. Oxidative stress: harms and benefits for human health. Oxid. Med. Cell. Longev. 13. https://doi.org/10.1155/2017/8416763. Article ID 8416763.

Quiros, P.M., Ramsay, A.J., Sala, D., Fernandez-Vizarra, E., Rodriguez, F., et al., 2012. Loss of mitochondrial protease OMA1 alters processing of the GTPase OPA1 and causes obesity and defective thermogenesis in mice. EMBO J. 31, 2117–2133.

Rafii, M.S., 2020. Alzheimer's disease in down syndrome: progress in the design and conduct of drug prevention trials. CNS Drugs 34, 785–794. https://doi.org/10.1007/s40263-020-00740-6.

Ristow, M., Schmeisser, S., 2011. Extending life span by increasing oxidative stress. Free Radic. Biol. Med. 51 (2), 327–336.

Ristow, M., Zarse, K., Oberbach, A., et al., 2009. Antioxidants prevent health-promoting effects of physical exercise in humans. Proc. Natl. Acad. Sci. U.S.A. 106 (21), 8665–8670.

Ritov, V.B., Menshikova, E.V., He, J., Ferrell, R.E., Goodpaster, B.H., et al., 2005. Deficiency of subsarcolemmal mitochondria in obesity and type 2 diabetes. Diabetes 54, 8–14.

Rotter, D., Peiris, H., Grinsfelder, D.B., et al., 2018. Regulator of Calcineurin 1 helps coordinate whole-body metabolism and thermogenesis. EMBO Rep. 19 (12), e44706. https://doi.org/10.15252/embr.201744706.

Ruegsegger, G.N., Creo, A.L., Cortes, T.M., Dasari, S., Sreekumaran Nair, K., 2018. Altered mitochondrial function in insulin deficient and insulin-resistant states. J. Clin. Invest. 128 (9), 3671–3681. https://doi.org/10.1172/JCI120843.

Schriner, S.E., Linford, N.J., Martin, G.M., et al., 2005. Medicine: extension of murine life span by overexpression of catalase targeted to mitochondria. Science 308 (5730), 1909–1911.

Seok, H.Y., Chen, J., Kataoka, M., Huang, Z.P., Ding, J., et al., 2014. Loss of MicroRNA-155 protects the heart from pathological cardiac hypertrophy. Circ. Res. 114, 1585–1595.

Sharma, P.K., Agrawal, V., Roy, N., 2011. Mitochondria-mediated hormetic response in life span extension of calorie-restricted *Saccharomyces cerevisiae*. Age 33 (2), 143–154.

Shi, W., Li, L., Ding, Y., Yang, K., Chen, Z., Fan, X., Jiang, S., Guan, Y., Liu, Z., Xu, D., Wu, L., 2018. The critical role of epigallocatechin gallate in regulating mitochondrial metabolism. Future Med. Chem. 10, 795–809. https://doi.org/10.4155/fmc-2017-0204.

Stringer, M., Abeysekera, I., Thomas, J., LaCombe, J., Stancombe, K., Stewart, R.J., Dria, K.J., Wallace, J.M., Goodlett, C.R., Roper, R.J., 2017. Epigallocatechin-3-gallate (EGCG) consumption in the Ts65Dn model of Down syndrome fails to improve behavioral deficits and is detrimental to skeletal phenotypes. Physiol. Behav. 177, 230–241. https://doi.org/10.1016/j.physbeh.2017.05.003.

Strydom, A., Dickinson, M.J., Shende, S., Pratico, D., Walker, Z., 2009. Oxidative stress and cognitive ability in adults with Down syndrome. Prog. Neuro Psychopharmacol. Biol. Psychiatr. 33 (1), 76–80.

Swatton, J.E., Sellers, L.A., Faull, R.L., Holland, A., Iritani, S., Bahn, S., 2004. Increased MAP kinase activity in Alzheimer's and Down syndrome but not in schizophrenia human brain. Eur. J. Neurosci. 19 (10), 2711–2719.

Taguchi, K., Motohashi, H., Yamamoto, M., 2011. Molecular mechanisms of the Keap1-Nrf2 pathway in stress response and cancer evolution. Gene Cell. 16 (2), 123–140.

Thompson, L.V., 2009. Age-related muscle dysfunction. Exp. Gerontol. 44, 106–111.

Vacca, R.A., Valenti, D., 2015. Green tea EGCG plus fish oil omega-3 dietary supplements rescue mitochondrial dysfunctions and are safe in a Down's syndrome child. Clin. Nutr. 34, 783–784. https://doi.org/10.1016/j.clnu.2015.04.012.

Vacca, R.A., Valenti, D., Caccamese, S., Daglia, M., Braidy, N., Nabavi, S.M., 2016. Plant polyphenols as natural drugs for the management of Down syndrome and related disorders. Neurosci. Biobehav. Rev. 71, 865–877. https://doi.org/10.1016/j.neubiorev.2016.10.023.

Vacca, R.A., Bawari, S., Valenti, D., Tewari, D., Nabavi, S.F., Shirooie, S., Sah, A.N., Volpicella, M., Braidy, N., Nabavi, S.M., 2019. Down syndrome: neurobiological alterations and therapeutic targets. Neurosci. Biobehav. Rev. 98, 234–255. https://doi.org/10.1016/j.neubiorev.2019.01.001.

Valenti, D., Manente, G.A., Moro, L., Marra, E., Vacca, R.A., 2011. Deficit of complex I activity in human skin fibroblasts with chromosome 21 trisomy and overproduction of reactive oxygen species by mitochondria: involvement of the cAMP/PKA signalling pathway. Biochem. J. 435, 679–688.

Valenti, D., Braidy, N., De Rasmo, D., et al., 2018. Mitochondria as pharmacological targets in Down syndrome. Free Radic. Biol. Med. 114, 69–83. https://doi.org/10.1016/j.freeradbiomed.2017.08.014.

Valenti, D., de Bari, L., de Rasmo, D., Signorile, A., Henrion-Caude, A., Contestabile, A., Vacca, R.A., 2016. The polyphenols resveratrol and epigallocatechin-3-gallate restore the severe impairment of mitochondria in hippocampal progenitor cells from a Down syndrome mouse model. Biochim. Biophys. Acta 1862, 1093–1104. https://doi.org/10.1016/j.bbadis.2016.03.003.

Valenti, D., Vacca, R.A., Moro, L., Atlante, A., 2021. Mitochondria can cross cell boundaries: an overview of the biological relevance, pathophysiological implications and therapeutic perspectives of intercellular mitochondrial transfer. Int. J. Mol. Sci. 22 (15), 8312. https://doi.org/10.3390/ijms22158312.

Wang, X., Hai, C.X., 2011. ROS acts as a double-edged sword in the pathogenesis of Type 2 diabetes mellitus: is Nrf2 a potential target for the treatment? Mini Rev. Med. Chem. 11 (12), 1082–1092.

Wong, H., Levenga, J., Cain, P., Rothermel, B., Klann, E., Hoeffer, C., 2015. RCAN1 overexpression promotes age-dependent mitochondrial dysregulation related to neurodegeneration in Alzheimer's disease. Acta Neuropathol. 130, 829–843.

Wust, R.C., de Vries, H.J., Wintjes, L.T., Rodenburg, R.J., Niessen, H.W., et al., 2016. Mitochondrial complex I dysfunction and altered NAD(P) H kinetics in rat myocardium in cardiac right ventricular hypertrophy and failure. Cardiovasc. Res. 111, 362–372.

Zhang, J.-C., Xu, H., Yuan, Y., Chen, J.-Y., Zhang, Y.-J., Lin, Y., Yuan, S.-Y., 2017. Delayed treatment with green tea polyphenol EGCG promotes neurogenesis after ischemic stroke in adult mice. Mol. Neurobiol. 54, 3652–3664. https://doi.org/10.1007/s12035-016-9924-0.

Zigman, W.B., 2013. Atypical aging in Down syndrome. Dev. Dis. Res. Rev. 18 (1), 51–67.

CHAPTER

10

Gene-dosage imbalance and altered neurogenesis in Down syndrome

1. Introduction

Besides myriads of DS-specific phenotypes, cognitive disability and mental retardation (MR) are potential weak points of life-quality of individuals with Down syndrome (DS). The on-time production of neurons from local pools of progenitor cells defines the development of the cerebral cortex, and migration of these new neurons then occurs within the growing tissue to position themselves before establishing appropriate connections with other cells. This mechanism forms functional circuits. Thus, disruptions in these early stages of brain development can contribute to intellectual disability, cognitive impairment, and epilepsy (Heng et al., 2010). Genetic abnormalities, such as trisomy of human chromosome 21 (HSA21), can direct abnormal brain development during fetal development for which human DS is the most appropriate and common example having diagnosed with invariant intellectual disability. Indeed, DS fetuses showed defective brain development at in utero age.

At birth, DS subjects present moderately reduced brain size and weight. In particular, alterations in hippocampal-dependent processes and cerebellar function deeply affect the visuospatial memory in DS. However, the underlying neuropathological features and the genes of molecular pathobiology of the decline in cognition processes across development are largely unelucidated. Despite reduced brain size and weight, the major morphological alterations in the DS brain specifically include the cerebellum and the hippocampus, defective cortical organization, decreased number and depth of sulci in the cerebral cortex, displacement of neurons within the cerebral cortex, aberrant gyrification, neuronal heterotopias, abnormal neuronal migration and differentiation, decreased neuronal densities, altered cell populations such as granule cells in cerebral cortices, and dendritic anomalies—altogether a disturbed neurogenesis (reviewed in Dierssen and Ramakers, 2006; Stachowicz, 2018). Indeed, neuronal density is reduced in distinct regions of DS brains, including the cochlear nuclei, cerebellum, hippocampus, basal forebrain, and the granular layers of the neocortex and brainstem regions. The neocortex is comprised of abnormal neuronal morphology and aberrant neuronal orientation. Notably, reduced volumes of cerebral and cerebellar hemispheres, mammillary bodies, ventral pons, and hippocampal formation have been demonstrated in young adults with DS (Raz et al., 1995).

Though the para-hippocampal gyrus was enlarged accompanied by disproportionately

Genetics and Neurobiology of Down Syndrome
https://doi.org/10.1016/B978-0-323-90456-8.00006-5

smaller frontal lobe volumes in DS, nevertheless, the contribution of the reduced brain size or neuronal density to cognitive alterations in DS is not a straightforward mechanism; rather the connection with the size of the basal dendritic arbors, their branching structure, and spine density may have specific pathogenesis and evolution in DS. Interestingly, similar neuropathological changes are manifested in brains of DS patients with Alzheimer disease (AD) at or beyond 35 years of age. The brains of AD patients appeared with progressive accumulation of senile plaques of Aβ-protein and neurofibrillary tangles and their identical region-specific distribution. The reduced brain size, reduced volumes of cerebral gray and white matter, larger subcortical and parietal gray matter, and temporal white matter may be responsible for verbal memory deficits and changes in short-term visuo-spatial memory (Pinter et al., 2001). Cognitive deficits in DS patients include long-term memory changes and difficulty in the assimilation of new skills. Importantly, unravelling the molecular signature underlying the neurobehavioral processes in DS would be a key challenge for identification of novel drug targets and set up new prevention strategies aimed at reducing the impact of metabolic disorders and cognitive decline in DS patients (Dierssen et al., 2020).

2. Developmental consequences of brain in DS

The advent of in vivo structural brain imaging techniques has demonstrated the neuroanatomical basis of brain malfunctioning and specific defects of neural development in DS brains. The data obtained have not only expanded knowledge of brain structures linked to specific neurobehavioral conditions but also demonstrated the interindividual variations among DS subjects carrying the same trisomic HSA21 as the genetic defect. Nevertheless, information on the fetal brain with DS is limited to some of the morphological reports. Abnormalities occur from 20th and 22nd weeks of gestation in neuronal differentiation, migration, and synaptogenesis, including immature gyral pattern, irregular neocortical laminar formation, and delayed myelination of cortical fibers (Wisniewski and Kida, 1994). Until the 40th week of gestation, reduction in neuronal density was observed in several brain regions, including cortical areas and basal nucleus with sprouting dendritic trees and unchanged dendritic spines (Becker et al., 1991). Further decrease in dendritic spines continues in the visual cortex from early infancy to about 4 months of age leading to detectable reduced dendritic branches at about 6 months of age.

The progressive postnatal degenerative process burdens the DS brains, exhibiting a trend of declining rate of cognitive development as children get older (reviewed in Nadel, 2003). In this direction, several studies have demonstrated reduced cortical width, abnormal cortical lamination patterns, altered dendritic arbors and dendritic spines, aberrant electrophysiological properties of membranes, reduced synaptic density, and abnormal synaptic morphology in DS patients (reviewed in Dierssen and Ramakers, 2006). Some of these changes might probably be related to abnormalities in pre- and postnatal development of the nervous system. However, no significant differences were reported in dendritic differentiation between normal and a 2.5-month-old DS infant in a quantitative analysis of dendrites in layer IIIc pyramidal neurons of the prefrontal cortex (prospective area 9) of the brains (Vuksic et al., 2002). Conversely, shorter than normal basilar dendrites of cortical pyramidal neurons were observed in DS subjects older than 4 months (Takashima et al., 1981). These features suggest that DS children begin their lives with morphologically normal layer III pyramidal neurons, but the pathologic changes begin from 2.5 months of postnatal age, with a steady decline of the brain features especially in apical dendrites of DS subjects older than 2 years, relative to that of age-matched normal children (reviewed in Dierssen and Ramakers, 2006).

In DS adults, significant decrease in dendritic branching and length and in spine density was reported both in the hippocampus and cortex with degenerative neuronal changes associated with dendritic abnormalities (Alves-Sampaio et al., 2010), and that was further supported by the presence of similar features of pyramidal neurons in parietal cortex and on nonpyramidal neurons in the motor cortex (reviewed in Dierssen and Ramakers, 2006). Interestingly, the non-AD brains of aged DS individuals are reported with a marked reduction in the number of spines in the middle and distal segments of apical dendrites of pyramidal neurons in cingulate cortex and in hippocampus, which were more severe in CA1 area of DS cases with AD (DSAD). Altogether, MR in DS could be attributed to a deficient neuronal network connectivity implicating a deficient information processing resulting in MR in DS. This theory was supported by reduced functional connections between brain circuit elements observed in functional imaging with positron emission tomography (Balogh et al., 2002), suggesting such abnormally connected brain might generally be affected with or might have disturbances in specific pathways or neurotransmitter systems (reviewed in Dierssen and Ramakers, 2006).

With respect to postnatal development of neurons in DS, a question remains whether the structural complexity of neurons undergo greater dendritic retraction compared to that of euploid brains during maturation. It is proposed that both abnormality in dendritic development of DS brain and the correct remodeling of the brain circuitry during postnatal development are responsible for maladaptive behaviors and cognitive performance of DS patients. However, exploration of the signaling pathways triggering the altered neuroplasticity and the molecules providing the information for this process may explain the mechanism of the complex processes of neural network formation and neural plasticity in DS, and that could be resolved from understanding the implication of specific dosage-sensitive genes and proteins responsible for aberrant neurodevelopment in DS.

The first trisomic mice of DS models, Ts16, with 129 of the 731 genes on MMU16 orthologous to HSA21 exhibit some characteristics of DS, including craniofacial malformations and alterations in the development of the cerebellum, hippocampus, and cortical plate (Benavides-Piccionea et al., 2004). The overall reduction in cortical neurons indicated abnormalities in neurogenesis, a failure of brain-derived neurotrophic factor-dependent neuron survival, and increased caspase-dependent apoptosis in Ts16 mice (reviewed in Benavides-Piccionea et al., 2004). The abnormalities in hippocampal neurons in these mice are demonstrated opposite in direction to abnormalities in Dorsal root ganglia (DRG) neurons, suggesting that all Ts16 neuronal types are not caused by a common membrane electrical abnormality. The hypothesis described that a common defect in signal transduction, possibly involving channel phosphorylation, might contribute to abnormalities in the hippocampus in this strain.

Altered corticogenesis appeared with slow tangential expansion of the neuroepithelium, resulting in a reduction of final telencephalic size and predicting a decrease in number of radial cortical units in the mature brain in Ts16 mice. Besides, delayed radial growth of the cortex was apparent due to reduction in the number of neocortical founder cells leading to a reduction in overall size of the cortex of a defective cortical neurogenesis in Ts16 strain. Indeed, altered proliferative rate delayed the generation of neocortical neurons, while the reduced founder cell generation leads to a proportional decrease in the overall number of neurons. Indeed, prenatal perturbations either in timing of neuron generation or the final number of neurons produced may contribute to significant neocortical abnormalities in this mouse model as observed in DS patients (Haydar and Reeves, 2012).

Transgenic mouse models such as Ts1Cje and Ts65Dn, that carry single candidate genes and chromosomal segments homologous to the Down syndrome critical region (DSCR) in excess, show some but not all the pathological, biochemical, and transcriptional changes seen in DS (reviewed in Cairns, 2001). The Ts65Dn mice containing three copies of MMU16 from App to Mx1, Ts1Cje mice having three copies from Sod1 to Mx1, and Ms1Ts65 having three copies from App to Sod1 have been considered in DS research (Sago et al., 1998). The Ts65Dn model includes most of the MMU16 region homologous to HSA21, but deficient in ~40% of HSA21 genes. This is the extensively studied model of DS. Interestingly, this strain does not reflect some of the gross abnormalities frequently observed in DS, such as cardiac valvular congenital abnormalities, despite many genes are overexpressed in the Ts65Dn model. Nevertheless, these mice exhibit analogous effects on cognitive deficits and craniofacial structures (reviewed in Benavides-Piccionea et al., 2004). Notably, neuroanatomical parallels of DS cerebellum was reported in Ts65Dn mice with decrease in granule cell number and cerebellar volume owing to reductions in both the internal granule layer and the molecular layer, though the brain size and histology appear normal in these mice. The cerebellar dysmorphology in the Ts65Dn mouse may be related to the global disruption of the cerebellar transcriptome (Saran et al., 2003). Furthermore, minor irregularities, including reduced volume of CA2 and decreased neuronal density in the dentate gyrus (DG), were noted following stereological morphometric studies in selected regions of the hippocampus of Ts65Dn mice (reviewed in Benavides-Piccionea et al., 2004). Diminished dendritic branching and spine density in the hippocampus and cortex, indicating alterations in the morphology and volume of these structures, was documented in Ts65Dn and Ts1Cje mice, those parallel to human DS. The hippocampus exhibited elevated levels of DS cell adhesion molecule (DSCAM) mRNA and protein and reduction of total dendrite length in DSCAM-overexpressing neurons in Ts1Cje mice (Alves-Sampaio et al., 2010).

The cortical pyramidal neurons showed reduced dendritic tree with markedly decreased spines, indicating abnormal circuit development whch could be responsible for behavioral deficits in these trisomic mice (Fig. 10.1) (Dierssen and Ramakers, 2006). Moreover, confocal microscopy in the hippocampus (fascia dentata, CA1) as well as the motor and somatosensory cortex, entorhinal cortex, and medial septum showed that both presynaptic and postsynaptic elements were noticeably enlarged and also showed abnormal internal membranes in Ts65Dn strain that parallels to human DS. Additionally, reduced spine density was noticed on the dendrites of dentate granule cells, with reorganization of inhibitory inputs, accompanied by a relative decrease in inputs to dendrite shafts and elevation of inputs to the necks of spines (Belichenko et al., 2004). Furthermore, specific presynaptic bouton enlargements and presence of a large number of vacuoles and multivesicular structures in spines were reported in Ts65Dn mice (Belichenko et al., 2004). Typical alterations in hippocampus inhibitory synapse position from shafts onto the necks of dendrites were also demonstrated in this model (Belichenko et al., 2004). Cognitive deficit was found to be connected with reduced hippocampal volumes in these tests (Cramer and Galdzicki, 2012). Interestingly, environmental enrichment displayed more branched and more spinous pyramidal cell structure in control animals; however, that had little effect in Ts65Dn mice and nonenriched controls (Dierssen, 2012; Dierssen et al., 2003).

At old ages, a combined dendritic labeling and synapse analysis revealed a markedly lower number (30%) of asymmetric (presumptive excitatory) synapses in Ts65Dn mice, while the number of symmetric (presumptive inhibitory) synapses appeared similar to that of controls (Benavides-Piccionea et al., 2004). Interestingly, the mean length of synaptic apposition of both asymmetric and symmetric synapses appeared significantly

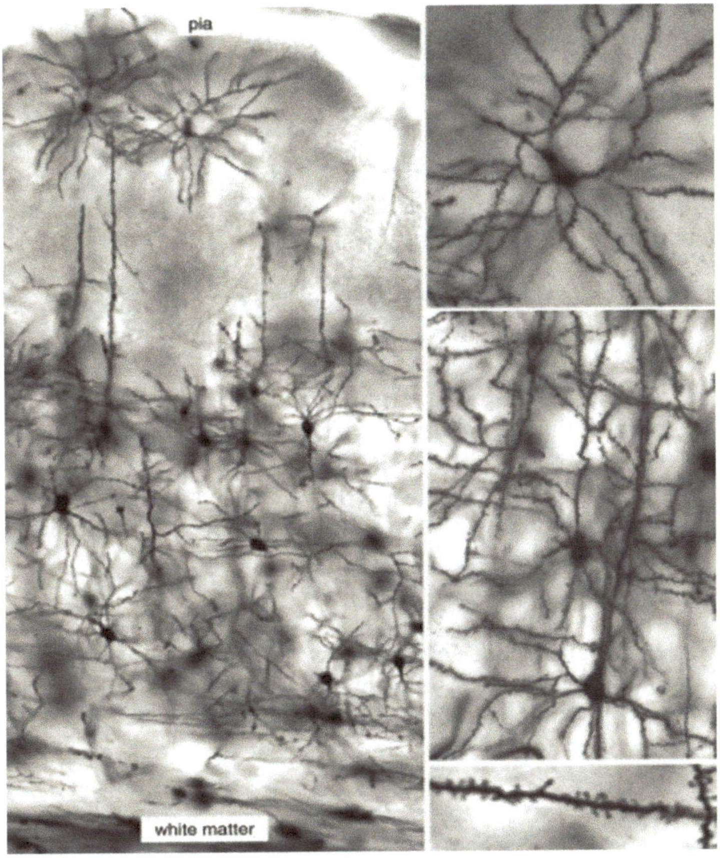

FIGURE 10.1 **An example of neuronal network structure in an adult mouse primary visual cortex as revealed by Golgi staining.** On the right are nonpyramidal and pyramidal neurons at a higher magnification, as well as part of a dendritic segment with spines. Alterations in neuronal morphology are usually not directly visible in brains of persons with MR or mouse models of MR and will only be revealed by quantitative analysis. Essential in this respect is a blind and random sampling procedure, to prevent various sources of bias. *Reprinted from Dierssen and Ramakers (2006) with permission from John Wiley and Sons ("John Wiley and Sons") and Copyright Clearance Center.*

larger in Ts65Dn mice (15% and 11%, respectively), indicating a preferential perturbation in the excitatory synapses and compensation of the deficits in asymmetric synapses by the increased contact zone area of the existing synapses. Neurochemical studies have revealed severely disturbed adenyl cyclase and phospholipase C signaling pathways, involved in activity-dependent synaptic stabilization and dendrite growth, in the brain of Ts65Dn mice and DS patients (Dierssen et al., 2003). Both basal and stimulated production of cyclic AMP (cAMP) appeared reduced in Ts65Dn mice suggesting that these alterations may affect the function of the central nervous system (CNS). In addition, severe deficits in the limbic system were corroborated with alterations of neurotransmitters, synaptic ultrastructure, and long-term potentiation (LTP) in Ts65Dn strain. The results of the cultured neurons derived from human DS fetuses and from DS animal fetuses suggested that malfunctioning of brain circuits may arise from abnormal neuronal characteristics of electrical and biochemical pathways in the DS brain because morphological and biochemical changes linked to experience-dependent plasticity of neuronal synapses indicate the long-term memory storage. Alterations in number and morphology of spine depict the mechanisms of conversion of the transient changes in synaptic activity into long-lasting memory.

Another trisomic model, designated as Ts{Rb [12.17(16)]}2Cje or Ts2Cj2, carrying a chromosomal rearrangement of the Ts65Dn genome, where the marker chromosome has been translocated to MMU12 forming a Robertsonian chromosome and retaining the dosage imbalance of HSA21 homologous genes from App to the telomere, showed expression levels similar to Ts65Dn within the triplicated region. Reduction in the spine density on the dendrites of dentate granule cells was reported in quantitative confocal microscopy of the fascia dentata accompanied by significantly enlarged dendritic spines disturbing the entire population in these mice compared with controls (Villar et al., 2005). Interestingly, the structural abnormalities of the dendritic spine appeared similar to those described in the Ts65Dn model (reviewed in Dierssen and Ramakers, 2006).

It is likely that impairment of global cognitive abilities and adaptive skills in individuals with DS have a causative link with the abnormalities in spine morphology, wherein the abnormal spines and associated abnormal neuronal network might be the primary cause of the cognitive deficit and MR. Alterations in spine morphology result in an increased proportion of immature spines and, subsequently, abnormal development or maturation of spine synapses in DS. In addition, dysregulation of the actin cytoskeleton and hippocampal functions appear to be more affected than prefrontal functions (reviewed in Dierssen and Ramakers, 2006). The trisomic mice exhibited DS-like cerebellar pathology with similar extent of significantly affected cerebellar volume in Ts1Cje and Ts65Dn, despite fewer triplicated genes present in Ts1Cje. However, little effect of dosage imbalance was demonstrated on the granule cell and Purkinje cell density in Ts1Cje mice. Noticeably, specific gene action in conjunction with the dosage imbalance defines the specific phenotypes in DS.

3. Genetic implication of neurologic alterations in DS

Dendritic alterations in DS may be a consequence of upregulated proteins coded by trisomic HSA21 genes. Specific genetic loci, especially the DSCR region, are primarily responsible for regulating the abnormal development and maturation of the DS brain (detailed in Chapter 6); however, nontrisomic and non-HSA21 genes are also involved in defining DS characteristics. Moreover, the phenotypic consequences of increased gene dosage also depend on the biological action of the gene product itself, such as an enzyme, structural protein, transcription factor, intracellular signaling molecule, cell surface marker, receptor subunit, etc. Indeed, multiple molecular cascades are likely to be involved in normal dendritic and spine maturation and the spatio-temporal expression patterns. Some of the most salient molecular mechanisms underlying the altered neurogenesis in DS individuals resulting in DS-specific neuropathology have been included here.

The anatomical and cellular abnormalities in the DS brain are orchestrated at birth. Impaired cognitive sense leading to moderate to severe degree of MR is the major disabling feature in individuals with DS and that is caused by significant abnormalities in the structure and function of brain matters (Raz et al., 1995; Guidi et al., 2008). Autopsy studies of brain specimens of human DS and orthologous mouse models have presented altered neural cells with perturbed manifestations, including rate of proliferation, oxidative stress, action of the glial cells, myelination, neurite overgrowth, and synaptic formation, which is contributed largely by the complex and global transcriptomic changes induced by the imbalance of gene dosage in triplicated HSA21 accompanied by non-HSA21-encoded transcription factors (Haydar and Reeves, 2012; Kleschevnikov et al., 2004; Larsen et al., 2008; Olmos-Serrano et al., 2016; Letourneau et al., 2014; Antonarakis, 2017). Notably, owing to the limited access to brain specimens for high-throughput investigation of the DS brain, use of neural derivatives from induced pluripotent stem cells (iPSCs) with trisomic HSA21 has been considered as an alternative in vitro model of DS. Myriads of in vitro studies on trisomic iPSC derivatives have demonstrated wide array of abnormalities in the DS brain on gene expression, neural progenitor cell proliferation, neurogenesis, fate and function of the glial cells, neurite outgrowth, synaptic morphology and mitochondrial function, amyloid deposition in cortical neurons and association with an AD-like pathology (reviewed in Sobol et al., 2019; Hibaoui et al., 2014; Murray et al., 2015).

3.1 Dysregulation of gene expression and alteration in neurogenesis

Molecular profiling of the brain specimens from human DS and iPSC-derived neural cells of DS origin has highlighted the insights into the mechanisms of abnormal neurogenesis in DS (Fig. 10.2) (Sobol et al., 2019); however, the studies have shortcomings by the differentiation protocols for generation of neuronal cells, developmental time points considered for analysis, and/or the methodological sensitivity and specificity for molecular profiling. Nevertheless, dysregulation of HSA21 genes, including perturbations of major functional clusters, an early upregulation of the hub gene APP, and the transcription factors OLIG1, OLIG2, and RUNX1, accompanied by cellular abnormalities are demonstrated to be involved in disturbed neurogenesis, deficient myelination and neuronal differentiation, and highlighted novel sequential and differentiation-dependent dynamics of disturbed functions, and pathways underlying the developmental abnormalities of the DS brain (Sobol et al., 2019).

Indeed, genome-wide transcriptional changes, mainly upregulation of 922 and 879 differentially expressed genes (DEGs), were observed in trisomic neural progenitor cells (NPCs) and differentiated NPCs (diffNPCs), respectively, which were 734 and 634 at the NPC stage and diffNPC stage, respectively (Fig. 10.2) (Sobol et al., 2019). This massive upregulation of transcriptomes and corresponding proteomes suggests extensive molecular alterations in trisomic cells with distinct temporal variations at differentiation stages. Notably, upregulation of 3.4% non-HSA21 genes was also reported along with downregulation of HSA21 genes, such as 6- to 7-fold downregulation of the transcription factors OLIG1 and OLIG2 in trisomic NPCs (Sobol et al., 2019). Six DEGs, such as EPB41L3, HEXB, SOD1, PMP22, NFASC, and LAMA2 of myelination gene ontology category (GO:0042552) and the general process of ensheathment of neurons (GO:0007272), were upregulated in diffNPCs but not in NPCs. This affirms an effect on oligodendrocyte differentiation and myelination pathways in the trisomic iPSC model.

Differential expression of several DSCR genes was reported contradictory based on the stage of development and differentiation and cell types (Olmos-Serrano et al., 2016). Marked downregulation of the transcription factors OLIG1 and OLIG2, detected in trisomic NPCs, was not detected in diffNPCs, suggesting its contribution on the fate of the glial cells at an early stage of fetal development leading to deficient myelination in DS brains (Sobol et al., 2019). However, OLIG1 and OLIG2 were conversely overexpressed in cells derived from the ventricular zone of DS fetuses (Lu et al., 2012) and in neural cells derived from iPSCs with trisomy 21 (Hibaoui et al., 2014). Overexpression of Olig1 and Olig2 was reported in the medial ganglionic eminence (MGE) of the brain of Ts65Dn mice compared to euploid littermates at E13.5 and E14.5 (Chakrabarti et al., 2010), suggesting Olig1/Olig2 overexpression as the genetic basis of the interneuron overproduction in brains of these mice. Subtraction of the gene dosage of Olig2 through cross-breeding rescued the phenotype of interneuron overproduction and brought down the frequency of spontaneous inhibitory postsynaptic currents (sIPSC) to normal level in Ts65Dn mice.

Overexpression of the DSCAM has been documented in human DS brains (Sterne et al., 2015), which may impose an inhibitory action on synaptogenesis and neurite outgrowth, indicating a direct connection to pathological

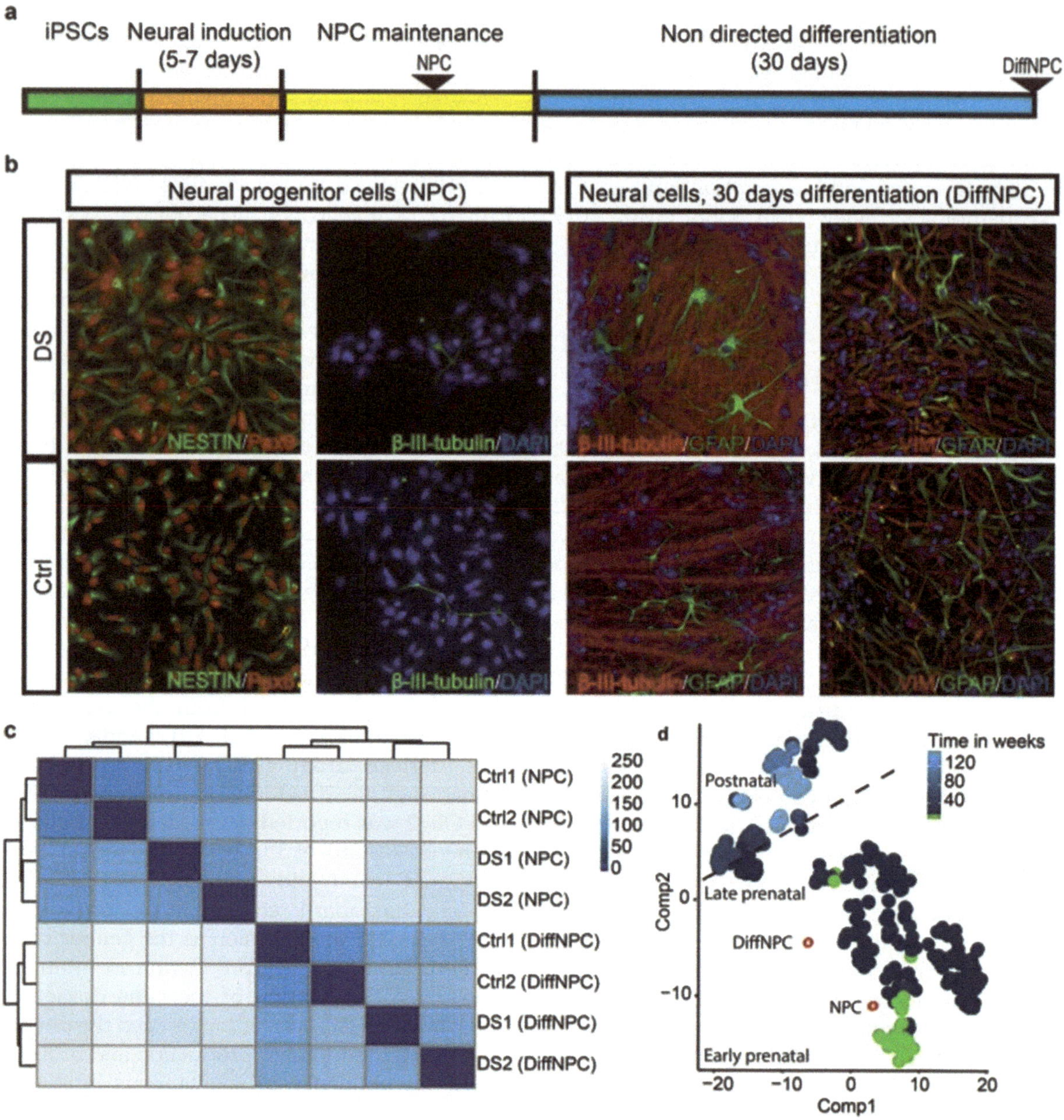

FIGURE 10.2 Generation and characterization of the iPSC model. (A) Schematic presentation of the protocol used to generate NPCs and diffNPCs from iPSCs. (B) Representative images of immunofluorescent stainings of cells with full trisomy 21 (DS) and euploid control (Ctrl) cells. Panels show neural progenitor cells (NPC) stained for NESTIN and Pax6 as well as differentiated neural progenitor cells (diffNPCs; 30 days of nondirected differentiation) stained for β-III-tubulin/TUJ1, GFAP, and Vimentin. (C) Heatmap of transcriptome sample-to-sample distances using the rlog-transformed values (DS1 and DS2: neural iPSC lines with T21; Ctrl1 and Ctrl2: euploid neural iPSC lines). (D) Clustering of transcriptome data from

changes in synaptic plasticity in DS brains. DSCAM is expressed in the hippocampus, amygdala, thalamus, caudate nucleus, and corpus callosum in human brains. Its presence in the cerebellar white matter of normal and DS brains suggested that DSCAM may be involved in regulating myelination (reviewed in Stachowicz, 2018). In Drosophila, a lethal mutation in Dscam specifically perturbed segregation of axonal branches (Wang et al., 2002). Axonal bifurcation is necessary for efficient innervation of multiple targets and that results in the formation and segregation of the sister branches. DSCAM demonstrated a functional role in dendritogenesis and synaptic plasticity in mouse model of DS (Alves-Sampaio et al., 2010). Dscam has been reported to regulate the inhibition of neocortical pyramidal neurons (PyNs) in a level-dependent fashion. Overexpression of Dscam resulted in increase in GABAergic innervation of cortical PyNs by chandelier and basket cells in the Ts65Dn mouse, which was rescued by genetic normalization of Dscam expression. Thus, overexpression of Dscam due to triplication was identified as the cause of excessive GABAergic innervation and inhibition in the neocortex of trisomic mouse model and also targeted as a potential pathogenic driver of neurological disorders in individuals with DS (Liu et al., 2020).

Another gene, EURL (also known as C21ORF91), a protein-coding gene located at the centromeric boundary of the DSCR (within 21q21.1), is generally expressed during cerebral cortex development. Altered EURL expression and its mRNA levels were described in the developing cerebral cortex of trisomic mice and DS brain tissue, and such perturbations of Eurl impaired the long-term positioning and dendritic spine densities of cortical projection neurons (Li et al., 2016). In trisomic mouse models of DS (Ts65Dn, Ts1Cje, Ts1Rhr and Tc1 strains), the representative EURL gene is not represented within the trisomic regions of HSA21; however, electroporation studies have established the first evidence that altered Eurl expression may result in alterations of cortical progenitor proliferation and neuronal differentiation. Altered EURL dosage may contribute to abnormal neuronal development, including reduced brain volume, disorganized cortical layers indicative of aberrant neuronal positioning, reduced dendritic spine densities on cortical projection neurons, and neurological deficits in DS. A mechanistic link was proposed between Eurl expression and enhancement of β-catenin signaling, suggesting altered Eurl levels are responsible for abnormal cortical organization and related brain disorders. Interaction of EURL with the coiled-coil domain-containing protein CCDC85B was hypothesized for modulation of β-catenin levels in cells (Li et al., 2016). Altogether, EURL has been identified as an important new factor of neuronal development that is likely to influence neuropathogenesis of DS and other HSA21-related disorders.

Overexpression of the CYYR1 (cysteine and tyrosine-rich protein 1) gene, encoding a highly conserved tyrosine-rich domain on 21q21.2, in trisomic iPSCs indicates its possible association with neurogenesis in DS (Sobol et al., 2019). The CYYR1 protein has been suggested to have a regulatory function on the intracellular growth factor receptor trafficking, and thus for receptor-mediated signaling (Justice et al., 2017). The implication of C2CD2 encoding a

neural iPSC lines derived from the two healthy donors and at two differentiation time points (NPC and diffNPC; *red circles*) together with transcriptomes from Brainspan brain samples of different postconceptional weeks (pcw) using the tSNE plot. Brainspan samples represent fetal ages up to 200 pcw. Early fetal samples at age between 8 and 10 pcw are colored green ($n = 30$), and remaining fetal samples are colored in different blue shades corresponding to their gestational ages >10 pcw. The *dashed line* indicates estimated position of clusters at full term. Clustering of samples was performed using the key gene set. *Reprinted from Sobol et al. (2019) under a CC-BY-NC-ND 4.0 International license.*

transmembrane protein is yet unclear. Another DSCR gene, RUNX1, encoding the transcription factor of AML1, was strongly upregulated in both NPCs and diffNPCs with trisomy 21 (Sobol et al., 2019). A crucial role of RUNX1 was described in neurogenesis and neurite outgrowth in DS (Halevy et al., 2016), indicating its involvement in altered neurite outgrowth and perturbed connectivity and synapse formation during DS brain development (Sobol et al., 2019).

In Ts2Cje mice, a bidirectional alteration to the synaptic inhibition on CA1 pyramidal cells and associated impairment in spatial memory was evidenced. The features were attributed to gene dosage of the kainate receptor (KAR) encoding gene Grik1. Grik1 is a glutamate receptor expressed in pyramidal cells of the cortex in fetal and early postnatal life and in adult. Interestingly, normalization of Grik1 dosage in these mice specifically restored spatial memory and reversed the bidirectional alterations to CA1 inhibition; however, the changes in synaptic plasticity or the other behavioral modifications could not be restored (Valbuena et al., 2019). The result proposed that some of the characteristic cognitive deficits in DS could be caused by the disturbed inhibitory tone rather than generalized overinhibition due to modified information gating.

Understanding the molecular basis of the DS brain development by employing the transcriptome analysis on protein coding sequences following deep RNA sequencing (RNA-seq) complemented with LC-MS/MS techniques has revealed 11 altered functional clusters and their temporal dynamics in neurogenesis of iPSC model with trisomy 21 (Fig. 10.3) (Sobol et al., 2019). The studies, particularly on iPSC-derived neural cell model, uncovered the transcriptomic similarities between two distinct differentiation stages of the fetal brain and temporal dynamic abnormalities during neurogenesis (Sobol et al., 2019; Murray et al., 2015; Jiang et al., 2013; Huo et al., 2018). Dysregulated DNA replication, pluripotency, synaptic formation, and neural signaling clusters were prominently appeared in the neural progenitor cells (NPCs) in early and critical stages of cell proliferation and transition from the stem cell stages of neurogenesis (Haydar and Reeves, 2012), and those were accompanied by decreased proliferative rate and disturbed neurite formation (Chen et al., 2014; Belichenko et al., 2007; Sobol et al., 2019). Interestingly, both trisomic and diffNPCs showed alteration in pathways for extracellular matrix (ECM)-receptor interactions, cell-to-cell communication, and integrin complex formation, indicating disrupted ECM-related functions and cell membrane–associated communication in trisomic cells during neuronal differentiation. In contrast, deficient TGF-β signaling, SMAD-associated signal transduction and different growth factor-receptor activities were observed at the later stage of NPC differentiation (Sobol et al., 2019).

Several other HSA21 genes that may be involved in development of the central nervous system or neural functions related to synaptic plasticity are indicated in the table below (Table 10.1):

Collectively, the triplicated gene dosage imbalance on HSA21 has significant downstream transacting effects on the entire genome, where the dysregulated expression of the DSCR and other genes of HSA21 have been demonstrated in defined differentiation stages, suggesting a direct association with abnormal neurogenesis and phenotypic outcome in DS. Notably, several alterations of functional clusters have already been reported in other nonneural trisomic cell types, including extracellular matrix receptor interactions, cell–cell adhesion, integrin complex, and glycolysis, establishing a general intrinsic defect in DS organogenesis (Conti et al., 2007; Mao et al., 2005). Nevertheless, many of the molecular and cellular perturbations in trisomic neural lines are consistent with altered cellular composition, impaired trophic interactions, and deficient functional neural

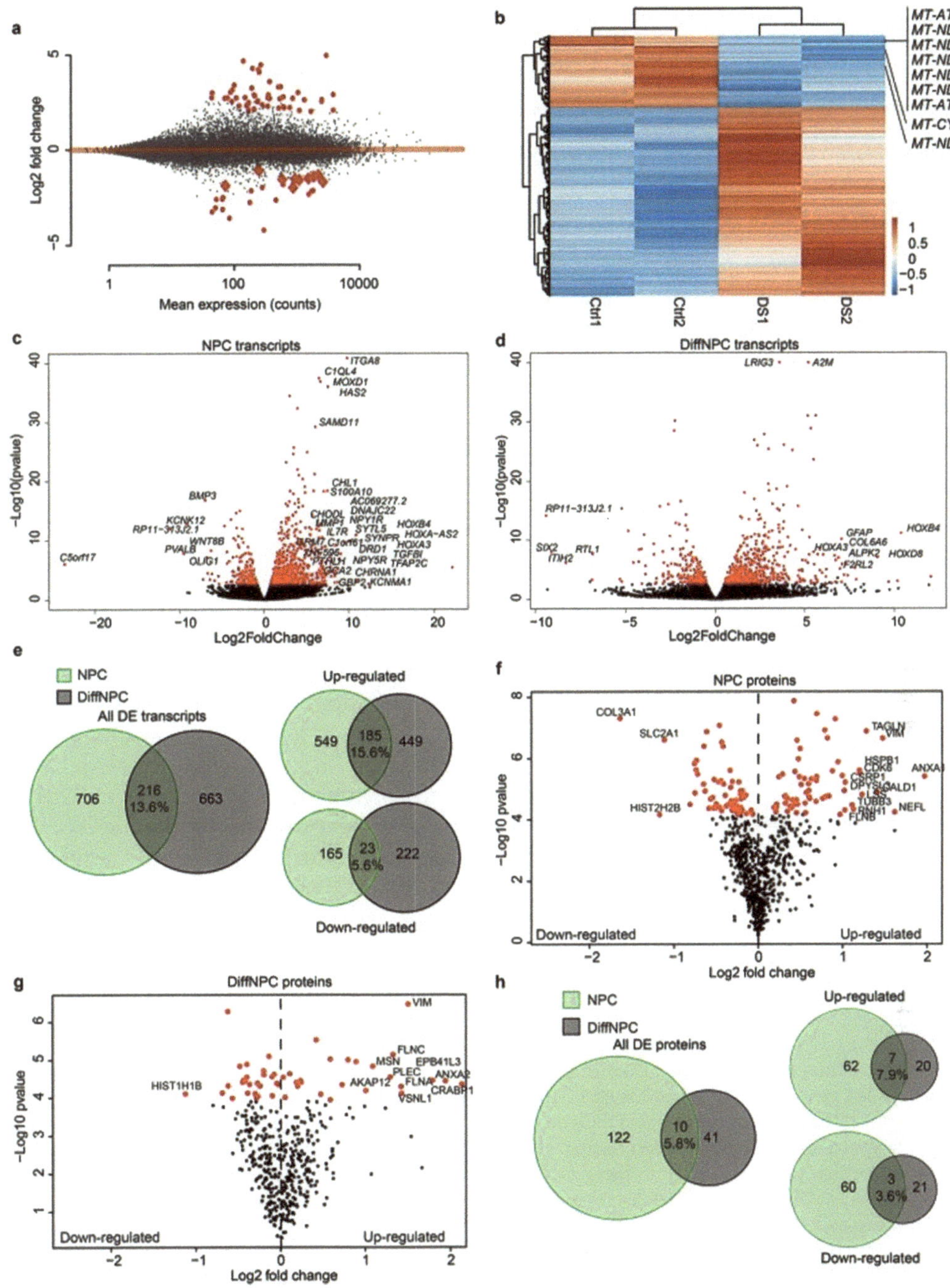

FIGURE 10.3 Differential levels of transcripts and proteins in T21 lines. (A) Overview of transcript levels in T21 lines (DS1 and DS2) relative that in euploid control lines (Ctrl1 and Ctrl2) at the diffNPC stage evaluated using DESeq2. The log2fold change of expression levels in T21 lines is shown versus the mean expression levels in control lines. *Red dots* correspond to

TABLE 10.1 HSA21 genes involved in brain development and synaptic plasticity.

Gene name	Gene function
SIM2 (single-minded homolog 2)	Synchronized cell division and establishment of proper cell lineage in *Drosophila*
GART (phosphoribosylglycinamide formyltransferase)	Expressed during prenatal development of the cerebellum
PCP4 (Purkinje cell protein 4)	Expressed exclusively in the brain most abundantly in the cerebellum
SOD1 (superoxide dismutase 1)	Role in apoptosis in fetal DS neural cultures
DSCR1 (Down syndrome critical region 1)	Role in LTP/LTD
ITSN (intersectin)	Role in endocytosis

Adapted from Benavides-Piccione et al. (2004).

circuits observed in brains of DS individuals (Larsen et al., 2008).

3.2 Deregulated pathways involved in neurogenesis in trisomic cells

Several aberrant pathways, including focal adhesion, extracellular matrix (ECM) function, and TGF-β signaling were demonstrated to be markedly associated with either the NPC or the diffNPC stage. Dysregulation of pathways of signaling to regulate the pluripotency and synapse formation (KEGG: 04550, 04724, and 04727) perturbed the transition from the neural stem cell stage leading to disrupted formation of more mature neuronal structures in trisomic NPCs (Sobol et al., 2019). At the diffNPC stage, glycolysis/gluconeogenesis (KEGG: 00010), axon guidance (KEGG: 04360) long-term depression (LTD) of cerebellum (KEGG: 04730), and a group of signaling pathways (KEGG: 04151, 04390, 04015, 04933) were affected, indicating the effects of these perturbed functions on neural differentiation. The fate choice of progenitor glia cells results in a predisposition for astrogliogenesis in DS neurogenesis that creates a toxic effect to neurons (reviewed in Sobol et al., 2019). An increase in expression of the GFAP, VIM, and S100B genes and the glia cell markers TAGLN2 and FABP7 observed in trisomic NPCs were

significant differential expression (DE; adjusted P value $< .05$) levels with a log2fold change of >1. *Gray dots* correspond to nonsignificant difference in expression levels. Mitochondrial (MT) transcript levels are annotated with *red diamonds,* all of them downregulated in T21 lines. (B) Heatmap and clustering of DE transcripts in diffNPC from control lines (Ctrl1 and Ctrl2) and the two T21 lines (DS1 and DS2). Data from MT transcript levels are annotated as reference. (C) Volcano plots of differentially expressed genes (DEGs; red; adjusted P value $< .05$) from comparison of T21 lines (DS1 and DS2) and euploid lines (Ctrl1 and Ctrl2) at the NPC stage. Selected DEGs are named. (D) Volcano plots of DEGs as in (C) but at the diffNPC stage. (E) Venn diagrams showing all DE genes when comparing T21 and euploid lines (left panel) at the NPC (green) and diffNPC (gray) stages of differentiation. The overlap in the Venn diagram therefore represents DE genes regardless of differentiation stages. Right panels show DE genes in T21 lines separated by up- and downregulated genes, respectively. (F) Volcano plots of differentially expressed (DE) proteins (red) from pairwise comparisons of T21 lines (DS1 and DS2) and euploid lines (Ctrl1 and Ctrl2) at the NPC stage. Selected DE proteins with log2 fold change < -1 and >1 are annotated with names. (G) DE proteins (red) from pairwise comparisons of T21 lines (DS1 and DS2) and euploid lines (Ctrl1 and Ctrl2) at the diffNPC stage. (H) Venn diagram showing all DE proteins when comparing T21 and euploid lines (left) at the NPC (green) and diffNPC (gray) stages of differentiation. The overlap in the Venn diagram therefore represents DE proteins regardless of differentiation stage. Right panels show DE proteins in T21 lines separated by up- and downregulated proteins, respectively. *Reprinted from Sobol et al. (2019) under a CC-BY-NC-ND 4.0 International license.*

consistent with increased formation of glia cells and astrocytes in DS (Fig. 10.3) (Colombo et al., 2005; Sobol et al., 2019). Conversely, a slight reduction in expression of LMX1B gene of neuronal markers was observed (Sobol et al., 2019). Notably, a significant increase in glia cells was evidenced in the trisomic diffNPCs. Taken together, the genetic profiles of trisomic neural cells exhibited an association of abnormalities in dynamics of pathways responsible for cell fate, differentiation, and energy metabolism with perturbation in brain development of DS individuals.

Notably, dysregulation of elements and pathways in DNA replication, ECM-, and growth factor receptor–mediated signaling were supported by marked reduction in growth and proliferation of NPCs observed in vitro (Fig. 10.4A and B) (Sobol et al., 2019). Impairment of dendritic and synaptic formations observed in DS brains was consistent with different distribution in numbers of neurite outgrowth in trisomic NPCs compared to euploid cell lines (Fig. 10.4C) (Sobol et al., 2019) and abnormal neurite formation and altered neural connectivity in brains of DS fetuses and orthologous mouse models (reviewed in Sobol et al., 2019). Collectively, decreased proliferative capacity and altered neurite formation were reported in trisomic neural cells at the NPC stage.

Besides predominant defects of synapse and neuronal clusters at early stage in trisomic neural cells, the influence of GABAergic, glutamatergic, and dopaminergic synapses and signaling may contribute to abnormalities in the early stage of development of specific cell progenitors and neural structures. In contrast, diffNPCs exhibited three perturbed clusters, such as TGF-β signaling, oxidative phosphorylation, and glycolysis, indicating their association with neuronal differentiation in trisomic cells (Fig. 10.5) (Sobol et al., 2019). Furthermore, the perturbation of collagen components, such as cell–cell adhesion, ECM–receptor interaction, and the integrin complex indicate their role during differentiation of trisomic neural cells. Importantly, the temporal dynamics of these perturbed functions affects the stages of growth, differentiation, and maturation of the developing brain.

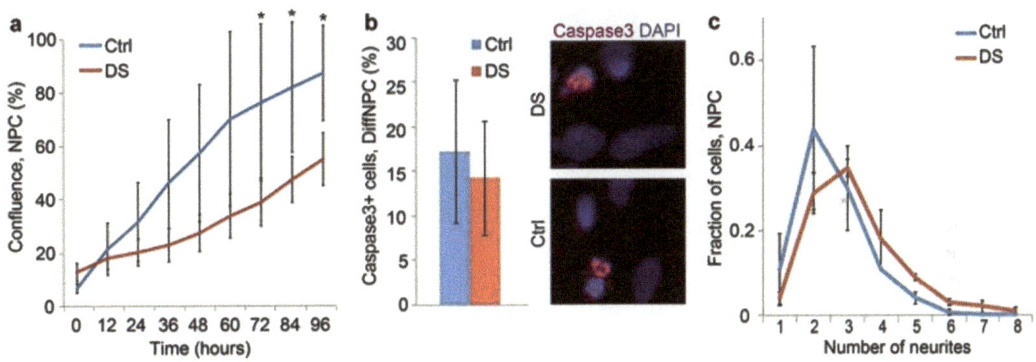

FIGURE 10.4 Functional characterization of T21 neural lines. (A) Growth rate of euploid (Ctrl; Ctrl1 and Ctrl2) and T21 NPCs (DS; DS1 and DS2) for 96 h illustrating reduced proliferation in trisomic lines (mean ± SD; *: $P < .05$). Live-cell imaging with IncuCyte was used for analysis. (B) Proportion of Caspase-3 positive diffNPCs in T21 (DS) and euploid (Ctrl) lines (left) with representative pictures from both lines (right). No significant difference was detected between the two groups from the analysis of 569 cells with T21 (DS1 and DS2) and 639 euploid cells (Ctrl1 and Ctrl2) (mean ± SD). Staining was visualized using confocal microscope and ImageJ software. (C) Analysis of neurite outgrowth in β–III–tubulin-positive NPCs. The trisomic NPCs show a different distribution in neurite numbers when compared to euploid NPCs (DS1 and DS2, $N = 2$, $n = 101$, Mo = 3; Ctrl1 and Ctrl2, $N = 2$, $n = 148$, Mo = 2; $F = 19.375$, mean ± SD, $P < .05$). *Reprinted from Sobol et al. (2019) under a CC-BY-NC-ND 4.0 International license.*

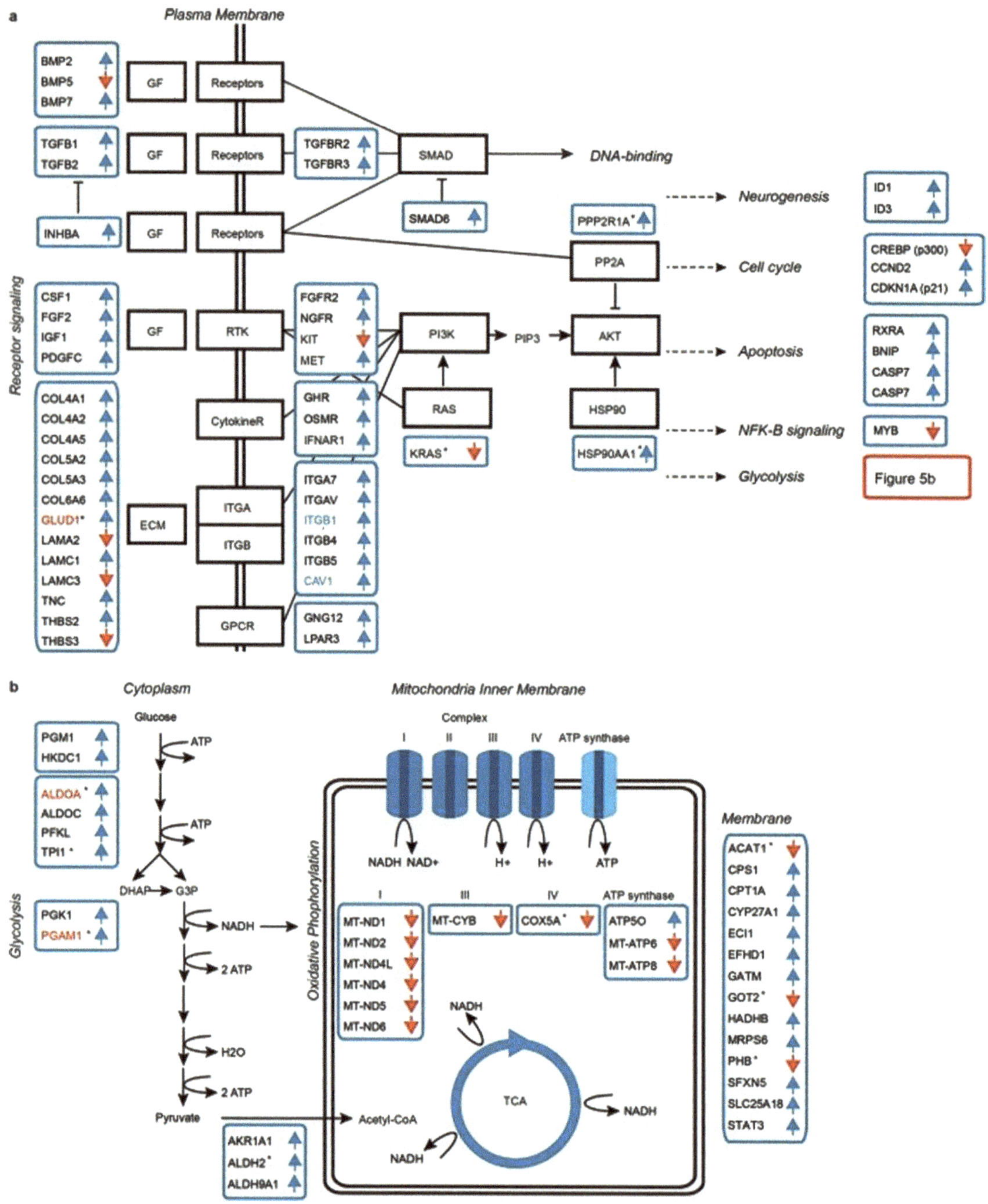

FIGURE 10.5 Graphical overview of selected and perturbed functional pathways in diffNPC with T21. (A) Schematic illustration of DE transcripts and proteins (black abbreviations within *blue boxes*) integrated in pathways. Elements are indicated as up- (*blue arrows*) or downregulated (*red arrows*). DE elements from protein analysis are indicated with a star (*asterisk*), and DE

Overexpression of DYRK1A has been reported to influence neurogenesis and dendritic development and associated with learning and memory defects observed in population with DS. In rodent, Dyrk1a is expressed in dividing neuronal progenitors of fetal adult brains and later in the adult cerebellum, hippocampus, and olfactory bulb (Hammerle et al., 2008). Notably, alterations in the phenotype of neocortical pyramidal cells were demonstrated in the Dyrk1Aþ/− mouse, along with smaller, less branched and less spinous basal dendritic arbors of layer III pyramidal cells in the trisomic mice compared with the control littermates. However, whether the pyramidal cells in juvenile mice of the Dyrk1Aþ/− strain cannot achieve the structural complexity that occurs in the wild-type mice or whether these mice undergo more intense dendrite retraction during maturation is yet to be ascertained. The involvement of DYRK1A might be apparent in both processes because overexpression of a kinase-deficient DYRK1A demonstrated perturbation of neurite outgrowth (Yang et al., 2001). In addition, a GTPase putative substrate of DYRK1A, Dynamin 1 (DYN1), has been attributed to synaptic vesicle recycling, clathrin-mediated endocytosis, intracellular membrane trafficking, and neurite outgrowth (Chen-Hwang et al., 2002). Finally, DYRK1A kinases are proposed to modulate the dendrite development by regulating DYN1-dependent vesicle trafficking. Thus, DYRK1A co-localizes with DYN1 and is transported through the growing dendritic tree (Hammerle et al., 2008). Interestingly, the pyramidal cells contain more than 70% of the neurons in the cortex, and thus, it is rational to implicate Dyrk1A while determining the complexity of circuits in the mature cerebral cortex (Dierssen et al., 2003). Moreover, it is imperative that the association of MR with disruption of the pyramidal phenotype could be the co-operating effects of various dosage-sensitive genes at the backdrop of non-HSA21 and nontrisomic genes. Taken together, Dyrk1A plays a fundamental role during neuronal development, and its proliferation and differentiation. Noticeably, lower Dyrk1A dosage may contribute to reduction in brain size, with no obvious changes in the lamination and anatomy of the cerebral cortex in mice (Fotaki et al., 2002).

DYRK1A contributes in brain development and neuronal differentiation through control of the AKT, MAPK/ERK, and STAT3-like critical signaling pathways or through the NFAT pathway in synaptic function (reviewed in Brault et al., 2021). The development of dendritic trees of neurons is regulated by Dyrk1A, while it modulates the action of the c-AMP response element binding protein (CREB), that is involved in signal transduction pathways related to synaptic plasticity and neuronal differentiation (Hammerle et al., 2008). An excess or haploinsufficiency of Dyrk1a in transgenic mice showed cognitive deficits similar to that observed in DS patients with specific hippocampal-dependent learning and memory impairment. The studies on the mechanisms of glutamatergic and GABAergic neurotransmitter dysfunction have presented excess GABAergic input and decreased activation of NMDA receptors and LTP in the hippocampal CA1 and dentate gyrus (DG) areas of Ts65Dn mice (reviewed in Brault et al., 2021). Additionally, enhanced hippocampal LTD was observed

elements from both transcriptome and proteome analysis are indicated in *red letters*. Hub genes are indicated in *blue letters*. Disturbed pathways for collagen, ECM—receptor interactions, integrin complex, receptor tyrosine kinase (RTK), and TGF-beta and growth factor (GF) receptor signaling are shown with their downstream effects on cell cycle regulation, apoptosis, glycolysis, and neurogenesis. *ECM*, extracellular matrix; *GF*, growth factors; *GPCR G*, protein coupled receptors; *ITGA/B*, integrin alpha and beta; *PM*, plasma membrane; *RTK*, receptor tyrosine kinases. (B) Highlight from (A) summarizing the molecular abnormalities associated with glycolysis and mitochondrial activity. The functional clusters are annotated using DAVID. *Reprinted from Sobol et al. (2019) under a CC-BY-NC-ND 4.0 International license.*

in the hippocampi in response to sustained activation of excitatory synapses and attributed to excessive signaling via NMDA receptors in Ts65Dn mice. A BAC transgenic mouse line overexpressing Dyrk1a showed alterations in glutamatergic synaptic proteins suggesting the contribution of DYRK1A to changes in glutamatergic neurotransmission. Interestingly, normalization of Dyrk1a from trisomy to disomy in Ts65Dn mice improved synaptic plasticity, GABAergic/glutamatergic balance, learning, and memory (reviewed in Brault et al., 2021). Notably, the heterozygous Dyrk1a-knockout mice revealed a decreased dendritic arborization and the spine density of glutamatergic pyramidal neurons of the cerebral cortex along with alterations in glutamatergic and GABAergic synaptic proteins. Thus, the dosage of Dyrk1a was hypothesized to alter the glutamatergic neurons of the hippocampus and cortex of DS, and altered development and/or normal working in adult brain underlie the cognitive deficits observed in DS, or Dyrk1a haploinsufficient MR, autosomal dominant 7 (MRD7), or mouse models.

Reduction of or conditional knockout allele of Dyrk1a gene dosage in Tg(Camk2a-Cre) transgenic line and trisomic Dp(16 Lipi-Zbtb21)1Yey mouse model (Dp1Yey; contains a segmental duplication of the 22.9 Mb Lipi-Zfp295 region including Dyrk1a) unraveled the consequence of Dyrk1a dosage from 0 to three on glutamatergic neurons (Brault et al., 2021). Molecular analyses did not present the impact of Dyrk1a dosage on locomotor activity, working memory, social behavior and epileptic susceptibility in postnatal glutamatergic neurons but indicated its involvement on long-term explicit memory, contextual fear memory, and exploratory behavior (Brault et al., 2021). Notably, a deregulation of transcriptional activity interacting with key postsynaptic proteins at the glutamatergic postsynapse implicated in the mechanism of long-term enhanced synaptic plasticity. Altogether, 297 up-regulated and 257 down-regulated genes were identified in Dyrk1aC/C (Dyrk1aC/C: $Dyrk1a^{Camk2aCre/Camk2aCre}$—recombination of both $Dyrk1a^{cKO}$ alleles with the Cre recombinase) mice compared with controls (Fig. 10.6) (Brault et al., 2021). The Dp1Yey mice include 65% of HSA21 mouse-orthologous gene and free from the 50 DS-irrelevant trisomic genes that are present on the mini chromosome of Ts65Dn but possess the deficits in working memory, long-term episodic memory, and learning. Indeed, genetic modification in homozygotes, heterozygotes, and the Dp(16Lipi-Zbtb21)1Yey trisomic mouse model extracted important information on the action of DYRK1A, which would further guide to identify DYRK1A-inhibitors for a better therapeutic approach (Brault et al., 2021; Nguyen et al., 2018). Treatment of the three DS models (Tg(Dyrk1a), Ts65Dn, Dp1Yey) with leucine L41, the pharmacological inhibitor of DYRK1A, normalized DYRK1A activity and corrected the novel object recognition (NOR) impairment in the three models (Nguyen et al., 2018).

Dyrk1a mRNA and protein are evidenced in the transition of neuroepithelial cells from proliferating to neurogenic divisions at early embryonic ages in mice. Its overexpression attenuates the neurite outgrowth influenced by a neurogenic factor in immortalized hippocampal cells. Interaction of DYRK1A with SYN1 (synapsin 1—a neuronal phosphoprotein associated with the cytoplasmic surface of the presynaptic vesicles and tethering them to the actin cytoskeleton) and with CAMK2 (a subunit of the calcium/calmodulin-dependent protein kinase II that phosphorylates SYN1 leading to the release of the vesicle pool) revealed DYRK1A-mediated phosphorylation of SYN1 on its S551 residue both in vitro and in vivo, highlighting the contribution of DYRK1A in SYN1-dependent presynaptic vesicle trafficking (Brault et al., 2021). Deregulated CAMK2 in proteomic analysis indicated its role in the glutamatergic postsynapse cascade leading to LTP. Importantly, postsynapse interaction of DYRK1A protein with

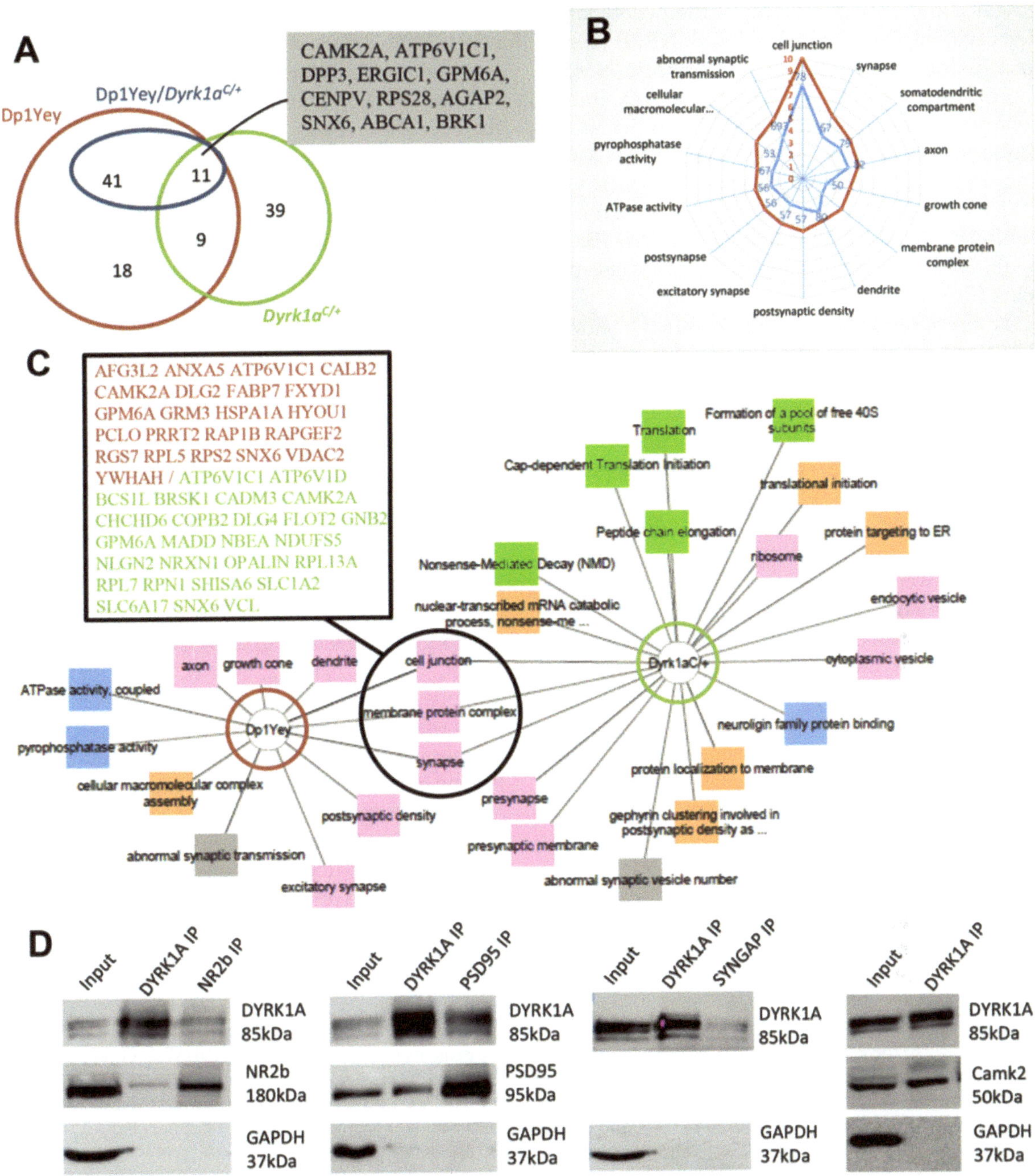

FIGURE 10.6 **Proteomic analysis.** (A) Venn diagram showing the numbers of deregulated proteins in the different mouse models. The numbers of proteins shown for the Dp1Yey/*Dyrk1a*$^{C/+}$ model (in dark blue) correspond to proteins that were deregulated in the Dp1Yey and back to normal levels in this model (proteins that are regulated by Dyrk1a in the trisomy). Proteins deregulated by both *Dyrk1a* up- and down-regulation (common to Dp1Yey, Dp1Yey/*Dyrk1a*$^{C/+}$, and *Dyrk1a*$^{C/+}$) are listed in the *gray-shaded box*. (B) Radar plots of GO terms that are mostly enriched in the Dp1Yey model (in red with scale bar

CAMK2 and critical proteins of the postsynaptic density complex (PSD), GLUN2B, NR2B (a subunit of the glutamatergic postsynaptic NMDA receptor, which plays a pivotal role in excitatory synaptic transmission), PSD95, and SYNGAP highlighted the presence of CAMK2A, NR2B, and PSD95 in the immunoprecipitates (IPs) of DYRK1A, while DYRK1A was found in the IPs of NR2B, PSD95 and SYNGAP (Fig. 10.6) (Brault et al., 2021), suggesting interaction of these proteins (Brault et al., 2021).

Transcriptomic analysis of Dyrk1a$^{C/C}$ showed down-regulation of genes for proteins associated with presynaptic vesicle cycle, implicating DYRK1A in neurotransmitter release (Brault et al., 2021). Deregulation of Amphiphysin (Amph—its protein is a known target of DYRK1A), and hyperphosphorylated Synapsin 2 (SYN2: paralog SYNAPSIN 1) by the overexpressed DYRK1A were reported in TgDyrk1a mice (reviewed in Brault et al., 2021). Additionally, DYRK1A-mediated phosphorylation of MUNC18-1 (which interacts with the SNARE complex protein Syntaxin 1A, whose transcripts are also decreased in the hippocampus of Dyrk1a$^{C/C}$ mice) deregulate the major subnetwork of biological cascades in DS models. Taken together, DYRK1A acts as a key factor for the glutamatergic synapse and thus, targeted for treatment of DYRK1A-linked diseases, such as AD.

Cystathionine-b-synthase (CBS), a key enzyme in the trans-sulfuration pathway, catalyzes the conversion of homocysteine into cystathionine. Impaired intelligence in DS was linked to CBS-derived overproduction of H_2S (Kamoun et al. (2003). J. Lejeune hypothesized in 1975 that the metabolic alterations and overall clinical feature of DS could be contributed by the increased expression of CBS based on the similarities in clinical features of homocystinurea and DS (Lejeune, 1975) and markedly increased circulating levels of sulf-hemoglobin (a product of the reaction of hemoglobin with H_2S) and ~2-fold increase (over CC) in urinary excretion of thiosulfate (a stable degradation product of H_2S) in individuals with DS (Kamoun et al., 2003; Caracausi et al., 2018).

3.3 Growth factor receptors, proteins, and altered neurogenesis in DS

Downregulation of the numerous components of the EGF signaling cascade, including the receptors (EGFR, ERBB3, ERBB4), and the ligand itself (HBEGF) has been reported in DS. Notably, 17 of the 75 growth factor signaling proteins, including EGFR, ERBB4, NTRK2, NTRK3, and RET, were demonstrated to be linked with regulation of neurogenesis (Sullivan et al., 2017). The two transmembrane receptors such as NTRK2 and NTRK3 for the brain-derived neurotrophic factor (BDNF) and neurotrophin 3, respectively, were downregulated in the plasma of people with DS. Selective failure of BDNF-dependent survival was accompanied by overexpression of a truncated, kinase-deficient isoform of the BDNF receptor NTRK2 in cultured hippocampal neurons of the Ts16

corresponding to the log of the *P*-value) and of the proportion of the proteins found deregulated in Dp1Yey, whose amount is normalized by the return in two copies in the Dp1Yey/*Dyrk1a*$^{C/+}$ model (in blue with scale bar corresponding to the % of Dp1Yey deregulated protein back to normal levels). (C) Visual representation of the GO enrichments for the deregulated proteins in Dp1Yey and *Dyrk1a*$^{C/+}$ hippocampi with connection between common terms. The different categories of GO are represented by different colors: pink for "cellular components," blue for "molecular functions," green for "pathways," orange for "biological processes," and gray for "phenotypes." The list of proteins in the box corresponds to proteins present in the common deregulated GO terms (in red deregulated proteins in Dp1Yey and in green proteins deregulated in *Dyrk1a*$^{C/+}$). (D) Western blots of DYRK1A, NR2B, PSD95, SYNGAP, and CAMK2 proteins following IPs of wild-type mice brain extracts. We found in NR2B, PSD95, and CAMK2 in the IPs of DYRK1A. We also detected DYRK1A in the IPs of NR2B, PSD95, and SYNGAP. *Reprinted from Brault et al. (2021). https://doi.org/10.1371/journal.pgen.1009777 under the terms of the Creative Commons Attribution License.*

mouse. Interestingly, introduction of adenovirus-mediated exogenous full-length NTRK2 into Ts16 neurons restored BDNF-dependent survival fully, whereas BDNF signaling failure was demonstrated in exogenous truncated NTRK2 expression in normal euploid neurons. This result has explained the dysregulation of NTRK2 isoform expression results in the failure of Ts16 neurons to respond to BDNF (reviewed in Benavides-Piccionea et al., 2004). Such a neurotrophin signaling defect could contribute to alteration in structural properties of the dendritic spines with consequences in neural plasticity and possibly in long-term memory storage.

Downregulation of a third neurotrophin receptor, SLITRK3, is essential for normal GABAergic synapse development (Takahashi et al., 2012). Also, low circulating levels of RET, a receptor tyrosine kinase of the cadherin superfamily with key roles in neural crest development has been reported in people with DS. Noticeably, loss-of-function mutations in RET cause Hirschsprung disease and that is significantly prevalent in people with DS (Tomuschat and Puri, 2015). The proteomic datasets of Sullivan et al. (2017) have indicated many hypothesis-generating results that may direct for future mechanistic investigations with a view to deciphering the molecular underpinnings of the developmental problems and cognitive deficits in patients with DS (Fig. 10.7) (Sullivan et al., 2017). The datasets have generated significant information on how trisomic HSA21 causes profound immune dysregulation, while many of the changes being highly important for those observed in type I interferonopathies and other autoinflammatory conditions.

The dynamic dendritic spine structure arises from the dendrite comprising of a spine shaft and spine head. Biochemically, it contains cytoskeleton proteins: β- and γ-isoforms of actin, spectrin, α-actinin, capping proteins, and drebrin (an actin-binding protein and regulator of the dendritic spine's cytoskeleton). The dendritic spines are extremely motile and provide a structural mechanism for synaptic plasticity. In dendritic spines, actin filaments constitute the major cytoskeletal components and configure the basic structural scaffolding of spines (Hayashi et al., 1996). Thus, alterations in actin filaments are likely responsible for changes in shape of the spines in DS.

The implication of specific MR-related dosage-sensitive genes in some single-gene transgenic models revealed overexpression of human S100B (stimulates glial proliferation) contributing stress in the dendritic pathology by changing the cytoskeletal markers such as the dendritic-associated protein, MAP-2, the growth-associated protein-43, and the dendritic spine marker, drebrin (reviewed in Dierssen and Ramakers, 2006). Such alterations resulted in an increased density of dendrites within the hippocampus of the transgenic animals. Co-localization of drebrin with actin filaments at dendritic spines has been detected in high-density long-term primary cultures of rat cerebral cortex neurons, suggesting it as an actin-regulatory component of spines. Thus, drebrin was postulated to regulate the assembly and disassembly of actin filaments and thereby change the shape of spines (reviewed in Benavides-Piccionea et al., 2004). Indeed, upon deletion of the actin-binding domain of drebrin, the distribution of protein in both spines and dendritic shafts suggested the requirement of its actin-binding activity for selective accumulation of drebrin in the spines. Reduced dendritic spine protein might impair and dismantle the functional assembly in the dendritic spine (reviewed in Weitzdoerfer et al., 2001).

Indeed, drebrin was remarkably reduced to not detectable level in several DS brain samples (Weitzdoerfer et al., 2001). Significant reduction in drebrin may be responsible for deterioration of dendritic spine structure in DS population. Since dendritic spines are the chief determinants of the brain morphogenesis, drebrin was hypothesized to be responsible for dendritic spine abnormalities and aberrant brain dysgenesis in DS patients. Indeed, adult DS brain showed dedifferentiation of neurons associated with

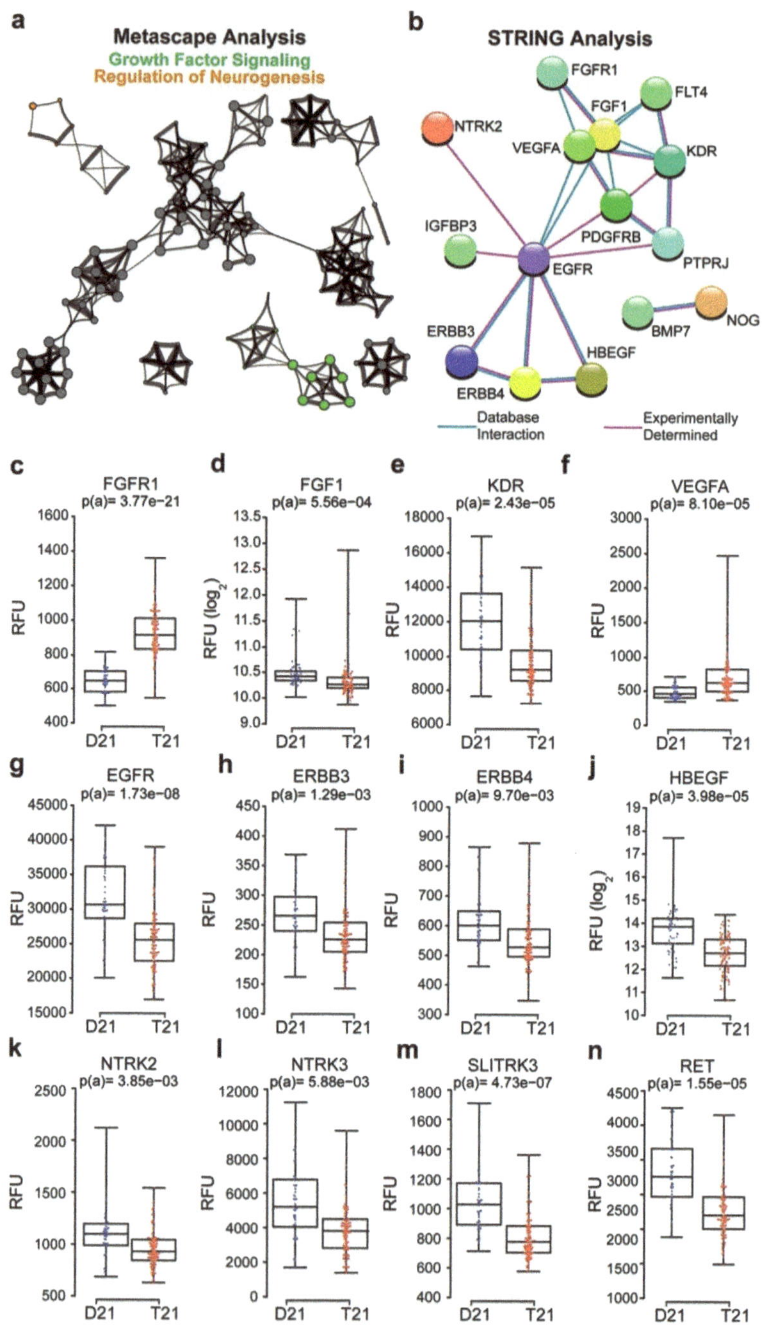

FIGURE 10.7 People with Down syndrome show different levels of proteins involved in growth factor signaling and control of neurogenesis. (A) Metascape analysis highlighting protein networks related to Growth Factor Signaling and Regulation of Neurogenesis among the differential proteins in the discovery study. (B) STRING analysis identifying interacting groups of deregulated proteins within the growth factor signaling category. (C–N) Box and whisker plots displaying SOMAscan data for individual proteins deregulated in individuals with T21. p(a) from KS test with Bonferroni correction. *Reprinted from Sullivan et al. (2017). https://doi.org/10.1038/s41598-017-13858-3 under the terms of the Creative Commons CC BY Attribution 4.0.*

reduced drebrin, suggesting drebrin a consequence of neuronal loss in AD and other dementing diseases (reviewed in Weitzdoerfer et al., 2001). Notably, reduced drebrin was reported in patients with AD. Interestingly, over 90% of the excitatory glutamatergic synapses occur on dendritic spines and cluster at actin-drebrin dendritic spines in mammalian brain. Thus, decreased drebrin could be assigned for deterioration of the excitatory amino acid transmission system in DS, which could explain the mechanisms of deficient nicotinergic functions shown in adult DS. The nicotinic acetylcholine receptor binds to actin- and drebrin-containing dendritic spines, and thus, spine retraction results in dispersion of this nicotinic receptor.

Expression of a presynaptic membrane–associated protein, SNAP25 of 25 kDa mass, is associated with synaptogenesis and neuronal maturation. This protein is expressed at the axonal growth cone—a sensory structure that receives directional information from surrounding and is also a motor structure for axonal elongation. Inhibition of SNAP25 protein by antisense oligonucleotides exhibited diminished axonal extension in developing neurons both in vitro and in vivo. Thus, SNAP25 has been hypothesized as the core component of the t-SNARE (target membrane SNAP receptor) complex containing syntaxin and SNAP25 and that plays a crucial role in vesicle docking and membrane fusion (reviewed in Weitzdoerfer et al., 2001). Evaluation of fetal DS cortex indicated impaired synaptogenesis at 19th week of gestation and that may also be responsible for shortened frontal occipital length of DS brain (Weitzdoerfer et al., 2001). Reduced SNAP25, α-SNAP in particular, was observed in adult brains of DS and AD.

A predominant expression of Chapsyn110—a membrane associated putative guanylate kinase, is associated with postsynaptic density (PSD) in brain. Chapsyn110 belongs to PSD-95 family and binds to Shaker K+ channel subunits and N-methyl-D-aspartate (NMDA) receptor directly, mediating their clustering in heterologous cells (reviewed in Weitzdoerfer et al., 2001). Increase in labeling of Chapsyn110 antibody was reported in synapses of rat hippocampus from postnatal day 2 to adult, which correlates to increase in NR2A—a subunit of NMDA receptor and suggests its importance in mature synapses (Sans et al., 2000). However, postsynaptic membrane density of Chapsyn110 at this stage of development of human DS fetus was unaffected (Weitzdoerfer et al., 2001).

Taken together, impaired dendritic spines, axonal growth, and function in the presynaptic terminal in the cerebral cortex was contributed by decreased level of drebrin, SNAP25, and α-SNAP in the DS fetal brain of the second trimester gestation (Weitzdoerfer et al., 2001). Reduction of these proteins may perturb spine motility and plasticity in DS brains, which may not be a simple consequence of neuronal loss. Reduced drebrin was reported causing preceding impaired arborization and atrophy of the dendritic tree, which is a morphological hallmark of DS brain. Also deficient drebrin was proposed to perturb the nicotinic and glutamatic receptor binding at the head of dendritic spine in DS.

Studies on protein–protein interaction (PPI) network demonstrated marked upregulation of 23 genes encoding hub proteins in trisomic NPCs and diffNPCs, of which upregulated proteins encoded by four genes: VCL (encoding vinculin), CAV1 (encoding caveolin-1), ITGB1 (encoding integrin subunit beta 1), and IL7R (encoding interleukin-7 receptor) dysregulated the functional modules, including focal adhesion, ECM interactions, receptor-mediated signaling, and cytoskeletal adaptation. In diffNPCs, increased expression of the membrane-bound ITGB1 protein dysregulated the ECM-receptor interaction, focal adhesion, cytoskeletal formation, migration, and neural-specific functions, including axon guidance and synapse formation (reviewed in Sobol et al., 2019). Additionally, ITGB1 activates the PI3K-Akt pathway of the cell cycle regulation. Critical importance of such interconnected functions perturbs the cell cycle regulation, cell morphology, and cytoskeleton plasticity, and those are associated with morphological and functional abnormalities of brains of human DS and DS models. Additionally, upregulation of the APP hub gene contributes to deposition of amyloid peptides (Aβ) and Alzheimer-like neurodegenerative pathomechanism as early as at 21 gestational weeks in DS brain (Wiseman et al., 2015) and trisomic iPSC models (Shi et al., 2012; Murray et al., 2015). Interestingly, upregulation of APP at an early stage of NPC suggests its contribution to promoting growth and plasticity of neural cells during neurodevelopment (Muller et al., 2017) and indicates a detrimental role of the APP protein in early stages of neurogenesis in DS (Sobol et al., 2019). The inflammatory responses resulted from Aβ deposition in senile plaques may trigger cytokine production, which may act as negative modulators of growth and maintenance of dendritic neurons in DS. However, vast majority of the hub proteins (~40) appeared unaltered in the disturbed functional networks, suggesting their more fine-tuned role in DS neurogenesis (Sobol et al., 2019). Altogether, dysregulation of the hub genes and associated perturbed pathways with dynamic variations in differentiation of trisomic neural cells indicate downstream effects on proliferation, maturation, energy metabolism, differentiation, and degeneration that collectively contribute to abnormalities in brains of human DS and orthologous mouse models, and trisomic iPSCs in vitro, to some extent (reviewed in Sobol et al., 2019).

Transcriptome and proteome data at either of the two stages of differentiation, revealed eight of the dysregulated clusters, such as DNA-replication, collagen, cell-adhesion, ECM–receptor interactions, integrin complex, TGF-β signaling, oxidative phosphorylation, and glycolysis, and that appeared consistent with incidences in fetal heart tissues and fibroblasts of DS patients (reviewed in Sobol et al., 2019). This report suggests regulatory abnormalities in DS organogenesis irrespective of tissue type. Nevertheless, five clusters are specific in trisomic NPCs, which include DNA replication, pluripotency, synaptic maturation, neuroactive signaling, and collagen binding clusters (Fig. 10.8) (Sobol et al., 2019).

4. Mitochondrial control of neural progenitor cells

Neurons are highly dependent on oxidative energy metabolism, and thus, dysfunction in mitochondrial energy metabolism has been suggested to contribute to a unified pathogenetic mechanism of neurodegeneration, though clinical features greatly vary in different neurodegenerative disorders. Mitochondria play the key role in oxygen metabolism, generate the majority of the cellular energy by the oxidative phosphorylation (OXPHOS) mechanism, regulate cytosolic Ca++ homeostasis and control cell death or apoptosis through the activation of the mitochondrial permeability transition pore (mtPTP), and also produce the reactive oxygen species (ROS) (described in detail in Chapter 8). Therefore, impairment in this organelle can install catastrophic consequences in cells resulting in mitochondrial dysfunction, which in turn may increase ROS production, reduce membrane potential, impair OXPHOS, and reduce ATP production. Collectively, mitochondrial dysfunction may cause havoc in the cells contributing to a number of metabolic and degenerative diseases, and that is more pronounced in the metabolically active tissues such as the heart and brain (Wallace and Chalkia, 2013).

The combined effect of loss of mitochondrial integrity and an increase in oxidative stress promote apoptosis (changes in caspase and respiratory chain protein expression) and gliosis (increase in GFAP). In the DS brain, free radicals perturb the mitochondrial respiratory system and stimulate apoptosis and gliosis, which collectively promote neuronal damage. Such neuronal losses during neurogenesis and neuronal degeneration contribute to dementia, early aging and AD in DS adults.

Mitochondrial dysfunction, altered mitochondrial morphology, and downregulation of the mitochondrial transcriptome are well-known features in trisomic cells (Arbuzova et al., 2002; Helguera et al., 2013). Notably, molecular

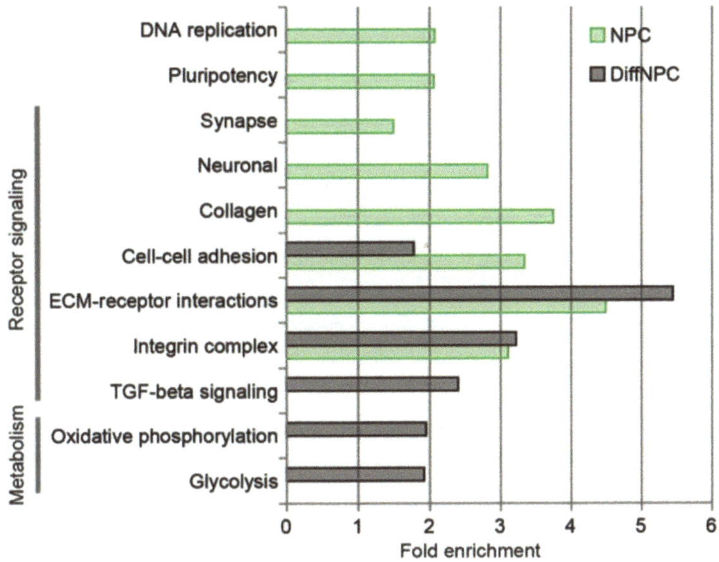

FIGURE 10.8 Dynamics of perturbed functional clusters in T21 neural lines based on integrated analysis of data from NPCs and diffNPCs. Eleven deficient functional clusters were identified based on annotations using DAVID considering all transcriptome and proteome data sets from NPCs and diffNPCs, respectively. Fold enrichment of clusters was obtained from separate data analysis of NPCs and diffNPCs, respectively. The clusters were named after representative annotations. Five dysregulated clusters (DNA replication, pluripotency, synapse, neuronal, and collagen) were characteristic for trisomic NPCs suggesting a significant impact of these clusters in early T21 neurogenesis. In diffNPCs, three dysregulated clusters are distinguished (TGF-beta signaling, oxidative phosphorylation, and glycolysis) suggesting their ascendant role with differentiation in T21 neurogenesis. Seven dysregulated clusters belong to receptor signaling pathways, and these were evident at both stages of differentiation implying a critical role for dysfunctional intercellular communication and ECM in T21. Fold enrichment scores of annotated clusters are shown for NPC (green) and diffNPC (gray). The complete clustering data are summarized in Supplemental Table S5 in Sobol et al. (2019). *Reprinted from Sobol et al. (2019) under a CC-BY-NC-ND 4.0 International license.*

perturbations indicated a reduced mitochondrial activity along with increased downregulation of mitochondrial transcription leading to distinctive dysregulated glycolysis and oxidative phosphorylation in trisomic differentiated NPCs. Reduced cellular growth in induced pluripotent stem cells (iPSCs) derived from DS population was neither associated with an increase in apoptosis as defined by caspase-3 staining nor with increased levels of ROS, despite H_2O_2 induction, indicating tolerance of the mixed neural cell populations with trisomic HSA21 to increased stress (Sobol et al., 2019; Weick et al., 2013). However, conflicting reports exist on enhanced ROS levels in trisomic skin fibroblasts and astroglial cells (Chen et al., 2014) and increased apoptosis and caspase-3 staining in NPCs with T21 (Hibaoui et al., 2014). Nevertheless, the discrepancies on the ROS levels, oxidative stress, and cellular function can be explained by the energy requirement of the cell types and culture conditions (Helguera et al., 2013).

Noticeably, the dysregulated mitochondrial transcription and increase in transcripts for glycolysis in trisomic cells led to decrease in the proliferative rate of NPCs and a significant increase in the fraction of glia cells along with a reduced proportion of neuronal cells. Interestingly, 10 of the 24 expressed mitochondrial (MT) genes were downregulated significantly in trisomic diffNPCs but not in trisomic NPCs (Fig. 10.9) (Sobol et al., 2019), which supports a

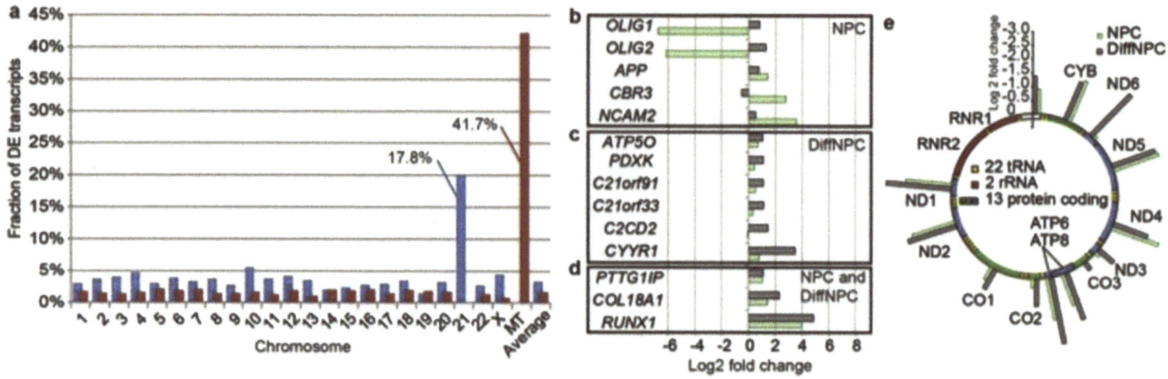

FIGURE 10.9 Temporal variation of differential expression from HSA21 genes and the MT genome in T21 lines. (A) Fraction of up- (blue) and downregulated (red) genes per chromosome in diffNPCs with T21. (B) DEGs on HSA21 that are specific for NPCs with T21 (cut-off $P < .01$, absolute log2fold ratio of >1). (C) DEGs on HSA21 specific for diffNPCs with T21. (D) DEGs on HSA21 shared at both differentiation stages. (E) Schematic presentation of the MT genome indicating the relative downregulation of MT gene expression (standing bars) in both NPCs (green) and diffNPCs (gray) with T21. *Reprinted from: Sobol et al.(2019) under a CC-BY-NC-ND 4.0 International license.*

ubiquitous downregulation of mitochondrial activity on cardiac tissues and fibroblasts of DS patients (illustrated in Chapter 8). Additionally, the downregulation was nonsignificant at the early NPC stage, while diffNPC showed a tendency of reduction of the mitochondrial transcripts encoding tRNAs with differentiation, suggesting a marked downregulation of mitochondrial transcription with differentiation in trisomic neural lines despite increase in mtDNA copy numbers (Sobol et al., 2019).

Upregulated storage of glycogen accompanied by an active glycolysis can take place in astroglia cells, while metabolism of neurons depends on the TCA cycle and oxidative phosphorylation (Magistretti and Allaman, 2015) and that is disrupted by dysregulation and/or downregulation of mitochondrial transcription in trisomic cell lines. Summarily, the downstream of regulatory factors for growth and cell fate decision in the mixed trisomic HSA21 neural cultures was implicated by the dysregulation of mitochondrial transcription and not by the increase in ROS levels. The downregulation of mitochondrial transcription leads to decrease in oxidative phosphorylation, which may lead to a consequence of reduced proliferative rate of varying cellular composition with variability in metabolic requirements (Sobol et al., 2019). Further studies on targeted candidates on trisomic HSA21 and its impact on the global expression of the disomic genes can explore the interference on brain development and improve the cognitive outcome in DS patients. Moreover, the abnormalities of anatomy, including cellularity and physiology of the DS brain, arise at birth. Therefore, targeting therapeutic strategies ideally should focus on improvement of cognitive development of DS through molecular understanding of neurogenesis in DS and tailoring druggable pathways accordingly (Antonarakis, 2017; Fernandez et al., 2007).

The trisomic diffNPCs exhibited an adaptation to cellular stress in line with the mitochondrial activity reported in DS individuals (Sobol et al., 2019). The diffNPCs under hydrogen peroxide (H_2O_2) stress appeared identical to that of the euploid neural cells in ROS production or apoptosis as measured by caspase-3 staining (Sobol et al., 2019); however, dysregulated genes responsible for ROS metabolism along with antioxidation and apoptosis were noticed in global gene expression analysis in both NPCs and diffNPCs with trisomic HSA21. Increased proportion of glia cells using glycolysis or via downregulation of energy metabolism has been hypothesized to render the tolerance to trisomic diffNPCs to increased cellular stress (Helguera et al., 2013). Nevertheless, contradictory findings exist in this aspect, which might be explained by the fact that the ROS levels, oxidative stress, and cellular function vary with energy requirements of the types and functions of cells as well as culture conditions (Helguera et al., 2013). Therefore, a collective effect of the impaired MT transcription, increase in transcripts for glycolysis, and reduced proliferative rate of NPCs contributed to increase the fraction of glia cells and to reduce the proportion of neuronal cells (Sobol et al., 2019). Interestingly, astroglia cells can store glycogen and possess an active and upregulated glycolysis, whereas neural cells are dependent on the TCA cycle and oxidative phosphorylation for metabolism. The downregulation of mitochondrial transcription may reflect the altered distribution of the 2 cell types and their different metabolism in trisomic lines (Sobol et al., 2019). Indeed, these findings are significant for identification of druggable pathways and elements, and tailoring therapeutic strategies can interfere with neurodevelopmental abnormalities and subsequently improve the cognition in DS population.

5. Concluding remarks

Functional imaging of cerebral metabolic rates in DS young adults exhibited fewer significant correlations in different brain regions and reduced functional connections between brain circuit elements leading to dysfunctional circuitry in adults (Schapiro et al., 1992). However, it may not be rational to attribute the cognitive deficits in DS only to reduced brain size or neuronal density. Notably, different aspects of neuronal cell structure, including the size of the basal dendritic arbors, their branching structure, and spine density may vary independently (reviewed in Elston, 2003). The neural mechanisms underlying MR may include abnormalities in the formation of neuronal networks and/or in information processing aspects of brain plasticity. Additionally, abnormalities in dendritic architecture and subsequent neuronal disruption determine the nature of a neuron's inputs and its role in the cortical circuitry, which are distinct in DS subjects and correlate with their cognitive profile to some extent. However, abnormal postnatal development and remodeling of the brain circuitry could correlate dendritic abnormalities and consequences on the cognitive abilities in DS patients. The question remains that if the degree of dendritic alterations correlates with cognitive function, then DS subjects with moderate level of MR should exhibit moderate level of dendritic abnormalities, and, if dendritic abnormalities indicate the global cognitive dysfunction, they should be observed in many other syndromes associated with MR, and thus, dendritic alterations should parallel the defects in neuronal maturation and/or decline in cognitive performance. Furthermore, neurons in DS may not achieve the level of structural complexity as that of normal children, or greater dendritic retraction may occur during maturation. In total, HSA21 gene dosage in DS underlies the alterations in neurogenesis and its consequences on cognitive ability.

References

Alves-Sampaio, A., Troca-Marín, J.A., Montesinos, M.L., 2010. NMDA-mediated regulation of DSCAM dendritic local translation is lost in a mouse model of Down's syndrome. J. Neurosci. 30, 13537–13548.

Antonarakis, S.E., 2017. Down syndrome and the complexity of genome dosage imbalance. Nat. Rev. Genet. 18 (3), 147–163. https://doi.org/10.1038/nrg.2016.154.

Arbuzova, S., Hutchin, T., Cuckle, H., 2002. Mitochondrial dysfunction and Down's syndrome. Bioessays 24 (8), 681–684. https://doi.org/10.1002/bies.10138.

Balogh, E.Z., Lengyel, Z., Emri, M., Szikszai, E., Esik, O., Kollar, J., Sikula, J., Tron, L., Olah, E., 2002. Cerebral glucose metabolism in Down syndrome using positron emission tomography. Orv. Hetil. 143, 1304–1307.

Becker, L., Mito, T., Takashima, S., Onodera, K., 1991. Growth and development of the brain in Down syndrome. Prog. Clin. Biol. Res. 373, 133–152.

Belichenko, P.V., Masliah, E., Kleschevnikov, A.M., Villar, A.J., Epstein, C.J., Salehi, A., Mobley, W.C., 2004. Synaptic structural abnormalities in the Ts65Dn mouse model of Down Syndrome. J. Comp. Neurol. 480, 281–298.

Belichenko, P.V., Kleschevnikov, A.M., Salehi, A., Epstein, C.J., Mobley, W.C., 2007. Synaptic and cognitive abnormalities in mouse models of Down syndrome: exploring genotype-phenotype relationships. J. Comp. Neurol. 504 (4), 329–345. https://doi.org/10.1002/cne.21433.

Benavides-Piccionea, R., Ballesteros-Yáñeza, I., de Lagránb, M.M., Elstonc, G., Estivillb, X., Fillatb, C., DeFelipea, J., Dierssen, M., 2004. On dendrites in Down syndrome and DS murine models: a spiny way to learn. Prog. Neurobiol. 74, 111–126. https://doi.org/10.1016/j.pneurobio.2004.08.001.

Brault, V., Nguyen, T.L., Flores-Gutiérrez, J., Iacono, G., Birling, M.-C., Lalanne, V., et al., 2021. *Dyrk1a* gene dosage in glutamatergic neurons has key effects in cognitive deficits observed in mouse models of MRD7 and Down syndrome. PLoS Genet. 17 (9), e1009777. https://doi.org/10.1371/journal.pgen.1009777.

Cairns, N.J., 2001. Molecular neuropathology of transgenic mouse models of Down syndrome. J. Neural. Transm. Suppl. 61, 289–301.

Caracausi, M., Ghini, V., Locatelli, C., Mericio, M., Piovesan, A., Antonaros, F., et al., 2018. Plasma and urinary metabolomic profiles of Down syndrome correlate with alteration of mitochondrial metabolism. Sci. Rep. 8, 2977. https://doi.org/10.1038/s41598-018-20834-y.

Chakrabarti, L., Best, T.K., Cramer, N.P., Carney, R.S.E., Isaac, J.T.R., Galdzicki, Z., Haydar, T.F., 2010. Olig1 and Olig2 triplication causes developmental brain defects in Down syndrome. Nat. Neurosci. 13, 927–934. https://doi.org/10.1038/nn.2600.

Chen, C., Jiang, P., Xue, H., Peterson, S.E., Tran, H.T., McCann, A.E., Parast, M.M., Li, S., et al., 2014. Role of astroglia in Down's syndrome revealed by patient-derived human-induced pluripotent stem cells. Nat. Commun. 5, 4430. https://doi.org/10.1038/ncomms5430.

Chen-Hwang, M.C., Chen, H.R., Elzinga, M., Hwang, Y.W., 2002. Dynamin is a minibrain kinase/dual specificity Yak1-related kinase 1A substrate. J. Biol. Chem. 277, 17597–17604.

Colombo, J.A., Reisin, H.D., Jones, M., Bentham, C., 2005. Development of interlaminar astroglial processes in the cerebral cortex of control and Down's syndrome human cases. Exp. Neurol. 193 (1), 207–217. https://doi.org/10.1016/j.expneurol.2004.11.024.

Conti, A., Fabbrini, F., D'Agostino, P., Negri, R., Greco, D., Genesio, R., D'Armiento, M., Olla, C., et al., 2007. Altered expression of mitochondrial and extracellular matrix genes in the heart of human fetuses with chromosome 21 trisomy. BMC Genom. 8, 268. https://doi.org/10.1186/1471-2164-8-268.

Cramer, N., Galdzicki, Z., 2012. From abnormal hippocampal synaptic plasticity in Down syndrome mouse models to cognitive disability in Down syndrome. Neural Plast. 2012, 101542.

Dierssen, M., 2012. Down syndrome: the brain in trisomic mode. Nat. Rev. Neurosci. 13, 844–858. https://doi.org/10.1038/nrn3314.

Dierssen, M., Fructuoso, M., Martínez de Lagrán, M., Perluigi, M., Barone, E., 2020. Down syndrome is a metabolic disease: altered insulin signaling mediates peripheral and brain dysfunctions. Front. Neurosci. 14, 670. https://doi.org/10.3389/fnins.2020.00670. Published 2020 Jul 8.

Dierssen, M., Benavides-Piccione, R., Martinez-Cue, C., Estivill, X., Florez, J., Elston, G.N., DeFelipe, J., 2003. Alterations of neocortical pyramidal cell phenotype in the Ts65Dn mouse model of Down syndrome: effects of environmental enrichment. Cerebr. Cortex 13, 758–764.

Dierssen, M., Ramakers, G.J.A., 2006. Dendritic pathology in mental retardation: from molecular genetics to neurobiology. Gene Brain Behav. 5 (Suppl. 2), 48–60.

Elston, G.N., 2003. Cortex, cognition and the cell: new insights into the pyramidal neuron and prefrontal function. Cerebr. Cortex 13, 1124–1138.

Fernandez, F., Morishita, W., Zuniga, E., Nguyen, J., Blank, M., et al., 2007. Pharmacotherapy for cognitive impairment in a mouse model of Down syndrome. Nat. Neurosci. 10, 411–413.

Fotaki, V., Dierssen, M., Alcántara, S., Martínez, S., Martí, E., Casas, C., Visa, J., Soriano, E., Estivill, X., Arbonés, M.L., 2002. Dyrk1A haploinsufficiency affects viability and causes developmental delay and abnormal brain morphology in mice. Mol. Cell Biol. 22, 6636–6647.

Guidi, S., Bonasoni, P., Ceccarelli, C., Santini, D., Gualtieri, F., Ciani, E., Bartesaghi, R., 2008. Neurogenesis impairment and increased cell death reduce total neuron number in the hippocampal region of fetuses with Down syndrome. Brain Pathol. 18 (2), 180–197. https://doi.org/10.1111/j.1750-3639.2007.00113.x.

Halevy, T., Biancotti, J.C., Yanuka, O., Golan-Lev, T., Benvenisty, N., 2016. Molecular characterization of Down syndrome embryonic stem cells reveals a role for RUNX1 in neural differentiation. Stem Cell Rep. 7 (4), 777–786. https://doi.org/10.1016/j.stemcr.2016.08.003.

Hammerle, B., Elizalde, C., Tejedor, F.J., 2008. The spatio-temporal and subcellular expression of the candidate Down syndrome gene Mnb/Dyrk1A in the developing mouse brain suggests distinct sequential roles in neuronal development. Eur. J. Neurosci. 27 (5), 1061–1074. https://doi.org/10.1111/j.1460-9568.2008.06092.x. EJN6092 [pii].

Hayashi, K., Ishikawa, R., Ye, L.H., He, X.L., Takata, K., Kohama, K., Shirao, T., 1996. Modulatory role of drebrin on the cytoskeleton within dendritic spines in the rat cerebral cortex. J. Neurosci. 16, 7161–7170.

Haydar, T.F., Reeves, R.H., 2012. Trisomy 21 and early brain development. Trends Neurosci. 35 (2), 81–91. https://doi.org/10.1016/j.tins.2011.11.001.

Helguera, P., Seiglie, J., Rodriguez, J., Hanna, M., Helguera, G., Busciglio, J., 2013. Adaptive downregulation of mitochondrial function in Down syndrome. Cell Metabol. 17 (1), 132–140. https://doi.org/10.1016/j.cmet.2012.12.005.

Heng, J.I., Chariot, A., Nguyen, L., 2010. Molecular layers underlying cytoskeletal remodelling during cortical development. Trends Neurosci. 33, 38–47. https://doi.org/10.1016/j.tins.2009.09.003.

Hibaoui, Y., Grad, I., Letourneau, A., Sailani, M.R., Dahoun, S., SantoniFA, G.S., Guipponi, M., et al., 2014. Modelling and rescuing neuro developmental defect of Down syndrome using induced pluripotent stem cells from monozygotic twins discordant for trisomy21. EMBO Mol. Med. 6 (2), 259–277. https://doi.org/10.1002/emmm.201302848.

Huo, H.-Q., Qu, Z.Y., Yuan, F., Ma, L., Yao, L., XuM, H.Y., Ji, J., et al., 2018. Modeling Down syndrome with patient iPSCs reveals cellular and migration deficits of GABAergic neurons. Stem Cell Rep. 10 (4), 1251–1266. https://doi.org/10.1016/j.stemcr.2018.02.001.

Jiang, J., Jing, Y., Cost, G.J., Chiang, J.C., Kolpa, H.J., Cotton, A.M., Carone, D.M., Carone, B.R., et al., 2013. Translating dosage compensation totrisomy 21. Nature 500 (7462), 296–300. https://doi.org/10.1038/nature12394.

Justice, E.D., Barnum, S.J., Kidd, T., 2017. The WAGR syndrome genePRRG4 is a functional homologue of the commissureless axonguidance gene. PLoS Genet. 13 (8), e1006865. https://doi.org/10.1371/journal.pgen.1006865.

Kamoun, P., Belardinelli, M.C., Chabli, A., Lallouchi, K., Chadefaux-Vekemans, B., 2003. Endogenous hydrogen sulfide overproduction in Down syndrome. Am. J. Med. Genet. A. 116A, 310–311.

Kleschevnikov, A.M., Belichenko, P.V., Villar, A.J., Epstein, C.J., Malenka, R.C., Mobley, W.C., 2004. Hippocampal long-term potentiation suppressed by increased inhibition in the Ts65Dn mouse, a genetic model of Down syndrome. J. Neurosci. 24, 8153–8160.

Larsen, K.B., Laursen, H., Graem, N., Samuelsen, G.B., Bogdanovic, N., Pakkenberg, B., 2008. Reduced cell number in the neocortical part of the human fetal brain in Down syndrome. Ann. Anat. 190 (5), 421–427. https://doi.org/10.1016/j.aanat.2008.05.007.

Lejeune, J., 1975. Réflexions sur la débilité de l' intelligence des enfants trisomiques 21. Pont. Acad. Sci. 3, 1–12.

Letourneau, A., Santoni, F.A., Bonilla, X., Sailani, M.R., Gonzalez, D., Kind, J., Chevalier, C., Thurman, R., et al., 2014. Domains of genome-wide gene expression dysregulation in Down's syndrome. Nature 508 (7496), 345–350. https://doi.org/10.1038/nature13200.

Li, S.S., Zhengdong, Q., Haas, M., Ngo, L., Jeong Heo, Y., Hyo Jung, K., Maria Britto, J., Daniella Cullen, H., Kate Vanyai, H., Tan, S.-S., Chan-Ling, T., Margaret Gunnersen, J., Ik-Tsen Heng, J., 2016. The HSA21 gene *EURL/C21ORF91* controls neurogenesis within the cerebral cortex and is implicated in the pathogenesis of Down Syndrome. Sci. Rep. 6, 29514. https://doi.org/10.1038/srep29514.

Liu, H., Caballero-Florán, R.N., Yang, T., Hull, J.M., Pan, G., Li, R., Veling, M.W., Isom, L.L., Kwan, K.Y., Huang, Z.J., Fuerst, P.G., Jenkins, P.M., Ye, B., 2020. Dscam gene triplication causes neocortical overinhibition in Down syndrome. bioRxiv. https://doi.org/10.1101/2020.01.03.894113.

Lu, J., Lian, G., Zhou, H., Esposito, G., Steardo, L., Delli-Bovi, L.C., Hecht, J.L., Lu, Q.R., et al., 2012. OLIG2 over-expression impairs proliferation of human Down syndrome neural progenitors. Hum. Mol. Genet. 21 (10), 2330–2340. https://doi.org/10.1093/hmg/dds052.

Magistretti, P.J., Allaman, I., 2015. A cellular perspective on brain energy metabolism and functional imaging. Neuron 86 (4), 883–901. https://doi.org/10.1016/j.neuron.2015.03.035.

Mao, R., Wang, X., Spitznagel Jr., E.L., Frelin, L.P., Ting, J.C., Ding, H., KimJW, Ruczinski, I., et al., 2005. Primary and secondary transcriptional effects in the developing human Down syndrome brain and heart. Genome Biol. 6 (13), R107. https://doi.org/10.1186/gb-2005-6-13-r107.

Muller, U.C., Deller, T., Korte, M., 2017. Not just amyloid: physiological functions of the amyloid precursor protein family. Nat. Rev. Neurosci. 18 (5), 281–298. https://doi.org/10.1038/nrn.2017.29.

Murray, A., Letourneau, A., Canzonetta, C., Stathaki, E., Gimelli, S., Sloan-Bena, F., Abrehart, R., Goh, P., et al., 2015. Brief report: isogenic induced pluripotent stem cell lines from an adult with mosaic Down syndrome model accelerated neuronal ageing and neurodegeneration. Stem Cell. 33 (6), 2077–2084. https://doi.org/10.1002/stem.1968.

Nadel, L., 2003. Down's syndrome: a genetic disorder in biobehavioral perspective. Gene Brain Behav. 2, 156–166.

Nguyen, T.L., Arnaud, D., Manousopoulou, A., Loaëc, N., Villiers, B., Pani, G., Karatas, M., Mechling, A.E., Harsan, L.-A., Limanton, E., Pierre Bazureau, J., Carreaux, F., Garbis, S.D., Meijer, L., Hérault, Y., 2018. Correction of cognitive deficits in mouse models of Down syndrome by a pharmacological inhibitor of DYRK1A. Dis. Models & Mech. 11, dmm035634. https://doi.org/10.1242/dmm.035634.

Olmos-Serrano, J.L., Kang, H.J., TylerWA, Silbereis, J.C., Cheng, F., ZhuY, Pletikos, M., Jankovic-Rapan, L., et al., 2016. Down syndrome developmental brain transcriptome reveals defective oligodendrocyte differentiation and myelination. Neuron 89 (6), 1208–1222. https://doi.org/10.1016/j.neuron.2016.01.042. 12.

Pinter, J.D., Eliez, S., Schmitt, J.E., Capone, G.T., Reiss, A.L., 2001. Neuroanatomy of Down's syndrome: a high-resolution MRI study. Am. J. Psychiatr. 158, 1659–1665.

Raz, N., Torres, I.J., Briggs, S.D., Spencer, W.D., Thornton, A.E., Loken, W.J., Gunning, F.M., McQuain, J.D., Driese, N.R., Acker, J.D., 1995. Selective neuroanatomic abnormalities in Down's syndrome and their cognitive correlates: evidence from MRI morphometry. Neurology 45, 356–366.

Sago, H., Carlson, E.J., Smith, D., Kilbridge, J., Rubin, E.M., Mobley, W., Epstein, C.J., Huang, T., 1998. Ts1Cje, a partial trisomy 16 mouse model for Down syndrome, exhibits learning and behavioral abnormalities. Proc. Natl. Acad. Sci. U.S.A. 95, 6256–6261.

Sans, N., Petralia, R.S., Wang, Y.X., Blahos 2nd, J., Hell, J.W., Wenthold, R.J., 2000. A developmental change in NMDA receptor-associated proteins at hippocampal synapses. J. Neurosci. 20 (3), 1260–1271. https://doi.org/10.1523/JNEUROSCI.20-03-01260.2000.

Saran, N.G., Pletcher, M.T., Natale, J.E., Cheng, Y., Reeves, R.H., 2003. Global disruption of the cerebellar transcriptome in a Down syndrome mouse model. Hum. Mol. Genet. 12, 2013–2019.

Schapiro, M.B., Haxby, J.V., Grady, C.L., 1992. Nature of mental retardation and dementia in Down syndrome: study with PET, CT, and neuropsychology. Neurobiol. Aging 13, 723–734.

Shi, Y., Kirwan, P., Smith, J., MacLean, G., Orkin, S.H., Livesey, F.J., 2012. A human stem cell model of early Alzheimer's disease pathology in Down syndrome. Sci. Transl. Med. 4 (124), 124ra129. https://doi.org/10.1126/scitranslmed.3003771.

Sobol, M., Klar, J., Laan, L., Shahsavani, M., Schuster, J., Annerén, G., Konzer, A., Mi, J., Bergquist, J., Nordlund, J., Hoeber, J., Huss, M., Falk, A., Dahl, N., 2019. Transcriptome and proteome profiling of neural induced pluripotent stem cells from individuals with Down syndrome disclose dynamic dysregulations of key pathways and cellular functions. Mol. Neurobiol. 56, 7113–7127. https://doi.org/10.1007/s12035-019-1585-3.

Stachowicz, K., 2018. The role of DSCAM in the regulation of synaptic plasticity: possible involvement in neuropsychiatric disorders. Acta Neurobiol. Exp. 78, 210–219. https://doi.org/10.21307/ane-2018-019.

Sterne, G.R., Kim, J.H., Ye, B., 2015. Dysregulated Dscam levels act through Abelson tyrosine kinase to enlarge presynaptic arbors. Elife 4, e05196.

Sullivan, K.D., Evans, D., Pandey, A., et al., 2017. Trisomy 21 causes changes in the circulating proteome indicative of chronic autoinflammation. Sci. Rep. 7, 14818. https://doi.org/10.1038/s41598-017-13858-3.

Takashima, S., Becker, L.E., Armstrong, D.L., Chan, F., 1981. Abnormal neuronal development in the visual cortex of the human fetus and infant with Down's syndrome. A quantitative and qualitative Golgi study. Brain Res. 225, 1–21.

Takahashi, H., et al., 2012. Selective control of inhibitory synapse development by Slitrk3-PTPδ trans-synaptic interaction. Nat. Neurosci. 15 (389–98), S1–S2.

Tomuschat, C., Puri, P., 2015. RET gene is a major risk factor for Hirschsprung's disease: a meta-analysis. Pediatr. Surg. Int. 31, 701–710.

Valbuena, S., García, A., Mazier, W., Paternain, A.V., Lerma, J., 2019. Unbalanced dendritic inhibition of CA1 neurons drives spatial-memory deficits in the Ts2Cje Down syndrome model. Nat. Commun. 10, 4991. https://doi.org/10.1038/s41467-019-13004-9.

Villar, A.J., Belichenko, P.V., Gillespie, A.M., Kozy, H.M., Mobley, W.C., Epstein, C.J., 2005. Identification and characterization of a new Down syndrome model, Ts [Rb(12.1716)]2Cje, resulting from a spontaneous Robertsonian fusion between T(171)65Dn and mouse chromosome 12. Mamm. Genome 16, 79–90.

Vuksic, M., Petanjek, Z., Rasin, M., Kostovic, I., 2002. Perinatal growth of prefrontal layer III pyramids in Down syndrome. Pediatr. Neurol. 27, 36–38.

Wang, J., Zugates, C.T., Liang, I.H., Lee, C.H., Lee, T., 2002. Drosophila Dscam is required for divergent segregation of sister branches and suppresses ectopic bifurcation of axons. Neuron 33, 557–559.

Wallace, D.C., Chalkia, D., 2013. Mitochondrial DNA genetics and the heteroplasmy conundrum in evolution and disease. Cold Spring Harbor Perspect. Biol. 5 (11), a021220.

Weick, J.P., Held, D.L., Bonadurer 3rd, G.F., Doers, M.E., Liu, Y., MaguireC, C.A., Knackert, J.A., et al., 2013. Deficits in human trisomy 21iPSCs and neurons. Proc. Natl. Acad.

Sci. U. S. A. 110 (24), 9962–9967. https://doi.org/10.1073/pnas.1216575110.

Weitzdoerfer, R., Dierssen, M., Fountoulakis, M., Lubec, G., 2001. Fetal life in Down Syndrome starts with normal neuronal density but impaired dendritic spines and synaptosomal structure. In: Lubec, G. (Ed.), Protein Expression in Down Syndrome Brain © Springer-Verlag/Wien.

Wiseman, F.K., Al-Janabi, T., Hardy, J., Karmiloff-Smith, A., Nizetic, D., Tybulewicz, V.L., Fisher, E.M., Strydom, A., 2015. A genetic cause of Alzheimer disease: mechanistic insights from Down syndrome. Nat. Rev. Neurosci. 16 (9), 564–574. https://doi.org/10.1038/nrn3983.

Wisniewski, K.E., Kida, E., 1994. Abnormal neurogenesis and synaptogenesis in Down syndrome brain. Dev. Brain Dysfunct. 7, 289–301.

Yang, E.J., Ahn, Y.S., Chung, K.C., 2001. Protein kinase Dyrk1 activates cAMP response element-binding protein during neuronal differentiation in hippocampal progenitor cells. J. Biol. Chem. 276, 39819–39824.

CHAPTER

11

Imbalance in glutamatergic and GABAergic transmission in Down syndrome and therapeutic targets

1. Introduction

Brains of individuals with Down syndrome (DS) are manifested with reduced volume and lower number of neurons and abnormal cellular morphology with decreased neuronal density in diverse regions of the brain, prominently in the cortex, hippocampus, and cerebellum (illustrated in Chapter 10). These alterations arise during early stage of brain development, due likely to defective proliferation of neuronal precursor cells during gestation (Contestabile et al., 2017). The anomalies extend through adulthood, resulting in genesis of GABAergic neurons, variation of the inhibitory drive, and modifications in the control of excitability in neural-network. As a result of such alterations, the DS individuals experience delayed cognitive progress in infancy and childhood, leading to mild-to-moderate degree of mental retardation (MR) with an intelligence quotient (IQ) ranging from 30 to 70 (Vicari et al., 2013). The situation is aggravated during adulthood by the progressive loss of cognitive abilities and the development of Alzheimer disease (AD) by the 4th decade of life. Alongside cognitive impairment, DS population exhibits other neurological and psychiatric features, including increased frequency of anxiety, clinically relevant sleep disorder, hyperactivity or movement disorders, and increased incidence of epileptic seizure through early life to adult age (reviewed in Contestabile et al., 2017). The possible mechanistic etiology of the cognitive impairment in DS has largely been studied on mouse models of DS, which indicated altered signaling of the neurotransmitter γ-aminobutyric acid (GABA) as one of the main determinants in reducing cognitive performance and memory functions. Noticeably, GABAergic dysfunctions perturb synaptic plasticity through alteration of optimal excitatory/inhibitory (E/I) synaptic balance and finally impair learning and memory (L/M) in DS and AD (Kleschevnikov et al., 2012b; Deidda et al., 2014; Li et al., 2016; Contestabile et al., 2017). GABA is produced by the GABAergic neurons in mammalian brain during the formation of GABAergic synapses between GABAergic neurons and their targets. The level of inhibitory action is measured by synaptic GABA release, which can be modulated by the synaptic E/I signals received by the GABAergic neurons.

The language of neuronal function and communication is translated through synaptic

Genetics and Neurobiology of Down Syndrome
https://doi.org/10.1016/B978-0-323-90456-8.00007-7

plasticity that depends on the structure of dendritic spines and the timing and strength of the pre- and postsynaptic dialogue (Cramer and Galdzicki, 2012). Such communication process was described as "fire together, wire together" (Hebb, 1949). Any disruption of synapse strengthening, including the formation of new dendritic spines, modification of the size or shape and levels of neurotransmitter release of the existing synapses perturbs the synaptic plasticity (Cramer and Galdzicki, 2012). The synapse dialogue is represented as long-term potentiation (LTP) and/or long-term depression (LTD), which acts as a learning and memory sensor and demonstrates the long-lasting synaptic efficacy and predicts synaptic plasticity. The cell adhesion molecules (CAMs), particularly in DS (DSCAM), together with other cell surface recognition molecules, are involved in the process of synapse strengthening and transmission of synaptic dialogue (Hattori et al., 2008), which are further regulated by glutamate (Glu) activation (Li et al., 2009). Unquestionably, DSCAM plays critical roles in neural circuit formation and synaptogenesis (Pérez-Núñez et al., 2016). However, understanding the DSCAM protein/gene is only an "observer" or an active player in synaptic plasticity, but its subsequent role in the pathomechanism of neurologic disorders has not been unzipped. Therefore, elucidation of the synaptic plasticity, dendritic branching, and spine density in the hippocampus and cortex may extract insights of learning potential in DS and guide to identify potential therapeutic targets.

DSCAM appeared essential for precise wiring during synapse formation in the learning process and for stabilization of the presynaptic structures at postsynaptic sites (Li et al., 2009; Stachowicz, 2018). DSCAM is overexpressed in DS individuals and was hypothesized to be the lead factor of cognitive impairment (Deidda et al., 2015b). Moreover, the neurotransmitter GABA appears as one of the key contributors to impairment in synaptic plasticity leading to cognitive and memory impairment via alteration in optimal E/I synaptic balance. GABA is produced from L-glutamate by L-glutamic acid decarboxylase (GAD). GABA plays an inhibitory function in the adult brain, whereas it acts as a fast excitatory neurotransmitter in the immature brain at the same time (Obrietan et al., 2002). Alterations in synthesis or release of GABA displayed a significant effect on normal brain function. A GABAergic imbalance was documented with GABAergic interneurons in the hippocampus and cortex of the Ts65Dn mice of DS models (Contestabile et al., 2017). Therapeutic efficacy of drugs on the molecular interactions and GABA-actions appeared extremely complex due to the structural heterogeneity of the receptors, existence of multiple allosteric binding sites, and a wide range of ligands that can bind to them (Ghit et al., 2021).

Association of the DSCAM-mediated signaling was reported with changes in glutamate (Glu) receptor. In particular, glutamatergic pathways and the glutamate (excitatory)/GABA (inhibitory) balance contribute to neuropsychotic illnesses in DS and many other disorders. In the prefrontal cortex of trisomic Ts65Dn mice, the dendritic synaptic inhibitory loop was formed by somatostatin-positive Martinotti cells (MCs) and pyramidal neurons (PNs) were strongly enhanced, with no alteration in their excitability. In contrast, perisomatic inhibition from parvalbumin-positive (PV) interneurons was unaltered, but PV cells of these DS mice lost their classical fast-spiking phenotype and exhibited increased excitability. These microcircuit alterations resulted in reduced pyramidal-neuron firing and increased phase locking to cognitive-relevant network oscillations in vivo, defining the important synaptic and circuit mechanisms underlying cognitive dysfunctions in DS (de San Martin et al., 2020). Hence, the glutamate/GABA balance is targeted to constitute pharmacological treatments, such as riluzole showed modulation of the glutamate and GABA balance (Vingerhoets et al., 2020). Essentially, the GABA interneurons are the main inhibitory neurons in the central nervous system (CNS), and the activity of the dysfunctional GABA interneuron can disrupt the E/I balance in the cortex, which underlie the pathophysiological mechanism of cognitive dysfunction. Thus, GABA-mediated synaptic inhibition and associated glutamatergic alterations can bring in the synaptic changes in DS. Furthermore, GABA was demonstrated to depolarize and excite targeted cells through an outwardly directed flux of chloride at late embryonic/early postnatal stages. Such depolarizing action of GABA and associated calcium influx regulate a variety of developmental processes from cell migration and differentiation to synapse formation (Pizzarelli and Cherubini, 2011). Downregulation of the hippocampal glutamate, dysregulation of N-methyl-D-aspartate receptors (NMDA-Rs), and both pre- and postsynaptic dysfunction at glutamatergic synapses exhibited behavioral impairments in DS mouse model (Kaur et al., 2014). Importantly, glutamatergic aberrations were suggested as the distinctive early pathology of cognitive disabilities in DS models, indicating a key role played by glutamatergic dysfunction on hippocampal pathogenesis of AD and DS. Significant glutamatergic dysregulation and neurotransmission deficits were observed in dementic or nondementic DS subjects and experimental models (Tan et al., 2014).

GABA has three receptors: GABA-A, GABA-B, and GABA-C. GABA-A receptors (GABA-ARs) are heteropentamers formed from a selection of 19 subunits: six α (alpha1-6), three β (beta1-3), three γ (gamma-3), three ρ (rho-3), and one each of the δ (delta), ε (epsilon), π (pi), and θ (theta), which result in the production of a considerable number of receptor isoforms. Each isoform exhibits distinct pharmacological and physiological properties. However, the majority of GABA-ARs are composed of two α subunits, two β subunits, and one γ subunit arranged as $\gamma2\beta2\alpha1\beta2\alpha1$ in counterclockwise manner around the center. The mature receptor has a central chloride ion channel gated by GABA neurotransmitter and modulated by a variety of different drugs. Therefore, understanding the relationship between GABA-AR deficits and CNS disorders thus has a significant impact on the discovery of disease pathogenesis and drug development. Blockade of GABA-A receptors in the hippocampus of DS mice may reverse LTP impairment and cognitive efficiency (Cramer and Galdzicki, 2012). Immunohistochemical study of the dentate gyrus in Ts65Dn mice showed a shift of GABAergic synapses to the spine necks, which may contribute to impairment of synaptic integration (Contestabile et al., 2017).

Alteration in GABAergic signaling system has been documented in several neurological and neurodevelopmental disorders (Braat and Kooy, 2015). The identification of possible mechanisms leading to cognitive deficits in DS has been largely conducted in DS mouse models. Interestingly, compelling studies in these animals indicate that altered signaling of excitatory and inhibitory mechanism of synaptogenesis is linked to human DS. In this chapter, alteration of GABAergic signaling in DS has been discussed with the current status of preclinical and clinical trials with GABAergic drugs for DS. Importantly, altered GABAergic signaling caused by trisomic HSA21 has been demonstrated in myriads of studies to be linked to the cognitive deficits in individuals with DS (Contestabile et al., 2017).

2. Neurotransmission in DS brain

Of the two principal inhibitory neurotransmitters (GABA and glycine), GABA has been demonstrated as the major inhibitory neurotransmitter in the CNS of the healthy adult vertebrates. GABA plays crucial roles during development of CNS, starting from the control of cell proliferation to the formation of synapses (reviewed in Lauder et al., 1998; Sernagor et al., 2010; Lujan et al., 2005). The two neurotransmitters are either released separately or co-released from presynaptic terminals of the inhibitory neurons. In the mature CNS, GABA acts as a major inhibitory neurotransmitter and binds to GABA receptors, opens a chloride channel, and reduces neuronal excitability through the ligand-gated GABA-A receptor channels and G protein coupled with GABA-B receptors. Both GABA-A and GABA-B receptors contribute to the inhibitory postsynaptic transmission all over the nervous system. The inhibitory action mediated by the GABA-A receptors is associated with hyperpolarization owing to an enhancement in conductance to chloride ions (Stein and Nicoll, 2003). Alterations in chloride (Cl^-) homeostasis may induce epileptic activities in tissues from patients with mesial temporal lobe epilepsy associated with hippocampal sclerosis (Huberfeld et al., 2007). It was postulated that GABA can have a depolarizing and excitatory action in mature cortical neurons leading to alteration in synaptic plasticity, potentially triggering learning and memory deficits in adults, and the excitatory activation of the GABA-A receptor was described in the neonatal neurons and spinal cord neurons.

2.1 Generation and migration of GABAergic neurons

The early stage of generation and migration of GABAergic neurons and their specific subtypes in the correct locations and in appropriate numbers and types would decide the expression of the final neural circuits, such as increase in the striatum and reduction in the globus pallidus occur in the number of paravalbumin (PV)-positive neurons of Tourette syndrome (Kataoka et al., 2010). GABAergic neurons generally arise from the ventral part of the telecephalon (subpallium); however, there are few exceptions in humans and nonhuman primates. The telecephalon or subpallium is subdivided into the lateral (LGE), medial (MGE), and caudal (CDE) ganglionic eminences, where GABAergic neurons in the olfactory bulb and striatum are derived from the progenitors from LGE, and cortical GABAergic neurons (often called cortical interneurons) originate from MGEs and CGEs (Xu et al., 2004).

The telecephalon has structurally less defined areas, including the anterior entopeduncula and preoptic area (POA). Different types of neural circuits containing cells called interneurons generate inhibitory activity in the brain and reduce the activity of the neurons. In Ts65dn mouse model of DS, interneurons called Martinotti cells help the brain to combine inputs from different sources, and interneurons called parvalbumin-positive basket cells help different areas of the brain to synchronize their activity, which consequently help in exchange of information in those areas (de San Martin et al., 2020).

The transcription factors for the mammalian achaete-scute homolog 1 (MASHI) and distalles homeobox (DLX) family members determine the fate of the GABAergic neurons in the subpallium (Fode et al., 2000). The Pax6 (paired box gene 6) produces the olfactory bulb interneurons in the dorsal LGE, and GSH2 (glutathione synthase 2) and low levels of Pax6 controls mainly striatal neurons, including most of the striatal projection neurons in the ventral LGE (Yun et al., 2001). The Nkx2.1 (NK2 homeobox 1) factor controls the origin of the cortical GABAergic neurons from MGE (Sussel et al., 1999). Expression of molecular markers for GABAergic progenitors in CGE has yet to be identified.

The GABAergic neurons are sub-typed by the diversified features, which includes morphology (basket, chandelier, bipolar, double bouquet cells, etc.), physiology (fast-spiking, regular firing, bursting, stuttering, etc.) and the molecular markers (Markram et al., 2004). The GABAergic subtypes are primarily determined at the progenitor stage and long before their functional integration into the neural circuits, which are revealed through the gene profiling of the cortical interneuron precursors (Batista-Brito et al., 2008). The specification of subtypes is genetically regulated by both spatial and temporal factors, which demonstrates that MGE neurons preferentially express parvalbumin (PV) and somatostatin (SST)/calbindin (CB), while CGE-derived neurons release calretinin (CR)/vasoactive intestinal peptide (VIP) or neuropeptide Y (NPY) (Batista-Brito et al., 2008). Temporally, SST neurons are developed before PV neurons, followed by VIP neurons (Miyoshi et al., 2007), and that indicates the neurogenesis in MGE (producing SST and PV neurons) occurs prior to CGE (producing VIP neurons) (Butt et al., 2005).

2.2 Synthesis and shunting of GABA transmission

GABA is synthesized from the main precursor glucose, and other precursors as amino acids and pyruvate in mammalian CNS, in a close-loop system that produces and conserves GABA (Fig. 11.1) (Ghit et al., 2021). The biomechanism of GABA shunt includes: 1. Production of L-glutamic acid through transamination of α-ketoglutarate produced from the metabolism of glucose in the Krebs cycle by GABA-α ketoglutarate transaminase (GABA-T); 2. production of GABA through decarboxylation of the L-glutamic acid by glutamic acid decarboxylase (GAD), which uses vitamin B6 (pyridoxine) as a cofactor and is only expressed in cells that use GABA as a neurotransmitter; 3. GABA is metabolized to succinic semialdehyde (SSA) by GABA-T when α-ketoglutarate is present, and it accepts the amino group extracted from GABA and reforms glutamic acid; 4. SSA is oxidized to succinate by the succinic semialdehyde dehydrogenase (SSADH) enzyme; and 5. succinate enters the Krebs cycle and completes the GABA-shunt loop.

The newly synthesized GABA is packaged into synaptic vesicles by a vesicular transporter, with subsequent transport of the vesicles into plasma membrane of cells (Südhof, 2013; Ghit et al., 2021). Depolarization of the presynaptic neuron discharges GABA to the synaptic cleft and diffuses toward postsynaptic receptors, which can then bind to postsynaptic GABA receptors and modulate ion channels, hyperpolarize the cell, and prevent action potential transmission. Despite binding to GABA-AR or GABA-BR, GABA acts as an inhibitor. GABA increases anionic conductance of chloride channel into the cell, where it binds to an ionotropic GABA-AR (Ghit et al., 2021). Eventually, the influx of increased chloride ion causes membrane hyperpolarization leading to decrease in neuronal excitability. GABA can then enter into three pathways: 1. GABA can be degraded

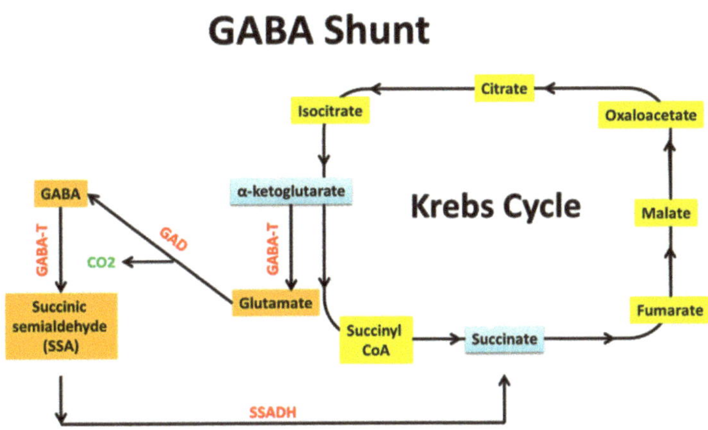

FIGURE 11.1 Schematic illustration of gamma aminobutyric acid (GABA) shunt. Transamination of α-ketoglutarate by GABA-α ketoglutarate transaminase (GABA-T) produce glutamate which is decarboxylated to GABA by glutamic acid decarboxylase (GAD). GABA-T metabolizes GABA to succinic semialdehyde, which is oxidized to succinate by succinic semialdehyde dehydrogenase (SSADH). Then, succinate can enter the Krebs cycle and complete the loop. *Reprinted from: Ghit et al. (2021) https://doi.org/10.1186/s43141-021-00224-0 under a Creative Commons Attribution 4.0 International License.*

extracellularly by GABA-T into SSA, which then enters the citric acid cycle; 2. GABA can be reuptaken to nerve terminals for utilization again; and 3. GABA can be reuptaken to the glial cell where it is metabolized to SSA by GABA-T or to glutamine which is then transported to neurons and converted to glutamate by glutaminase and re-enters the GABA shunt. In absence of GAD in glia, GABA cannot be synthesized again from glutamate (Ghit et al., 2021).

2.3 Molecular ontogeny of glutamic acid decarboxylase

Synthesis of GABA occurs by activation of the amino acid glutamate by the GAD, which has two isoforms having molecular weights of 65-kDa and 67-kDa produced from two different genes. GAD65 gene (GAD2) is located on HSA10 (10p11.23) and on mouse chromosome 2 (2 9.0 cM), whereas GAD67 gene (GAD1) is mapped onto 2q31 and in mouse chromosome 2 (2 43.0 cM) (Bu et al., 1992). The 67-kDa GAD isoforms diffusely distributed in the cellular cytoplasm and the 65-kDa GAD isoform gets attached to the synaptic vesicles (Hsu et al., 2000). The GAD67 gene produces 25-kDA leader (GAD25) and the enzymatically active protein GAD44 during development of CNS in mouse and rat embryos (reviewed in Varju et al., 2001). They are abundant in proliferating progenitors (Varju et al., 2001) and downregulated during neuronal differentiation in conjunction with an upregulation of GAD67 expression (Katarova et al., 2000). The diversity in GABAergic signaling, owing to several pre- and postsynaptic factors, is the target of many compelling drugs that are being widely considered in current clinical use (Mohler, 2011).

The molecular markers used for examination of the autopsy tissues of patients with neurodevelopmental disorders, commonly include calcium-binding proteins such as parvalbumin (PV), calretinin (CR), and calbindin (CB) and neuropeptides, such as somatostatin (SST), vasoactive intestinal peptide (VIP), neuropeptide Y (NPY), and cholecystokinin (CCK) (reviewed in Ramamoorthi and Lin, 2011). Boutons from PV neuron subtypes are distinguishable by differences in protein levels of the GABA-synthesizing enzymes GAD65 and GAD67. CB and CR neurons and their respective subtypes, SST and VIP, gave rise to three subpopulations of GAD-containing boutons: GAD65+, GAD67+, and GAD65/GAD67+ in macaque monkey. The result suggested that the GAD65+ and GAD65/GAD67+ boutons arise from neuron-expressing GAD65 and GAD67 mRNA, while the GAD67+ boutons arise from GAD67-only mRNA-expressing neurons. At the transcript level, the two unique subtypes of CB and CR neurons indicated distinct subpopulations of CB/GAD+ and CR/GAD + boutons, where GABA synthesis from CB/GAD67+ and CR/GAD67+ neurons was thought to be more vulnerable to disease-associated deficits in GAD67 expression, as observed in schizophrenia (Rocco et al., 2016). Among the PV neurons, basket cells give rise to GAD65/GAD67+ boutons, whereas chandelier cells give rise to GAD67+ boutons (Glausier et al., 2014); both subtypes display similar short-term depression across species and cortical regions.

2.4 Migration and settlement of GABAergic neurons

Migration and settlement of the GABAergic neurons displayed that the newly born cortical neurons first migrate tangentially away from the ganglionic eminence and then migrate through the two migratory streams into the regions above and below the developing cortical plate (Lavdas et al., 1999). Following that, the cortical GABAergic neurons follow various radial migration modes to inhabit into the specific cortical layers in an inside-out manner according to their birth order, where the older cells settle in the deeper layers and younger cells occupy the superficial layers (Anderson et al., 2002).

The peripheral or lateral migration of GABAergic neurons are regulated by many factors, including the sets of attractant and repellent guidance molecules (similar to those used by glutamatergic neurons), growth factors and their receptors, and neurotransmitters (Hernandez-Miranda et al., 2010). Dysregulation of many of these factors are implicated in neurological disorders, such as altered migration of GABAergic neurons may result from defects in Neuregulin 1 (NRG1)-ErbB4 (receptor of NRG1) signaling leading to pathology of a subset of patients with schizophrenia (Corfas et al., 2004). NRG1, secreted in the cortical regions, may serve as a potent chemoattractant and guide the ErbB4-expressing GABAergic neurons to drift tangentially into the cortex (Flames et al., 2004). The fate and direction of migration of the GABAergic neurons are controlled by the expression of a number of transcription factors. Downregulation of the semaphoring receptor neuropilin 2 (NRP2) is controlled by Nkx2.1. The repellent ligands of NRP2, SEMA3A, and SEMA3F are expressed at striatum; however, neurons with low levels of Nkx2.1 migrate to the cortex instead of striatum and high levels of Nkx2.1 migrate to the striatum (Nobrega-Pereira et al., 2008).

The tangential migration of GABAergic neurons distributes them longitudinally around the cortical plate, whereas the radial migration sorts them into the correct cortical layers, where they are connected to glutamatergic projection neurons within each layer that can directly control the E/I balance of the cortical circuits. However, the migration mechanisms of different subtypes of GABAergic neurons can cause alterations in distribution of certain subtypes in development of neurodevelopmental disorders (Kalanithi et al., 2005). Involvement of the signaling pathway of the chemokine CXCL12 and its receptors CXCR4 and CXCR7 is associated with the timing of switching from tangential to radial migration of GABAergic neurons (Lopez-Bendito et al., 2008); Wang et al. (2011). Disturbances in cortical radial migration of glutamatergic neurons result in lissencephaly causing mental retardation and seizures in humans (Olson and Walsh, 2002). However, involvement of radial migration of GABAergic neurons in such condition could exists, since gross anatomical changes, similar to those seen in lissencephaly, are implicated by abnormal laminar positioning of GABAergic neurons, which could also have a significant effect on the function of cortical circuits. Altogether, involvement of the laminar distribution of GABAergic neurons in neurodevelopmental disorders, which do not involve gross anatomical abnormalities, may extract new insights into their etiology.

3. The GABAergic receptors and ontogenic changes

GABA acts through Cl^- permeable ionotropic bicuculline-sensitive type A (GABA-A) receptors (GABA-AR) and metabotropic baclofen-sensitive GABA-B receptors (GABA-BR), which are coupled to K^+ and Ca^{2+} channels through G-proteins in neuronal membranes. Another Cl^- permeable bicuculline- and baclofen-insensitive receptor subtype, GABA-C receptors (GABA-CR), is also activated by GABA in retina and lower level in other CNS areas (Rozzo et al., 2002). All GABA receptors are seemingly located pre- and postsynaptically, such as GABA-BRs are presynaptically located, and GABA-AR and GABA-CR are postsynaptically located on the cell membrane (Kullmann et al., 2005). GABA-ARs are of particular interest because of the possibility of marked postsynaptic alteration of GABAergic signaling and modification of expression in their subunits. In addition, association of the subtypes of receptors has been described with significant physiological and specific cognitive outcomes (Rudolph and Mohler,

2013). GABA-CRs are now considered as part of GABA-AR isoforms and entirely made up of rho (ρ) subunits (Olsen and Sieghart, 2008).

3.1 GABA-A receptors

GABA-ARs are composed of five subunits of oligomeric proteins, and several variants of these subunits are reported in mammals, including six α (α1−6), three β (β1−3), three γ (γ1−3), three ρ (ρ1−3), and one each of δ, ε, π, and θ (Fig. 11.2A) (Ghit et al., 2021), which render variable pharmacological and kinetic properties to the receptors (Fritschy and Panzanelli, 2014). The alternative splicing of several genes underlie the diversity of GABA-ARs. The genes of GABA-AR subunits are arranges as: α2, α4, β1, and γ1 on chromosome four; α1, α6, β2, and γ2 on chromosome five; α5, β3, and γ3 on chromosome 15; and α3, ε, and θ on the X chromosome. Notably, functional and pharmacological properties are implicated on the composition and arrangement of GABA-ARs (reviewed in Ghit et al., 2021). Although the amino acid sequence, level of expression, and localization in brain tissues of each subunit have been investigated, their interaction and formation of many different isoforms are yet to be elucidated. Remarkably, a single cell can possess variety of isoforms. Nevertheless, an adult isoform is apparently composed of stoichiometric 2α, 2β, and 1γ subunits and arranged in γ2β2α1β2α1 counterclockwise pattern around a central pore (CR) (Fig. 11.2B) (Ghit et al., 2021). For example, the adult brain, predominantly containing the α1−3, β2−3, and γ2 isoforms, constitutes the synaptic GABA-ARs and leads to synaptic neurotransmission mediating phasic inhibition in the brain. Notably, these subunits release fast kinetics with speedy onset and desensitization and generate phasic currents with spatial and temporal accuracy following GABA release from presynaptic vesicles. On the other hand, substitution of α5-for α1-3 or δ-containing subunits for a γ2 of GABA-ARs appears extrasynaptic, displays higher affinity for GABA and desensitizes GABA-ARs slowly and less extensively than postsynaptic GABA-ARs containing the β and/or the γ-subunits (Storustovu and Ebert, 2006). Remarkably, GABA-ARs containing the α5 subunit display a decrease in desensitization kinetics compared to that of the receptors containing other α subunits (Gingrich et al., 1995). Thereby, the α5- or δ-containing receptors are capable to detect ambient GABA released by the synaptic cleft and cause long-lasting tonic inhibition mediated by the extrasynaptic receptors (Luo et al., 2013). GABA-ARs containing α5 or δ subunit cannot anchor to postsynaptic scaffold protein complex and thus, most likely, cannot accumulate at postsynaptic site (Crestani et al., 2002). Thus, the subunit composition is extensively targeted for allosteric modulators like benzodiazepines or neurosteroids in pharmaceutical developments (Luscher et al., 2011).

In general, GABA-ARs belong to the cystein-loop receptor family and a mature GABA-AR subunit is composed of ~450 amino acid residues containing N-terminal, a large hydrophilic extracellular domain (ECD), and four hydrophobic transmembrane domains (TMD: TM1−TM4). The TM2 domain forms the pore of the anionic chloride channel, and intracellular domain (ICD) between TM3 and TM4 and that is the site of protein interactions and posttranslational modifications that defines receptor activity (Fig. 11.2C) (Ghit et al., 2021; Jacob et al., 2008). The agonist binding site has mainly α subunits, while γ subunits are involved in linking GABA-ARs to the postsynaptic cytoskeleton. Moreover, combination of subunits varies depending on the synaptic and extrasynaptic location of GABA-ARs, and the isoform expression of the subunits and composition of receptors change with subcellular localization and functional properties, brain area, and development, assuring that receptor properties are set to fulfill the specific neuronal needs (Luscher et al. 2011).

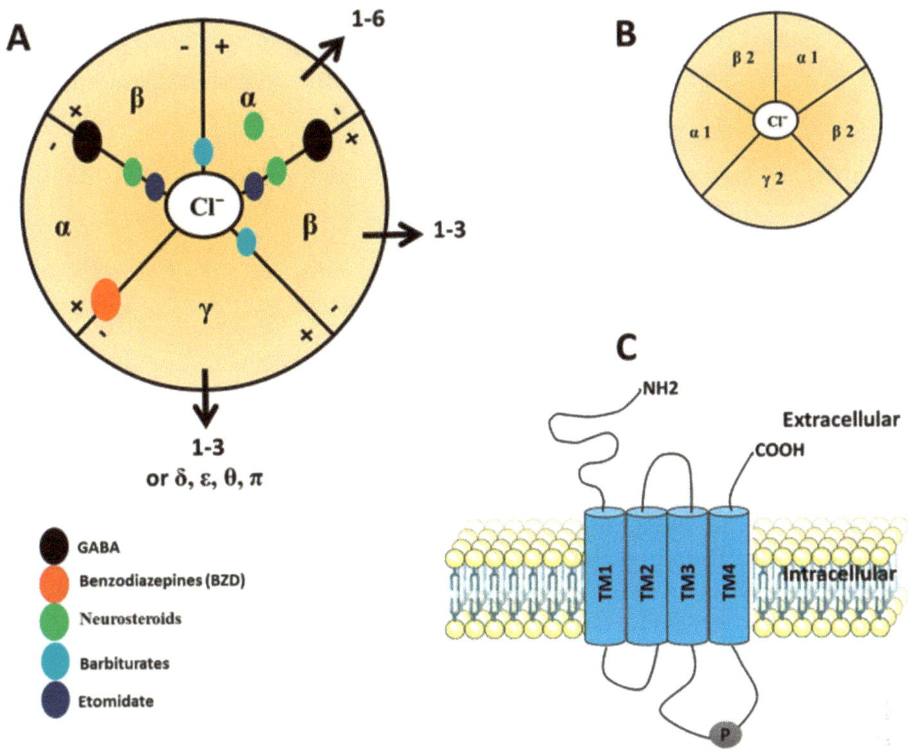

FIGURE 11.2 Schematic representation of gamma aminobutyric acid (GABA)$_A$ receptor structure. (A) GABA$_A$ receptors are heteropentamers that form a chloride-ion-permeable channel. They are formed by 19 subunits: α1–6, β1–3, γ1–3, δ, ε, θ, π, and ρ1–3. The GABA binding sites are located at the junction of β+/α–, whereas benzodiazepines (BZDs) are located at α+/γ– interface. Anesthetics are located at different sites where barbiturates bind to α+/β– and γ+/β– interfaces, while etomidate binds to β+/α– interface. The binding site of the neurosteroids is located at α subunit as well as the β+/α– interface. (B) The most popular GABA$_A$R isoform is composed of α1, β2, and γ2 subunits arranged γ2β2α1β2α1 counterclockwise around the central pore. (C) The mature subunit contains a large hydrophilic extracellular N-terminal, four hydrophobic transmembrane domains (TMD: TM1–TM4), and a small extracellular C terminus. TM1 and TM2 are connected by a short intracellular loop, while a short extracellular loop connects TM2 and TM3. Besides, TM3 and TM4 are connected by a lengthy intracellular loop that can be phosphorylated. *Reprinted from Ghit et al. (2021); https://doi.org/10.1186/s43141-021-00224-0 under a Creative Commons Attribution 4.0 International License.*

The neurotransmitter GABA and psychotropic drugs, such as benzodiazepines (BZDs), bind to the N-terminal at α-β and α-γ interfaces, respectively, whereas neurosteroids and anesthetics, such as barbiturates, are located within the TMD of α and β subunits (Fig. 11.2A) (Ghit et al., 2021).

In the CNS, variable expression of the GABA-AR subunits indicates α6 expression only in the cerebellum while ρ expression mainly, but not exclusively, in the retina. The postsynaptic location of GABA-ARs in the adult brain is mainly composed of the α1–3, β1–3, and γ2 where a GABA neurotransmitter can bind with and open chloride channels and thus, increases the anion conductance for a short period, leading to hyperpolarization of a depolarized membrane. This type of GABA inhibition has been demonstrated as phasic inhibition. While a tonic inhibition has been demonstrated when the extrasynaptic location of GABA-ARs is composed of the α4–6, β2/3, and δ subunits and the low GABA concentration can open these receptors for a longer period (reviewed in Ghit et al., 2021). The most prominent isoforms mediating tonic inhibition in extrasynaptic GABA-ARs are α4βδ, α6βδ, and α1βδ receptors in the forebrain, cerebellum, and hippocampus, respectively. Notably, the concentration of the α2, α3, and β3 subunit-containing receptors is ~100 times higher at synapses than in the extrasynaptic membrane. Interestingly, all γ2-containing receptors are not concentrated postsynaptic, such as α5βγ2 receptors are found at extrasynaptic sites and involved in tonic inhibition. Besides phasic and tonic inhibition, the γ2 subunit is necessary for postsynaptic clustering of the GABA-ARs, and it is substituted by the γ3 subunit for development of the postnatal brain. Moreover, apart from the CNS, GABA-ARs have been located in other organs, such as different types of immune cells, liver cells, pancreatic islet β-cells, and airway smooth muscle cells (reviewed in Ghit et al., 2021). However, regulation of assembly and distribution of GABA-ARs isoforms are largely unknown.

Notably, little is known about the changes in expression of the GABA-AR subunits during development of the spinal cord motoneurons (MNs). A synchronized peak of mRNAs for α2, α3, β2, β3, γ2, and γ3 subunits is demonstrated at neonatal stages in rat, while the cervical spinal cord of adult rats revealed mRNAs for α1, α4, α5, β1-2, γ1, and γ3 subunits in relatively few cells scattered in the gray matter, while the mature MNs displayed α2β3γ2 transcripts (Ma et al., 1994). Remarkably, mRNA level of the α3 detected at early six neural plasticity developmental stage in the lateral motor column, reduces around birth and disappears in the adult (Ma et al., 1994). Therefore, prenatal to postnatal development of MNs did not display apparent switch for expression of GABA subunits. Since brainstem and spinal motoneurons are largely targeted by GABAergic interneurons, an optimum level of maturation is essential for their GABA receptors and GABA innervations. Several studies have documented the depolarizing GABA effects on the maturation of neurite outgrowth, promoting both E/I synaptogenesis and in controlling the switch for depolarizing to hyperpolarizing (Akerman and Cline, 2007; Represa and Ben-Ari, 2005; Sernagor et al., 2010).

The expression of the different GABA receptors was not significantly different in the hippocampus of Ts65Dn mice of DS model compared to wild-type mice for α1, α2, α3, α5, and γ2 subunits of GABA-AR (Belichenko et al., 2009a; Best et al., 2012). However, age-related changes in β2/3 subunits of GABA-AR have been described, as reduced immunostaining was observed in the dentate gyrus (DG) in 3-month-old Ts65Dn mice, and that disappears at later age (Belichenko et al., 2009a). Similar expression of GABA-ARs containing the α5 subunit was demonstrated in Ts65Dn mice through in vivo binding of the selective α5 subtype radioligand [3H]RO0154513 (Martínez-Cué et al., 2013). Decrease in γ-oscillation power was reported in the hippocampal slices and layer four of somatosensory cortex slices of Ts65Dn strain (Hanson et al., 2013; Cramer et al., 2015). Nevertheless, GABA confers depolarizing effects during development of the immature neuronal cells, whereas most of the mature CNS regions experience its hyperpolarizing effects (Ben-Ari et al., 2007). Such GABA-mediated depolarizing effects lead to Ca^{2+} influx and generate spontaneous electrical activities in the immature structures of the CNS, in association with conventional excitatory effect of glutamate and other classical neurotransmitters such as acetylcholine.

3.2 GABA-B receptors

GABA-BRs, the $G_{i/o}$-protein-coupled receptors (GPCRs), are mostly localized extrasynaptically and activated by ambient GABA spilled out of synapses owing to their high affinity for GABA despite of their distance of location (Contestabile et al., 2017). In particular, The $G\alpha_{i/o}$ subunit of GABA-BR can lead to inhibition of adenylate cyclase resulting in lowering of cAMP levels and PKA pathway activity (Padgett and Slesinger, 2010), while the $G\beta\gamma_{i/o}$ subunit inhibits the voltage-gated calcium channels leading to opening of the G-protein coupled inwardly rectifying potassium channels (GIRK/Kir3) and formation of tetramers by different compositions of GIRK1–4 subunits (Koyrakh et al., 2005). The inhibition of the voltage-gated calcium channels leads to decrease in vesicle release by presynaptic GABA-BRs following a calcium-independent mechanism (Rost et al., 2011), whereas GABA-BR subunits of the dendrite and spine necks regulate neuronal excitability by coupling to the GIRK channels (Koyrakh et al., 2005). The GIRK2-containing homo- and heterotetramers contribute to GABA-BR-activated potassium currents (Koyrakh et al., 2005). Notably, the protein complexes formed with GABA-BR subunits in heterologous systems can virtually be activated by all GIRK subunits (Ciruela et al., 2010). The GABA-BRs inhibit neuronal excitability by shunting excitatory currents, hyperpolarizing the membrane with slow inhibitory postsynaptic potentials (PSPs), and contributing in maintaining resting potential (VREST) through GIRK channels because GIRK channel opening facilitates an outward hyperpolarizing K^+ current and a decrease in the input resistance (Koyrakh et al., 2005; Gassmann and Bettler, 2012). This mechanism prevents back-propagation in dendrites, and thus, GABA-BR activation prevents dendritic calcium spikes through inhibition of postsynaptic voltage-gated calcium channels.

4. Glutamatergic and GABAergic systems during development

A balance between excitatory glutamatergic input and inhibitory GABAergic transmission (E/I) maintains the neuronal excitability in mature neurons and that is established during development. The glutamate receptors are expressed prior to the glutamatergic synaptogenesis in neurons. Interestingly, glutamate receptor acts to adjust the developing GABAergic synapse, and GABA then acts as the main excitatory transmitter during early stages of development, not only at the synaptic and network level, but also on cell cycle and migration (Ben-Ari et al., 2007; Hampe et al., 2017). Both GABAergic and glutamatergic neurons are widely diversified in the CNS, and their expressions are widespread in the embryonic stage. Thus, the mechanism of GABAergic signaling is not always inhibitory at the synaptic level. At the early stages of development, the GABAergic excitatory function is influenced by the increased level of the intracellular chloride concentration in immature neurons, which triggers sodium spikes and activates voltage-gated Ca^{2+} channels (Ben-Ari et al., 2007). Eventually, chloride efflux is progressively reduced indicating a transition from a depolarizing to a hyperpolarizing effect in the spinal cord.

Notably, neuronal chloride homeostasis is maintained by channels, exchangers, and cotransporters. The developmental changes of potassium-chloride transporter 2 (KCC2), the principal chloride extruder with higher expression in mature neurons, and sodium-potassium-chloride cotransporter isoform 1 (NKCC1), that ensures chloride uptake with higher expression in immature neurons, play the master role for the alterations in chloride efflux associated with maturation of neurons (Hampe et al., 2017).

4.1 Chloride-mediated inhibition of motoneuron maturation

The effects of GABA binding to GABA-AR mainly depend on the chloride equilibrium potential (ECl). A chloride influx is induced through the activation of the chloride permeable channels in mature neurons, where the intracellular Cl^- concentration $[Cl^-]i$ is lower than extracellular Cl^- concentration $[Cl^-]o$. In contrast, the intracellular Cl^- concentration is higher in the immature neurons that manifest GABA as an excitatory neurotransmitter. Thus, switching of excitatory to inhibitory effects of GABA was evidenced during CNS development (Allain et al. 2011). In the adult brain, the chloride gradient across the neuronal cell membrane is mediated by the exporter K–Cl co-transporter (KCC2), which maintains low $[Cl^-]i$. The influx of negative chloride ions hyperpolarizes the cell membrane potential resulting in inhibition of the neuronal activity. Interestingly, opening of GABA-AR ion channels may enhance the concurrent excitatory currents that are driven by the neurotransmitter glutamate and thus, prevent the action potential of the membrane. Such GABA-AR opening short-circuits the depolarizing synaptic currents by reducing the local input resistance (Khazipov et al., 2015). This mechanism of GABA-AR-mediated transmission causes the adult membrane physiologically hyperpolarizing and inhibitory. On the other hand, low KCC2 expression, being coupled with high expression of the chloride importer NKCC1, generates high $[Cl^-]i$ leading to an outward chloride flux that depolarizes the cell membrane potential through the GABA-AR system in the early stage of neurodevelopment (Fiumelli et al., 2005).

The strength of GABAergic synapses is regulated by glutamatergic transmission (Rosato-Siri et al., 2002). Moreover, GABA signaling, in most immature CNS regions, is established before glutamatergic transmission, indicating that GABA is the primary excitatory transmitter during early development of the brain (Ben-Ari et al., 2007). This has been further supported by the pharmaceutical investigation of the spontaneous GABA activity with acetylcholine as the earliest spontaneous motor activity in the spinal cord, followed by glutamate interference (Myers et al., 2005). The maturation of synaptic development in the spinal MNs of rat occurs during postnatal stages, if started in the embryonic life, following a developmental shift from primarily long-duration GABAergic miniature inhibitory postsynaptic currents (mIPSCs) to short-duration glycinergic mIPSCs after birth (Gao et al., 2001). During postnatal development, commissural connections, mostly mediated by glycine and GABAergic inhibition, regulate the onset and duration of neurochemically induced locomotor activity (Hinckley et al., 2005).

Integration processes of excitatory synaptic inputs depend on postsynaptic potentials (PSPs) generated from GABA-ARs, shunting inhibition, and also from the localization of GABA-ARs, such as GABAergic synapses are mostly located on the dendritic shaft and on spines on the dendrites of CA1 PNs (Megías et al., 2001). This localization facilitates participation of synaptic integration of two different sets of inputs, which include: one input is already partially processed along the way from their synaptic origin to the dendritic shaft, and the other is originated in neighborhood synapses of the same spine. Importantly, varying [Cl−]i gradient in different cellular compartments critically determines the GABA-AR participation for input integration in a compartment specific fashion, highlighting maximum [Cl−]i in the axon and axonal initial segment, lower in the soma, and gradually decreases along the dendrites (Khirug et al., 2008; Waseem et al., 2010).

4.2 Interplay of KCC2 and NKCC1 and mechanism of GABA polarization

The effects of the switching of excitatory to inhibitory mechanism of GABA-AR is closely linked to decrease in [Cl−]i during the course of development, which primarily depends on the variation in ontogenic expression of the NKCC1. NKCC1 uptakes chloride ions, while KCC2 extrudes chloride ions (reviewed in Allain et al., 2011). However, other exchangers can control the chloride gradient by anion (Cl^-−$HCO3^-$) exchangers (AE), which can be Na^+ independent or Na^+ driven (Farrant and Kaila, 2007). NKCC1, which maintains a high [Cl−]i and is predominant in early stages of development, disappears at later stages with increase in KCC2 by lowering intracellular chloride levels (Ben-Ari, 2001). However, the relationship of KCC2, NKCC1, and EGABA-AR (GABA-AR reversal potential) with neuronal development during the embryonic life through the depolarizing to hyperpolarizing mechanism of GABA on spinal MNs is yet to be fully elucidated. Nevertheless, understanding the inhibitory effects of shift of GABA from depolarizing to hyperpolarizing in spinal MNs may involve AEs in association with KCC2 and NKCC1 (Gonzalez-Islas et al., 2009).

GABA-AR-mediated depolarizing PSPs, though not sufficient in most cases to reach the threshold for generation of action-potential, are able to activate voltage-gated calcium channels (VGCCs), remove the voltage-dependent magnesium block from the N-methyl-D-aspartate receptors (NMDA-Rs) (Ben-Ari, 2001), and most likely add up to concurrent excitatory inputs (Gulledge and Stuart, 2003). As a result, an initial mild depolarizing effect of GABA is able to contribute to a significant excitation for generating an action potential firing, suggesting that GABA-A-mediated transmission is generally depolarizing and possibly excitatory in the immature brain. However, depolarizing GABA-A currents may shunt the concurrent excitatory inputs. Noticeably, in different cell compartments of mature neurons, a depolarizing GABA-AR response may also occur depending on timing and location of GABA-ARs at a relatively small difference in [Cl−]i gradient (Cellot and Cherubini, 2014; Chiang et al., 2012). Therefore, even small changes in [Cl−]i are sufficient to alter the polarity of GABAergic responses because the reversal potential for GABA-AR-mediated chloride currents (ECl) occurs very near to the resting membrane potential (VREST). Notably, the inward chloride flow will be hyperpolarizing when ECl is more negative than VREST, and conversely, the chloride flow occurs in the opposite direction by generating depolarizing PSPs, when ECl is less negative than VREST (Contestabile et al., 2017). [The NMDA receptor, a glutamate and ion channel protein receptor, is activated upon binding of glycine and glutamate to it. The NMDA-R is a heteromeric complex that interacts with multiple intracellular proteins by three different subunits: GluN1, GluN2, and GluN3].

It is understood that during neuronal maturation, GABA-AR-mediated transmission undergoes a developmental switch from depolarizing (excitatory) to hyperpolarizing (inhibitory) state, and inhibition and excitation can shift under some circumstances and/or in specific cellular compartments (Gulledge and Stuart, 2003; Ruusuvuori et al., 2013). Additionally, neuronal physiology is regulated by the mechanism of ionic movements and that is determined by GABA-AR signaling and chloride transporters (Deidda et al., 2014). Consequently, upregulated NKCC1 was observed in the brain of both Ts65Dn mice and DS patients and increase in[Cl−]i in CA1 PNs of Ts65Dn mice and depolarized ECl in neurons by 7.7 mV compared to WT mice (Deidda et al., 2015b). Thus, ECl was less negative in trisomic neurons than VREST, predicting an outward Cl^- depolarizing currents upon GABA-AR activation. As a result, trisomic hippocampal slices showed an increase in the mean firing frequency of individual CA1 neurons upon exogenous application of GABA (Fig. 11.3) (Contestabile et al., 2017); however, it was reduced by blockade of GABA-AR-mediated transmission with bicuculline. Therefore, GABA-A signaling appeared excitatory in neurons of the Ts65Dn mice following prolonged bath-application of the drugs in acute brain slices, though GABA presents the complex dual excitatory and shunting/inhibitory effect

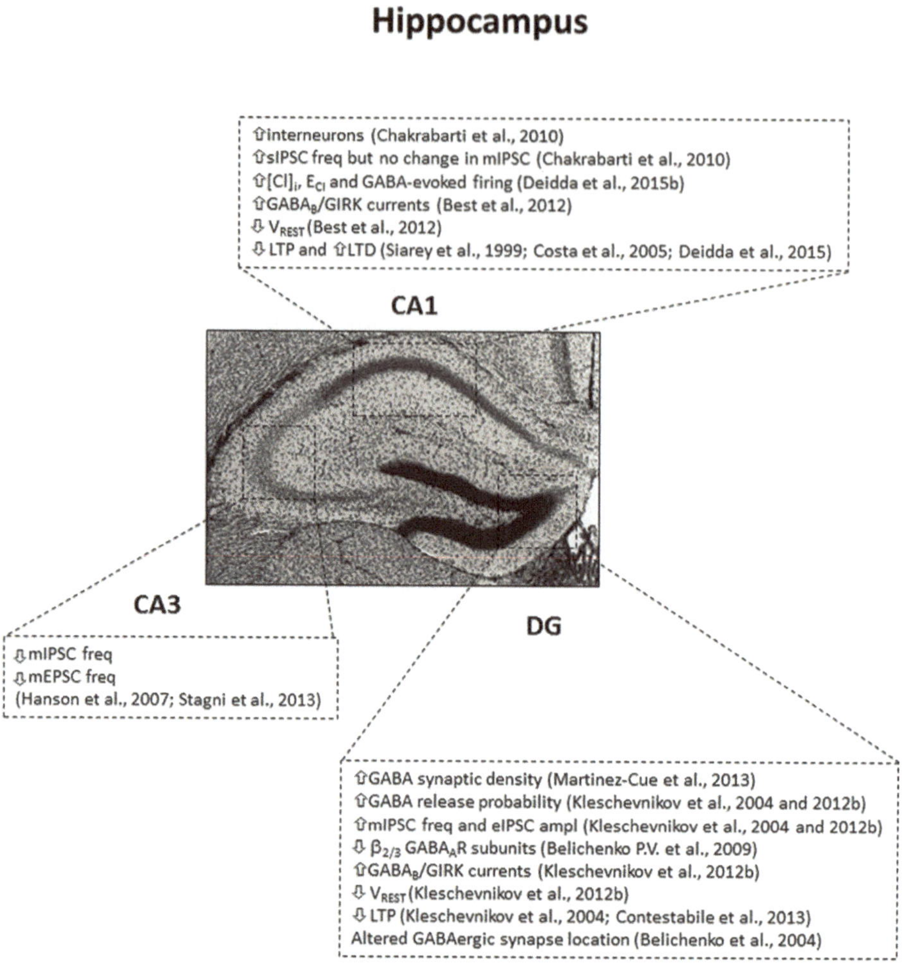

FIGURE 11.3 Subregion-specific gamma aminobutyric acid (GABA)-related dysfunction in the hippocampus of the Ts65Dn mouse model. *[Cl]i*, intracellular chloride concentration; *ampl*, amplitude; E_{Cl}, reversal potential of GABA-$_A$R-mediated Cl currents; *eIPSC*, evoked IPSC; *freq*, frequency; *IPSC*, inhibitory postsynaptic currents; *mEPSC*, miniature excitatory postsynaptic currents; *mIPSC*, miniature IPSC; *sIPSC*, spontaneous IPSC; *VREST*, resting membrane potential. *Arrows* indicate increases or decreases of the reported measures. *Reprinted from Contestabile et al. (2017), https://doi.org/10.3389/fncel.2017.00054, under the terms of the Creative Commons Attribution License (CC BY).*

following ECI depolarization (Contestabile et al., 2017). Moreover, reduced response to bicuculline application on network burst-amplitude and duration in hippocampal cultures from Ts65Dn mice and also decrease in network bursts upon GABA application in both WT and Ts65Dn culture were reported (Stern et al., 2015). However, the association of evoked synaptically released or tonic GABA signaling to the depolarizing effect of GABA-ARs and the strength and direction of GABA-A transmission has not been explored in Ts65Dn slices and in individuals with DS. Notably, knowledge is limited on the contribution of GABAergic signaling on network dynamics in trisomic neurons.

GABA-AR-driven depolarization favors excitation, while GABA-AR-driven shunting favors inhibition, in spite of the fact that GABA-AR-induced depolarization may or may not reach the threshold for neuronal excitation. Remarkably, the shunting effect of GABA-AR opening in Ts65Dn neurons may be very mild on dendritic signal processing because most synapses are located on the spine neck. Conversely, the GABA-AR depolarizing PSPs may generate the depolarized ECl and disseminate to spine heads and dendritic shaft, affecting NMDA receptor- and voltage-gated channel openings consequently, resulting in add up of the concurrent excitatory PSPs in the spine neck (Contestabile et al., 2017). Thus, the concomitant hyperpolarized shift of VREST in Ts65Dn neurons having triplication of GIRK2 (Best et al., 2012) may also cause depolarizing of GABA. Nevertheless, the specific antagonist of the pharmacological inhibition of NKCC1 transport activity with bumetanide completely re-established the hyperpolarizing and inhibitory action of GABA in trisomic neurons of Ts65Dn mice (Deidda et al., 2015b).

4.3 Phasic and tonic inhibition

Upon binding to respective receptors at postsynaptic membrane, GABA-A and glycine elicit inhibitory postsynaptic currents (IPSCs). GABA-ARs mediate two distinct forms of inhibition involved in inhibitory processing in the CNS: phasic and tonic. GABA-AR-mediated phasic versus tonic currents regulate neuronal physiology. Phasic inhibition is mediated by both GABAergic and glycinergic inhibitory postsynaptic currents in the spinal cord dorsal horn, whereas, tonic inhibitory currents were mediated by GABA-A receptors only and not by glycine receptors in dorsal horn lamina II region (Ataka and Gu, 2006). IPSCs provide phasic inhibition in neuronal network and are important for information processing. At synaptic sites, low concentrations of ambient GABA can activate high-affinity GABA-ARs that are expressed at extrasynaptic sites to elicit a sustained inhibitory current (Nusser and Mody, 2002). Phasic GABA transmission participates in synaptic-input integration. Notably, fast feed-forward inhibition mechanism acts as a coincidence detector of excitatory inputs, which ensures that a second input being desynchronized when a first input is canceled out, suggesting that excitatory inputs integrates only when they are perfectly synchronized (reviewed in Contestabile et al., 2017; Pouille and Scanziani, 2001). A similar but opposite integration occurs when GABA is depolarizing. Small excitatory potentials generated through GABA-AR opening last after the channel closure and recapitulate with concurrent excitatory inputs, leading to achievement of action potential threshold (Gulledge and Stuart, 2003).

Tonic depolarizing GABA-A responses negatively regulate adult neurogenesis in the dentate gyrus (DG) of the hippocampus by GABAergic Parvalbumin (PV) interneurons (Pontes et al., 2013; Pallotto and Deprez, 2014). On the other hand, phasic GABA-A-mediated transmission ensures simultaneous and temporally limited inhibition, able to synchronize network activity, and generate network oscillations, observed in computational models as well as experiments with in vitro slices from juvenile (P14–27) and adult mice. Additionally, GABA-BRs can synchronize hippocampal network activity at low oscillation frequency and be activated during cortical up-states, contributing to their termination (Mann et al., 2009). Remarkably, GABA plays crucial roles in neurodevelopment, and thus, alterations in the development of the GABAergic system result in onset of neurodevelopmental disorders (Ramamoorthi and Lin, 2011; Deidda et al., 2014) such as epilepsy, anxiety, and AD-like neurodegenerative diseases (Li et al., 2016), indicating the critical contribution of the inhibition on the neural circuit development and function. Noticeably, mutations in the transcriptional regulator MeCP2 (methyl-CpG binding protein 2) in GABAergic neurons implicates most of the phenotypes of Rett syndrome and some of the autism spectrum disorders (ASD) (Rubenstein and Merzenich, 2003).

Summarily, the phasic inhibition acts as a coincidence detector and undergoes a tight spatial and temporal integration with excitatory inputs in a compartment-specific manner, whereas tonic inhibition sets the neuronal excitability background by affecting excitatory input–output relation. Moreover, phasic and tonic GABA transmissions are essential for synchronization of neuronal firing and maintaining the physiological network oscillation. Notably, tonic inhibitory currents are persistently exposed to low concentration of ambient GABA and generate relatively persistent changes of input conductance, leading to rightward shifting of the input–output relationship. Thereby, tonic inhibition decreases the magnitude, duration, and propagation length of excitatory postsynaptic potential (PSPs), limiting their temporal/spatial summing up possibilities and inhibiting neuronal excitability (Contestabile et al., 2017). These effects are mainly mediated by shunting, which is specific for the inhibitory action of tonic GABA transmission and less relevant for the phasic inhibition, at a situation of longer opening of tonic GABA-ARs (Khazipov et al., 2015).

4.4 Transition of GABA expression during embryonic life

In mouse, GABA-ir somata was first detected at embryonic day 11.5 (E11.5), exclusively at brachial level, in the ventral horn, with sharp increase both at the brachial and lumbar level of the ventral horn at E13.5 (Allain et al., 2011). The transient expression of GABA immunoreactivity exhibited reduced number in the ventral gray matter, and GABA-ir fibers contacting MNs at E15.5 in stained perikarya. The first synaptic activity was observed at E12.5 in mouse spinal MNs when the GABAergic phenotype appears to be mostly expressed by interneurons located in the ventral gray matter, with a glycinergic synaptic transmission following increase in GABAergic synaptic activity (reviewed in Allain et al., 2011). Interestingly, GABA effects vary depending on the developmental stage. At early stages, the GABA effects appear excitatory contributing with cholinergic inputs to the genesis of spontaneous network activity, when MNs are still growing to their peripheral targets with the support of GABA-mediated spontaneous activity for correct motor axon guidance (reviewed in Allain et al., 2011). Principally, GABA signaling is established as the chief excitatory transmitter before glutamatergic transmission during early development of the immature CNS regions. The development of the glutamatergic synaptic transmission was reported around E14.5 after the GABAergic expression in the embryonic spinal cord (Allain et al., 2011). The pharmacologic testing has demonstrated GABA-generated acetylcholine suggesting GABA as the spontaneously expressed first ubiquitous signal of motor activity, which then interfered by glutamate expression. Notably, glutamatergic transmission then regulates the strength of GABAergic synapses (reviewed in Allain et al., 2011).

The synaptic transmission from the embryonic life through maturation during postnatal stages indicates a developmental shift from primarily long-duration GABAergic miniature inhibitory postsynaptic currents (mIPSCs) to short-duration glycinergic mIPSCs ending at birth in spinal MNs (reviewed in Allain et al., 2011). In rat, at E15.5, commissural GABAergic connections get involved in synchronous excitatory action on rhythm-generating networks, while these GABAergic commissural connections contribute to reciprocal inhibition during left and right alternation in the ventral spinal cord at E18.5, whereas these inhibitory commissural inputs are mediated by glycine and not by GABA anymore at E20.5. This transition of the primordial role of GABA-AR for ensuring spontaneous activity to reciprocal inhibition between left and right sides of the ventral spinal cord may explain the reason of such a huge expression of GABA in ventral spinal networks at E15.5 in the mouse. Notably, at postnatal developmental stages, the commissural connections are mostly mediated by glycine, while GABAergic inhibition regulates the onset and duration of neurochemically induced locomotor activity (reviewed in Allain et al., 2011).

4.5 Maturation of the GABAergic neurotransmission

During the first 3 weeks of postnatal development, inhibitory synaptic transmission changes in multiple ways in different brain areas with varying glycinergic and GABAergic transmission to the overall inhibitory message to the postsynaptic neurons. Notably, the predominant GABAergic transmission switches to the main glycinergic neurotransmission depending on the development of the lumbar spinal cord and the lateral superior olive of the young rodents; however, the GABAergic neurotransmission controls development of collicular neurons (reviewed in Allain et al., 2011). Interestingly, corelease of glycine and GABA can occur from the same presynaptic vesicle in neonatal spinal MNs causing a mixed glycinergic/GABAergic synaptic event, while the mixed inhibitory synapses have also been functionally detected in MNs of the hypoglossal nucleus. However, the mixed synapses are not specific to the inhibitory input on MNs because these synapses are also present on the spinal interneurons. The ratio of mixed synapses reduces during development in Renshaw cells and other spinal cord interneurons reflecting an intermediate stage of maturation of glycinergic synapses; however, the mixed inhibitory synapses remain functional in the adult (reviewed in Allain et al., 2011).

Interestingly, although the developmental shift from glycinergic/GABAergic to pure glycinergic neurotransmission depends majorly on the maturation of the presynaptic elements, the association of postsynaptic GlyRs and GABA-ARs was demonstrated with the same postsynaptic density at all ages tested. Thus, mIPSCs are effectively mixed glycinergic/GABAergic at P3–P5 and predominantly glycinergic at P15. Nevertheless, since GABAergic synapses are unlikely to contain postsynaptic GABA-ARs, newly formed "nonfunctional" GABAergic synaptic contacts may be developed as observed in the cerebellum and in neurons of the rat lateral superior olive (reviewed in Allain et al., 2011). A hypothesis on the GABAergic presynaptic terminals could face postsynaptic glycine receptor (GlyR) clusters was proposed to explain the mismatch between the pre- and the postsynaptic element of inhibitory synapses. Collectively, the maturation of inhibitory synapses was postulated to occur through differential regulation of GlyT2 and GAD65 expression at the level of a single synaptic terminal and not from a redistribution of GlyRs and GABA-ARs at postsynaptic site. However, such a discrepancy between the pre- and the postsynaptic element in other CNS area during development remains unexplained. Moreover, the glycinergic and GABAergic synapses control the development of MNs in a region-specific manner during programmed cell death as reported in gephyrin-deficient mice that lack all postsynaptic GlyRs and some GABA-AR clusters (Allain et al.,2011; Banks et al., 2005). Notably, region-specific expression of different GlyR and GABA-AR subunits was demonstrated in amyotrophic lateral sclerosis (ALS).

Summarily, GABA-AR activation plays a pivotal role in the development of spinal cord and brainstem MNs as well as in some pathological conditions, but it is unclear to what extent such a diversity is important involving functionally different GABA-ARs and the communication pathway between GABAergic/glycinergic pre-synaptic neurons and postsynaptic receptors in DS pathobiology.

4.6 Functions of GABA-AR and GABA-BR on synaptic plasticity

Owing to having the potential of variety of signaling dynamics by GABA-receptors (A- and B-receptors), GABA plays a pivotal role in the processes of healthy neurodevelopment at all ages and also under several pathological conditions. In early life, depolarizing and possibly excitatory GABA-A-mediated transmission activity specifically play key roles in promoting neurodevelopment. At this stage, when α5-or δ-containing extrasynaptic GABA-ARs are predominant, and synapses are not formed yet, high affinity subunits can detect ambient GABA released from growth cones and astrocytes (Cellot and Cherubini, 2013). Since depolarizing tonic currents facilitate spiking, it increases the probability of firing coincidence from different cells and importance of synaptic wiring for synaptogenesis. Importantly, migration and maturation of PNs, axon growth, and synaptic plasticity are mediated by these tonic currents regulated by GABA-BRs (Ben-Ari et al., 2007; Luhmann et al., 2015). Notably, activation of GABA-BR cannot elicit the GIRK-mediated (hyperpolarizing) currents in early neurodevelopment, though GABA-BRs maintain the presynaptic inhibitory function on neurotransmitter release and regulate neuronal developments, synapse formation and circuit development. Interestingly, binding of postsynaptic GABA-BRs to inwardly rectifying potassium (Kir) channels is delayed during development (Contestabile et al., 2017).

Therefore, both GABA-A- and GABA-B-mediated transmission controls synaptic plasticity during neurodevelopment, wherein GABA-A receptors interfere with the opening and closure of the visual cortex at the time when brain plasticity can be evoked by environmental stimuli and experimental conditions in slices and in vivo for induction and monocular deprivation (Levelt and Hübener, 2012). Consequently, GABA-ARs can exert a negative regulation on hippocampal plasticity and cognition in the adult brain. Notably, GABA-AR-mediated inhibition has been reported to suppress the LTP both in vitro and in vivo (reviewed in Contestabile et al., 2017). Thus, the positive regulators of GABA-ARs such as benzodiazepines or GABA-AR activation can impair the memory function (Raccuglia and Mueller, 2013). Interestingly, LTP can be enhanced by GABA-AR blockade. Notably, LTD facilitated by GABA-AR activation suggested that GABA-A-mediated inhibition balances the ratio between LTP and LTD in vitro. In contrast, GABA-BR-mediated membrane hyperpolarization, inhibition of voltage gated calcium channels, back-propagating spikes, and lowering of the cAMP/PKA activity usually contribute to prevention of synaptic plasticity. Thus, GABA-BRs suppress LTP and memory performance; though a dual role mediated by auto- and hetero-GABA-BRs was reported in plasticity regulation (reviewed in Contestabile et al., 2017). Collectively, GABAergic involvement regulates brain plasticity in adult neurogenesis.

5. GABA signaling in trisomic condition

During development of the DS brain, alterations in neurogenesis and reduction in neuronal number in the cortex are well documented (Becker et al., 1991). Such alterations are specific to area, cell type, and age as evidenced with the small primary foci with granular, presumably GABAergic neurons in layer II and layer IV of the cortex. Interestingly, cortical neuron density appears normal in early gestation; however, a continual decrease in the number of neurons occurs from late gestation to throughout early life (Weitzdoerfer et al., 2001). In consistence with this, in vitro studies of human neuronal progenitor cells (hNPCs) derived from DS cortex (DS-hNPCs) demonstrated normal neuron numbers initially but reduces with time indicating reduced neurogenesis and selective apoptosis of DS neurons (Busciglio and Yankner, 1995). Gene dosage imbalance in DS-hNPCs might underlie the defects in development of interneuron progenitor cells. Alteration of the expression of three subtypes of GABA-AR was observed, including upregulation of the α2 subunit, while α5 and β3 subunits were downregulated (Bhattacharyya et al., 2009). Collectively, gene-dosage expression indicates an inherent decreased GABAergic interneuron neurogenesis in DS. Indeed, autopsy in DS tissue revealed unbalanced excitatory and inhibitory systems (Seidl et al., 2001). The mouse models of DS showed similar results with increased inhibition in the hippocampus indicating failed induction of LTP (Kleschevnikov et al., 2004). Notably, the hippocampal slices displayed reduced synaptic plasticity with significantly decreased LTP in DS transgenic mice (Kleschevnikov et al., 2004; Belichenko et al., 2007), and treating the slices with GABA-A antagonists rescued these deficits, which suggested an increased inhibition manifests the imbalance of neurotransmission. Remarkably, a reduction in inhibitory neurotransmission resulted in an enhanced hippocampal-mediated cognitive tasks in trisomic mouse (Fernandez et al., 2007). In addition, GABA-AR composition alters in trisomic mice with increase in localization of GAD and vesicular GABA transporter, while the glutamate transporters remain unaltered (Belichenko et al., 2009a). A significant overall decrease in β2/3 subunit of GABA-AR has been demonstrated in the dentate gyrus in early stage of DS-pathology in trisomic mice, followed by a significant increase during postnatal 3–8 months. At 3 months of age, an altered ratio of β2/3 to α1 was observed in several areas of the hippocampus, though no significant changes occur in α1 subunit (Belichenko et al., 2009a), indicating an age-related increase in inhibitory neurotransmission in DS, including AD in DS (DSAD) (Rissman and Mobley, 2011).

The hippocampus showed region-specific mechanisms underlying the GABAergic signaling deficits in Ts65Dn mice. A decrease in the number of excitatory synapses in conjunction with a relative increase in the inhibitory synaptic markers was demonstrated in the cortex and hippocampus of adult Ts65Dn mice (Belichenko et al., 2009a; Perez-Cremades et al., 2010), suggesting an increase in the number of inhibitory interneurons in these trisomic mice (Chakrabarti et al., 2010). In consistence with the increase in synaptic markers, the rate of spontaneous inhibitory postsynaptic currents (sIPSCs) appeared higher in the pyramidal cell layer of CA1 in Ts65Dn strain. Notably, the frequency of mIPSC or the paired pulse ratio remained unaltered and that was consistent with an increase in the number of functional interneurons but inconsistent with the enhanced number of inhibitory synapses or enhanced probability of GABA release in CA1 (Chakrabarti et al., 2010; Best et al., 2012; Mitra et al., 2012). Stimulation of the stratum radiatum or pyramidale showed a transient increase in evoked inhibition during the second postnatal week but not in the first or third postnatal week, in mIPSC and in Ts65Dn mice (Mitra et al., 2012). Such transient inhibitory synaptic transmission was hypothesized to occur during critical period of neuronal network development with a subsequent potential to cause network dysfunction in adult mice (Smith-Hicks, 2013). The diffusion-time (td)-dependent diffusion MRI (dMRI) showed reduced granule cell density and abnormal dendritic arborization of Purkinje cells in Ts65Dn mice compared with euploid mice (Wu et al., 2021).

5.1 Trisomy HSA21 and GABA-A signaling

The number of GABAergic interneurons increased during embryonic development of neural progenitor cells in both the cortex and the hippocampus of Ts65Dn mice model (Chakrabarti et al., 2010; Pérez-Cremades et al., 2010; Hernández-González et al., 2015). Notably, amplified neurogenesis appeared in the medial ganglionic eminence (MGE), where most interneurons develop during embryonic development (Marín and Müller, 2014), with prominence of Parvalbumin and Somatostatin-positive GABAergic interneurons (Chakrabarti et al., 2010). Additionally, spontaneous increase in GABAergic postsynaptic events in CA1 PNs was also reported in the same DS model. However, alterations in the frequency of mIPSC, which represents activity-independent quantal release of GABA, GABAergic synapses, and evoked GABA-A transmission in the hippocampal CA1 region of adult Ts65Dn mice were not evidenced in electrophysiology study of GABAergic transmission (Chakrabarti et al., 2010; Best et al., 2012). Therefore, the enhancement of spontaneous GABAergic events observed in Ts65Dn mice was speculated to be a general enhancement in interneuron excitability, instead of specific increase in the number of GABAergic synapses (Fig. 11.3) (Contestabile et al., 2017). The temporal cortex and hippocampus showed the density of symmetric synapses (putative GABAergic) unaffected in the adult Ts65Dn mice (Kurt et al., 2004; Belichenko et al., 2009a). Additionally, selective redistribution of GABAergic synapses from the dendrite shaft and the spine heads to the spine neck was reported in trisomic neurons in Ts65Dn mice (Belichenko et al., 2004; Belichenko et al., 2009a; Kleschevnikov et al., 2012a,b) and in the dentate gyrus (DG) of Ts1Cje mouse model (Belichenko et al., 2007), indicating that the impairment of synaptic input integration could be a possibility due to shifting of location of GABAergic synapses in DS mice. Alteration of contacts on the spine head and dendritic shaft may respectively integrate inputs at local synaptic and dendritic levels, while on the spine neck, the GABAergic contacts may facilitate integration of spine-converging inputs. Nevertheless, the elevated number of GABAergic neurons in the DS brain is seemingly not directly associated with increased synaptic contacts. Moreover, the GABAergic synaptic density may normalize in adulthood for compensatory mechanisms, following defective brain development. Indeed, both mIPSC frequency and evoked GABA-A transmission (but not release probability) were increased differently in the CA1 region of 2 weeks old Ts65Dn mice than in adult Ts65Dn mice (Mitra et al., 2012), which was further supported by the increased GABAergic synaptic density observed in the inner molecular layer and granular layer of the hippocampal DG specifically in the adult Ts65Dn mice (Martínez-Cué et al., 2013; Mojabi et al., 2016). Thus, the brain areas highly innervated by GABAergic fibers could face a more difficult mechanism for development. However, GABAA-mediated dysfunctions due to age-related changes could not be ruled out because of its variability in different areas of the trisomic brain. Interestingly, increase in mIPSC frequency and evoked GABAergic transmission were observed in the DG of adult Ts65Dn mice, though such alteration was mainly attributed to the increased release probability at GABAergic terminals, instead of increase in synapse number (Kleschevnikov et al., 2004, 2012b). This suggests a possibility of sub-regional differences in the density and action of the GABAergic synapses (reviewed in Contestabile et al., 2017). In addition, reduction in mIPSC frequency was observed in the hippocampal CA3 region of Ts65Dn mouse model (Hanson et al., 2007; Stagni et al., 2013). Altogether, an increase in excitability, enhanced release probability, and decrease in tonic GABA inhibition were found in cerebellar granule cells of the Ts65Dn mice (Das et al., 2013; Szemes et al., 2013).

5.2 Trisomy HSA21 and GABA-B signaling

Triplication of the KCNJ6 gene encodes the subunit two of the GIRK channel (GIRK2/Kir3.2) of HSA21 caused overproduction of G-protein-coupled inward-rectifying potassium (GIRK2) mRNA and protein in hippocampus, cortex, and midbrain in Ts65Dn mice (Harashima et al., 2006). Additionally, the brains of the Ts65Dn mice revealed overexpression of the GIRK1 protein with normal levels of mRNA; however, GIRK1 gene (KCNJ3 mapped onto 2q24.1) is not triplicated in DS. Therefore, overexpression of the GIRK1 protein was linked to enhanced hetero-dimerization with GIRK2 subunit, which drives GIRK subunit trafficking and decrease the protein turnover (reviewed in Contestabile et al., 2017; Harashima et al., 2006). As a result, the primary hippocampal neuron cultures, CA1 PNs and the DG granule cells from acute slices of the Ts65Dn hippocampus exhibited an increase in GABA-B/GIRK currents (Best et al., 2012; Kleschevnikov et al., 2012a, b). Such upregulated GABA-B/GIRK signaling might disturb the neuronal excitability and plasticity in Ts65Dn neurons through the enhanced GIRK-mediated shunting inhibition, decrease in excitatory PSPs, back-propagating action potentials, and alteration in neuronal passive properties (Contestabile et al., 2017; Koyrakh et al., 2005). Interestingly, subtraction of GIRK2 gene dosage from trisomy to disomy through cross breeding of GIRK2$^{+/-}$ mice rescued the GABA-B-triggered increased currents in CA1 PNs in Ts65Dn mice (Joshi et al., 2016). The stratum lacunosum molecule (SLM) displayed more intensely expressed GABA-B receptors than the other layers of the CA1 (Harashima et al., 2006). The SLM is likely to serve as primary modulators of the direct perforant pathway inputs into CA1 due to its larger synaptically evoked inhibitory currents (reviewed in Smith-Hicks, 2013) and thus, plays pivotal roles in learning and memory (Brun et al., 2002). Interestingly, stimulation of SLM enhanced the ratio of monosynaptic GABA-B/GIRK to GABA-A currents significantly in Ts65Dn mice without changing the expression or function GABA-A subunit.

In consistence with increased GABA-B/GIRK signaling activated by the trisomic KCNJ6, hyperpolarization of VREST was observed by 2.3 and 6.2 mV in CA1 PNs and DG granule cells (DGGC), respectively (Best et al., 2012; Kleschevnikov et al., 2012b); however, the incidence in the hippocampus of the Ts65Dn mice for GABA-BR subunit 1 was not statistically significant compared to the euploid mice. A compensatory adaptation mechanism toward the increase of GIRK2 levels was hypothesized to lower the GABA-BR subunit 2 (Best et al., 2012; Kleschevnikov et al., 2012b).

Beside GABA-BRs, GIRK channels may also mediate signaling from many other neurotransmitters, such as acetylcholine, dopamine, opioid, serotonin, somatostatin, and adenosine through their metabotropic GPCRs (reviewed in Contestabile et al., 2017). Therefore, triplicated GIRK2 may result in a wide spectrum of versatile effects in DS depending on the involvement of the neurotransmitter systems, such as enhancement of GIRK2 channels mediated induction of hypothermic response observed following administration of an agonist 8-OH-DPAT of serotonin 5-HT1A and 5-HT7 receptors in Ts65Dn mice (Costa et al., 2005; Stasko et al., 2006).

6. Alterations of GABAergic circuits in DS model and neurodevelopmental disorders

The basic wiring of the human nervous system is configured during the early developmental period, when the key events such as birth and migration of neurons, synapse formation and maturation occur, and experience-dependent refinement of circuit connections create functional neural circuits with well-balanced E/I components. Hence, disruption of E/I balance can cause myriads of neurodevelopmental disorders, including cognitive and emotional disabilities, due to lack of effective treatments (Rubenstein and Merzenich, 2003). However, although E/I imbalance is often observed in neurodevelopmental disorders such as epilepsy, autism, and schizophrenia, the underlying evidence directly associating the

E/I imbalance in such diseases is not fully understood.

Cognitive deficit, including altered working memory, impaired long-term memory, language, and reading incompetence, and spatial skills is one of the serious phenotypic expressions in DS. Studies on human patients and mouse models have described a plethora of anatomical, cellular, and physiological alterations that might contribute to the cognitive impairment. However, a direct causal link between genetic, physiological, and intellectual alterations remains largely unexplored. Understanding the specific contribution of the triplicated HSA21 genes to specific alterations in brain of DS may disentangle its multifaceted expression in cognitive performance in DS, and thus development of effective therapeutics to overcome the memory issue of DS patients. Interestingly, physiological association between alterations in the E/I balance and the characterization of the cognitive deficit phenotype in DS has governed an important aspect in pharmaceutical research.

The contribution of GABAergic activity on cognitive disabilities in DS has been tested in Ts2Cje mouse model of DS and suggested that alterations in GABAergic systems may increase inhibitory tone, underlying deficits in synaptic plasticity and cognitive functions in DS. DS-mouse models present reduced excitatory drive and significant enhancement of GABAergic activity in different brain regions, especially in the hippocampal formation. However, a precise explanation of the mechanisms on how cognitive deficits are influenced by the excessive inhibition and delineation of the link between the triplication of specific HSA21 genes and the increased inhibitory tone in mouse models of DS is still lacking. Thus, the unbalanced excitatory and GABAergic activity needs explanation on alterations in synaptic plasticity and behavior in mouse models of DS, especially as pharmacological blockade of GABA receptors restores anomalous physiological and cognitive phenotypes. Notably, the existence of bidirectional alterations to the synaptic inhibition on CA1 pyramidal cells and its association with the impaired spatial memory were reported in the Ts2Cje mouse model of DS (Valbuena et al., 2019). The alterations of these phenotypes were assigned to the triplication of the kainate receptor (KAR) encoding gene Grik1, wherein normalization of Grik1 dosage in this model specifically restored spatial memory with reversal of the bidirectional alterations to CA1 inhibition. However, the changes in synaptic plasticity and/or other behavioral modifications could not be reverted. This information led to hypothesize that disturbed inhibitory tone contributes to cognitive deficits via modified information gating in DS population (Valbuena et al., 2019).

In DS, intellectual deficits caused by an excessive activity of inhibitory circuits have been hypothesized based on the recovery of behavioral and neurophysiological deficits through the use of GABA-A receptor blockers (de San Martin et al., 2018). However, direct evidence of over-inhibition, involvement of specific GABAergic circuits, and the anatomical, molecular, and functional diversity of cortical inhibitory neurons toward understanding the cognitive deficits in DS are obscure (Tremblay et al., 2016). Notably, the integrity of the prefrontal cortex (PFC) plays the executive role for the synchronization of task-relevant and large-scale neuronal activity in brain (Helfrich and Knight, 2016). The network of this synchronization is characterized by the neuronal oscillations distributed at different brain regions, including rhythmic fluctuations of the electrical activity of single neurons, local neuronal populations, and multiple neuronal assemblies (Buzsa'ki and Wang, 2012). Such neuronal oscillations are controlled by a balanced and coordinated activity of the excitatory PNs and different types of inhibitory neurons that use GABA as the neurotransmitter. Specifically, the PV-positive inhibitory interneurons control PN spiking activity tightly by forming synapses onto the perisomatic region of PNs, and drive network oscillations in the range of

30–100 Hz as γ-frequency (Buzsa'ki and Wang, 2012). The γ-oscillations control many of the PFC cognitive functions such as cognitive flexibility (Cho et al., 2015) and sustained attention (Kim et al., 2016). In contrast, the MCs, the somatostatin (SST)-positive interneurons, inhibit distal dendrites of PNs, and thus, control the integration of distal dendritic glutamatergic synaptic inputs originating from different regions across the brain (Tremblay et al., 2016). Dendritic integration of multipathway inputs is essential for working memory (Abbas et al., 2018; Kim et al. 2016). Therefore, the two forms of inhibition, such as PV interneurons and MCs, are involved in the entrainment of network oscillations as characterized by the two major cortical inhibitory circuits and rendering a precise division of labor during cortical activity (Chen et al., 2017; Veit et al., 2017). Thus, PVs and MCs, especially inhibition from SST interneurons, contribute to medial-(m)PFC-dependent memory functions and cognitive performance (Abbas et al., 2018; Cho et al., 2015; Cummings and Clem, 2020). The enhancement of β-γ-band power and neural synchronization in the frequency ranges observed in the Ts65Dn mice correlates well with augmented hippocampal-PFC synchronization and LFP γ-band power during natural non-REM sleep in Ts65Dn mice (AlemanyGonza'lez et al., 2020; de san Martin et al., 2020). Such alterations of γ-oscillations observed in Ts65Dn mice were postulated to be involved in the pathophysiology of sleep apnea of DS individuals (Fernandez et al., 2017; Gime'nez et al., 2018).

The evaluation of the synaptic and intrinsic properties of two major inhibitory circuits of the prefrontal cortex in Ts65Dn strain of DS model displayed strong potentiation of the dendritic inhibitory synaptic loop involving MCs and PNs in trisomic mice compared to euploid control mice; however, the perisomatic synaptic inhibitory control of PNs by PV cells remained unaffected in trisomic mice, though the excitability of PV cells was significantly altered. These GABAergic circuit-specific alterations were consistent with reduced PN spiking and enhanced coupling with β-γ-activity network in these trisomic mice in vivo (de San Martini et al., 2020). Notably, α5-mediated dendritic inhibition in the hippocampus, which is known to control the dendritic integration, NMDA-R activation, nonlinear dendritic integration, and action potential (AP) firing strongly (Schulz et al., 2019), was also increased in Ts65Dn mice. In addition, the specific synaptic strengthening of the dendritic inhibitory loop involving MCs, suggested a major impact on PN dendritic integration and electrogenesis in trisomic mice (de San Martini et al., 2020). Elevated dendritic inhibition by MCs could cause the deficits of long-term plasticity of glutamatergic synapses in DS patients similar to that appeared in the trisomic hippocampus of mouse models. Pharmacologic approach for recovering the LTP deficits was thought of by using allosteric modulators of α5-GABA-ARs (Duchon et al., 2020; Martı'nez-Cue' et al., 2013; Schulz et al., 2019). Similarly, the enhanced MC-PN inhibitory loop observed in trisomic mice may indicate the underlying mechanism for the rescue of cognitive deficits in these mice.

The dendritic inhibitory loop and increased PV-cell excitability are associated with the changes of PN spiking activity as observed in Ts65Dn mice in vivo (de san Martin et al., 2020). Remarkably, the suppressed spike rates and enhanced tuning in the β-γ-frequency range appeared consistent with increased activity of inhibitory neurons (Atallah et al., 2012; Cardin et al., 2009; Chen et al., 2017; Veit et al., 2017). The over-inhibition was resulted from highly specific synaptic and intrinsic alterations of dendritic and somatic inhibitory circuits in Ts65Dn mice.

6.1 GABA signaling in DS patient–derived stem cells

In DS, the hallmark of GABAergic system has been documented with reduced brain size and decreased density of neurons, which mainly arise from disturbed neurogenesis during brain development. However, the phenomena have been more extensively studied in trisomic mice than human DS. It is less detailed for individuals with DS, which is limited to availability of DS patients with partial trisomy 21 (reviewed in Contestabile et al., 2017). The two histological studies on autopsy of DS brain samples have demonstrated increased cellularity of bonafide GABAergic interneurons. Reduction in Golgi-stained aspinous stellate (putative GABAergic) cells in the somatosensory, visual, and auditory cortices, and Parvalbumin and Calbindin-positive non-PNs in the frontal and temporal cortices was evidenced in DS patients (reviewed in Contestabile et al., 2017). Nevertheless, such decrease of GABAergic neurons may arise as a secondary effect due to Alzheimer-like degeneration in DS (GABA concentration is specifically decreased in AD brain) but not in aging DS population. In this line, microarray-based evaluation of altered gene expression in human DS, cortical hNPCs showed indicative changes in gene dosage for reduced genesis of GABAergic interneurons in vitro (reviewed in Contestabile et al., 2017). Interestingly, the same study displayed increased expression of the $\alpha2$ subunit of the GABA-AR but downregulation of the $\alpha3$ and $\alpha5$ subunits in trisomic cells relative to controls (Bhattacharya et al., 2009). This finding on GABA-AR composition favoring the $\alpha2$ subunit over the $\alpha5$ subunits may indicate a fastened GABA-AR opening, possibly by reducing the chance of shunting inhibition and generating quick PSPs, a situation that may be of particular relevance of depolarizing GABAA-mediated transmission in humans. Indeed, overexpressed NKCC1 in the DS brains, which also indicates a direct parallel with the Ts65Dn model (Deidda et al., 2015b), possibly suggests parallelism of human depolarized ECl to animal models. Nevertheless, decreased or unaltered GABA levels were observed in neurochemical and ^{1}H-MRS studies on trisomic human brains (reviewed in Contestabile et al., 2017).

Interestingly, reprogramming facility of somatic cells has excited the study of trisomic neurons derived from DS-patients, and attracted as a valuable tool for modeling neurological diseases in vitro (detailed in Chapter 14) (Takahashi et al., 2007, 2012). This approach has facilitated production of different induced pluripotent stem cell (iPSCs) lines of DS origin. Indeed, studies using trisomic iPSCs have highlighted a general impairment in synaptogenesis in iPSC neurons derived from DS patients accompanied by reduced frequency of both inhibitory and excitatory spontaneous postsynaptic currents (sPSCs). The GABAergic neurogenesis and synaptogenesis were induced in iPSCs for neuronal differentiation. Notably, the frequency of neurons expressing GABAergic markers, the fraction of GABAergic synapses, and the ratio of glutamatergic to GABAergic sPSCs were largely unaffected (Weick et al., 2013; Hibaoui et al., 2014).

7. Perspectives on therapeutics of GABAergic transmission

The GABAergic system plays a crucial role in maintaining the E/I balance required for normal neural circuit function in brain. The E/I imbalances caused by disturbances in the development of this GABAergic system, starting from the generation of inhibitory neurons to the formation of their synaptic connections, have been associated with several neurodevelopmental disorders, and such imbalances are thought to be the underlying mechanism of epileptogenesis. Perturbation of action of the GABAergic interneurons—the primary inhibitory neurons, results in hyperactivity in the epileptic circuitry. The mechanisms of the optogenetic excitation of GABAergic interneurons are widely used for suppression of seizure (Ye and Kaszuba, 2017). These mechanisms include: excitation of postsynaptic neurons by GABAergic interneurons via the enhanced reversal potential of GABA receptors in the postsynaptic cells; transient GABA depletion due to continuous activity in

GABAergic interneurons, which prevents their inhibitory effect on pyramidal cells; synchronization of the network activity during seizure by GABAergic interneurons; and inhibition of other interneurons by some GABAergic interneurons leading to dis-inhibition of PNs and network hyperexcitability. Therefore, understanding the impairments at different stages of GABAergic development can identify the potential modulators of the GABAergic system, which can suggest therapeutic strategies for treatment of neurodevelopmental disorders.

Notably, recent research suggests that selective modulation of the GABAergic system is a promising intervention for the treatment of DS- and schizophrenia-associated cognitive defects. Evidence derived from postmortem and animal studies has described how abnormal GABA interneurons and GABAergic neurotransmission could perturb the neuronal oscillations and suggested many agonists, antagonists, and positive and negative allosteric modulators (including dual allosteric modulators) for $\alpha5/\alpha3/\alpha2$ GABA-A and GABA-B receptors (Gardiner, 2015). There are various potential GABA-associated therapies, including histone deacetylases (HDAC) inhibitors that target GABA system through epigenetic modulations, GABA prodrug, and presynaptic GABA transporter inhibitors (reviewed in Xu and Wong, 2018). However, pharmacological therapeutics for ameliorating cognitive deficits in DS has not been established yet, though several preclinical and clinical trials with GABAergic drugs for DS are underway.

Research and investigation mostly include mouse models to find a compelling treatment for improvement of learning and memory in DS. Mice have orthologs of the HSA21 genes map to many mouse chromosomes (mouse chromosomes 10 (MMU10), 16 (MMU16), and 17 (MMU17)), while Ts65Dn trisomic mice consisting 88 orthologs of HSA21 protein coding genes and five microRNA genes is used as the most preferred DS model (Rueda et al., 2012). For improving the overall quality of life of DS, especially for rescuing learning and memory deficits, more than 20 drugs having diverse properties, such as N-methyl-D-aspartate receptor (NMDA-R) antagonist, GABA-AR antagonists, acetylcholinesterase inhibitors, and the green tea component have been tested and appeared to be effective (Braudeau et al., 2011; Block et al., 2017; Costa, 2008; Das et al., 2013; Busciglio et al., 2013).

7.1 Modulation of synaptic plasticity and GABA signaling in DS

Altered GABA-functioning has been linked to DS-specific cognitive phenotypes, majorly based on studies of hippocampal synaptic plasticity in trisomic mice. Synaptic plasticity has been defined on widely accepted LTP and LTD models, which are illustrated by the strengthening or weakening (respectively) of synaptic efficiency measured from specific stimulation protocols. LTP depicts the cellular contexts of memory and that is essential for different cognitive actions, including learning (Nabavi et al., 2014). Indeed, altered hippocampal slices demonstrated the synaptic plasticity paradigms (both LTP and LTD) in Ts65Dn mice and thus, indicated the impairment of cognitive characteristics parallels to human DS and trisomic mouse models.

Given the association of the GABAergic inhibition in regulating synaptic plasticity and the defective GABAergic transmission in DS animals, potential and targeted GABAergic therapeutics have been evaluated on different forms of hippocampal LTP in trisomic mice. Notably, LTP of Shaffer collateral-CA1 synapses (CA3-CA1 LTP) was rescued by acute application and recording bath of the GABA-AR antagonist picrotoxin (PTX) in Ts65Dn slices. Similarly, application of either PTX or the GABA-BR antagonist CGP55845 restored potentiation at perforant path-DGGC synapses (DG-LTP) (Kleschevnikov et al., 2012b), suggesting a causal link between plasticity deficits and GABAergic transmission in DS. Indeed, application of the NKCC1 inhibitor bumetanide (a FDA-approved loop diuretic) rescued learning and memory, specifically discriminative memory in the object recognition test, spatial memory in the object location test and associative memory in the contextual fear conditioning task in Ts65Dn mice (Deidda et al., 2015b). The result indicated the importance of depolarizing GABAergic signaling in the impairment of synaptic plasticity in DS. In this line, chronic treatment with the GABA-AR blocker pentylenetetrazole (PTZ) rescued DG-LTP in acute slices in Ts65Dn mice. The effect persisted up to 1–3 months following cessation of treatment (Fernandez et al., 2007). Similarly, chronic treatment with RO4938581, a selective negative allosteric modulator (NAM) of α5-containing GABA-ARs, rescued CA3-CA1 LTP in Ts65Dn mice both in vivo and in vitro; however, the effect was not assessed after drug withdrawal (reviewed in Contestabile et al., 2017; Martínez-Cué et al., 2013).

Interestingly, the Ts1Cje, Ts1Rhr, and triple trisomic (Dp10/Dp16/Dp17) mice showed deficits in synaptic plasticity with similarly impaired DG-LTP, which was rescued by PTX treatment (Belichenko et al., 2009b). In contrast, CA3-CA1 LTP was reduced in Ts1Cje, Dp16 and triple trisomic mice; but unchanged in Ts1Rhr and Dp10 mice; and markedly increased in Dp17 mice. However, these studies were not evaluated for the contribution of GABAergic signaling (Yu et al., 2010). Interestingly, other electrophysiological investigations on alteration of GABAergic signaling were demonstrated in acute brain slices of DS mouse model. Thus, further evaluation of GABAergic dysfunctions would be meaningful in primary neuronal cultures in vitro and in the intact brain in vivo (Best et al., 2007).

7.2 Therapeutics for GABAergic signaling in DS

The LTP was markedly reduced in the CA1 and DG in trisomic mice due to enhancement of GABA-mediated inhibition in several hippocampal sub-regions. Interestingly, rescue of LTP could be achieved following acute or chronic treatment with GABA antagonist in Ts65Dn hippocampus (Fernandez and Garner, 2008). Importantly, despite GABA signaling, alternative mechanisms have been shown to underlie LTP and/or cognitive deficits in Ts65Dn mice. Interestingly, a variety of strategies, including: acute application of the GluN2B-selective antagonist Ro25–6981 (Hanson et al., 2013) or of the green tea polyphenolic compound epigallocatechin-3-gallate (EGCG, an inhibitor of the DS triplicated kinase DYRK1A) (Xie et al., 2008), chronic treatments with EGCG (De la Torre et al., 2016), the monoacylglycerol lipase inhibitor JZL184 (Lysenko et al., 2014), the neuro-hormone melatonin (Corrales et al., 2013), the Sonic Hedgehog agonist SAG1.1 (Das et al., 2013), an antidepressant antagonist of the serotonin reuptake inhibitor fluoxetine (Guidi et al., 2014), or also exposure to an enriched environment (EE) (Begenisic et al., 2015) rescued CA3-CA1 LTP and/or behavioral performances in Ts65Dn mice (reviewed in Contestabile et al., 2017). Moreover, most of these tests have exhibited direct or indirect effect on modulation of the GABAergic mechanism. However, some of these treatments did not establish such link as observed with Ro25–6981 that selectively reduced the activation of GABAergic interneurons in the hippocampal Stratum Radiatum (Hanson et al., 2013), while JZL184 decreased GABAergic transmission by possible modulation of the presynaptic cannabinoid receptors (Zhang et al., 2009; Lee et al., 2015).

Treatment of Ts65Dn mice both with fluoxetine and exposure to EE showed beneficial effects on LTP and memory, probably by reducing release from GABAergic terminals (Begenisic et al., 2015). Indeed, fluoxetine exhibited reduction of GABAergic neurotransmission in the hippocampus independently from the inhibition of serotonin reuptake (Méndez et al., 2012) and decrease in the levels of extracellular GABA in vivo (reviewed in Contestabile et al., 2017).

Remarkably, fluoxetine may inhibit GIRK channels, possibly by normalizing the enhanced GABA-BR signaling in Ts65Dn mice. In this context, it is noteworthy that GIRK channels are also coupled with serotonin receptors 5-HT1A (Llamosas et al., 2015; Montalbano et al., 2015), and thus, triplicated GIRK2 can impact on serotoninergic signaling in DS. In reality, stimulation of 5-HT1A receptors can potentially lower the neuronal firing frequency and gamma oscillations through activation of GIRK channels in the hippocampus (Johnston et al., 2014). The therapeutic efficacy of fluoxetine in Ts65Dn mice was speculated from decrease in GIRK-mediated 5-HT1A and GABA-BR signaling in the dorsal Raphe induced by fluoxetine treatment (Cornelisse et al., 2007); however, a possible link between overexpressed GIRK2 mediated changes in serotonin signaling and the impact of fluoxetine on learning and memory has not yet been discerned in DS mice. Interestingly, prenatal treatment of Ts65Dn mice with fluoxetine exhibited normal brain neuronal proliferation and dendritic growth at 45 days (Guidi et al., 2014). The fluoxetine molecule interacts with the histone deacetylase class of enzymes that catalyze a reduction of the acetyl group. Chronic postnatal treatment with fluoxetine rescued abnormalities in neurogenesis and stimulated the production of neurons and their incorporation into functional networks. However, higher level of daily fluoxetine consumption was hypothesized to provoke seizures (reviewed in Rondal, 2020).

Use of an EGCG enriched extract, the major catechin present in green tea, during gestation of transgenic mice overexpressing Dyrk1a (mBACtgDyrk1a), was planned for inhibition of DYRK1A function (Souchet et al., 2019). Chronic prenatal treatment revealed reduced levels of inhibitory markers, restored VGAT1/VGLUT1 balance, and rescued density of GAD67 interneurons in the cortex of adult Ts65Dn mice, which was similar to GABAergic and glutamatergic markers and interneuron density observed in Dp(16)1Yey mice, trisomic for 140 HSA21 orthologs. Prenatally treated adult mice of this strain showed improvement in novel object recognition memory in cognitive and behavior testing but did not show improvement in Y maze paradigm. Prenatal EGCG treatment modulated trisomic neural crest cells deficiencies at embryonic age and normalized some craniofacial phenotypes, including cranial vault in adult Ts65Dn mice (McElyea et al., 2016). In addition, postnatal (P) EGCG treatment during P3–P15 restored neurogenesis at P15, but not at P45, and did not restore normal cognitive performance evaluated either in Y maze or Morris water maze tests (Stagni et al., 2016). Thus, prenatal EGCG treatment displayed an empirical support for a prenatal intervention to target specific neural circuitries, even with higher efficacy in a more complex DS model (Souchet et al., 2019), though adult mice treated with EGCG for 1 month did not rescue levels of inhibitory interneuron marker GAD67. Acute treatment with EGCG positively modulated GABAA-mediated transmission (Park et al., 2011). In this direction, a clinical trial named TESDAD (NCT01699711) on evaluation of the efficacy of long-term green tea extract on young adults with DS exhibited some behavioral improvements and partial correction of cognitive deficits from a dietary supplement of EGCG (de la Torre et al., 2016), but that was below the threshold of clinical relevance. Nevertheless, mouse models exhibited varying results with EGCG treatment.

Indeed, an IC50 of 0.3 μM of EGCG rescued LTP deficits in CA1-CA3 of Ts65Dn and in neocortex of transgenic mBACtgDyrk1a, and thus, EGCG has been established as a noncompetitive inhibitor of DYRK1A. Interestingly, treatment of a transgenic mouse model for human DYRK1A treated from gestation to adulthood showed strong correction of molecular brain derived neurotrophic factor (BDNF) levels and cognitive deficits (novel object recognition tests). In trisomic mice, Dyrk1a is overexpressed as early as E11.5 (telencephalon) and E15.5 (brain). Since the prenatal period is critical for GABAergic interneurons, prenatal treatment EGCG appears relevant for investigation of the adults for persistence of changes induced at fetal age (Brault et al., 2021; Benoit et al., 2019).

The positive effect of melatonin, which acts as a positive allosteric modulator of GABA-A receptors, is unlikely to depend on reduced GABAergic signaling (Cheng et al., 2012).

7.3 GABA transmission as a pharmacological target to rescue cognitive deficits in DS

Cognitive pharmacotherapy is a very active field of DS research. Although a number of molecules have been tested, conclusive outcomes have yet to be ascertained. A number of molecules are under testing, and hopefully some of them could prove effective as adjunct treatment for boosting brain development and neurotransmission for DS population. Nevertheless, short- and long-term negative side effects need to be controlled, and also possible toxicity of the drugs needs to be assessed stringently.

Owing to involvement of GABA signaling in neurophysiology, cognition, and synaptic plasticity in DS, GABAergic transmission has been targeted in many pharmacological studies aiming improvement of learning and memory processes in DS mice. Alleviation of the neuronal inhibition using an antagonist named pentylenetetrazol (PTZ) or two different inverse agonists selective for the α5-subunit containing receptor was reported in the Ts65Dn mouse of DS. The cognitive function measured by novel object recognition (NOR) was rescued by using the noncompetitive GABA-AR antagonists PTZ, picrotoxin (PTX), and bilobalide in a seminal study on long-term declarative memory at nonepileptic doses (Fernandez et al., 2007). Strikingly, the positive effect of PTZ (a drug previously used in the clinic) persisted after another 2 months of cessation of treatment, indicating stability of the cognitive performance that was induced by long-term neuronal-circuit rearrangements. In addition, the Morris water maze test further indicated rescued learning and memory in Ts65Dn mice through chronic administration of PTZ (Rueda et al., 2008; reviewed in Contestabile et al., 2017). Conversely, the same study showed a negative effect of PTZ in a test for equilibrium in Ts65Dn mice, indicating a probability of side effect of the drug. Nevertheless, the inverse antagonists reduced the inhibitory GABAergic transmission and thus, could rescue the learning and memory deficits in Ts65Dn mice (Martínez-Cué et al., 2014). However, therapeutic intervention with PTZ as GABA-AR antagonists has been controversial for possible clinical application in humans with DS because of its proepileptic and anxiogenic side effects. Thus, seizures induced by GABA antagonists via modulation of the E/I balance remains a particular safety concern in research for reversal of cognitive deficits in DS population. Importantly, a follow-up study with 10 times lower dose of PTZ appeared effective without compromising its effectiveness, defining a safer therapeutic window of PTZ keeping its proepileptic and anxiogenic side effects (Colas et al., 2013). Given the positive response of PTZ treatment on improved learning and memory performance and the observation on higher expression of α5 subunit of GABA-AR in the hippocampus in knock-out mice (Collinson et al., 2002), efficacy of two other inverse agonists selective for the α5 subunit of the GABA-AR and acting as negative allosteric modulators was evaluated in a second series of experiment. Both acute and chronic treatments with these two drugs such as α5IA and RO4938581, developed by different pharmaceutical companies, as cognitive enhancers proved their effectiveness in ameliorating cognitive performance in Ts65Dn mice in both the NOR and Morris water maze tests (reviewed in Contestabile et al., 2017). Interestingly, these drugs did not cause any proepileptic or anxiogenic side effects that may be linked to GABA-AR antagonism (Braudeau et al., 2011; Martínez-Cué et al., 2013).

A clinical trial named COMPOSE (Cognition and Memory in People with DS) has been registered in the Australian-New Zealand clinical trial registry (ACTRN12612000652875) for testing efficacy of PTZ (BTD-001) (reviewed in Contestabile et al., 2017). This phase IB trial with two different doses has been targeted for evaluating the safety, tolerability, preliminary efficacy, and pharmacodynamics of BTD-001 in adults and adolescents with DS.

Two clinical trials for testing the efficacy of the selective negative allosteric modulator of the α5-containing GABA-AR subtype, Basmisanil (RG1662/RO5186582)—a derivative of RO4938581, rescued learning and memory in Ts65Dn mice (Martínez-Cué et al., 2013). The first one (CLEMATIS) was designed as a phase II placebo-controlled study (NCT02024789) for evaluating the efficacy and safety of RG1662 at two different doses in individuals with DS. However, Hoffmann-La Roche has announced through a media release that the CLEMATIS study did not satisfy its primary and secondary endpoints as no significant difference between the placebo and the treated groups was observed on improving cognitive functions in DS patients. Thus, the second placebo-controlled study (NCT02484703) on evaluation of the pharmacokinetics, pharmacodynamics, efficacy, and safety of RG1662 in children with DS has been discontinued. Nevertheless, the drug was well-tolerated with no apparent side effects (reviewed in Contestabile et al., 2017).

Additionally, use of the selective GABA-A α5 NAM, RO4938581 (3-bromo-10-(difluoromethyl)-9H-benzo[f]imidazo[1,5-a][1,2,4]triazolo [1,5-d][1,4]diazepine), rescued hippocampal LTP and hippocampus dependent cognitive deficits in Ts65Dn mice (Martinez-Cue et al., 2013). The GABA-A-benzodiazepine receptor, a selective agonist for the α5-subtype (aαIA), restored learning and memory in NOR and the Morris water maze test in trisomic mice (Braudeau et al., 2011). Indeed, a GABA-B receptor antagonist, CGP55845, improved the contextual memory in fear-conditioning task in Ts65Dn mice (Kleschevnikov et al., 2012b). Since the preclinical data highlighted a strong link between GABA signaling and DS cognitive impairment, the pharmacological interventions have been translated into a number of clinical trials on evaluation of the effect of GABA antagonism on cognitive phenotypes in individuals with DS and in DS mouse models.

Furthermore, given the fact that the chloride importer NKCC1 is upregulated in DS and GABA-A transmission is depolarizing in adult Ts65Dn mice, bumetanide (an FDA-approved loop diuretic) was assessed for its efficacy as an NKCC1 inhibitor on learning and memory in trisomic mice (Contestabile et al., 2017). The test on chronic NKCC1 inhibition revealed rescue of discriminative memory in the object recognition test, spatial memory in the object location test, and associative memory in the contextual fear conditioning task in Ts65Dn mice (Deidda et al., 2015b). Interestingly, similar effect of bumetanide was also achieved following an acute treatment but quickly lost upon treatment cessation. This situation suggested that bumetanide outcome directly relied on NKCC1 inhibition rather than on neuronal-circuit rearrangements. These reports are in consistence with previous results on the positive effect of GABA-AR antagonists on learning and memory in trisomic mice. In fact, aberrant depolarizing GABA-A signaling can be reduced by GABA-AR antagonists in Ts65Dn mice in spite of enhancement of GABA-AR signaling in DS.

Conversely, since modulation of $[Cl^{-}]i$ is proposed to render only small effect on shunting inhibition, bumetanide would modulate the GABA-A currents to shunt the concurrent excitatory inputs. Thus, it would reduce the potential for proepileptic side effects. On the other hand, GABA-AR antagonists in conjunction with reducing receptor transmission may also lessen the shunting inhibition, which may likely to increase the risk of seizures and alter neuronal input integration to a greater extent. As regards the increased GABA-BR signaling in DS, a specific GABA-BR antagonist CGP55845 restored cognitive performance in the NOR test and associative memory in the contextual fear conditioning test in Ts65Dn mice in both acute and chronic treatment (reviewed in Contestabile et al., 2017), indicating the probable involvement of metabotropic GABA signaling in cognitive impairment in DS.

Interestingly, infusion of the bicuculline (BIC; 50 ng in 0.5 μL saline/hemisphere), a GABA-A-receptor antagonist, in the medial prefrontal cortex (PFC) of rats revealed that intra-PFC BIC decreased risky choice and reward sensitivity during probabilistic discounting and reduced the preference for larger rewards associated with a greater effort cost, similar to the behavioral squeal observed in schizophrenia. BIC infusion enhanced decision latencies and altered the ability to "stay on task" as recorded by the reduced rates of instrumental responding. Altogether, these results implicate prefrontal GABAergic dysfunction in decision-making process in schizophrenia and other neuropsychiatric conditions having similar neurobiological and behavioral alterations (Piantadoshi et al., 2016).

Piracetam has shown an increase in brain oxygenation and improvement in GABA neurotransmission (reviewed in Rondal, 2020). A double-blind study with 80–100 mg/kg piracetam per day for 15 weeks on DS children could not exhibit a significant benefit on learning, attention, and memory, as well as on the adaptive scales. Rather, side effects of irritability and poor sleep were observed in seven out of 18 children.

Antioxidant therapy with Vitamin E showed positive results on memory and basal forebrain pathology in mouse models of DS but no positive response in humans with DS (reviewed in Rondal, 2020). Treatment with minocycline, an antibiotic commonly used to treat acne, exhibited neuroprotective effects and inhibited neuron apoptosis. Treatment of 10-month old Ts65Dn mice for 3 months significantly improved cognitive tasks (reviewed in Rondal, 2020). Supply of nutrients to the neurons and detoxification of the extracellular milieu for neutralizing excess glutamate by the glial cells (astrocytes) was induced in vitro from stem cells derived from fibroblasts of persons with DS (Chen et al., 2014); however, the same did not favor in vivo neurogenesis when transplanted into iPSCs derived from mice brains. Interestingly, minocycline corrected the deficiency by modulating the expression of S100B gene that regulates various cellular processes, including the glial function. Testing with hydroxyurea for 3 months, a derivative of urea, for activation of neural pathways produced significant improvement of memory retention of spatial information in 3-month old Ts65Dn mice (reviewed in Rondal, 2020). Indeed, there are series of biochemical agents that have exhibited nootropic (nonspecific brain boosting) and antioxidant effects.

Collectively, experiments with different drugs indicated improvement in learning and memory, cognitive performance and also neurogenesis in Ts65Dn mice. However, there was variability in experimental protocols in chronic or acute treatments, duration of treatment, and sex and age of mice (prenatal to adult age), which might draw a variable outcome. Thus, standardization of harmonized protocols and experiments could draw a convincing outcome for targeting therapeutics. The presence of non-HSA21 orthologous genes (~50 genes in Ts65Dn mice) is a challenging issue for using the genetically modified mice. Overexpression of these genes might alter the molecular basis of drug responses, and thus, protocols of the preclinical and clinical trials shall be defined accordingly. Moreover, the neurobiological differences between mice and humans cannot be ignored. *Per se*, cognitive pharmacotherapy in DS is in infancy, and thus, the results obtained from preliminary testing should be dealt with caution before envisaging on DS patients. Maximum caution must be considered on drug safety and the occurrence of adverse events. Taken together, these data indicate that a large number of effective drug treatments restoring LTP and/or memory in Ts65Dn mice may rely on a common ground of action through the modulation of the GABAergic system.

7.4 Modulation of GABA/BDNF mechanism

A positive feedback loop for GABA/BDNF has been proposed during early development (Obrietan et al., 2002). An age-related increase in BDNF levels has been demonstrated in individuals with DS (Dogliotti et al., 2010). The impaired hippocampal synaptic plasticity coupled with impaired local translation in DS mouse models has been implicated with loss of synaptic sensitivity as a result of changes in Glutamate/BDNF and basally saturated NMDA signaling (Troca-Marín et al., 2012). In this direction, enhancement of local translation of dendritic mRNA such as Dscam was attributed to long-term synaptic plasticity (LTP) (Li et al., 2011; Troca-Marín et al., 2012). In addition, locally induced synthesis of new proteins and structural modifications during synapse formation are important attributes in the back-modulation of synaptic plasticity (Kossut, 2007).

The neural cell adhesion molecules (NCAMs) influence GABA-ergic synapse specificity (Sassoè-Pognetto et al., 2011). IgSF9b, a highly expressed synaptic adhesion molecule in inhibitory synapses of the hippocampus, cortical neurons, and GABA-ergic interneurons has been described to play a crucial role in major depressive disorders (Woo et al., 2013). Downregulation of IgSF9b was hypothesized to cause high GABAergic transmission to pyramidal cells resulting in an overall inhibition. Over the years, research has established the association of upregulation of DSCAM and cognitive dysregulation and mental retardation in DS. The association of NMDA action and DSCAM expression has been demonstrated in mice, where treatment with NMDA in wild-type mice exhibited increased locally translated dendritic DSCAM protein levels in hippocampal neurons (Fig. 11.4) (Stachowicz, 2018); however, that was not the case in trisomic Ts1Cje hippocampal neurons having inherently increased levels of DSCAM (Alves-Sampaio et al., 2010). Interestingly, treatment with an NMDA antagonist (APV) abolished the increase in DSCAM protein level in both wild-type and Ts1Cje hippocampal neurons (Alves-Sampaio et al., 2010). In succession, DSCAM-mediated mechanisms played important roles during changes in synaptic plasticity, and thus, revitalization of Dscam-mediated signaling by induction of LTF was required during learning-related synapse formation (Li et al., 2009). In the context of learning and memory formation, presence of mTOR signaling in the dendritic elements appeared important (Jernigan et al., 2011), while hyperactivation of the Akt-mTOR pathway has been demonstrated as a characteristic outcome of DS disability (Troca-Marín et al., 2012). The Ts1Cje model showed enhanced phosphorylation of Akt-mTOR as a result of elevation of pro-BDNF and BDNF (Troca-Marín et al., 2012). Thus, it is collectively hypothesized that DSCAM is regulated by the NMDA pathway and is connected with dysregulation of m-TOR signaling to cause intellectual disability. Furthermore, the association between the glutamatergic pathways with DSCAM expression may suggest that this gene/protein is seemingly involved in the regulation of dendrite morphology and synaptic plasticity in DS and other mood disorders.

In Ts65Dn mice, presynaptic terminals and dendritic spines appeared enlarged along with thickening of the postsynaptic density of the CA1 region of the hippocampus (Belichenko et al., 2004; Kurt et al., 2004). Taken together, DSCAM-ran transsynaptic signaling, and interactions between PSD-95/Dscam may highlight a wide range of DSCAM-dependent signaling at the level of modulation of glutamate receptors, alongside NMDA-R, and metabotropic

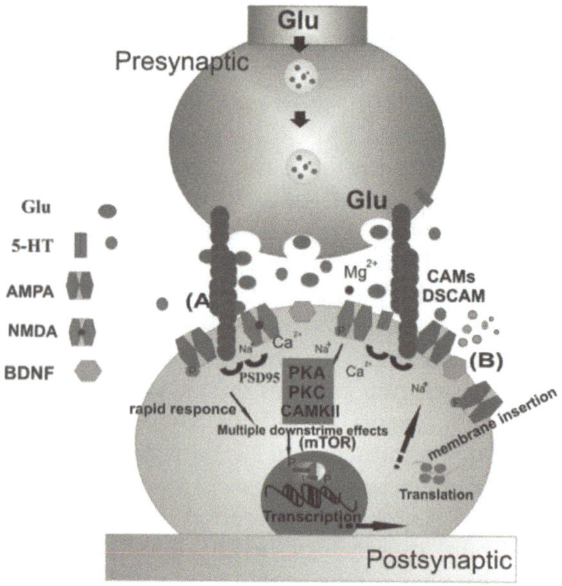

FIGURE 11.4 Schematic drawing of DSCAM localization on a glutamate synapse. Following Li et al. (2009), Dscam is a first-step safeguard in a whole cascade of events during glutamate synapse communication (A). It not only stabilizes the presynaptic part to the postsynaptic but is a guarantor of precise timing during this process. When presynaptically Glu is released, timed with depolarization, Mg^{2+} ions are released and binding of Glu to the postsynaptic part occurs, allowing permeation of Ca^{2+} ions. NMDA is composed of three main groups of subunits: GluN1, GluN2 (GluN2A, GluN2B, GluN2C, GluN2D), and GluN3 (GluN3A, GluN3B). It was documented, that during LTP a larger involvement of GluN1/GluN2A in total activity takes part (Paoletti et al., 2013). In opposition, LTD-engages mostly GluN1/GluN2B subunits. Ca^{2+} entry through NMDA receptors, or release by activation of metabotropic receptors, activates intracellular signaling pathways: inositol triphosphate (IP3), activation of CaMKII and insertion of GluR-1 AMPA receptor into the postsynaptic membrane. This process involves Dscam, as Li et al. (2009) documented its role in AMPA receptor stabilization (B). Furthermore, DSCAM protein level is regulated by NMDA-dependent synaptic activity and regulation is lost in trisomic neurons. The whole cascade of events is engaged in signaling, from the synapse to the nucleus, if gene transcription has to occur (mitogen activated protein kinase (MAPK), cAMP-responsive element binding protein (CREB)). CREB activation seems to be most important step in this cascade, initiating protein synthesis and leading to occurrence of LTP. One of the early genes in the process of synaptic plasticity is brain-derived neurotrophic factor (BDNF). Increased BDNF release is connected with activation of AMPA and mTOR pathways. These pathways are disrupted in DS persons, as Troca-Marin et al. (2012) showed increased local translation of BDNF, with greatest AMPA receptor delivery and occurrence of an excitatory loop. *Reprinted from Stachowicz K., 2018 The role of DSCAM in the regulation of synaptic plasticity: possible involvement in neuropsychiatric disorders. Acta Neurobiologiae Experimentalis. 78(3), 210–219, DOI: https://doi.org/10.21307/ane-2018-019, under CC-BY-4.0.*

glutamate receptors (mGluRs). The contribution of ketamine, a NMDA antagonist, is capable of producing rapid antidepressant effects in bipolar depression and patients with suicidal ideation (Banasr et al., 2011; Grady et al., 2017). Notably, elevated DSCAM level was reported in bipolar patients (Banasr et al., 2011). The therapeutic target of ketamine and rapamycin (mTOR) has been explained through autopsy studies of depressed patients exhibiting a potential link between mTOR signaling activity and deficits in synaptic proteins (Jernigan et al., 2011; Banasr et al., 2011). Indeed, Ketamine's antidepressant-like effect of was displayed with elevated hippocampal and prefrontal cortical mTOR and BDNF levels (Zhou et al., 2013). The increase in synthesis of BDNF was explained as a mechanism to obstruct the NMDA receptor activity (Szewczyk et al., 2012). Importantly, enhancement of BDNF levels was observed from chronic antidepressant treatment (Duric and Duman, 2013).

7.5 Modulators of excitatory deficits in DS

Several studies have indicated the key role of GABAergic signaling causing neuronal network deficits in DS mice. Notably, deficits in excitatory inputs and glutamatergic transmission could also result in an imbalance in E/I transmission in the trisomic brain, which are associated with delayed development and diminished production of excitatory glutamatergic neurons in the cortex, DG, and cerebellum of the Ts65Dn mice (Chakrabarti et al., 2007, 2010; Guidi et al., 2014; reviewed in Contestabile et al., 2017). In this line, the cortex and hippocampus of Ts65Dn mice showed a reduced density of glutamatergic synapses in electron microscopy and immunohistochemistry (Hibaoui et al., 2014). However, 1H MRS evaluation of different metabolites and neurotransmitters in the Ts65Dn hippocampus in vivo showed no difference in the concentration of both GABA and glutamate (Santin et al., 2015). Given the evidence in the published reports, a possible decrease of excitatory inputs is likely, while electrophysiological evaluation of the functionally glutamatergic signaling revealed reduced frequency of miniature excitatory postsynaptic currents (mEPSC) in the CA3 hippocampal region in trisomic mice (Stagni et al., 2013). Notably, a diminished ratio of postsynaptic NMDA/AMPA-evoked responses was observed in the CA1 region of these mice (Das et al., 2013).

The NMDA receptors (NMDA-Rs) in association with glutamate play essential roles in excitatory transmission and learning and memory processes. NMDA-Rs contribute to synaptic dysfunction by generating free radicals when it binds to excessive amounts of glutamate (Kamat et al., 2013). Interestingly, pharmacological manipulations of NMDA-Rs in Ts65Dn mice showed increase in electrophysiological and behavioral responses (Costa et al., 2008). Inhibition of NMDA transmission with a noncompetitive antagonist memantine rescued learning and memory performance using different behavioral tests in Ts65Dn mice following both acute and chronic administration; however, a complete mechanistic explanation of such outcome needs further evaluations (Costa et al., 2008; Costa, 2014). Memantine, an NMDA-R antagonist (approved by the FDA), binds the N-methyl D-aspartate 2A (NR2A) and N-methyl D-aspartate 2B (NR2B) subunits and modulates the excitatory neurotransmission through antagonizing the activity intermittently with high rates (Costa et al., 2008; Lipton, 2007). This antagonist is reported to prevent glutamate binding by binding to NMDA-Rs and thus, protects cognitive and memory deficits (Olivares et al., 2012). It regulates the activity of neurotransmitter glutamate, and acts on the NMDA receptors for regulating the quantity of calcium ions entering neurons in circulation of the neural influx. Glutamate overload enhances the activity of NMDA receptors leading to an excess of calcium penetrating the cells resulting in alteration in neural transmission. Chronic administration of memantine produced side effects similar to those observed with cholinergic medication (transient nausea, vomiting, loss of appetite, and diarrhea). The protein profiles of normal and trisomic mice with and without memantine treatment have demonstrated the impact of memantine on learning capability in Ts65Dn trisomic mice (Nguyen et al., 2011; Costa et al., 2008). Summarily, a comprehensive assessment of the involvement of glutamatergic signaling in E/I imbalance in DS mouse models could explain the glutamatergic functions in the trisomic brain in more detail.

A study on administration of memantine to human young adults with DS showed encouraging clinical signs (Capone, 2011). Interestingly, the efficacy of memantine was further evaluated in two clinical trials for improving cognitive functions in DS patients. The drug was well-tolerated; however, the first study (NCT00240760) did not exhibit any improvement on cognitive functions (Hanney et al., 2012), which might have been compromised by the advanced age of participants. Indeed, the irreversible pathological and degenerative changes of AD-related degeneration might have already occurred at the time of treatment, since DS patients are at increased risk of developing AD at early age (Costa, 2014). The second trial (NCT01112683) produced no significant difference in the primary outcome on a relatively small number of young adults with DS, though some encouraging improvement in verbal memory was detected in a secondary measure (Boada et al., 2012). Thus, the clinical trials with memantine have presented limited success for improving cognitive functioning in persons with DS and AD (Boada et al., 2012).

7.6 Modulation of GABA in DS neurodevelopment and critical period plasticity

Majority (but not all) of the brain defects in DS originate during embryonic and early life; hence, DS is indeed recognized as a neurodevelopmental disorder. Notably, both GABA-AR and GABA-BR actions significantly contribute to brain development and alterations in ambient GABA. Notably, the increased number of GABAergic interneurons may underlie some of the major brain alterations from fetal age to adult life (Gaiarsa and Porcher, 2013). Nevertheless, evaluation of the timing of the depolarizing/hyperpolarizing GABA switch (Ben-Ari, 2002) and related developmental changes in expression of GABA-AR subunits in trisomic mice may indicate the effects of aberrant GABAergic signaling in neural circuit formation in DS.

Additionally, GABA-BR signaling may contribute in perturbation of DS brain development through modulation of calcium channels and adenylate cyclase. However, trisomic GIRK2 is unlikely to contribute directly to GABAB-mediated alterations in early development of DS brain because co-operating effects of GABA-BRs and GIRK channels occur during the second postnatal week of life in rats (reviewed in Contestabile et al., 2017). Nonetheless, overexpression of GIRK2 may favor its premature coupling to GABA-BRs in DS. Moreover, modulation GABAergic signaling and/or intracellular Cl^- accumulation during particular developmental stages when brain circuits are likely to be more prone to plastic changes may improve learning and memory through adulthood.

Therefore, testing of the long-term effects of GABA-AR inhibitors (i.e., PTZ or α5 negative allosteric modulator treatments) or lowering $[Cl^-]$i with bumetanide during brain development was thought of, especially during important stages of DS brain development. Interestingly, prenatal or early postnatal treatment with Fluoxetine, SAG1.1 and choline exhibited rescue of memory performances later in life in Ts65Dn mice (Das et al., 2013; Guidi et al., 2014; Velazquez et al., 2013; Ash et al., 2014). Nevertheless, early pharmacological interventions may exert side effects of the drugs during developmental phase, and despite the beneficial effects, adverse effects of the drugs may trigger some organ-specific or systemic functions at adult age. Indeed, the NEMO trial (NCT01434225) on use of bumetanide on newborns caused hypoxic ischemic encephalopathy and hearing loss as adverse effects, which led to interruption of the trial (Pressler et al., 2015).

Interestingly, the outcome of prenatal and/or perinatal treatment with fluoxetine of Ts65Dn offsprings prompted a clinical trial aiming the assessment of the efficacy of Fluoxetine treatment during second trimester gestation and postnatal age up to 2 years (Guidi et al., 2014). However, prenatal administration of fluoxetine has resulted in birth defects (Reefhuis et al., 2015), and treatment in adult patients increased the risk of seizure susceptibility (Pisani et al., 1999). Treatment at a later age also increased aggression and changed emotional and social behaviors in males (reviewed in Contestabile et al., 2017). Therefore, the dosage and timing of administration should be critically analyzed. Moreover, fluoxetine treatment during embryonic and early postnatal life altered the migration of cortical GABAergic interneurons in rodents.

Aberrant GABAergic transmission was hypothesized to affect the critical period plasticity in the visual cortex. Indeed, this critical period plasticity depends on the depolarizing action of GABA during early development and its persistence can be extended by lowering the depolarizing GABA-A signaling with bumetanide treatment (Deidda et al., 2015a). Therefore, an early bumetanide intervention is expected to provide an extended window for neuronal plasticity in DS. Interestingly, the Ts65Dn mice showed deficits in cortical visual evoked potentials (VEPs), though critical-period plasticity in visual cortical circuits was not studied in trisomic animals (Scott-McKean et al., 2010). Additionally, exposure to EE either during early development or adulthood restored cortical VEP responses in trisomic mice, possibly through modulation of GABAergic transmission (Begenisic et al., 2011, 2015).

7.7 GABA alteration for other DS symptoms

Besides cognitive impairments, myriads of DS symptoms such as epilepsy, sleep disorders, and anxiety have been associated with defective GABAergic transmission (reviewed in Contestabile et al., 2017). Since the shift in GABA-AR-mediated responses renders impact on the E/I balance and support brain neuronal circuit hyperexcitability, increase in seizures might be consistent with depolarizing GABA-A signaling in DS. However, the effect of increased GABA signaling may reduce the overall neuronal network activity and thus, may decrease the incidence of seizures in Ts65Dn mice. In this direction, administration of a prodrug, γ-butyrolactone (GBL), for the GABA-BR agonist γ-hydroxybutyrate (GHB) enhanced episodes of seizures in Ts65Dn mice, which was otherwise rescued by normalization of GIRK2 gene dosage (Joshi et al., 2016).

However, the molecular mechanism of GHB action is yet to be established (Bay et al., 2014; Venzi et al., 2015), and thus, it may be speculated that in the backdrop of GABA dysregulation in DS, co-operating effect of GABA-BR-GIRK2 signaling and depolarizing GABA-A signaling may contribute to GBL-induced epileptic symptoms occurred in Ts65Dn mice. Notably, GHB may further enhance the already overexpressed GIRK2-mediated signaling and thus, facilitate drifting of VREST abnormally toward hyperpolarization. This situation would accentuate the depolarizing GABA-A signaling with augmentation of all depolarizing inputs and consequently trigger the epileptic incidence. In addition, the trisomic mice would also increase the incidence of audiogenic seizure, which can be controlled by administration of Fenobam, a metabotropic glutamate receptor subtype mGluR5 antagonist (reviewed in Contestabile et al., 2017), but not by bumetanide induced inhibition of NKCC1 (Deidda et al., 2015b). Furthermore, treatment with the α5-containing GABA-AR negative modulator RO4938581 showed decrease in the hyperactivity phenotype in the trisomic Ts65Dn mice (Martínez-Cué et al., 2013), but not with bumetanide treatment (Deidda et al., 2015b). The hypothesis based on these findings indicated that altered GABA signaling may increase susceptibility to seizure and hyperactive phenotypes in Ts65Dn mice.

The high incidence of sleep apnea in DS patients and Ts65Dn mice have been demonstrated with increased awaking and higher theta power in sleep EEG (Colas et al., 2013). However, these trisomic mice showed little or no differences in circadian rhythms, which could not be modulated by PTZ treatment (Ruby et al., 2010). Noticeably, PTZ was effective in restoring memory functions in Ts65Dn mice only when it was administrated during the diurnal phase, but not during the nocturnal phase, suggesting a likely different circadian effect of the GABAergic system on learning and memory in DS (Colas et al., 2013). Nevertheless, recently identified link between GABA signaling and memory function in circadian arrhythmic animals necessitated further studies to unravel the possible circadian effects in DS (Ruby et al., 2008). It may be apparent that alterations of GABA signaling and/or expression of NKCC1 may play important role in sleep disorders in DS because GABA-A signaling and possibly Cl^- homeostasis have been described as key regulator of sleep and circadian rhythms (reviewed in Contestabile et al., 2017).

In addition, the GABA-A signaling has been reported to be intricately associated with anxiety disorders (Vicari et al., 2013), that has not yet been assessed in trisomic mice. Interestingly, GIRK2 knockout mice exhibited low extent of anxiety, but anxiety was expressed in GABA-BR-knockout mice. Thus, coupling of GIRK channels to metabotropic neurotransmitter receptor mechanisms but not to GABA-BRs, was proposed to regulate the anxiety-related symptoms in DS. Indeed, modulation of signal transduction of dopamine receptors was favored by the overexpression of GIRK2 in the midbrain in Ts65Dn mice, which may control anxiety and other related features of DS (Harashima et al., 2006). Summarily, association of altered GABAergic transmission, impairment of adult neurogenesis and AD, and its collective contribution on impairment of cognition have not yet been explored fully in the field of DS research (Li et al., 2016; Rissman and Mobley, 2011).

8. Concluding remarks

The gene dosage of triplicated HSA21 has unequivocally been identified as the key element in pathomechanisms associated with DS population. Knowledge on contribution of genes has been expanded through creation of DS mouse models and derivation of iPSCs from DS patients. Still no therapeutic has been defined for DS-specific clinical features, especially for cognitive deficits, neurodegeneration, and dementia. Understanding the abnormal signaling from GABA-A and GABA-B receptors and E/I balance on the neuronal defects in DS have been expanded using the Ts65Dn mouse model. The genesis of defective GABAergic transmission tracks back into fetal onset of defective neurodevelopment, with possible E/I imbalance and neuronal and plasticity impairments that percolates to adulthood. However, the gravity of the GABAergic inhibitory versus the glutamatergic excitatory transmission in this irregularity is still less understood. Indeed, cooperation of depolarizing GABA-A signaling and glutamatergic excitatory alterations appeared to be brain-region specific in trisomic Ts65Dn mice. In this direction, expanded studies involving DS patient–derived iPSCs may contribute to the future understanding of DS neuropathology to a greater extent. Indeed, a low predictive potential of the Ts65Dn mouse model could be replaced by use of trisomic iPSCs in pharmaceutical investigation of potential drug candidates. Discordant results of DS mice, such as changes in GABA-AR subunit composition specific for human DS, were lacking in the Ts65Dn mice (Bhattacharyya et al., 2009) and could be responsible for the failure of therapeutic outcome with α5 selective antagonists in DS patients. Moreover, creation of the "perfect" replica of trisomic HSA21 in DS models has come out with unintended rearrangements. Thus, future assays on the efficacy of each new pharmacological treatment may focus on more than one DS model both in vitro and in vivo and that can control the failures, inadequacy, and discrepancies in clinical trials. Overall, although the available data from human studies are limited and seem not to support the data derived from animal research on increased GABA-mediated transmission due to the overproduction of GABAergic interneurons, further electrophysiological studies on iPSCs-derived neurons are needed. On the other hand, no study has yet conclusively assessed the occurrence of depolarizing GABA-A signaling or increased GABA-B-mediated transmission in human DS neurons.

The present chapter has illustrated the multifaceted consequences of DS brain abnormalities. The alterations in neurotransmission, especially the impaired GABAergic transmission, represent the collective expression of different molecular mechanisms exerted by gene-dosage imbalance. Seemingly, the abnormalities of neurotransmission originate (at least in part) during neurodevelopment at gestation and that goes through complex synaptic, physiological, and circuit changes through adulthood, ultimately causing cognitive impairment and other neurological manifestations. As a matter of fact, the unmet identification of effective pharmacological interventions for alleviation of cognitive deficits in DS constitutes an incredible complex challenge for extraction of the insights about underlying GABA-related impairments and other neurodevelopmental disorders related to defective GABAergic transmission.

References

Abbas, A.I., Sundiang, M.J.M., Henoch, B., Morton, M.P., Bolkan, S.S., Park, A.J., Harris, A.Z., Kellendonk, C., Gordon, J.A., 2018. Somatostatin interneurons facilitate Hippocampal-Prefrontal synchrony and prefrontal spatial encoding. Neuron 100, 926–939. https://doi.org/10.1016/j.neuron.2018.09.029.

Akerman, J., Cline, H.T., 2007. Refining the roles of-GABAergic signaling during neural circuit formation. Trends Neurosci. 30 (8), 382–389.

Alemany-Gonza´lez, M., Gener, T., Nebot, P., Vilademunt, M., Dierssen, M., Puig, M.V., 2020. Prefrontal-hippocampal functional connectivity encodes recognition memory and is impaired in intellectual disability. Proc. Natl. Acad. Sci. U. S. A. 117, 11788–11798. https://doi.org/10.1073/pnas.1921314117.

Allain, A.-E., Le Corronc, H., Delpy, A., Cazenave, W., Meyrand, P., et al., 2011. Maturation of the GABAergic transmission in normal and pathologic motoneurons. Neural Plast. 905624. https://doi.org/10.1155/2011/905624.

Alves-Sampaio, A., Troca-Marín, J.A., Montesinos, M.L., 2010. NMDA-mediated regulation of DSCAM dendritic local translation is lost in a mouse model of Down's syndrome. J. Neurosci. 30, 13537–13548.

Anderson, S.A., et al., 2002. Distinct origins of neocortical projection neurons and interneurons in vivo. Cerebr. Cortex 12, 702–709.

Ash, J.A., Velazquez, R., Kelley, C.M., Powers, B.E., Ginsberg, S.D., Mufson, E.J., et al., 2014. Maternal choline supplementation improves spatial mapping and increases basal forebrain cholinergic neuron number and size in aged Ts65Dn mice. Neurobiol. Dis. 70, 32–42. https://doi.org/10.1016/j.nbd.2014.06.001.

Ataka, T., Gu, J.G., 2006. Relationship between tonic inhibitory currents and phasic inhibitory activity in the spinal cord lamina II region of adult mice. Mol. Pain 2, 36. https://doi.org/10.1186/1744-8069-2-36.

Atallah, B.V., Bruns, W., Carandini, M., Scanziani, M., 2012. Parvalbumin-expressing interneurons linearly transform cortical responses to visual stimuli. Neuron 73, 159–170. https://doi.org/10.1016/j.neuron.2011.12.013.

Banasr, M., Dwyer, J.M., Duman, R.S., 2011. Cell atrophy and loss in depression: reversal by antidepressant treatment. Curr. Opin. Cell Biol. 23, 730–737.

Banks, G.B., Kanjhan, R., Wiese, S., et al., 2005. Glycinergic and GABAergic synaptic activity differentially regulate motoneuron survival and skeletal muscle innervation. J. Neurosci. 25 (5), 1249–1259.

Batista-Brito, R., et al., 2008. Gene expression in cortical inter neuron precursors is prescient of their mature function. Cerebr. Cortex 18, 2306–2317.

Bay, T., Eghorn, L.F., Klein, A.B., Wellendorph, P., 2014. GHB receptor targets in the CNS: focus on high-affinity binding sites. Biochem. Pharmacol. 87, 220–228. https://doi.org/10.1016/j.bcp.2013.10.028.

Becker, L., Mito, T., Takashima, S., Onodera, K., 1991. Growth and development of the brain in Down syndrome. Prog. Clin. Biol. Res. 373, 133–152.

Begenisic, T., Sansevero, G., Baroncelli, L., Cioni, G., Sale, A., 2015. Early environmental therapy rescues brain development in a mouse model of Down syndrome. Neurobiol. Dis. 82, 409–419. https://doi.org/10.1016/j.nbd.2015.07.014.

Begenisic, T., Spolidoro, M., Braschi, C., Baroncelli, L., Milanese, M., Pietra, G., et al., 2011. Environmental enrichment decreases GABAergic inhibition and improves cognitive abilities, synaptic plasticity, and visual functions in a mouse model of Down syndrome. Front. Cell. Neurosci. 5, 29. https://doi.org/10.3389/fncel.2011.00029.

Belichenko, P.V., Masliah, E., Kleschevnikov, A.M., Villar, A.J., Epstein, C.J., Salehi, A., Mobley, W.C., 2004. Synaptic structural abnormalities in the Ts65Dn mouse model of Down syndrome. J. Comp. Neurol. 480, 281–298.

Belichenko, P.V., Kleschevnikov, A.M., Salehi, A., Epstein, C.J., Mobley, W.C., 2007. Synaptic and cognitive abnormalities in mouse models of Down syndrome: exploring genotype-phenotype relationships. J. Comp. Neurol. 504 (4), 329–345. https://doi.org/10.1002/cne.21433.

Belichenko, P.V., Kleschevnikov, A.M., Masliah, E., Wu, C., Takimoto-Kimura, R., Salehi, A., et al., 2009a. Excitatory-inhibitory relationship in the fascia dentata in the Ts65Dn mouse model of Down syndrome. J. Comp. Neurol. 512, 453–466. https://doi.org/10.1002/cne.21895.

Belichenko, N.P., Belichenko, P.V., Kleschevnikov, A.M., Salehi, A., Reeves, R.H., Mobley, W.C., 2009b. The "Down syndrome critical region" is sufficient in the mouse model to confer behavioral, neurophysiological and synaptic phenotypes characteristic of down syndrome. J. Neurosci. 29, 5938–5948. https://doi.org/10.1523/JNEUROSCI.1547-09.2009.

Ben-Ari, Y., Gaiarsa, J.L., Tyzio, R., Khazipov, R., 2007. GABA: a pioneer transmitter that excites immature neurons and generates primitive oscillations. Physiol. Rev. 87, 1215–1284. https://doi.org/10.1152/physrev.00017.2006.

Ben-Ari, Y., 2001. Developing networks play a similar melody. Trends Neurosci. 24 (6), 353–360.

Ben-Ari, Y., 2002. Excitatory actions of gaba during development: the nature of the nurture. Nat. Rev. Neurosci. 3, 728–739. https://doi.org/10.1038/nrn920.

Best, T.K., Siarey, R.J., Galdzicki, Z., 2007. Ts65Dn, a mouse model of Down syndrome, exhibits increased GABAB-induced potassium current. J. Neurophysiol. 97, 892–900. https://doi.org/10.1152/jn.00626.2006.

Benoit, S., Arnaud, D., Gu, Y., et al., 2019. Prenatal treatment with EGCG enriched green tea extract rescues GAD67 related developmental and cognitive defects in Down syndrome mouse models. Sci. Rep. 9, 3914. https://doi.org/10.1038/s41598-019-40328-9.

Best, T.K., Cramer, N.P., Chakrabarti, L., Haydar, T.F., Galdzicki, Z., 2012. Dysfunctional hippocampal inhibition in the Ts65Dn mouse model of Down syndrome. Exp. Neurol. 233, 749–757. https://doi.org/10.1016/j.expneurol.2011.11.033.

Bhattacharyya, A., McMillan, E., Chen, S.I., Wallace, K., Svendsen, C.N., 2009. A critical period in cortical interneuron neurogenesis in down syndrome revealed by human neural progenitor cells. Dev. Neurosci. 31, 497–510. https://doi.org/10.1159/000236899.

Block, A., Ahmed, M.M., Rueda, N., Hernandez, M.C., Martinez-Cué, C., Gardiner, K.J., 2017. The GABAA α5-selective modulator, RO4938581, rescues ProteinAnomalies in the Ts65Dn mouse model of down syndrome. Neuroscience 372.

Boada, R., Hutaff-Lee, C., Schrader, A., et al., 2012. Antagonism of NMDA receptors as a potential treatment for Down syndrome: a pilot randomized controlled trial. Transl. Psychiat. 2, e141.

Braat, S., Kooy, R.F., 2015. The GABA receptor as a therapeutic target forneurodevelopmental disorders. Neuron 86, 1119–1130. https://doi.org/10.1016/j.neuron.2015.03.042.

Braudeau, J., Delatour, B., Duchon, A., Pereira, P.L., Dauphinot, L., de Chaumont, F., et al., 2011. Specific targeting of the GABA-A receptor alpha5 subtype by a selective inverse agonist restores cognitive deficits in Down syndrome mice. J. Psychopharmacol. 25, 1030–1042. https://doi.org/10.1177/0269881111405366.

Brault, V., Nguyen, T.L., Flores-Gutiérrez, J., et al., 2021. *Dyrk1a* gene dosage in glutamatergic neurons has key effects in cognitive deficits observed in mouse models of MRD7 and Down syndrome. PLoS Genet. 17 (9), e1009777. https://doi.org/10.1371/journal.pgen.1009777.

Brun, V.H., Otnass, M.K., Molden, S., Steffenach, H.A., Witter, M.P., Moser, M.B., et al., 2002. Place cells and place recognition maintained by direct entorhinal-hippocampal circuitry. Science 296, 2243–2246. https://doi.org/10.1126/science.1071089.

Bu, D.F., Erlander, M.G., Hitz, B.C., et al., 1992. Two human glutamate decarboxylases, 65-kDa GAD and 67-kDa GAD, are each encoded by a single gene. Proc. Natl. Acad. Sci. U. S. A. 89 (6), 2115–2119.

Busciglio, J., Yankner, B.A., 1995. Apoptosis and increased generation of reactive oxygen species in Down's syndrome neurons in vitro. Nature 378, 776–779.

Busciglio, J., Capone, G., O'Bryan, J., Gardiner, K.J., 2013. Down syndrome: genes, model systems, and progress towards pharmacotherapies and clinical trials forcognitive deficits. Cytogenet. Genome Res. 141, 260–271.

Butt, S.J., et al., 2005. The temporal and spatial origins of corticalinterneurons predict their physiological subtype. Neuron 48, 591–604.

Buzsa´ ki, G., Wang, X.J., 2012. Mechanisms of gamma oscillations. Annu. Rev. Neurosci. 35, 203–225. https://doi.org/10.1146/annurev-neuro-062111-150444. PMID: 22443509.

Capone, G., 2011. Pharmacotherapy for children with Down syndrome. In: Rondal, J.A.P., Spiker, D.A. (Eds.), Neurocognitive Rehabilitation of Down Syndrome. Cambridge University Press, Cambridge, pp. 96–116.

Cardin, J.A., Carle´ n M., Meletis, K., Knoblich, U., Zhang, F., Deisseroth, K., Tsai, L.H., Moore, C.I., 2009. Driving fast-spiking cells induces gamma rhythm and controls sensory responses. Nature 459, 663–667. https://doi.org/10.1038/nature08002.

Cellot, G., Cherubini, E., 2013. Functional role of ambient GABA in refining neuronal circuits early in postnatal development. Front. Neural Circ. 7, 136. https://doi.org/10.3389/fncir.2013.00136.

Cellot, G., Cherubini, E., 2014. GABAergic signaling as therapeutic target for autism spectrum disorders. Front. Pediatr. 2, 70. https://doi.org/10.3389/fped.2014. 00070.

Chakrabarti, L., Galdzicki, Z., Haydar, T.F., 2007. Defects in embryonic neurogenesis and initial synapse formation in the forebrain of the Ts65Dn mouse model of Down syndrome. J. Neurosci. 27 (43), 11483–11495. https://doi.org/10.1523/jneurosci.3406-07.2007.

Chakrabarti, L., Best, T.K., Cramer, N.P., Carney, R.S.E., Isaac, J.T.R., Galdzicki, Z., et al., 2010. Olig1 and Olig2 triplication causes developmental brain defects in Down syndrome. Nat. Neurosci. 13, 927–934. https://doi.org/10.1038/nn.2600.

Chen, C., Jiang, P., Xue, H., Peterson, S.E., Tran, H.T., McCann, A.E., ParastMM, Li, S., et al., 2014. Role of astroglia in Down's syndrome revealed by patient-derived human-induced pluripotent stem cells. Nat. Commun. 5, 4430. https://doi.org/10.1038/ncomms5430.

Chen, G., Zhang, Y., Li, X., Zhao, X., Ye, Q., Lin, Y., Tao, H.W., Rasch, M.J., Zhang, X., 2017. Distinct inhibitory circuits orchestrate cortical beta and gamma band oscillations. Neuron 96, 1403–1418. https://doi.org/10.1016/j. neuron.2017.11.033.

Cheng, X.P., Sun, H., Ye, Z.Y., Zhou, J.N., 2012. Melatonin modulates the GABAergic response in cultured rat hippocampal neurons. J. Pharmacol. Sci. 119, 177–185. https://doi.org/10.1254/jphs.11183fp.

Chiang, P.-H., Wu, P.-Y., Kuo, T.-W., Liu, Y.-C., Chan, C.-F., Chien, T.-C., et al., 2012. GABA is depolarizing in hippocampal dentate granule cells of the adolescent and adult rats. J. Neurosci. 32, 62–67. https://doi.org/10.1523/JNEUROSCI.3393-11.2012.

Cho, K.K., Hoch, R., Lee, A.T., Patel, T., Rubenstein, J.L., Sohal, V.S., 2015. Gamma rhythms link prefrontal interneuron dysfunction with cognitive inflexibility in dlx5/6+/-mice. Neuron 85, 1332–1343. https://doi.org/10.1016/j. neuron.2015.02.019. PMID: 25754826.

Ciruela, F., Fernández-Dueñas, V., Sahlholm, K., Fernández-Alacid, L., Nicolau, J.C., Watanabe, M., et al., 2010. Evidence for oligomerization between GABAB receptors and GIRK channels containing the GIRK1 and GIRK3 subunits. Eur. J. Neurosci. 32, 1265–1277. https://doi.org/10.1111/j.1460-9568.2010.07356.x.

Colas, D., Chuluun, B., Warrier, D., Blank, M., Wetmore, D.Z., Buckmaster, P., et al., 2013. Short-term treatment with the GABAA receptor antagonist pentylenetetrazole produces a sustained pro-cognitive benefit in a mouse model of Down's syndrome. Br. J. Pharmacol. 169, 963–973. https://doi.org/10.1111/bph.12169.

Collinson, N., Kuenzi, F.M., Jarolimek, W., Maubach, K.A., Cothliff, R., Sur, C., et al., 2002. Enhanced learning and memory and altered GABAergic synaptic transmission in mice lacking the α5 subunit of the GABAA receptor. J. Neurosci. 22, 5572–5580.

Contestabile, A., Magara, S., Cancedda, L., 2017. The GABAergic hypothesis for cognitive disabilities in down syndrome. Front. Cell. Neurosci. 11. https://doi.org/10.3389/fncel.2017.00054. Article 54.

Corfas, G., et al., 2004. Neuregulin 1-erbB signaling and the-molecular/cellular basis of schizophrenia. Nat. Neurosci. 7, 575–580.

Cornelisse, L.N., van der Harst, J.E., Lodder, J.C., Baarendse, P.J., Timmerman, A.J., Mansvelder, H.D., et al., 2007. Reduced 5-HT1A- and GABAB receptor function in dorsal raphé neurons upon chronic fluoxetine treatment of socially stresse drats. J. Neurophysiol. 98, 196–204. https://doi.org/10.1152/jn.00109.2007.

Corrales, A., Vidal, R., García, S., Vidal, V., Martínez, P., García, E., et al., 2013. Chronic melatonin treatment rescues electrophysiological and neuromorphological deficits in a mouse model of Down syndrome. J. Pineal Res. 56, 51–61. https://doi.org/10.1111/jpi.12097.

Costa, A.C., Grybko, M.J., 2005. Deficits in hippocampal CA1 LTP induced by TBS but not HFS in the Ts65Dn mouse: a model of Down syndrome. Neurosci. Lett. 382, 317–322. https://doi.org/10.1016/j.neulet.2005.03.031.

Costa, A.C., Scott-McKean, J.J., Stasko, M.R., 2008. Acute injections of the NMDA receptor antagonist memantine rescue performance deficits of the Ts65Dn mousemodel of Down syndrome on a fear conditioning test. Neuropsychopharmacology 33, 1624–1632.

Costa, A.C., 2014. The glutamatergic hypothesis for Down syndrome: the potential use of N-methyl-D-aspartate receptor antagonists to enhance cognition and decelerate neurodegeneration. CNS Neurol. Disord. - Drug Targets 13, 16–25. https://doi.org/10.2174/18715273113126660183.

Cramer, N., Galdzicki, Z., 2012. From abnormal hippocampal synaptic plasticity in down syndrome mouse models to cognitive disability in down syndrome. Neural Plast. 101542, 2012.

Cramer, N.P., Xu, X.F., Haydar, T., Galdzicki, Z., 2015. Altered intrinsicand network properties of neocortical neurons in the Ts65Dn mouse model of Down syndrome. Phys. Rep. 3, e12655. https://doi.org/10.14814/phy2.12655.

Crestani, F., Keist, R., Fritschy, J.M., et al., 2002. Trace fear conditioning involves hippocampal α5 GABAA receptors. Proc. Nat. Acad. Sci. U. S. Am. 99 (13), 8980–8985.

Cummings, K.A., Clem, R.L., 2020. Prefrontal somatostatin interneurons encode fear memory. Nat. Neurosci. 23, 61. https://doi.org/10.1038/s41593-019-0552-7.

Das, I., Park, J.-M., Shin, J.H., Jeon, S.K., Lorenzi, H., Linden, D.J., et al., 2013. Hedgehog agonist therapy corrects structural and cognitive deficits in a Down syndrome mouse model. Sci. Transl. Med. 5. https://doi.org/10.1126/scitranslmed.3005983, 201ra120.

De la Torre, R., De Sola, S., Hernandez, G., et al., 2016. Safety and efficacy of cognitive training plus epigallocatechin-3-gallate in young adults with Down's syndrome (TESDAD): a doubleblind, randomized, placebo-controlled, phase 2 trial. Lancet Neurol. 15, 801–810.

de San Martin, J.Z., Delabar, J.M., Bacci, A., Potier, M.C., 2018. GABAergic over-inhibition, a promising hypothesis for cognitive deficits in down syndrome. Free Radic. Biol. Med. 114, 33–39. https://doi.org/10.1016/j.freeradbiomed.2017.10.002. PMID: 28993272.

de San Martin, J.Z., Donato, C., my Peixoto Je're', Aguirre, A., Choudhary, V., De Stasi, A.M., Lourenço, J., Marie-Claude, P., Bacci, A., 2020. Alterations of specific cortical GABAergic circuits underlie abnormal network activity in a mouse model of Down syndrome. Elife 9, e58731. https://doi.org/10.7554/eLife.58731. 23.

Deidda, G., Bozarth, I.F., Cancedda, L., 2014. Modulation of GABAergic transmission in development and neurodevelopmental disorders: investigating physiology and pathology to gain therapeutic perspectives. Front. Cell. Neurosci. 8, 119. https://doi.org/10.3389/fncel.2014.00119.

Deidda, G., Allegra, M., Cerri, C., Naskar, S., Bony, G., Zunino, G., et al., 2015a. Early depolarizing GABA controls critical-period plasticity in the rat visual cortex. Nat. Neurosci. 18, 87–96. https://doi.org/10.1038/nn.3890.

Deidda, G., Parrini, M., Naskar, S., Bozarth, I.F., Contestabile, A., Cancedda, L., 2015b. Reversing excitatory GABAAR signaling restores synaptic plasticity and memory in a mouse model of Down syndrome. Nat. Med. 21, 318–326. https://doi.org/10.1038/nm.3827.

Dogliotti, G., Galliera, E., Licastro, F., Corsi, M.M., 2010. Age-related changes in plasma levels of BDNF in Down syndrome patients. Immun. Ageing 7, 2.

Duchon, A., Gruart, A., Albac, C., Delatour, B., Zorrilla de San Martin, J., Delgado-García, J.M., Hé rault, Y., Potier, M.C., 2020. Long-lasting correction of in vivo LTP and cognitive deficits of mice modelling down syndrome with an a5-selective GABAA inverse agonist. Br. J. Pharmacol. 177, 1106–1118. https://doi.org/10.1111/bph.14903.

Duric, V., Duman, R.S., 2013. Depression and treatment response: dynamic interplay of signaling pathways and altered neural processes. Cell. Mol. Life Sci. 70, 39–53.

Farrant, M., Kaila, K., 2007. The cellular, molecular and ionic basis of GABAA receptor signalling. Prog. Brain Res. 160, 59–87.

Fernandez, F., Garner, C.C., 2008. Episodic-like memory in Ts65Dn, a mouse model of Down syndrome. Behav. Brain Res. 188, 233–237. https://doi.org/10.1016/j.bbr.2007.09.015.

Fernandez, F., Morishita, W., Zuniga, E., Nguyen, J., Blank, M., Malenka, R.C., et al., 2007. Pharmacotherapy for cognitive impairment in a mouse model of Down syndrome. Nat. Neurosci. 10, 411–413. https://doi.org/10.1038/nn1860.

Fernandez, F., Nyhuis, C.C., Anand, P., Demara, B.I., Ruby, N.F., Spanò, G., Clark, C., Edgin, J.O., 2017. Young children with Down syndrome show normal development of circadian rhythms, but poor sleep efficiency: a cross-sectional study across the first 60 months of life. Sleep Med. 33, 134–144. https://doi.org/10.1016/j.sleep.2016. 12.026.

Fiumelli, H., Cancedda, L., Poo, M.M., 2005. Modulation of GABAergic transmission by activity via postsynaptic Ca2C-dependent regulation of KCC2 function. Neuron 48, 773–786. https://doi.org/10.1016/j.neuron.2005.10.025.

Flames, N., et al., 2004. Short- and long-range attraction of cortical GABAergic interneurons by neuregulin-1. Neuron 44, 251–261.

Fode, C., et al., 2000. A role for neural determination genes inspecifying the dorsoventral identity of telencephalic neurons. Genes Dev. 14, 67–80.

Fritschy, J.M., Panzanelli, P., 2014. GABAA receptors and plasticity of inhibitory neurotransmission in the central nervous system. Eur. J. Neurosci. 39, 1845–1865. https://doi.org/10.1111/ejn.12534.

Gaiarsa, J.-L., Porcher, C., 2013. Emerging neurotrophic role of GABAB receptors in neuronal circuit development. Front. Cell. Neurosci. 7, 206. https://doi.org/10.3389/fncel.2013.00206.

Gao, B.X., Stricker, C., Ziskind-Conhaim, L., 2001. Transition-from GABAergic to glycinergic synaptic transmission innewly formed spinal networks. J. Neurophysiol. 86 (1), 492–502.

Gardiner, K.J., 2015. Pharmacological approaches to improving cognitive function in Down syndrome: current status and considerations. Drug Design Dev. Ther. 9, 103–125.

Gassmann, M., Bettler, B., 2012. Regulation of neuronal GABAB receptor functions by subunit composition. Nat. Rev. Neurosci. 13, 380–394. https://doi.org/10.1038/nrn3249.

Ghit, A., Assal, D., Al-Shami, A.S., et al., 2021. GABAA receptors: structure, function, pharmacology, and related disorders. J. Genet. Eng. Biotechnol. 19, 123. https://doi.org/10.1186/s43141-021-00224-0.

Gimé nez, S., Videla, L., Romero, S., Benejam, B., Clos, S., Ferná ndez, S., Martínez, M., Carmona-Iragui, M., Antonijoan, R.M., Mayos, M., Fortuna, A., Peñ acoba, P., Plaza, V., Osorio, R.S., Sharma, R.A., Bardé, s I., Rebillat, A.S., Lleó, A., Blesa, R., Videla, S., et al., 2018. Prevalence of sleep disorders in adults with down syndrome: a comparative study of SelfReported, Actigraphic, and polysomnographic findings. J. Clin. Sleep Med. 14, 1725–1733. https://doi.org/10.5664/jcsm.7382.

Gingrich, K.J., Roberts, W.A., Kass, R.S., 1995. Dependenceof the GABAA receptor gating kinetics on the α-subunitisoform: implications for structure-function relations andsynaptic transmission. J. Physiol. 489 (2), 529–543.

Glausier, J., Fish, K., Lewis, D., 2014. Altered parvalbumin basket cell inputs in the dorsolateral prefrontal cortex of schizophrenia subjects. Mol. Psychiatr. 19, 30–36. https://doi.org/10.1038/mp.2013.152.

Gonzalez-Islas, C., Chub, N., Wenner, P., 2009. NKCC1 andAE3 appear to accumulate chloride in embryonic motoneurons. J. Neurophysiol. 101 (2), 507–518.

Grady, S.E., Marsh, T.A., Tenhouse, A., Klein, K., 2017. Ketamine for the treatment of major depressive disorder and bipolar depression: a review of the literature. Mental Health Clinician 7, 16–23.

Guidi, S., Stagni, F., Bianchi, P., et al., 2014. Prenatal pharmacotherapy rescues brain development in a Down's syndrome mouse model. Brain 137, 380–401.

Gulledge, A.T., Stuart, G.J., 2003. Excitatory actions of GABA in the cortex. Neuron 37, 299–309. https://doi.org/10.1016/s0896-6273(02)01146-7.

Hampe, C.S., Mitoma, H., Manto, M., December 20, 2017. GABA and glutamate: their transmitter role in the CNS and pancreatic islets. In: GABA and Glutamate - New Developments in Neurotransmission Research. Janko Samardzic, IntechOpen. https://doi.org/10.5772/

intechopen.70958. Available from: https://www.intechopen.com/chapters/57103.

Hanney, M., Prasher, V., Williams, N., Jones, E.L., Aarsland, D., Corbett, A., et al., 2012. Memantine for dementia in adults older than 40 years with Down's syndrome (MEADOWS): a randomised, double-blind, placebo-controlled trial. Lancet 379, 528–536. https://doi.org/10.1016/S0140-6736(11)61676-0.

Hanson, J.E., Blank, M., Valenzuela, R.A., Garner, C.C., Madison, D.V., 2007. The functional nature of synaptic circuit ryisalteredinarea CA3 of the hippocampus in a mouse model of Down's syndrome. J. Physiol. 579, 53–67. https://doi.org/10.1113/jphysiol.2006.114868.

Hanson, J.E., Weber, M., Meilandt, W.J., Wu, T., Luu, T., Deng, L., et al., 2013. GluN2B antagonism affects interneurons and leads to immediate and persistent changes in synaptic plasticity, oscillations, and behavior. Neuropsychopharmacology 38, 1221–1233. https://doi.org/10.1038/npp.2013.19.

Harashima, C., Jacobowitz, D.M., Witta, J., Borke, R.C., Best, T.K., Siarey, R.J., et al., 2006. Abnormal expression of the G-protein-activated inwardly rectifying potassium channel 2 (GIRK2) in hippocampus, frontal cortex and substantia nigra of Ts65Dn mouse: a model of Down syndrome. J. Comp. Neurol. 494, 815–833. https://doi.org/10.1002/cne.20844.

Hattori, D., Millard, S., Wojtowicz, W.M., Zipursky, S.L., 2008. Dscam-mediated cell recognition regulates neural circuit formation. Annu. Rev. Cell Dev. Biol. 24, 597–620.

Hebb, D.O., 1949. The Organization of Behavior: A Neuropsychological Theory. Wiley and Sons, New York, USA.

Helfrich, R.F., Knight, R.T., 2016. Oscillatory dynamics of prefrontal cognitive control. Trends Cognit. Sci. 20, 916–930. https://doi.org/10.1016/j.tics.2016.09.007. PMID: 27743685.

Hernández-González, S., Ballestin, R., López-Hidalgo, R., Gilabert-Juan, J., BlascoIbáñez, J.M., Crespo, C., et al., 2015. Altered distribution of hippocampal interneurons in the murine Down syndrome model Ts65Dn. Neurochem. Res. 40, 151–164. https://doi.org/10.1007/s11064-014-1479-8.

Hernandez-Miranda, L.R., et al., 2010. Molecules and mechanisms involved in the generation and migration of cortical interneurons. ASN Neuro 2, e00031.

Hibaoui, Y., Grad, I., Letourneau, A., Sailani, M.R., Dahoun, S., SantoniFA, G.S., Guipponi, M., et al., 2014. Modelling and rescuing neuro developmental defect of Down syndrome using induced pluripotentstem cells from monozygotic twins discordant for trisomy21. EMBO Mol. Med. 6 (2), 259–277. https://doi.org/10.1002/emmm.201302848.

Hinckley, C., Seebach, B., Ziskind-Conhaim, L., 2005. Distinctroles of glycinergic and GABAergic inhibition in coordinatinglocomotor-like rhythms in the neonatal mousespinalcord. Neuroscience 131 (3), 745–758.

Hsu, C.C., Davis, K.M., Jin, H., et al., 2000. Association of Lglutamic acid decarboxylase to the70-k Daheatshock protein as a potential anchoring mechanism to synaptic vesicles. J. Biol. Chem. 275 (27), 20822–20828.

Huberfeld, G., Wittner, L., Clemenceau, S., Baulac, M., Kaila, K., Miles, R., Rivera, C., 2007. Perturbed chloride homeostasis and GABAergic signaling in human temporal lobe epilepsy. J. Neurosci. 27 (37), 9866–9873.

Jacob, T.C., Moss, S.J., Jurd, R., 2008. GABAA receptor trafficking and its role in the dynamic modulation of neuronal inhibition. Nat. Rev. Neurosci. 9, 331–343. https://doi.org/10.1038/nrn2370.

Jernigan, C.S., Goswami, D.B., Austin, M.C., Iyo, A.H., Chandran, A., Stockmeier, C.A., Karolewicz, B., 2011. The mTOR signaling pathway in the prefrontal cortex is compromised in major depressive disorder. Prog. Neuro-Psychopharmacol. Biol. Psychiat. 35, 1774–1779.

Johnston, A., McBain, C.J., Fisahn, A., 2014. 5-Hydroxytryptamine1A receptor-activation hyperpolarizes pyramidal cells and suppresses hippocampal gamma oscillations via Kir3 channel activation. J. Physiol. 592, 4187–4199. https://doi.org/10.1113/jphysiol.2014.279083.

Joshi, K., Shen, L., Michaeli, A., Salter, M., Thibault-Messier, G., Hashmi, S., et al., 2016. Infantile spasms in down syndrome: rescue by knockdown of the GIRK2 channel. Ann. Neurol. 80, 511–521. https://doi.org/10.1002/ana.24749.

Kalanithi, P.S., et al., 2005. Altered parvalbumin-positive neuron distribution in basal ganglia of individuals with Tourette syndrome. Proc. Natl. Acad. Sci. U.S.A. 102, 13307–13312.

Kamat, P.K., Rai, S., Swarnkar, S., Shukla, R., Ali, S., Najmi, A.K., et al., 2013. Okadaic acidinduced Tau phosphorylation in rat brain: role of NMDA receptor. Neuroscience 238, 97–113.

Kataoka, Y., et al., 2010. Decreased number of parvalbumin and cholinergic interneurons in the striatum of individuals with Tourette syndrome. J. Comp. Neurol. 518, 277–291.

Katarova, Z., Sekerkova, G., Prodan, S., Mugnaini, E., Szabo, G., 2000. Domain-restricted expression of two glutamic acid decarboxylase genes in midgestation mouse embryos. J. Comp. Neurol. 424 (4), 607–627.

Kaur, G., Sharma, A., Xu, W., et al., 2014. Glutamatergic transmission aberration: a major cause of behavioral deficits in a murine model of Down's syndrome. J. Neurosci. 34 (15), 5099–5106. https://doi.org/10.1523/JNEUROSCI.5338-13.2014.

Khazipov, R., Valeeva, G., Khalilov, I., 2015. Depolarizing GABA anddevelopmental epilepsies. CNS Neurosci. Ther. 21, 83–91. https://doi.org/10.1111/cns.12353.

Khirug, S., Yamada, J., Afzalov, R., Voipio, J., Khiroug, L., Kaila, K., 2008. GABAergic depolarization of the axon initial segment in cortical principalneurons is caused by the Na-K-2Cl cotransporter NKCC1. J. Neurosci. 28, 4635–4639. https://doi.org/10.1523/JNEUROSCI.0908-08.2008.

Kim, D., Jeong, H., Lee, J., Ghim, J.W., Her, E.S., Lee, S.H., Jung, M.W., 2016a. Distinct roles of parvalbumin- and Somatostatin-Expressing interneurons in working memory. Neuron 92, 902–915. https://doi.org/10.1016/j. neuron.2016.09.023.

Kim, H., A¨ hrlund-Richter, S., Wang, X., Deisseroth, K., Carle', n M., 2016b. Prefrontal parvalbumin neurons in control of attention. Cell 164, 208–218. https://doi.org/10.1016/j.cell.2015.11.038. PMID: 26771492.

Kleschevnikov, A.M., Belichenko, P.V., Villar, A.J., Epstein, C.J., Malenka, R.C., Mobley, W.C., 2004. Hippocampal long-term potentiation suppressed by increased inhibition in the Ts65Dn mouse, a genetic model of Down syndrome. J. Neurosci. 24, 8153–8160.

Kleschevnikov, A.M., Belichenko, P.V., Faizi, M., et al., 2012a. Deficits in cognition and synaptic plasticity in a mouse model of Down syndrome ameliorated by GABAB receptor antagonists. J. Neurosci. 32, 9217–9227. https://doi.org/10.1523/JNEUROSCI.1673-12.2012.

Kleschevnikov, A.M., Belichenko, P.V., Gall, J., George, L., Nosheny, R., Maloney, M.T., et al., 2012b. Increased efficiency of the GABAA and GABAB receptor-mediated neurotransmission in the Ts65Dn mouse model of Down syndrome. Neurobiol. Dis. 45, 683–691. https://doi.org/10.1016/j.nbd.2011.10.009.

Kossut, M., 2007. In: Przewłocka, B. (Ed.), Synaptogenesis in Learning and Memory (In Polish), Script of XXIV Winter School of IF PAN, pp. 25–30.

Koyrakh, L., Luján, R., Colón, J., Karschin, C., Kurachi, Y., Karschin, A., et al., 2005. Molecular and cellular diversity of neuronal G-protein-gated potassium channels. J. Neurosci. 25, 11468–11478. https://doi.org/10.1523/JNEUROSCI.3484-05.2005.

Kullmann, D.M., Ruiz, A., Rusakov, D.M., Scott, R., Semyanov, A., Walker, M.C., 2005. Presynaptic, extrasynaptic andaxonal GABAA receptors in the CNS: where and why? Prog. Biophys. Mol. Biol. 87 (1), 33–46.

Kurt, M.A., Kafa, M.I., Dierssen, M., Davies, D.C., 2004. Deficits of neuronal density in CA1 and synaptic density in the dentate gyrus, CA3 and CA1, in a mouse model of Down syndrome. Brain Res. 1022, 101–109.

Lauder, J.M., Liu, J., Devaud, L., Morrow, A.L., 1998. GABA as a trophic factor for developing monoamine neurons. Perspect. Dev. Neurobiol. 5 (2–3), 247–259.

Lavdas, A.A., et al., 1999. The medial ganglionic eminence gives rise toa population of early neurons in the developing cerebral cortex. J. Neurosci. 19, 7881–7888.

Lee, S.H., Ledri, M., Tóth, B., Marchionni, I., Henstridge, C.M., Dudok, B., et al., 2015. Multiple forms of Endocannabinoid and endovanilloid signaling regulate the tonic control of GABA release. J. Neurosci. 35, 10039–10057. https://doi.org/10.1523/JNEUROSCI.4112-14.2015.

Levelt, C.N., Hübener, M., 2012. Critical-period plasticity in the visual cortex. Annu. Rev. Neurosci. 35, 309–330. https://doi.org/10.1146/annurev-neuro-061010-113813.

Li, H.L., Huang, B.S., Vishwasrao, H., Sutedja, N., Chen, W., Jin, I., Hawkins, R.D., Bailey, C.H., Kandel, E.R., 2009. Dscam mediates remodeling of glutamate receptors in Aplysia during de novo and learning-related synapse formation. Neuron 61, 527–540.

Li, S., Jin, M., Koeglsperger, T., Shepardson, N., Shankar, G., Selkoe, D., 2011. Soluble A βoligomers inhibit long-term potentiation through a mechanism involving excessive activation of extrasynaptic NR2B-containing NMDA receptors. Neuroscience 31, 6627–6638.

Li, Y., Sun, H., Chen, Z., Xu, H., Bu, G., Zheng, H., 2016. Implications of GABAergic neurotransmission in Alzheimer's disease. Front. Aging Neurosci. 8, 31. https://doi.org/10.3389/fnagi.2016.00031.

Lipton, S.A., 2007. Pathologically-activated therapeutics for neuroprotection: mechanism of NMDA receptor block by memantine and S-nitrosylation. Curr. Drug Target. 8, 621–632.

Llamosas, N., Bruzos-Cidón, C., Rodríguez, J.J., Ugedo, L., Torrecilla, M., 2015. Deletion of GIRK2 Subunit of GIRK channels alters the 5-HT1A receptor-mediated signaling and results in a depression-resistant behavior. Int. J. Neuropsychopharmacol. 18, pyv051. https://doi.org/10.1093/ijnp/pyv051.

Lopez-Bendito, G., et al., 2008. Chemokine signaling controls intra cortical migration and final distribution of GABAergic inter neurons. J. Neurosci. 28, 1613–1624.

Luhmann, H.J., Fukuda, A., Kilb, W., 2015. Control of cortical neuronal migration by glutamate and GABA. Front. Cell. Neurosci. 9, 4. https://doi.org/10.3389/fncel.2015.00004.

Lujan, R., Shigemoto, R., Lopez-Bendito, G., 2005. Glutamate and GABA receptor signalling in the developing brain. Neuroscience 130 (3), 567–580.

Luo, R., Partridge, J.G., Vicini, S., 2013. Distinct roles of synaptic and extra synaptic GABAA receptors in striatal inhibition dynamics. Front. NeuralCirc. 7, 186. https://doi.org/10.3389/fncir.2013.00186.

Luscher, B., Shen, Q., Sahir, N., 2011. The GABAergic deficit hypothesis of major depressive disorder. Mol. Psychiatry 16 (4), 383–406. https://doi.org/10.1038/mp.2010.120.

Lysenko, L.V., Kim, J., Henry, C., Tyrtyshnaia, A., Kohnz, R.A., Madamba, F., et al., 2014. Monoacylglycerol lipase inhibitor JZL184 improves behavior and neural properties in Ts65Dn mice, a model of down syndrome. PLoS One 9, e114521. https://doi.org/10.1371/journal.pone.0114521.

Ma, W., Behar, T., Chang, L., Barker, J.L., 1994. Transient increase in expression of GAD65 and GAD67 mRNAs during postnatal development of rat spinal cord. J. Comparat. Neurol. 346 (1), 151–160.

Mann, E.O., Kohl, M.M., Paulsen, O., 2009. Distinct roles of GABAA and GABAB receptors in balancing and terminating persistent cortical activity. J. Neurosci. 29, 7513–7518. https://doi.org/10.1523/JNEUROSCI.6162-08.2009.

Marín, O., Müller, U., 2014. Lineage origins of GABAergic versus glutamatergic neurons in the neocortex. Curr. Opin. Neurobiol. 26, 132–141. https://doi.org/10.1016/j.conb.2014.01.015.

Markram, H., et al., 2004. Interneurons of the neocortical inhibitory system. Nat. Rev. Neurosci. 5, 793–807.

Martínez-Cué, C., Martínez, P., Rueda, N., Vidal, R., García, S., Vidal, V., Corrales, A., Montero, J.A., Pazos, Á., Flórez, J., Gasser, R., Thomas, A.W., Honer, M., Knoflach, F., Trejo, J.L., Wettstein, J.G., Hernández, M.C., 2013. Reducing GABAA a5 receptor-mediated inhibition rescues functional and neuromorphological deficits in a mouse model of down syndrome. J. Neurosci. 33, 3953–3966. https://doi.org/10.1523/JNEUROSCI.1203-12.2013.

Martínez-Cué, C., Delatour, B., Marie-Claude, P., 2014. Treating enhanced GABAergic inhibition in Down syndrome: use of GABA α5-selective inverse agonists. Neurosci. Biobehav. Rev. 46 (Pt 2), 46. https://doi.org/10.1016/j.neubiorev.2013.12.008.

McElyea, S.D., et al., 2016. Influence of prenatal EGCG treatment and Dyrk1a dosage reduction on craniofacial features associated with Down syndrome. Hum. Mol. Genet. https://doi.org/10.1093/hmg/ddw309.

Megías, M., Emri, Z., Freund, T.F., Gulyás, A.I., 2001. Total number and distribution of inhibitory and excitatory synapses on hippocampalCA1 pyramidal cells. Neuroscience 102, 527–540. https://doi.org/10.1016/s0306-4522(00)00496-6.

Méndez, P., Pazienti, A., Szabó, G., Bacci, A., 2012. Direct alteration of a specific inhibitory circuit of the hippocampus by antidepressants. J. Neurosci. 32, 16616–16628. https://doi.org/10.1523/JNEUROSCI.1720-12.2012.

Mitra, A., Blank, M., Madison, D.V., 2012. Developmentally altered inhibition inTs65Dn, a mouse model of Down syndrome. Brain Res. 1440, 1–8. https://doi.org/10.1016/j.brainres.2011.12.034.

Miyoshi, G., et al., 2007. Physiologically distinct temporal cohorts of cortical interneurons arise from telencephalic Olig2-expressing precursors. J. Neurosci. 27, 7786–7798.

Mohler, H., 2011. The rise of a new GABA pharmacology. Neuropharmacology 60, 1042–1049. https://doi.org/10.1016/j.neuropharm.2010.10.020.

Mojabi, F.S., Fahimi, A., Moghadam, S., Moghadam, S., Windy McNerneny, M.W., Ponnusamy, R., et al., 2016. GABAergichyperinnervation of dentate granule cells in the Ts65Dn mouse model of down syndrome exploring the role of app. Hippocampus 26, 1641–1654. https://doi.org/10.1002/hipo.22664.

Montalbano, A., Corradetti, R., Mlinar, B., 2015. Pharmacological characterization 5-HT1A autoreceptor-coupled GIRK channels in rat dorsal raphe 5-HT neurons. PLoS One 10, e0140369. https://doi.org/10.1371/journal.pone.0140369.

Myers, C.P., Lewcock, J.W., Hanson, M.G., et al., 2005. Cholinergic is required during embryonic development to mediate proper assembly of spinal locomotor circuits. Neuron 46 (1), 37–49.

Nabavi, S., Fox, R., Proulx, C.D., Lin, J.Y., Tsien, R.Y., Malinow, R., 2014. Engineering a memory with LTD and LTP. Nature 511, 348–352. https://doi.org/10.1038/nature13294.

Nguyen, C.D., Costa, A.C., Cios, K.J., Gardiner, K.J., 2011. Machine learning methods predict locomotor response to MK-801 in mouse models of down syndrome. J. Neurogenet. 25 (1–2), 40–51.

Nobrega-Pereira, et al., 2008. Postmitotic Nkx2-1 controls the migration of telencephalic interneurons by direct repression of guidance receptors. Neuron 59, 733–745.

Nusser, Z., Mody, I., 2002. Selective modulation of tonic and phasic inhibitions in dentate gyrus granule cells. J. Neurophysiol. 87, 2624–2628.

Obrietan, K., Gao, X.B., Van Den Pol, A.N., 2002. Excitatory actions of GABA increase BDNF expression via a MAPK-CREB-dependent mechanism – a positive feedback circuit in developing neurons. J. Neurophysiol. 88, 1005–1015.

Olivares, D., Deshpande, V.K., Shi, Y., Lahiri, D.K., Greig, N.H., Rogers, J.T., et al., 2012. N-methyl D-aspartate (NMDA) receptor antagonists and memantine treatment for Alzheimer's disease, vascular dementia and Parkinson's disease. Curr. Alzheimer Res. 9 (6), 746–758.

Olsen, R.W., Sieghart, W., 2008. International Union of Pharmacology. LXX. Subtypes of γ-aminobutyric acidA receptors: classification on the basis of subunit composition, pharmacology, and function. Pharmacol. Rev. 60, 243–260. https://doi.org/10.1124/pr.108.00505. Update.

Olson, E.C., Walsh, C.A., 2002. Smooth, rough and upside-down neocortical development. Curr. Opin. Genet. Dev. 12, 320–327.

Padgett, C.L., Slesinger, P.A., 2010. GABAB receptor coupling to G-proteins and ion channels. Adv. Pharmacol. 58, 123–147. https://doi.org/10.1016/S1054-3589(10)58006-2.

Pallotto, M., Deprez, F., 2014. Regulation of adult neurogenesis by GABAergic transmission: signaling beyond GABAA-receptors. Front. Cell. Neurosci. 8, 166. https://doi.org/10.3389/fncel.2014.00166.

Paoletti, P., Bellone, C., Zhou, Q., 2013. NMDA receptor subunit diversity: impact on receptor properties, synaptic plasticity and disease. Nat. Rev. Neurosci. 14, 383–400.

Park, K.S., Han, J.Y., Moon, D.C., Hong, J.T., Oh, K.W., 2011. (-)-Epigallocatechin-3-O-gallate augments pentobarbital-induced sleeping behaviors through Cl− channel activation. J. Med. Food 14, 1456–1462. https://doi.org/10.1089/jmf.2010.1529.

Perez-Cremades, D., Hernandez, S., Blasco-Ibanez, J.M., Crespo, C., Nacher, J., Varea, E., 2010. Alteration of inhibitory circuits in the somatosensory cortex of Ts65Dn mice, a model for Down's syndrome. J. Neural. Transm. 117, 445–455. https://doi.org/10.1007/s00702-010-0376-9.

Pérez-Núñez, R., Barraza, N., Gonzalez-Jamett, A., Cárdenas, A.M., Barnier, J.V., Caviedes, P., 2016. Overexpressed Down syndrome cell adhesion molecule (DSCAM) deregulates P21-activated kinase (PAK) activity in an in vitro neuronal model of Down syndrome: consequences on cell process formation and extension. Neurotox. Res. 30, 76–87.

Piantadoshi, P.T., Khayambashi, S., Magdalen, G., Schluter, M.G., Kutarna, A., Floresco, S.B., 2016. Perturbations in reward-related decision-making induced by reduced prefrontal cortical GABA transmission: relevance for psychiatric disorders. Neuropharmacology 101, 279–290. https://doi.org/10.1016/j.neuropharm.2015.10.007.

Pisani, F., Spina, E., Oteri, G., 1999. Antidepressant drugs and seizure susceptibility: from in vitro data to clinical practice. Epilepsia 10, S48–S56. https://doi.org/10.1111/j.1528-1157.1999.tb00885.x.

Pizzarelli, R., Cherubini, E., 2011. Alterations of GABAergic signaling in autism spectrum disorders. Neural Plast. 12. https://doi.org/10.1155/2011/297153. Article ID 297153.

Pontes, A., Zhang, Y., Hu, W., 2013. Novel functions of GABA signaling in adult neurogenesis. Front. Biol. 8, 496–507. https://doi.org/10.1007/s11515-013-1270-2.

Pouille, F., Scanziani, M., 2001. Enforcement of temporal fidelity in pyramidal cells by somatic feed-forward inhibition. Science 293, 1159–1163. https://doi.org/10.1126/science.1060342.

Pressler, R.M., Boylan, G.B., Marlow, N., Blennow, M., Chiron, C., Cross, J.H., et al., 2015. Bumetanide for the treatment of seizures in newborn babies with hypoxic ischaemic encephalopathy (NEMO): an open-label, dose finding and feasibility phase 1/2 trial. Lancet Neurol. 14, 469–477. https://doi.org/10.1016/s1474-4422(14)70303-5.

Raccuglia, D., Mueller, U., 2013. Focal uncaging of GABA reveals a temporally defined role for GABAergic inhibition during appetitive associative olfactory conditioning in honeybees. Learn. Mem. 20, 410–416. https://doi.org/10.1101/lm.030205.112.

Ramamoorthi, K., Lin, Y., 2011. The contribution of GABAergic dysfunction to neurodevelopmental disorders. Trends Mol. Med. 17, 452–462. https://doi.org/10.1016/j.molmed.2011.03.003.

Reefhuis, J., Devine, O., Friedman, J.M., Louik, C., Honein, M.A., National Birth Defects Prevention Study., 2015. Specific SSRIs and birth defects: bayesian analysis to interpret new data in the context of previous reports. BMJ 351, h3190. https://doi.org/10.1136/bmj.h3190.

Represa, Ben-Ari, Y., 2005. Trophic actions of GABA onneuronal development. Trends Neurosci. 28 (6), 278–283.

Rissman, R.A., Mobley, W.C., 2011. Implications for treatment: GABAA receptors in aging, Down syndrome and Alzherimer's disease. J. Neurochem. 117, 613–622. https://doi.org/10.1111/j.1471-4159.2011.07237.x.

Rocco, B.R., Sweet, R.A., Lewis, D.A., Fish, K.N., 2016. GABA-synthesizing enzymes in calbindin and calretinin neurons in monkey prefrontal cortex. Cerebr. Cortex 26, 2191–2204. https://doi.org/10.1093/cercor/bhv051.

Rondal, J.A., 2020. Down syndrome: a curative prospect? AIMS Neurosci. 7 (2), 168–193. https://doi.org/10.3934/Neuroscience.2020012.

Rosato-Siri, M., Grandolfo, M., Ballerini, L., 2002. Activity dependent modulation of GABAergic synapses in developing rat spinal networks in vitro. Eur. J. Neurosci. 16 (11), 2123–2135.

Rost, B.R., Nicholson, P., Ahnert-Hilger, G., Rummel, A., Rosenmund, C., Breustedt, J., et al., 2011. Activation of metabotropic GABA receptors increases the energy barrier for vesicle fusion. J. Cell Sci. 124, 3066–3073. https://doi.org/10.1242/jcs.074963.

Rozzo, M.A., Franzot, J., Chiaruttini, C., Nistri, A., Tongiorgi, E., 2002. Expression and dendritic mRNA localization of GABAC receptor ρ1 and ρ2 subunits in developing rat brain and spinal cord. Eur. J. Neurosci. 15 (11), 1747–1758.

Rubenstein, J.L., Merzenich, M.M., 2003. Model of autism: increased ratio of excitation/inhibition in key neural systems. Gene Brain Behav. 2, 255–267.

Ruby, N.F., Hwang, C.E., Wessells, C., Fernandez, F., Zhang, P., Sapolsky, R., et al., 2008. Hippocampal-dependent learning requires a functional circadian system. Proc. Natl. Acad. Sci. U. S. A. 105, 15593–15598. https://doi.org/10.1073/pnas.0808259105.

Ruby, N.F., Fernandez, F., Zhang, P., Klima, J., Heller, H.C., Garner, C.C., 2010. Circadian locomotor rhythms are normal in Ts65Dn "Down syndrome" mice and unaffected by pentylenetetrazole. J. Biol. Rhythm. 25, 63–66. https://doi.org/10.1177/0748730409356202.

Rudolph, U., Mohler, H., 2013. GABAA receptor subtypes: therapeutic potential in down syndrome, affective disorders, schizophrenia, and autism. Annu. Rev. Pharmacol. Toxicol. 54, 483–507. https://doi.org/10.1146/annurev-pharmtox-011613-135947.

Rueda, N., Flórez, J., Martínez-Cué, C., 2008. Chronic pentylenetetrazole but not donepezil treatment rescues spatial cognition in Ts65Dn mice, a model for Down syndrome. Neurosci. Lett. 433, 22–27. https://doi.org/10.1016/j.neulet.2007.12.039.

Rueda, N., Flórez, J., Martínez-Cué, C., 2012. Mouse models of Down syndrome as a tool to unravel the causes of mental disabilities. Neural Plast. 2012, 584071.

Ruusuvuori, E., Huebner, A.K., Kirilkin, I., Yukin, A.Y., Blaesse, P., Helmy, M., et al., 2013. Neuronal carbonicanhydrase VII provides GABAergic excitatory drive to exacerbate febrile seizures. EMBO J. 32, 2275–2286. https://doi.org/10.1038/emboj.2013.160.

Santin, M.D., Valabrèguea, R., Rivalsc, I., Pénagera, R., Paquina, R., Dauphinota, L., et al., 2015. Invivo 1HMRS study in microlitre voxels in the hippocampus of a mouse model of Down syndrome at 11.7 T. NMR Biomed. 27, 1143–1150. https://doi.org/10.1002/nbm.3155.

Sassoè-Pognetto, M., Frola, E., Pregno, G., Briatore, F., Patrizi, A., 2011. Understanding the molecular diversity of GABAergic synapses. Front. Cell. Neurosci. 5, 4.

Schulz, J.M., Knoflach, F., Hernandez, M.C., Bischofberger, J., 2019. Enhanced dendritic inhibition and impaired NMDAR activation in a mouse model of down syndrome. J. Neurosci. 39, 5210–5221. https://doi.org/10.1523/JNEUROSCI.2723-18.2019.

Scott-McKean, J.J., Chang, B., Hurd, R.E., Nusinowitz, S., Schmidt, C., Davisson, M.T., et al., 2010. The mouse model of Down syndrome Ts65Dn presents visual deficits as assessed by pattern visual evoked potentials. Invest. Ophthalmol. Vis. Sci. 51, 3300–3308. https://doi.org/10.1167/iovs.09-4465.

Seidl, R., Cairns, N., Singewald, N., Kaehler, S.T., Lubec, G., 2001a. Differences between GABA levels in Alzheimer's disease and Downsyndrome with Alzheimer-like neuropathology. Naunyn-Schmiedeberg's Arch. Pharmacol. 363, 139–145. https://doi.org/10.1007/s002100000346.

Sernagor, E., Chabrol, F., Bony, G., Cancedda, L., 2010. GABAergic control of neurite outgrowth and remodeling during development and adult neurogenesis: general rules and differences in diverse systems. Front. Cell. Neurosci. 4 (11), 11.

Smith-Hicks, C.L., 2013. GABAergic dysfunction in pediatric neuro-developmental disorders. Front. Cell. Neurosci. 7, 269. https://doi.org/10.3389/fncel.2013.00269.

Souchet, B., Duchon, A., Gu, Y., et al., 2019. Prenatal treatment with EGCG enriched green tea extract rescues GAD67 related developmental and cognitive defects in Down syndrome mouse models. Sci. Rep. 9, 3914. https://doi.org/10.1038/s41598-019-40328-9.

Stachowicz, K., 2018. The role of DSCAM in the regulation of synaptic plasticity: possible involvement in neuropsychiatric disorders. Acta Neurobiol. Exp. 78 (3), 210–219. https://doi.org/10.21307/ane-2018-019.

Stagni, F., Magistretti, J., Guidi, S., Ciani, E., Mangano, C., Calzà, L., et al., 2013. Pharmacotherapy with fluoxetine restores functional connectivity from the dentate gyrus to field CA3 in the Ts65Dn mouse model of down syndrome. PLoSOne8 e61689. https://doi.org/10.1371/journal.pone.0061689.

Stagni, F., et al., 2016. Short- and long-term effects of neonatal pharmacotherapy with epigallocatechin-3-gallate on hippocampal development in the Ts65Dn mouse model of Down syndrome. Neuroscience 333, 277–301. https://doi.org/10.1016/j.neuroscience.2016.07.031.

Stasko, M.R., Scott-McKean, J.J., Costa, A.C., 2006. Hypothermic responses to 8-OH-DPAT in the Ts65Dn mouse model of Down syndrome. Neuroreport 17, 837–841. https://doi.org/10.1097/01.WNR.0000220129.78726.bb.

Stein, V., Nicoll, R.A., 2003. GABA generates excitement. Neuron 37 (3), 375–378. https://doi.org/10.1016/s0896-6273(03)00056-4.

Stern, S., Segal, M., Moses, E., 2015. Involvement of potassium and cation channels in hippocampal abnormalities of embryonic Ts65Dn and Tc1 trisomic mice. EBioMedicine 2, 1048–1062. https://doi.org/10.1016/j.ebiom.2015.07.038.

Storustovu, S.I., Ebert, B., 2006. Pharmacological characterization agonists at delta-containing GABAA receptors: functional selectivity for extrasynaptic receptors is dependent on the absence of gamma2. J. Pharmacol. Experiment. Therapeut. 316 (3), 1351–1359.

Südhof, T.C., 2013. Neurotransmitter release: the last millisecond in the life of a synaptic vesicle. Neuron 80, 675–690. https://doi.org/10.1016/j.neuron.2013.10.022.

Sussel, L., et al., 1999. Loss of Nkx2.1 homeobox gene function results in a ventral to dorsal molecular respecification within the basal telencephalon: evidence for a transformation of the pallidum in to the striatum. Development 126, 3359–3370.

Szemes, M., Davies, R.L., Garden, C.L.P., Usowicz, M.M., 2013. Weaker control of the electrical properties of cerebellar granule cells by tonicallyactive GABAA receptors in the Ts65Dn mouse model of Down's syndrome. Mol. Brain 33. https://doi.org/10.1186/1756-6606-6-33.

Szewczyk, B., Pałucha-Poniewiera, A., Poleszak, E., Pilc, A., Nowak, G., 2012. Investigational NMDA receptor modulators for depression. Expet Opin. Invest. Drugs 21, 91–102.

Takahashi, H., et al., 2012. Selective control of inhibitory synapse development by Slitrk3-PTPδ trans-synaptic interaction. Nat. Neurosci. 15 (1–2), 389–398.

Takahashi, K., Tanabe, K., Ohnuki, M., Narita, M., Ichisaka, T., Tomoda, K., et al., 2007. Induction of pluripotent stem cells from adult human fibroblasts by defined factors. Cell 131, 861–872. https://doi.org/10.1016/j.cell.2007.11.019.

Tan, G.M., Beacher, F., Daly, E., et al., 2014. Hippocampal glutamate-glutamine (Glx) in adults with Down syndrome: a preliminary study using *in vivo* proton magnetic resonance spectroscopy (^{1}H MRS). J. Neurodev. Disord. 6, 42. https://doi.org/10.1186/1866-1955-6-42.

Tremblay, R., Lee, S., Rudy, B., 2016. GABAergic interneurons in the neocortex: from cellular properties to circuits. Neuron 91, 260–292. https://doi.org/10.1016/j.neuron.2016.06.033. PMID: 27477017.

Troca-Marín, J.A., Alves-Sampaio, A., Montesinos, M.L., 2012. Deregulated mTOR-mediated translration in intellectual disability. Prog. Neurobiol. 96, 268–282.

Valbuena, S., García, A.', Mazier, W., Paternain, A.V., Lerma, J., 2019. Unbalanced dendritic inhibition of CA1 neurons drives spatial-memory deficits in the Ts2Cje down syndrome model. Nat. Commun. 10, 4991. https://doi.org/10.1038/s41467-019-13004-9.

Varju, P., Katarova, Z., Madarasz, E., Szabo, G., 2001. GABA signalling during development: new data and old questions. Cell Tissue Res. 305 (2), 239–246.

Veit, J., Hakim, R., Jadi, M.P., Sejnowski, T.J., Adesnik, H., 2017. Cortical gamma band synchronization through somatostatin interneurons. Nat. Neurosci. 20, 951–959. https://doi.org/10.1038/nn.4562.

Velazquez, R., Ash, J.A., Powers, B.E., Kelley, C.M., Strawderman, M., Luscher, Z.I., et al., 2013. Maternal choline supplementation improves spatial learning and adult hippocampal neurogenesis in the Ts65Dn mouse model of Down syndrome. Neurobiol. Dis. 58, 92–101. https://doi.org/10.1016/j.nbd.2013.04.016.

Venzi, M., Di Giovanni, G., Crunelli, V., 2015. A critical evaluation of the g-hydroxybutyrate (GHB) model of absence seizures. CNS Neurosci. Ther. 21, 123–140. https://doi.org/10.1111/cns.12337.

Vicari, S., Pontillo, M., Armando, M., 2013. Neurodevelopmental and psychiatric issues in Down's syndrome: assessment and intervention. Psychiatr. Genet. 23, 95–107. https://doi.org/10.1097/YPG.0b013e32835fe426.

Vingerhoets, C., Tse, D.H., van Oudenaren, M., Hernaus, D., van Duin, E., Zinkstok, J., Ramaekers, J.G., Jansen, J.F., McAlonan, G., van Amelsvoort, T., August, 2020. Glutamatergic and GABAergic reactivity and cognition in 22q11.2 deletion syndrome and healthy volunteers: a randomized double-blind 7-Tesla pharmacological MRS study. J. Psychopharmacol. 34 (8), 856–863. https://doi.org/10.1177/0269881120922977.

Wang, Y., et al., 2011. CXCR4 and CXCR7 have distinct functions in regulating interneuron migration. Neuron 69, 61–76.

Waseem, T., Mukhtarov, M., Buldakova, S., Medina, I., Bregestovski, P., 2010. Genetically encoded Cl-Sensor as a tool for monitoring of Cl-dependent processes in small neuronal compartments. J. Neurosci. Methods 193, 14–23. https://doi.org/10.1016/j.jneumeth.2010.08.002.

Weick, J.P., Held, D.L., Bonadurer 3rd, G.F., Doers, M.E., Liu, Y., MaguireC, C.A., Knackert, J.A., et al., 2013. Deficits in human trisomy 21iPSCs and neurons. Proc. Natl. Acad. Sci. U. S. A. 110 (24), 9962–9967. https://doi.org/10.1073/pnas.1216575110.

Weitzdoerfer, R., Dierssen, M., Fountoulakis, M., Lubec, G., 2001. Fetal life in Down syndrome starts with normal neuronal density but impaired dendritic spines and synaptosomal structure. J. Neural. Transm. Suppl. 61, 59–70.

Woo, J., Kwon, S.K., Nam, J., Choi, S., Takahashi, H., Krueger, D., Park, J., Lee, Y., Bae, J.Y., Lee, D., Ko, J., Kim, H., Kim, M.H., Bae, Y.C., Chang, S., Craig, A.M., Kim, E., 2013. The adhesion protein IgSF9b is coupled to neuroligin 2 via S-SCAM to promote inhibitory synapse development. J. Cell Biol. 201, 929–944.

Wu, D., Zhang, Y., Cheng, B., Mori, S., Reeves, R.H., Feng, J., Gao, 2021. Time-dependent diffusion MRI probes cerebellar microstructural alterations in a mouse model of Down syndrome. Brain Communicat. https://doi.org/10.1093/braincomms/fcab062.

Xu, M., Wong, A., 2018. GABAergic inhibitory neurons as therapeutic targets for cognitive impairment in schizophrenia. Acta Pharmacol. Sin. 39, 733–753. https://doi.org/10.1038/aps.2017.172.

Xie, W., Ramakrishna, N., Wieraszko, A., Hwang, Y.W., 2008. Promotion of neuronal plasticity by (−)-epigallocatechin-3-gallate. Neurochem. Res. 33, 776–783. https://doi.org/10.1007/s11064-007-9494-7.

Xu, Q., et al., 2004. Origins of cortical interneuron subtypes. J. Neurosci. 24, 2612–2622, 12.

Ye, H., Kaszuba, S., 2017. Inhibitory or excitatory? Optogenetic interrogation of the functional roles of GABAergic interneurons in epileptogenesis. J. Biomed. Sci. 24, 93. https://doi.org/10.1186/s12929-017-0399-8.

Yu, T., Liu, C., Belichenko, P., Clapcote, S.J., Li, S., Pao, A., et al., 2010. Effects of individual segmental trisomies of human chromosome 21 syntenic regions on hippocampal long-term potentiation and cognitive behaviors in mice. Brain Res. 1366, 162–171. https://doi.org/10.1016/j.brainres.2010.09.107.

Yun, K., et al., 2001. Gsh2 and Pax6 play complementary roles in dorsoventral patterning of the mammalian telencephalon. Development 128, 193–205.

Zhang, S.Y., Xu, M., Miao, Q.L., Poo, M.M., Zhang, X.H., 2009. Endocannabinoid-dependent homeo static regulation of inhibitory synapses by miniature excitatory synaptic activities. J. Neurosci. 29, 13222–13231. https://doi.org/10.1523/JNEUROSCI.1710-09.2009.

Zhou, W., Wang, N., Yang, C., Li, X.M., Zhou, Z.Q., Yang, J.J., 2013. Ketamine-induced antidepressant effects are associated with AMPA receptors-mediated upregulation of mTOR and BDNF in rat hippocampus and prefrontal cortex. Eur. Psychiatr. 29, 419–423.

CHAPTER

12

Alzheimer disease and neuroinflammation in Down syndrome: Effects of gene dosage and therapeutics

1. Introduction

The brain in Down syndrome (DS) autopsies showed abnormal distribution of neurons in some regions of brain along with diminished neurogenesis, neuronal hypocellularity, and hypoplasia, which resulted in reduction in the number of neurons and synaptic transmission (Contestabile et al., 2010; Delabar et al., 1993; Lev and Melamed, 2002). It is likely that genetic imbalance is responsible for development of neurodegenerative factors, including Alzheimer disease (AD). Overexpressed APP, due to trisomy of chromosome 21 (HSA21), causes overproduction, deposition, and accumulation of amyloid-β (Aβ) peptide in the brain, resulting in activation of microglial cells and generation of cascade events, including abnormal release of cytokines and chemokines. Prolonged activation and exacerbation of Aβ deposition in microglial cells of brain influence production of reactive oxygen species (ROS), oxidative stress, neurotoxicity, and neuronal death leading to a neuroinflammatory response favoring neuronal loss and an early onset of Alzheimer disease (EOAD) in DS (Wilcock and Griffin, 2013; Bagyinszky et al., 2014). As a result, dysregulation of pro- and anti-inflammatory molecules favors neuropathology and AD-pathogenesis in DS population (Wilcock and Griffin, 2013).

Upregulation of HSA21-encoded APP gene influences abnormalities in cytokine levels and immune deficiencies in DS and is responsible for the development of neurodegenerative disorders, including AD (Contestabile et al., 2010; Carta et al., 2002; Gomes and Otten, 2021; Nateghi et al., 2012). As such, people with DS are intrinsically deficient in immune systems and DS phenotypes and clinical expressions hypothesized DS as an immune disorder (discussed in detail in Chapters 2 and 7). Treatment with interferon consistently activated the cells obtained from individuals with DS indicating insights of the pathomechanism of autoimmune disorders, AD, and leukemia in DS. It was postulated that an increase in interferon signaling triggers the protective defences of the immune system in DS (Sullivan et al., 2017).

A specific genetic profile implicating dysregulation of genes involved in Aβ and tau biology, and other fundamental mechanisms associated

Genetics and Neurobiology of Down Syndrome
https://doi.org/10.1016/B978-0-323-90456-8.00014-4

with intracellular signaling, cell homeostasis, and death, and the cholinergic and glutamatergic systems indicate cortical dysfunction in the frontal cortex in demented DS individuals (Perez et al., 2019). Interestingly, the dysregulation of immune-related genes caused an inflammatory state in the adult brain; however, these genes remained unaltered in fetal brains, indicating their direct link to AD pathogenesis in DS (Lockstone et al., 2007). A direct correlation between inflammatory and other AD-related biomarkers and also an increase in the circulating levels of inflammatory cytokines was reported in blood in adult DS (Naude et al., 2015; Rodrigues et al., 2014). Autopsy studies of DS people revealed a different inflammatory phenotype from age-matched neurotypical controls indicating an increase in proinflammatory expression of genes such as interleukin (IL)-6 and IL-12. Such increase in neuroinflammation was consistent in the brain of both DS with AD (DSAD) neuropathology and sporadic AD (Wilcock et al., 2015a,b); however, elevated CD86 and FcgR1 observed in DSAD was different from sporadic AD, and that feature has suggested extravasation of serum proteins into the brain, which may promote the formation of immune complexes (Martini et al., 2020). Altogether, persistent systemic inflammation activates delivery of cytokines and other soluble molecules directly by involving the glial cells, especially microglia or from peripheral immune cells, such as mast cells. Neuroinflammation contributes to the pathogenesis of AD and thus, characterizes AD as a neurocentric disease.

Microglial cells interact with amyloid plaques and result in production of activated macrophage phenotype (M1), induction of a proinflammatory cascade, and finally production of chemokines and cytokines responsible for clearance of Aβ and prevention of neuronal death (Wang et al., 2015). In addition, an alternative activated macrophage phenotype (M2) is involved in tissue reparation through the production of anti-inflammatory cytokines and reduction in proinflammatory cytokine concentrations, where a balance between "classical" (M1) and "alternative" (M2) activations is essential for release or inhibition of inflammatory mediators (Ferrante and Leibovich, 2012). Some of the HSA21 genes are involved in the inflammatory response in association with M1 phenotype and also cause neuropathological manifestations in people with DS. These genes contribute to upregulation of glial activation (Wilcock and Griffin, 2013). Interestingly, an increase in markers of M2 phenotypes (CD64 and CD86) was reported in DSAD but not in sporadic AD. Hypocomplementia due to trisomy 21 was postulated to modulate AD progression in DSAD individuals through microglia-mediated synapse destruction during AD progression (Hong et al., 2016), wherein both β-amyloid deposits and soluble B2M are known to activate the classic complement cascade (Nybo et al., 1999).

Taken together, a distinct microglial phenotype is characterized in individuals with DS and more particularly individuals with DSAD. DSAD is further characterized with the presence of increased number of dystrophic microglia and fewer ramified microglial cells, which indicate the presence of fewer resting state microglia with more damage, and an increase in ameboid microglial cells. Furthermore, individuals with DS appear to have age-related gradual progress of specific microglial changes that could lead to the development and progression of AD. These changes reinforce the differences in the neuroinflammation in DS individuals. Moreover, expression of the large variety of inflammatory genes located on HSA21 could contribute to the morphological alterations of microglia cells leading to the onset and evolution of AD pathology. However, more exploration of the neuropathological changes from autopsy of DS cases would allow understanding the early events in DSAD and address a critical gap in clinical management of AD pathogenesis, also in familial and sporadic AD, which could finally lead to identify targetable drug candidates.

Neuroinflammation is described as an inflammatory process of the nervous system. It is implicated with the presence of neuritic plaques and neurofibrillary tangles where inflammation can directly influence plaques and tangles, and plaques and tangles can also directly influence inflammation in a similar way in the brains of both sporadic AD and DSAD patients. However, identification and characterization of early events that contribute to and/or regulate the expression and interaction of the triplicated HSA21 gene products is essential for understanding of the neurodegenerative cycles and the mechanisms of evolution of the neuro-patho-physiological progression of DSAD and others. Studies of DS fetuses have established three such early events, which are symbiotic for induction and expression by cytokines subsequent to neuroinflammatory changes. In particular, these include overexpression of two HSA21 gene products such as APP and S100β, and the resultant co-operating overexpression of the pluripotent neuroinflammatory cytokine IL-1, which is encoded by HSA2 genes: IL-1A and IL-1B (reviewed in Wilcock and Griffin, 2013). Therefore, complex interactions among APP, glial activation, S100B, and IL-1 are influenced by the upregulation of expression of IL-1α and β. Such interactions result in multiple neural insults, each of which is leading to gliosis-related neuroinflammation and increasing risk for development of neuropathological changes in AD.

Neuroinflammation plays critical role in AD development. Overexpression of the astrocyte-derived cytokine S100β and the neuroinflammatory cytokine IL-1β have been identified as early events in DS (Wilcock and Griffin, 2013). Barger and Harmon (1997). The first evidence of the link between neuronal stress, APP expression and neuroinflammation, demonstrating that an increased release of the α-secretase-cleaved fragment, sAPPα (from the nonamyloidogenic pathway), activates microglia, and influences the expression of IL-1β. The association of the overexpression of APP, S100β, and IL-1β collectively influenced the activation of glia resulting in elevation of IL-1β and S100β in the nervous system and onset of neuropathogenesis of both AD and DS. The overexpression of APP in DS might promote self-propagating cycles of neuroinflammatory cytokines IL-1β and S100β, whose overproduction in turn, increases APP expression (Wilcock and Griffin, 2013). Interestingly, a histopathological study showed vast microglial activation in AD compared to "mismatched" patients (who had Aβ accumulation in brains but with no AD symptoms) (reviewed in Gomez et al., 2020).

In addition, the domain of exosome application, especially BDEs, appears relatively novel in understanding the disease mechanisms and biomarkers of the course of AD, and their ability to mediate cell-to-cell communication in the processes associated with AD. Furthermore, the mechanisms of bi-directional transport of cargo-carrying exosomes across the blood-brain-barrier (BBB), changes in number or size of exosomes secreted, alterations in cargo constituents, and identification of differences in specific cell types may help in reversal of physiological changes in nervous system disorders (Song et al., 2020). The present chapter has addressed the genetic and biochemical expressions contributing to AD in DS. The markers of AD and neuroinflammation have been discussed in detail, along with therapeutic aspects of AD.

2. Preclinical changes of cognitive performance

A predictable order of changes in biomarkers occurs over a period of more than 2 decades, and thus, AD in patients with DS has been hypothesized with a long preclinical phase (Fortea et al., 2020). Onset of AD-related neuropathological changes at middle age of DS is a virtual certainty; however, timing of the onset of dementia is less certain, though estimates of approximately 50%–70% of DS individuals developing dementia by age 60–70 years are reported (Lott and Head, 2001). A recent report described the onset of pathophysiology of AD occurs in more than half of DS adults by the age of 55 years and 2/3rd in their 60–70 years of age (Hartley et al., 2020). Tracking the preclinical events of AD in this population may include a temporal progression from cognitively stable to detectable level of Aβ plaques and onset of prodromal mild cognitive impairment (MCI) (mild functional declines) and then AD (substantial cognitive and functional declines) (Firth et al., 2018; Startin et al., 2019). Such tracking would help to introduce preventive therapeutic intervention through clinical trials, which may delay or prevent DSAD and/or sporadic and autosomal dominant type of AD of the general population.

Assessment of the Cued Recall Test, including episodic memory (listing of learning tasks), visual attention and executive functioning, and motor planning and coordination was conducted to track performance on cognitive measures in relation to neocortical and striatal Aβ via [C-11] Pittsburgh Compound-B (PiB) positron emission tomography (PET) imaging in nondemented DS. The assay revealed that changes in PiB were associated with the level and rate of decline in cognitive performance involving episodic memory, visual attention, executive functioning, and visuo-spatial ability at two to five times over 10-year period (Hartley et al., 2020). Among the three PiB groups (consistently PiB negative (PiB-), transition from PiB- to PiB positive (PiB-/+) and consistently PiB+), the Cued Recall Test varied and appeared significantly prominent in the consistent PiB + group with elevated Aβ for longer and greater cognitive decline across time. The study indicated that the Cued Recall Test could be considered as a promising indicator for studying the preclinical to prodromal cognitive changes in DSAD population.

It is noteworthy to record the population-level variability in transition to the preclinical and prodromal stages for conducting clinical trials for AD in DS population, such as biological sex posit - an age-related variation in the estrogen level. However, the effect of age-related changes between two sexes on pathomechanism of AD has conflicting reports (Schupf et al., 2006; Temple et al., 2008). In addition, variation in DS phenotypes and severity of clinical expression and associated comorbidities may contribute to the variability of preclinical to prodromal transition of AD.

3. Dementia in AD

People with DS are at higher risk for AD due to triplication of HSA21. Many of the HSA21 genes are associated with AD-related changes, including increased production of APP, β-amyloid cleavage enzyme 2 (BACE2), and so on. AD is clinically characterized by a progressive loss of memory and cognition, and pathological diagnosis is associated with two signature lesions, including neuritic plaques composed of aggregated amyloid-β (Aβ) peptides, neurofibrillary tangles composed of hyperphosphorylated, and aggregated tau protein in the brain tissue (Newell et al., 1999). Aβ peptide is a cleavage product of amyloid precursor protein (APP). APP is a transmembrane protein and is differentially cleaved by secretase enzymes. Nevertheless, AD is a disease of aging, where currently one of eight Americans over the age of 65 years and half of those over 85 years is affected with AD (www.alz.org accessed December 12, 2020). Aβ-deposits in DS begin to accumulate in childhood and progresses with increase in age (Leverenz and Raskind, 1998), with acceleration between the ages of 35 and 45 years when other

AD symptoms such as neurofibrillary tangles and inflammation begin to appear (Wisniewski et al., 1985). However, onset of dementia occurs by ages 60–70 years in 50%–70% of DS individuals (Janicki and Dalton, 2000). Primarily, triplication of APP gene leads to EOAD and dementia in individuals with DS. As such, people with DS are intrinsically deficient in immune systems; thus, upregulation of APP favors abnormalities in cytokine levels in DS and is likely responsible for the development of neurodegenerative disorders, including DSAD.

AD is a key factor of early aging in DS population with progressive loss of memory and early course of dementia in adults (Fig. 12.1). Severe cognitive deterioration, including acquired apraxia and agnosia, has been reported at age 30 years in 28% of individuals with DS, with a higher propensity in the subsequent years (Head et al., 2012). The dementia in DS is manifested with changes in personality and behavior, pragnosia, or socially deficient communication from previous well-developed social capacities. The severity of the preexisting cognitive disability may also predict the rate of cognitive deterioration in DS. Interestingly, women with DS may develop dementia 10–20 years earlier than non-DS women of the general population, which is further linked to a genetic polymorphism in the hydroxysteroid-17β-dehydrogenase gene that modifies the circulating levels of neuroprotective estrogen through conversion of estrone to estradiol. The onset of dementia in DS

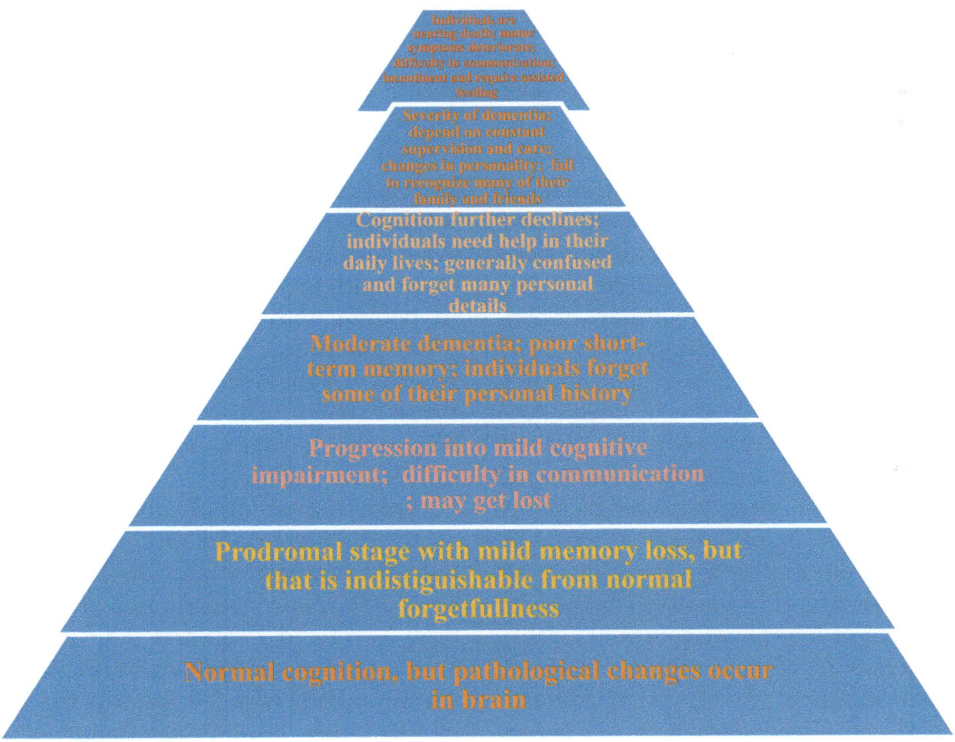

FIGURE 12.1 Early stages of clinical manifestation and dementic consequences of Alzheimer disease. *Modified version of Table 1 of Sheppard and Coleman (2020).*

women was linked to three of five single nucleotide polymorphisms (SNPs) in this gene. Importantly, 2–3 times of higher risk of AD development was evidenced in DS women with high-risk SNPs (Head et al., 2012).

Studies of plasma Aβ have provided variable results with respect to distinguishing demented from nondemented individuals with DS and aging (reviewed in Head et al., 2016). Conflicting reports on the age related levels of Aβ1-42 and Aβ1-40 and implication on dementia in DS have been resolved with increasing trend of plasma Aβ1-40 and decreased Aβ1-42 as predictors of conversion to dementia in DS (Schupf et al., 2010). Indeed, DS adults with reduced plasma Aβ1-42 appeared 5 times more likely to become demented within 4 years (Head et al., 2016).

4. Bio-markers of AD

In the present era, DS patients are experiencing longer life span (exceeding 60 years) than before, due mainly to increasing awareness among the parents, patients, and public about the clinical consequences of the disorder and early onset of the therapeutic intervention, and improvement of medical care (Bittles et al., 2007; Hithersay et al., 2019). The impact of clinical complications of AD, along with other comorbidities, is affecting 5·8 million people worldwide and reported as the leading cause of death for ~90% adults with DS (Ballard et al., 2016). Overexpression of the gene-dosage effect of APP gene has been established as the primary genetic basis of DSAD (McCarron et al., 2017). Recent studies are focusing more on the natural history of biomarker changes for the development of DSAD (Dubois et al., 2014). The characteristics of the three markers, including deposition of Aβ, hyperphosphorylated tau, and neurodegeneration of AD pathogenesis, are noticeably similar to those observed in the sporadic and autosomal dominant forms of AD. The onset of the neurodegenerative changes, as measured by increased plasma level of neurofilament light chain (NFL) or brain hypo-metabolism, might occur at early age in subjects with DS, even before detectable fibrillar Aβ deposition by neuroimaging.

Neuroimaging by positron emission tomography (PET) provides assessment of fibrillar Aβ plaque deposition and brain metabolism and structural MRI assessed brain atrophy, extracting insight into the pathogenesis and progression of AD and enables drug discovery and clinical trials. These imaging facilities have been augmented with further advancements for precise underpinnings of DSAD, such as accumulation of tau and related cognitive decline, glucose hypometabolism, reduction in gray matter volume, and loss of the frontal white matter integrity that correlates with decline in cognitive function. DS people with mild cognitive impairment (MCI) are characterized with low levels of cerebrospinal fluid (CSF) Aβ-42 (soluble) and elevated levels of total tau (t-tau) and phosphorylated tau (p-tau). Indeed, elevated levels of Aβ-42 and other amyloid species are apparent in early childhood with low levels of CSF tau; however, CSF Aβ-42 levels decline with increase in CSF tau levels with increasing age of the DS individuals. In people with DS, plasma level of NFL increases with age and correlates with decreased adaptive behavior scores. Plasma NFL can further distinguish between DS and DSAD for its predictive role as a biomarker of dementia in DS.

Order and timing of alterations in biomarkers in adult DSAD population has been estimated with the apolipoprotein E (APOE) allele carrier status, plasma concentrations of amyloid β peptides 1–42 and 1–40 and their ratio (Aβ1–42/1–40), total tau protein, and NFL, tau phosphorylation at threonine 181 (p-tau), and one or more of PET with ^{18}F-fluorodeoxyglucose, PET with amyloid tracers, and MRI (Fortea et al., 2020). Studies of early clinicopathological and biomarkers, including patterns of cerebral Aβ-deposition, atrophy, and hypometabolism, indicated similar cortical regions in the AD-affected sporadic and autosomal dominant forms and in DSAD (Neale et al., 2018). However, a clear difference in cognitive performance and plasma Aβ1−42 concentrations or hippocampal volumes was reported from the onset in the youngest individuals with DS through their entire life span (Fortea et al., 2020).

4.1 Mechanism of Aβ processing, cleavage Aβ, degradation, and clearance in the brain

APP is cleaved by three enzymes, α-secretase, β-secretase, and γ-secretase. Competition of α- and β-secretases for cleavage drives processing of APP through amyloidogenic and nonamyloidogenic pathways (Fig. 12.2) (Chow et al., 2010; Coskun et al., 2010). In the nonamyloidogenic pathway, APP is cleaved by the α-secretase enzyme and generates fragments of amyloid precursor-protein-α (sAPPα), which are neuroprotective in functions, whereas in the amyloidogenic pathway, APP cleavage is mediated by β-secretase (BACE1 and BACE2) enzymes and generates APP-β (sAPPβ) (Cole and Vassar, 2007). Later, sAPPα and sAPPβ are cleaved by γ-secretase enzyme and produce shorter P3 fragments ($A\beta_{17\text{-}x}$) of unknown biological function and Aβ peptides for AD initiation, respectively (Chow et al., 2010; Siegel et al., 2017). Abnormalities in APP cleavage processing and/or imbalances between Aβ generation and clearance lead to excess production of long Aβ peptides (Aβ40/42) that can facilitate accumulation of amyloid plaques and onset of AD-related neuropathology (MacLeod et al., 2015; Siegel et al., 2017). DS individuals, over 40 years of age, display decrease in α-secretase and increase in β-secretase activities suggesting an association between alterations of secretase

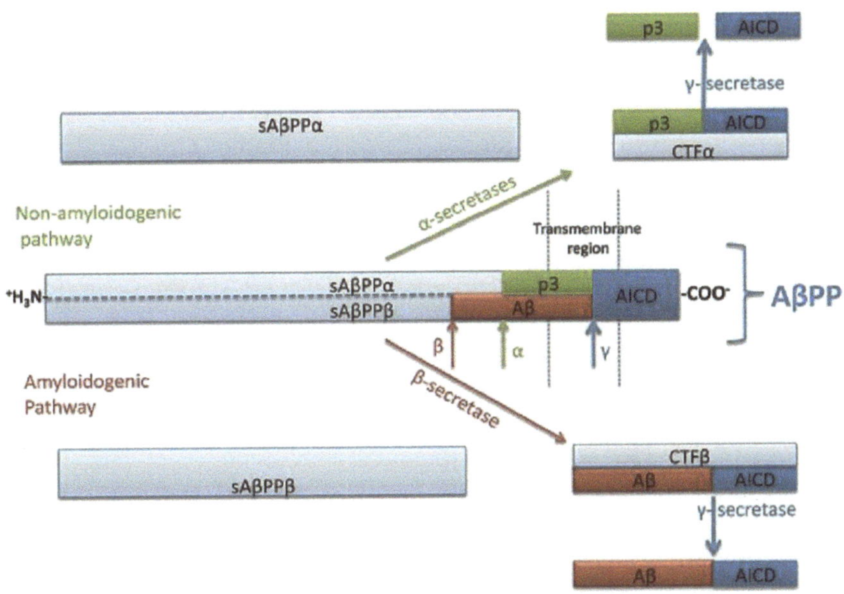

FIGURE 12.2 Schematic representation of amyloid β-precursor protein (Aβ-PP) processing for amyloidogenic and nonamyloidogenic pathways. *Reprinted from Coskun et al. (2010), with permission from, IOS press.*

activity and production of amyloidogenic fragments and accumulation of Aβ (Nistor et al., 2007).

The inherent mechanism involves several enzymes, including insulin-degrading enzyme (IDE), neprilysin, and tissue plasminogen activator for subsequent degradation and clearance of accumulated Aβ, either in soluble or insoluble form, in the brain (reviewed in Head et al., 2016). Despite triplication of APP and BACE2 and lifelong overexpression of these proteins primarily contribute to increased production of Aβ in DS. Overt AD neuropathology is detectable only after 40 years of age. Elevation of neprilysin protein concentration correlates with levels of insoluble Aβ in DS brain. This situation indicates that degradation and clearance of Aβ occur in the DS brain by Aβ-degrading enzymes, and thus, the therapeutic potential of enhancing Aβ degradation and clearance has been thought of for targeted drug development.

4.2 Soluble Aβ oligomers in DS

Once Aβ is cleaved from APP, it may appear first as a soluble isoform within neurons or in the extracellular space, including oligomers and protofibrils. Aβ-derived diffusible ligands, which accumulate inside the neurons, induce neurotoxicity and may cause more neuronal dysfunction than extracellular Aβ. Aβ oligomers are abundant in AD brain. Soluble Aβ may accumulate prior to extracellular Aβ and may be more critical in causing neuronal dysfunction than extracellular Aβ in DS. Notably, Aβ oligomers are crucial for causing neuronal dysfunction, even prior to overt neuronal loss. Despite the extensive extracellular Aβ accumulation feature of all ADs, including DSAD, inconsistent link between Aβ and severity of dementia has been demonstrated (reviewed in Head et al., 2016). Indeed, transgenic mice with overexpressing human mutant APP displayed cognitive dysfunction before the accumulation of insoluble and extracellular Aβ, indicating the role of the soluble forms of Aβ. Importantly, oligomeric Aβ renders neuronal dysfunction before neuronal loss occurs during development as well as aging in DS (Cenini et al., 2012); however, its mechanism of action in DS brain is yet to be elucidated.

Additionally, oligomeric Aβ may cause mitochondrial dysfunction through interactions with dynamin-related protein 1 (DRP1), and exacerbating the oxidative defect in DS (Manczak et al., 2011). Autopsy of DS brains highlighted the importance of soluble and insoluble Aβ, oligomers, and intracellular Aβ associated with clinical signs of dementia. The Aβ deposits in DS brain may play a critical role in the generation of oxidative stress, similar to sporadic AD patients. In turn, neuronal oxidative stress may enhance Aβ production and vice versa, which was evidenced in frontal and entorhinal cortex in AD and DS subjects with neuritic plaques, suggesting that oxidation of Aβ is an important event in plaque biogenesis (reviewed in Cenini et al., 2012). Indeed, increased oxidative damage, elevated ROS level, and mitochondrial dysfunction may trigger a variety of pathways, including aberrant APP processing in DS (Busciglio et al., 2002). Collectively, a positive correlation between oxidative stress and the levels of Aβ40 peptide has been demonstrated. Additionally, overexpression of SOD-1 in DS may influence oxidative damage and degradation of Aβ since enzymes responsible for Aβ-degradation and clearance may themselves be altered by oxidative damage. An elevation of the oxidized Aβ-degrading enzyme, neprilysin, may explain the mechanism underlying the accumulation of extracellular Aβ in AD brain (Wang et al., 2003).

4.3 Intracellular Aβ in DS

Insoluble Aβ deposits progressively increases over time, despite a larger amount of Aβ exists in the soluble form. However, conflicting reports exist on intracellular Aβ1-40 and Aβ1-42 levels. Nevertheless, Aβ1-40 degrades more rapidly within lysosomes than the longer and higher toxic Aβ1-42/43 (reviewed in Head et al., 2016). Interestingly, a "solid phase" pathway has described a nonspecific proteolytic cleavage, which leads to heterogeneous Aβ production, including Aβ1-40/42/43 from APP. APP processing occurs within the endosomal/lysosomal system, and thus, intracellular Aβ deposits likely appear in endosomal/lysosomal location in DS. The intracellular Aβ is an event of early age and that accumulates in both infants and children with DS. Notably, accumulation of intracellular Aβ is consistently observed prior to the appearance of extracellular Aβ deposits, which is recapitulated in transgenic mouse models of AD. Therefore, it is apparent that accumulation of intracellular Aβ occurs within neurons prior to age-associated extracellular Aβ deposition at much earlier age in DS than in the general population. The accumulation of neuronal Aβ has been hypothesized to be associated with caspase cleavage products leading to increased apoptosis, which may subsequently be partially responsible for brain atrophy and neuronal loss in DS (reviewed in Head et al., 2016).

4.4 Senile Aβ plaques

Virtually, all adults with DS exhibit an age-dependent increased risk for developing dementia and appear with significant increase in neuritic plaques and neurofibrillary tangles leading to a neuropathological diagnosis of AD at ≥40 years of age. Senile plaques are derived from the longer precursor protein, β-APP and thus, contain the β-amyloid (Aβ) peptide. APP is upregulated due to trisomic HSA21 in DS and thus, myriads of neurobiological studies have focused primarily on APP processing and the temporal events in Aβ pathogenesis in DS. Aβ is thought to be a crucial factor in AD pathogenesis and overexpression of APP increase the levels of Aβ in DS.

Cerebral Aβ deposition occurs decades earlier in DS compared to brains of age matched control and AD. Although extracellular Aβ accumulation appears consistently with deposits of diffuse plaques after the age of 30 years, widespread-diffused Aβ42 plaques have been reported in brain sections of teenagers and DS individuals in their 20s (Figs. 12.3 and 12.4) (Head et al., 2016). It is noteworthy that diffuse plaques herald the neuritic plaques in the cortex with age. Indeed, the diffuse Aβ-42 plaques precede fibrillar senile plaques containing dystrophic neurites and formation of neurofibrillary tangles. Interestingly, Aβ-42 plaques are more prevalent over Aβ-40 plaques at all ages of DS. Notably, diffuse Aβ-42 is deposited in DS cerebellum and striatum in the third decade of life, while fibrillar plaques are not often observed in these brain areas in older DS brain, suggesting provincial effects on plaque maturation in DS. On the whole, senile plaques increase in the same brain regions and cortical layers as in AD; however, DS brains display higher plaque density. Moreover, an ApoE-ε4 allele doubles the burden of cerebral amyloid plaque in DS and shortens their life span. The ApoE protein is described early in plaques in DS brain, which acts as a possible chaperone for Aβ deposition (reviewed in Head et al., 2016). Interestingly, family history of dementia has been linked with white matter degradation in non-DS women with at least one apolipoprotein E4 allele.

Furthermore, an exponential rise in plaque accumulation between 30 and 40 years of age hallmark the neuropathology of diagnosis of AD in DS. An acceleration phase of disease development continues in DSAD and consistently increases the levels of plaques and tangles. Interestingly, younger individuals with AD pathology also exhibit rapid accumulation of plaques and tangles, but not sufficient level for a neuropathologic diagnosis of AD. Intracellular Aβ deposits have been reported in autopsy of

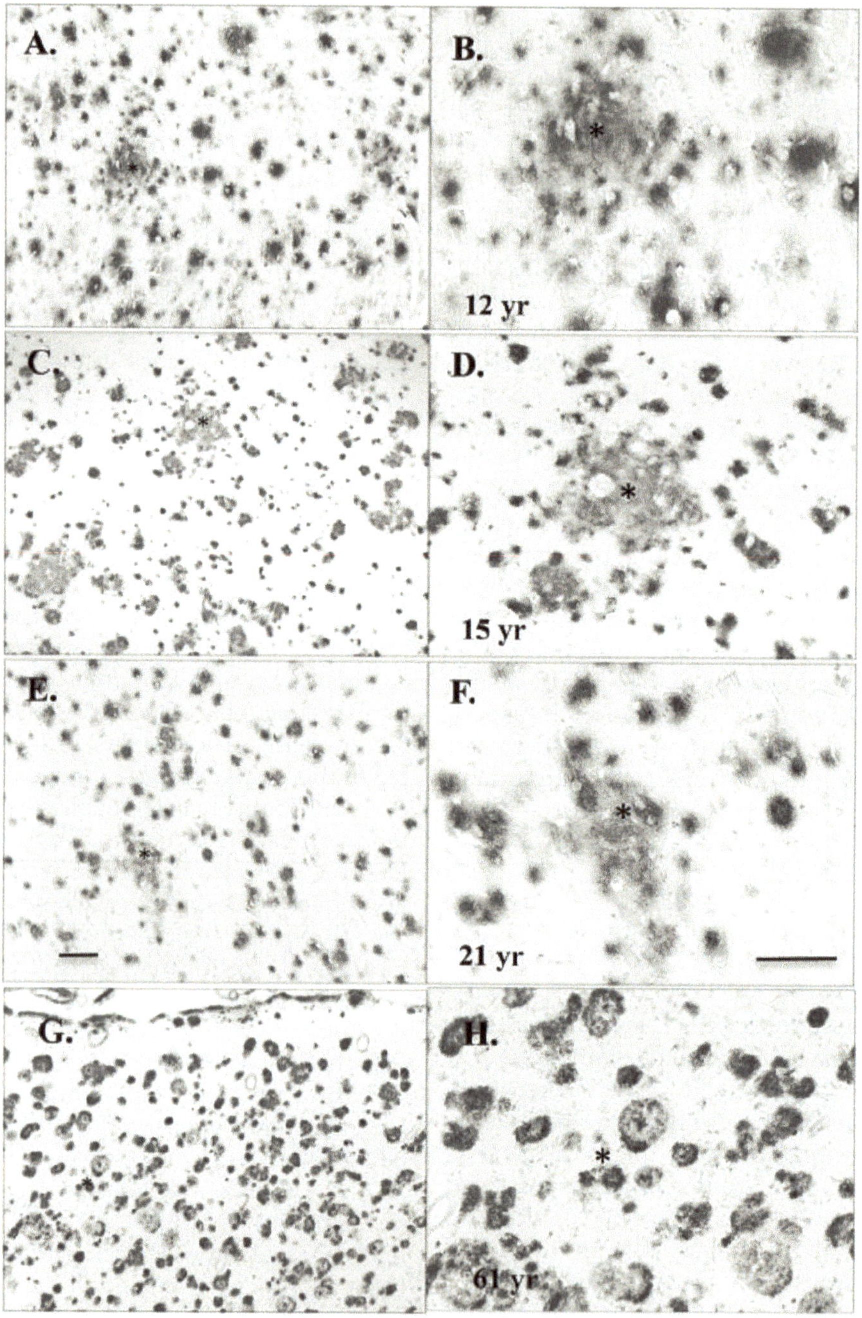

FIGURE 12.3 Aβ42 immunostaining in the frontal cortex of Down syndrome autopsy cases ranging in age from (A, B) 12 years, (C, D) 15 years, (E, F) 21 years, and (G, H) 61 years. Note the extensive diffuse plaque deposition in younger cases compared with cored and compacted plaques, subpial plaques, and vascular amyloid in the older DS brain (C42 pAb kindly provided by Dr. Takaomi Saido at the RIKEN Brain Institute, Japan and described in reference #38 (Mori et al., 2002). Plaques denoted by an asterisk at low magnification on the left (A, C, E, G) are shown at high magnification on the right (B, D, F, H). Scale bars, 100 microns. *Reprinted from Head et al. (2016). https://doi.org/10.2174/1567205012666151020114607, with permission from EUREKA SCIENCE (FZC) (Bentham Science Publishers, Ltd) and Copyright Clearance Center.*

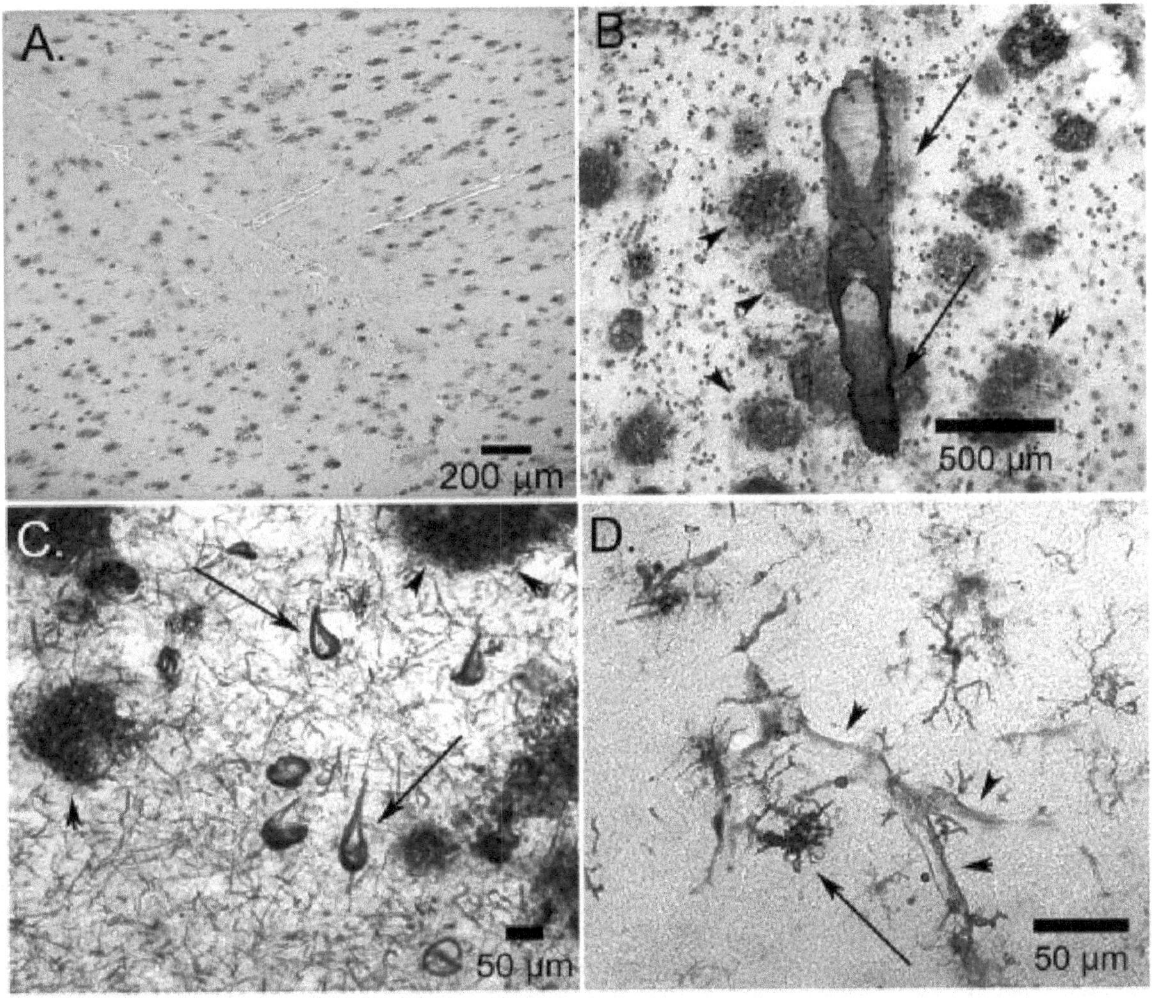

FIGURE 12.4 Neuropathological features of Down syndrome brain. (A) Diffuse plaque labeling using an antibody against Aβ (R1282, gift from Dr. D. Selkoe, Boston, MA) in the temporal cortex of an individual with Down syndrome aged 21 years. (B) Significant cerebral amyloid angiopathy in a 46-year-old male with Down syndrome and Alzheimer disease detected by immunostaining of the frontal cortex with Aβ1-40 antibody (*arrows*), which can be distinguished from extracellular plaques (*arrowheads*). (C) Neurofibrillary tangles labeled using the PHF-1 antibody (provided by Dr. Peter Davies-arrows) in the frontal cortex of a 46-year-old male with Down syndrome and Alzheimer disease. (D) Immunostaining using ionized calcium binding adaptor molecule (IBA-1) for microglial cells shows significant hypertrophy (*arrows*) and association with the cerebrovasculature (*arrowheads*) suggesting neuroinflammation. *Reprinted from Head et al. (2016). https://doi.org/10.2174/1567205012666151020114607, with permission from EUREKA SCIENCE (FZC) (Bentham Science Publishers, Ltd) and Copyright Clearance Center.*

younger DS individuals and as young as 3 years. The extracellular Aβ is thought to be mainly of neuronal origin. In cases with intracellular Aβ localized to endosomes, intracellular organelles are involved in degradation and turning over the Aβ proteins within cells (reviewed in Head et al., 2016). The extracellular Aβ can be post-translationally modified by isomerization, racemization, and oxidation in DS over time. Formation of pyroglutamate-3 Aβ by glutaminyl cyclase and its deposition into plaques and blood vessel walls have been reported in AD and DS brains. In addition, overexpression of BACE2 caused elevation of unmodified Aβ11-40/42 in DS brain. Also, deposits of APP P3 peptide at the nonamyloidogenic α-secretase site (Aβ17leu) have been described in DS brain extracts and cerebellar plaques (reviewed in Head et al., 2016). The mechanisms of modifications of Aβ production may possibly serve as potential chronobiological age markers of onset and progression of AD.

4.5 Oxidative stress and AD in Down syndrome

Although Aβ-42 is regarded as more toxic than Aβ-40, neurotoxicity of Aβ-40 has been reported in neuronal cultures. Furthermore, the neurotoxicity of Aβ-40 may perturb the survival and differentiation of the neuronal progenitor cells by generating oxidative damage. Aβ-42 is capable of causing oxidative stress through free radical reactions. Interestingly, the amount of Aβ-40 is consistently higher than Aβ-42 in the frontal cortex in DS, particularly in DSAD. This observation led to the hypothesis that oxidative stress of the DS brains may be more strongly associated with Aβ-40 than Aβ-42 (Cenini et al., 2012). Furthermore, elevated plasma Aβ-40 level in DS adults is associated with onset of dementia. Notably, age-related elevation of oxidized DNA/RNA occurs in DS prior to Aβ deposition; however, once Aβ-42 deposition begins in their temporal cortex, the level of oxidized DNA/RNA is reduced (reviewed in Cenini et al., 2012). Thus, it is likely that Aβ acts as an antioxidant in DS and also in AD (Cenini et al., 2012). Nevertheless, oxidative stress is significantly correlated with Aβ-40 in DS brain, and increased oxidative stress may increase risk of the DS brain to subsequent AD neuropathology.

Furthermore, enhanced lipid peroxidation leads to various aldehydic products, such as 4-hydroxy-2-nonenal (HNE) is responsible for the damaging effects of oxidative stress and neurodegenerative diseases such as AD. Lipoperoxidation is increased in prenatal DS brains (reviewed in Cenini et al., 2012). The embryonic brains of Ts1Cje mice exhibited HNE-bound proteins (Ishihara et al., 2009). Significant increase in another marker of lipid peroxidation, 8,12-isoiPF2α-VI, was reported in the urine of young DS subjects. Additionally, intracellular accumulation of ROS and increased lipid peroxidation observed in cortical neurons cultured from prenatal DS cases caused neuronal apoptosis (Busciglio et al., 2002). Also the amniotic fluid from mothers carrying DS fetuses showed significant increase in markers of oxidation (Perluigi et al., 2011). Summarily, lipid peroxidation was hypothesized as an early and possibly a chronic event in DS. Additionally, overexpression of SOD1 may increase lipid peroxidation in DS. In general, SOD1 converts superoxide radical into H_2O_2, which is subsequently neutralized by glutathione peroxidase or catalase. In absence of catalase or glutathione peroxidase activity in DS brains due to increased SOD1 level, accumulation of H_2O_2 may lead to increased oxidative stress. Consistent with this concept, gene transfection of SOD1 into two different cell lines displayed increased lipid peroxidation, elevated protein carbonyls, and a trend toward increased levels of 8-hydroxyguanine and 3-NT (Lee et al., 2001).

The oxidative stress primarily contributes to neurodegeneration in DS patients at a much earlier age and AD in adult patients (Lu et al., 2011; Bambrick and Fiskum, 2008). Therefore, oxidative stress–induced neurodegeneration could alter the neuronal numbers and synaptic plasticity through mitochondrial dysfunction and also affect the generation of neurons during development. Thus, it is most likely that the effects of overexpression of HSA21 genes promote the cognitive dysfunction in DS patients throughout their lifetime. Additionally, synergistic interaction of HSA21 genes, including APP and S100β, increases H_2O_2 levels and decreases membrane potential in the mitochondria of neuroprogenitor cells of DS and underlies the pathomechanism of neurodegeneration in DS brain. Moreover, several HSA21 genes have been demonstrated to enhance APP-dependent toxic effects on the mitochondria. Network prediction analysis has indicated several HSA21 proteins interacting with components of the JAK/STAT pathway (Sturgeon et al., 2012).

4.6 Mitochondrial dysfunction and Aβ processing

The cultures of astrocytes and neuronal cells derived from fetal DS brain showed modification in the processing of amyloid β-precursor protein (Aβ-PP), indicating increased levels of Aβ-PP and C99, decreased levels of secreted Aβ-PPs and C83, and intracellular accumulation of insoluble Aβ-42 (Busciglio et al., 2002). Such pattern of Aβ-PP processing is recapitulated by inhibition of mitochondrial metabolism in normal astrocytes and that was consistent with impaired mitochondrial function in astrocytes derived from DS. Intracellular Aβ-42 and reduced Aβ-PPs are demonstrated in brains of both DS and AD. The recombinant or astrocyte-produced Aβ-PPs significantly enhanced the survival of DS neurons, suggesting Aβ-PPs as a neuronal survival factor. Thus, intracellular deposition of Aβ-42 and reduced levels of Aβ-PPs caused by mitochondrial dysfunction may establish a chronic state of increased neuronal vulnerability in DS (Busciglio et al., 2002). Moreover, enhancement of Aβ deposition induces cellular toxicity causing decrease in mitochondrial biogenesis, which in turn influences dementia in DS (Coskun et al., 2010). Interestingly, a compensatory upregulation of mitochondrial biogenesis was reported in DS, but non-AD brains showing a slight increase in mtDNA levels; however, that was not consistent in DSAD (Coskun et al., 2010). Similar Aβ-accumulation was described following poisoning the mitochondria of normal cells, suggesting influence of the defective mitochondrial function to upsurge APP and Aβ-42 in DS cells. In general, the secreted form of APP appears neuroprotective from various toxins, including Aβ-42. Thus, impairment of mitochondrial function influenced by overexpression of the triplicated APP exerts less secretion of the protective form of APP in brain cells, enhancing the cellular vulnerability to the deadly cycle of damage from Aβ-42 accumulation (Busciglio et al., 2002).

4.7 Tau phosphorylation and neurofibrillary tangles in DS

Tau phosphorylation and aggregation has been considered as an AD-associated biomarker evidenced in DS brain. In particular, the hyperphosphorylated microtubule−associated tau protein wasconsistentinplaque-associateddystrophicneurites and intraneuronal neurofibrillary tangles (NFTs)(reviewed in Head et al., 2016) (Fig. 12.4). Overexpression of the DYRK1A gene potentially contributes to phosphorylation of tau protein making it a better substrate for GSK3β phosphorylation. Notably, DYRK1A phosphorylates the alternate splicing factors leading to an increase ratio of 3R:4R tau and that causes neurodegeneration. Therefore, brains of the middle-aged and older DS exhibit an elevated number of DYRK1A-positive and 3R-positive NFTs compared to sporadic AD. Another possible contributing factor is the overexpression of the regulator of calcineurin-1 (RCAN1) or calcipressin1 gene, which inhibits calcineurin, and may enhance tau phosphorylation by reducing calcineurin phosphatase activity and increasing GSK3β levels. Overexpression of APP may also contribute to tau phosphorylation as upregulation of DYRK1A and RCAN1 is influenced by Aβ-42 in DS brain. Thus, overexpression of DYRK1A, RCAN1, and APP genes are involved in driving tau pathology and neurodegeneration in DS.

Notably, early signs of tau pathological changes occur in the outer layer of the hippocampus of middle-aged (30–40 years) DS individuals, and later by NFTs in the hippocampal CA1 region and subiculum, leading to neuronal loss in the entorhinal cortex (reviewed in Head et al., 2016; Sheppard and Coleman, 2020). In general, AD-like similar distribution of NFTs (Fig. 12.5) occurs in DS, with a beginning in entorhinal cortex to spreading neocortex through the hippocampus, but density of NFTs is higher in DS than AD brain.

4.8 Neuronal loss in DS

Neuronal loss is evidenced starting in the subcortical areas of DS brain, such as locus coeruleus, basal forebrain nuclei, and entorhinal cortex, which further appear in hippocampus, temporal cortex, and other cortical areas. Cholinergic neuronal loss in the basal forebrain nuclei that extends to hippocampus and cerebral cortex, including the nucleus basalis of Meynert, increases with age in DS and DSAD pathology. Thus, it is likely that early synapse loss in AD and DS cortex may retrospectively reflect neurodegeneration in subcortical nuclei (Head et al., 2016). Moreover, 42% reduction in the overall volume of the entorhinal cortex was reflected in older DS brain, and as high as 90% decrease in the number of intact, NFT-free neurons was reported compared to controls, indicating neurofibrillary degeneration as the main factor of neuronal loss in DS (reviewed in Head et al., 2016). However, overall neuronal loss in older DS brain was lesser than that of sporadic AD brain, which was explained by the lower number of neurons present in these brain regions in young DS individuals compared to age-matched controls, implicating a smaller neuronal reserve in DS. In addition, a greater risk for AD is associated with degeneration of the white matter.

4.9 Exosomal factors of AD

Exosomes, one type of extracellular vesicles (EVs), are intricately involved in the trans-synaptic spread of pathogenic factors underlying

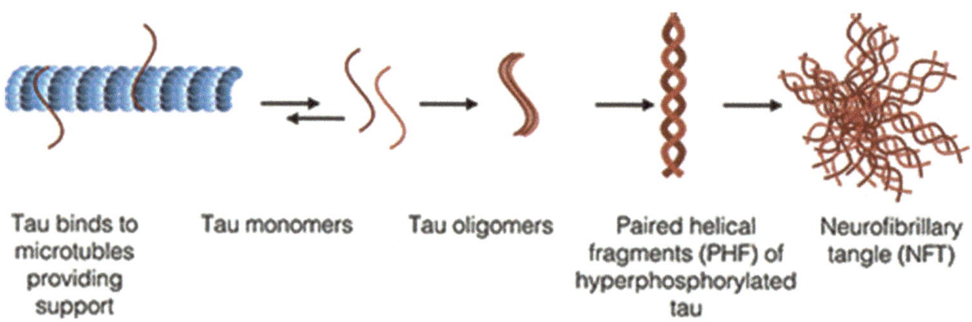

FIGURE 12.5 Microtubule-associated protein tau aggregation results in the accumulation of neurofibrillary tau tangles (NFTs). Tau is believed to play a role in the stabilization of microtubules. Hyperphosphorylated tau polymerization leads to the creation of insoluble paired helical fragments (PHFs), which further aggregate into NFTs. *Reprinted from Sheppard and Coleman (2020). https://doi.org/10.36255/exonpublications.alzheimersdisease.2020.ch1, under Creative Commons Attribution-Non Commercial 4.0 International (CC BY-NC 4.0) https://creativecommons.org/licenses/by-nc/4.0/.*

AD and other neurological diseases. Normal exosomes are nano-sized vesicles, produced in endosomal multivesicular bodies (MVBs), and secreted by most cells in the body. Importantly, exosomes mediate intercellular communication and genetic exchange, and thus, they have drawn a major focus in research and depicted a novel biological sequelae involved in situations bothering the DS population. Studies of exosomes have demonstrated complex multicellular processes underlying DS etiologies of abnormalities in formation and secretion of these extracellular vesicles or exosomes (Hamlett et al., 2020).

It has been established that an extra copy of APP promotes early-onset AD in DS individuals; even mild overexpression (20%) and partial triplication of APP (in mosaic DS) contribute to development of AD. Interestingly, structural and functional abnormalities of early endosomes, which are seen as the earliest neuronal pathology in AD patients, depend on the increased APP expression in trisomic fibroblasts derived from DS individuals (Jiang et al., 2010). Notably, an association of APP gene dosage and the C-terminal APP fragment generated by BACE1 (βCTF) contribute to pathological endosomal effects, neuronal vulnerability, and disease; however, that is not mediated by Aβ (Nixon, 2007; Jiang et al., 2010). Indeed, treatment of DS fibroblasts with shRNA effectively reduced APP expression and reversed the AD-related early endosome phenotype. However, endosomal pathology was not reversed when several other trisomic and/or disomic genes were silenced, such as SOD1 and calpastatin (non-HSA21). Similar reversal of both neuronal endosomal pathology and degenerative changes in basal forebrain cholinergic neuron (BFCN) was observed in Ts65Dn mice through a selective deletion of one copy of App or a small portion of the MMU16 containing App (reviewed in Jiang et al., 2010).

Aβ production is a direct consequence of increased APP copy number evidenced in DS population. Although, endosome dysfunction depends on elevated APP levels in DS fibroblasts, it could not be blocked by reducing the Aβ production with a γ-secretase inhibitor, which supported the previous observations of Aβ-independent endosome pathology in human AD and mouse models of AD. In addition, neuronal endosomes were not abnormally enlarged despite abundant Aβ deposition in brain of familial AD with presenilin mutations (PSFAD), indicating involvement of APP trafficking pathways in pathogenic mechanisms of PSFAD other than the endocytic pathway (Nixon, 2007). Indeed, endosomes are abnormally enlarged in Ts65Dn mice in absence of Aβ elevation. Notably, multiple lines of evidence suggest that enhancement of βCTF is the single common APP-related alteration associated with abnormal enlargement and proliferation of endosome leading to accelerated endocytosis (Jiang et al., 2010).

Accumulation of βCTF has been documented in early endosomes in cell models of AD and in enlarged endosomes of AD brain. βCTF appears more efficient for influencing the membrane conductances and forming channels in the plasma membrane than Aβ in a concentration-dependent manner, and thus, βCTF can disrupt neuronal ionic homeostasis. Interestingly, the neurodegenerative and behavioral deficits could be lowered by reducing the BACE1 levels with siRNA in App transgenic mice. Importantly, both βCTF and BACE1 are elevated in the DS brains and also abundant in early endosomes (reviewed in Jiang et al., 2010). Additionally, endocytosis maintains the structure of dendrites and axons and supports the synaptic transmission and regulation of receptor molecules those are critical for neuronal survival and repair as well as efficient uptake and intracellular delivery of lipids in neurons. Taken together, endosomal dysfunction is dependent on APP gene dosage and elevation of βCTF but independent of Aβ or αCTF levels. and that is responsible for EOAD in DS. Thus, BACE1 cleaved βCTF of APP, and multiple APP metabolites, including but not limited to Aβ, are crucial for the development of endosomal pathology of EOAD in DS (Jiang et al., 2010).

5. Age-related consequences of AD pathogenesis

Appearance of Aβ plaques has been demonstrated in DS patient as young as 8 years of age and found to have been consistently accumulating throughout life (Leverenz and Raskind, 1998). Symptomatically, prevalence of AD in individuals with DS progresses exponentially with age. Alterations in the first biomarkers such as CSF Aβ1−42/1−40 and plasma NFL values appear as early as the 3rd decade (28−30 years) of life in DS individuals, that is more than 20 years before the prodromal stage of diagnosis of AD (Fortea et al., 2020). In the 4th decade of life, changes occur in amyloid PET uptake and in ^{18}F-fluorodeoxyglucose PET and CSF p-tau, followed by hippocampal atrophy and deterioration of cognition in the 5th decade of life. The prodromal AD appeared at a median age of ∼50·2 years, and AD-related dementia at 53·7 years, while symptomatic AD progressed with age in DS patients, reaching 90%−100% in the 7th decade of life resulting in death (Table 12.1). Notably, the phases of AD evolution in DS include a long preclinical phase when the biomarkers follow a predictable order of changes over a period of more than 2 decades. Summarily, a long phase of preclinical changes begins more than 2 decades before the onset of AD-related dementia. Neurodegenerative changes, as indicated by plasma NFL increase or brain hypometabolism, might occur much earlier than previously thought in individuals with DS, even before detectable fibrillar Aβ deposition (Fortea et al., 2020). Interestingly, a significant negative correlation was demonstrated between CSF Aβ1-42 and age (from 21 to 60 years), which is implicated with a relatively consistent decrease in CSF Aβ1-42 in older DS adults and that is consistent with sporadic AD. Reduction of CSF Aβ-42 has been correlated with formation of plaque in brain parenchyma.

Age-related (21−56 years, average age 41 years) increase in CSF tau levels was demonstrated in DS relative to controls and to sporadic AD. In DS children, CSF tau did not discriminate children under the age of 56 months. Indeed, older DS patients (>40 years) exhibited significantly higher total tau compared to younger DS individuals (reviewed in Head et al., 2016). Altogether, increase in CSF tau was observed in DS individuals as a function of age.

6. Genes involved in neuropathologic events in DSAD

AD is a chronic neurodegenerative disease, progressively increases in advancing age, characterized by decline in memory, cognitive abilities, motor skills and so on, in both EOAD (30−60 years) and late-onset AD (LOAD: >60 years). In people with DS, AD appears 2−3 decades earlier from the onset of AD symptoms by the age of ≤40 years. Notably, the obligatory pattern of cognitive deficits in DS makes the neuropathological recognition of AD difficult (Wiseman et al., 2015). Individuals with DS display severe cognitive deterioration by the age of 30 years, including difficulties with verbalizing, apraxia, and other dementia-related dysfunctions of the frontal lobe (Head et al., 2016), and rapid increase in neuropathology may diagnose DSAD by 40 years of age (Lott and Head, 2019). Neuritic plaques developed from accumulation of insoluble Aβ deposition in brain parenchyma initially forms oligomers and subsequently forms fibrils and senile plaques slowly from the aggregation of abundance of Aβ (Murphy and LeVine, 2010; Gomez et al., 2020). Amyloid fibrils are generated from soluble proteins and that subsequently agglomerate to form insoluble aggregates, which are not easily degradable. Such aggregation

TABLE 12.1 Age-related consequences of AD development in DS.

Age of DS patients	Changes in AD markers	Consequence of changes	Disease stage
28–30 years	CSF Aβ1–42/1–40 and plasma NFL, increases in CSF p-tau concentrations, and reductions in brain metabolism, caused by mutation in Presenilin-1	Onset of fibrillar amyloid deposition; however, detectable by PET after ~10 years	More than 20 years before prodromal AD diagnosis
30–40 years	CSF p-tau and NFL concentrations; decrease in hippocampal volume due to atrophy		20 years before prodromal AD diagnosis
30 years	Early decrease in CSF Aβ1–42/1–40 values was accompanied by increase in plasma NFL concentrations, increase in CSF p-tau concentrations, and reductions in brain metabolism	Plasma NFL concentrations also increased visually from the earliest observed ages and were significantly different at age 30 years	20 years before prodromal AD diagnosis, similar to autosomal dominant Alzheimer's disease
37.5 years	Decrease in cerebral metabolism		12· 7 years before expected symptom. Onset of AD
40 years	Changes in Centiloid Aβ scale scores and brain atrophy occurred together with cognitive impairment	CSF p-tau and NFL concentrations began to increase at this age	Hippocampal volumes decreased faster
50 years	Prodromal AD and AD dementia	Dementia diagnosis occurred approximately 3 years after prodromal AD diagnosis.	Strikingly similar to that described in autosomal dominant and sporadic AD
55· 8 years	AD and AD dementia		
60 years	Progress of AD and dementia		
70 years	Death		

The table is prepared from the information available in Fortea et al. (2020).

eventually results in neurotoxicity and influences kinase/phosphatase activity by enhancing hyperphosphorylation of tau protein and formation of NFTs, and that consequently affects the synaptic and neuronal functions leading to neuronal death and the eventual appearance of AD (Lee et al., 2017a,b; Gomez et al., 2020). The tau protein, formed from the intraneuronal aggregations of hyperphosphorylated NFTs, is a microtubule-associated protein, and is generally present in axon and plays a crucial role in stabilization of microtubules in neurons (Brion, 1998; Maccioni et al., 2010). An abnormal accumulation and aggregation of hyperphosphorylated tau filaments in neuronal cells and dendrites eventually induces neurotoxicity and initiates AD pathophysiology (Šimić et al., 2016; Fonseca-Santos et al., 2015).

Imbalance in gene dosage and noncoding RNAs (ncRNAs), located at 21q22.11–21q22.3 in particular, are underlying factors of EOAD in DS (Chakrabarti et al., 2010; Mekkawy et al., 2016). Database for Annotation Visualization and Integrated Discovery (DAVID) demonstrated this region harboring ~464 genes to be the critical region for most of the neuropathological signs in DSAD, which was further refined to 441 using Gene Entrez database (Gomes et al., 2021). Of these, 19 genes have been ascribed with neurological dysfunctions, especially genes of APP processing (APOE, BACE1, BACE2, DYRK1A, PSEN1, PSEN2, ETS2, SUMO3, Mir-155), Aβ-deposition and plaque formation (CST3) and tau hyperphosphorylation (CXADR, DYRK1A, RCAN1, MARK4). In addition, SOD1, RUNX1, and ABCG1 (lipid transporter) have been demonstrated in DSAD. However, contribution of many of the genes of HSA21 critical region remains unelucidated for neurological diseases. Nonetheless, impact of dosage imbalances of HSA21 genes and/or interaction with non-HSA21 genes might affect the development of tissue and cellular differentiation causing neuronal reductions and neurological complications in DSAD (Contestabile et al., 2010). Upregulation of the triplicated RCAN1 and DYRK1A kinase dysregulates tau phosphorylation in brains of DS and that leads to EOAD (Jung et al., 2011). Notably, overexpression of DYRK1A suppressed the activity of a major Aβ-degrading enzyme in the brain named neprilysin (NEP) through phosphorylation at the NEP cytoplasmic domain. Significant decrease in NEP activity was evidenced in trisomic fibroblasts relative to that of euploid cells (Asai et al., 2017). Thus, this DYRK1A-driven suppression of NEP activity was targeted to look for DYRK1A inhibitors, which could be effective against DSAD as well as sporadic AD.

DYRK1A is responsible for cognitive deficits, which is a common factor between DS and AD. Overexpression of DYRK1A is reflected with: 1. increased levels of DYRK1A mRNA and DYRK1A protein in autopsy of human AD brains; 2. calpain 1-induced cleavage of DYRK1A in AD brains; 3. DYRK1A-mediated phosphorylation of the critical AD biomarkers, such as APP, presenilin 1, Tau (also known as MAPT), septin, and neprilysin; 4. DYRK1A regulates splicing of tau mRNA; and 5. inhibition of DYRK1A reversed cognitive defects in three widely used mice models of AD: 3xTG-AD, APP/PS1 and Aβ25-35 peptide-injected wt mice (reviewed in Nguyen et al., 2018). Thus, investigation of the regulation and substrates of brain DYRK1A may direct development of potent and selective DYRK1A inhibitors to treat cognitive deficits observed in different clinical indications.

Besides contribution of RCAN1 and DYRK1A, triplication and upregulation of APP were hypothesized as the primary cascade of amyloid deposition to cause dementia in people with DS (Castro et al., 2017; Lott and Head, 2005). Overexpression of APP results in elevation of Aβ peptide level in the brain following accumulation and development of neuritic plaques (also known as senile, dendritic, or amyloid plaques) and neurofibrillary tangles overtime, and that is evidenced in human DS, DS-mouse models, and induced pluripotent stem cells (iPSCs) derived from human DS (Wiseman et al., 2015). Such deposition occurs in the intracellular (neuron) and extracellular spaces (around the blood vessel and neurons) of the brain leading to development of dementia (Hithersay et al., 2017; Annus et al., 2016; Lott and Head, 2019). APP was 4- to 5-fold overexpressed in DS than euploid humans (Beyreuther et al., 1993). Reducing APP expression in axonal growth cones of DS with siRNA normalized the phenotypes, suggesting APP overexpression may perturb axonal pathfinding and circuit formation in the complex microenvironment of developing DS brain (Sosa et al., 2014). Normal APP protein is not neurotoxic; however, its overproduction and accumulation cause neurotoxicity (MacLeod et al., 2015). A genomic duplication of the APP locus observed in familial EOAD in general population supports the association of increased copy number of APP and an AD-like dementia, even in the absence of clinical findings of DS (Rosenberg et al., 2016). Interestingly, lack of triplication of APP in a DS patient with partial trisomy of HSA21 resulted in absence of dementia, low plasma Aβ-42, very low brain amyloid levels detectable by functional imaging, and the absence of AD neuropathology (Doran et al., 2017), indicating the compulsory involvement of the increased APP copy number in development of DSAD. However, the mechanism of upregulated APP action on generalized transcriptional dysregulation might interact with other trisomic and nontrisomic genes causing disruption of endosomal systems, mitochondrial dysfunction, and immune system dysfunction (Wiseman et al., 2015). Nevertheless, the central role of APP in AD and dementia pathogenesis in DSAD would seem well supported by myriads of studies.

The mechanism of alterations in APP cleavage and dysregulation of production and degradation of Aβ has been explained by dominant autosomal mutations or genetic polymorphisms of APP, Presenilin 1 and 2 (PSEN 1/2 or PS1/2), ABCG1, and apolipoprotein E (APOE) genes (Wiseman et al., 2015; Gomez et al., 2020). In normal conditions, PSEN 1/2 proteins are components of the γ-secretase complex that regulates APP processing; however, mutations in PSEN1 and PSEN2 genes can enhance production of long Aβ peptides (Aβ-40/42), which are known to be the marked pathogenic form of Aβ and associated with onset of EOAD and advanced grades of AD (Alonso Vilatela et al., 2012; Ryu et al., 2010). In addition, DYRK1A activates phosphorylation of PSEN1 in DS, indicating its contribution to increased γ-secretase activity and elevated amounts of Aβ40/42 in the brain consequently (Wiseman et al., 2015; Ryu et al., 2010).

With regards to contribution of APOE in neurodegeneration, it is known that APOE protein participates in neuronal signaling, regulation of the cholinergic neurotransmitter system, and clearance of Aβ (Mohandas et al., 2009). It is also considered as a biomarker of LOAD (Ryu et al., 2010), showing an association between APOE gene, particularly alleles μ4 (apoE4), and the development of dementia in DS adults (Zigman and Lott, 2007). Another HSA21 gene, ABCG1, contributes to regulation of cholesterol and induces processing of APP leading to clearance of Aβ from brain (Wiseman et al., 2015). Other HSA21 genes, such as BACE1, BACE2, and SUMO3, are involved in posttranscriptional modification of APP, indicating some change in Aβ production (Wiseman et al., 2015). Another HSA21 factor, the RUNX1, is involved as transcriptional regulator in neural progenitor cells and also implicated for the development of neuronal phenotypes in DS and neuropathogenic phenotype in DSAD (Halevy et al., 2016; Hithersay et al., 2017).

Furthermore, additional HSA21 factors such as microRNA-155 (miR-155) and ETS2 transcription factor are involved in modulating and processing of APP. The functional mechanisms of these two genes in APP processing are different; however, both are postulated to induce Aβ generation (Chatterjee et al., 2011; Wang et al., 2014). Importantly, non-HSA21 genes such as CST3 (single-nucleotide polymorphisms in cystatin C) and MARK4 (microtubule affinity-regulating kinase 4) were associated with dementia in DS adults (Lee et al., 2017a). The CST3 encodes an inhibitory protein CysC that binds with Aβ, inhibiting its aggregation in the brain and subsequently, preventing the formation of fibrils (Perlenfein et al., 2017). The MARK4 proteins are phosphorylated microtubule-associated proteins, wherein expression of aberrant MARK4 relates to increase in phosphorylation of tau and may lead to initiation of EOAD (Lund et al., 2014). Therefore, genes of HSA21 and non-HSA21 independently or co-operatively and/or their interaction may facilitate aggregation of Aβ and may contribute to neuroinflammation and DSAD pathogenesis (Guidi et al., 2017; Wilcock and Griffin, 2013).

Humanin (HN) has been demonstrated to attenuate the Aβ-peptide induced memory deficits, tau phosphorylation, dendritic degeneration, neuronal apoptosis, and neuronal dysfunction and to ameliorate the senile dementia present in individuals with AD (Chai et al., 2014). A humanin isoform, MTRNR2L12, has been described as a biomarker of severe cognitive disability (Bik-Multanowski et al., 2015) and early dementia in DS patients (Chai et al., 2014). Altogether, antiapoptotic and neuroprotective effects of HN may direct further studies on its possible therapeutic use to treat DS-associated clinical/phenotypic features, especially during the early developmental period.

7. Ubiquitinylation and proteasome system of AD development

One of the putative mechanisms proposed on impairments of the proteasome system and decrease in autophagosome function in DS, before and after the development of AD, suggested that accumulation of oxidized/misfolded protein could be an early hallmark in DS brain (reviewed in Tramutola et al., 2017). Notably, reduced protein degradation contributes to the increase in polyubiquitinated proteins, which virtually disturb other cellular functions, because polyubiquitinylation is essential for both proteasomal and endosomal—lysosomal degradation. Thus, disrupted ubiquitin (Ub) signaling might have significant consequences for neuronal functions: regulation of degradative pathways, endocytic trafficking, inflammation, and DNA repair. Interestingly, increased oxidative stress levels perturb protein degradation leading to accumulation of polyubiquitinated proteins (Tramutola et al., 2017). An altered pattern of polyubiquitinylation was observed in DS cases before and after the development of AD neuropathology. Increased polyubiquitinylation was combined with elevated oxidation for PRDX2, HSP60, UCH-L1, V0-ATPase, ATP synthase, CK, and DRP-2. Interestingly, increased polyubiquitinylation of UCH-L1 and PRDX2 was reported in DSAD compared to DS and age-matched controls. In addition, diminished level of PRDX2 protein was demonstrated in DS fetal brains (reviewed in Tramutola et al., 2017). Moreover, antioxidant enzymes, including glutathione peroxidase (GPX) and thioredoxins, were reduced with an altered SOD-1/GPX activity ratio in DS brains. In addition, increased oxidative stress coupled with reduced proliferation and increased apoptosis cause neuronal death and intellectual disability in DS.

Characterization of molecular chaperones (HSPs) for recognition of the oxidized/misfolded proteins showed oxidatively modified and increasingly polyubiquitinated HSP60 causing impairment of its function in AD brain. Interestingly, proteasomes degrade the unfolded/damaged proteins that are not refolded by chaperones. UCH-L1, a member of the proteasomal system, is abundant in the brain and maintains free Ub levels in neurons. It hydrolyzes the Ub chains from the carboxyl terminus to allow degradation of the targeted protein for gaining access to the proteasome. Interestingly, UCH-L1 itself undergoes oxidative damage and error-prone polyubiquitinylation during neurodegeneration (Tramutola et al., 2017). Decreased proteasomal activity of UCH-L1 was consistent with its increased polyubiquitinylation/carbonylation observed both in DS with and without AD. Additionally, endosomal—lysosomal system of removal of damaged proteins is affected with proteasomal and endosomal—lysosomal degradation. Notably, perturbation of autophagosome formation was reported in DS brain, before and after development of AD pathology, in association with the hyperactivation of the mTOR pathway (reviewed in Tramutola et al., 2017). Decreased acidification of lysosomes in and mutation of lysosomal ATPase contribute to autophagy-related neurodegenerative diseases, including AD. *Per se*, mitochondrial dysfunction and altered glucose metabolism lower ATP synthesis in DS brain (discussed in detail in Chapter 8).

Accelerated and early aging in DS cases have been characterized with increased polyubiquitinylation in proteins such as DRP-2, ATP synthase, and V0-ATPase as putative markers of accelerated aging phenotype. Interestingly, reduced level of DRP-2 was documented in trisomic fetal brain, which appears increasingly polyubiquitinated both in aged DS and controls (Tramutola et al., 2017). Moreover, DRP-2 level is reduced in adult DS brain at protein or mRNA level.

Polyubiquitinylation of gelsolin, a cytoskeletal protein present in the cytoplasm and also secreted in biological fluids, showed an antiamyloidogenic role in AD. Gelsolin is increasingly carbonylated in the cerebral spinal fluid (CSF) of cases with AD and MCI, indicating a critical role of gelsolin in AD, which was further supported by the evidence of its alteration in young DS brain before the onset of AD neuropathology (Tramutola et al., 2017). An association between cleaved gelsolin forms and levels of Aβ-peptide was reported in the frontal cortex of DS subjects (reviewed in Tramutola et al., 2017). Interestingly, a comparison of DSAD and DS without AD reflected an enhanced polyubiquitinylation of PRDX2, UCH-L1, DRP-2, and V0-ATPase along with proteins exhibiting low levels of poly-Ub chains, including tubulin (different isoforms), α-internexin, Ser/Thr PP2A phosphatase, ubiquilin, succinyl CoA transferase, D-phosphoglycerate dehydrogenase, and programmed cell death-interacting protein (Tramutola et al., 2017). Thus, it has been established that polyubiquitinated proteins accumulate with aging and neurodegeneration; however, the role played by ubiquitinylating/deubiquitinylating enzymes remains unzipped. Ubiquilin, markedly involved in protein degradation, was observed with reduced polyubiquitinylation with an N-terminal sequence that terminates with a conserved 50 amino acid sequence known as ubiquitin-associated domain (UBA) (reviewed in Tramutola et al., 2017).

Ubiquilin-1, reduced in both DS and DSAD compared with their respective age-matched controls, delivers polyubiquitinated proteins to the proteasome or to autophagosomes or influences aggresome formation to evade accumulation of misfolded proteins. Notably, ubiquilin-1 gene is associated with increased risk of LOAD (reviewed in Tramutola et al., 2017). Overexpression of ubiquilin-1 decreases the Aβ-42/40 ratio and avoids APP-induced cellular toxicity. Such relationship of ubiquilin-1 protein levels in AD patients may target some specific therapeutic approach for AD.

Thus, alteration of ubiquitinylation of proteins involved in energy metabolism, antioxidant responses, the cytoskeleton, and synaptic plasticity could serve as markers of AD neuropathology. Interestingly, polyubiquitination of some proteins was observed before the onset of AD neuropathology, while some were expressed in developed AD. Summarily, the disturbance of polyubiquitinylation was postulated to favor AD neuropathology. Altogether, reduction in proteosomal activity and impaired autophagy, oxidized/misfolded proteins tagged by poly-Ub chains may collectively create a toxic environment for neurons in DS. In addition, protein aggregation triggered by oxidation/polyubiquitinylation indicates possibility of deposition of other protein aggregates that may or may not colocalize or interact with plaques and tangles (Tramutola et al., 2017). Thus, there may be many other active protein aggregates alongside Aβ and tau, which could be resolved from understanding of the composition and chemical properties of polyubiquitinated proteins and provide novel insights into the pathogenesis and potential treatments of DS and DSAD, including other protein deposition disorders. Moreover, interaction of trisomic HSA21 genes with other global disomic genes may complicate the scenario.

8. Perspectives of neuroinflammation in DSAD

Neuroinflammation is a complex process with many phenotypic variations and was thought to be the causative factor for the onset of

AD-related dementia in individuals with DS (Wilcock, 2012; Wilcock and Griffin, 2013; Wilcock et al., 2015a,b). Nevertheless, neuroinflammation, dementia, and DSAD are primarily contributed by the triplicated HSA21 genes. The science of neuroinflammation in the brain and its disorders has progressed over time and continues to evolve as capabilities of understanding of the system develops. The evolution from the concept of "immunologically privileged" to exhibit an almost complete spectrum of inflammatory responses to the correct stimuli and environment in the brain has been established. From considering brain as a cytotoxic loop, current understanding of harnessing the inflammatory system of brain is being studied to ameliorate AD pathologies and improve clinical outcomes. Noticeably, significant neuroinflammation with activation of microglia, astrocytes, and other neurons participating in this inflammatory response was observed in humans with DS as well as DS mouse models (Griffin et al., 1998; Stoltzner et al., 2000). However, microglial cells are the primary inflammatory cells identified as a unique cell subtype in the brain observed by Del Rio Hortega in the 1920s. Microglia cells are dynamic and can adapt and change its morphology between ramified and ameboid-like cells under stress and pathological situations (Wolf et al., 2017). The oligodendrocytes and vascular pericytes also participate in the neuroinflammatory process (Wang et al., 2017).

8.1 Neuroinflammation in AD

An accentuated inflammatory response is likely before appearance of symptoms of neuropathology in DS. Triplication of HSA21 genes emphasizes the inflammatory response, influencing high expression and release of several proinflammatory cytokine markers (Table 12.2) (reviewed in Gomes et al., 2021; Wilcock and Griffin, 2013). Overexpression, upregulation, or production of IL-1β results in increase in APP function, which finally contributes to deposition of Aβ and activation of pathways involved in the formation of senile plaques and NFTs. HSA21 genes involved in the inflammatory process, formation of senile plaques, and/or NFT include: S100B (S100 calcium binding protein astrocyte-derived) and SOD1 (superoxide dismutase) that participle in oxidative stress; and ADAMTS1 (ADAM metalloproteinase with thrombospondin type 1 motif, 1), ADAMTS5 (ADAM metalloproteinase with thrombospondin type 1 motif, 5), and CXADR (Coxsackie virus and adenovirus receptor) that activate pathways such as mitogen-activated protein kinase (MAPK)-p38 (Wilcock and Griffin, 2013). The inflammatory markers present in blood, brain, and CSF may indicate signs of AD-related dementia in persons with DS and draw a diagnosis (Hartley et al., 2015; Startin et al., 2019; Alhajraf et al., 2019); however, investigations of an early diagnosis would be beneficial. Identification of biomarkers of neuroinflammation is also important for development of drugs and new therapeutic methods that may facilitate a treatment of EOAD in DS population (Castro et al., 2017).

The contribution of the three major interferon receptors such as IFNAR1, IFNAR2, and IFNGR2 in triplet condition includes enhancement of interferon signaling followed by subsequent increase in release of the M1 (activated macrophage phenotype) markers, specifically IL-1β, TNFα, and IL-6. These markers will exacerbate the mechanism of oxidative stress in the DS brain owing to triplication and overexpression of TIAM1, SOD1, and PRMT2, through induction of NADPH oxidase by TIAM1, enhancement of oxidation by high dosage of SOD1, and inhibition of the production of the antioxidant, nitric oxide, by PRMT2. Collectively, these factors enhance neurodegeneration in the DS brain in conjunction with the primary changes of pathogenetic markers such as formation of Aβ-plaques and NFTs. Therefore, the M1 inflammatory response was postulated to be the main state of enhancement, primarily associated

TABLE 12.2 Alteration of cytokines in DS.

Cytokine markers	Samples studied	Type and expression level
IL-1	Serum	Increased level of IL-1α
	Brain	Upregulated expression of IL-1Ra at 40 years; increased immunoreactivity in astrocytes of IL-1 in postnatal DS; decreased level of IL-1β in Ts65Dn mouse
	Plasma	Increased level of IL-1β, after stimulation of influenza A
	GCF	Increased level of IL-1β
	PBMCs	Decreased level of IL-1β in aged DS after stimulation with LPS
IL-2	Serum	Increased in DS; diminished production of in aged DS after stimulation with phytohemagglutinin (PHA)
	PBMCs	Reduced in DS when stimulated with Aβ1-42
IL-4	Serum	Increased in DS
	GCF	Increased in DS
IL-6	Brain	Elevated gene expression in DS < 40 years age; Elevated gene expression in DSAD >40 years age
	Plasma or serum	Increased in DSAD; increased in DS with severe AD; increased in DS
	PBMCs	Decreased level in DS; reduced gene expression in DS
	GCF	Increased in DS
IL-7	Plasma or serum	Increased in DS
IL-8	Plasma or serum	Increased in DS; increased in DSAD
IL-10	Brain	Increased gene expression in DS < 40 years age; increased gene expression in DSAD >40 years age
	Plasma or serum	Increased in DS and DSAD; increased in DS after stimulation with *Streptococcus pneumonia*
	Gingival	Reduced gene expression in DS with periodontal disease
	GCF	Increased in DS
	PBMCs	Increased in DS after PHA stimulation; increased in DS compared to intellectual disability and control; increased gene expression in DS
IL-12	Brain	Increased gene expression in DS < years age
	GCF	Increased in DS
IL-15	Plasma	Increased in DS
IL-17A	PBMCs	Increased gene expression of IL-17A in DS

(*Continued*)

TABLE 12.2 Alteration of cytokines in DS.—cont'd

Cytokine markers	Samples studied	Type and expression level
IFN-α	Plasma	Increased in DS
IFN-γ	Plasma or serum	Increased in DS compared to control and intellectual disability; increased in DSAD; increase in intracellular production in DS versus MR and HC
	GCF	Increased in DS
	PBMCs	Increased in DS after stimulation with PHA or Cytomegalovirus; increased in DS; decreased in DS when stimulated with Aβ1-42

DS, Down syndrome; *DSAD*, DS with Alzheimer disease; *IL*, interleukin; *GCF*, gingival cervicular fluid; *IFN-γ*, interferon-γ; *TNF-α*, tumor necrosis factor-α.
The table has been prepared from the information available in Gomes et al. (2021).

with neurodegeneration in DS brain. The preponderance of AD in DS can be explained with the contribution of triplicated set of the genes for creating a unique inflammatory environment; however, further studies will be worthy for better understanding of the inflammatory response and its clinical management with the concomitant approach of developing targeted dugs for sporadic and DSAD.

In young people with DS (postnatal to 41 years of age), microglia in brain tissue appeared more IL-1 positive compared to age-matched controls (Griffin et al., 1998). However, autopsies of DS over 40 years of age were consistent with a reduced number of resting microglia with increased dystrophic morphology of the microglial cells in the gray matter of the brain, possibly due to microglial dysfunction as a result of aging (Xue and Streit, 2011). Moreover, microglial morphology in DS, particularly in people with DS-AD, remains largely unexplored.

8.2 Neuroinflammatory markers of DSAD

Expression of active markers such as MHC-II (associated with antigen presentation), CD68 (a lysosomal protein), and CD36 (a class B scavenger receptor) surrounding amyloid plaques or neurofibrillary tangle has been reported in microglia of AD (Xiang et al., 2006). Such markers could induce toxicity leading to onset of AD pathogenesis or help microglia to remove the abnormal protein deposits from the brain; however, no consensus has been drawn on the negative or positive effects of the expression of these markers. Thus, the studies were focused on the effects of the cytokine markers on the behavior of the microglia contributing to development of AD pathogenesis.

Among the most common cytokines altered in AD were described as IL-1β, IL-6, TNF-α, and TGF-β. The association of IL-1β was first demonstrated with the development of neuritic amyloid plaques from diffused deposits in human autopsy tissues (Griffin et al., 1995), which led to hypothesize a "cytokine cycle" suggesting that the IL-1β production may initiate consequences of increased APP production and its processing by neurons. Recruitment and activation of astrocytes lead to signaling in the microglia in response to amyloid deposits, and induce more expression of IL-1β (Griffin et al., 1998). In series, IL-1β influences S100β production in astrocytes (Sheng et al., 1996) and promotes neurite growth (Kligman and Marshak, 1985). Elevation of serum IL-1β has been reported in cases with MCI increasing the likeliness of development of dementia, which indicated that serum IL-1β could be considered as an early marker of dementia and AD pathology (Forlenza et al., 2009). Additionally, genome-wide association studies (GWASs) have identified AD-associated polymorphisms of IL-1β (reviewed in Di Bona et al., 2008). However, IL-1β overexpression resulted in amelioration of amyloid pathology in the hippocampus of APP/PS1 transgenic mice with a 50% reduction in plaque load with concomitant increase in IL-1β in the hippocampus (Shaftel et al., 2007). Summarily, further studies are essential to explore the variable antagonistic or synergistic effects of genetic polymorphisms of IL-1β on plaque development or removal in the hippocampus.

The microglia are the main source of IL-6, another cytokine that mediates immune response and inflammatory reaction in CNS, astrocytes, neurons, and endothelial cells, which are all capable of producing IL-6 (Hampel et al., 2005). In brain tissue of AD, elevated IL-6 has been detected in the pathologically relevant regions (Strauss et al., 1992). In a mouse model with amyloid deposition, microglial cells, influenced by overexpression of IL-6, phagocytosized the amyloid deposits and that resulted in amelioration of the burden of amyloid deposits (Chakrabarty et al., 2010). However, conflicting reports are listed on its negative or positive effects, including the fact that IL-6 may enhance the acute-phase proteins, increasing vascular permeability, activation of lymphocytes, and antibody synthesis in AD (reviewed in Gadient and Otten, 1997) or may enhance neuronal survival (Kushima et al., 1992) and suppress demyelination in a model of multiple sclerosis (Rodriguez et al., 1994).

Another cytokine, TNF-α, has been reported with both beneficial and detrimental effects in the CNS, which includes its potent proinflammatory and cytotoxic effects in certain conditions, trophic effects on hippocampal neurons, and protection against free-radical damage of the primary neurons (Barger and Harmon, 1997). Such dichotomous effect of TNFα was dependent on its receptor subtypes: TNF-α receptor 1 (TNFR1) and TNF-α receptor 2 (TNFR2) through which the TNF-α works (Quintana et al., 2007). TNFR1 enhances neuronal death through the TNF-receptor-associated death domain protein and caspase-8-activated apoptosis (Depuydt et al., 2005), whereas TNFR2 mediates the prosurvival action of TNF-α through the nuclear factor-κB- (NFκB)-mediated antiapoptotic pathway (Hehlgans and Pfeffer, 2005). Elevation of expression of TNFR1 is described in the brain of AD with concomitant decrease in the levels of TNFR2 (Zhao et al., 2003). The diversified actions of TNF-α through its receptors have many subtleties of these effects described in the CNS. A fusion protein, etanercept, acts as a distracting receptor for TNF to reduce its effects at the biologically active receptors. Earlier reports on a small number of AD patients demonstrated improved cognition through perispinal delivery of etanercept (Tobinick and Gross, 2008). Clinical trials were conducted with etanercept combining TNFR2 for the treatment of AD, and the Fc-portion of IgG was used to treat Crohn's disease, arthritis, and other autoimmune disorders (Silva et al., 2010). Thalidomide was also considered for clinical trials for treatment of AD based on its anti-TNF-α effects, which showed improvement of learning and memory in transgenic mice (Alkam et al., 2008).

The cytokine TGFβ, a growth factor, plays a significant role in tissue development, repair, and homeostasis and is mostly associated with repair mechanisms and more importantly, it is not known for its detrimental or cytotoxic effects in the CNS. It mostly acts for the formation of a glial scar and upregulation of extracellular matrix proteins. Elevation of TGF-β levels are reported in the brain, but its decrease was observed in serum of patients with AD (De Servi et al., 2002). Notably, progression of AD pathology and associated neurodegeneration were attributed to deficiencies of TGF-β signaling in human brain

(Tesseur et al., 2006). Overproduction of TGF-β by astrocytes resulted in lowering of parenchymal amyloid deposits but increased deposition of amyloid in the cerebrovasculature in APP transgenic mice.

There is an increasing concern on better understanding of the inflammatory state in the brain of AD patients. A hypothesis on special mechanism of "classical activation" and "alternative activation" of microglia has been demonstrated in the brain (Colton et al., 2006). The "classical activation" entails the effects of type 1 T helper (Th1) cytokines, such as IFN-γ, IL-1β, TNF-α, and IL-6. The "alternative activation" was described as a state associated with anti-inflammatory, repair, wound healing effects mediated by IL-10, TGF-β, IL-4, IL-13, arginase 1 (AG1), and tissue remodeling factors: Found in inflammatory zone 1 (FIZZ1) and chitinase 3-like 3 (YM1). The hypothesis established the expression of both classical and alternative activation markers in the cultured microglial cells, transgenic AD mouse models, and autopsy of brain tissues from human AD, while the alternative activation markers were expressed to the same degree, sometimes more than the classical activation markers commonly associated with an inflammatory response.

Inflammatory response and cytokine stimulation have led to understand the basis of neuroinflammation, AD pathogenesis, and therapeutic development. Of the four distinct inflammatory states in the peripheral macrophage of the brain (Mantovani et al., 2004), the M1 response is stimulated by IFN-γ and/or TNF-α and is characterized by actions of IL-1β, IL-6, IL-12, and low degree of IL-10 inflammatory cytokines. The M2 response is mediated by the alternative activation state and is further classified into M2a, M2b, and M2c sub-types, each of which has specific stimuli and responses. Initiation of the M2a response is intervened by IL-4 and/or IL-13 and is characterized by the tissue remodeling factors: FIZZ, YM1, arginase 1, IL-1Ra, low degree of IL-12, and mannose receptor C1 (MRC1). M2b response is stimulated by immune complexes, specifically including CD86, IL-10, TNF-α, IL-1β, IL-6, low degree of IL-12, which are also involved in stimulation of M1 and M2a states. The sub-type M2c response is excited by IL-10, TGF-β, MRC1, Sphk1, and low level of IL-12. M2c response is characterized by a series of markers that actively antagonize M1 signaling pathways and thus, is also known as an acquired deactivation state.

8.3 Microglial changes associated with neuroinflammation in DSAD

Studies using HLA-DR immune-staining reported higher microglial loads in DS-AD and sporadic AD, which was consistent with increased inflammation in brain (Hopperton et al., 2018). The role played by the microglial cells in AD pathogenesis has been evidenced from the interaction with plaques and tangles leading to an increased risk of AD due to polymorphisms in genes expressed in microglial cells such as TREM2 and CD33 as well as changes in their morphology in the presence of AD pathology (Serrano-Pozo et al., 2013; Calderon et al., 2017; Davies et al., 2017). The ramified to ameboid alterations in morphology of microglial cells in human accompanied by changes in motility can be exacerbated in AD and other pathological situations, which may reflect the functional state of the cells that potentially contribute to the development and progression of AD (Torres-Platas et al., 2014; Franco-Bocanegra et al., 2019). The ramified shape of the microglial cells appeared as small and round body with extended long branching processes, while ameboid microglia are large and round body rich in cytoplasm and lack branching processes, which may occur in response to acute destruction of the central nervous system tissue. The ameboid-like cells can be labeled immuno-cytochemically using macrophage cell surface markers.

The microglial morphology exhibited the posterior cingulate cortex (PCC) in the brains of people with DS with or without AD across the lifespan. The significant characteristics of PCC include: lower glucose metabolic rates relative to their control groups (Azari et al., 1994), reduced glucose metabolism in the precuneus and posterior parieto-temporal lobes (observed through fluoro-deoxyglucose PET), which can lead to neuronal and synaptic loss before the appearance of symptoms in sporadic AD, and significantly higher expression of the glial biomarkers along with lower expression of other neuronal health biomarkers in this region as measured by magnetic resonance spectroscopy, in the demented DS adults (Lin et al., 2016). The Iba1 immune staining of the microglial cells and their processes allowed a better characterization of the morphology and distribution of different microglial phenotypes in the brains of DS people with and without AD (Figs. 12.6 and 12.7) (Martini et al., 2020). The study revealed a significant genotype effect (DS vs. non-DS) in the number of hypertrophic microglia in the white matter of the brain, which have enlarged, short processes, and thicker bodies, in response to injury (Bachstetter et al., 2015; Streit et al., 2014), and contrasts with other reports having fewer hypertrophic microglia in DS than controls (Xue and Streit, 2011). However, those studies considered the changes in the gray matter of the temporal cortex. The differences highlighted the importance of the alterations in both the white and gray matter.

A significant reduction in white matter and vascular pathology with increasing age of DS could potentially elicit hypertrophied microglial cells observed in the white matter (Head et al., 2017; Helman et al., 2019). Age-related changes in white matter, response of the microglia, and increase in microglial proteins were reported in brain tissue of autopsy with early onset of AD (Raj et al., 2017). Increase in activated microglia in the white matter and atrophy in the gray matter in other dementias (Ohm et al., 2019), indicated abnormalities in white matter is an early event preceding gray matter atrophy, especially in EOAD (Caso et al., 2015). A significant reduction in the number of ramified microglia in the gray matter was associated with increasing age, dementia, hippocampal sclerosis of aging, hippocampal sclerosis of aging with AD, and dementia with Lewy bodies and in the AD group and that was predominant in DS individuals (Xue and Streit, 2011). An increase in rod-shaped microglia with age was also observed in the autopsy of controls and DS cases, which was higher in DSAD compared to sporadic AD. These rod-shaped microglia were thought to provide protection to neurons and trophic support (Ziebell et al., 2012). The precise role of rod-shaped microglia in aging and AD pathogenesis in DS has yet been deciphered.

Ameboid microglial cells have higher motility owing to their reduced and retracted processes. These cells are most commonly found in the developing brain, before transforming into ramified microglia (Kettenmann et al., 2011; Harry and Kraft, 2012), and are phenotypically very similar to macrophages found in the rest of the body (Karperien et al., 2013). Therefore, it was hypothesized that these microglia could arise from monocytes entering into the brain in response to injury or stress (Boche et al., 2013), or they are senescing in the brain. The monocytes may enter the brains of DS people from the blood as vascular neuropathology may lead to small microhemorrhages in the brain leading to increase in ameboid microglial cells (Head et al., 2017; Helman et al., 2019). A significantly increased number of ameboid microglia were reported in DSAD cases than age-matched controls or young individuals with DS (Martini et al., 2020). However, the clinical significance of ameboid microglia in DSAD is yet to be explored fully.

A significant increase of dystrophic microglia preceded neurofibrillary pathology in brains of AD and in autopsy of older DS individuals

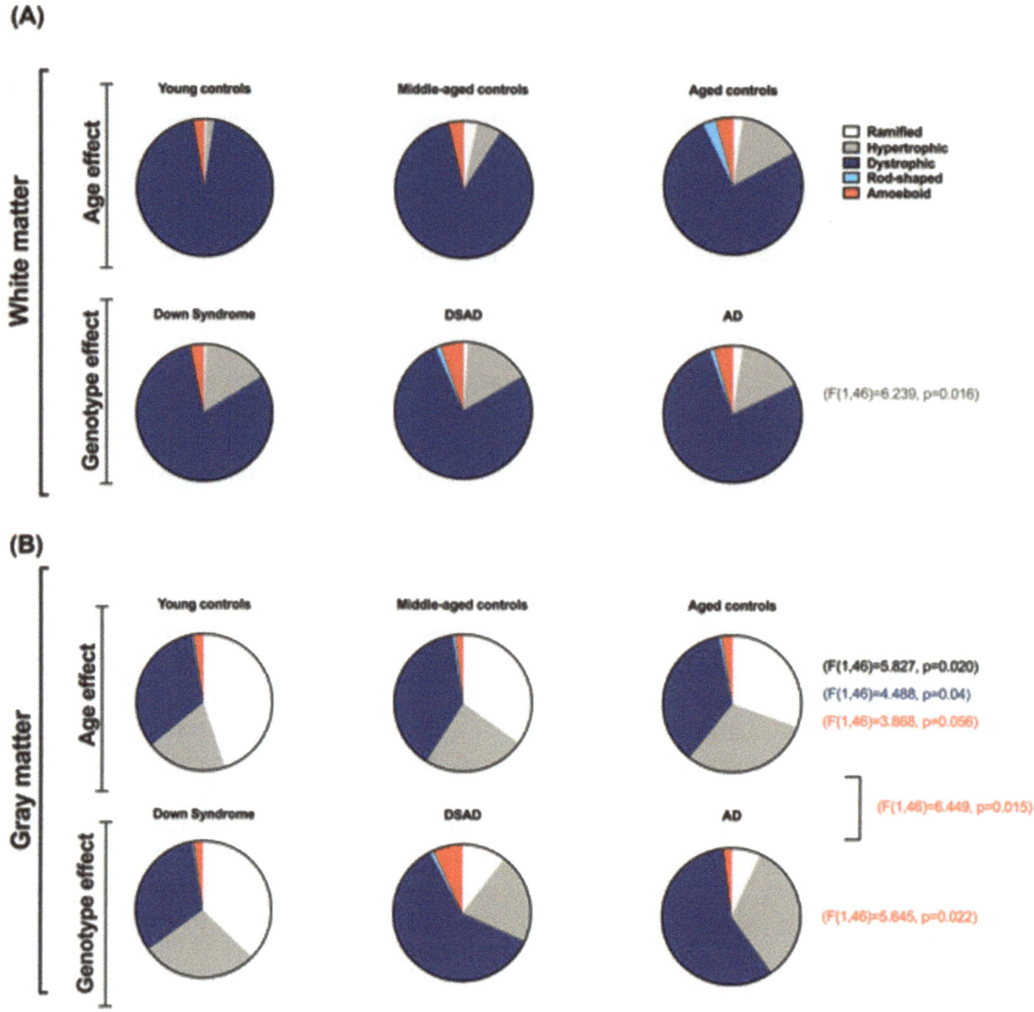

FIGURE 12.6 Distribution of microglia phenotypes across different groups in white and gray matter. The average number of microglia in each of the five classifications according to each group in the white (A) and gray (B) matter, demonstrating the age and genotype effects related to the morphology of cells. Data are presented as means. (A), In the white matter, hypertrophic microglia increase as a main effect of genotype. (B), In the gray matter, the age effect is significant for ramified and dystrophic cells. Ameboid cells have an effect of age, genotype, and a significant interaction between these factors. *Reprinted from Martini et al. (2020). https://doi.org/10.1002/dad2.12113, with permission from John Wiley and Sons and Copyright Clearance Center.*

with a preponderance in individuals with DSAD (Fig. 12.7) (Martini et al., 2020; Bachstetter et al., 2015; Streit et al., 2014; Paasila et al., 2019) and that suggested the importance of the loss of microglial function and activation of toxic and reactive cells. Importantly, presence of increased numbers of dystrophic microglia in the white matter of young non-AD individuals with DS indicated the vulnerability of the brain white matter to degeneration, and that could be exacerbated by the biological changes of aging (Augé et al., 2019). Dystrophic microglia owing to

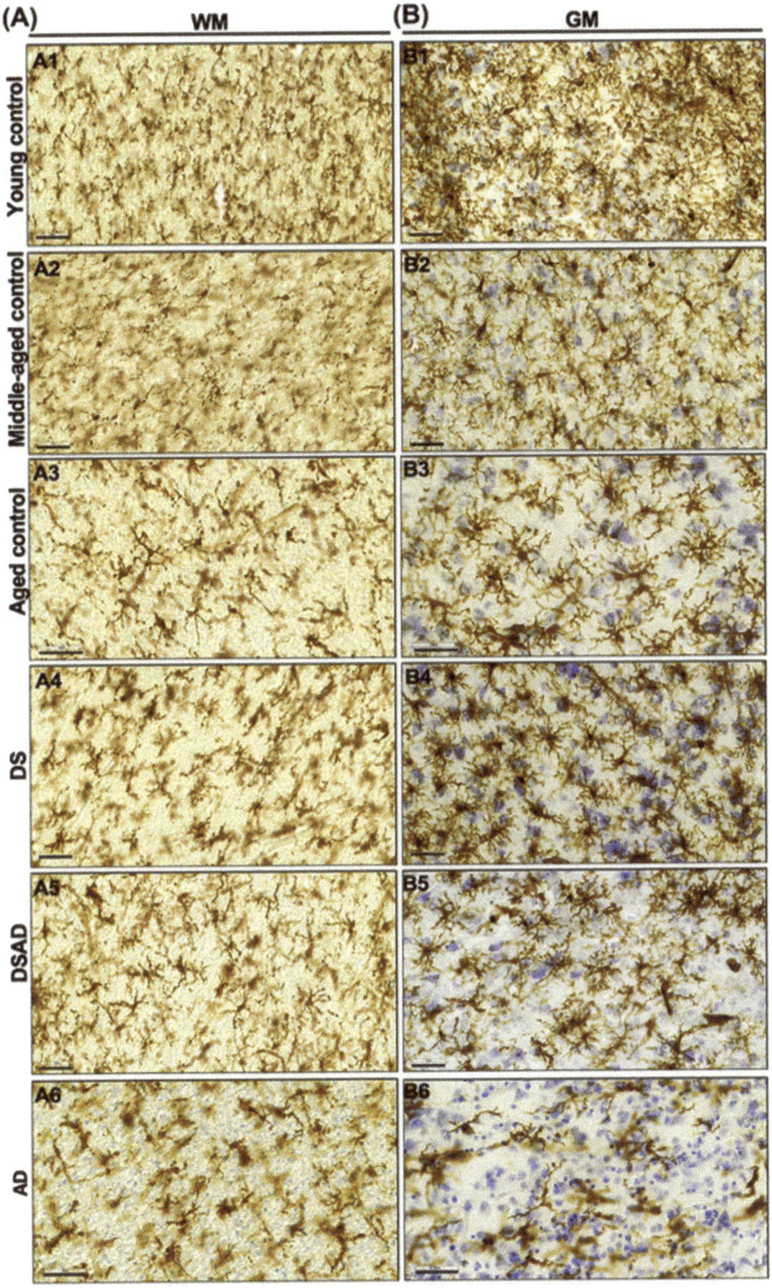

FIGURE 12.7 Group specific patterns in Iba1 microglia staining. Representative images of microglia morphologies observed in the white (A) and gray (B) matter of young controls (A1, B1), middle-aged controls (A2, B2), aged controls (A3, B3), Down syndrome (A4, B4), Down syndrome with neuropathological Alzheimer's disease (A5, B5), and Alzheimer disease (A6, B6), respectively. Scale: 50 μm. *Reprinted from Martini et al. (2020). https://doi.org/10.1002/dad2.12113, with permission from John Wiley and Sons and Copyright Clearance Center.*

microglial dysfunction or age-related senescence have fragmented or "beaded" processes, which is supported by the fact that human microglial processes shorten with age leading to a reduced arborization area and reduced capacity for gray matter surveillance and thus, were thought to be important markers for AD neuropathology (Davies et al., 2017; Wolf et al., 2017). Additionally, tau-positive neurons are associated with dystrophic microglia, suggesting its fundamental role in disease progression (Streit et al., 2009).

9. Genes of neuroinflammation in Down syndrome

The neuroinflammation pathways responsible for AD are directly involved in DSAD; however, gene dosage of the triplicated HSA21, more specifically the inflammatory genes of HSA21 (Table 12.3), may further aggravate the inflammatory state of the DS brain.

CXADR, a gene encodes for a protein called coxsackie virus and adenovirus receptor (herein abbreviated CXADR), has dual function such as a viral receptor and an adhesion molecule linked to functions of tight junctions. Its overexpression is noted in brain and the systemic secretory organs (pancreas, small intestine, and testis). Additionally, an inherent role of this protein has been reported in the models of myocardial inflammation and cardiac injury in absence of viral infection suggesting its inflammatory response in heart (Ito et al., 2000). It has also been demonstrated that increased CXADR can influence stress-induced mitogen-activated protein kinase (MAPK) pathways in the heart through enhanced production of IFN-γ, IL-12, IL-1β, TNF-α, and IL-6 (Yuen et al., 2011). In DS, CXADR-induced inflammatory response of M1 state can be activated by all of these inflammatory cytokines. Noticeably, CXADR's function in tight junctions leading to transendothelial migration of neutrophils may facilitate its altered expression in endothelial cells of the cerebrovasculature in DS patients, which may result in infiltration of peripheral inflammatory cells into the brain and influence the inflammatory response (Zen et al., 2005).

A significant overexpression of the protein encoded by ADAMTS1, a subtype of the ADAMTS (a disintegrin and metalloproteinase with thrombospondin motif) family, is influenced by the inflammatory cytokine IL-1β in DS, whereas another member ADAMTS5 was not increased to that extent (Miguel et al., 2005). Both ADAMTS1 and ADAMTS5 are located on HSA21 and dependent on IL-1β, suggesting their association with inflammatory response. The secretion of ADAMTS1 has been reported to contain a signal peptide in the N-terminal region. Both ADAMTS1 and ADAMTS5 act as proteinase and degrade the extracellular matrix proteoglycans such as aggrecan and versican (Kuno et al., 2000). Triplication of these proteinases in DS and further enhancement by the cytokine IL-1β may lead to exacerbate the degradation of extracellular matrix proteins in response to inflammatory insults. Moreover, DS brain with higher level of IL-1β immunoreactivity indicated presence of more of IL-1β to stimulate the ADAMTSs in the DS brain (Griffin et al., 1998).

An increase in the T-cell lymphoma invasion and metastasis 1 (TIAM1), a guanine nucleotide exchange factor for Rac1, was reported in the fetal brain and suggested that overexpression of Tiam1 could lead to increase oxidative stress in response to an inflammatory insult that involves IL-1β in the DS brain (Subasinghe et al., 2011). The Tiam1 was reported in pancreatic β-cells as a critical regulatory factor in cytokine IL-1β-induction of NADPH oxidase, indicating the necessity of IL-1β-induction for the activation of NADPH oxidase (Subasinghe et al., 2011).

The members of the interferon receptors IFNAR1, IFNAR2, and IFNGR2, located on HSA21, are triplicated in DS and thus, overexpressed. Both IFNAR1 and IFNAR2 respond to

TABLE 12.3 List of the inflammation related genes located on chromosome 21.

Gene	Protein	Function
CXADR	Coxsackie virus and adenovirus receptor	Activation of JNK and p38-MAPK pathways leading to production of M1 cytokines
ADAMTS1	ADAM metalloproteinase with thrombospondin type 1 motif 1	Secreted protease known to be induced by IL-1β
ADAMTS5	ADAM metalloproteinase with thrombospondin type 1 motif, 5	Secreted protease known to be induced by IL-1β and TGFβ.
TIAM1	T-cell lymphoma invasion and metastasis 1	Necessary for cytokine-mediated generation of oxidative species through NADPH oxidase
SOD1	Superoxide dismutase 1	Scavenges superoxide radicals producing H_2O_2 and O_2
BACE2	β-site APP-cleaving enzymes-2	Cleaves APP for less Aβ and increases IL-1R2, a decoy protein for excess IL-1 capture
βAPP	Amyloid beta precursor protein >5 fold overexpression in DS	Neuronal acute phase protein precursor of fragments Aβ in Alzheimer plaques and sAPP for induction of IL-1β
IFNAR2	Interferon (alpha, beta, and omega) receptor 2	Activates JAK/STAT-mediated anti-inflammatory pathway in response to IFNα/β
IFNAR1	Interferon (alpha, beta, and omega) receptor 1	Activates JAK/STAT-mediated anti-inflammatory pathway in response to IFNα/β
IFNGR2	Interferon gamma receptor 2	Activates JAK/STAT-mediated anti-inflammatory pathway in response to IFN-γ
RIPK4	Receptor-interacting serine-threonine kinase 4	Necessary for signaling through TNFR1
CBS	Cystathione-beta-synthase	Catalyzes production of hydrogen sulfide (H2S); a bimodal regulator of inflammation
S100B	S100 calcium binding protein B (astrocyte-derived Cytokine)	Constitutive expression by astrocytes, upregulates IL-1β and βAPP expression, released in response to TNFα
PRMT2	Protein arginine methyltransferase 2	Blocks the actions of NFκB in the nucleus

APP, precursor protein for β-amyloid; *Aβ*, β-amyloid; *IL*, interleukin; *JAK/STAT*, Janus kinase signal transducer and activator of transcription; *MAPK*, mitogen-activated protein kinase; *NFκB*, nuclear factor-kappa B; *TGF*, transforming growth factor; *TNF*, tumor necrosis factor; *TNFR*, tumor necrosis factor receptor.

Adapted from Wilcock and Griffin (2013). http://www.jneuroinflammation.com/content/10/1/84, under the terms of the Creative Commons Attribution License (http://creativecommons.org/licenses/by/2.0).

cytokine IFN-α, IFN-β, and activate the JAK/STAT signaling pathway upon ligand binding leading to induction of proinflammatory gene expression such as IL-1β, TNF-α, and IL-6. IFNGR2, though uses the same signaling pathway, specifically responds to IFN-γ. In DS patients, hyper-responsiveness to IFN is likely and that may lead to an increased inflammatory response in brain. A mouse model of DS with trisomy 16 showed triplication of Ifngr2 and Ifnar2, and these mice developed significant complications in utero and hardly survived till birth (Maroun et al., 2000). Improvement in the phenotype of these mice following treatment of these fetuses with anti-IFN IgG suggested that the triplication of the IFN receptors significantly contributed to the severe pathological complication in these mice. The same group later showed that studies following a partial knockout of the IFNAR2 and IFNGR2 in cultured neurons derived from the trisomy 16 mouse fetuses revealed an improvement of growth and viability (Maroun et al., 2000).

One of the multiple cell–signaling pathways of the receptor-interacting serine-threonine kinase 4 (RIPK4) is its involvement as a protein kinase for the activation of NFkB. RIPK4 is also involved in the signaling cascade of the TNF-α receptor TNFR1 (Rountree et al., 2010). It is noteworthy that the TNFR1 is significantly associated with the toxic effects of TNF-α; thus, it was speculated that overexpression of RIPK4 due to triplication in DS may increase responsiveness of TNFR1 to TNF-α aggravating the effects of TNFR1; however, further studies are essential for better understanding of this mechanism.

Cystathione beta synthase (CBS), a cytosolic enzyme, catalyzes the desulfhydration of cysteine-producing hydrogen sulfide (H_2S), and the activity of the enzyme is altered when it binds with nitric oxide (NO) or carbon monoxide (CO) in its heme pocket (Taoka and Banerjee, 2001). H_2S is characterized as an atypical cellular messenger having many normal physiological functions (Barañano et al., 2001). Having an apparent bimodal action, H_2S is recognized as a complicated signaling molecule, which upon inflammation appeared to be anti-inflammatory at low levels, while high H_2S levels may exacerbate inflammation in some instances (Whiteman et al., 2011). However, how the overexpression of CBS influences the DS phenotype and whether the amount of H_2S produced in DS patients is of the anti-inflammatory or proinflammatory concentrations remain unsolved.

The S100β gene encodes the β-subunit of the calcium-binding S100 proteins present in the mammalian brain in the form of a homodimer. This protein is involved in glial cell activation, neuroinflammation, and in the regulation and maintenance of the serotonergic nervous system, particularly on the hippocampus (reviewed in Shapiro et al., 2010). Elevation in S100 β expression in brain is associated with sporadic AD and DSAD. The S100β protein is secreted by the astrocytes in response to IL-1β and cyclic-AMP in the brain since it is primarily located at and expressed by astrocytes. S100β dichotomously acts as another inflammatory mediator: at low concentrations, it appears to enhance survival of neurons and stimulate neurite outgrowth, whereas its high concentration influences cell death and apoptosis (Sorci et al., 2004). Overexpression of S100β and its higher concentrations would place the S100β in the toxic category in DS brain, indicating its negative role in the aging-related pathological consequences (Griffin et al., 1998).

The neural precursors extracted from DS fetus demonstrated a constitutive overexpression of S100β in vitro, which led to increase in ROS formation, activation of stress response kinases, and increased apoptosis (Shapiro et al., 2010). Further studies have reported that such neural precursors in DS adopt more gliocentric progenitor phenotypes than euploid controls, with a consequent reduction in neurogenesis. Induction of S100β can occur through receptor for advanced glycation end products (RAGE) and consequent activation of JNK/p38 and JAK/STAT signaling, which are known to serve as potential downstream effectors to reactive gliosis, induction of S100β and glial-associated aquaporin 4 (Esposito et al., 2008; Lu et al., 2011). Increased levels of S100β and APP enhance RAGE activation and inflammation with reactive gliosis via cyclical cascade.

Protein arginine methyltransferase 2 (PRMT2), an enzyme, catalyzes the methylation of arginine that regulates the JAK-STAT signaling pathway, which is also key for many inflammatory processes, including IFN-γ, IFN-α, and IL-6 (Mowen et al., 2001). In addition, production of the asymmetric dimethylarginine (ADMA) through the natural degradation of proteins containing methylated arginine facilitates normal cell signaling and inflammation, where the ADMA acts as an endogenous inhibitor of nitric oxide synthase (NOS) (Vallance et al., 1992). Elevated concentration of ADMA was reported in DS with pulmonary hypertension compared to non-DS patients having pulmonary hypertension (Cua et al., 2011) and that led to hypothesize that if similar events take place in DS brain, then there could be reduced production of NO and enhanced activation of JAK-STAT pathway, where both factors will influence the inflammatory state of the brain. However, whether the triplication of PRMT2 results in significant changes in ADMA concentrations in the DS brain is still unclear.

SOD1 is a soluble cytoplasmic and mitochondrial interspace enzyme that converts superoxide radicals to molecular oxygen and H_2O_2 (Rakhit and Chakrabartty, 2006), and a potent endogenous antioxidant. Mutations in the SOD1 gene are commonly associated with genetic affinity to anterolateral sclerosis (ALS). Its potential loss of function mutations and overexpression increase oxidative stress, especially in absence of a peroxide-detoxifying enzyme (reviewed in Wilcock and Colton, 2008; Wilcock, 2012). Thus, triplication of SOD1 results in overexpression of the protein, which subsequently causes oxidative stress in DS.

Summarily, neuroinflammation, dementia, and AD in DS are primarily contributed by triplication of HSA21 genes. The triplets of the three major interferon receptors, such as IFNAR1, IFNAR2, and IFNGR2, influence interferon signaling followed by subsequent increase in release of the M1 markers, especially IL-1β, TNF-α, and IL-6. These markers may augment the mechanism of oxidative stress in the DS brain owing to triplication and overexpression of TIAM1, SOD1, and PRMT2, through induction of NADPH oxidase by TIAM1, enhancement of oxidation by high dosage of SOD1, and inhibition of the production of the antioxidant nitric oxide by PRMT2. Collectively, these factors will enhance neurodegeneration in the DS brain in conjunction with the primary changes of pathogenetic markers such as amyloid plaques and neurofibrillary tangles. Therefore, the M1 inflammatory response was postulated to be the main state of enhancement of neurodegeneration in the DS brain having triplicated HSA21. The preponderance of AD in DS can be explained with the contribution of triplicated set of the genes for creating a unique inflammatory environment; however, further studies will be worthy for better understanding of the inflammatory response and its clinical management with the concomitant approach of developing targeted drugs for sporadic AD and DSAD.

10. Therapeutics of AD management in DS

The resemblances of sporadic and autosomal dominant AD with DSAD make the DS population a suitable target for preventive treatments of AD. Therapeutic approaches of AD have been focusing more on immunotherapy by either active vaccination or passive immunization to introduce anti-Aβ antibodies in the DS population and AD patients (reviewed in Wilcock and Colton, 2008). The strong rationale for testing anti-Aβ therapies lies on the efficacy in this population, mainly due to less age-related co-occurring conditions and fewer nonamyloid vascular risk factors.

10.1 Clinical trials

A NIH-funded phase Ib study (NCT02738450) has tested an anti-Aβ vaccine named **ACI-24** in DS adults. The safety, tolerability, and immunogenicity of ACI-24 were evaluated as the primary endpoints in the multicenter (double-blind, randomized, placebo-controlled, dose-escalation) study on 16 DS adult patients aged 25–45 years. The secondary parameters included effects on AD-biomarkers as well as cognitive and clinical function of AD pathology. Another study showed antiaggregation effects of scyllo-inositol, an endogenous myoinositol isomer, on lowering amyloid deposition in CSF and brain. The study demonstrated an improving trend on cognitive ability in mild cases with AD (reviewed in Snyder et al., 2020). In addition, a phase II randomized, double-blind, placebo-controlled study on orally administered scyllo-inositol has addressed pharmacokinetic studies in 24 DS adults.

Another target on the inflammatory response in AD followed administration of nonsteroidal anti-inflammatory drugs (NSAIDs), which showed great outcome on significant protection from AD with its long-term use in retrospective epidemiological studies (Szekely et al., 2004). The NSAID-result shifted APP cleavage to promote Aβ-38 production, as opposed to Aβ-40 or Aβ-42 due to presence of γ-secretase modifying properties (Weggen et al., 2001). However, the prospective clinical trial named ADAPT conducted by NIA/NIH failed to produce any significant benefit (Meinert et al., 2009), rather a subset of ADAPT trial patients were benefited from NSAID use. Attenuation of cognitive decline was reported in a subgroup of Naproxen-treated AD patients, labeled as "slow decliners," whereas "fast decliners" experienced accelerated cognitive decline (Leoutsakos et al., 2012). It is thus hypothesized that different inflammatory states may lead to differential treatment outcome in different AD patients and that directs the future exploration of understanding of inflammation and its control in AD.

The clinical trials of the FDA-approved drugs have met with moderate or little success on the treatment of AD in the general population and DSAD (reviewed in Head et al., 2012; discussed in detail in the Chapter 11). Notably, the disease-modifying approaches were thought to be critically important in this regard. Although prevention of age-related complexities might be the most promising approach, both pharmacological and nonpharmacological interventions appear highly complex for this complex genetic disorder.

10.2 Practical concerns of clinical trials

Clinical trials in the DSAD population raise many phenotypic concerns. Given the fact of the wide variation in the baseline of intellectual capabilities, capacity of memory, attention, language, and communication shall be taken into consideration for accurate cognitive assessment and interpretation of outcome. Notably, clinical diagnosis of AD-associated dementia appears to be a tremendous challenge (Snyder et al., 2020). In the absence of the baseline knowledge of cognitive function, assessment of exponential development of subtle AD dementia would be complicated in DS. Indeed, an astute observer may identify the earliest signs of dementia in DS adults. In addition, baseline assessment of other contributing factors, including current concomitant medications and medical illnesses, status of TSH, and vitamin B12, or recent psychosocial events might build the evidence of declining cognition. Nevertheless, there are currently no standard clinical guidelines and instruments for assessment of AD-related dementia in DS adults. Furthermore, other challenging factors such as developing inclusion and exclusion criteria for clinical trials in DSAD and incidence of sensory impairments for responding to the questionnaires might limit the generalizability of the participation. In fact, recognition of prodromal symptoms or MCI might be inaccurate due to lack of verbal communication abilities of the trial participants and that might limit the registration of the potential representative of the larger DS population in trials.

On top, trial design also consider participant's home verses an assisted living facility with a view to identifying the behavioral changes and response to the trail by the care giver. However, currently there is no validated battery of test for monitoring the effects of medication in trials on DS and assess the treatment outcome from trial participation. A working group of the NICHD has identified measures that may assess the multiple cognitive and behavioral domains; however, that requires further evaluation and validation for demonstrating the sequence of decline events and onset of symptoms during the prodromal stages of DSAD (Esbensen et al., 2017; Snyder et al., 2020). In this regard, CAMDEX-DS questionnaire appeared useful for the standardized informant ratings of symptoms and explore the changes in behaviors related to frontal lobe dysfunction, and comprehensive batteries for cognitive test such as the LonDownS battery for identification of the early decline in memory and attention (reviewed in Snyder et al., 2020). Additionally, use of neuroimaging biomarkers appeared more problematic as AD symptoms progress in DS individuals, which might truly be a difficult challenge while scanning the disease progress. Thus, complication of selection and validation of biomarkers must be taken into account for potential use in clinical trials. Moreover, evaluation of trial interventions and cross-target complexities depends on the cerebral amyloid angiopathy (CAA) and neuroinflammation in the DSAD population.

Therefore, further understanding of the underlying pathophysiological and genetic mechanisms is essential in drug discovery for DSAD. Development of suitable animal and/or cellular models that reflect the gene dosage and regulatory sequences of HSA21 are required for better understanding of genotype–phenotype association, identification, and exploration of the potential therapeutic targets and assessment of new potential pharmacological and nonpharmacological approaches for DSAD population. Indeed, successful trials on non-DS population may readily translate their applicability in DS with AD population.

10.3 Therapeutic intervention in DS models

10.3.1 *Mouse models*

Use of mouse models of DS for the development of novel therapeutics or prevention strategies has enabled careful dissection of the role of individual or groups of genes to this DS phenotype. However, translation of the outcomes from mouse models to human clinical trials is not necessarily direct approach, though some of the mouse studies have demonstrated important proof of concept on promising new pharmacological or nonpharmacological therapeutics. In this direction, use of the norepinephrine precursor L-threo-DOPS (Droxidopa) improved learning and memory in DS mice, suggesting it as a possible target for clinical trials targeting improvement of age- and AD-related cognitive dysfunction and mitigation of neuroinflammation in human DS. Modulation of the Aβ production mediated by the overexpression of the APP could be a promising pharmacological approach. The treatment with anti-Aβ antibodies is expected to lower the load of Aβ in the brain and that has been evidenced in transgenic mouse models with improvement of memory and learning skill (Wilcock and Colton, 2008; Morgan et al., 2000). Noticeably, direct anti-Aβ injection into the brain of transgenic mice revealed a dependence of amyloid removal on microglial activation (Wilcock et al., 2006). However, conflicting report exists on anti-Aβ antibody induced transient activation of microglia, and a reduced efficacy with the deglycosylated antibody that incapacitated the IgG molecule for interacting with effector microglia cells (Wilcock et al., 2006).

Clinical trials on γ-secretase inhibitors for reduction of BACE activity have indicated adverse effects (reviewed in Head et al., 2012). However, BACE activity in young animals of the DS mouse model exhibited decrease in learning and memory deficits, which suggested that both overexpression of APP and overproduction of Aβ are responsible for intellectual disability in younger individuals and age-related AD development (Netzer et al., 2010).

Normalization of APP gene dosage from trisomy to disomy was thought of a potential therapeutic strategy for further exploration, which could potentially be accomplished by targeting mRNA levels of APP with antisense oligonucleotide activation, translation of APP mRNA, processing of APP with γ-secretase modulators, removal of Aβ with immunotherapy, or reducing a key regulator of intracellular trafficking and Ras-related protein a4 (Rab5) (reviewed in Snyder et al., 2020). Notably, reduction of one copy of APP in a mouse model of DS decreased endosomal pathology.

Non-Aβ treatment approaches targeting tau, microglial function, interferon-related signal transduction, oxidative stress, or inflammatory events were proposed to be effective in DSAD. In that direction, use of a granulocyte-macrophage colony-stimulating factor (GMCSF), a cytokine produced by the innate immune system, displayed reduction of Aβ deposition in vivo with reversal of memory deficits in aged DS mice compared to normal mice. However, side effects are likely in therapeutic use of GMCSF and that should be counted in therapeutic development with GMCSF (reviewed in Snyder et al., 2020). Nevertheless, reproduction of the pathological lesions of human DS and AD in the DS animal model remains a major challenge for the DS and AD research on therapeutics.

The implication of overexpressed DYRK1A in the development of cognitive deficits in DS and AD was treated with pharmacological inhibitors of brain DYRK1A in mouse models of DS. Interestingly, treatment with leucettine L41, a DYRK1A inhibitor, demonstrated normalization of DYRK1A activity in Tg(Dyrk1a), Ts65Dn, Dp1Yey mice with trisomic Dyrk1a (Nguyen et al., 2018), suggesting the possibility of pharmacological correction of cognitive impairment in DS. However, similar outcome is yet to be tested in human DS.

10.3.2 Cellular models

Use of induced pluripotent stem cells (iPSCs) and directly induced neurons (iNs) derived from human DS individuals have emerged as useful DS models for the study of age-related and neurodegenerative diseases in DS, especially DSAD (discussed in detail in Chapter 14). Indeed, iPSCs serve as pure genetic models; however, the scope of modeling for age-related changes is limited in their ability, in particular, those linked with epigenetic changes. Interestingly, iNs are able to maintain aging-specific features of aging neurons (Mertens et al., 2018).

10.4 Therapeutics on mitochondrial dysfunction and oxidative stress in DSAD

Control of the inevitable feature of oxidative damage by supplementing with antioxidants has demonstrated little or no clinical benefit in demented DS adults (Lott et al., 2011). Indeed, such compensatory mechanism on memory function was targeted by antioxidant supplementation in individuals of younger ages (Perluigi and Allan Butterfield, 2012), suggesting antioxidant supplementation would be a multi-targeted preventative approach (detailed in Chapter 9).

Production of ROS mediated by mitochondrial dysfunction and higher rate of mtDNA mutations exacerbate the mitochondrial dysfunction in DS. In this regard, several possible interventions, including dietary changes and/or supplementation with mitochondrial cofactors, are expected to protect age-associated cognitive declines. In addition, combined therapy of supplementation with antioxidants (such as vitamins E and C) and mitochondrial cofactors (such as lipoic acid, acetylcarnitine) may be of particular benefit for DS adults; however, such approaches are suggested for preventive measures rather than treatment of AD in DS (Lott et al., 2011). Moreover, interaction among the Aβ, oxidative damage and neuroinflammation, and the relative scarcity of data on neuroinflammation in the aging process of DS brains has indicated unexplored potential of new interventional targets (Wilcock, 2012). Several key lifestyle measures such as protection from gait and cataract issues, dietary supplements, etc. may prevent AD-related functional decline to some extent.

10.5 Exosomes as novel therapeutics for AD

The blood–brain barrier (BBB) seals the cell-to-cell contacts with a continuous endothelial membrane to prevent transport of the potentially toxic compounds between the brain and blood (Matsumoto et al., 2017) but facilitates selective movement of some compounds such as small (<400 Da) lipid soluble molecules into and out of the brain (Sanchez-Covarrubias et al., 2014). Since exosomes have an inherent capacity to cross the BBB and remain active in the brain, systemic injection of naked exosomes is considered ideal vehicles for delivery of small-molecule drugs to the brain in mice. Studies on transmitting exosomes through intranasal injection into the mouse brain also appeared successful (reviewed in Song et al., 2020). Expression of a fluorescently tagged forebrain astrocyte protein aldolase C in the brain tissue indicated the communication transmitted by exosomes from the brain to the rest of the body (Gomez-Molina et al., 2019). Indeed, use of nonpathogenic exosomes may change the duration and progression of AD disease. However, mechanism of such bi-directional transmission of exosomes for crossing the BBB remains unelucidated.

Interestingly, therapeutic effect of exosomes demonstrated significant downregulation of age-associated signaling molecules such as Insulin-like growth factor-1 receptor (IGF1-R) and upregulation of telomerase-related genes, including Men1, Mre11a, Tep1, Terf2, Tert, and Tnks in aged mice (Lee et al., 2018). In addition, intracerebral administration of exosomes to transgenic AD mice resulted in lowering the toxic oligomers and fibrils in a microglial-dependent manner, indicating clearance of Aβ in vivo (Yuyama and Igarashi, 2017). Notably, mesenchymal stromal–derived exosomes were considered as therapeutic targets in vivo for improvement of neurovascular plasticity in other neurodegenerative diseases such as stroke (reviewed in Song et al., 2020). Importantly, exosomal transmission systems have attracted extensive interest across the fields of pharmaceutical sciences, molecular cell biology, and nanoengineering for directing precise delivery of nanomedicine (Zhu et al., 2018). Thus, exosomes are exceedingly considered as promising type of vehicle for delivery of novel drugs owing to their capacity of crossing the BBB, shielding their cargo from enzymatic and chemical degradation, and their innate secretion of enzymes for degrading toxic fibrils (reviewed in Song et al., 2020; Tran et al., 2019). Indeed, introduction of exogenous exosomes into the CNS appears to be potential novel strategy for AD therapies. However, use of exosomes and the field of research on use of extracellular vesicles (EV) in the CNS for therapeutic applications are still at the initial stage and yet to be developed for treatment of DSAD.

Absence of unique region-specific markers for circulating exosomes and the inaccessibility of specific brain tissue from living patients make quantification of changes in EV cargo extremely difficult (Song et al., 2020). Interesting lines of research have explored the development of AD using pathogenic EVs and studied the sequestration of toxic plaques using exogenous healthy EVs; however, distinguishing the EV types on the basis of protein markers or size alone is difficult. Use of brain derived exosomes (BDEs) may resolve the issues of brain-associated disorders using blood biopsies and enrichment of BDEs from exosomes isolated from plasma (reviewed in Song et al., 2020). A pilot study on BDEs from plasma has examined biomarkers, such as tau, Aβ-42, and IL-10 for depressive patients and in military personnel with mild traumatic brain injuries. Studies on the cargo proteins and miRNA from astrocytic-derived exosomes and other cell-derived exosomes have analyzed the mechanistic insight into AD and other brain-related disorders. Interestingly, BDEs in plasma and other biological body fluids such as CSF and urine showed their potential for clinical use in treating nervous system disorders (reviewed I Song et al., 2020).

11. Conclusion and future direction

Imaging studies display similar pathologic features between sporadic AD and DSAD, but DSAD begins at an earlier age. Reunderstanding of the genomics, metabolomics, lipidomics, and other (−)omics, and many mechanistic lines of investigation may define the DSAD pathophysiology. AD in DS population is genetically driven with higher gene-dosage effects and interaction with nontrisomic genes in ways specific to DSAD unlike sporadic AD. Prominently, overexpression of APP and chronic oxidative stress are critical for DSAD pathology. Studies of exosomes, metabolomics, TREM2, DYRK1A, neurotransmitters, NFL, novel biomarkers in DS, and modifiable risk factors including sleep are contributing significantly. However, gaps exist in the knowledge of the best biomarkers for DSAD, specifically markers of vascular, inflammatory, oxidative stress, neuro-excitation, calcification, etc. across the continuum of disease. Establishment of a consensus research framework and development of a data-driven assessment battery for DSAD may advance clinical research and clinical trials to identify recruitment priorities, and empower the trials appropriately. In addition, establishment of coordinated research and medical care centers (centers of excellence) within the framework of existing DS and/or AD centers may contribute in better understanding of the progression and phenoconversion of DSAD and thus, accelerate development of effective treatments.

The substantial gap in the present understanding of aging process in people with DS and development of AD and dementia may continue to persist, but are critically important to address. Studies on monitoring the changes in overall health, cognitive function, development of dementia, and AD, mortality, and so on will be critically essential for development of policies and therapeutic interventions with a view to ensuring healthy aging among DS population. The social and medical infrastructure is not sufficiently prepared for providing care to sporadic or familial AD; thus, management of DSAD remains a key challenge. Notably, limited success has been achieved with FDA-approved treatments for AD in the general population. Therefore, exploration of novel therapeutic interventions with potential prophylactic approaches in the context of aging, dementia, and AD for individuals with DS and AD of the general population stands critically important.

Several international research consortia and collaborations, including the Horizon 21 Genetics Consortium (funded by J-L Institute), AD Biological correlates in DS (NIH, PI Granholm), Inflammation and NGF Dysfunction in the Evolution of AD Pathology in DS (multicenter), Improving AD care in adults with DS (submitted to GBHI and Alzheimer's Association Pilot Award), Horizon 21 Cognitive Project, HEROES consortium funded by the JPND in Europe, NIH Alzheimer's Biomarker Consortium for DS (ABC-DS), and the Crnic Institute's Human Trisome Project (NCT02864108) are advancing the understanding of DSAD. In addition, ABC-DS, LonDownS Consortium, Down Syndrome Biomarker Initiative (DSBI), and DABNI have demonstrated critical data on the natural mechanism of AD in DS and enabled clinical trials. Further, there is a growing need for the autopsy study of brain tissues from well-characterized

DSAD patients of different ages by the international research community (Snyder et al., 2020). The initial network, including 10 participating universities in the United States and Europe, and the Down Syndrome Biobank Consortium (DSBC) is developing the standardized protocols for harmonized procurement of DS tissues from the different sites of DS brains.

Development of clinical trial networks and infrastructure for multicenter collaborations using harmonized protocols have been considered as another important priority, which is currently underway with LuMIND IDSC Foundation Down Syndrome Clinical Trials Network (DS-CTN), the NIH-funded Alzheimer's Clinical Trial Consortium-Down Syndrome (ACTC-DS) and the Longitudinal Investigation for Enhancing Down Syndrome Research (LIFE-DSR). Expanded development and utilization of brain banking, collaborative data sharing, and evaluation across studies using pharmacological and nonpharmacological intervention as well as combination therapy approaches are also essential for prevention of dementia in the DS population.

References

Alhajraf, F., Ness, D., Hye, A., Strydom, A., 2019. Plasma amyloid and tau as dementia biomarkers in Down syndrome: systematic review and meta-analyses. Dev. Neurobiol. 79, 684–698.

Alkam, T., Nitta, A., Mizoguchi, H., et al., 2008. Restraining tumor necrosis factor-alpha by thalidomide prevents the Amyloid beta-induced impairment of recognition memory in mice. Behav. Brain Res. 189 (1), 100–106.

Alonso Vilatela, M.E., López-López, M., Yescas-Gómez, P., 2012. Genetics of Alzheimer's disease. Arch. Med. Res. 43, 622–631.

Annus, T., Wilson, L.R., Hong, Y.T., Acosta-Cabronero, J., Fryer, T.D., Cardenas-Blanco, A., et al., 2016. The pattern of amyloid accumulation in the brains of adults with Down syndrome. Alzheimers Dement. 12, 538–545.

Asai, M., Kawakubo, T., Mori, R., Iwata, N., 2017. Elucidating pathogenic mechanisms of early onset Alzheimer's disease in Down syndrome patients. Yakugaku Zasshi 137 (7), 801–805. https://doi.org/10.1248/yakushi.16-00236-2.

Augé, E., Bechmann, I., Llor, N., Vilaplana, J., Krueger, M., Pelegrí, C., 2019. Corpora amylacea in human hippocampal brain tissue are intracellular bodies that exhibit a homogeneous distribution of neo-epitopes. Sci. Rep. 9, 2063.

Azari, N.P., Pettigrew, K.D., Pietrini, P., Horwitz, B., Schapiro, M.B., 1994. Detection of an Alzheimer disease pattern of cerebral metabolism in Down syndrome. Dementia 5, 69–78.

Bachstetter, A.D., Van Eldik, L.J., Schmitt, F.A., et al., 2015. Disease-related microglia heterogeneity in the hippocampus of Alzheimer's disease, dementia with Lewy bodies, and hippocampal sclerosis of aging. Acta Neuropathol. 3, 32.

Bagyinszky, E., Youn, Y.C., An, S.S., Kim, S., 2014. The genetics of Alzheimer's disease. Clin. Interv. Aging 9, 535–551.

Ballard, C., Mobley, W., Hardy, J., Williams, G., Corbett, A., 2016. Dementia in Down's syndrome. Lancet Neurol. 15, 622–636.

Bambrick, L.L., Fiskum, G., January 10, 2008. Mitochondrial dysfunction in mouse trisomy 16 brain. Brain Res. 1188, 9–16. https://doi.org/10.1016/j.brainres.2007.10.045.

Barañano, D.E., Ferris, C.D., Snyder, S.H., 2001. Atypical neural messengers. Trends Neurosci. 24 (2), 99–106.

Barger, S.W., Harmon, A.D., 1997. Microglial activation by Alzheimer amyloid precursor protein and modulation by apolipoprotein E. Nature 388, 878–881. https://doi.org/10.1038/42257.

Beyreuther, K., Pollwein, P., Multhaup, G., Mönning, U., König, G., Dyrks, T., et al., 1993. Regulation and expression of the Alzheimer's beta/A4 amyloid protein precursor in health, disease, and Down's syndrome. Ann. N. Y. Acad. Sci. 695, 91–102.

Bik-Multanowski, M., Pietrzyk, J.J., Midro, A., 2015. MTRNR2L12: A candidate blood marker of early Alzheimer's disease-like dementia in adults with Down syndrome. J. Alzheim. Dis. 46, 145–150. https://doi.org/10.3233/JAD-143030.

Bittles, A.H., Bower, C., Hussain, R., Glasson, E.J., 2007. The four ages of Down syndrome. Eur. J. Publ. Health 17, 221–225.

Boche, D., Perry, V.H., Nicoll, J.A., 2013. Review: activation patterns of microglia and their identification in the human brain. Neuropathol. Appl. Neurobiol. 39, 3–18.

Brion, J.P., 1998. Neurofibrillary tangles and Alzheimer's disease. Eur. Neurol. 40, 130–140.

Busciglio, J., Pelsman, A., Wong, C., Pigino, G., Yuan, M., Mori, H., Yankner, B.A., 2002. Altered metabolism of the amyloid b precursor protein is associated with mitochondrial dysfunction in Down's syndrome. Neuron 33, 677–688.

Calderon, D., Bhaskar, A., Knowles, D.A., et al., 2017. Inferring relevant cell types for complex traits by using single-cell gene expression. Am. J. Hum. Genet. 101, 686–699.

Carta, M.G., Serra, P., Ghiani, A., Manca, E., Hardoy, M.C., Del Giacco, G.S., et al., 2002. Chemokines and pro-inflammatory cytokines in Down's syndrome: an early marker for Alzheimer-type dementia? Psychother. Psychosom. 71, 233–236.

Caso, F., Agosta, F., Mattavelli, D., et al., 2015. White matter degeneration in a typical Alzheimer disease. Radiology 277, 162–172.

Castro, P., Zaman, S., Holland, A., 2017. Alzheimer's disease in people with Down's syndrome: the prospects for and the challenges of developing preventative treatments. J. Neurol. 264, 804–813.

Cenini, G., Amy, L., Dowling, S., Beckett, T.L., Barone, E., Mancuso, C., Murphy, M.P., LeVine III, H., Lott, I.T., Schmitt, F.A., Allan Butterfield, D., Head, E., February 2012. Association between frontal cortex oxidative damage and beta amyloid as a function of age in Down syndrome. Biochim. Biophys. Acta 1822 (2), 130–138. https://doi.org/10.1016/j.bbadis.2011.10.001.

Chai, G.-S., Duan, D.-X., Ma, R.-H., Shen, J.-Y., Li, H.-L., Ma, Z.-W., et al., 2014. Human in attenuates Alzheimer-like cognitive deficits and pathological changes induced by amyloid β-peptide in rats. Neurosci. Bull. https://doi.org/10.1007/s12264-014-1479-3.

Chakrabarti, L., Best, T.K., Cramer, N.P., et al., 2010. Olig1 and Olig2 triplication causes developmental brain defects in Down syndrome. Nat. Neurosci. 13 (8), 927–934. https://doi.org/10.1038/nn.2600.

Chakrabarty, P., Jansen-West, K., Beccard, A., et al., 2010. Massive gliosis induced by interleukin-6 suppresses Aβ deposition in vivo: evidence against inflammation as a driving force for amyloid deposition. FASEB J. 24 (2), 548–559.

Chatterjee, A., Dutta, S., Sinha, S., Mukhopadhyay, K., 2011. Exploratory investigation on functional significance of ETS2 and SIM2 genes in Down syndrome. Dis. Markers 31, 247–257.

Chow, V.W., Mattson, M.P., Wong, P.C., Gleichmann, M., 2010. An overview of APP processing enzymes and products. NeuroMol. Med. 12, 1–12.

Cole, S.L., Vassar, R., 2007. The Alzheimer's disease beta-secretase enzyme. BACE1. Mol Neurodegen. 2–22.

Colton, C.A., Mott, R.T., Sharpe, H., Xu, Q., Van Nostrand, W.E., Vitek, M.P., 2006. Expression profiles for macrophage alternative activation genes in AD and in mouse models of AD. J. Neuroinfl. 3. Article no. 27.

Contestabile, A., Benfenati, F., Gasparini, L., 2010. Communication breaks-down: from neurodevelopment defects to cognitive disabilities in Down syndrome. Prog. Neurobiol. 91, 1–22.

Coskun, P.E., Wyrembak, J., Derbereva, O., Melkonian, G., Doran, E., Lott, I.T., Head, E., Cotman, C.W., Wallace, D.C., 2010. Systemic mitochondrial dysfunction and the etiology of Alzheimer's disease and Down syndrome dementia. J. Alzheimers Dis. 20 (Suppl. 2(0 2)), S293–S310. https://doi.org/10.3233/JAD-2010-100351.

Cua, C.L., Rogers, L.K., Chicoine, L.G., et al., 2011. Down syndrome patients with pulmonary hypertension have elevated plasma levels of asymmetric dimethylarginine. Eur. J. Pediatr. 170 (7), 859–863.

Davies, D.S., Ma, J., Jegathees, T., Goldsbury, C., 2017. Microglia show altered morphology and reduced arborization in human brain during aging and Alzheimer's disease. Brain Pathol. 27, 795–808.

De Servi, B., La Porta, C.A.M., Bontempelli, M., Comolli, R., 2002. Decrease of TGF-β1 plasma levels and increase of nitric oxide synthase activity in leukocytes as potential biomarkers of Alzheimer's disease. Exp. Gerontol. 37 (6), 813–821.

Delabar, J.M., Theophile, D., Rahmani, Z., Chettouh, Z., Blouin, J.L., Prieur, M., et al., 1993. Molecular mapping of twenty-four features of Down syndrome on chromosome 21. Eur. J. Hum. Genet. 1, 114–124.

Depuydt, B., Van Loo, G., Vandenabeele, P., Declercq, W., 2005. Induction of apoptosis by TNF receptor 2 in a T-cell hybridoma is FADD dependent and blocked by caspase-8 inhibitors. J. Cell Sci. 118 (3), 497–504.

Di Bona, D., Plaia, A., Vasto, S., et al., 2008. Association between the interleukin-1β polymorphisms and Alzheimer's disease: a systematic review and meta-analysis. Brain Res. Rev. 59 (1), 155–163.

Doran, E., Keator, D., Head, E., Phelan, M.J., Kim, R., Totoiu, M., Barrio, J.R., Small, G.W., Potkin, S.G., Lott, I.T., 2017. Down syndrome, partial trisomy 21, and absence of Alzheimer's disease: the role of APP. J. Alzheimers Dis. 56 (2), 459–470. https://doi.org/10.3233/JAD-160836.

Dubois, B., Feldman, H.H., Jacova, C., et al., 2014. Advancing research diagnostic criteria for Alzheimer's disease: the IWG-2 criteria. Lancet Neurol. 13, 614–629.

Esbensen, A.J., Hooper, S.R., Fidler, D., et al., 2017. Outcome measures for clinical trials in Down syndrome. Am. J. Intellect. Dev. Disabil. 122 (3), 247–281.

Esposito, G., Jaime, I., Lu, J., De Filippis, D., Scuderi, C., Ganesh, V.S., Folkerth, R., Hecht, J., Shin, S., Iuvone, T., Chesnut, J., Steardo, L., Sheen, V., February 1, 2008. Genomic and functional profiling of human Down syndrome neural progenitors implicates S100B and aquaporin 4 in cell injury. Hum. Mol. Genet. 17 (3), 440–457. https://doi.org/10.1093/hmg/ddm322.

Ferrante, C.J., Leibovich, S.J., 2012. Regulation of macrophage polarization and wound healing. Adv. Wound Care 1, 10–16.

Firth, N.C., Startin, C.M., Hithersay, R., et al., 2018. Aging related cognitive changes associated with Alzheimer's disease in Down syndrome. Annals Clin. Translat. Neurol. 5, 741–751.

Fonseca-Santos, B., Gremião, M.P., Chorilli, M., 2015. Nanotechnology-based drug delivery systems for the treatment of Alzheimer's disease. Int. J. Nanomed. 4 (10), 4981–5003.

Forlenza, O.V., Diniz, B.S., Talib, L.L., et al., 2009. Increased serum IL-1β level in Alzheimer's disease and mild cognitive impairment. Dement. Geriatr. Cognit. Disord. 28 (6), 507–512.

Fortea, J., Vilaplana, E., Carmona-Iragui, M., Benejam, B., Videla, L., Barroeta, I., Fernández, S., Altuna, M., Pegueroles, J., Montal, V., Valldeneu, S., Giménez, S., González-Ortiz, S., Muñoz, L., Estellés, T., Illán-Gala, I., Belbin, O., Camacho, V., Wilson, L.R., Annus, T., Osorio, R.S., Videla, S., Lehmann, S., Holland, A.J., Alcolea, D., Clarimón, J., Zaman, S.H., Blesa, R., Lleó, A., June 27, 2020. Clinical and biomarker changes of Alzheimer's disease in adults with Down syndrome: a cross-sectional study. Lancet 395 (10242), 1988–1997. https://doi.org/10.1016/S0140-6736(20)30689-9.

Franco-Bocanegra, D.K., McAuley, C., Nicoll, J.A.R., Boche, D., 2019. Molecular Mechanisms of microglial motility: changes in ageing and Alzheimer's Disease. Cells 8.

Gadient, R.A., Otten, U.H., 1997. Interleukin-6 (IL-6)—a molecule with both beneficial and destructive potentials. Prog. Neurobiol. 52 (5), 379–390.

Gomes, F.C., Mattos, M.F., Goloni-Bertollo, E.M., Pavarino, É.C., January-February. Alzheimer's disease in the Down syndrome: an overview of genetics and molecular aspects. Neurol. India 69 (1), 32–41. https://doi.org/10.4103/0028-3886.310062.

Gomez, W., Morales, R., Maracaja-Coutinho, V., Parra, V., Nassif, M., 2020. Down syndrome and Alzheimer's disease: common molecular traits beyond the amyloid precursor protein. Aging 12 (1), 1011–1033. https://doi.org/10.18632/aging.102677.

Gomez-Molina, C., Sandoval, M., Henzi, R., Ramirez, J.P., Varas-Godoy, M., Luarte, A., et al., 2019. Small extracellular vesicles in rat serum contain astrocyte-derived protein biomarkers of repetitive stress. Int. J. Neuropsychopharmacol. 22, 232–246. https://doi.org/10.1093/ijnp/pyy098.

Griffin, W.S.T., Sheng, J.G., Roberts, G.W., Mrak, R.E., 1995. Interleukin-1 expression in different plaque types in Alzheimer's disease: significance in plaque evolution. JNEN (J. Neuropathol. Exp. Neurol.) 54 (2), 276–281.

Griffin, W.S.T., Sheng, J.G., Royston, M.C., et al., 1998. Glial neuronal interactions in Alzheimer's disease: the potential role of a 'cytokine cycle' in disease progression. Brain Pathol. 8 (1), 65–72.

Guidi, S., Stagni, F., Bartesaghi, R., 2017. Targeting APP/AICD in Down syndrome. Oncotarget 8 (31), 50333–50334. https://doi.org/10.18632/oncotarget.18860.

Halevy, T., Biancotti, J.C., Yanuka, O., Golan-Lev, T., Benvenisty, N., 2016. Molecular characterization of Down syndrome embryonic stem cells reveals a role for RUNX1 in neural differentiation. Stem Cell Rep. 7, 777–786.

Hamlett, E.D., LaRosa, A., Mufson, E.J., Fortea, J., Ledreux, A., Granholm, A.C., 2020. Exosome release and cargo in Down syndrome. Develop. Neurol. https://doi.org/10.1002/dneu.22712.

Hampel, H., Haslinger, A., Scheloske, M., et al., 2005. Pattern of interleukin-6 receptor complex immunoreactivity between cortical regions of rapid autopsy normal and Alzheimer's disease brain. Eur. Arch. Psychiatr. Clin. Neurosci. 255 (4), 269–278.

Harry, G.J., Kraft, A.D., 2012. Microglia in the developing brain: a potential target with lifetime effects. Neurotoxicology 33, 191206.

Hartley, D., Blumenthal, T., Carrillo, M., DiPaolo, G., Esralew, L., Gardiner, K., et al., 2015. Down syndrome and Alzheimer's disease: common pathways, common goals. Alzheimers Dement. 11, 700–709.

Hartley, S.L., Handen, B.L., Devenny, D., Tudorascu, D., Piro-Gambetti, B., Zammit, M.D., Laymon, C.M., Klunk, W.E., Zaman, S., Cohen, A., Christian, B.T., 2020. Cognitive indicators of transition to preclinical and prodromal stages of Alzheimer's disease in Down syndrome. Alzheimer's Dement. 12 e12096.

Head, E., Silverman, W., Patterson, D., Ira, T., 2012. Lott. Aging and Down syndrome. Curr. Gerontol. & Geriatr. Res. 2012, 6. https://doi.org/10.1155/2012/412536. Article ID 412536.

Head, E., Lott, I.T., Wilcock, D.M., Lemere, C.A., 2016. Aging in Down syndrome and the development of Alzheimer's disease neuropathology. Curr. Alzheimer Res. 13 (1), 18–29. https://doi.org/10.2174/1567205012666151020114607.

Head, E., Phelan, M.J., Doran, E., et al., 2017. Cerebrovascular pathology in Down syndrome and Alzheimer disease. Acta Neuropathol. Commun. 5, 93.

Hehlgans, T., Pfeffer, K., 2005. The intriguing biology of the tumour necrosis factor/tumour necrosis factor receptor superfamily: players, rules and the games. Immunology 115 (1), 1–20.

Helman, A.M., Siever, M., McCarty, K.L., et al., 2019. Microbleeds and cerebral amyloid angiopathy in the brains of people with Down syndrome with Alzheimer's disease. J. Alzheimers Dis. 67, 103–112.

Hithersay, R., Hamburg, S., Knight, B., Strydom, A., 2017. Cognitive decline and dementia in Down syndrome. Curr. Opin. Psychiatr. 30, 102–107.

Hithersay, R., Startin, C.M., Hamburg, S., et al., 2019. Association of dementia with mortality among adults with Down syndrome older than 35 years. JAMA Neurol. 76, 152–160.

Hong, S., Beja-Glasser, V.F., Nfonoyim, B.M., Frouin, A., Li, S., Ramakrishnan, S., Merry, K.M., Shi, Q., Rosenthal, A., Barres, B.A., Lemere, C.A., Selkoe, D.J., Stevens, B., May 6, 2016. Complement and microglia mediate early synapse loss in Alzheimer mouse models. Science 352 (6286), 712–716. https://doi.org/10.1126/science.aad8373.

Hopperton, K.E., Mohammad, D., Trepanier, M.O., Giuliano, V., Bazinet, R.P., 2018. Markers of microglia in post-mortem brain samples from patients with Alzheimer's disease: asystematic review. Mol. Psychiatr. 23, 177198.

Ishihara, K., Amano, K., Takaki, E., Ebrahim, A.S., Shimohata, A., Shibazaki, N., Inoue, I., Takaki, M., Ueda, Y., Sago, H., Epstein, C.J., Yamakawa, K., 2009. Increased lipid peroxidation in Down's syndrome mouse models. J. Neurochem. 110, 1965–1976.

Ito, M., Kodama, M., Masuko, M., et al., 2000. Expression of coxsackievirus and adenovirus receptor in hearts of rats with experimental autoimmune myocarditis. Circ. Res. 86 (3), 275–280.

Janicki, M.P., Dalton, A.J., 2000. Prevalence of dementia and impact on intellectual disability services. Ment. Retard. 38 (3), 276–288.

Jiang, Y., Mullaney, K.A., Peterhoff, C.M., Chea, S., Schmidt, S.D., Boyer-Boiteau, A., Ginsberg, S.D., Cataldo, A.M., Mathews, P.M., Nixon, R.A., 2010. Alzheimer's-related endosome dysfunction in Down syndrome is Aβ-independent but requires APP and is reversed by BACE-1 inhibition. Proc. Natl. Acad. Sci. USA 107 (4), 1630–1635. www.pnas.org/cgi/doi/10.1073/pnas.0908953107.

Jung, M.S., Park, J.H., Ryu, Y.S., Choi, S.H., Yoon, S.H., Kwen, M.Y., et al., 2011. Regulation of RCAN1 protein activity by Dyrk1A protein-mediated phosphorylation. J. Biol. Chem. 286, 40401–40412.

Karperien, A., Ahammer, H., Jelinek, H.F., 2013. Quantitating the subtleties of microglial morphology with fractal analysis. Front. Cell. Neurosci. 7, 3.

Kettenmann, H., Hanisch, U.K., Noda, M., Verkhratsky, A., 2011. physiology of microglia. Physiol. Rev. 91, 461–553.

Kligman, D., Marshak, D.R., 1985. Purification and characterization of a neurite extension factor from bovine brain. Proc. Natl. Acad. Sci. USA 82 (20), 7136–7139.

Kuno, K., Okada, Y., Kawashima, H., et al., 2000. ADAMTS-1 cleaves a cartilageproteoglycan, aggrecan. FEBS Lett. 478 (3), 241–245.

Kushima, Y., Hama, T., Hatanaka, H., 1992. Interleukin-6 as a neurotrophic factor for promoting the survival of cultured catecholaminergic neurons in a chemically defined medium from fetal and postnatal rat midbrains. Neurosci. Res. 13 (4), 267–280.

Lee, M., Hyun, D., Jenner, P., Halliwell, B., 2001. Effect of overexpression of wild-type and mutant Cu/Zn superoxide dismutases on oxidative damage and antioxidant defences: relevance to Down's syndrome and familial amyotrophic lateral sclerosis. J. Neurochem. 76, 957–965.

Lee, J.H., Lee, A.J., Dang, L.H., Pang, D., Kisselev, S., Krinsky-McHale, S.J., et al., 2017. Candidate gene analysis for Alzheimer's disease in adults with Down syndrome. Neurobiol. Aging 56, 150–158.

Lee, N.C., Chien, Y.H., Hwu, W.L., 2017. A review of biomarkers for Alzheimer's disease in Down syndrome. Neurol. Ther. 6, 69–81. https://doi.org/10.1007/s40120-017-0071-y.

Lee, B.R., Kim, J.H., Choi, E.S., Cho, J.H., Kim, E., 2018. Effect of young exosomes injected in aged mice. Int. J. Nanomed. 13, 5335–5345. https://doi.org/10.2147/IJN.S170680.

Leoutsakos, J.M., Muthen, B.O., Breitner, J.C., Lyketsos, C.G., April 2012. ADAPT Research Team. Effects of non-steroidal anti-inflammatory drug treatments on cognitive decline vary by phase of pre-clinical Alzheimer disease: findings from the randomized controlled Alzheimer's Disease anti-inflammatory prevention trial. Int. J. Geriatr. Psychiatr. 27 (4), 364–374. https://doi.org/10.1002/gps.2723.

Lev, N., Melamed, E., 2002. Neurological complications in Down's syndrome. Harefuah 141, 820–823.

Leverenz, J.B., Raskind, M.A., 1998. Early amyloid deposition in the medial temporal lobe of young Down syndrome patients: a regional quantitative analysis. Exp. Neurol. 150 (2), 296–304.

Lin, A.L., Powell, D., Caban-Holt, A., et al., 2016. (1)H-MRS metabolites in adults with Down syndrome: effects of dementia. Neuroimag. Clin. 11, 728–735.

Lockstone, H.E., Harris, L.W., Swatton, J.E., Wayland, M.T., Holland, A.J., Bahn, S., 2007. Gene expression profiling in the adult Down syndrome brain. Genomics 90, 647–660.

Lott, I.T., Head, E., 2001. Down syndrome and Alzheimer's disease: a link between development and aging. Ment. Retard. Dev. Disabil. Res. Rev. 7, 172–178.

Lott, I.T., Head, E., March 2005. Alzheimer disease and Down syndrome: factors in pathogenesis. Neurobiol. Aging 26 (3), 383–389. https://doi.org/10.1016/j.neurobiolaging.2004.08.005.

Lott, I.T., Head, E., 2019. Dementia in Down syndrome: unique insights for Alzheimer disease research. Nat. Rev. Neurol. 15, 135–147.

Lott, I.T., Doran, E., Nguyen, V.Q., Tournay, A., Head, E., Gillen, D.L., 2011. Down syndrome and dementia: a randomized, controlled trial of antioxidant supplementation. Am. J. Med. Genet. 155 (8), 1939–1948.

Lu, J., Esposito, G., Scuderi, C., Steardo, L., Delli-Bovi, L.C., Hecht, J.L., et al., 2011. S100B and APP promote a gliocentric Shift and impaired neurogenesis in Down syndrome neural progenitors. PLoS One 6 (7), e22126. https://doi.org/10.1371/journal.pone.0022126.

Lund, H., Gustafsson, E., Svensson, A., Nilsson, M., Berg, M., Sunnemark, D., et al., 2014. MARK4 and MARK3 associate with early tau phosphorylation in Alzheimer's disease granulovacuolar degeneration bodies. Acta Neuropathol. Commun. 2, 22.

Maccioni, R.B., Farías, G., Morales, I., Navarrete, L., 2010. The revitalized tau hypothesis on Alzheimer's disease. Arch. Med. Res. 41, 226–231.

MacLeod, R., Hillert, E.-K., Cameron, R.T., Baillie, G.S., 2015. The role and therapeutic targeting of α-, β- and γ-secretase in Alzheimer's disease. Future Sci. OA 1, FSO11.

Manczak, M., Calkins, M.J., Reddy, P.H., 2011. Impaired mitochondrial dynamics and abnormal interaction of amyloid beta with mitochondrial protein Drp1 in neurons from patients with Alzheimer's disease: implications for neuronal damage. Hum. Mol. Genet. 20, 2495–2509.

Mantovani, A., Sica, A., Sozzani, S., Allavena, P., Vecchi, A., Locati, M., 2004. The chemokine system in diverse forms of macrophage activation and polarization. Trends Immunol. 25 (12), 677–686.

Maroun, L.E., Heffernan, T.N., Hallam, D.M., 2000. Partial IFN-α/β and IFN-γ receptor knockout trisomy 16 mouse fetuses show improved growth and cultured neuron viability. J. Interferon Cytokine Res. 20 (2), 197–203.

Martini, A.C., Helman, A.M., McCarty, K.L., Lott, I.T., Doran, E., Schmitt, F.A., Head, E., October 14, 2020. Distribution of microglial phenotypes as a function of age and Alzheimer's disease neuropathology in the brains of people with Down syndrome. Alzheimers Dement. 12 (1), e12113. https://doi.org/10.1002/dad2.12113.

Matsumoto, J., Stewart, T., Sheng, L., Li, N., Bullock, K., Song, N., et al., 2017. Transmission of alpha-synuclein-containing erythrocyte-derived extracellular vesicles across the blood-brain barrier via adsorptive mediated transcytosis: another mechanism for initiation and progression of Parkinson's disease? Acta Neuropathol. Commun. 5, 71. https://doi.org/10.1186/s40478-017-0470-4.

McCarron, M., McCallion, P., Reilly, E., Dunne, P., Carroll, R., Mulryan, N., 2017. A prospective 20-year longitudinal follow-up of dementia in persons with Down syndrome. J. Intellect. Disabil. Res. 61, 843–852.

Meinert, C.L., McCaffrey, L.D., Breitner, J.C., 2009. Alzheimer's disease anti-inflammatory prevention trial: design, methods, and baseline results. Alzheimer's Dement. 5 (2), 93–104.

Mekkawy, M.K., Mazen, I.M., Kamel, A.K., Vater, I., Zaki, M.S., 2016. Genotype/phenotype correlation in a female patient with 21q22.3 and 12p13.33 duplications. Am. J. Med. Genet. 170A, 1050–1058.

Mertens, J., Reid, D., Lau, S., Kim, Y., Gage, F.H., 2018. Aging in a dish: iPSC derived and directly induced neurons for studying brain aging and age-related neurodegenerative diseases. Annu. Rev. Genet. 52, 271–293.

Miguel, R.F., Pollak, A., Lubec, G., 2005. Metalloproteinase ADAMTS-1 but not ADAMTS-5 is manifold overexpressed in neurodegenerative disorders as Down syndrome, Alzheimer's and Pick's disease. Mol. Brain Res. 133 (1), 1–5.

Mohandas, E., Rajmohan, V., Raghunath, B., 2009. Neurobiology of Alzheimer's disease. Indian J. Psychiatr. 51, 55–61.

Morgan, D., Diamond, D.M., Gottschall, P.E., et al., 2000. A β peptide vaccination prevents memory loss in an animal model of Alzheimer's disease. Nature 408 (6815), 982–985.

Mori, C., et al., 2002. Intraneuronal Abeta42 accumulation in Down syndrome brain. Amyloid 9 (2), 88–102.

Mowen, K.A., Tang, J., Zhuetal, W., 2001. Argininemethylation of STAT1 modulates IFNα/β-induced transcription. Cell 104 (5), 731–741.

Murphy, M.P., LeVine, H., 2010. Alzheimer's disease and the amyloid-beta peptide. J. Alzheimers Dis. 19, 311–323.

Nateghi Rostami, M., Douraghi, M., Miramin Mohammadi, A., Nikmanesh, B., 2012. Altered serum pro-inflammatory cytokines in children with Down's syndrome. Eur. Cytokine Netw. 23, 64–67.

Naude, P.J., Dekker, A.D., Coppus, A.M., et al., 2015. Serum NGAL is Associated with distinct plasma amyloid-beta peptides according to the clinical diagnosis of dementia in Down syndrome. J. Alzheimers Dis. 45, 733–743.

Neale, N., Padilla, C., Fonseca, L.M., Holland, T., Zaman, S., 2018. Neuroimaging and other modalities to assess Alzheimer's disease in Down syndrome. NeuroImag. Clin. 17, 263–271.

Netzer, W.J., Powell, C., Nong, Y., et al., 2010. Lowering beta-amyloid levels rescues learning and memory in a Down syndrome mouse model. PLoS One 5 (6). Article ID e10943.

Newell, K.L., Hyman, B.T., Growdon, J.H., Hedley-Whyte, E.T., 1999. Application of the National Institute on aging (NIA)-Reagan Institute criteria for the neuropathological diagnosis of Alzheimer disease. JNEN (J. Neuropathol. Exp. Neurol.) 58 (11), 1147–1155.

Nguyen, T.L., Duchon, A., Manousopoulou, A., et al., September 27, 2018. Correction of cognitive deficits in mouse models of Down syndrome by a pharmacological inhibitor of DYRK1A. Dis. Model Mech. 11 (9), dmm035634. https://doi.org/10.1242/dmm.035634.

Nistor, M., Don, M., Parekh, M., Sarsoza, F., Goodus, M., Lopez, G.E., et al., 2007. Alpha- and beta-secretase activity as a function of age and beta-amyloid in Down syndrome and normal brain. Neurobiol. Aging 28, 1493–1506.

Nixon, R.A., 2007. Autophagy, amyloidogenesis and Alzheimer disease. J. Cell Sci. 120, 4081–4091.

Nybo, M., Nielsen, E.H., Svehag, S.E., December 1999. Complement activation by the amyloid proteins A beta peptide and beta 2-microglobulin. Amyloid 6 (4), 265–272. https://doi.org/10.3109/13506129909007338.

Ohm, D.T., Kim, G., Gefen, T., et al., 2019. Prominent microglial activation in cortical white matter is selectively associated with cortical atrophy in primary progressive aphasia. Neuropathol. Appl. Neurobiol. 45, 216–229.

Paasila, P.J., Davies, D.S., Kril, J.J., Goldsbury, C., Sutherland, G.T., 2019. The relationship between the morphological subtypes of microglia and Alzheimer's disease neuropathology. Brain Pathol. 29, 726–774.

Perez, S.E., Miguel, J.C., He, B., et al., 2019. Frontal cortex and striatal cellular and molecular pathobiology in individuals with Down syndrome with and without dementia. Acta Neuropathol. 137, 413–436.

Perlenfein, T.J., Mehlhoff, J.D., Murphy, R.M., 2017. Insights into the mechanism of cystatin C oligomer and amyloid formation and its interaction with β-amyloid. J. Biol. Chem. 292, 11485–11498.

Perluigi, M., Allan Butterfield, D., 2012. Oxidative stress and Down syndrome: a Route toward Alzheimer-like dementia. Curr. Gerontol. & Geriatr. Res. 10. https://doi.org/10.1155/2012/724904. Article ID 724904.

Perluigi, M., di Domenico, F., Fiorini, A., Cocciolo, A., Giorgi, A., Foppoli, C., Butterfield, D.A., Giorlandino, M., Giorlandino, C., Schininà, M.E., Coccia, R., April 2011. Oxidative stress occurs early in Down syndrome pregnancy: a redox proteomics analysis of amniotic fluid. Proteonomics Clin. Appl. 5 (3–4), 167–178. https://doi.org/10.1002/prca.201000121.

Quintana, A., Molinero, A., Florit, S., et al., 2007. Diverging mechanisms for TNF-α receptors in normal mouse brains and in functional recovery after injury: from gene to behavior. J. Neurosci. Res. 85 (12), 2668–2685.

Raj, D., YinZ, B.M., Doorduin, J., et al., 2017. Increased white matter inflammation in aging- and Alzheimer's disease brain. Front. Mol. Neurosci. 10, 206.

Rakhit, R., Chakrabartty, A., 2006. Structure, folding, and misfolding of Cu,Zn superoxide dismutase in amyotrophic lateral sclerosis. Biochim. Biophys. Acta 1762 (11–12), 1025–1037.

Rodrigues, R., Debom, G., Soares, F., et al., 2014. Alterations of ectonucleotidases and acetylcholinesterase activities in lymphocytes of Down syndrome subjects: relation with inflammatory parameters. Clin. Chim. Acta 433, 105–110.

Rodriguez, M., Pavelko, K.D., McKinney, C.W., Leibowitz, J.L., 1994. Recombinant humanIL-6suppresses demyelination in a viral model of multiple sclerosis. J. Immunol. 153 (8), 3811–3821.

Rosenberg, R.N., Lambracht-Washington, D., Yu, G., Xia, W., 2016. Genomics of Alzheimer disease: a review. JAMA Neurol. 73, 867–874.

Rountree, R.B., Willis, C.R., Dinh, H., et al., 2010. RIP4 regulatesepidermaldifferentiation andcutaneousinflammation. J. Invest. Dermatol. 130 (1), 102–112.

Ryu, Y.S., Park, S.Y., Jung, M.-S., Yoon, S.-H., Kwen, M.-Y., Lee, S.-Y., Choi, S.-H., Radnaabazar, C., Kim, M.-K., Kim, H., Kim, K., Song, W.-J., Chung, S.-H., 2010. Dyrk1A-mediated phosphorylation of Presenilin 1: a functional link between Down syndrome and Alzheimer's disease. J. Neurochem. 115, 574–584. https://doi.org/10.1111/j.1471-4159.2010.06769.x.

Sanchez-Covarrubias, L., Slosky, L.M., Thompson, B.J., Davis, T.P., Ronaldson, P.T., 2014. Transporters at CNS barrier sites: obstacles or opportunities for drug delivery? Curr. Pharmaceut. Des. 20, 1422–1449. https://doi.org/10.2174/13816128113199990463.

Schupf, N., Winsten, S., Patel, B., et al., 2006. Bioavailable estradiol and age at onset of Alzheimer's disease in postmenopausal women with Down syndrome. Neurosci. Lett. 406, 298–302.

Schupf, N., et al., 2010. Change in plasma Aβ peptides and onset of dementia in adults with Down syndrome. Neurology 75 (18), 1639–1644.

Serrano-Pozo, A., Gomez-Isla, T., Growdon, J.H., Frosch, M.P., 2013. Hyman BT A phenotypic change but not proliferation underlies glial responses in Alzheimer disease. Am. J. Pathol. 182, 2332–2344.

Shaftel, S.S., Kyrkanides, S., Olschowka, J.A., Miller, J.N., Johnson, R.E., O'Banion, M.K., 2007. Sustained hippocampal IL-1 beta overexpression mediates chronic neuroinflammation and ameliorates Alzheimer plaque pathology. J. Clin. Invest. 117, 1595–1604.

Shapiro, L.A., Bialowas-McGoey, L.A., Whitaker-Azmitia, P.M., 2010. Effects of S100B on serotonergic plasticity and neuroinflammation in the Hippocampus in Down syndrome and Alzheimer's disease: studies in an S100B overexpressing mouse model. Cardiovasc.

Psychiatr. & Neurol. 13. https://doi.org/10.1155/2010/153657. Article ID 153657.

Sheng, J.G., Mrak, R.E., Rovnaghi, C.R., Kozlowska, E., Van Eldik, L.J., Griffin, W.S.T., 1996. Human brain S100β and S100β mRNA expression increases with age: pathogenic implications for Alzheimer's disease. Neurobiol. Aging 17 (3), 359–363.

Sheppard, O., Coleman, M., 2020. Alzheimer's disease: etiology, neuropathology and pathogenesis. In: Huang, X. (Ed.), Alzheimer's Disease: Drug Discovery. Exon Publications, Brisbane, Australia, ISBN 978-0-6450017-0-9. https://doi.org/10.36255/exonpublications.alzheimersdisease.2020.ch1.

Siegel, G., Gerber, H., Koch, P., Bruestle, O., Fraering, P.C., Rajendran, L., 2017. The Alzheimer's disease γ-secretase generates higher 42:40 ratios for β-amyloid than for p3 peptides. Cell Rep. 19, 1967–1976.

Silva, L.C.R., Ortigosa, L.C.M., Benard, G., 2010. Anti-TNF-α agents in the treatment of immune-mediated inflammatory diseases: mechanisms of action and pitfalls. Immunotherapy 2 (6), 817–833.

Šimić, G., Leko, M.B., Wray, S., Harrington, C., Delalle, I., Jovanov-Milošević, N., et al., 2016. Tau protein hyperphosphorylation and aggregation in Alzheimer's disease and other tauopathies, and possible neuroprotective strategies. Biomolecules 6, 6.

Snyder, H.M., Bain, L.J., Brickman, A.M., Carrillo, M.C., Esbensen, A.J., Espinosa, J.M., Fernandez, F., Fortea, J., Hartley, S.L., Head, E., Hendrix, J., Kishnani, P.S., Lai, F., Lao, P., Lemere, C., Mobley, W., Mufson, E.J., Potter, H., Zaman, S.H., Granholm, A.C., Rosas, H.D., Strydom, A., Whitten, M.S., Rafii, M.S., July 2020. Further understanding the connection between Alzheimer's disease and Down syndrome. Alzheimers Dement. 16 (7), 1065–1077. https://doi.org/10.1002/alz.12112.

Song, Z., Xu, Y., Deng, W., Zhang, L., Zhu, H., Yu, P., Qu, Y., Zhao, W., Han, Y., Qin, C., May 29, 2020. Brain derived exosomes are a double-edged sword in Alzheimer's disease. Front. Mol. Neurosci. 13, 79. https://doi.org/10.3389/fnmol.2020.00079.

Sorci, G., Riuzzi, F., Agneletti, A.L., Marchetti, C., Donato, R., 2004. S100B causes apoptosis in a myoblast cell line in a RAGE-independent manner. J. Cell. Physiol. 199 (2), 274–283.

Sosa, L.J., Postma, N.L., Estrada-Bernal, A., Hanna, M., Guo, R., Busciglio, J., Pfenninger, K.H., 2014. Dosage of amyloid precursor protein affects contact guidance in Down syndrome. FASEB J. 28 (1), 195–205. https://doi.org/10.1096/fj.13-232686:10.1096/fj.13-232686.

Startin, C.M., Lowe, B., Hamburg, S., Hitherway, R., Strydom, A., Consoritum, L.D.S., 2019. Validation of the cognitive scale for Down syndrome (CSDS) to detect longitudinal cognitive decline in adults with Down syndrome. Front. Psychiatr. 10, 158.

Stoltzner, S.E., Grenfell, T.J., Mori, C., et al., 2000. Temporal accrual of complement proteins in amyloid plaques in Down's syndrome with Alzheimer's disease. Am. J. Pathol. 156, 489–499.

Strauss, S., Bauer, J., Ganter, U., Jonas, U., Berger, M., Volk, B., 1992. Detection of interleukin-6 and α2-macroglobulin immunore activity in cortex and hippocampus of Alzheimer's disease patients. Lab. Invest. 66 (2), 223–230.

Streit, W.J., Braak, H., Xue, Q.S., Bechmann, I., 2009. Dystrophic(senescent)rather than activated microglial cells are associated with tau pathology and likely precede neurodegeneration in Alzheimer's disease. Acta Neuropathol. 118, 475–485.

Streit, W.J., Xue, Q.S., Tischer, J., Bechmann, I., 2014. Microglial pathology. Acta Neuropathol. 2, 142.

Sturgeon, X., Le, T., Mahiuddin Ahmed, M., Gardiner, K.J., 2012. Chapter 5 - pathways to cognitive deficits in Down syndrome. Prog. Brain Res. 197, 73–100. https://doi.org/10.1016/B978-0-444-54299-1.00005-4.

Subasinghe, W., Syed, I., Kowluru, A., 2011. Phagocyte-like NADPH oxidase promotes cytokine-induced mitochondrial dysfunction in pancreatic β-cells: evidence for regulation by Rac1. Am. J. Physiol. 300 (1), R12–R20.

Sullivan, K.D., Evans, D., Pandey, A., Hraha, T.H., Smith, K.P., Markham, N., et al., 2017. Trisomy 21 causes changes in the circulating proteome indicative of chronicauto inflammation. Sci. Rep. 7 (1), 14818 pmid: 29093484.

Szekely, C.A., Thorne, J.E., Zandi, P.P., et al., 2004. Nonsteroidal anti-inflammatory drugs for the prevention of Alzheimer's disease: a systematic review. Neuroepidemiology 23 (4), 159–169.

Taoka, S., Banerjee, R., 2001. Characterization of NO binding to human cystathionine β-synthase: possible implications of the effects of CO and NO binding to the human enzyme. J. Inorg. Biochem. 87 (4), 245–251.

Temple, V., Jozsvai, E., Konstantareas, M.M., Hewitt, T.A., 2008. Alzheimer dementia in Down's syndrome: the relevance of cognitive ability. Intellect Disabil. Res. 45, 47–55.

Tesseur, I., Zou, K., Esposito, L., et al., 2006. Deficiency in neuronal TGF-β signaling promotes neurodegeneration and Alzheimer's pathology. J. Clin. Invest. 116 (11), 3060–3069.

Tobinick, E.L., Gross, H., 2008. Rapid cognitive improvement in Alzheimer's disease following perispinal etanercept administration. J. Neuroinfl. 5. Article no. 2.

Torres-Platas, S.G., Comeau, S., Rachalski, A., et al., 2014. Morphometric characterization of microglial phenotypes in human cerebral cortex. J. Neuroinfl. 11, 12.

Tramutola, A., Di Domenico, F., Barone, E., Arena, A., Giorgi, A., di Francesco, L., Schinina, M.E., Coccia, R., Head, E., Butterfield, D.A., Perluigi, M., 2017. Polyubiquitinylation profile in Down syndrome brain before and after the development of Alzheimer neuropathology. Antioxid. Redox Signal. 26 (7). https://doi.org/10.1089/ars.2016.6686.

Tran, P., Xiang, D., Tran, T., Yin, W., Zhang, Y., Kong, L., et al., 2019. Exosomes and nanoengineering: a match made for precision therapeutics. Adv. Mater. 18, e1904040. https://doi.org/10.1002/adma.201904040.

Vallance, P., Leone, A., Calver, A., Collier, J., Moncada, S., 1992. Accumulation of an endogenous inhibitor of nitric oxide synthesis in chronic renal failure. Lancet 339 (8793), 572–575.

Wang, D.-S., Iwata, N., Hama, E., Saido, T.C., Dickson, D.W., 2003. Oxidized neprilysin in aging and Alzheimer's disease brains. Biochem. Biophys. Res. Commun. 310, 236–241.

Wang, X., Huang, T., Zhao, Y., Zheng, Q., Thompson, R.C., Bu, G., et al., 2014. Sorting nexin 27 regulates Aβ production through modulating γ-secretase activity. Cell Rep. 9, 1023–1033.

Wang, W.Y., Tan, M.S.1, Yu, J.T.1, Tan, L., 2015. Role of pro-inflammatory cytokines released from microglia in Alzheimer's disease. Ann. Transl. Med. 3, 136.

Wang, J., Gu, B.J., Masters, C.L., Wang, Y.J., 2017. A systemic view of Alzheimer disease - insights from amyloid-β metabolism beyond the brain. Nat. Rev. Neurol. 13, 612–623. https://doi.org/10.1038/nrneurol.2017.111. Internet.

Weggen, S., Eriksen, J.L., Das, P., et al., 2001. A subset of NSAIDs lower amyloidogenic Aβ42 independently of cyclooxygenase activity. Nature 414 (6860), 212–216.

Whiteman, M., Le Trionnaire, S.L., Chopra, M., Fox, B., Whatmore, J., 2011. Emerging role of hydrogen sulfide in health and disease: critical appraisal of biomarkers and pharmacological tools. Clin. Sci. 121 (11), 459–488.

Wilcock, D.M., 2012. Neuroinflammation in the aging Down syndrome brain; lessons from Alzheimer's disease. Curr. Gerontol. & Geriatr. Res. 10. https://doi.org/10.1155/2012/170276. Article ID 170276.

Wilcock, D.M., Colton, C.A., 2008. Anti-amyloid-β immunotherapy in Alzheimer's disease: relevance of transgenic mouse studies to clinical trials. J. Alzheim. Dis. 15 (4), 555–569.

Wilcock, D.M., Griffin, W.S., July 16, 2013. Down's syndrome, neuroinflammation, and Alzheimer neuropathogenesis. J. Neuroinfl. 10, 84. https://doi.org/10.1186/1742-2094-10-84.

Wilcock, D.M., Alamed, J., Gottschall, P.E., et al., 2006. Deglycosylated anti-amyloid-β antibodies eliminate cognitive deficits and reduce parenchymal amyloid with minimal vascular consequences in aged amyloid precursor protein transgenic mice. J. Neurosci. 26 (20), 5340–5346.

Wilcock, D.M., Hurban, J., Helman, A.M., Sudduth, T.L., McCarty, K.L., Beckett, T.L., Ferrell, J.C., Murphy, M.P., Abner, E.L., Schmitt, F.A., Head, E., September 2015. Down syndrome individuals with Alzheimer's disease have a distinct neuroinflammatory phenotype compared to sporadic Alzheimer's disease. Neurobiol. Aging 36 (9), 2468–2474. https://doi.org/10.1016/j.neurobiolaging.2015.05.016.

Wilcock, D.M., Hurban, J., Helman, A.M., et al., 2015b. Down syndrome individuals with Alzheimer's disease have a distinct neuroinflammatory phenotype compared to sporadic Alzheimer's disease. Neurobiol. Aging 36, 2468–2474.

Wiseman, F.K., Al-Janabi, T., Hardy, J., Karmiloff-Smith, A., Nizetic, D., Tybulewicz, V.L., Fisher, E.M., Strydom, A., 2015. A genetic cause of Alzheimer disease: mechanistic insights from Down syndrome. Nat. Rev. Neurosci. 16, 564–574.

Wisniewski, K.E., Wisniewski, H.M., Wen, G.Y., 1985. Occurrence of neuropathological changes and dementia of Alzheimer's disease in Down's syndrome. Ann. Neurol. 17 (3), 278–282.

Wolf, S.A., Boddeke, H.W., Kettenmann, H., 2017. Microglia in physiology and disease. Annu. Rev. Physiol. 79, 619–643.

Xiang, Z., Haroutunian, V., Ho, L., Purohit, D., Pasinetti, G.M., 2006. Microglia activation in the brain as inflammatory biomarker of Alzheimer's disease neuropathology and clinical dementia. Dis. Markers 22 (1–2), 95–102.

Xue, Q.S., Streit, W.J., 2011. Microglial pathology in Down syndrome. Acta Neuropathol. 122, 455–466.

Yuen, S., Smith, J., Caruso, L., Balan, M., Opavsky, M.A., 2011. The coxsackie-adenovirus receptor induces an inflammatory cardiomyopathy independent of viral infection. J. Mol. Cell. Cardiol. 50 (5), 826–840.

Yuyama, K., Igarashi, Y., 2017. Exosomes as carriers of Alzheimer's amyloidss. Front. Neurosci. 11, 229. https://doi.org/10.3389/fnins.2017.00229.

Zen, K., Liu, Y., McCall, I.C., et al., 2005. Neutrophil migration across tight junctions is mediated by adhesive interactions between epithelial coxsackie and adenovirus receptor and a junctional adhesion molecule-like protein on neutrophils. Mol. Biol. Cell 16 (6), 2694–2703.

Zhao, M., Cribbs, D.H., Anderson, A.J., et al., 2003. The induction of the TNFα death domain signaling pathway in Alzheimer's disease brain. Neurochem. Res. 28 (2), 307–318.

Zhu, Q., Heon, M., Zhao, Z., He, M., 2018. Microfluidic engineering of exosomes: editing cellular messages for precision therapeutics. Lab Chip 18, 1690–1703. https://doi.org/10.1039/c8lc00246k.

Ziebell, J.M., Taylor, S.E., Cao, T., Harrison, J.L., Lifshitz, J., 2012. Rod microglia: elongation, alignment, and coupling to form trains across the somatosensory cortex after experimental diffuse brain injury. J. Neuroinfl. 9, 247.

Zigman, W.B., Lott, I.T., 2007. Alzheimer's disease in Down syndrome: Neurobiology and risk. Ment. Retard. Dev. Disabil. Res. Rev. 13, 237–246.

CHAPTER

13

MicroRNAs and epigenetic signatures in Down syndrome

1. Introduction

Trisomy of the chromosome 21 (HSA21) in individuals with Down syndrome (DS) contributes to various neuroanatomical abnormalities, including reduction in brain size, brain weight, and neuronal density; neuronal distribution; and dendritic abnormalities at the cellular level. Many hypotheses such as gene dosage imbalance, amplified developmental instability, and DS critical region were proposed to relate the extra HSA21 to DS phenotypes (Delabar et al., 1993, detailed in Chapter 7). The growing knowledge of the contribution of the microRNAs (miRNAs/miRs), epigenetic modifications, allelic differences, gene dosage compensation, and also additive, subtractive, and epistatic mechanisms to DS phenotypes is not only deciphering the complexity of genetic association but also directing therapeutic development, especially for the improvement of the intellectual status and Alzheimer disease (AD) associated features.

HSA21, the smallest human chromosome, has been demonstrated to contain approximately 700 genes. Alongside, it harbors approximately 1050 small noncoding RNAs (ncRNAs), including 30 miRNAs that do not translate into protein but are associated with regulation of transcription of protein coding genes (Table 13.1). In the whole mammalian genome, approximately, 2000 miRNAs are present with conserved sequences. Interestingly, miRNAs constitute only 3% of the whole genome. However, the bioinformatics annotations have uncovered only some of the HSA21-encoded miRNAs (HSA21-miRNAs) in DS, such as miR-99a, let-7c, miR-125b-2, miR-155, and miR-802 miRNA genes with the 1.5 ratio of transcription level encoded, due to triplication of HSA21. The miRNAs are comprised of a large group of small RNAs with 21–25 nucleotides and act as posttranscriptional regulators through complementary-based silencing or degradation of the distinctive messenger RNAs (mRNAs), to avoid their further translation (Matamala et al., 2018).

The miRNAs modulate gene expression by binding to their target mRNAs; thus, affecting their stability or preventing translation. However, each of miRNA can target hundreds of mRNAs, and modulate the expression of about 90% of the known genes. Indeed, approximately 60% of the mRNAs possess one or more sequences, which are more likely to be silenced with miRNAs (reviewed in Mahernia et al., 2020). The expression of miRNAs is tightly regulated both at the transcriptional and the

Genetics and Neurobiology of Down Syndrome
https://doi.org/10.1016/B978-0-323-90456-8.00011-9

TABLE 13.1 RNAs, including microRNAs (miRNAs) and long noncoding RNAs (lncRNAs) of HSA21.

Symbol	Location	Description
RNA18SN1/2/3	21p11.2	RNA, 18S ribosomal N1/2/3
RNA45SN2	21p11.2	RNA, 45S pre-ribosomal N2
18S preribosomal N4/5	21p11.2	RNA
5.8S ribosomal N4 RNA28SN2	21p11.2	RNA
28S ribosomal N2 RNA		RNA
45S preribosomal N3 RNA		RNA
miR6724-1/4	21p11.2	microRNA 6724-1/4
miR 3648-1/2	21p11.2	microRNA 3648-1/2
miR 8069-1/2	21p12	microRNA 8069-1/2
miR 3156-3	21q11.2	microRNA 3156-3
miR 3118-1	21q11.2	microRNA 3118-1
miR 548X	21q21.1	microRNA 548x
miR 548XHG	21q21.1	miR548X host gene
miR 99AHG	21q21.1	mir-99a-let-7c cluster host gene
miR 99A	21q21.1	microRNA 99a
miR LET7C	21q21.1	microRNA let-7c
miR R125B2	21q21.1	microRNA 125b-2
miR 6130	21q21.2	microRNA 6130
miR 4759	21q21.3	microRNA 4759
miR 155	21q21.3	microRNA 155
miR 4327	21q22.11	microRNA 4327
miR 6501	21q22.11	microRNA 6501
miR 802	21q22.12	microRNA 802
miR 155HG		miR155 host gene
miR5692B	21q22.3	microRNA 5692b
miR 6508	21q22.2	microRNA 6508
miR 4760	21q22.2	microRNA 4760
miR 3197	21q22.2	microRNA 3197
miR 6814	21q22.3	microRNA 6814
miR 6070	21q22.3	microRNA 6070
miR 6815	21q22.3	microRNA 6815

TABLE 13.1 RNAs, including microRNAs (miRNAs) and long noncoding RNAs (lncRNAs) of HSA21.—cont'd

Symbol	Location	Description
lnc01667	21p11.2	long intergenic nonprotein coding RNA 1667
lnc 01669	21p12	long intergenic nonprotein coding RNA 1669
lnc01674	21q11.2	long intergenic nonprotein coding RNA 1674
lnc 02246	21q11.2	long intergenic nonprotein coding RNA 2246
lnc00308	21q21.1	long intergenic nonprotein coding RNA 308
lnc 01687/1683	21q21.1	long intergenic nonprotein coding RNA 1687/1683
lnc01425	21q21.1	long intergenic nonprotein coding RNA 1425
lnc 00317	21q21.1	long intergenic nonprotein coding RNA 317
lnc00320	21q21.1	long intergenic nonprotein coding RNA 320
lnc02573	21q21.1	long intergenic nonprotein coding RNA 2573
lnc01692	21q21.2	long intergenic nonprotein coding RNA 1692
lnc01684/1689	21q21.2	long intergenic nonprotein coding RNA 1684/1689
lnc00515	21q21.3	long intergenic nonprotein coding RNA 515
lnc01697/1695/161	21q21.3	long intergenic nonprotein coding RNA 1697/1695/161
lnc00113	21q21.3	long intergenic nonprotein coding RNA 113
lnc00159	21q22.11	long intergenic nonprotein coding RNA 159
lnc00307	21q22.11	long intergenic nonprotein coding RNA 307
lnc01690	21q22.11	long intergenic nonprotein coding RNA 1690
lnc00945	21q22.11	long intergenic nonprotein coding RNA 945
lnc01548	21q22.11	long intergenic nonprotein coding RNA 1548
lnc00160	21q22.12	long intergenic nonprotein coding RNA 160
lnc01426	21q22.12	long intergenic nonprotein coding RNA1426
lnc01436	21q22.12	long intergenic nonprotein coding RNA 1436
lnc01423	21q22.2	long intergenic nonprotein coding RNA 1423
lnc01700	21q22.2	long intergenic nonprotein coding RNA 1700
lnc00323	21q22.2	long intergenic nonprotein coding RNA 323
lnc00111	21q22.3	long intergenic nonprotein coding RNA 111
lnc00112	21q22.3	long intergenic nonprotein coding RNA 112
lnc01671	21q22.3	long intergenic nonprotein coding RNA 1671
lnc01668	21q22.3	long intergenic nonprotein coding RNA 1668

(Continued)

TABLE 13.1 RNAs, including microRNAs (miRNAs) and long noncoding RNAs (lncRNAs) of HSA21.—cont'd

Symbol	Location	Description
lnc00322	21q22.3	long intergenic nonprotein coding RNA 322
lnc01679	21q22.3	long intergenic nonprotein coding RNA 1679
lnc01678	21q22.3	Long intergenic nonprotein coding RNA 1678
lnc02575	21q22.3	long intergenic nonprotein coding RNA 2575
lnc01424	21q22.3	long intergenic nonprotein coding RNA 1424
lnc00163	21q22.3	long intergenic nonprotein coding RNA 163
lnc00334	21q22.3	long intergenic nonprotein coding RNA 334
lnc00316	21q22.3	long intergenic nonprotein coding RNA 316
lnc01694	21q22.3	long intergenic nonprotein coding RNA 1694
PICSAR	21q22.3	P38 inhibited cutaneous squamous cell carcinoma associated lincRNA
PICSAR	21q22.3	P38 inhibited cutaneous squamous cell carcinoma associated lincRNA

Modified from: file:///H:/DOWN%20SYNDROME/All%20Gene%20list%20by%20Location%20(Kb).pdf. Accessed June 24, 2020.

posttranscriptional level during development of cells, and transcriptional regulation is lost when miRNAs are translocated or ectopically expressed (reviewed in Farroni et al., 2018). The miRNAs are important regulators in biological processes, including development, adaptation, response to stress, cell fate determination, proliferation and differentiation of cells, immune reaction, and apoptosis. Stressed primary human neuronal-glial cells exhibited an increase in miR-155 (reviewed in Mahernia et al., 2021a).

Epigenetic alterations define the transmissible changes in the context of gene expression that are not contributed by alterations in the DNA structure itself. Such epigenetic alterations, consequences of disease development and possible drug targets have been described in different contexts (reviewed in Ganguly and Kadam, 2016a,b; Ganguly, 2017; Ganguly et al., 2018). A global hypermethylation in all genomic regions and all autosomes in DS placenta demonstrated genes with promoter hypermethylation, enriched for functions relevant to DS phenotypes, and suggested that epigenetic perturbation may be one important mechanism linking the triplication of HSA21 with DS phenotypic expression (Ganguly and Kadam, 2016a). Interestingly, functionally Dnmt3a can promote nonpromoter-mediated DNA methylation of genes responsible for neurogenesis, such as Dlx2, Gbx2, and Sp8 in neural progenitor cells by antagonizing polycomb repression, leading to increased expression of these genes (reviewed in Lu et al., 2016). It was postulated that elevated expression of DYRK1A in HSA21 might induce global epigenetic changes by down-regulating the repressor element one silencing transcription factor (REST) (reviewed in Ganguly and Kadam, 2016a).

In addition, the enhancer of zeste homolog 2 (EZH2), the main catalytic member of the polycomb repressive complex 2 (PRC2), encodes a histone methyltransferase that is responsible for mono-, di-, and trimethylation of lysine 27 of histone 3 (H3K27me1, me2, m3, respectively), was previously described on 21q22, and was thought to be the candidate gene for some of the DS phenotypes. The PRC2 complex contains EZH1 and EZH2 as its catalytic subunits. Subsequently, sequencing of exons mapped the EZH2 to 7q35 and demonstrated that the sequence previously isolated from 21q22 corresponds to a "pseudogene." However, further investigation was suggested for considering the small-molecule inhibitors of EZH2 as potential drug targets for DS leukemia, in cooperation with other gene expression. Another missense, possibly dominant negative, or gain of function, mutation has been recorded in the mRNA-splicing factor U2AF1 (U2 small nuclear RNA auxillary factor 1) mapped onto 21q22.3 and reported to cause myeloproliferative disorders being enriched in patients with del20q11 and ASXL1 mutations (Ganguly and Kadam, 2016a,b). Interaction of triplicated U2AF1 may thus increase the risk of leukemogenesis in DS. Indeed, research on identification of targeted small-molecule inhibitors may direct novel therapeutics for treatment of DS leukemia. Nevertheless, since global hypermethylation seen in DS with trisomic HSA21 appeared higher than that observed in trisomic HSA13 (Patau syndrome) and HSA18 (Edward syndrome), overexpression of HSA21-encoded DNMT3L and many other HSA21 genes and/or oxidative changes were proposed to be responsible for this incidence. However, further examination of additional cases was recommended for extraction of greater insights into the contributions and mechanistic associations.

The present chapter has illustrated the miRNAs (located on HSA21 and other chromosomes) and their possible role in development of DS phenotypes, particularly in AD. Additionally, the epigenetic alterations have been discussed with highlights on their impact on DS features in detail. Interestingly, epigenetic expression in DS varies depending on the tissue type and developing age. Moreover, discussion on small-molecule inhibitors of epigenetic targets is an important concern of therapeutic development for DS individuals. The discussion on miRNAs and epigenetic expression has been presented in two independent sections with respective and relevant sub-topics pertaining to DS phenotypes and therapeutics.

2. MicroRNAs and biogenesis

RNAs, known to primarily act in the "DNA → RNA → protein" central dogma as an intermediate between DNA and protein, have gained attraction in recent years for their involvement in a wide variety of biological processes, including epigenetic control of gene expression. Several noncoding regions of the human genome, so called "junk DNA" that produce RNA transcripts but not translated into protein are defined as ncRNAs. ncRNAs have been recognized and classified as small nuclear RNAs (snRNAs, involved in splicing), small nucleolar RNAs (snoRNAs, modify nucleotides in rRNA and other RNAs, and RNase P that cleaves pre-tRNAs), miRNAs, piwi-interacting RNAs (piRNAs), short interfering RNAs (siRNAs), and long ncRNAs (lncRNAs). Many of the ncRNAs are grouped as RNA-interference (RNAi) and expressed in cells targeting the expression of other biomolecules, instead of coding a protein (reviewed in Ganguly, 2017).

The biogenesis of miRNA starts with production of pri-miRNA by the transcription of miRNA genes activated by RNA polymerase II. The pri-miRNAs with huge transcripts of multiple miRNA sequences fold into hairpin structures, which subsequently get converted into pre-miRNA with microprocessing of RNase III enzyme Drosha and the DGCR8 protein in the nucleus. The pre-miRNA is transported to the cytoplasm from the nucleus with the involvement of exportin-5 complex, which is then cleaved off the stem loop with the cytoplasmic endonuclease DICER, in a complex with TAR RNA-binding protein (TRBP) and PKR-activating protein (PACT) (reviewed in Mahernia et al., 2021a). The mature miRNA is released by unwinding of the duplex miRNA and degradation of the passenger strand. Then, mature miRNA is furnished with matched base-pairs by the RNA-induced silencing complex (RISC) containing Argonaute (AGO) protein. Following matching and exact pairing or approximately near complementary to each other of the miRNA and the target mRNA, the target mRNA is degraded, and the target gene is silenced (reviewed in Mahernia et al., 2021a). The miRNA genes are present as a single unit or organized in gene clusters in the genome.

2.1 Noncoding RNAs

The miRNAs are also known as ncRNAs. Indeed, thousands of ncRNA families constitute the most representative transcriptional units of the mammalian genome for rendering crucial roles in key molecular mechanisms in different domains of life (Perez-Nievas et al., 2013). Triplication of these ncRNAs indicates potential overexpression in DS. Notably, upregulation of different miRNAs detected in both AD and DS brains could potentially downregulate the expression of regulatory and anti-inflammatory genes in both DS and AD patients. These ncRNAs might interfere with the expression of protein coding genes and/or regulatory sequences encoded in HSA21 (HSA21-ncRNA) and be involved in Aβ-aggregation triggering the early onset of AD in DS (Gomez et al., 2020). Strikingly, each miRNA can downregulate expression of several proteins simultaneously through base-pair interactions with the target mRNA, and subsequently alter multiple signaling pathways (Mahernia et al., 2021a). ncRNAs may form DNA:RNA triplex through interaction with many other genes resulting in triplex mediated up- or down-regulation of expression, silencing of translation, or guiding epigenetic methylation (Ganguly, 2017). Nevertheless, multiple interactive associations between ncRNAs and their mutations result in dysregulation of miRNAs, which in turn leads to various disease pathogenesis, including oncogenesis. However, the role of ncRNAs is not fully explored and understood in DS research. Therefore, further study is suggested to surface the mechanistic involvement of ncRNAs in disease development, which will also allow identification of new therapeutic candidates in single or in combination, and many more insights pertaining to functions and silencing of ncRNAs, especially in DS and other chromosomal aneuploidies.

2.2 Long noncoding RNAs

All over the human genome, there are over 170,000 lncRNAs, and these noncoding transcripts are composed of over 200 nucleotides and originate from intergenic regions or with some level of overlapping with protein coding genes. Over 1000 lncRNAs are mapped onto HSA21, and thus, triplication of HSA21 in DS obviously indicates overexpression of lncRNAs, which may perturb myriads of systemic functions and contribute to pathogenesis in DS. Moreover, their tissue-specific abundance and increased expression in neuronal tissues may indicate a significant cross-talk between lncRNAs and neurodegenerative diseases, including DS and AD. A microarray analysis of autopsy of elderly AD patients revealed upregulation of 24 lncRNAs and downregulation of 84 lncRNAs compared to healthy controls (reviewed in Gomez et al., 2020). However, the mechanisms of their action and subsequent impact have not been fully deciphered, and thus, limited information is available on lncRNA transcripts of HSA21 (HSA21-lncRNA) and their interaction with other genes of disomic status in DS.

Noticeably, association of some of the HSA21-lncRNAs with AD in DS (DSAD) has been indicated with their expression levels in triplicated condition (Table 13.2). The abundance of expression of the Down Syndrome Critical Region 9

TABLE 13.2 Long noncoding RNAs (lncRNAs) associated with Down syndrome (DS) and Alzheimer disease (AD).

LncRNAs	Features	Affected system	Association with diseases	References
DSCR9 lncRNA	Tissue-specific expression; abundant in the heart and brain with higher abundance in hippocampus and amygdala	Correlated with genes important for development and function of the nervous system	Regulates proteins contribute to DS and other neurological diseases	Chen et al. (2018)
BACE1-Antisense Strand (BACE-AS) lncRNA	A conserved lncRNA; transcribed from the antisense strand of the BACE1 gene; BACE1-AS confers moderating effect BACE1-mRNA expression and influence the formation of Aβ	Nervous system in AD; cardiac dysfunction	Causes overexpression of the BACE1 mRNA that encodes the protein involved in proteolysis of the APP protein	Wan et al. (2017); Bacalini et al. (2015); Li et al. (2019); Greco et al. (2017)
Brain Cytoplasmic RNA 1 (BCYRN1) lncRNA	modulates the protein synthesis in dendrites	Nervous system	Increased in AD compared to healthy individuals; associated with synaptodendritic deterioration	Wan et al. (2017)
n341006 lncRNA	Its downregulation was associated with the protein ubiquitination pathway		Significantly upregulated n336934 lncRNA showed to be linked to cholesterol homeostasis, which is dysregulated in AD	Luo and Chen (2016)

(DSCR9), a HSA21-lncRNA, was evidenced in the heart, hippocampus, and amygdala of the DS brain. The report has highlighted the importance of DSCR9 in the development and functions of the nervous system, suggesting its contribution on regulation of proteins associated with DS-specific and other neurodegenerative diseases (Chen et al., 2018). BACE1 encodes the protein involved in cleaving the APP and generation of the neurotoxic Aβ peptides in AD patients. The BACE1-antisense strand (BACE-AS), a lncRNA, is transcribed from the antisense strand (AS) of the BACE1 gene. BACE1-AS is responsible for overexpression of BACE1 mRNA required for proteolysis of the APP protein (Gomez et al., 2020; Wan et al., 2017; Bacalini et al., 2015). Another lncRNA, the brain cytoplasmic RNA 1 (BCYRN1) is involved in modulation of the protein synthesis in dendrites. Its overexpression in AD was demonstrated to be responsible for synaptodendritic deterioration compared to healthy controls (Wan et al., 2017). A gene set enrichment study has reported association of the downregulated n341006 lncRNA with the protein ubiquitination pathway, whereas significantly upregulated n336934 lncRNA was related to cholesterol homeostasis, a dysregulated pathway observed in AD (Luo and Chen, 2016).

An RNA expression study in disomic and trisomic iPSCs revealed a significant alteration in expression of lncRNAs relative to the protein-coding genes in trisomic iPSCs (Qiu et al., 2017). In addition, lncRNAs have been associated with different mitochondrial activities, such as mitochondrial organization, electron transport, ATP synthesis, and mitochondrial membrane organization. Remarkably, most of the protein-coding genes responsible for mitochondrial functions were downregulated in trisomic iPSCs, indicating likeliness of an association of lncRNA expression with mitochondrial dysfunction observed in individuals with DS (Qiu et al., 2017). Altogether, further studies can unravel the association of HSA21-lncRNAs with DS phenotypes and direct development of targeted therapeutics.

3. HSA21-MicroRNAs in Down syndrome

Of the 19 precursors and 26 mature miRNAs encoded on HSA21, only five, including miR-155, miR-802, miR-125b-2, let-7c, and miR-99a, have been characterized with DS phenotypes (Table 13.3). Overexpression of these miRNAs observed in DS fetal hippocampus (10–15 fold) and heart samples indicated their exclusive expression in neurons but not in microglial cells or astrocytes (Siew et al., 2013). Trisomic overexpression of HSA21-miRNAs has been partly linked with the defective neurodevelopment in DS individuals. Overexpression of miR-155 and miR-802 was observed in trisomic Ts65Dn mouse model. Triplication of these miRNAs showed hippocampal deficits and that might disturb the learning, memory, long-term synaptic plasticity, and neurogenesis in DS (Keck-Wherley et al., 2011). Interestingly, both miR-155 and miR-802 target the methyl CpG binding protein2 (MeCP2) and confer deleterious effects on development. Following degradation and repressed expression of MeCP2, overexpression of miR-155 and miR-802 caused developmental defects in the trisomic iPS-derived neuronal progenitor cells (iPS-NPCs) with decreased number of neurons compared to diploid cells (Lu et al., 2012). Repressed expression of MeCP2 protein was hypothesized to cause neurochemical abnormalities detected in brains of DS individuals (Kerkel et al., 2010). The miR-155 DNA has been proposed to downregulate the polymerase beta and MeCP2 expression (Ahmed et al., 2013). Silencing of miR-155 and miR-802 in trisomic Ts65Dn mice restored the MeCP2 expression and its target genes (reviewed in Siew et al., 2013). Assessment of hippocampal neurons from mice that either lacked expression or expressed twice the normal levels of MeCP2 has proposed that the regulation of glutamatergic synapse number by MeCP2 might be a mechanism for the altered synaptic strength of neurons in DS (reviewed in Brás et al., 2018; Chao et al., 2007).

Interestingly, MeCP2 repressed miR-199b expression in brain of a mouse model of Rett syndrome (reviewed in Siew et al., 2013). In contrast, miR-199b has been reported to suppress sirtuin1 (SIRT1), a key factor of dendritic development. Since overexpression of miR-155 and miR-802 leads to repression of MeCP2 expression, the repressed MeCP2 is expected to subsequently relieve the suppression of miR-199b and decrease SIRT1 expression (reviewed in Siew et al., 2013). This interconnected cascade of complex regulation of miRNA expression has indicated disruption of a series of complex events that may be involved in brain development in DS. Thus, modalities of selective inactivation of these HSA21-miRNAs have been targeted for identification of novel therapeutic candidates for treatment of DS.

Brain deficits and systemic immune pathology such as activated microglia and increased inflammatory signaling have been reported in DS. The autopsy of DS brain tissue and sporadic AD patients exhibited overexpression of miRNA-155, while repression of the complement factor H (CFH) (a soluble inherent immune regulatory glycoprotein was observed in brain tissues of AD, DS, and experimental AD models. It acted as an essential repressor of the innate immune response) in brains and peripheral tissues of DS individuals

TABLE 13.3 The HSA21-encoded microRNAs (miRNAs) and Down syndrome (DS) symptoms.

miRNAs	Target genes/signaling pathways	Model/organisms of DS	Expression level	Associated DS patho-physiology
miR-155	MeCP2 (Keck-Wherelay et al., 2011); CFH (Li et al., 2012); PICALM (Paschou et al., 2012)	DS patients	Overexpressed in fetal hippocampus and heart tissues (reviewed in Siew et al., 2013); overexpressed in prefrontal cortex at all ages (fetal age to adulthood); overexpressed in adult temporal lobe, spleen, liver (Li et al., 2012); overexpressed in autopsy of DS and AD brains	Lower the complement factor H mRNA (Brás et al., 2018); downregulate MeCP2 expression (Ahmed et al., 2013); neuronal loss of protection against complete opsonization and leukocyte infiltration into the penumbra; ancient regulator of the immune system (Vigorito et al., 2013); Angiotensin II type 1 receptor (a marker of cardiac pathology); less cardiovascular disease among DS patients
		Ts65Dn mice	Overexpressed in the adult hippocampus (reviewed in Siew et al., 2013); Overexpressed in adult hippocampus and blood samples (Keck-Wherelay et al., 2011); Anti-miR treatment restored MeCP2 in adult hippocampus	
		iPSCs derived from human amniotic fluid with DS	Overexpressed in trisomic iPSCs (Lu et al., 2012)	
		TNFα-stressed human neuronal glial cells	Anti-miR treatment restored CFH level (Li et al., 2012)	
		Superior temporal neocortex of AD	Overexpressed in neocortical samples (Lukiw and Alexandrov, 2012)	
miRNA let-7c	TLP7 (Lehmann et al., 2012)	iPSCs derived from human amniotic fluid with DS	Overexpressed in trisomic iPSCs (Lu et al., 2012)	Target protein Ezh2; acts as tumor suppressor; CHD in DS (Coppola et al., 2014); lower the cell proliferation of lung adenocarcinoma (Wang et al., 2013)
miR-99a	TGFβ pathway (Turcatel et al., 2012)	DS patients	Overexpressed in fetal hippocampus and heart tissues (reviewed in Siew et al., 2013); overexpressed in prefrontal cortex at all ages (fetal age to adulthood) (reviewed in Siew et al., 2013)	Nucleosome-remodeling factor Smarca5; acts as tumor suppressor; weakening of the Nodal/Smad2 signaling, CHD in DS (Coppola et al., 2014)

Continued

TABLE 13.3 The HSA21-encoded microRNAs (miRNAs) and Down syndrome (DS) symptoms.—cont'd

miRNAs	Target genes/signaling pathways	Model/organisms of DS	Expression level	Associated DS patho-physiology
miR-802	MeCP2 (Keck-Wherelay et al., 2011)	DS patients	Overexpressed in fetal hippocampus and heart tissues (reviewed in Siew et al., 2013); overexpressed in prefrontal cortex at all ages (fetal age to adulthood) (reviewed in Siew et al., 2013)	Ship1 (inositol phosphatase); malfunctioning of the hippocampal synaptic plasticity
		Ts65Dn mice	Overexpressed in the adult hippocampus (reviewed in Siew et al., 2013); Overexpressed in adult hippocampus and blood samples (Keck-Wherelay et al., 2011); Anti-miR treatment restored MeCP2 in adult hippocampus (reviewed in Siew et al., 2013)	Hippocampal deficits; responsible for learning, memory and long-term synaptic plasticity (Keck-Wherelay et al., 2011)
		iPSCs derived from human amniotic fluid with DS	Overexpressed in trisomic iPSCs (Lu et al., 2012)	
miR-125b	CDKN2A (Pogue et al., 2010); NR2A and EphA4 (Edbauer et al., 2010); Nestin (Cui et al., 2012)	IL-6-stressed normal human astrocytes (Siew et al., 2013)	Not available	
		Hippocampus of newborn Sprague–Dawley rat pups	Overexpressed miR-125b reduced Nestin expression and inhibited neural stem/progenitor cells proliferation (Cui et al., 2012)	Acts as tumor suppressor for solid tumors (affects the growth potential and invasive capabilities of human breast cancer cells); oncogenic for AMKL
		E19 rat hippocampal neurons in culture	FMRP was partially depending on miR-125b in suppressing NR2A (Edbauer et al., 2010). Overexpressed miR-125b led to longer and thinner dendritic protrusions with reduced synaptic strength (Edbauer et al., 2010).	
	Tau hyperphosphorylation		Synaptic protein deficiency in AD brains (Bacalini et al., 2015); synapsin II mRNA could be a potential target	
miR-125b-2		DS patients	Overexpressed in fetal hippocampus and heart tissues (reviewed in Siew et al., 2013); overexpressed in prefrontal cortex at all ages (fetal age to adulthood) (reviewed in Siew et al., 2013)	DICER1, ST18; oncogenic for DS-AMKL in pediatric DS; oncogenic miRNA (Klusmann et al., 2010)

Adapted from Siew et al. (2013); Mahernia et al. (2021a).

(Li et al., 2012). The interaction of miR-155 and CFH are demonstrated to be centrally involved in the pathogenesis causing inflammatory neurodegeneration and susceptibility to infection in DS patients (Wan et al., 2017). An interaction of miR-155 was predicted with presynaptic mRNAs such as phosphatidyl inositol-binding clathrin assembly protein (PICALM) marker of AD (Paschou et al., 2012). Since DS patients are at higher risk for early onset of AD, further investigations on the association between miR-155 and brain pathology are essential to decipher its role in regulating inflammatory and innate immune responses.

The miR-125b plays a crucial role in survival of neural cells and development of spine morphology (Schaefer et al., 2007). A marked upregulation of miR-125b was described in interleukin-6-stressed normal human astrocytes (Pogue et al., 2010). Overexpression of HSA21-miR-125b has been implicated on the neuropathology of DS patients. Attenuation of astrogliosis and increase in cyclin-dependent kinase inhibitor 2A (CDKN2A), a negative regulator of cell growth, were demonstrated with anti-miR-125b treatment. Thus, suppression of CDKN2A mediated by miR-125b suggested that astrogliosis in DS brains could be attributed to the glial proliferative effect (Pogue et al., 2010). Furthermore, miR-125b has been targeted for glutamate receptor, ionotropic, N-methyl D-aspartate subunit 2A (NR2A), Ephrin type A receptor 4 (EPHA4), and Nestin (Edbauer et al., 2010; Cui et al., 2012). The miR-125b influences neurite outgrowth and dendritic branching (Le et al., 2009), resulting in generation of longer and thinner dendritic spine with reduced spine width by targeting NR2A (Edbauer et al., 2010). Suppression of NR2A and EphA4 expression has been demonstrated in the process of long-term potentiation (LTP) and/or long-term depression (LTD) (reviewed in Siew et al., 2013). The miR-125b expression was reported to target 3_UTR of Nestin and reduce levels of both Nestin mRNA and associated protein. Overexpression of miR-125b was attributed to inhibit the proliferation of neural stem/progenitor cells via suppression of Nestin (Cui et al., 2012). Summarily, overexpression of miR-125b was observed to cause striking changes in neuronal development. Therefore, association of the miR-125b with impairment of synaptogenesis and LTP in DS patients may direct targeting of relevant transcripts for potential drug development.

Another HSA21-miR miR-99a has been ascribed to regulate the TGF-β pathway, which is directly involved in retrograde synaptic signaling and the Wnt signaling in stem cell differentiation (reviewed in Siew et al., 2013). In addition, the miR-let-7c is demonstrated to be highly expressed in the hippocampal region of adult mice. It exerts neuronal loss in a dose- and time-dependent manner through interaction with toll-like receptor7 (TLR7) (reviewed in Siew et al., 2013). However, further investigations are necessary to unravel the association of miR-99a and let-7c on regulation of the neuronal function in DS individuals.

The learning disability in children with DS is a collective expression of auditory deficits, language difficulties, reading impairment, limited verbal short-term memory capacity, hearing impairments, and so on. Notably, this phenotype has been attributed to HSA21-miRNAs such as MiR-125b-2, let-7c, and miR-99a (Weston et al., 2006; Siew et al., 2013). However, further investigation is suggested in human DS for exploration of their roles in regulating auditory development and/or function contributing to learning disability in DS individuals.

Taken together, miRNAs of HSA21 are likely to be directly linked with development and differentiation of neuronal cells, synaptic plasticity, and more. Further refinement of involvement of these miRNAs in manifestation of DS phenotypes may elucidate better understanding of the underlying mechanisms of development of intellectual disability and neurodegeneration in DS population. More importantly, miRNA-based therapeutics have shown to be effective to restore expression of targeted genes and recover their downstream functions. Therefore, delineation of further understanding of the contribution of the defective neurogenesis, gliogenesis, and cell cycle events on intellectual impairment in DS patients would help in identification of therapeutic candidates for rescuing the impaired phenotypes linked to quality of life of DS patients.

4. Disease consequences of HSA21-miRNAs

MiRNAs play an important role in stem cell self-renewal and differentiation and reprogramming of multiple oncogenic cascades (reviewed in Ganguly, 2017). The ncRNAs have been ascribed to a number of human diseases, such as cancer, cardiovascular, and neurodegenerative disorders, inflammatory abnormalities, AD, and so on, though the mechanism of action has not been fully elucidated (Sethupathy et al., 2007; Chen et al., 2018; Luo and Chen, 2016). In MDS and other diseases, ncRNAs are involved in a number of processes via transcription from intronic regions of genes and posttranscriptional regulation of exon skipping and intron retention. Of the several ncRNA families, the miRNAs and the lncRNAs are well characterized with their association with disease consequences that result from several key molecular events. RNA-sequencing has identified several unannotated lncRNAs that are differentially expressed in normal and malignant T-lymphocytes. Germline mutation and dysregulation of lncRNAs have also been attributed to a number of inherited diseases (reviewed in Ganguly, 2017). In DS, their overexpression due to triplication could be deemed in part to contribute to onset and progress of several DS-specific phenotypes (Hattori et al., 2000). Evidently, miR-155 and miR-802 contribute to cognitive disability in DS. The significant patho-physiological role of global miRNAs of the human genome has been delineated; however, the pathophysiologic link of HSA21-miRNAs in DS is not fully understood. DS phenotypes resemble the pathological hallmarks of cell senescence and early aging. The biological mechanisms and related signaling of miR-99a, let-7c, miR-125b-2, miR-155, and miR-802 of HSA21 (Table 13.3) contributing to DS symptoms need precise understanding for clinical management and therapeutic development.

DS people are more susceptible to AD and other neurodegenerative diseases. Overexpression of the HSA21-miRNAs may contribute to this pathomechanism and also affect specific haploinsufficiency-related genes (Brás et al., 2018). Triplication of DYRK1A and APP is responsible for aggregation of Aβ peptide and hyperphosphorylation of tau in brain leading to genesis of AD in DS. Mitigation of tau and Aβ pathology was evidenced in the hippocampus of Ts65Dn mouse model of DS (Chaves et al., 2020), indicating an association between miRNAs, such as miR-17, -20a, -101, -106b, -199b, -26a, and 26b, and some of their target mRNAs, such as APP, DYRK1A, and BDNF, with hyperphosphorylated protein levels. The results demonstrated the importance of miR-17, -20a, -101, -106b, -199b, -26a, and 26b for lowering the tau and Aβ damage in DS (Chaves et al., 2020). Indirect perturbation of Aβ metabolism is also described through the action on 3′UTR of BACE1, ABCA1, APP, and other related genes. In addition, miRNA-mediated regulation of synaptic function is linked to the pathophysiology of plasticity-related diseases, including AD and fronto-temporal dementia. Notably, miRNAs play important roles in adjusting synaptic plasticity and pathobiology of AD (reviewed in Mahernia et al., 2021a). Aberrant expression of let-7 and miR-155 miRNAs has been reported to contribute to Aβ formation and tau phosphorylation.

Overexpression of miR-125b has been demonstrated to contribute to tau hyperphosphorylation and the upregulation of p35, cdk5, and p44/42-MAPK signaling in initial neurons leading to the AD pathology in DS. Importantly, knockdown of DUSP6 and PPP1CA phosphatases and the Bcl-W antiapoptotic factor influenced tau hyperphosphorylation, indicating involvement of these phosphatases on mediating the action of miR-125b on tau (Banzhaf-Strathmann et al., 2014). In addition, mild overexpression of miR-125b resulted in synaptic protein deficiency in brains of AD suggesting synapsin II mRNA could be a potential target of miR-125b (Bacalini et al., 2015). A computational analysis has described that the phosphoprotein synapsin II mRNA encodes neuron-specific neurotransmitter, contains 14 potential binding sites for different miRNAs in its 3′UTR mRNA, and is markedly downregulated in AD (Gomez et al., 2020).

Overexpression of miR-155 leads to alterations in levels of specific target proteins in DS and AD (Brás et al., 2018). Ubiquitous expression of miR-155 has been demonstrated in active B cells, T cells, macrophages, and dendritic cells. Regulation of the T helper cell differentiation is mediated by miR-155 (reviewed in Mahernia et al., 2020). Additionally, upregulation of both miR-125b and miR-155 were evidenced in memory B cells (MBCs), activated B cells, and plasma blasts (PBs)/plasma cells highlighting the critical role of HSA21-miRNAs in the regulation of antigen-experienced B cells and immune response (Vigorito et al., 2013). With regards to cancer incidence in DS, the miR-125b-2 was hypothesized as an oncogenic miRNA that contributes to acute megakaryocytic leukemia (AMKL) in DS children (Klusmann et al., 2010). In pediatric acute myeloid leukemia (AML), the miRNAs, including miR-100, miR-125b, miR-335, miR-146, and miR-99a, are highly expressed (O'Connell et al., 2008). Noticeably, acquired mutations in GATA1, an important hematopoietic transcription factor, are evident in leukemic cells of AMKL in DS children. In this regard, cooperative effects between GATA1 and miR-486-5p has been proposed to enhance the survival of leukemic cells by miR-486-5p in DS patients (Shaham et al., 2015). Interestingly, the overexpression of miR-155 is documented in human cancers of the thyroid, cervix, pancreas, lung, breast, colon, and other solid malignancies (reviewed in Mahernia et al., 2020). Notably, assessment of miR-155 has been considered as an important tool for diagnosis and treatment of tumors.

Interestingly, overexpression of miR-125a or miR-125b affected the growth potential and decreased the motility and invasive capabilities of Erb-B2 receptor tyrosine kinase 2-dependent human breast cancer cells in vitro. In addition, overexpression of the miRNA let-7c identified in breast cancer has been demonstrated to regulate the tumorigenicity of breast cancer cells. Indeed, let-7c suppressed the tumor formation capacity of breast cancer stem cells, suggesting low rate of breast cancer is likely among women with DS. Moreover, overexpression of let-7 displayed low rate of cell proliferation in lung cells and proliferation of lung adenocarcinoma (Wang et al., 2013) and also affected cell cycle progression in a malignant liver cell line (reviewed in Brás et al., 2018). Furthermore, the incidence of prostate cancer is also less in DS patients compared to the general population. Overexpression of the HSA21-encoded miR-99 family of miRNAs has been documented to inhibit the proliferation of prostate cancer cells and reduce the level of prostate-specific antigen (PSA) (Sun D et al., 2011). Taken together, overexpression of these two (let-7 and miR-99) miRNAs was thought to be protective toward development of solid tumors in DS patients, since these miRNAs were observed to inhibit proliferation of cells in lung, breast, and prostate cancers (Sun et al., 2011, 2016; reviewed in Brás et al., 2018). Notably, the HSA21-miR-99a, let-7c, and miR-125b miRNAs act as tumor suppressors and thus, are capable of reducing the incidence of solid tumors in DS patients (Xu et al., 2013a). However, inhibition of vascular endothelial growth factor–mediated angiogenic signaling mediated by overexpression of the calcineurin inhibitor DSCR1 may also contribute to lower the risk of solid tumors in DS (Baek et al., 2009).

In DS, overexpression of the miR-99a/let-7c cluster was hypothesized to be associated with congenital heart defects (CHD) (Coppola et al., 2014). Interestingly, the expression levels of BACH1 and miRNA-155 were not increased in hearts of individuals with DS (Hefti et al., 2016), though increased expression of BACH1 was observed in trisomic fibroblasts (1.6 relative fold) but not in trisomic lymphoblasts (1.2 relative fold). Interindividual genetic variation was proposed to likely contribute to expression of miR-155 and BACH1 in different DS individuals. Notably, inhibition of BACH1 protein translation by miR-155 conferred a plausible protection against oxidative stress in euploid endothelial cells, indicating a positive association between BACH1 mRNA and miR-155 expression (Pulkkinen et al., 2011; reviewed in Hefti et al., 2016). However, similar association has not yet been studied in DS cells.

Interestingly, overexpression of miR-155 was reported to regulate complement factor H (CFH) expression in AD (Lukiw and Alexandrov, 2012). Since CFH protects neurons from axonal injury, complement opsonization, and leukocyte infiltration in the brain parenchyma, overexpression of miR-155 was hypothesized to be involved in the brain pathology of DS patients (Brás et al., 2018). The miR-155 lowered the level of CFH in DS tissues. Strikingly, repression of the abundance of the miR-155 mRNA perturbs the CFH. Furthermore, miR-155 interacts with the angiotensin II type 1 (AGNII-TI) receptor, a significant marker of the cardiac pathology. Overexpression of miRNA-155 is expected to lower the risk of cardiovascular disease among DS patients (reviewed in Mahernia et al., 2021a; Brás et al., 2018). Overexpression of miR-146a (a key factor of neuronal development) was noticed higher in astroglial cells within the white matter of the hippocampus in DS than normal fetuses, indicating postnatal persistence of its elevated expression (Arena et al., 2017). Indeed, deregulation of miR-146a and miR-155 has been reported in the APP/PS1 mouse of AD model and Ts65Dn of DS mouse model (Della et al., 2014).

Identification of two novel HSA21-miRNAs, such as miR-nov1 and miR-nov2, has explained their location at up- and downstream of the annotated miR-802 loci. Notably, miR-nov1 and miR-nov2 were reported in cord blood mononuclear cells of DS fetuses. The overexpression of miR-nov2, located in the "DSCR," has been described in DS lymphocytes (Xu et al., 2013b). Interestingly, the study has predicted 97 mRNA targets of miR-nov2, which are associated with cellular growth and death, cellular localization, and transport of proteins. Remarkably, downregulation of miR-99a, let-7c, miR-125b-2, and miR-155 were demonstrated in cord blood mononuclear cells derived from DS (Xu et al., 2013a; reviewed in Brás et al., 2018).

4.1 Association of HSA21-miRNAs on mitochondrial output in DS

The miRNAs were predicted to play important roles in the cardiac phenotype in DS since they are overexpressed in the heart and regulate cardiac development. Of the five HSA21-miRNAs such as miR-99a-5p, miR-125b-2-5p, let-7c-5p, miR-155-5p, and miR-802-5p studied, the expression of these miRNAs in heart tissues from DS fetuses showed overexpression of miR-99a-5p, miR-155-5p, and let-7c-5p in trisomic hearts. Indeed, different databases and cross-validated studies using the gene expression profiling dataset have demonstrated expression and downregulation of eighty-five targets of let-7c-5p, 33 of miR-155-5p, and 10 of miR-99a-5p in trisomic fetal heart. Noticeably, nuclear encoded mitochondrial genes were found downregulated in trisomic hearts, where mitochondrial dysfunction is a hallmark of DS phenotypes. In this regard, special attention to let-7c-5p and miR-155-5p revealed their downregulation in trisomic fetal hearts and involvement in mitochondrial function (Izzo et al., 2017). The overexpression strategy of MMU-let-7c and MMU-miR-99a in cardiomyogenesis was investigated in murine embryonic stem cells. At early stage of differentiation, overexpression of these miRNAs demonstrated let-7c-induced cardiogenesis indicating their involvement in heart development, whereas the effect was repressed by miR-99a through alteration of the Smad2 signaling (Coppola et al., 2014). Thus, the overexpression of HSA21-miRNAs was predicted to downregulate specific target genes responsible for DS phenotypes.

In addition, significantly lower expression of five mRNAs, including ATP8, CO3, ND2, ND3, and ND4 was reported in DS fibroblasts, of which ND2, ND3, and ND4 encode the complex I, CO3 encodes complex IV, and ATP8 encodes the complex V. The result suggested that down-regulation of these mitochondrial mRNAs might increase the susceptibility of DS to AD (Salemi et al., 2018). However, further investigation in this line can establish the stated association.

The SLC25A4 (solute carrier family 25 member 4) or ANT1 (adenine nucleotide translocator 1) acts as a gated pore for translocation of ADP and ATP between the cytoplasm and mitochondria and regulate the intracellular energetic balance. Interestingly, the let-7c-5p predicted SLC25A4/ANT1 as a possible therapeutic target for both mitochondrial dysfunction and cardiac anomalies in individuals with DS. Thus, its dysregulation has been associated to mitochondrial cardiomyopathies (reviewed in Izzo et al., 2017). ANT has four isoforms encoded by four different genes and are distributed in a tissue-specific pattern in humans. The SLC25A4/ANT1 gene is primarily expressed in the heart and skeletal muscle; however, it is down-regulated in human DS fetal hearts and fibroblasts with concomitant upregulation of let-7c-5p (reviewed in Izzo et al., 2017).

Interestingly, SLC25A4/ANT1 is one of the 37 nuclear encoded mitochondrial genes (NEMGs) downregulated after NRIP1 overexpression. Notably, silencing of NRIP1 increased the expression of SLC25A4/ANT1 in trisomic cells, suggesting a more efficient exchange of ATP and an improvement of mitochondrial activity in DS samples. As a result, genetic inactivation of the heart isoform caused damage in mtDNA and increased reactive oxygen species (ROS) leading to perturbation of oxidative phosphorylation (OXPHOS) and mitochondrial function (reviewed in Izzo et al., 2017). These findings lead to select targets for HSA21-miRNAs, including miR-155-5p for improvement of mitochondrial function and heart development. Indeed, 33 predicted target genes of miR-155-5p appeared downregulated in trisomic fetal hearts relative to euploid ones; eight of them, including MECP2 gene, is a validated target (reviewed in Izzo et al., 2017). MECP2 is also included among the let-7c-5p predicted targets.

The HSA21-miR-155-5p has been described to regulate mitochondrial biogenesis by targeting mitochondrial transcription factor A (TFAM) at the posttranscriptional level in fibroblasts and heart samples with diploid genome but not in the DS setting (Quiñones-Lombraña and Blanco, 2015). This result was explained by the fact that mitochondrial biogenesis was already markedly decreased in DS cells; thus, interaction of miR-155-5p with TFAM could not exert significant changes in mitochondria-specific parameters, such as mtDNA content, citrate synthase activity, and number of mitochondria in trisomic cells. Involvement of additional factors during mitochondrial biogenesis in the DS setting was speculated. Furthermore, miR-155-5p has been hypothesized to be an inducer of cardiac hypertrophy, suggesting clinical potential of its possible inhibition to counteract this pathology (reviewed in Izzo et al., 2017). These results have recommended investigation of both let-7c-5p and miR-155-5p miRNAs for understanding the molecular mechanisms underlying the cardiac malformations in DS patients.

5. Association of other non-HSA21 miRNAs with HSA21-miRNAs

Despite overexpression of the trisomic HSA21 genes, expression of and/or interaction with non-HSA21 encoded disomic genes was proposed to contribute to the variability observed in DS phenotypes. Similar explorations on the implication of non-HSA21-miRNAs on HSA21 targeted genes on DS features have demonstrated inconsistent findings on the association of non-HSA21 miRNAs with the triplicated HSA21-miRNAs (reviewed in Brás et al., 2018). Significant difference in miRNA expression was reported in trisomic blood and placenta samples (Lim et al., 2015). Up-regulation of miR-10b (on 2q31 between HOXD4 and HOXD8), miR-542-5p (on Xq26.3), miR-654 (on 14q32), and miR-615 (on 12q13, within the intron of the Hoxc5 gene) was reported in placentas with DS along with HSA21-encoded miR-99a, miR-125b, and let-7c (Svobodová et al., 2016). In addition, up-regulation of miR-1973 (on 4q26) and miR-3196 (on chromosome 20) in trisomic placentas exhibited perturbation of the regulation of genes contributing to the development of the nervous system (Lim et al., 2015). However, the synergistic or antagonistic interaction of the non-HSA21-miRNAs with HSA21-miRNAs has not been fully elucidated. Notably, overexpression of three miRNAs including miR-27b (on 9q22.32), miR-27a (on 19p13), and miR-329 (on chromosome 14) and overproduction of seven proteins, including growth factor receptor-bound protein 2, thymosin β10, RuvB-like 2, mitogen-activated protein kinase 1, tyrosine-protein phosphatase nonreceptor type 11, α -actin-2, and protein tyrosine kinase 2 were reported in cord blood samples from normal and DS fetuses (Lin et al., 2016; reviewed in Brás et al., 2018). Additionally, astroglial cells within the hippocampal white matter of DS fetuses exhibited a higher level of miR-146a (on 5q33.3) expression compared to normal fetuses, and notably, this elevated expression persisted during postnatal development (Arena et al., 2017). This expression level of miR-146a was thought to be an important determinant of neuronal development, especially in DS children (Nguyen et al., 2016). However, type of tissue and developmental stage at study point could underlie the variable expression of miRNAs. Nevertheless, differential expression of non-HSA21 miRNAs and associated proteins are expected to contribute to DS pathogenesis.

Overexpression of other non-HSA21-miRNAs such as miR-214 (on 1q24.3, in intron 14 of the Dynamin-3 gene (DNM3)) and miR-223 (on Xq12) was observed in the hippocampus of adult Ts65Dn mice (Keck-Wherley et al., 2011); however, expression profiles of these miRNAs have not been fully characterized with DS phenotypes. Nevertheless, miR-214 was proposed to regulate the synaptic proteins, such as cortactin-binding protein 2 (CTTNBP2) and ankyrin2 (ANK2) (Paschou et al., 2012). The miR-124 has also been reported to repress the expression of sodium-calcium exchanger1 (Ncx1) (Aurora et al., 2012). The Ncx1 is demonstrated to be crucial for compartmentalization of the spine-dendrites and in the recovery of presynaptic calcium mechanism in cerebellar parallel fiber (Roome et al., 2013; Siew et al., 2013). The miR-223 has been reported to regulate the glutamate receptor subunits NR2B and GluR2. Interestingly, its overexpression caused inhibition of NMDA-induced calcium influx in hippocampal neurons (Harraz et al., 2012). Therefore, targeting downstream gene of both miR-214 and miR-223 was hypothesized to have critical roles in regulation of neuronal development and function; however, a direct link of these miRNAs with neuronal dysfunction in DS and their targets could not be ascertained yet.

Dysregulation of the nuclear factor of activated T-cells (NFAT) pathway by the HSA21-encoded DYRK1A has been an established incidence in DS patients (Arron et al., 2006, detailed in Chapter 6). Remarkably, both mouse chromosome 9-derived miR-199b and HSA2-derived miR-1246 were predicted responsible for down-regulation of DS orthologous MMU16-encoded Dyrk1a and human DYRK1A leading to induction of apoptosis through NFAT pathway (Da Costa Martins et al., 2010). Dysregulated NFAT signaling pathway has been observed in AD (Hudry et al., 2012; Siew et al., 2013). Considering the association of DYRK1A and NFAT signaling with AD, further investigations on the miR-199b:Dyrk1a and miR-1246:DYRK1A interactions are expected to further enlighten the pathomechanism of AD mediated by these miRNAs in individuals with DS.

Furthermore, microarray-based expression profile of miRNA in hippocampal tissues from DS fetuses reported expression of miR-138-5p (on 3p21. 3), and the report hypothesized that the function of miR-138-5p in the hippocampus and the down-regulation of its target, EZH2 (enhancer of zeste homolog 2), might be involved in neurological deficiency resulting in impairment of intellectual ability in DS patients (Shi et al., 2016). In addition, possible involvement of interleukin (IL)-1β, IL-12 receptor subunit β2, autism susceptibility candidate 2 (AUTS2), and KIAA2022 have been reported to cause atrioventricular septal defect (ASD) in DS patients (Wang et al., 2016); thus, their expression was proposed to be targeted by miR-518a, miR-518e, miR-518f, miR-528a, and miR-96. Further investigations, in this direction, may be useful for better understanding of the expression of DS phenotypes and their management accordingly.

6. Drug targets of miRNAs

Individuals with DS are cognitively impaired, which progresses to dementia following development of AD. However, wide variability in the level of cognitive ability creates a real problem in therapeutic assessment of potential druggable candidates. Therefore, an attempt on development of a validated devise could produce the Arizona Cognitive Test Battery (ACTB) for reproducible assessment of the cognitive and language skills in DS patients of different ages. The ACTB instrument was capable to measure the overall cognitive ability in the context of hippocampal and cerebellar function and its cross-site usage, data-driven consistency, and precise phenotypic profiling (reviewed in Mahernia et al., 2021a,b). ACTB was endorsed for a clinical trial with 74 DS participants and 50 mental age-matched controls (Kreutzer et al., 2013). The study had recommended inclusion of DS patients of less than 40 years age for cognitive assessment in preclinical anti-AD drug therapy because cognitive ability progressively declines with age after the age of 40 years.

Restoration of intellectual abilities has been the major focus while exploring the promising strategies for development of therapeutics for DS individuals. Importantly, HSA21-miRNAs play crucial roles in DS-associated dementia and leukemia. Thus, HSA21-miRNAs are considered as putative therapeutic targets, possibly using chemically engineered oligonucleotides, named "antagomiRs" as useful pharmacological tool for the treatment of DS. In this direction, both advantages and drawbacks of miRNA-targeted drug discovery have been debated with variable outcome (Mahernia et al., 2021a). Overexpression of HSA21-miRNAs was thought to disturb several important signaling pathways in a complex situation of dysregulation via the co-operating effects of HSA21-encoded overexpressed genes and non-HSA21 disomic genes, leading to silencing of several mutual or nonmutual proteins. Thus, modulation of HSA21-encoded miRNAs was targeted for amelioration of the translation of several underexpressed proteins linked to DS-related intellectual performance (Van Meter et al., 2020).

Selective inactivation of HSA21-miRNAs was considered for reducing clinical complications or treatment of DS. Silencing of miR-155 or miR-802 through intraventricular injection of selective mir-antagonists (antagomiR) resulted in normalization of appropriate miRNA, MeCP2, CREB1, and MEF2C expressions in vivo. The result suggested that incorrect repression of MeCP2 might cause the neurochemical abnormalities in the DS brains, which might be secondary to trisomic overexpression of HSA21-derived miRNAs (reviewed in Mahenia et al., 2021a).

Gene edition of the HSA21 by CRISPR-Cas9 technology has hypothesized the possibility of using blood-brain-barrier permeable nanoparticles for encapsulation of the CRISPR-Cas9 protein cocktails (Tafazoli et al., 2019). Indeed, recent approaches of exploitation of nanoparticles have become the central issue of drug development. Noticeably, nano-formulation of HSA21-miRNAs could be the future modality of drug development and delivery for assessing the AD factors in clinical trials on DS population.

Overexpression of miR-155 and miR-125b was reported in peripheral blood mononuclear cells (PBMCs). Antagonistic modulation of miR-155 and miR-125b with antagomiRs was targeted on tonsillar memory B cells in DS children (Farroni et al., 2018). Use of activation-induced cytidine deaminase (AID) protein, a miR-155 target, on sorted trisomic cells revealed significant decrease of miR-155 in memory B cells in vitro. miR-155 was significantly overexpressed in CpG-activated PBMCs, where miR-125b appeared constitutively higher than normal. In addition, use of an anti-miR-155-regulated miR-155 abundance could return the complement factor H (CFH) to homeostatic levels (Li et al., 2012). Indeed, the significant correlation between miR-155 upregulation and CFH downregulation regulates the innate immune response. Interestingly, alteration of the expression of both miR-155 and miR-125b was observed in B cells of DS individuals both in vivo and in vitro; however, further research was suggested on identification of potential anatagomiRs for treatment of DS.

Decrease in sorting nexin 27 (SNX27), a vital protein for memory and learning, leads to synaptic dysfunction in the brain. Interestingly, inhibition of overexpressed miR-155 could upregulate the SNX27 protein in the brains of trisomic mice resulting in rescue of cognitive and synaptic deficits (reviewed in Mahernia et al., 2021a). Additionally, therapeutic application of the Argonaute 2 (Ago2) protein facilitated RNA-mediated gene silencing by silencing of the target mRNA. The bi-lobed structure of Ago2 contains the amino-terminal and PIWI-Argonaute-Zwille (PAZ) domains in one lobe, while the other lobe contains the middle (MID) and P-element-induced wimpy testis (PIWI) domains (reviewed in Mahernia et al., 2020). The PAZ and MID domains bind to the guide RNAs in their 3′ and 5′-phosphate ends, respectively. The limited paring interactions between the "seed region" of miRNA and the target mRNAs in the 3′-UTR end help Ago2 to recognize the complementary sequence. The seed region of miRNA includes nucleotides two to seven or eight from the 5′-end, where the PAZ domain acts for holding the 3′-end of miRNA. The mid-domain binds to the 5′ side and anchors the miRNA (reviewed in Mahernia et al., 2020). Ago2 is found in the nucleus of all postnatal mammalian cells and correlates with other processes beyond posttranslational regulation (Benhamed et al., 2012). Notably, Ago2 interacts with chromatin modifiers and also participates in the DNA repair process (reviewed in Mahernia et al., 2020). In addition, these proteins can increase the stability of miRNAs and also acts as a Dicer-independent RNA slicer. Interestingly, overexpression of the Ago2 protein was described in several types of carcinoma and tumorigenesis (Ye et al., 2015).

The interaction of Ago2 has been reported with the nucleoside (uridine) of miR-155 (Mahernia et al., 2020). A potent small molecule inhibitor was rationally selected following the virtual screening of the OTAVA_CNS_library against the miR-155/Ago2 complex. The molecular dynamic simulations have demonstrated the pharmacological activity of the selected small molecule under the influence of virtual mutations, including Thr526 to isoleucine and Gln545 to alanine. Involvement of Thr526 and Gln545 appeared as essential contacts between the miR-155 and Ago2. Thus, Ago2/miR-155 complex structure acts in the main cellular processes such as proliferation, differentiation, and cellular death (Masciarelli et al., 2014); it was suggested as a model for identification of feasible and approachable drug targets (Mahernia et al., 2020). Indeed, designing small molecules via inhibition of the RISC loading process has been attempted in different ways (Tan et al., 2011). Noticeably, the compounds mimicking the mRNA structure are able to bind the seed region of miRNA within the active site of Ago2 (Schmidt et al., 2013). In addition, another strategy to inhibit the activity of Ago2 has been targeted with cell-permeable small molecules (Masciarelli et al., 2014). Further study revealed the binding and inhibition of pre-miRNAs with few peptides and linear peptides (poly-N-substituted glycine). A diverse library was designed against affinity binding to pre-miRNA, synthesized with combinatorial macrocyclic γ-AA-peptides and screened using the fluorescence polarization (FP) assay (Teng et al., 2016).

The pharmacological output of the targeted small molecules was examined through the influence on the downstream regulations of miR-155, for identification of a pre-miR-155 binder that can inhibit oncogenic miR-155 maturation in vitro and in cell, and induce apoptosis in cancer cells (Yan et al., 2019). The investigation demonstrated the potential of the macrocyclic peptidomimetics to serve as new scaffold for RNA targeting. Another study reported that the transcription of a cluster of let-7c/99a/125b-2 miRNAs could be suppressed through the androgen-induced repression at the transcriptional level, showing binding of the androgen to the cluster host gene as a receptor and then recruitment of the polycomb protein EZH2 (Sun et al., 2014). In addition, inhibition of the overexpressed miRNAs, including the miR-155, let-7c, miR-99a, and miR-125b-2, was tested with small molecules accompanied with the double-verified upregulation of downstream proteins (Mahernia et al., 2021b). Therefore, this pharmacological testing has directed translational research for seeking new therapeutic targets with less off-site toxicity and more specificity and ability. Identification of specific targets during initiation of biogenesis of miRNAs through transcription of their gene in the nucleus may lead to more impactful therapeutic outcome on phenotypic manifestation. However, due to lack of information regarding mechanism of action tailored to DS model, both in vitro and in vivo, the pharmacological inhibition of these HSA21-miRNAs challenges the translatability of these tests, and thus, further complicates the task of therapeutic developments.

Overexpression of the miR-155 contributes directly and indirectly to the onset of AD or its progression in DS. Since the miR-155 has potential to simultaneously reduce the translation of several genes at posttranscriptional levels, targeting the miR-155 was thought of for identification of anti-AD therapeutic candidates. In this respect, recognition of novel small molecules and computational drug design approaches and in vitro studies for selective miR-155 processing were aimed for inhibiting the miR-155 function or its biogenesis and maturation.

Interestingly, a series of imidazo[1,2-a]pyridine derivatives was used as new nonpeptic candidates for their BACE1 and miR-155 binder inhibitory potential in vitro and in vivo for the treatment of DS with AD (Mahernia et al., 2021b). The compounds exhibited potential as an inhibitor of BACE1 and miR-155. Truly, systematic computational approaches, including 3D modeling, docking-based virtual screening, and molecular dynamics simulation would help in identification of small molecule inhibitors of pre-miR-155 maturation.

7. Perspectives of microRNAs in prenatal diagnosis of DS

Detection of the encapsulated miRNAs in exosomes circulating in extracellular medium is possible in multitude of tissues, including in prenatal diagnostic samples (Mahernia et al., 2021a). Most of the placental miRNAs are released in the blood circulation of pregnant woman via exosomes (Luo et al., 2009). Exosomes may play a significant role in the intercellular communication in placentation; the formation of the vascular system between the mother and fetus and the induction of maternal immune tolerance to the fetus (Ludwig and Giebel, 2012). Therefore, exosomes were targeted as the potential miRNA biomarkers for early NIPT of various pregnancy complications, including those connected to placentation and preeclampsia (Salomon et al., 2017). Interestingly, promising results extracted from the circulating nanoparticles were considered for assessment of miRNAs in young individuals with DS and their siblings (Salvi et al., 2019). Since DS pregnancies exhibit abnormal placentation, study of miRNA in maternally circulating exosomes released from such impaired placenta could be considered as promising markers for NIPT of DS pregnancies (Zedníková et al., 2020).

Upregulated miR-99a, miR-542-5p, miR-10b, miR-125b, miR-615, let-7c, and miR-654 miRNAS were demonstrated in placentas with DS, suggesting perturbation of several genes responsible for intercellular communication (e.g., connexin 1), cytotrophoblast cell adhesion (cadherin 11) or syncytiotrophoblast differentiation (syncytin-1 and 2) and, thus, affect physiological consequences of placental development in DS (Svobodova et al., 2016). There are several reports on comparison of miRNA levels in the plasma of pregnant women bearing euploid and DS pregnancies; however, variability in technologies used and lack of a harmonized protocol have yielded variable results on number and type of miRNA expression in plasma of DS pregnancies (reviewed in Svobodova et al., 2016). Nevertheless, the technique and the results collected on miRNA determination in plasma from pregnant women is not recommended for noninvasive prenatal testing (NIPT) of fetal DS.

The screening of specific miRNAs markers attributing to DS pathology would be helpful in prenatal diagnosis of DS and its management accordingly. The miRNAs are being increasingly considered as a novel noninvasive maternal biomarker for prenatal diagnosis of DS by employing methods such as microarray, deep sequencing, and quantitative RT-PCR in maternal whole blood. Three mechanisms were proposed on the entry of the fetal miRNAs into the maternal blood circulation: 1. cellular release of exosomes, apoptotic bodies, and microvesicles that contain waste materials and extracellular nucleic acids; 2. the intervillous space sustains the placenta with nutrients and oxygen; and 3. the connection between placenta, mother, and fetus (reviewed in Mahernia et al., 2021a). Noticeably, among the extracellular nucleic acids, miRNAs appear to be more stable in maternal blood due to the resistance against various pH conditions and enzymatic degradation. Thereby, the HSA21-miRNAs are proposed to constitute a potential biomarker for diagnosis of DS and other diseases. Evidently, the maternal plasma level of miR-99a and miR-3156 was significantly higher in pregnancies with DS (Kamhieh-Milz et al., 2014). Of the 695 miRNAs identified, 36 differentially expressed mature miRNAs were identified as potential biomarkers. The differentially expressed miRNAs were enriched in signaling pathways such as mucin type-O-glycans, ECM-receptor interactions, TGF-beta, and endocytosis, which have been previously associated with DS (Kamhieh-Milz et al., 2014). By evaluating all 36 miRNAs using the Diana mirPath tool, the detected KEGG pathways accurately reflected many known pathomechanisms of DS. Additionally, miRNAs may characterize the severity of DS disabilities and may recruit antagoMirs for treating DS features in the future. However, increased levels of miR-99a, let-7c, miR-125b-2, and miR-155 in pregnant women compared to nonpregnant women could not display marked difference in these miRNAs between euploid and DS pregnancies (Kotlabova et al., 2013). Therefore, HSA21-miRNAs are not recommended as reliable biomarkers for noninvasive diagnosis of pregnancies with DS as of now.

8. Spectrum of epigenetic expression in DS

Among the most important epigenetic regulations, DNA methylation is the best-known mechanism. Evidently, hypermethylation of cytosine phosphate guanine (CpG) sites within gene promoter regions can reduce specific gene expression, which consequently contributes to development of specific clinical phenotypes such as Angelman and Prader–Willi syndromes, Beckwith–Wiedemann syndrome, and so on. In this context, constitutional DNA hyper/hypomethylation in DS can alter the expression of the complex DS genome during embryogenesis and also can affect organogenesis even if not closely matched with embryogenesis-specific transcriptome, proteome, and metabolome alterations. Noticeably, some methylation patterns can persist after birth. Consistent with this, methylation patterns detectable in blood mononuclear cells may indicate the biological consequences of development of CHD in DS. Importantly, various tissues of individuals with DS appeared with aberrant DNA methylation at different ages (Lu et al., 2016; Henneman et al., 2018).

The interface of environmental factors and the individual genome depicts the epigenetic mechanisms and its influence on regulation of development, differentiation, and establishment of cellular identity, and consequently on various diseases. In the postgenomic era, *cis* (involves short-range effects of single-nucleotide polymorphisms (SNPs) and haplotypes) and *trans* (involves the effects of chromosomal aneuploidy on DNA methylation of CpG sites) genetic–epigenetic interactions are gaining increasing attention in DS research. In presence of trisomic HSA21, the functional impact of epigenetic consequences may significantly contribute to the clinical phenotypic variation in DS. The epigenome-wide profiling of DNA methylation in several tissues of DS, including leukocytes, skin fibroblasts, buccal cells, blood, placenta, and brain, demonstrated evenly distributed differentially methylated regions (DMRs) on all chromosomes, instead of enrichment of only HSA21 genes (El Hajj et al., 2016; Jin et al., 2013; Jones et al., 2013; Kerkel et al., 2010; Hovrath et al., 2015). Remarkably, global DNA hypermethylation, with only a partial overlap of DMRs, was observed in the cortex and placenta of DS (El Hajj et al., 2016; Jin et al., 2013). Indeed, both hypermethylation and hypomethylation events were described in DS samples, with higher hypermethylation in brain regions and placenta (Bacalini et al., 2015; Mendioroz et al., 2015; Jin et al., 2013). Notably, cell-free fetal DNA (cffDNA) in maternal blood represents the chorionic villi cells (e.g., trophoblasts), and therefore, DNA methylation profiling in cffDNA via analysis of fetal DS-specific DNA methylation patterns is being increasingly considered for NIPT of DS in the first trimester of pregnancy in maternal blood (Ganguly and Kadam, 2019). Importantly, epigenetic patterning in DS is cell type-specific, and hence, methylation profiles of blood cells will not represent the other organs.

Mechanistically, DNA methylation involves addition of a methyl group to the 5′ carbon of the cytosine pyrimidine ring from a methyl donor S-adenosyl methionine that usually occurs in a CpG dinucleotide site. The DNA methyl transferase (DNMTs) enzymes, including DNMT1, DNMT3A, DNMT3B, and DNMT3L, activate this reaction. Interestingly, this DNA methylation reaction can be reversed by DNA demethylation enzymes, such as AID, APOBEC, and TETs. DNA methylation itself physically obstructs binding of transcriptional proteins to the gene, facilitating the binding of the methylated DNA with proteins as methyl-CpG-binding domain (MBD) proteins. Practically, several proteins are responsible for the regulation of DNA methylation dynamics such as chromatin readers of 5-methyl cytosine (5mC), methylation-sensitive transcription factors, and Ten-Eleven Translocation (TET) family of DNA hydroxylases. Notably, oxidation of 5mC to 5-hydroxymethylcytosine (5hmC), 5-formylcytosine (5fC), and 5-carboxylcytosine (5caC) is catalyzed by the TET enzymes (TET1, TET2, and TET3), which can act as distinct epigenetic marks or represent intermediates of DNA demethylation. Interestingly, TET-mediated active DNA demethylation is involved in DNA repair pathway mediated by thymine DNA glycosylase (TDG) to remove the unnecessary oxidized 5mC residues (Schübeler, 2015). With recruitment of additional proteins such as histone deacetylases and other chromatin remodeling proteins to the locus, MBD can modify histones. Alterations in DNA methylation, substantially occur during stem-cell differentiation, are generally explained as epigenetic effects. Notably, global DNA methylation observed in placental villi and adult buccal epithelial cells has been attributed to cognitive impairment in DS (Kerkel et al., 2010; Jones et al., 2013; Jin et al., 2013). However, involvement of genes in ubiquitination pathway for neurological phenotypes remains unelucidated.

Various CpG sites were differentially methylated between euploid and DS fetuses, which were mostly hypermethylated in DS across the whole genome. A DNA methylation profiling at single-base resolution covering approximately 3.2 million CpG sites of the human genome in the chorionic villi cells of human DS and euploid fetuses identified new DMRs across the whole epigenome at the early developmental stage in DS (Lim et al., 2019). Notably, most differentially methylated CpG sites (DMCs) were hypermethylated in DS, while hypomethylation accounts for only ~4%. Remarkably, the functional annotation of identified DMR genes indicated significant association of the hypermethylated DMRs with various congenital DS phenotypes and DS pathophysiology, including chromosomal aberration of DS (Lim et al., 2019). This finding highlighted the possibility of fetal epigenetic therapy through inhibition of the global DNA methylation in DS.

Assessment of CpG regions and promoters in trisomic neural induced pleuripotent stem cells (iPSCs) derived from DS revealed a total of 500 stable and differentially methylated CpGs that were annotated to CpG islands of 151 genes across the genome (Laan et al., 2020). The enriched genes of the DNA binding category recognized 37 important factors involved in transcriptional regulation and chromatin structure, including the provincial epigenetic changes of the transcription factor genes such as ZNF69, ZNF700, ZNF763, and also the HOXA3, HOXB3, and HOXD3 genes. The chromatin remodeling was evidenced through the similar clustering of differential methylation in the CpG islands of the HIST1 genes. Thus, a set of genes in neural iPSCs of DS origin displayed differential methylation, which is linked to DS brain development (Laan et al., 2020). However, further studies appeared essential for providing a novel framework of epigenetic changes and transcriptional dysregulation during neurogenesis in DS.

Interestingly, global hypermethylation appeared conserved in relation to different tissue types and developmental stages in DS, indicating variable number and location of DMCs in the whole genome (El Hajj et al., 2016; Jin et al., 2013; Jones et al., 2013; Kerkel et al., 2010; Hovrath et al., 2015; Lu et al., 2016). Seemingly, variability of hypermethylation in DS was documented as following (Lim et al., 2019): eight genes (TMEM131, TCF7, CD3Z/CD247, SH3BP2, EIF4E, PLD6, SUMO3, and CPT1B) in peripheral blood leukocytes of adults; nine genes (C1orf35, CPT1B, DECR2, FAM83H, GLI4, LRRC14, LRRC24, STK19, and TNXB) in the fetal cortex; three genes (CPT1B, TCF7, and FAM62C) in the fetal placenta; and hypermethylation in the 5′UTR region of CPT1B gene in chorionic villi cells at an early developmental stage of fetuses. Indeed, hypermethylation of the CPT1B (located on HSA22; one of the carnitine palmitoyltransferase 1 genes) is conserved as documented in various tissue samples of different developmental stage of DS, and thus, this may be considered as a common epigenetic marker in DS (Lim et al., 2019). Since this enzyme is required for the transport of long chain fatty acyl-CoAs from the cytoplasm to the mitochondria (El Hajj et al., 2016) and profound mitochondrial dysfunction has been reported in DS (illustrated in Chapters 8 and 9), placental mitochondrial dysfunction during gestation could also be critically contributing to health of the mother and fetus/offspring. Therefore, hypermethylation of CPT1B may be considered as an epigenetic marker of mitochondrial dysfunction in the developing DS fetus.

In addition, more than 40 hypermethylated DMCs were demonstrated indicating the methylation degree of ZNF837, C11orf95, KIAA1875, SHROOM1, and CASZ1 genes by methylation-specific real-time PCR and methyl-capture sequencing (MC-Seq) (Lim et al., 2019). Notably, SHROOM1 and CASZ1 are involved in pathophysiology of DS. SHROOM family members are linked to the development of nervous system and other tissues and may also be involved in the assembly of microtubule arrays during cell elongation. SHROOM1 contributes to atrial septal defect (Lim et al., 2019). A marked hypermethylation of SHROOM1 in all differentially methylated sites was reported in individuals with knee osteoarthritis. Indeed, the incidence of heart and joint problems was higher in DS patients. Therefore, it is apparent that hypermethylation of SHROOM1 may be potentially associated with the pathophysiology of the CHD and osteoarthritis in DS (Lim et al., 2019). Collectively, the multifaceted involvement of SHROOM1 indicated hypermethylated SHROOM1 as a promising candidate gene for deciphering its functional involvement in osteoarthritis pathology, which in turn can be considered as a potential target for therapeutic development.

Furthermore, adult and fetal brain and the fetal placenta in DS displayed hypermethylation of ZNF837 (mapped onto chromosome 19, encodes zinc finger protein 837). Notably, the chorionic villi cells of DS fetuses exhibited ZNF837 as the DMR gene with the largest DMCs in the early developmental stage (Lim et al., 2019). ZNF837 gene is located in the nucleus and may be involved in transcriptional regulation. Additionally, the C11orf95 (chromosome 11 open reading frame 95) has been characterized with ependymoma as tumors of the brain and spinal cord (Lim et al., 2019). Spontaneous translocation of C11orf95–RELA fusion protein activates the nuclear factor-kappa B (NF-κB) target genes in the nucleus and rapidly transforms the neural stem cells, and the precursor cells of ependymoma tumors (Parker et al., 2014). However, ependymoma has been reported in two patients with DS (reviewed in Lim et al., 2019). Noticeably, the epigenetic function, protein-coding potential, and mechanism of KIAA1875 gene (previous HGNC symbols for WDR97 and encode WD Repeat Domain 97) have yet to be discerned with further studies on DS population. Collectively, it is imperative that continuous epigenetic changes may occur in the whole genome from embryos early in development in DS because of the chromosomal aneuploidy forms the basis of the developmental defects. Moreover, alterations in potential pathways owing to epigenetic changes remain to be further elucidated in DS patients.

In general, gain of methylation in the promoter/enhancer region of the RUNX1 gene (HSA21 encoded, a hematopoietic transcription factor) has been reported with DMRs in whole blood leukocytes, including T cells of DS patients compared to control. An epigenome wide association study (EWAS) in dried bloodspots of newborns with DS has described 710 CpGs, of which CpGs in the RUNX1, and FLI1 of HSA11 (plays a critical role in megakaryopoiesis) were significantly hypermethylated (Wiemels et al., 2019). Additionally, of the 847 DMRs identified, the RUNX1 DMR containing 16 CpGs displayed an overlap with a superenhancer in hematopoietic stem cells (HSCs). Notably, addition of sex combs like transcriptional regulator 1 (ASXL1, located on 20q11.21) encodes a chromatin-binding protein and is involved in epigenetic regulation of hematopoietic cells, primarily for development and differentiation of normal hematopoietic cells, recruits PRC2 to specific loci and in maintenance of activation and silencing of development-related Hox genes through chromatin conformational changes (Ganguly, 2017). Interestingly, ASXL1 mutations are frequent in patients with RUNX1-RUNX1T1 rearrangement and seem to be correlated to the presence of RUNX1 mutations resulting in adverse prognosis of diseases. In this direction, a complex association of ASXL1 with triplicated RUNX1 gene is likely, which can contribute to DS leukemia. Interestingly, the blood components with lower B-cells, CD4 T-cells, and monocytes, and higher CD8 T-cells and nucleated red blood cells (nRBCs) in DS compared to non-DS newborns may indicate transient abnormal myelopoiesis in DS neonates (Wiemels et al., 2019); however, further investigation is suggested for better understanding of the problem. A functional enrichment and network analysis revealed that RUNX1, inhibitor of DNA binding 4 (ID4) (involved in neuronal differentiation and transcriptional suppression), and interaction among other genes, including nuclear receptor subfamily four group A member 2 (NR4A2)-early growth response (EGR)2 and NR4A2-EGR3 might play crucial role in DS pathogenesis (Zhang et al., 2016).

8.1 Epigenetics of ncRNAs

The ncRNAs involved in epigenetic processes are broadly named as short ncRNAs (with <30 nucleotides (nts); miRNAs, siRNAs, piRNAs) and lncRNAs (>200 nts), which are involved in heterochromatin formation, DNA methylation targeting, histone modification, and finally silencing of gene expression. Approximately, 35,000 lncRNA transcripts have been reported in >10,000 loci in human genome that is involved in 5′ capping, splicing, and polyadenylation, but have little or no role in open reading frame (ORF) (reviewed in Ganguly, 2017). Understanding involvement of ncRNAs in other nondevelopmental mechanisms and oncogenesis is at an incipient stage.

Identification of novel pathways will be challenging to help in manipulation of lncRNAs by targeted therapeutics for improvement of erythropoiesis and other hematopoietic lineages. Interactive modulation of expression of oncogenes and enhancement of tumor suppressors, cell cycle regulators, and the resultant proteins may be targeted for designed development of epigenetic modifiers. LncRNAs are targeted for its specific folding conformations, which could facilitate surface manipulation by other RNAs, peptides, and small molecules. Taken together, further understanding of ncRNA functions may eventually identify pharmacologic targets for modulation of pathological gene expression.

9. DS phenotypes associated with epigenetic modulation

Methylation of cytosine residues in the CpG sites can be influenced by myriads of environmental stressors such as pollution, food habits, body mass index (BMI), periconceptional maternal supplementation, in utero environment, maternal genotype and variants, lifestyle factors, etc., and physiological mechanism (Bik-Multanowski et al., 2018). In pregnancies with DS, deficiency in maternal folic acid and genetic variations of methylene tetrahydrofolate reductase gene (MTHFR) folate pathway genes, and DNA methylation is implicated in development of DS phenotypes, including CHD (reviewed in Babić Božović et al., 2019; Ganguly and Kadam, 2016b). Maternal hypomethylation of the long interspersed nucleotide element-1 (LINE-1), a surrogate measure of global DNA methylation and also a potential biomarker of several diseases, has been reported to increase the risk of CHD (particularly septal defects) with preponderance in DS children (Babić Božović et al., 2019).

Reduced DNA methylation of LINE1 was remarkably associated with the MTHFR 677T genotype, low-folate diet, and high BMI in mothers of DS pregnancies affected with CHD (Cai et al., 2014; Babić Božović et al., 2019). Higher BMI and/or obesity are linked to maternal risk of low folate level in the blood and undiagnosed diabetes, which in turn is responsible for CHD (Hobbs et al., 2010). Consistently, higher risk of CHD in individuals with DS has been linked to folate metabolism and the resulting epigenetic modifications (Elsayed et al., 2014; Coppedè, 2015). Periconceptional maternal folate supplementation is suggested to neutralize the impact of the MTHFR C677T polymorphism and reinstate normal enzyme activity, thus, protect CHD in offspring. Notably, a significant association with LINE-1 hypomethylation level was reported with maternal smoking, medication, alcohol use, age, dietary folate intake, and MTHFR C677T genotype (Babić Božović et al., 2019). However, accurate data on pre- and postnatal incidence of CHD in pregnancies with DS may alleviate the inconsistent findings on the association of maternal LINE-1 methylation with CHD in DS children. In this direction, multifactorial analysis of myriads of endogenous and exogenous maternal factors could discern the association of maternal LINE-1 DNA methylation with the events of pathologies in their children.

The NRG1 gene is important for heart development. Its elevated methylation and the corresponding reduced activity of three functionally connected genes, such as ERBB3, SHC3, and SHC4, were expected to be associated with CHD in children with DS (Dobosz et al., 2019). Notably, all of the above genes are key components of the ErbB/EEGF pathway that is crucial for endocardial cushions development (Rupert and Coulombe, 2015). However, the role of constitutive hypermethylation of NRG1 gene promoter and the corresponding altered expression of ERBB3, SHC3, and SHC4 genes on development of endocardial cushion defect-related heart defects in individuals with DS requires further exploration and replication in trisomic mouse models.

CASZ1 is expressed in several developing tissues, including neural, endothelial, and cardiac tissues, and aberrant regulation of CASZ1 could potentially contribute to cognitive defects, CHD, hypertension, and many others (reviewed in Lim et al., 2019). Reportedly, the CASZ1 function interacts with HSA21-located resident-congenital heart disease 5 (CHD5) protein, and thus, CHD5-CASZ1 interaction was demonstrated to be necessary for cardiac morphogenesis (Liu et al., 2014). Nevertheless, the precise mechanism of CASZ1 and CHD5 interaction has yet to be unraveled. To summarize, a potential link of the misregulation of CASZ1 with one or more DS pathologies was suggested. CASZ1 is also involved in regulating the cellular adhesion in multiple tissues and retroperitoneal sarcoma and retroperitoneum carcinoma. Hypermethylation of CASZ1 in DS may extract additional inputs relating to the DS-phenotypes contributed by its misregulation. CASZ1, encoded on chromosome 1, is a protein coding gene of castor zinc finger 1 (a zinc finger transcription factor).

Constitutive overexpression of HSA21-harbored DNMT3L, which encodes a methyltransferase, has been characterized in several investigations. The DNMT3L can stimulate the methylation activity of DNMT3A/3B, which has been associated with regulation of gliogenesis, neurogenesis, and synaptic plasticity through regulation of expression of related genes, such as Bdnf, Reln, Dlx2, Gbx2, Sp8, and Stat1 (reviewed in Lu et al., 2016). Therefore, such overexpression in DS was expected to cause abnormal neurodevelopment. Indeed, DNMT3L overexpression in neuroprogenitors caused potential increase in methylation of the ubiquitin associated TSPYL5 gene in the promoter region through DNMT3B. DNMT3L downregulates its mRNA expression, and consequently upregulates mRNA expressions of TP53 and APP (effectors of TSPYL5) presumably via TP53 (Lu et al., 2016). Thus, TP53 has been implicated to alterations in APP and PSD95, and consequently, alter the associated phenomena such as proliferation, survival, and synaptic connectivity of neural cells, cell death, and DS associated dementia and mental retardation. Interestingly, TSPYL5-mediated suppression of TP53 expression occurs by reducing the activity of a deubiquitylase for TP53, USP7, in ubiquitination mechanism (Epping et al., 2011). Thus, down-regulation of TSPYL5 mRNA by DNMT3L overexpression will enhance TP53 expression and influence apoptosis (a crucial endophenotype factor during cortical development in DS) during early cortical development (Lu et al., 2016). Notably, the mechanism of potential influence of DNMT3L overexpression on increasing APP could be mediated directly or indirectly by ubiquitination/deubiquitination genes, which is not deciphered yet, but it increases the toxic effects of this gene and gliosis. DNMT3L overexpression led to elevation of APP and PSD95 expression in differentiating neurons; however, DNMT3L-shRNA could partially save the up-regulation of APP and PSD95 in DS cells. Interestingly, inhibition of DNMT3L rescued DS-associated impairments. These results explained the mechanistic insights into epigenetic changes in DS, and modification of associated genes of DS neural endophenotype. Notably, the mechanism of this action might help in identification of potential drug candidates. The PBMCs obtained from DS showed a downregulation of DNMT3A (Ciccarone et al., 2018). A decrease in expression of DNMT3A and DNMT3B has been reported in DS brain (El Hajj et al., 2016); however, no difference was observed for DNMT3A in iPSCs derived from the discordant twins with DS (Sailani et al., 2015). Furthermore, deregulation of DNMT3A and other DNMTs has been demonstrated in several trisomic tissues during aging process (Ciccarone et al., 2018).

The Dnmt3l gene is present in three copies in the Dp(10)1Yey mice but only in the normal two copies in the Dp(16)1Yey and Ts65Dn lines of DS models. Notably, hypermethylation of the differentially CpG methylated genes appeared stronger in Dp(10)1Yey than in Dp(16)1Yey mice of DS model (Mendioroz et al., 2015). Differentially methylated CpGs in Ts65Dn and Dp(16)1Yey mice showed a significant enrichment of binding sites for the MMU16-associated transcription factors Olig2, Gabpa, Erg, and Ets2 (Do et al., 2017).

Remarkably, the spine dysgenesis in cortex, which is characterized with sparse, small, short stalks combined with unusually long spines in early development, demonstrated that both increased and/or decreased expression of PSD95 might cause cognitive impairment in DS (reviewed in Lu et al., 2016). Similar effect of cognitive impairment was implicated to decreased expression of PSD95 in Ts65Dn mice (reviewed in Lu et al., 2016). However, PSD95 expression was increased in both the cell bodies and synapses of neurons in human DS (Lu et al., 2016). The discrepancy was explained from the fact that the Ts65Dn model is trisomic with extra copies of chromosome 16 and 17 but not with chromosome 10, which contains the Dnmt3l gene. Thus, the Ts65Dn mice do not have the extra copy of Dnmt3l, unlike human DS. In addition, increased expression of PSD95 and spine density and inhibition of LTP were described in the Dyrk1a transgenic mice, which also exhibit DS-associated cognitive deficits. In addition, overexpression of Dnmt3a, an effector of Dnmt3l, can potentially increase spine density and that could be correlated with increase in PSD95 (reviewed in Lu et al., 2016). Taken together, DNMT3L is involved in regulation of aberrant gene expression in DS; however, it is likely that other factors might also be involved in mediating the epigenetic changes such as mitochondrial dysfunction and oxidative stress in DS. Notably, fluctuation in methylation such as more hypermethylated probes being hypomethylated and more hypomethylated probes being hypermethylated was suggested to be caused by proteotoxic stress from trisomic chromosomes or oxidative stress (reviewed in Lu et al., 2016).

Aberrant DNA methylation patterns were hypothesized to contribute to the heterogeneity of the DS disease. The peripheral blood mononuclear cells (PBMCs) obtained from DS subjects displayed reduced levels of 5-hydroxymethylcytosine, TET1, and other components of the DNA methylation/demethylation machinery in the MARK-AGE study (Ciccarone et al., 2018). Association of DNA hypermethylation was suggested with downregulation of TET genes and/or of REST transcription factor in placenta with DS (Jin et al., 2013). Whole blood of DS patients displayed hypomethylation of a TET1 gene regulator region (TET1 CpG island 3′-shore) compared to their siblings (Bacalini et al., 2015). Moreover, similar level of 5hmC, TET1, and TET3 in DS and the elderly population supported the premature aging features of DS (Ciccarone et al., 2018). In addition, a report has described that 5mC and 5hmC account for most of the alterations of DNA methylation in DS cerebellum, while the level of 5hmC was reduced in DS samples (Ciccarone et al., 2018). Remarkably, a positive correlation between TET1 expression and 5hmC levels has led to propose a causative role of TET1 downregulation in DNA hypermethylation in DS patients (Ciccarone et al., 2018). Indeed, many neurodegenerative diseases, including AD and multiple sclerosis, were encountered with global alterations in 5hmC levels. Practically, loss of 5hmC is a hallmark of cancer. Likewise, blood cells acquire a gradual loss of 5hmC in aging process (Valentini et al., 2016). Interestingly, an association of concomitant reduction of 5hmC and expression of TET1 and TET3 was demonstrated with aging (Valentini et al., 2016). The contribution of 5hmC was interpreted with the highest levels of DNA hydroxymethylation reported in brain tissues where the genes, which are largely related to brain development and function, exhibited differentially hydroxymethylated regions (Mendioroz et al., 2015). Additionally, trisomic PBMCs exhibited hyper-hydroxymethylation that characterizes the internal promoter region of the PR domain containing 8 (PRDM8) gene, accompanying with up-regulation of the associated transcript variant 2 (Lu et al., 2016). Thus, the result demonstrated aberrant DNA methylation patterns, which might disturb the transcriptional status of immune cells and render a global disturbance of methylation control in DS subjects. Taken together, deregulation of 5hmC was postulated to contribute to the establishment of disease-associated DNA methylation patterns, and a downstream of 5hmC alteration in PBMCs of DS can potentially influence immune-senescence (age-related clinical impairment of immune system) with higher rate of infections, hematological malignancies, and autoimmune diseases in DS patients.

10. Therapeutic targets of epigenetic expression

Knowledge has been rapidly expanding on the contribution of epigenetic dysregulation in cancer and myriads of other diseases, through the technological advancement during the past several years. Primarily, higher priority was paid to histone methylation and acetylation and targeting somatic alterations in histone-modifying enzymes or epigenetic proteins as the starting point in drug discovery. The recurrent point mutations that activate or alter protein function are targeted as the subset of early biological validation for drug development. Notably, the high-throughput investigative result on epigenetic genes, which encode enzymes, is highly likely to represent novel tractable therapeutic targets. Abnormal epigenome disturbs several pathways such as DNA modifications, cytosine methylation, modifications of histone proteins, or RNA-associated gene silencing (Ganguly and Kadam, 2016b). Indeed, somatic mutations in coding regions and enhancer elements of several genes are responsible for key epigenetic regulation, which perturb normal regulation of gene expression. The epigenetic consequences in DS predict involvement of several mutually interactive pathways, including neurological perturbations. Interestingly, epigenetic modifications are frequently reversible due to their inherent plasticity and thus, dysregulation of epigenetic mechanisms is targeted by small-molecule inhibitors as potential therapeutic approach.

Since global hypermethylation in DS neuroprogenitors is established, inhibition of DNMT or DNA deamination was targeted to rescue or treat the DS phenotypes. The plethora of knowledge gained from the therapeutic targets of myelodysplastic syndromes (MDS) and the outcome of the "first-generation" hypomethylating agents such as Azacitidine (Vidaza) and Decitabine (Dacogen), and histone deacetylase (HDAC) inhibitors (HDACi) (Vorinostat, Romidepsin) has sensitized drug discovery and development of "second-generation" targeted epigenetic agents (reviewed in Ganguly and Kadam, 2016b; Ganguly, 2017). Many other DNMT3A inhibitors, such as ASTX727, E7727 and so on, were considered for testing as potential druggable candidates (Ganguly, 2017). Indeed, combinatorial therapy of Vorinostat with standard chemotherapy (Idarubicin and Cytarabine) was proposed to yield improved response in newly diagnosed MDS and AML patients. Combination of HDACi and DNMT3A inhibitor has also been tested (Kirschbaum et al., 2014).

Epigenetic drug discovery for DS-phenotypes is at its incipient stage. Understanding the complexities of the "histone code" and associated scientific and pragmatic challenges are daunting task in the recent emerging field of drug discovery and identification of small molecules that directly and specifically silence the coding genes or increase expression of tumor suppressors. A mechanistic understanding of epigenetic dysregulation and subsequent perturbation of biological function needs to be elucidated for development of DS therapeutics. There are underlying genetic and/or epigenetic mechanisms that predispose some tumors to EZH2 enzyme inhibition. The clinical utility of DNMT3A inhibitors and pan-HDAC inhibitors as well as the rapid preclinical advancement of the second generation of epigenetic modulators guides for future epigenetic drug discovery and development.

The epigenetic regulation of PRC2 that catalyzes trimethylation of lysine 27 on histone H3 via EZH2 methyltransferase, confers stemness, and regulates differentiation during embryonic development. Given these activities of EZH2 and H3K27me3, plastic and dynamic features of cancer cells, especially cancer stem cells, were thought to be closely associated with this epigenetic mechanism (reviewed in Ganguly and Kadam, 2016b). Similar mechanism is likely for leukemogenesis in DS patients. Indeed, EZH2 has been considered as a novel target for identification of small candidate compounds that can inhibit EZH2 activity. PRC2 complexes have been used on a variety of substrates for identification of direct EZH2 inhibitors those bind directly to the protein and inhibit its enzyme activity through biochemical assays. Some of EHZ2 inhibitors have displayed greater biochemical potency and a longer enzyme-inhibitor residence time when H3K27me3 peptide is bound to PRC2. Some of the EZH2 inhibitors, such as GSK126, GSK343, EPZ005687, EPZ-6438, EPZ-5676 UNC1999, UNC2400, DZNep, EI1, E7438, and so on, were tested for therapeutic potential, primarily targeting MDS (reviewed in Ganguly and Kadam, 2016b). Nowadays, there are a number of small molecule inhibitors in consideration for clinical trial for establishing them as compelling drug targets.

In addition, development of more specific inhibitor/activator of methylation or deamination may target specific promoters of HSA21 genes of important pathways. Dysfunction of the ubiquitin proteasome/lysome system (UPLS) system causes accumulation of proteins and/or overdegradation in cellular organelles (Lu and Sheen, 2013). Thus, developing activator or inhibitor of proteasomes of the ubiquitin-proteasome pathway may directly target the subunits of proteasome (the core of the proteolysis machinery), instead of targeting upstream ubiquitination and recognition of ubiquitinated protein substrates. The proteasome inhibitors such as Bortezomib (Velcade) are approved for clinical treatment for multiple myeloma (reviewed in Lu and Sheen, 2013). Investigation on the proteasome activators, such as Blm10/PA200, is underway. Maintenance of mitochondrial function and OXPHOS pathway has been targeted with an investigational drug EPI-743 for inherited mitochondrial respiratory chain disease. EPI-743 is a small molecule, orally absorbable, and can easily reach the central nervous system. It synchronizes energy generation in mitochondria by targeting the NADPH quinone oxidoreductase 1 (NQO1) enzyme for reducing the cellular redox stress (reviewed in Lu and Sheen, 2013).

Targets with small-molecule inhibitors have brought an important breakthrough in drug discovery in the present era. Indeed, the identification of new efficacious inhibitors is a great challenge targeting molecular mutations involved in pathway machineries and disease consequences. Thus, better understanding of epigenetic modification, splicing mechanism, and genes/enzymes involved would be important with pharmaceutical value. The sensory activity of TP53 in splicing mechanism demonstrates its importance in mRNA splicing. Further extensive investigation on new pathway genes would be essential for extraction of information toward pharmaceutical development for refinement of treatment modalities of DS phenotypes in complex genetic scenario with coexistence of multiple miRNA and epigenetic expressions. However, more information on interplay of mutations on larger database would provide newer modalities of drug development with specific target agents. Interindividual variability in genetic make-up in a scenario of complex trisomic and disomic gene-dosage and multiple genetic alterations further necessitates reunderstanding of genotype—phenotype association for development of DS-specific therapeutics.

11. Concluding remarks

Down syndrome is broadly characterized with typical cranio-facial features, intellectual disability, dementia, and AD. Constitutive overexpression of miRNAs, including miR-155, miR-802, miR-99, miR-125b and let-7c, and the consequent production of haploinsufficiency of their specific target proteins are potentially involved in manifestation of DS phenotypes. Collectively, involvement of HSA21-miRNAs in DS pathobiology was demonstrated as: miR-155 and miR-802 with neuropathology; the cluster of miR-99/let-7c with CHD; miR-125b-2, miR-155, and miR-99a with AMKL; miR-155-5p with mitochondrial biogenesis; and protective role of miR-125b, miR-99, and let-7c on breast, prostrate, and lung cancers. The miR125b-2 has been considered as an oncogenic miRNA for contributing to leukemia in DS patients. In addition, interaction of HSA21-miRNAs with non-HSA21-miRNAs has a significant role in DS pathologic expression. Overexpression of miR-99a-5p, miR-155-5p, and let-7c-5p in the heart of trisomic fetuses as well as other miRNAs in several other organs in human DS, DS models, and DS-derived iPSCs have directed development of miRNA-targeted drugs for alleviating the DS features and/or reversal of the consequences of overexpressed miRNAs. Indeed, identification of antagomiRs is increasingly being considered for testing in different systems for deciphering their druggable potential.

In addition, global hypermethylation was observed in the DS genome, and expression of methylation status varies depending on the tissue of origin and developmental stage. Targeted gene analysis has demonstrated the implication of overexpressed DNMT3L on regulation of tumor suppressor TP53 in DS patients. Interestingly, the epigenetic studies in DS have indicated silencing of pathways and expression of proteins. Thus, expression profiling of epigenetic regulation of pathway genes contributes to pre-existing knowledge of oxidative stress, gliosis, and mitochondrial dysfunction in DS. In this context, identification of small molecule inhibitors has yielded promising results, in particular for management of hematopoietic malignancies. Similar target is expected to work for leukemogenesis in DS population. However, DS patients have many more clinical consequences of gene-dosage effect, and consequent expression of miRNAs and epigenetic markers. Thus, screening of "methylomes" (methylation markers) contributing to pathways, including OXPHOS, ubiquitination, neurodevelopment and so on raised the high likelihood of significant effects of DNA methylation on DS. Moreover, precise understanding of gene-dosage expression of the triplicated HSA21 in the context of miRNAs and epigenetic regulation necessitates further exploration of the independent expression and synergistic or antagonistic interaction in the complex genetic environment.

Nevertheless, the pharmacological mechanism of action of the proposed small molecule inhibitors of miRNAs and epigenetic signatures of DS genome would require appropriate test systems for targeting specific phenotypic features of DS. The most noticeable challenge in the pharmacology of identification of small molecule inhibitors is their unpredictable specificity of action in the cellular microenvironments, both in vivo and in vitro. Thus, future outreach must take into account such hurdles of targeted drug developments. Indeed, studies on preclinical DS model have achieved limited success while investigating the efficacy and safety of the niche of small molecules. Moreover, seeking the answers of insightful questions by taking care of the overlooked factors might precisely establish the translational and clinical breakthroughs related to development of DS-therapeutics for management of clinical consequences of trisomic HSA21.

References

Ahmed, A., Simon, K., Cabelof, D., 2013. Mirna-155 as modulator of DNA polymerase beta and base excision repair (ber). Environ. Mol. Mutagen. 54, S32.

Arena, A., Iyer, A.M., Milenkovic, I., Kovács, G.G., Ferrer, I., Perluigi, M., Aronica, E., 2017. Developmental expression and dysregulation of miR-146a and miR-155 in Down's syndrome and mouse models of Down's syndrome and Alzheimer's disease. Curr. Alzheimer Res. 14, 1305–1317. https://doi.org/10.2174/1567205014666170706112701.

Arron, J.R., Winslow, M.M., Polleri, A., Chang, C.-P., Wu, H., Gao, X., et al., 2006. NFATdysregula- tionbyincreaseddo-sageofDSCR1 and DYRK1Aonchromosome21. Nature 441, 595–600.

Aurora, A.B., Mahmoud, A.I., Luo, X., Johnson, B.A., Van Rooij, E., Matsuzaki, S., et al., 2012. MicroRNA-214 protects the mouse heart from ischemic injury by controlling Ca^{2+} overload and cell death. J. Clin. Invest. 122, 1222–1232.

Babić Božović, I., Stanković, A., Živković, M., Vraneković, J., Mahulja-Stamenković, V., Brajenović -Milić, B., 2019. Maternal LINE-1 DNA methylation and congenital heart defects in down syndrome. Front. Genet. 10, 41. https://doi.org/10.3389/fgene.2019.00041.

Bacalini, M.G., Gentilini, D., Boattini, A., Giampieri, E., Pirazzini, C., Giuliani, C., Fontanesi, E., Scurti, M., Remondini, D., Capri, M., Cocchi, G., Ghezzo, A., Del Rio, A., et al., 2015. Identification of a DNA methylation signature in blood cells from persons with Down Syndrome. Aging 7, 82–96. https://doi.org/10.18632/aging.100715.

Baek, K.H., Zaslavsky, A., Lynch, R.C., Britt, C., Okada, Y., Siarey, R.J., Lensch, M.W., Park, I.H., Yoon, S.S., Minami, T., et al., 2009. Down's syndrome suppression of tumour growth and the role of the calcineurin inhibitor DSCR1. Nature 459, 1126–1130.

Banzhaf-Strathmann, J., Benito, E., May, S., Arzberger, T., Tahirovic, S., Kretzschmar, H., et al., 2014. MicroRNA-125b induces tau hyperphosphorylation and cognitive deficits in Alzheimer's disease. EMBO J. 33 (15), 1667–1680. https://doi.org/10.15252/embj.201387576.

Benhamed, M., Herbig, U., Ye, T., Dejean, A., Bischof, O., 2012. Senescence is an endogenous trigger for microRNA-directed transcriptional gene silencing in human cells. Nat. Cell Biol. 14 (3), 266. https://doi.org/10.1038/ncb2443.

Bik-Multanowski, M., Revhaug, C., Grabowska, A., Dobosz, A., Madetko-Talowska, A., Zasada, M., et al., 2018. Hyperoxia induces epigenetic changes in newborn mice lungs. Free Radic. Biol. Med. 121, 51–56.

Brás, A., Rodrigues, A.S., Gomes, B., Rueff, J., 2018. Down syndrome and microRNAs. Biomed. Rep. 8 (1), 11–16. https://doi.org/10.3892/br.2017.1019.

Cai, G., Sun, X., Zhang, L., Hong, Q., 2014. Association between maternal body mass index and congenital heart defects in offspring: a systematic review. Am. J. Obstet. Gynecol. 211, 91–117. https://doi.org/10.1016/j.ajog.2014.03.028.

Chao, H.T., Zoghbi, H.Y., Rosenmund, C., 2007. MeCP2 controls excitatory synaptic strength by regulating glutamatergic synapse number. Neuron 56, 58–65.

Chaves, J., Felippe, M., Almeida, M., Bacovsky, T., Ferrari, M., 2020. microRNAs expression correlates with levels of APP, DYRK1A, hyperphosphorylated Tau and BDNF in the hippocampus of a mouse model for Down syndrome during ageing. Neurosci. Lett. 714, 134541. https://doi.org/10.1016/j.neulet.2019.134541.

Chen, M., Wang, J., Luo, Y., Huang, K., Shi, X., Liu, Y., Li, J., Lai, Z., Xue, S., Gao, H., Chen, A., Chen, D., 2018. Identify Down syndrome transcriptome associations using integrative analysis of microarray database and correlation-interaction network. Hum. Genom. 12, 2. https://doi.org/10.1186/s40246-018-0133-y.

Ciccarone, F., Valentini, E., Malavolta, M., Zampieri, M., Bacalini, M.G., Calabrese, R., Guastafierro, T., Reale, A., Franceschi, C., Capri, M., Breusing, N., Grune, T., Moreno-Villanueva, M., Bürkle, A., Caiafa, P., May 9, 2018. DNA hydroxymethylation levels are altered in blood cells from down syndrome persons enrolled in the MARK-AGE project. J. Gerontol. A Biol. Sci. Med. Sci. 73 (6), 737–744. https://doi.org/10.1093/gerona/glx198.

Coppedè, F., 2015. The genetics of folate metabolism and maternal risk of birth of a child with down syndrome and associated congenital heart defects. Front. Genet. 6, 223. https://doi.org/10.3389/fgene.2015.00223.

Coppola, A., Romito, A., Borel, C., Gehrig, C., Gagnebin, M., Falconnet, E., et al., 2014. Cardiomyogenesis is controlled by the mir-99a/let-7c cluster and epigenetic modifications. Stem Cell Res. 12 (2), 323–337. https://doi.org/10.1016/j.scr.2013.11.008.

Cui, Y., Xiao, Z., Han, J., Sun, J., Ding, W., Zhao, Y., et al., 2012. MiR-125b orchestrates cell proliferation, differentiation and migration in neural stem/progenitor cells by targeting Nestin. BMC Neurosci. 13, 116. https://doi.org/10.1186/1471-2202-13-116.

Da CostaMartins, P.A., Salic, K., Gladka, M.M., Armand, A.-S., Leptidis, S., ElAzzouzi, H., et al., 2010. MicroRNA-199b targets the nuclear kinase Dyrk1a in an auto- amplification loop promoting cal- cineurin/NFATsignalling. Nat. Cell Biol. 12, 1220–1227.

Delabar, J.M., Theophile, D., Rahmani, Z., Chettouh, Z., Blouin, J.L., Prieur, M., et al., 1993. Molecular mapping of twenty-four features of Down syndrome on chromosome 21. Eur. J. Hum. Genet. 1, 114–124.

Della Ragione, F., Gagliardi, M., D'Esposito, M., Matarazzo, M.R., 2014. Non-coding RNAs in chromatin disease involving neurological defects. Front. Cell. Neurosci. 8, 54. https://doi.org/10.3389/fncel.2014.00054.

Do, C., Xing, Z., Yu, Y.E., Tycko, B., 2017. Trans-acting epigenetic effects of chromosomal aneuploidies: lessons from Down syndrome and mouse models. Epigenomics 9 (2), 189–207. https://doi.org/10.2217/epi-2016-0138.

dobosz, A., grabowska, agnieszka, bik-multanowski, miroslaw, 2019. Hypermethylation of NRG1 gene correlates with the presence of heart defects in Down's syndrome. J. Genet. 98, 110. https://doi.org/10.1007/s12041-019-1152-8.

Edbauer, D., Neilson, J.R., Foster, K.A., Wang, C.-F., Seeburg, D.P., Batterton, M.N., etal, 2010. Regulation of synaptic structure and functionbyFMRP-associated microRNA smiR-125b and miR- 132. Neuron 65, 373–384.

El Hajj, N., Dittrich, M., Böck, J., Kraus, T.F., Nanda, I., Müller, T., et al., 2016. Epigenetic dysregulation in the developing Down syndrome cortex. Epigenetics 11, 563–578.

Elsayed, G.M., Elsayed, S.M., Ezz-Elarab, S.S., 2014. Maternal MTHFR C677T genotype and septal defects in offspring with down syndrome: a pilot study. Egypt. J. Med. Hum Genet. 15, 39–44. https://doi.org/10.1016/j.ejmhg.2013.09.003.

Epping, M.T., Meijer, L.A., Krijgsman, O., Bos, J.L., Pandolfi, P.P., Bernards, R., 2011. TSPYL5 suppresses p53 levels and function by physical interaction with USP7. Nat. Cell Biol. 13, 102–108.

Farroni, C., Marasco, E., Marcellini, V., et al., 2018. Dysregulated miR-155 and miR-125b are related to impaired B-cell responses in down syndrome. Front. Immunol. 9, 2683. https://doi.org/10.3389/fimmu.2018.02683. Published 2018 Nov 20.

Ganguly, B.B., Kadam, N.N., 2016a. Down syndrome: from the age of characterization to the era of curative approach. Nucleus. https://doi.org/10.1007/s13237-016-0187-y. Published online: October 21.

Ganguly, B.B., Kadam, N.N., 2016b. Mutations of myelodysplastic syndromes (MDS): an update. Mutat. Res. Rev. 769, 47–62. https://doi.org/10.1016/j.mrrev.2016.04.009.

Ganguly, B.B., Kadam, N.N., 2019. Genetic race: prevalence of diseases and guidelines for prevention. MGM J. Med. Sci. 6 (1), 32–41. https://doi.org/10.5005/jp-journals-10036-1228.

Ganguly, B.B., Banerjee, D., Agarwal, M.B., 2018. Impact of chromosome alterations, genetic mutations and clonal hematopoiesis of indeterminate potential (CHIP) on the classification and risk stratification of MDS. Blood Cell Mol. Dis. 69, 90–100. https://doi.org/10.1016/j.bcmd.2017.10.001.

Ganguly, B.B., 2017. Small molecule inhibitors of epigenetic mutations as compelling drug targets for myelodysplastic syndromes. Curr. Canc. Drug Target. 17 (7), 586–602. https://doi.org/10.2174/15680096170330145002. Impact factor:3.5.

Gomez, W., Morales, R., Maracaja-Coutinho, V., Parra, V., Nassif, M., 2020. Down syndrome and Alzheimer's disease: common molecular traits beyond the amyloid precursor protein. Aging 12 (1), 1011–1033.

Greco, S., Zaccagnini, G., Fuschi, P., Voellenkle, C., Carrara, M., Sadeghi, I., Bearzi, C., Maimone, B., Castelvecchio, S., Stellos, K., Gaetano, C., Menicanti, L., Martelli, F., 2017. Increased BACE1-AS long noncoding RNA and b-amyloid levels in heart failure. Cardiovasc. Res. 113, 453–463. https://doi.org/10.1093/cvr/cvx013.

Harraz, M.M., Eacker, S.M., Wang, X., Dawson, T.M., andDawson, V.L., 2012. MicroRNA-223 is neu- roprotective by targeting glutamate receptors. Proc. Natl. Acad. Sci. U. S. A 109, 18962–18967.

Hattori, M., Fujiyama, A., Taylor, T., Watanabe, H., Yada, T., Park, H.-S., et al., 2000. The DNA sequence of human chromosome 21. Nature 405 (6784), 311–319. https://doi.org/10.1038/35012518.

Hefti, E., Quiñones-Lombraña, A., Redzematovic, A., Hui, J., Javier, G., 2016. Blanco. Analysis of mtDNA, miR-155 and BACH1 expression in hearts from donors with- and without- Down syndrome. Mitochondr. DNA A DNA MappSeq Anal. 27 (2), 896–903. https://doi.org/10.3109/19401736.2014.926477.

Henneman, P., Bouman, A., Mul, A., Knegt, L., van der Kevie-Kersemaekers, A.M., Zwaveling-Soonawala, N., et al., 2018. Widespread domain-like perturbations of DNA methylation in whole blood of Down syndrome neonates. PLoS One 13, e0194938.

Hobbs, C.A., Cleves, M.A., Karim, M.A., Zhao, W., MacLeod, S.L., 2010. Maternal folate-related gene environment interactions and congenital heart defects. Obstet. Gynecol. 116, 316–322. https://doi.org/10.1097/AOG.0b013e3181e80979.

Horvath, S., Garagnani, P., Bacalini, M.G., Pirazzini, C., Salvioli, S., Gentilini, D., et al., 2015. Accelerated epigenetic aging in Down syndrome. Aging Cell 14, 491–495.

Hudry, E., Wu, H.-Y., Arbel-Ornath, M., Hashimoto, T., Matsouaka, R., Fan, Z., et al., 2012. Inhibition of the NFAT pathway alleviates amyloid β neurotoxicity in a mouse model of Alzheimer's disease. J. Neurosci. 32, 3176–3192.

Izzo, A., Manco, R., Cristofaro, T.de, Bonfiglio, F., Cicatiello, R., Mollo, N., Martino, M.De, Genesio, R., Zannini, M., Conti, A., Nitsch, L., 2017. Overexpression of chromosome 21 miRNAs may affect mitochondrial function in the hearts of down syndrome fetuses. Int. J. Genom. 10. https://doi.org/10.1155/2017/8737649. Article ID 8737649.

Jin, S., Lee, Y.K., Lim, Y.C., Zheng, Z., Lin, X.M., Ng, D.P., Holbrook, J.D., Law, H.Y., Kwek, K.Y., Yeo, G.S., et al., 2013. Global DNA hypermethylation in down syndrome placenta. PLoS Genet. 9, e1003515.

Jones, M.J., Farré, P., McEwen, L.M., Macisaac, J.L., Watt, K., Neumann, S.M., Emberly, E., Cynader, M.S., Virji-

Babul, N., Kobor, M.S., 2013. Distinct DNA methylation patterns of cognitive impairment and trisomy 21 in Down syndrome. BMC Med. Genom. 6, 58.

Kamhieh-Milz, J., Moftah, R.F.H., Bal, G., Futschik, M., Sterzer, V., Khorramshahi, O., et al., 2014. Differentially expressed micrornas in maternal plasma for the noninvasive prenatal diagnosis of down syndrome (trisomy 21). BioMed Res. Int. 2014, 402475. https://doi.org/10.1155/2014/402475.

Keck-Wherley, J., Grover, D., Bhattacharyya, S., Xu, X., Holman, D., Lombardini, E.D., et al., 2011. Abnormal microRNA expression in Ts65Dn hippocampus and whole blood: contributions to down syndrome phenotypes. Dev. Neurosi. 33 (5), 451–467. https://doi.org/10.1159/0003300884.

Kerkel, K., Schupf, N., Hatta, K., Pang, D., Salas, M., Kratz, A., Minden, M., Murty, V., Zigman, W.B., Mayeux, R.P., Jenkins, E.C., Torkamani, A., Schork, N.J., et al., 2010. Altered DNA methylation in leukocytes with trisomy 21. PLoS Genet. 6, e1001212. https://doi.org/10.1371/journal.pgen.1001212.

Kirschbaum, M., Gojo, I., Goldberg, S.L., et al., 2014. A phase 1 clinical trial of vorinostat in combination with decitabine in patients with acute myeloid leukaemia or myelodysplastic syndrome. Br. J. Haematol. 167 (2), 185–193.

Klusmann, J.-H., Li, Z., Böhmer, K., Maroz, A., Koch, M.L., Emmrich, S., et al., 2010. miR-125b-2 is a potential oncomir on human chromosome 21 in megakaryoblastic leukemia. Gene. Dev. 24 (5), 478–490. https://doi.org/10.1101/gad.1856210.

Kotlabova, K., Doucha, J., Chudoba, D., Calda, P., Dlouha, K., Hromadnikova, I., 2013. Extracellular chromosome 21-derived micrornas in euploid & aneuploid pregnancies. Indian J. Med. Res. 138 (6), 935–943.

Kreutzer, J.S., DeLuca, J., Caplan, B., 2013. Encyclopedia of Clinical Neuropsychology. Springer, New York, pp. 2618–2621.

Laan, L., Klar, J., Sobol, M., Jan, H., Shahsavani, M., Kele, M., Fatima, A., Zakaria, M., Annerén, G., Falk, A., Schuster, J., Dahl, N., 2020. DNA methylation changes in Down syndrome derived neural iPSCs uncover codysregulation of ZNF and HOX3 families of transcription factors. Clin. Epigenet. https://doi.org/10.1186/s13148-019-0803-1.

Le, M.T.N., Xie, H., Zhou, B., Chia, P.H., Rizk, P., Um, M., et al., 2009. MicroRNA-125b promotes neuronal differentiation in human cells by repressing multiple targets. Mol. Cell Biol. 29, 5290–5305.

Lehmann, S.M., Krüger, C., Park, B., Derkow, K., Rosenberger, K., Baumgart, J., et al., 2012. An unconventional role for miRNA: let-7 activates Toll-like receptor 7 and causes neurodegeneration. Nat. Neurosci. 15, 827–835.

Li, Y.Y., Alexandrov, P.N., Pogue, A.I., Zhao, Y., Bhattacharjee, S., Lukiw, W.J., 2012. miRNA-155 upregulation and complement factor H (CFH) deficits in Down's Syndrome. Neuroreport 23 (3), 168–173. https://doi.org/10.1097/WNR.0b013e32834f4eb4.

Li, F., Wang, Y., Yang, H., Xu, Y., Zhou, X., Zhang, X., Xie, Z., Bi, J., 2019. The effect of BACE1-AS on β-amyloid generation by regulating BACE1 mRNA expression. BMC Mol. Biol. 20, 23. https://doi.org/10.1186/s12867-019-0140-0.

Lim, J.H., Lee, D.E., Kim, S.Y., Kim, H.J., Kim, K.S., Han, Y.J., et al., 2015. MicroRNAs as potential biomarkers for noninvasive detection of fetal trisomy 21. J. Assist. Reprod. Genet. 32 (5), 827–837. https://doi.org/10.1007/s10815-015-0429-y.

Lim, J.H., Kang, Y.-J., Lee, B.Y., Han, Y.J., Chung, J.H., Kim, M.Y., Kim, M.H., Kim, J.W., Cho, Y.-H., Ryu, H.M., 2019. Epigenome-wide base-resolution profiling of DNA methylation in chorionic villi of fetuses with Down syndrome by methyl-capture Sequencing. Clin. Epigenet. 11, 180. https://doi.org/10.1186/s13148-019-0756-4.

Lin, H., Sui, W., Li, W., Tan, Q., Chen, J., Lin, X., Guo, H., Ou, M., Xue, W., Zhang, R., et al., 2016. Integrated microRNA and protein expression analysis reveals novel microRNA regulation of targets in fetal down syndrome. Mol. Med. Rep. 14, 4109–4118.

Liu, Z., Li, W., Ma, X., Ding, N., Spallotta, F., Southon, E., et al., 2014. Essential role of the zinc finger transcription factor Casz1 for mammalian cardiac morphogenesis and development. J. Biol. Chem. 289, 29801–29816.

Lu, J., Sheen, V., 2013. Genetic and Epigenetic Mechanisms in Down Syndrome Brain. In: Down Syndrome. INTECH. https://doi.org/10.5772/52807.

Lu, J., Lian, G., Zhou, H., Esposito, G., Steardo, L., Delli-Bovi, L.C., Hecht, J.L., Lu, Q.R., Sheen, V., 2012. OLIG2 overexpression impairs proliferation of human Down syndrome neural progenitors. Hum. Mol. Genet. 21, 2330–2340.

Lu, J., Mccarter, M., Lian, G., Esposito, G., Elena, C., Laurent, C., Delli-Bovi, Hecht, J., Sheen, V., 2016. Global hypermethylation in fetal cortex of Down syndrome due to DNMT3L overexpression. Hum. Mol. Genet. 25 (9), 1714–1727. https://doi.org/10.1093/hmg/ddw043.

Ludwig, A.K., Giebel, B., 2012. Exosomes: small vesicles participating in intercellular communication. Int. J. Biochem. Cell Biol. 44 (1), 11–15. https://doi.org/10.1016/j.bioce l.2011.10.005.

Lukiw, W.J., Alexandrov, P.N., 2012. Regulation of complement factor H(CFH) by multiplemiR- NAsin Alzheimer's-disease (AD) brain. Mol. Neurobiol. 46, 11–19.

Luo, Q., Chen, Y., 2016. Long noncoding RNAs and Alzheimer's disease. Clin. Interv. Aging 11, 867–872. https://doi.org/10.2147/CIA.S107037.

Luo, S.S., Ishibashi, O., Ishikawa, G., Ishikawa, T., Katayama, A., Mishima, T., Takizawa, T., Shigihara, T., Goto, T., Izumi, A., Ohkuchi, A., Matsubara, S., Takeshita, T., Takizawa, T., 2009. Human villous trophoblasts express and secrete placenta-specific microRNAs into maternal circulation via exosomes. Biol. Reprod. 81 (4), 717–729. https://doi.org/10.1095/biolr eprod .108.07548 1.

Mahernia, S., Hassanzadeh, M., Sarvari, S., Amanlou, M., 2020. Targeting the microRNA binding domain of argonaute 2: rational inhibitor design and study of mutation effects on protein-ligand interaction. J. Biomol. Struct. Dyn. 38 (16), 4710–4717. https://doi.org/10.1080/07391102.2019.1688688.

Mahernia, S., Sarvari, S., Fatahi, Y., Amanlou, M., 2021a. The role of HSA21 encoded mirna in down syndrome pathophysiology: opportunities in miRNA-targeted pharmacotherapy and diagnosis of the down syndrome. Pharmaceut. Sci. 27 (3), 302–312. https://doi.org/10.34172/PS.2020.97.

Mahernia, S., Hassanzadeh, M., Adib, M., Peytam, F., Haghighijoo, Z., Iraji, A., Mahdavi, M., Edraki, N., Amanlou, M., 2021b. The possible effect of microRNA-155 (miR-155) and BACE1 inhibitors in the memory of patients with down syndrome and Alzheimer's disease: design, synthesis, virtual screening, molecular modeling and biological evaluations. J. Biomol. Struct. Dyn. 1–13. https://doi.org/10.1080/07391102.2021.1873861.

Masciarelli, S., Quaranta, R., Iosue, I., Colotti, G., Padula, F., Varchi, G., et al., 2014. A small-molecule targeting the microRNA binding domain of argonaute 2 improves the retinoic acid differentiation response of the acute promyelocytic leukemia cell line NB4. ACS Chem. Biol. 9 (8), 1674–1679. https://doi.org/10.1021/cb500286b.

Matamala, J.M., Arias-Carrasco, R., Sanchez, C., Uhrig, M., Bargsted, L., Matus, S., Maracaja-Coutinho, V., Abarzua, S., van Zundert, B., Verdugo, R., Manque, P., Hetz, C., 2018. Genome-wide circulating microRNA expression profiling reveals potential biomarkers for amyotrophic lateral sclerosis. Neurobiol. Aging 64, 123–138. https://doi.org/10.1016/j.neurobiolaging.2017.12.020.

Mendioroz, M., Do, C., Jiang, X., et al., 2015. Trans effects of chromosome aneuploidies on DNA methylation patterns in human Down syndrome and mouse models. Genome Biol. 16, 263. https://doi.org/10.1186/s13059-015-0827-6.

Nguyen, L.S., Lepleux, M., Makhlouf, M., Martin, C., Fregeac, J., Siquier-Pernet, K., Philippe, A., Feron, F., Gepner, B., Rougeulle, C., et al., 2016. Profiling olfactory stem cells from living patients identifies miRNAs relevant for autism pathophysiology. Mol. Autism. 7 (1).

O'Connell, R.M., Rao, D.S., Chaudhuri, A.A., Boldin, M.P., Taganov, K.D., Nicoll, J., et al., 2008. Sustained expression of microrna-155 in hematopoietic stem cells causes a myeloproliferative disorder. J. Exp. Med. 205 (3), 585–594. https://doi.org/10.1084/jem.20072108.

Parker, M., Mohankumar, K.M., Punchihewa, C., et al., 2014. C11orf95-RELA fusions drive oncogenic NF-κB signalling in ependymoma. Nature 506 (7489), 451–455. https://doi.org/10.1038/nature13109.

Paschou, M., Paraskevopoulou, M.D., Vlachos, I.S., Koukouraki, P., Hatzigeorgiou, A.G., Doxakis, E., 2012. miRNA regulons associated with synaptic function. PLoS One 7, e46189. https://doi.org/10.1371/journal.pone.0046189.

Perez-Nievas, B.G., Stein, T.D., Tai, H.C., Dols-Icardo, O., Scotton, T.C., Barroeta-Espar, I., Fernandez-Carballo, L., de Munain, E.L., Perez, J., Marquie, M., Serrano-Pozo, A., Frosch, M.P., Lowe, V., et al., 2013. Dissecting phenotypic traits linked to human resilience to Alzheimer's pathology. Brain 136, 2510–2526. https://doi.org/10.1093/brain/awt171.

Pogue, A.I., Cui, J.G., Li, Y.Y., Zhao, Y., Culicchia, F., Lukiw, W.J., 2010. Micro RNA-125b (miRNA125b) function in astrogliosis and glial cell proliferation. Neurosci. Lett. 476, 18–22.

Pulkkinen, K.H., Yla-Herttuala, S., Levonen, Al, 2011. Heme oxygenase 1 is induced by miR-155 via reduced BACH1 translation in endothelial cells. Free Radic. Biol. Med. 51, 2124–2131.

Qiu, J.J., Liu, Y.N., Ren, Z.R., Yan, J.B., 2017. Dysfunctions of mitochondria in close association with strong perturbation of long noncoding RNAs expression in down syndrome. Int. J. Biochem. Cell Biol. 92, 115–120. https://doi.org/10.1016/j.biocel.2017.09.017.

Quiñones-Lombraña, A., Blanco, J.G., 2015. Chromosome 21-derived hsa-miR-155-5p regulates mitochondrial biogenesis by targeting Mitochondrial Transcription Factor A (TFAM). Biochim. Biophys. Acta 1852 (7), 1420–1427. https://doi.org/10.1016/j.bbadis.2015.04.004.

Roome, C.J., Knöpfel, T., andEmpson, R.M., 2013. Functional contri- bution sof the plasma membrane calcium ATPase and the sodium- calcium exchange rat mouse parallel fibre to Purkinjeneu- ronsynapses. PflügersArch. 465, 319–331.

Rupert, C.E., Coulombe, K.L., 2015. The roles of neuregulin-1 in cardiac development, homeostasis, and disease. Biomark. Insight. 10 (Suppl. 1), 1–9.

Sailani, M.R., Santoni, F.A., Letourneau, A., et al., 2015. DNA-methylation patterns in trisomy 21 using cells from monozygotic twins. PLoS One 10, e0135555. https://doi.org/10.1371/journal.pone.0135555.

Salemi, M., Giambirtone, M., Barone, C., Grazia Salluzzo, M., Russo, R., Lo Giudice, M., Cutuli, S., Ridolfo, F., Romano, C., 2018. Mitochondrial mRNA expression in fibroblasts of Down syndrome subjects. Hum. Cell 31, 179–181. https://doi.org/10.1007/s13577-018-0205-2.

Salomon, C., Guanzon, D., Scholz-Romero, K., Longo, S., Correa, P., Illanes, S.E., Rice, G.E., 2017. Placental exosomes as early biomarker of preeclampsia: potential role of exosomal MicroRNAs across gestation. J. Clin. Endocrinol. Metab. 102 (9), 3182–3194. https://doi.org/10.1210/jc.2017-00672.

Salvi, A., Vezzoli, M., Busatto, S., Paolini, L., Faranda, T., Abeni, E., Caracausi, M., Antonaros, F., Piovesan, A., Locatelli, C., Cocchi, G., Alvisi, G., De Petro, G., Ricotta, D., Bergese, P., Radeghieri, A., 2019. Analysis of a nanoparticle enriched fraction of plasma reveals miRNA candidates for down syndrome pathogenesis. Int. J. Mol. Med. 43 (6), 2303–2318. https://doi.org/10.3892/ijmm.2019.4158.

Schaefer, A., O'Carroll, D., Tan, C.L., Hillman, D., Sugimori, M., Llinas, R., et al., 2007. Cerebellar neurodegeneration in the absence of microRNAs. J. Exp. Med. 204, 1553–1558.

Schmidt, M.F., Korb, O., Abell, C., 2013. MicroRNA-specific argonaute 2 protein inhibitors. ACS Chem. Biol. 8 (10), 2122–2126. https://doi.org/10.1021/cb400246k.

Schübeler, D., 2015. Function and information content of DNA methylation. Nature 517, 321–326. https://doi.org/10.1038/nature14192.

Sethupathy, P., Borel, C., Gagnebin, M., Grant, G.R., Deutsch, S., Elton, T.S., Hatzigeorgiou, A.G., Antonarakis, S.E., 2007. Human microRNA-155 on chromosome 21 differentially interacts with its polymorphic target in the AGTR1 3′ untranslated region: a mechanism for functional single-nucleotide polymorphisms related to phenotypes. Am. J. Hum. Genet. 81, 405–413. https://doi.org/10.1086/519979.

Shaham, L., Vendramini, E., Ge, Y., Goren, Y., Birger, Y., Tijssen, M.R., McNulty, M., Geron, I., Schwartzman, O., Goldberg, L., et al., 2015. MicroRNA-486-5p is an erythroid oncomiR of the myeloid leukemias of Down syndrome. Blood 125, 1292–1301.

Shi, W.L., Liu, Z.Z., Wang, H.D., Wu, D., Zhang, H., Xiao, H., Chu, Y., Hou, Q.F., Liao, S.X., 2016. Integrated miRNA and mRNA expression profiling in fetal hippocampus with Down syndrome. J. Biomed. Sci. 23 (48).

Siew, W.-H., Tan, K.-L., Babaei, M.A., Cheah, P.-S., Ling, K.-H., 2013. MicroRNAs and intellectual disability (ID) in Down syndrome, X-linked ID, and Fragile X syndrome. Front. Cell. Neurosci. https://doi.org/10.3389/fncel.2013.00041. Volume7 | Article41.

Sun, D., Lee, Y.S., Malhotra, A., Kim, H.K., Matecic, M., Evans, C., et al., 2011. Mir-99 family of micrornas suppresses the expression of prostate-specific antigen and prostate cancer cell proliferation. Canc. Res. 71 (4), 1313–1324. https://doi.org/10.1158/0008-5472.CAN-10-1031.

Sun, D., Layer, R., Mueller, A.C., Cichewicz, M.A., Negishi, M., Paschal, B.M., et al., 2014. Regulation of several androgen-induced genes through the repression of the mir-99a/let-7c/mir-125b-2 mirna cluster in prostate cancer cells. Oncogene 33 (11), 1448–1457. https://doi.org/10.1038/onc.2013.77.

Sun, X., Xu, C., Tang, S.-C., Wang, J., Wang, H., Wang, P., et al., 2016. Let-7c blocks estrogen-activated wnt signaling in induction of self-renewal of breast cancer stem cells. Canc. Gene. Ther. 23 (4), 83. https://doi.org/10.1038/cgt.2016.3.

Svobodová, I., Korabečná, M., Calda, P., Břešťák, M., Pazourková, E., Pospíšilová, Š., et al., 2016. Differentially expressed miRNAs in trisomy 21 placentas. Prenat. Diagn. 36 (8), 775–784. https://doi.org/10.1002/pd.4861.

Tafazoli, A., Behjati, F., Farhud, D., Abbaszadegan, M.R., 2019. Combination of genetics and nanotechnology for Down syndrome modification: a potential hypothesis and review of the literature. Iran. J. Publ. Health 48 (3), 371–378. https://doi.org/10.18502/ijph.v48i3.878.

Tan, G.S., Chiu, C.-H., Garchow, B.G., Metzler, D., Diamond, S.L., Kiriakidou, M., 2011. Small molecule inhibition of RISC loading. ACS Chem. Biol. 7 (2), 403–410. https://doi.org/10.1021/cb200253h.

Teng, P., Shi, Y., Sang, P., Cai, J., 2016. Γ-aapeptides as a new class of peptidomimetics. Chemistry 22 (16), 5458–5466. https://doi.org/10.1002/chem.201504936.

Turcatel, G., Rubin, N., El-Hashash, A., Warburton, D., 2012. MIR-99a and MIR-99b modulate TGF-β induced epithelial to mesenchymal plasticity in normal murine mammary gland cells. PLoS One 7, e31032. https://doi.org/10.1371/journal.pone.0031032.

Valentini, E., Zampieri, M., Malavolta, M., et al., 2016. Analysis of the machinery and intermediates of the 5hmC-mediated DNA demethylation pathway in aging on samples from the MARK-AGE Study. Aging 8, 1896–1922. https://doi.org/10.18632/aging.101022.

Van Meter, E.N., Onyango, J.A., Teske, K.A., 2020. A review of currently identified small molecule modulators of microRNA function. Eur. J. Med. Chem. 188, 112008. https://doi.org/10.1016/j.ejmech.2019.112008.

Vigorito, E., Kohlhaas, S., Lu, D., Leyland, R., 2013. miR-155: an ancient regulator of the immune system. Immunol. Rev. 253, 146–157. https://doi.org/10.1111/imr.12057.

Wan, P., Su, W., Zhuo, Y., 2017. The role of long noncoding RNAs in neurodegenerative diseases. Mol. Neurobiol. 54, 2012–2021. https://doi.org/10.1007/s12035-016-9793-6.

Wang, P.Y., Sun, Y.X., Zhang, S., Pang, M., Zhang, H.H., Gao, S.Y., Zhang, C., Lv, C.J., Xie, S.Y., 2013. Let-7c inhibits A549 cell proliferation through oncogenic TRIB2 related factors. FEBS Lett. 587, 2675–2681.

Wang, L., Li, Z., Song, X., Liu, L., Su, G., Cui, Y., 2016. Bioinformatic analysis of genes and microRNAs associated with atrioventricular septal defect in Down syndrome patients. Int. Heart J. 57, 490–495.

Weston, M.D., Pierce, M.L., Rocha- Sanchez, S., Beisel, K.W., Soukup, G.A., 2006. MicroRNA gene expression in the mouse inner ear. Brain Res. 1111, 95–104.

Wiemels, J.L., Muskens, I.S., Li, S., Pandey, P., Roy, R., Hansen, H.M., Siegmund, K.D., Mueller, B.A., Ma, X., Metayer, C.atherine, Smith, A.J de, 2019. The genome-wide impact of trisomy 21 on DNA methylation and its implications for hematologic malignancies. Blood 134 (Suppl. ment_1), 2510. https://doi.org/10.1182/blood-2019-131455.

Xu, Y., Li, W., Liu, X., Ma, H., Tu, Z., Dai, Y., 2013a. Analysis of microRNA expression profile by small RNA sequencing in Down syndrome fetuses. Int. J. Mol. Med. 32, 1115–1125.

Xu, Y., Li, W., Liu, X., Chen, H., Tan, K., Chen, Y., Tu, Z., Dai, Y., 2013b. Identification of dysregulated microRNAs in lymphocytes from children with Down syndrome. Gene 530, 278–286.

Yan, H., Zhou, M., Bhattarai, U., Song, Y., Zheng, M., Cai, J., et al., 2019. Cyclic peptidomimetics as inhibitor for mir-155 biogenesis. Mol. Pharm. 16 (2), 914–920. https://doi.org/10.1021/acs.molpharmaceut.8b01247.

Ye, Z., Jin, H., Qian, Q., 2015. Argonaute 2: a novel rising star in cancer research. J. Canc. 6 (9), 877–882. https://doi.org/10.7150/jca.11735.

Zedníková, I., Chylíková, B., Šeda, O., Korabečná, M., Pazourková, E., Břešťák, M., Krkavcová, M., Calda, P., Hořínek, A., 2020. Genome-wide miRNA profiling in plasma of pregnant women with down syndrome foetuses. Mol. Biol. Rep. 47, 4531–4540. https://doi.org/10.1007/s11033-020-05545-w.

Zhang, J., Zhou, W., Liu, Y., Li, N., November, 2016. Integrated analysis of DNA methylation and RNA-sequencing data in Down syndrome. Mol. Med. Rep. 14 (5), 4309–4314. https://doi.org/10.3892/mmr.2016.5778.

CHAPTER

14

In vivo and in vitro models for research on Down syndrome

1. Introduction

Animal models have become prominent tools for understanding genotype–phenotype association of the triplicated HSA21 genes and biological implication of specific genes, individually and/or in co-operation, in individuals with Down syndrome (DS). Impact of gene dosage has been delineated using trisomic and knockout mice, separately and through cross breeding by making the specific genes disomic. Thus, more knockout models may accelerate the pace of DS research for understanding each single gene present in trisomic status in mice but are syntenic to HSA21 in human DS. Generation of transgenic mice has further advantage of studying gene expression of trisomic genes, which are hypothesized to be associated with particular disease phenotypes in DS patients. Additionally, genetic manipulation of MMU10 and MMU17 genes syntenic to HSA21 would strongly be justified for understanding the contribution of these genes, which have otherwise appeared with milder phenotypic effects in DS individuals. Studies on mice with segmental trisomy revealed phenotypes relevant to DS, which might have caused by genes located in these regions. However, interaction of these trisomic genes with trisomy of MMU16 syntenic region might be responsible for development of complex phenotypes and might not be a straightforward mechanism for rendering expression of gene-specific characteristics of DS pathology.

Transgenic mouse models of DS carrying single candidate genes and chromosomal segments have contributed enormous knowledge in DS research but not on all the DS-associated pathological, biochemical, and transcriptional consequences. Prominently, the Ts65Dn mice (with three copies of MMU16 from App to Mx1), Ts1Cje mice (with trisomic segment from Sod1 to Mx1), Ms1Ts65 (with trisomic region from App to Sod1), and many more mouse models have been generated for understanding the impact of dosage imbalance on DS biology. Notably, most of the MMU16 region in Ts65Dn mice is homologous to HSA21 but deficient in ~40% of HSA21 genes; however, it is an extensively studied model of DS (Ishihara, 2021; Gough et al., 2020; Roper et al. 2020). The Ts1Cje with partial trisomy of MMU16 exhibits learning and behavioral abnormalities of DS. The Ts1Cje mouse contains a translocated segment of MMU16 and is at dosage imbalance for 67% of the genes triplicated in Ts65Dn model. Hence, phenotypic contribution of the region of difference between Ts65Dn and Ts1Cje appeared attractive. The Ms1Ts65 model and its

Genetics and Neurobiology of Down Syndrome
https://doi.org/10.1016/B978-0-323-90456-8.00013-2

littermates were studied for the region of difference (APP to Sod1) in relevance to DS phenotypes. Interestingly, the performance of the Morris water maze test of Ts1Cje mice was somewhat similar to that of Ts65Dn mice, but Ts65Dn was more severely affected in the reverse probe tests. Notably, the deficits of Ms1Ts65 mice were less severe than those of Ts65Dn.

Use of Ts65Dn and other models having segmental trisomy in pharmaceutical development for amelioration of some crucial feature of DS has allowed clinical trials. Various treatment models established using the mouse strain with variable gene dosage of particular genes are on the rise for preclinical research. Moreover, a substantial improvement of some deleterious phenotype reminiscent to DS individuals is the judging factor for proceeding clinical trial in humans. However, the power of contribution of gene dosage in different mouse strains having different genetic background may not lead to straightforward understanding of effects of therapeutic agents. To this perspective, syntenic genes of MMU10 or MMU17 to HSA21 was thought to influence the drug effect, and it was demonstrated that DS patients, who developed AML, have presented with high incidence of event-free survival facilitated by their increased sensitivity to cytosine arabinoside. Interestingly, increased sensitivity was attributed to the CBS gene, located in MMU17 in mice. A similar outcome was described in DS patients treated with methotrexoate (one of the most frequently used anticancer drug), and that was attributed to trisomy of the Slc19a1 gene (a reduced folate carrier (RFC) mapped onto MMU10). To this effect, trisomic Ts65Dn strain or other transgenic mice with trisomic gene for CBS or RFC might escalate the sensitivity to cytosine arabinoside or methotrexoate-like drugs. Collectively, more such studies would accumulate more effective information for understanding the mechanism of drug action by creating more appropriate mouse models.

It is worthwhile to understand the value of mouse models in DS research for attending crucial features of DS therapeutically, such as intellectual disability, stigmatic facial dysmorphism, early onset of aging, etc., which discriminates DS individuals from their peers. Mouse models have already established its critical use in understanding the genetic etiology of DS pathology and that will certainly continue to grow further with promising outcome. Thus, mouse models would have significant contribution to the lives of DS individuals. And most importantly, research in these genetically tailored models may help in deciphering the genes responsible for features common in DS and non-DS population. Researchers using different models concluded that mouse models have been of vital importance to study and understand many aspects of DS. However, primary differences exist at the molecular and systems levels between rodents and primates, mainly due to variation in considering the portions of mouse chromosomes (MMU10, MMU16, and MMU17) syntenic to HSA21 to a copy of HSA21 itself. Moreover, introducing the full range of human-specific genes triplicated in DS into the mouse may not accurately reproduce the effect of those genes expressed in the human environment. In addition, the triplication of mouse genes orthologous to HSA21 may express differently and yield different phenotypes than those found in DS patients. Furthermore, mouse husbandry method of generation of the mouse breeds also implicates the variation as observed in Ts1Rhr bred on the inbred C57BL/6 background versus mixed C57BL/6-C3H background. The first breed of trisomic animals displayed increased volume of neural crest in first pharyngeal arch (PA1) region during embryonic development and a reduced body weight during adulthood, while the latter ones showed no difference in the volume of PA1, rather body weight increased in the adult trisomic animals compared to controls. Interestingly, changing phenotypes of Ts65Dn were chronicled over generations, illustrating that even animals of the same trisomic line carrying same genetic background but of different historic times and generation exhibit significant variation in well-documented phenotypes, such as cortical expansion during embryonic development, the trajectory of oligodendrogenesis and maturation, and learning and memory performance with a drift in wild-type mice. The findings suggested that the phenotypes of trisomic animals may not be preserved from generation to generation (reviewed in Dierssen et al., 2021; Shaw et al., 2020), and variability between animals may be a representation of the variability within the DS population.

Thus, human models were thought to recapitulate human-specific gene expression and brain development as alternatives to mouse models. The cell lineages with induced pluripotent stem cells (iPSCs) were created with cells from DS patients (DS-iPSCs), with a view to achieving closer replicates of DS phenotypes, but in vitro. In this direction, several iPSCs and embryonic stem cell (ESC) lineages were validated, specifically for understanding the perturbation of neurogenesis in neural progenitor cells (NPCs) to recapitulate cognitive imbalance seen in DS individuals. A great deal of research has been carried out in this context, which has been discussed in detail in this chapter. However, analytical differences and nonhomogeneous sources of iPSC lines debarred drawing a satisfactory resolution on understanding the genotype–phenotype association reminiscent to DS patients. Hence, organoids are being created from iPSC lines with an attempt to achieve precise information about the gene-dosage effects. Currently, cerebral organoids (COs) have yielded satisfactory results for understanding neurogenesis as a cumulative and/or single gene effect in trisomic doses. The value of DS iPSCs and the use of brain organoids have been emphasized for accurate study of the neural development and that represents the human genetic environment because the in vitro experiments on the parent iPSCs do not recapitulate the environment of a developing brain. Additionally, interlab differences in selecting cell lines and in culturing techniques can also affect the development of cells in vitro. Thus, harmonization of iPSC techniques used, DS-iPSCs selected, and sharing of isogenic DS cells can mitigate variations in experimental outcomes.

In this chapter, a few of the several mouse models are discussed, which have trisomic genetic complex or chromosomal regions syntenic to various regions of HSA21, selected transgenic, and knockout genes. In addition, the emphasis has been paid on genes examined in both transgenic and knockout models and combination of knockout and trisomic complex for elucidation of expression by restoring disomy. Also focus has been paid on genes involved in functional pathways and associated with expression of DS phenotypes, such as APP, RCAN1, SYNJ1, and so on for their involvement in synaptic plasticity and/or function and also likely to be responsible for intellectual disabilities in DS patients, along with RUNX1, ETS2, and ERG group of genes because of their contribution to cancer in DS. Several studies based on trisomic iPSCs, in particular with NPCs and hematopoietic progenitors, have been discussed. In addition, studies in DS iPSC–derived COs have drawn serious attention in the present era as a potential model for DS research and targeting DS therapeutics. Taken together, the chapter highlights the current scenario of DS research in terms of modeling for trisomic HSA21 gene-expression in vivo and in vitro on one hand and on the other, potential uses of DS models in pharmaceutical testing of targeted drug candidates.

2. Models of DS research

A great deal of variability in phenotypic expression was observed among the individuals with DS, which was thought to be due to incomplete penetrance occurring in only some people with DS and variable expressivity (differing in severity among people with DS). Given that fact, the principal goal of research was focused on understanding the causative relationship of imbalance of gene dosage and phenotypic expression in DS individuals. Initial emphasis was paid on partial trisomies of HSA21. However, since mouse models offer access to every tissue at all stages of development, highly amenable to genomic and chromosome manipulation and precise quantification of phenotypic expression, different strains of mouse models of DS were created with segmental trisomy or knock down of genes. Additionally, mice are higher organisms with requisite biological characteristics, which enable studies of neurological, behavioral, cardiac, hematological, musculoskeletal and several other manifestations reminiscent to humans. Also, mouse models are well characterized for testing of various drugs, including ones associated with DS. Most importantly, mice are small, short lived, rapidly reproducible with ease, and highly economic to maintain and breed. Alongside, there were several other models developed for understanding the complexity of DS phenotype, including computer models, in vitro models, iPSCs of specific lineages, iPSC-derived organoids, lower organisms, and many others; however, mouse models are widely considered in DS research.

3. Mouse models in DS research

With the advent of advanced genetic engineering technologies, there is tremendous improvement in understanding the gene dosage effect on DS phenotypes. However, little is known about the interrelation and/or interaction of the triplicated HSA21 genes with hundreds of thousands of the neighboring disomic genes. In addition, the environmental impact on the global disomic and trisomic HSA21 genes, and the effect of DNA methylation and alteration in chromatin regulation cannot be ruled out (Antonarakis, 2017; Letourneau et al., 2014; Mendioroz et al., 2015). A few non-HSA21 genes have been characterized to not contributing to individual phenotypic variation (Roper and Reeves, 2006).

Most of the knowledge gathered on alterations of the GABAergic signaling in the DS brain has been extracted from the creation and study of mouse models of DS (Fig. 14.1) (Herault et al., 2017). The recent study on the Vertebrate Genome Annotation (VEGA) database1 has demonstrated that from the total of 222 protein coding genes of the HSA21, 218 map to the long arm of HSA21 (21q). The mouse genes ortholog to those mapped on the long arm of HSA21 is distributed on three syntenic regions present on mouse chromosomes 10 (MMU10), 16 (MMU16), and 17 (MMU17). Remarkably, the distal portion of the MMU16 encompasses a large (~28 Mb) region that represents ~55% of HSA21 orthologous protein-coding genes (Antonarakis et al., 2004). Therefore, genetic manipulation of the MMU16 region has been considered for creation of many of the available mouse models of DS, of which the Ts65Dn mouse is used widely among all murine models of DS.

3.1 Ts16

With the advent of chromosome engineering techniques, trisomic mice could be created for any chromosomal region. Ts16 was the first mouse model of DS developed in 1979 with trisomy for almost all of MMU16, including the region syntenic to HSA21. The strain was produced by selection of centric fusions using breeding scheme (Gropp et al., 1974). Owing to having some features reminiscent to DS, the mouse was thought to be a suitable model for certain aspects of DS phenotypes and thus, indicated its potential use in DS research (Epstein, 1981). Indeed, the hypothesis was supported

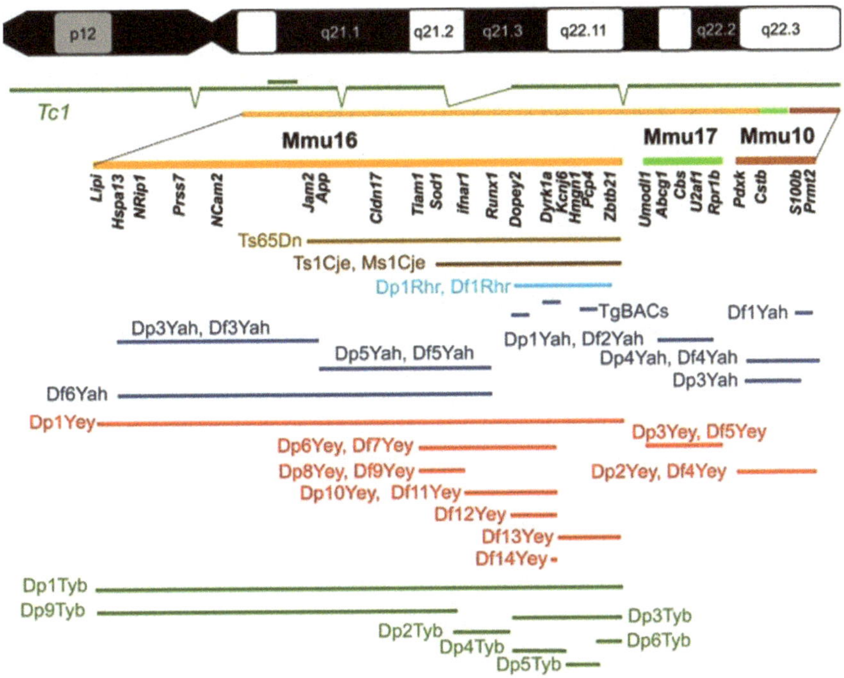

FIGURE 14.1 Syntenic/orthologous regions of HSA21 in different mouse models created for DS research. *Reprinted from Herault et al., 2017. https://doi.org/10.1242/dmm.029728 with permission from Herault Y and under the terms of the Creative Commons Attribution License (http://creativecommons.org/license/by/3.0).*

with discovery of syntenic SOD1, IFNAR, and GART genes of HSA21 (reviewed in Vaccano et al., 2012). The result had established first landmark evidence of mouse model of human DS and indicated the possibility of generation of more such suitable mouse models for DS research. However, Ts16 embryos rarely survive until birth or die shortly after birth. Hence, the model was useful for the study of embryonic or fetal development of DS pregnancies such as edema and fetal cardiac, neurologic, and thymic abnormalities but limiting their use as a model for studying behavior, postnatal developments relevant to achieve adulthood, and several other aspects that happen during lifetime of a DS patient, including aging (Rueda et al., 2012).

Nevertheless, cell lines developed from Ts16 mice were useful to study the relevant biological processes in human DS and had potential to study the biology of aging and age-related disorders (Arriagada et al., 2010; Bambrick and Fiskum, 2008). The major drawback of this model was the presence of many of the trisomic MMU16 genes, which are not present in HSA21 but orthologous to HSA3, HSA8, HSA12, HSA6, and HSA22, indicating additional imbalance in gene dosage, and that might have caused embryonic lethality suggesting it as an unsuitable model for DS research. Such disadvantage of this strain led to produce more specific and suitable mouse models.

3.2 Ts43H (Ts(16^{17})43H)

This mouse is trisomic for 30 MB of proximal MMU17 and presented behaviors reminiscent to human DS phenotypes (Vacik et al., 2005). The model is trisomic for at least nine HSA21 genes orthologous to MMU17. The breakpoint on the segmental chromosome of Ts43H occurs in HSA21 homologous region. Since Ts43H model has trisomy of a large segment of MMU17, a dosage imbalance with a higher number of genes was envisaged compared to Ts65Dn or Tc1 mouse models. Trisomic HSA21 orthologs and trisomy of large nonorthologous genomic regions and/or trisomic heterogeneity might have contributed to DS-related phenotypes and altered distinct pathways in this model (Table 14.1).

3.3 Ts65Dn (Ts(17^{16})65Dn)

The Ts65Dn mouse model significantly resembles HSA21 than the Ts16 model and was first introduced in 1993 (Davisson et al., 1993). The model was produced by breeding with males with irradiated testes and screened the litters for chromosomal alterations involving MMU16 and appeared as one of the best characterized mouse models. In Ts65Dn, the cells possess an extra copy of only a segment of genes on MMU16 and a segment of MMU17 genes. Practically, the Ts65Dn mouse carries a freely segregating extra translocation chromosome configured with fusion of the entire genomic region distal to Mir-155 on MMU16 with the centromeric portion of MMU17. Thus, importantly, the Ts65Dn mice is disomic for about 16 HSA21/MMU16 genes and also trisomic for a centromere proximal region of MMU17 (Yu et al., 2010a). The trisomic segment of this chromosomal rearrangement between MMU16 and MMU17 carries 60 MMU17 genes, which are mostly overexpressed in heart. Importantly, the two trisomic genes such as Synj2 and Tiam2 of MMU17 in these mice are related to Synj1 and Tiam1 genes encoded by HSA21/MMU16. This extra chromosome contains 90 non-KRTAP—HSA21 protein coding orthologs, and 35 protein-coding genes derived from the MMU17, which are not triplicated in DS (Gupta et al., 2016). This model has exhibited various DS phenotypes, including behavioral abnormalities and cognitive defects (Rueda et al., 2012). As the Ts65Dn model is the most well characterized and extensively studied DS model, creation of this model is considered as a seminal achievement in DS research. A number of DS features of this model have become standard to compare the incidence and extent of phenotypes in other trisomic mouse models (Richtsmeier et al., 2002).

Phenotypically, the Ts65Dn mice carry similar features with that of other trisomic mice that carry the entire region of MMU16 syntenic to HSA21, and also the mice trisomic for all mouse chromosome regions syntenic to HSA21. The freely segregating extra chromosome in Ts65Dn model is unique and different from other models created by chromosome engineering technique (Ts1Cje and Ts2Cje) or serendipitously with duplication of the relevant chromosome region (Vaccano et al., 2012). Thus, this model may facilitate extensive investigation for understanding the genetic consequences in human DS. On the other hand, the other models can be somewhat correlated with DS carrying Robertsonian translocation involving one of the trisomic 21 meaning including two centromeres—one or both of the HSA21. There are several DS individuals carrying apparently balanced translocations between HSA21 and other chromosomes, which may not involve the centromeres of the two chromosomes but can implicate all the classical features of DS individuals. This led to the hypothesis that expression of DS phenotypes, particularly that are not related to reproduction, may not require the extra chromosome (Vacano et al., 2012).

Ts65Dn mice recapitulate several dysfunctions present in DS individuals, such as low birth weight, male sterility, abnormal craniofacial appearance, cardiovascular malformations, structural deficiencies in neurologic system, and several cognitive and behavioral impairments, including executive functions, such as working memory and cognitive flexibility

TABLE 14.1 Mouse models of Down syndrome.

Common name (MGI-name)	Salient features	Genetic drawback	No. of HSA21 orthologues	Gene-dosage	Trisomic offspring at weaning (%)	Background genetics of mice	References
Ts16	Created by breeding with centric fusions; HSA21 syntenic genes: SOD1, IFNAR, GART; 100% embryonic lethality; suitable for embryonic development, but not for postnatal growth	Many trisomic MMU16 genes (~75.5 MB) were not present in HSA21, but orthologous to HSA3, 8, 6, 12, and 22	144	Trisomic for almost all of MMU16 (~98 MB)	NA	NR	Gropp et al. (1974), Epstein (1981)
Ts43H (Ts(16^{17}) 43H)	Trisomic for 30 MB of proximal MMU17, and presented behaviors reminiscent to DS-related phenotypes	Trisomy of large nonorthologous genomic regions and/or trisomic heterogeneity	9–15	Trisomy of a large segment of MMU17	11	NR	Vacik et al. (2005)
Ts65Dn (Ts(17^{16}) 65Dn)	Irradiating the testes of male mice and subsequent breeding; a freely-segregating extra translocation chromosome configured with fusion of the entire genomic region distal to miR-155 on MMU16 with the centromeric end of MMU17; the segment is homologous to HSA21 from Mrp/39 to the end of MMU16	~35 protein-coding genes derived from the MMU17, which is not triplicated in DS	104	Segmental trisomy of MMU16, contains ~ half of the HSA21 orthologs	20–36	B6C3B or B6EiC3Sn	Davisson et al. (1993), Reeves et al. (1995), Hattori et al. (2000)
Ms1Rhr/Ts65Dn			70		NR		Olson et al. (2004)

Continued

TABLE 14.1 Mouse models of Down syndrome.—cont'd

Common name (MGI-name)	Salient features	Genetic drawback	No. of HSA21 orthologues	Gene-dosage	Trisomic offspring at weaning (%)	Background genetics of mice	References
Ts1Cje	Translocation between MMU16 and MMU12; trisomic segment contain ~75 genes syntenic to HSA21; both female and male Ts1Cje mice appeared equally fertile; exhibited milder form of DS-related phenotypes and/or appear with some discrete characteristics	Complex genetic recombination with trisomy; the translocation caused deletion of 7 genes located in the most telomeric segment of MMU12, which are monosomic; disomic for 19 genes, including APP and SOD1, probably due to insertion of *neo* sequence	81	Trisomic for a smaller segment (interval Sod1 to Mx1) of MMU16, but covers almost 2/3rd of trisomic region in Ts65Dn	50	B6EiC3Sn (or B6 congenic)	Sago et al. (1998)
Ms1Cje/Ts65Dn			22		23		Sago et al. (1998)
Ts2Cje (Ts [Rb(12.Ts17^{16}65Dn)] 2Cje)	Robertsonian translocation between the trisomic MMU16 (marker chromosome) of Ts65Dn model and MMU12; similar features relative to Ts65Dn models; conferred reproductive advantages with enhanced male fertility; female mice appeared with 40 chromosomes, but 41 arms due to Robertsonian translocation	Was an equivalent model to Ts65Dn; however, it had increased male fertility and frequency of transmission of the trisomic segment through female germline	104	Trisomic for small segment of MMU16	43	B6EiC3Sn	Villar et al. (2005)

Ts1Rhr	Carries a tandem duplication of a tiny MMU16 region comprising of 33 HSA21 protein-coding orthologs from the so-called "DS critical region" (DSCR); behavioral deficits observed in T-maze and novel object recognition test, but not in Morris water test; increased body and brain weight and expanded posterior hippocampal region comparable to diploid mice	DSCR was necessary but not sufficient to cause DS-related phenotypes	33	Trisomic for the mouse equivalent of the human DSCR region	NR	NR	Delabar et al. (1993), Korenberg et al. (1994), Olson et al. (2004)
Ts1 Yah	Trisomic for the HSA21 syntenic region on MMU17 containing 12 genes between abcg1 and U2af1; exhibited learning and memory deficits and longer long term potentiation (LTP)	NR	12	trisomic for MMU17 syntenic to HSA21		NR	Pereira et al. (2009)
Tc1 (Tc(HSA21) 1TybEmcf)	Transchromosomal mice created by microcell-mediated chromosome engineering technique using embryonic stem cells and insertion of the HSA21; contains almost the entire freely segregating HSA21 (91%); duplication of S100β and PRMT2 genes; Tc1 mice are mosaics; cross breeding of Ms4Yah strain, which carries a deletion of the region of MMU10 that is orthologous to HSA21, with the Tc1 strain demonstrated that trisomic genes in this interval are not responsible for learning defects; ~70% of brain nuclei carry HSA21 though tissue-specific variability exists in cellular trisomy	This chromosome has several deletions with loss of a small number of HSA21 genes; irradiation of HSA21 prior to transfer to mouse cells resulted in genetic alterations, including deletions, duplications and other rearrangements. The strain has the advantage of allowing study of the HSA21 genes, including those which are not found in the orthologous regions of MMU10, MMU16 and MMU17	~212	Almost, the entire HSA21 was introduced to mouse embryonic stem cells to generate Tc1	>40	Breeding of F1 of 129S8 and C57BL/6J	O'Doherty et al. (2005)

Continued

TABLE 14.1 Mouse models of Down syndrome.—cont'd

Common name (MGI-name)	Salient features	Genetic drawback	No. of HSA21 orthologues	Gene-dosage	Trisomic offspring at weaning (%)	Background genetics of mice	References
Dp10/Dp16/Dp17 (Dp(10)1Yey/+, Dp(16)1Yey/+, Dp(17)1Yey/+)	Cre/lox-mediated chromosome engineering technique created tandem duplication on the corresponding chromosomes MMU10, MMU16 and MMU17 respectively; symptoms, including increased rate of heart defects and learning and memory deficits were comparable to symptoms of human DS; Dp16 model genetically more accurate representation of human DS; behavioral and synaptic plasticity deficits in Dp16 and triple trisomic (Dp10/Dp16/Dp17) mice was comparable to Ts65Dn mice	Crossing of the Dp10, Dp16 and Dp17 lines enabled creation of fully trisomic mice with all HSA21 orthologs, these triple trisomic mice appeared with compromised viability and poor breeding capacity, and that has restricted their use in several other experimental studies		Mice are individually trisomic for the three complete syntenic regions on MMU10, 16 and 17 respectively		Mixed bred	Yu et al. (2010a), Li et al. (2007)
Dp(16Cbr1-Fam3b)1Rhr[Ts1Rhr; Dp1Rhr]	Cre-mediated rearrangement on MMU16 between loxP sites in areas proximal to Cbr1 and distal to Fam3b causing duplication of Cbr1 and Fam3b genes			Duplication of genes, including Cbr1 and Fam3b		B6.129S6	Olson et al. (2004)
Dp(16Lipi-Zbtb21)1Yey[Dp(16)1Yey; Dp1Yey]	Cre-mediated rearrangement on MMU16 between loxP sites in areas proximal to Lipi (D930038D03Rik) and distal to Zbtb21 (Zfp295) causing duplication of Lipi and Zbtb21 genes			Duplication of genes, including Lipi and Zbtb21		B6N5	Li et al. (2007)

Dp(16Tiam1-Kcnj6)6Yey[Dp(16)2Yey; Dp6Yey]	Cre-mediated rearrangement on MMU16 between loxP sites in areas proximal to Tiam1 and distal to Kcnj6 causing duplication of Tiam1 and Kcnj6 genes	Duplication of genes, including Tiam1and Kcnj6	Mixed 129S7 and 129S1	Liu et al. (2011)
Dp(16Tiam1-Il10rb)8Yey[Dp(16)3Yey; Dp8Yey]	Cre-mediated rearrangement on MMU16 between loxP sites in areas proximal to Tiam1 and distal to Il10rb causing duplication of Tiam1 and Il10rb genes	Duplication of genes, including Tiam1and Il10rb	Mixed 129S7 and 129S1	Liu et al. (2014)
Dp(16Ifnar1-Kcnj6)10Yey[Dp(16)4Yey; Dp4Yey]	Cre-mediated rearrangement on MMU16 between loxP sites in areas proximal to Ifnar1 and distal to Kcnj6 causing duplication of Ifnar1 and Kcnj6 genes	Duplication of genes, including Ifnar1 and Kcnj6	Mixed 129S7 and 129S1	Liu et al. (2014)
Dp(16Hspa13-App)2Yah[Ts2Yah; Dp2Tah]	Cre-mediated rearrangement on MMU16 between loxP sites inserted in Hspa13 and App causing inactivation of segment homologous to HSA21q11-q21	Duplication of Hspa13 and App genes	C57BL/6J(>N6)	Brault et al. (2015)
Dp(16App-Runx1)5Yah[Ts5Yah; Dp5Yah]	Cre-mediated rearrangement on MMU16 between loxP sites inserted in App and Runx1 caused inactivation of both flanking genes and duplication of HSA21 homologous segment	Duplication of HSA21 homologous segment	C57BL/6J(>N6)	Raveau et al. (2012)

Continued

TABLE 14.1 Mouse models of Down syndrome.—cont'd

Common name (MGI-name)	Salient features	Genetic drawback	No. of HSA21 orthologues	Gene-dosage	Trisomic offspring at weaning (%)	Background genetics of mice	References
Dp(16Lipi-Zbtb21) 1TybEmcf[Dp1Tyb]	Cre-mediated rearrangement on MMU16 between two loxP sites inserted proximal to Lipi and distal to Zbtb21 for duplication of segment homologous to HSA21			Duplication of HSA21 homologous segment		C57BL/ 6J.129P2	Lana-Elola et al. (2016)
Dp(16Mis18a-Runx1)2TybEmcf [Dp2Tyb]	Cre-mediated rearrangement on MMU16 between two loxP sites inserted proximal to Mis18a and distal to Runx1 for duplication of segment homologous to HSA21			Duplication of HSA21 homologous segment		C57BL/ 6J.129P2	Lana-Elola et al. (2016)
Dp(16Mir802-Zbtb21)3TybEmcf [Dp3Tyb]	Cre-mediated rearrangement on MMU16 between two loxP sites inserted proximal to miR-802 and distal to Zbtb21 for duplication of segment homologous to HSA21			Duplication of HSA21 homologous segment		C57BL/ 6J.129P2	Lana-Elola et al. (2016)
Dp(16Mir802-Dscr3)4TybEmcf [Dp4Tyb]	Cre-mediated rearrangement on MMU16 between two loxP sites inserted proximal to miR-802 and distal to Dscr3 for duplication of segment homologous to HSA21			Duplication of HSA21 homologous segment		C57BL/ 6J.129P2	Lana-Elola et al. (2016)
Dp(16Dyrk1a-B3galt5)5TybEmcf [Dp5Tyb]	Cre-mediated rearrangement on MMU16 between two loxP sites inserted proximal to Dyrk1a and distal to B3galt5 for duplication of segment homologous to HSA21			Duplication of HSA21 homologous segment		C57BL/ 6J.129P2	Lana-Elola et al. (2016)

Dp(16Igsf5-Zbtb21)6TybEmcf[Dp6Tyb]	Cre-mediated rearrangement on MMU16 between two loxP sites inserted proximal to Igsf5 and distal to Zbtb21 for duplication of segment homologous to HSA21	Duplication of HSA21 homologous segment	C57BL/6J.129P2	Lana-Elola et al. (2016)
Dp(16Lipi-Hunk)9TybEmcf[Dp9Tyb]	Cre-mediated rearrangement on MMU16 between two loxP sites inserted proximal to Lipi and distal to Hunk for duplication of segment homologous to HSA21	Duplication of HSA21 homologous segment	C57BL/6J.129P2	Lana-Elola et al. (2016)
Dp(10Prmt2-Pdxk)2Yey[Dp(10)1Yey; Dp2Yey]	Cre-mediated rearrangement on MMU10 between loxP sites in areas proximal to Prmt2 and distal to Pdxk causing duplication of Prmt2 and Pdxk genes	Duplication of genes, including Prmt2 and Pdxk	B6N4	Yu et al. (2010b)
Dp(10Cstb-Prmt2)3Yah[Ts3Yah; Dp3Yah]	Cre-mediated rearrangement on MMU10 between two loxP sites inserted proximal to Cstb and distal to Col6a1 caused inactivation of both flanking genes and duplication of segment homologous to HSA21	Duplication of segment homologous to HSA21	C57BL/6J(>N6)	Duchon et al. (2008)
Dp(10Cstb-Prmt2)4Yah[Ts4Yah; Dp4Yah]	Cre-mediated rearrangement on MMU10 between two loxP sites inserted proximal to Cstb and distal to Prmt2 caused inactivation of both flanking genes and duplication of segment homologous to HSA21	Duplication of segment homologous to HSA21	C57BL/6J(>N6)	Duchon et al. (2008)

Continued

TABLE 14.1 Mouse models of Down syndrome.—cont'd

Common name (MGI-name)	Salient features	Genetic drawback	No. of HSA21 orthologues	Gene-dosage	Trisomic offspring at weaning (%)	Background genetics of mice	References
Dp(17Abcg1-Rrp1b) 3Yey[Dp(17)1Yey; Dp3Yey]	Cre-mediated rearrangement on MMU17 between loxP sites in areas proximal to Abcg1 and distal to Rrp1b causing duplication of Abcg1 and Rrp1b genes			Duplication of genes, including Abcg1 and Rrp1b		B6N4	Yu et al. (2010b)
Dp(17Abcg1-Cbs) 1Yah[Ts1Yah; Dp1Yah]	Cre-mediated rearrangement on MMU17 between two loxP sites inserted in Abcg1 and downstream of Cbs (before U2af1) caused deletion of Abcg1 and duplication of the segment homologous to HSA21			Duplication of the segment homologous to HSA21 and deletion of Abcg1		C57BL/6J(>N6)	Pereira et al. (2009)
Df(16Cbr1-Fam3b) 1Rhr: (Ms1Rhr; Df1Rhr)	Cre-mediated rearrangement on MMU16 between loxP sites in areas proximal to Cbr1 and distal to Fam3b causing deletion of Cbr1 and Fam3b genes			Deletion of genes, including Cbr1 and Fam3b		B6.129S6	Olson et al. (2004)
Df(16Tiam1-Kcnj6) 7Yey[Df(16)2Yey; Df7Yey]	Cre-mediated rearrangement on MMU16 between loxP sites in areas proximal to Tiam1 and distal to Kcnj6 caused deletion of Tiam1 and Kcnj6 genes			Deletion of genes, including Tiam1 and Kcnj6		Mixed 129S7 and 129S1	Liu et al. (2011)

Df(16Tiam1-Il10rb) 9Yey[Df(16)3Yey; Df9Yey]	Cre-mediated rearrangement on MMU16 between loxP sites in areas proximal to Tiam1 and distal to Il10rb caused deletion of Tiam1 and Il10rb genes		Deletion of genes, including Tiam1 and Il10rb	Mixed 129S7 and 129S1	Liu et al. (2014)
Df(16Ifnar11- Kcnj6) 11Yey[Df(16)4Yey; Df11Yey]	Cre-mediated rearrangement on MMU16 between loxP sites in areas proximal to Ifnar1 and distal to Kcnj6 caused deletion of Tiam1 and Kcnj6 genes		Deletion of genes, including Ifnar1 and Kcnj6	Mixed 129S7 and 129S1	Liu et al. (2014)
Df(16Setd4- Kcnj6) 12Yey[Df(16)5Yey; Df12Yey]	Cre-mediated rearrangement on MMU16 between loxP sites in areas proximal to Setd4 and distal to Kcnj6 caused deletion of Setd4 and Kcnj6 genes		Deletion of genes, including Setd4 and Kcnj6	C57BL/6J(>N6)	Liu et al. (2014)
Df(16Kcnj15-Mx2) 13Yey[Df(16)6Yey; Df13Yey]	Cre-mediated rearrangement on MMU16 between loxP sites in areas proximal to Kcnj15 and distal to Mx2 caused deletion of Kcnj15 and Mx2 genes		Deletion of genes, including Kcnj15 and Mx2	C57BL/6J(>N6)	Jiang et al. (2015)
Df(16Dyrk1a-Kcnj6)14Yey[Df(16) 7Yey; Df14Yey]	Cre-mediated rearrangement on MMU16 between loxP sites in areas proximal to Dyrk1a and distal to Kcnj6 caused deletion of Dyrk1a and Kcnj6 genes		Deletion of genes, including Dyrk1a and Kcnj6	C57BL/6J(>N6)	Jiang et al. (2015)

Continued

TABLE 14.1 Mouse models of Down syndrome.—cont'd

Common name (MGI-name)	Salient features	Genetic drawback	No. of HSA21 orthologues	Gene-dosage	Trisomic offspring at weaning (%)	Background genetics of mice	References
Df(16App- Runx1) 5Yah[Ms5Yah; Df5Yah]	Cre-mediated rearrangement on MMU16 between loxP sites inserted in App and Runx1 caused inactivation of both flanking regions and deletion of the HSA21q22 homologous segment			Deletion of segment homologous to HSA21q22		C57BL/6J(>N6)	Raveau et al. (2012)
Df(16Hspa13- App) 3Yah[Ms3Yah; Df3Yah]	Cre-mediated rearrangement on MMU16 between loxP sites inserted in Hspa13 and App caused inactivation of both flanking regions and deletion of the HSA21q22 homologous segment			Deletion of segment homologous to HSA21q22		C57BL/6J(>N6)	Brault et al. (2015)
Df(10Prmt2-Pdxk) 4Yey[Df(10)1Yey; Df2Yey]	Cre-mediated rearrangement on MMU10 between loxP sites in areas proximal to Prmt2 and distal to Pdxk caused deletion of Prmt2 and Pdxk genes			Deletion of genes, including Prmt2 and Pdxk		Mixed 129S7 and 129S1	Yu et al. (2010a)
Df(10Prmt2-Col6a1) 1Yah[Ms1Yah; Df1Yah]	Cre-mediated rearrangement on MMU10 between loxP sites inserted in Prmt2 and Col61a caused inactivation of both flanking genes and deletion of the HSA21q22 homologous segment			Deletion of segment homologous to HSA21q22		C57BL/6J(>N6)	Besson et al. (2007)

Df(10Prmt2-Cstb)4Yah[Ms4Yah;Df4Yah]	Cre-mediated rearrangement on MMU10 between loxP sites inserted in Prmt2 and Cstb caused inactivation of both flanking genes and deletion of the HSA21q22 homologous segment		Deletion of segment homologous to HSA21q22	C57BL/6J(>N6)	Duchon et al. (2008)
Df(17Abcg1-Rrp1b)5Yey[Df(17)1Yey;Df5Yey]	Cre-mediated rearrangement on MMU17 between loxP sites in areas proximal to Abcg1 and distal to Rrp1b caused deletion of Abcg1 and Rrp1b genes		Deletion of genes, including Abcg1 and Rrp1b	Mixed 129S7 and 129S1	Yu et al. (2010a)
Df(17Abcg1-Cbs)2Yah[Ms2Yah;Df2Yah]	Cre-mediated rearrangement on MMU17 between loxP sites inserted in Abcg1 and Cbs caused inactivation of both flanking genes and deletion of the HSA21q22 homologous segment		Deletion of segment homologous to HSA21q22	C57BL/6J(>N6)	Pereira et al. (2009)

Df, deletion; *Dp*, duplication; *NR*, not reported.
The table has been prepared with the information available in Herault et al. (2017), and Moore and Roper (2007).

(Richtsmeier et al., 2000). Interestingly, Ts65Dn model of DS has been extensively used for understanding the cognitive and electrophysiological abnormalities in DS because it represents many of the phenotypic features of the human DS (Dierssen, 2012; Rueda et al., 2012). The DS reminiscent features, including small hippocampus and cerebellum and abnormal skull shape, CHD, myeloproliferative disorders, reduced bone density, and altered incidence and response to certain cancers. Most interestingly, Ts65Dn model appeared with learning and memory deficits, behavioral disorders, and early onset of Alzheimer disease (AD)-like features and neurodegeneration, including aging (Holtzman et al., 1996; Roper et al., 2020). Early neurodegeneration, including early loss of forebrain cholinergic neurons, could be linked to learning and memory deficits in these mice. The study on the neurological basis of its mental impairment in this model exhibited an inhibition in the dentate gyrus (DG), where the GABA-A antagonists were able to resolve some of this impairment (Fernandez et al., 2007). These mice were seen with a delay in development and unusual behaviors reminiscent to retardation in human DS and eventually encountered astrocytic hypertrophy and other forms of neurodegeneration (Holtzman et al., 1996). They also contained abnormally large neural synapses and other structural changes (Belichenko et al., 2004). Videofluoroscopic swallow studies (VFSS) conducted on adults of Ts65Dn model showed markedly slower swallow rates, longer inter-swallow intervals, and greater numbers of jaw excursion cycles preceding each swallow (Glass et al., 2019). However, quantitative analyses of the intrinsic (transverse muscle) and extrinsic (genioglossus (GG), styloglossus (SG), and hyoglossus (HG)) tongue muscles revealed no significant difference between genotype groups of myosin heavy chain isoform profiles.

The trisomic condition of this strain syntenic to HSA21 has an obvious overexpression of APP gene and age-related increase in levels of the products of APP protein metabolism (Choi et al., 2009). However, these mice do not exhibit development of plaques or tangles of AD characteristics demonstrated in human DS and transgenic mouse models of DS. Aged Ts65Dn mice exhibited accumulation of tau/reelin containing clusters in the CA1 region of the hippocampus along with extracellular granular deposits of tau/reelin, which are also reported in mouse models of AD (reviewed in Vaccano et al., 2012). Enhancement of age-related disorders, including increased risk of lymphoma has been observed in systemic studies of the Ts65Dn model (Levine et al., 2009; Vaccano et al., 2012). Collectively, the triplicated genes in the Ts65Dn mouse appeared most consistent with DS-related cognitive abnormalities (Rueda et al., 2012), severe behavioral deficits in different learning and memory tasks, electrophysiological alterations in both synaptic transmission, and hippocampal synaptic plasticity, in comparison with other DS mouse models. In addition, Ts65Dn mouse exhibit altered dendritic spine morphology and impaired neurogenesis in the developing brain and in neurogenic niches of the adult brain (Belichenko et al., 2004; Reeves et al., 1995; Rueda et al., 2012; reviewed in Vaccano et al., 2012). This strain also exhibits some sleep alterations and hyperactivity in locomotor behavior similar to DS patients (Reeves et al., 1995; reviewed in Vaccano et al., 2012).

Although this model exhibits highly similar features reminiscent to human DS, there are weaknesses emanating from the genetic point of view (Gardiner et al., 2003), such as the gene for retinal degeneration which leads to blindness, and thus, restricts use of this model for tests requiring vision. However, that gene has been bred out of the Ts65Dn model, making the model to be otherwise equivalent to the original TS65Dn strain. Another drawback of this model is the reproductive potential; the male mice of this model are functionally infertile, and poor transmission of the small marker chromosome. T(17(16))65Dn, carrying the distal end of the MMU16 takes place, thereby limiting efficient production of these animals for experimental purposes. Furthermore, Ts65Dn mouse, though not spontaneously epileptic, shows increased incidence of seizures in some experimental epilepsy paradigms (Cortez et al., 2009; Westmark et al., 2010; Joshi et al., 2016).

Ts65Dn mouse muscle stem cells exhibit accumulation of DNA damage and overexpression of a histone deubiquitinating enzyme, Usp16, that regulates response to the DNA damage (Pawlikowski et al., 2018). Such dysfunctions of the muscle stem cells may impair muscle regeneration, including signs of early aging and contribute to pathomechanism in DS. These mice also have significantly reduced numbers of hematopoietic stem cells (HSCs) along with an increase in production of reactive oxygen species (ROS) compared to euploid cells of the wild-type littermates (Lorenzo et al., 2013). HSCs from Ts65Dn mice exhibited elevated level of spontaneous DNA double-strand breaks, which correlates with significantly reduced HSC clonogenic activity compared to controls (Wang et al., 2016). HSCs from Ts65DN mice are also less proficient in repair of DNA double-strand breaks than cells of the normal euploid mice. These observations were hypothesized to be caused by additional copy of genes syntenic to HSA21, which may selectively impair the ability of HSCs to repair double-strand breaks and that contributes to hematological abnormalities and malignancies in human DS. Additional DS mouse models have been created to overcome the shortcomings of Ts65Dn model, such as the Ts1Cje and the Ts1Rhr, which carry a smaller triplication of the MMU16 syntenic region. Nonetheless, it is the only mouse model extensively used for preclinical identification of pharmacological interventions, especially those which are relevant to cognitive impairment in DS (Gardiner, 2014).

3.4 Ts1Cje (Ts12; 16C-tel)1Cje)

The Ts1Cje mouse is partially trisomic for a smaller segment of MMU16 compared to the triplicated region in Ts65Dn model and that is translocated to the distal portion of MMU12. The translocation between MMU16 and MMU12 in Ts1Cje model was created with an attempt to inactivate Sod1 by targeted mutagenesis. Thus, the Ts1Cje mice have three copies of the distal portion of MMU16 from the genes Sod1 to Mx1, and the Sod1 gene is inactivated in this model. The MMU16 in the Ts1Cje mouse carries a duplicated segment that contains 75 well-curated genes of HSA21 protein coding orthologs (Sago et al., 1998; Gupta et al., 2016). Notably, the translocation caused deletion of seven genes located in the most telomeric segment of MMU12, which are monosomic in Ts1Cje model (Duchon et al., 2011). Two of these seven deletions, namely, Abcb5 and Itgb8, are related to ABCG1 (located on MMU17) and CD18 (located on MMU10) of HSA21; however, these are not trisomic in either Ts1Cje or Ts65Dn mice (Vaccano et al., 2012). Additionally, the Ts1Cje mice are disomic for 19 genes, including APP and SOD1, while these genes are trisomic in Ts65Dn or Ts2Cje mice (which are genetically equivalent). In Ts1Cje mice, expression of genes involved in Jak-STAT signaling pathway has been characterized throughout development (Lee et al., 2019). Nevertheless, the DS features, including oxidative stress, mitochondrial dysfunction, and tau hyperphosphorylation linked to neurodegeneration and premature aging, were present in Ts1Cje mice, even though this model was diploid for App and Sod1 genes (Shukkur et al., 2006). These findings indicate that other trisomic genes may play contributing role for expression of these features in Ts1Cje mice. The Ts1Cje mice also exhibited deficits in synaptic plasticity and behavior in the T-maze test, similar to that of Ts65Dn mice; however, memory performances were insignificant in the

Morris water maze, which was substantially spared in the novel object recognition (NOR) test (Sago et al., 1998; Siarey et al., 2005; Belichenko et al., 2007; Fernandez and Garner, 2007). Additionally, Ts1Cje mice appeared hypoactive (Sago et al., 1998). Both female and male Ts1Cje mice appeared equally fertile. Altogether, the Ts1Cje mice showed some of the DS-related phenotypes, which were present in somewhat milder form and/or appeared with some discrete characteristics (Rueda et al., 2012).

Moreover, owing to presence of DS-reminiscent features and elevated levels of BDNF and phosphorylated Akt-mTOR in Ts1Cje mice, this strain was considered for pharmaceutical testing of rapamycin. Interestingly, the treatment in these mice repaired the neuronal defects such as abnormally high local dendritic protein translation that plays a key role in memory formation (Troca-Marín et al., 2011). The result has indicated the potential use of rapamycin and other inhibitors of mTOR for mitigating learning and memory loss and intellectual disability in human DS. The mTOR (mammalian target of rapamycin) acts as a key regulator of metabolism, dendritic morphology, and synaptic plasticity, whereas rapamycin acts an inhibitor of mTOR and is potential to improve the health and expand lifespan of the diploid and noninbred mice (Miller et al., 2011).

3.5 Ts2Cje (Ts[Rb(12.Ts17^{16}65Dn)] 2Cje)

This model was identified after a fortuitous translocation of the Ts65Dn marker chromosome. Indeed, the Ts2Cje model carries the same segment of MMU16 triplicated in the Ts65Dn mouse but is translocated to chromosome 12. The male sterility and transmission of genetic element of Ts65Dn mice have been overcome by creating this as a derivative model of Ts65Dn, named as Ts2Cje, where the extra marker chromosome of Ts65Dn has undergone a Robertsonian translocation with MMU12. The stable chromosomal rearrangement conferred enhancement of breeding potential, though diminished in male mice of this strain and increased in incidence of transmitted segmental trisomy through the female germline (Villar et al., 2005; Moore et al., 2011). The dosage imbalance of HSA21-homologous genes was retained from App to the telomere and expression level within the triplicated region was similar to Ts65Dn. Reduced spine density on the dendrites of dentate granule cells was accompanied by significantly enlarged dendritic spines in the Ts2Cje model compared to euploid mice. The structural dendritic spine abnormalities appeared similar to those observed in Ts65Dn mice. Summarily, the Ts2Cje model offered reproductive advantages without sacrificing the integrity of the Ts65Dn model of DS.

3.6 Ts1Rhr and related mice

The Ts1Rhr mouse was generated by Cre/lox chromosome engineering technique that conferred a tandem duplication of a tiny MMU16 region comprising of 33 HSA21 protein-coding orthologs from the so-called "DS critical region" (DSCR; Delabar et al., 1993; Korenberg et al., 1994). This mouse is trisomic for 33 genes, equivalent of the hypothetical human DSCR region (Deitz and Roper, 2011). DSCR was postulated to be critical for development of DS characteristics (see Chapters 6 and 7). The initial report demonstrated that the human DSCR in this model was not sufficient to produce the craniofacial phenotype of Ts65Dn model or human DS (Olson et al., 2004). It was postulated that DSCR was necessary but not sufficient to produce structural, learning, and memory deficits assessed by Morris water maze in Ts65Dn mice, as seen in human DS (Olson et al., 2007). A comprehensive behavioral testing using the same mice revealed that the trisomic DSCR was sufficient to cause DS-specific behavioral, neurophysiological, and synaptic characteristics (Belichenko et al., 2009).

Behavioral deficits were observed in the T-maze and NOR tests but not in the Morris water maze test (Olson et al., 2007; Belichenko et al., 2009). The test has expressed alteration of 20 of 48 DS-related features, but some were in milder form than that of Ts65Dn or Ts1Cje models. Notably, the Ts1Rhr mice showed characteristics, including increased body and brain weight and expanded posterior hippocampal region comparable to diploid mice; however, these features were not evident in human DS and other mouse models, such as Ts65Dn and/or Ts1Cje. These mice showed impaired synaptic plasticity in the hippocampal dentate gyrus (DG), but plasticity was unaffected in the cornu ammonis (CA1) hippocampal region. In addition, DS-like craniofacial dysmorphology and the small dysmorphic mandible appeared to be developed from larger embryonic craniofacial precursors (Deitz and Roper, 2011). However, variability of these phenotypes was reported in subsequent studies with the Ts1Rhr model, which was explained by differences in the genetic background. Notably, increased expression of Dyrk1a and differential expression of Ets2 could underlie mandibular precursor (Deitz and Roper, 2011). Altogether, the phenotypic expression involving the locomotor behavior in Ts1Rhr mice was almost similar to that of wild-type animals (Belichenko et al., 2009). The result suggested that there could be variable implications of partial trisomy 21 and full trisomy on development of DS features as observed in Ts65Dn mice with reduced bone density (Blazek et al., 2011), whereas the feature was not seen in the Ts1Rhr model (Olson and Mohan, 2011). Collectively, the Ts1Rhr mice though showed some of the DS-related phenotypes, appeared with somewhat milder features and/or appear with some discrete characteristics (Rueda et al., 2012; Vaccano et al., 2012).

3.7 Ts1Yah

The Ts1Yah mice were trisomic for the HSA21 syntenic region on MMU17 containing 12 genes between abcg1 and U2af1 (Pereira et al., 2009). Interestingly, these mice exhibited learning and memory deficits on open field, Y arm maze, and NOR tests. Additionally, these mice were more efficient in learning in the Morris water maze test and showed larger and long term potentiation (LTP) responses than the diploid mice. Their efficient LTP could be related to improved performance in the Morris water maze test. Such provocative findings have hypothesized that DS phenotypes are caused by interaction of many regions of HSA21, indicating certain genes or regions of trisomic HSA21 may improve some aspects of cognition. Such compensatory mechanism has been postulated to exist in human DS also (Head et al., 2007).

Ts1Yah mice indicated necessary precautions for examining the chromosomally engineered mice. The genetic expression of the trisomic region revealed that the expression of the two genes, Abcg1 and U2af1, at the triplicated or deleted ends remained unaltered in monosomic, disomic, and/or trisomic mice, while Ubash3a, Tff2, Tff3, and Tmprss3 were equally expressed in monosomic and trisomic segments. The other genes appeared expressed according to their gene dosage in the region. The Umodl1 gene, adjacent to the Abcg1 but not in the engineered region, showed apparent overexpression in thalamus but not in the hippocampus and cerebellum. Thus, the chromosomally engineered mice have illuminated the unexpected expression of genes located near the rearranged chromosomal region (Vacano et al., 2012).

3.8 The relational context of the Ts65Dn, Ts1Cje, and Ts2Cje models

The genetic conditions in Ts1Cje and Ts2Cje mice may indicate a useful approach of comparison to associate the genetic combination and phenotypic expression. Presumably, the two strains have difference in copy number of genes, which indicates differences in features, such as less severe learning deficits in Ts1Cje compared to Ts2Cje mice, and absence of degeneration of basal forebrain cholinergic neurons (BFCN) (Sago et al., 1998). On the other hand, the similarities in phenotype such as enlarged brain ventricles and decreased neurogenesis observed in the two strains could be contributed by the common set of trisomic genes (Ishihara et al., 2010, 2021). Notably, many of the neuroanatomical features reminiscent to human DS, including regionally selective decrease in dendritic spines, are reported in Ts1Cje, though it was less severe (Belichenko et al., 2007). Ts1Cje mice do not display the age related decline in BFCN neurons typical of Ts65Dn mice (Sago et al., 1998). Also, unlike Ts65Dn mice, Ts1Cje mice show more deficits in spatial than nonspatial learning. Moreover, while comparing the phenotypes, trisomy of MMU17 genes in Ts65Dn and monosomy of MMU12 genes in Ts1Cje shall be kept in mind. The Ts65Dn mice are aneuploid having one freely segregating extra chromosome, which is absent in Ts1Cje, Ts2Cje, and many other mouse models. The extra chromosome in Ts65Dn mice may exert some additional phenotypic effects (Duchon et al., 2011).

3.9 Other mouse models of DS with trisomy or monosomy for the entire chromosomal regions of MMU16, MMU17, and MMU10 syntenic to HSA21

Mouse models of DS research have heralded the understanding of the genotype–phenotype association of human DS (Table 14.1). However, elucidation of phenotypic expression and contribution of the triplet set of genes of HSA21 could not be achieved fully even after developing so many genetically modified mouse models of DS. Thus, a continuous effort and evolving genetic engineering techniques have produced enormous number of models for DS research. Ideally, the mouse model of DS shall have trisomy of all mouse genes syntenic to HSA21 (the relevant segments of MMU16, MMU17, and MMU10). The use of chromosome engineering technology and selective breeding has resulted in seminal accomplishment of developing trisomic mice for the relevant regions of each mouse chromosome (Yu et al., 2010a). Development of such mice was aimed to allow evaluation of the intriguing effect of trisomy of each individual syntenic chromosomal region in different strains.

Undoubtedly, the laboratory mouse, especially Ts65Dn model, has made tremendous contributions to understanding the association of gene dosage imbalance and DS phenotypes. Despite its utilization in DS research, it has limitations, including its trisomy being only ~60% of HSA21-orthologous mouse genes, and its translocation chromosome has a third copy of several genes which are not HSA21 orthologs (reviewed in Moyer et al., 2021). Therefore, several (more than 20) new mouse models have been developed since the creation of Ts65Dn strain, to study the consequences of trisomic HSA21 genes or their mouse orthologs (Herault et al., 2017; Gupta et al., 2016). Notably, newer models such as mouse artificial chromosome models, iPSC-based chimeric organoid models, and rat models are being increasingly considered in DS research.

In the recent era, Cre-/lox-mediated chromosome engineering technique has created three new "triple trisomic" murine lines, such as Dp10, Dp16, and Dp17 also termed as Dp(10)1Yey/+, Dp(16)1Yey/+, and Dp(17)1Yey/+, respectively, through tandem duplication on the corresponding chromosomes (MMU10, MMU16, and MMU17). These mice are individually trisomic for the three complete syntenic regions on MMU10, MMU16, and MMU17, respectively (Li et al., 2007; Yu et al., 2010a,b). Dp10 mice did not exhibit any detected alterations in hippocampal LTP or in learning and memory behaviors. Dp17 mice showed increased hippocampal LTP; however, Morris water maze or contextual fear conditioning tests did not reveal any significant change in learning and memory behavior. The Dp16 mice exhibited abnormalities in hippocampal LTP, and both Morris water maze and fear conditioning tests were similar to that of Ts65Dn mice. This model presents an array of symptoms, including an increased rate of heart defects and learning and memory deficits, which are comparable to symptoms seen in human DS. These mice also showed an enhancement of birth defects in the pancreas (annuler pancreas) and intestinal malrotation. Interestingly, unlike the Ts65Dn model, Dp16 mice contain a duplication of only the parts of MMU16 that is homologous to HSA21. This makes the Dp16 model a genetically more accurate representation of human DS. Dp16 mice are not trisomic for MMU17 genes, which are trisomic in Ts65Dn mice, suggesting that these genes are unlikely to be important for the measured phenotypic changes. Also, the measured phenotypes indicate that genes of MMU10 and MMU17 are not major contributors of phenotypes of human DS but confer minimal effect. Therefore, the study of these mice indicates involvement of various regions, and their interactions with each other is important for expression of DS-like phenotypes. However, further extensive investigation in these mice may draw different interpretations regarding the genotype–phenotype association relevant to human DS.

Interestingly, the dissection of the different triplicated regions in these Dp10, Dp16, and Dp17 mice was useful for understanding the relative contribution to the diverse phenotypes of DS pathology. Moreover, crossing of the Dp10, Dp16, and Dp17 lines enabled creation of fully trisomic mice with all HSA21 ortholog genes, though these triple trisomic mice appeared with compromised viability and poor breeding capacity, and that has restricted their use in several other experimental studies for delineating the gene-expression in DS (Yu et al., 2010a; Belichenko et al., 2015). Nevertheless, the behavioral and synaptic plasticity deficits in Dp16 and triple trisomic (Dp10/Dp16/Dp17) mice were comparable to that of Ts65Dn mice (Yu et al., 2010a,b; Belichenko et al., 2015). Additionally, normal or somewhat enhanced performances of the single trisomic Dp10 and Dp17 strains highlighted the significance of the MMU16 syntenic region on the most features of DS phenotypes (Yu et al., 2010b).

These compound strains of DS mouse model are currently available as the most complete model for DS research. One of the important DS phenotypes is CHD observed in ~50% of the individuals with DS, which was thought to be caused by the triplicated HSA21. Studies conducted in the three mouse models delineated that only MMU16 region is responsible for CHD in mice. Further investigation in Dp(16Tiam1-Kcnj6)Yey/+ and Df(16)1Yey/+ (produced by chromosome engineering) has refined the specific region responsible for CHD, which might be either a 5.43 Mb duplication or a corresponding deletion of an area extending from and including the Tiam1 and Kcnj6 genes (Liu et al., 2011; Vaccano et al., 2012). Breeding of these two mice strains restored disomy of genes of this region and identified the necessity of trisomy of Tiam1-Kcnj6 region to produce heart defects in these experimental mice. The attempt was followed for breeding trisomic mice with knockouts of individual genes for studying genotype–phenotype association in various DS mouse models. Interestingly, these three mouse lines of DS model appeared ideal for having genetic triplication; however, these lines did not have the extra freely segregating chromosome comparable to that found in most individuals with DS as well as in Ts65Dn mouse model. Notably, the phenotypic consequences have been demonstrated to be caused by the co-operating effects of the extra unpaired chromosome and interaction with the disomic chromosomes because the trisomic condition could have significant impact on the global gene expression and/or chromatin structure (Reinholdt et al., 2009; Dierssen, 2012).

3.10 Transchromosomal model: Tc1

Understanding of the variable phenotypic expression caused by the increased gene dosage of HSA21 in trisomic mouse models exhibited some or the other drawbacks; thus, introduction of the HSA21 was attempted into the mouse genome for delineation of the precise expression of the trisomic HSA21 in the mouse system. The Tc1 mouse was produced using microcell-mediated chromosome transfer and embryonic stem (ES) cell technologies with the freely segregating copy of HSA21 with 91% of the full complement of HSA21 genes (Fig. 14.1) (O'Doherty et al., 2005). Notably, almost the entire HSA21 was introduced into the mouse ES cells to generate the Tc1 mutant strain. Thus, this model was suitably known as "transchromosomal model." This mouse expresses HSA21 sequences at mRNA, protein, and functional level (O'Doherty et al., 2005; Reynolds et al., 2010). Importantly, vascular endothelial growth factors (VEGF) -dependent angiogenic response was restored in the Tc1 mice indicating overexpression of transcripts encoded by four genes (Reynolds et al., 2010). However, irradiation of HSA21 prior to transfer to mouse cells resulted in genetic alterations, including deletions, indicated loss of a small number of HSA21 genes, leaving ~83% of the genes of HSA21 (Reynolds et al., 2010). Notably, an apparent duplication of S100β and PRMT2 genes was demonstrated in the inserted chromosome in Tc1 mice (Duchon et al., 2011).

Interestingly, complete sequencing of the HSA21 demonstrated that γ−irradiation in Tc1 mice induced de novo rearrangements, such as mosaicism, meaning incomplete trisomy, wherein the chromosome is present along with other clone (different chromosomal complement) in variable number of cells in a tissue (Gribble et al., 2013). In addition to rearrangements of ~20% of HSA21 genes, random loss of the human chromosome in cells of this model depicts every individual as a developmentally unique mosaic of trisomic and euploid cell (O'Doherty et al., 2005; Gribble et al., 2013). Despite the presence of secondary molecular aberrations, the Tc1 mice expressed a number of DS-like phenotypes, including learning and memory impairment, cardiac defects, neurologic and craniofacial abnormalities, and decreased tumor angiogenesis (O'Doherty et al., 2005; Galante et al., 2009; Reynolds et al., 2010). Thus, the Tc1 strain has the advantage of allowing the study of the HSA21 genes, including those which are not found in the orthologous regions of MMU10, MMU16, and MMU17 of mice strains (Gardiner et al., 2003). However, phenotypic features of these mice were largely dependent on the inherent genetic background of the mouse. The additional caution on using this mouse strain was loss of the extra HSA21 chromosome on inbred genetic background. Cross breeding of these mice with mouse strains carrying small deletions resulted in a decrease in copy number of trisomic genes indicating precise association of genes with DS characteristics from the deleted intervals. Furthermore, the cross breeding of Ms4Yah strain, which carries a deletion of the region of MMU10 that is orthologous to HSA21 (Fig. 14.1), with the Tc1 strain demonstrated that trisomic genes in this interval are not responsible for learning defects (Duchon et al., 2010). Such cross mapping has indicated the necessity of cross-breeding with even smaller deletions with a "knockout" for the relevant gene, which might eventually identify the individual dosage-sensitive genes in the mouse model. Tc1 mice have exhibited germline transmission of almost the intact HSA21 (proximal and distal gaps eliminate ~10% of the HSA21 sequence and 8% of HSA21 genes); however, the possibility of tissue-specific variability exists in cellular trisomy.

Fusion of a mouse artificial chromosome with the long arm (q) of HSA21 enabled development of the TcMAC21 model (Kazuki et al., 2020). The strain possessed almost the intact copy of the HSA21q, appeared nonmosaic, displayed several features of trisomic HSA21 of DS, and thus, was represented as a better model compared to other trisomic mouse models (Moyer et al., 2021). However, the possibility of loss of the human chromosome or masking of its features by the mouse artificial chromosome in mouse cells remains. Nonetheless, a continuous effort is necessary for creating newer models with full or segment of HSA21 for establishing singe gene-single phenotype links.

3.11 Transgenic mice

Transgenic mice are produced with an artificially introduced additional foreign genetic material, preferably a single gene, indicating gain of function of the gene or overexpression of corresponding proteins. The transgenic mice are increasingly used in DS research to study the biochemical and phenotypic implications of the individual extra gene in vivo. Transgenic mice have been developed with single HSA21 orthologous genes or regions that result in dosage imbalance (Altafaj et al., 2001; Roubertoux et al., 2006). The features of the transgenic mice often concentrate on copy number of genes, the site of integration of the transgene, level of transcription and protein expression, tissue specificity and time of expression, and obviously, the genetic make up of the mice. Studies involving molecular cloning of individual HSA21 encoded genes facilitated understanding of the expression and organization of their proteins and their possible contribution to DS characteristics. Since single trisomic genes may pose significant impact on specific phenotype, a comparison of spatial and temporal gene expression with other segmental trisomic models and/or with DS individuals can extract critical information on genotype–phenotype association. Also, emphasis shall be bestowed on differences in quantitative gene expression, limited splice variants and absence of potential regulatory regions in transgenic mice. Several transgenic mice strains have been studied for delineating the possible association of several triplicated genes of HSA21 with phenotypic consequences.

3.11.1 Transgenic mice to study intellectual disability and altered brain function in DS individuals

A number of HSA21 genes and their function have been demonstrated to be associated with neurodegenerative disorders such as AD and amyotrophic lateral sclerosis (ALS or Lou Gehrig disease), synaptic function, and development and degeneration of neurological system. The following HSA21-encoded genes are altered due to triplication, which have been studied in transgenic mice for deciphering the effect of single gene on specific phenotypic expression related to DS.

3.11.1.1 APP

Several studies have demonstrated that APP, encodes amyloid β precursor protein, and is overexpressed in some trisomic mouse models and in DS individuals (discussed in detail in Chapters 6, 7, and 12). Briefly, abnormalities in posttranslation processing and cleavage of the APP by the β-secretase result in aggregation of β-amyloid peptide following cleavage by γ-secretase (Vaccano et al., 2012). Neurological feature of AD is expressed by the formation of amyloid plaques. The initial transgenic models with wild-type overexpression of APP did not present neurodegenerative features of AD pathology or amyloid plaques, except in one transgenic line (reviewed in Vaccano et al., 2012). Later, transgenic mice developed for wild type (WT) human APP displayed an increase in APP expression that caused a modest but significant reduction in nerve growth factor (NGF) transport (Salehi et al., 2006). Further research has reported multiple pathological signs in mice due to overexpression of WT APP, including significant histopathological abnormalities in cytoskeleton, cognitive deficits, synaptic dysfunction, and also cell loss in the hippocampus and entorhinal cortex (Simón et al., 2009). The alterations were associated with an early enhancement of phosphorylated tau protein and elevation of APP-derived carboxy terminal fragments and almost undetectable levels of Aβ peptide. These observations strongly indicated the involvement of Aβ peptide as an independent pathomechanism in AD.

Interestingly, mutations leading to overexpression of mutant APP and induction of plaque pathology in familial AD (FAD) have been observed in transgenic mice. These mice exhibited early onset of DS-like AD (DSAD) features. DSAD has been characterized with overexpression of WT APP (reviewed in Vaccano et al., 2012). Overexpression caused by duplication of the APP gene has been postulated to lead to FAD (reviewed in Vaccano et al., 2012).

The level of APP overexpression in the mouse models was much higher (often up to 10 times) than physiological levels suggesting that such high levels of APP or any other protein could be toxic for the cells. Amyloidopathy-linked progressive neurodegeneration observed in humans was infrequent in mice. Notably, amyloidopathy is characterized with cognitive decline and an established feature of normal aging and/or a predementia stage, rather than that of a complete neurodegenerative disease (Ashe and Zahs, 2010). Notably, the amount of plaque is not correlated well with severity of cognitive decline in people with AD (Head et al., 2007). Moreover, although several studies have been conducted on transgenic mice of APP models, further investigations have been suggested for establishing a one-on-one genotype–phenotype association.

3.11.1.2 RCAN1

The regulator of calcineurin one gene (RCAN1), encodes the calcipressin-1 protein, interacts with calcineurin A and inhibits calcineurin-dependent signaling pathways, such as transcription factors for activation of nuclear factor of activated T-cells (NFAT) (illustrated in previous chapters in detail). RCAN1 is up-regulated by calcineurin signaling indicating its regulation through a negative feedback loop. Its expression is enhanced by Ca^{2+} in a calcineurin-dependent manner causing sustained calcineurin repression through a negative feedback mechanism. Calcineurin is a Ca^{2+}–dependent phosphate, ubiquitously expressed in the heart, skeletal muscle, developing and adult brain, and in endocrine tissue in abundance. Such Ca^{2+}–dependent neuronal functions are responsible for cytoskeletal stabilization, neurite outgrowth, release of neurotransmitters, and also apoptosis. DSAD is significantly contributed by the regulation of calcineurin by RCAN1 gene. RCAN1 is overexpressed in brain in DS fetuses and Ts65Dn trisomic mice (1.8 fold) and thought to affect their CNS development (Ermak et al., 2001; Baek et al., 2009). Also, its overexpression is described in the hippocampus and cerebral cortex in individuals with sporadic AD, wherein its chronic overexpression might cause neurofibrillary tangles of AD pathology (Ermak et al., 2001).

Transgenic mouse models developed with a human RCAN1 cDNA splice variant exhibited a 4-fold enhancement of expression under the control of endogenous promoter. A noncalcineurin-dependent expression was reported in chromaffin cells obtained from these transgenic mice, responsible for disruption in exocytosis and vesicle trafficking mechanisms (Keating et al., 2008). Such overexpression was much higher than the apparent 1.5- to 2-fold increase observed normally in individuals with DS.

Another RCAN1 transgenic mouse model was produced under the control of the platelet-derived growth factor β (PDGF-β) promoter, which revealed a disruption in visuo-spatial learning. The Morris water maze test exhibited 1.3- to 1.5-times increase in expression in the hippocampus and cerebral cortex accompanied by poor performance in these mice. However, performance on memory tasks was not different, indicating unimpaired retention after learning a task (Dierssen, 2012). The result led to the conclusion that overexpression of RCAN1 may interfere in learning, but not memory, and disrupt the balance in hippocampus following calcineurin-dependent phosphorylation and/or dephosphorylation.

3.11.1.3 SYNJ1

HSA21-encoded SYNJ1 (synaptojanin 1) gene is trisomic in individuals with DS. It acts through dephosphorylation of phosphotidylinositol-4,5-bisphosphate (PtdIns(4,5)P_2) and may stabilize the phosphate kinase type-1γ (PIPK1) (illustrated in previous chapters). The PtdIns(4,5)P_2 acts as an intracellular signaling phospholipid and plays key roles in cytoskeletal dynamics, membrane trafficking, and signal transduction mechanism. Thus, its balance is crucial for cellular signaling events, and that is critically maintained by PIPK1 and SYNJ1 at neuronal synapses. On the basis of its crucial role in cellular dynamics through regulation of PtdIns(4,5)P_2, and contribution of overexpressed SYNJ1 for dysfunction of PtdIns(4,5)P_2 metabolism in neurophysiological changes observed in DS, it is hypothesized to cause early onset of AD pathology (reviewed in Vaccano et al., 2012; Voronov et al., 2008). Overexpression of SYNJ1 may increase sensitivity of the neurons to β-amyloid plaque insults, which are developed in DS due to overexpression of APP.

Transgenic mouse models were developed for both human and mouse SYNJ1 gene with BAC constructs, and these mice appeared with a 2.5-fold elevation in transcript levels along with 59% and 38% increase in protein levels in human and mouse models, respectively. Alteration in PtdIns(4,5)P_2 metabolism evidenced in brains of these mice was rendered by the overexpressed SYNJ1, in association with poor performance in the Morris water maze test indicating cognitive and learning deficits (Voronov et al., 2008). It was hypothesized on the basis of pleiotropic nature of the PtdIns(4,5)P_2, that abnormalities in PtdIns(4,5)P_2 metabolism might affect several cellular activities.

3.11.2 Transgenic mouse models to study the cancer spectrum in people with DS

Children with DS are at increased risk of childhood leukemia. Conversely, development of solid tumors in these individuals is seldom, and the occurrence has been demonstrated to be age related. Similar features have been described in trisomic mice of DS models. Further investigation has been carried out in transgenic and knockout mice with manipulated genes syntenic to HSA21 and relevant to cancer.

3.11.2.1 ETS2

The v-ets erythroblastosis virus E26 oncogene homolog 2 (avian; murine homolog Ets2) has been mapped onto HSA21, encoding a protein called C-ets-2. This protein acts as a transcription factor of the ETS family. The conserved ETS domain appears as a winged helix-turn-helix DNA-binding domain within these proteins and binds with GGAA/T (the core consensus DNA sequence) of the target genes (Dwyer et al., 2007). The ETS family of transcription factors is involved in activation or regression of genes responsible for cellular proliferation and differentiation, development of stem cells, transformation and tumorigenesis, senescence and apoptosis (Dwyer et al., 2007). Ets2 is involved in the development of the anterior-posterior (AP) axis and paraxial mesoderm (PM) at developmental stage and also essential for development of the trophoblast. Overexpression of ETS2 has been linked to increase in apoptosis and contributes to DS pathology (illustrated in previous chapters). Ets2 transcription factors are present in neurons and seemingly appeared to be critical in formation of neuromuscular junction in mice (reviewed in Vaccano et al., 2012). Indeed, a nearly 2-fold overexpression of Ets2 caused abnormalities in cervical skeletal, and neurocranial and vicerocranial development in mouse models, which was similar to trisomy 16 mouse models and DS patients. A ubiquitous overexpression of Ets2 was evidenced in this mouse model under the control of a metallothionein promoter, resulting in physiological conditions reminiscent to DS individuals and trisomic mice (reviewed in Vaccano et al., 2012). Transgenic mice developed for Ets2 studies revealed that App gene is activated by the Ets2 transcription factors via specific Ets-binding sites and co-operating with the API transcription factor. The cultured cells from brains and primary neuronal cultures developed from Ets2 transgenic mice and fibroblasts exhibited abnormalities, including elevation of App protein and β-amyloid plaques reminiscent to DS patients suggesting the exacerbating adverse effects of APP overexpression (reviewed in Vaccano et al., 2012).

3.11.2.2 RUNX1

RUNX1 (also known as AML1; murine homolog is Runx1) is a hematopoietic transcription factor and is essential for normal hematopoiesis and megakaryopoiesis (illustrated in previous chapters). RUNX1 is essential for terminal differentiation of megakaryocytic progenitors. The heterodimeric transcription complex formed by RUNX1 and core-binding factor β (CBF-β) is frequently observed in leukemia-associated translocations. DS children are prone to leukemia, wherein 10% DS children are born with transient myeloproliferative disorder (TMPD), which often progresses to acute megakaryocytic leukemia (AMKL). Interestingly, triplicated RUNX1 is not directly involved in development of TMPD or progression of AMKL in DS children, though mutations and translocations of RUNX1 have frequently been demonstrated in acute myeloid leukemias (AMLs) (Ganguly and Kadam, 2017; Ganguly et al., 2017; reviewed in Vaccano et al., 2012). Interaction of the trisomic RUNX1 with GATA1, a non-HSA21 gene, may cause GATA1 dysfunctional and favor development of AMKL in DS children. Notably, GATA1 was dysfunctional in DS children and development of AMKL. GATA1 acts as a hematopoietic transcription factor.

For understanding the role of RUNX1 in leukemic development, transgenic mouse models were developed under the control of the GATA1 hematopoietic regulatory domain. The whole bone marrow of these mice showed nearly 5-fold overexpression of Runx1 transcript and its protein; however, such overexpression did not influence the leukemic development. Later, these Runx1 transgenic mice were crossed with BXH2 mouse model of myeloid leukemia, which led to gain of an extra copy of Runx1, that is, trisomic Runx1. These Runx1-BXH2 mice showed early onset of myeloid leukemia, elevation of TMPD, and AMKL risk reminiscent to DS children (Yanagida et al., 2005).

3.12 Mouse models with genes inactivated by targeted mutagenesis

Deletion of a gene could be a powerful tool for evaluation of function and necessity of a gene for viability and contribution to expression of features. Such manipulation of gene dosage has hypothesized that the gene dosage effect may identify the responsible gene specific for DS phenotypes and also developmental instability caused by chromosomal imbalance and/or abnormal expression of several genes (collectively gene-dosage effect).

3.13 Gene-knockout models to study intellectual disability/brain function

3.13.1 APP

A vector designed to replace the App promoter exon one and part of the first intron with a neomycin phosphotransferase gene (PGKneo) cassette was used to generate an App null mutant mice. These mice remained viable without any overt abnormalities, particularly neuroanatomical anomalies of the brain tissue, and did not present detectable App mRNA or protein in App null animals (Zheng et al., 1996). However, function of one member of the amyloid β (A4) precursor-like protein (Aplp1 and Aplp2) genes indicated that the null mice for App and Aplp2 caused perinatal lethality, and similar incidence was observed in null mice for both Aplp1 and Aplp2. Conversely, null mice for only Aplp2 had no apparent abnormalities suggesting the critical role of Aplp2 and that was further supported by the findings that null mice for App and APLP1 remained viable. Nevertheless, no noticeable histopathological aberrations were reported in brain along with viability of the cortical neurons in basal culture in mice null for both App and Aplp2 (Heber et al., 2000).

A significant reduction in transport of NGF in the hippocampus was demonstrated indicating downregulation of gene expression of NGF receptor (p75NTR) affecting the basal forebrain cholinergic neurons (BFCN) in Ts65Dn mice (Salehi et al., 2006). That effect was milder in Ts1Cje mice compared to normal euploid mice and that led to the hypothesis that Ts65Dn mice were trisomic with triplicated App gene but not in Ts1Cje mice. Cross breeding of Ts65Dn mice with null mice for App gene partially rescued NGF action by making App trisomic to disomic. Transgenic mice generated with human APP allele revealed expression of this gene comparable to the level seen in individuals with DS and showed decreased NGF transport.

Therefore, App was major, but not an exclusive contributor for reduction of NGF transport and subsequent BFCN reduction in Ts65Dn mice. The hypothesis was further supported by the observation in transgenic mice developed for both App and presenilin 1 (PSEN1 or PS1), which showed a synergistic effect in lowering NGF transport and the BFCN number. The endosomal alterations detected in Ts65Dn mice (not in disomic Ts1Cje or Ts65Dn mice for App developed by cross breeding: Ts65Dn: App+/+/−) indicated requirement of App for endosomal alterations. Such effects are demonstrated in DS at fetal age, and as an early sign in sporadic and DSAD, even before deposition of β-amyloid and as soluble Aβ levels. Interestingly, endosomal alterations are not observed in transgenic mice developed for either human APP London (APP670/671 and APP717) or Swedish (APP670/671) mutations (reviewed in Vaccano et al., 2012). These mutations caused marked overexpression (2-fold for London and 7-fold for Swedish transgenic) of App suggesting the necessity of overexpression of App; however, overexpression is not sufficient, and thus, needs high expression of one or more genes of MMU16 trisomic in Ts65Dn mice. Decrease in NGF transport might partly be responsible for endosomal alterations (Harrington et al., 2011). Endosomal alterations have been detected in the hippocampus, neocortex, and basal forebrain.

3.13.2 *RCAN1*

Knockout mice was generated for an Rcan1 null allele by using a vector designed for replacing exon 5 and 6 with β-galactosidase (Vega et al., 2003). The null Rcan1 allele did not produce any detectable transcript, and the mice appeared viable and fertile without expressing any overt abnormalities. The knockout mice produced for calcineurin A appeared with diminished response to hypertrophic stimuli rendered by its β-catalytic subunit. Calcineurin is essential and adequate for cardiac hypertrophy to respond to pathological and/or physiological stimuli. Notably, increased expression of calcipressin-1, an inhibitor of calcineurin-dependent signaling, decreases the hypertrophic response, and in contrast, its reduced expression increases the hypertrophic response. Mice null for Rcan1 and transgenic for a muscle-specific expression that activates calcineurin exhibited an aggravated hypertrophic response such as acute fibrosis. However, cardiac hypertrophic response was decreased in these null mice against aortic banding or chronic adrenergic stimulation, which indicated a dual role of calcipressin-1 for cardiac hypertrophy and depends on differences in hypertrophic stimulation (Vacano et al., 2012). The Ts16 mice showed increased expression of Rcan1 isoforms in developing heart and brain compared to diploid littermates and decreased expression of NFAT transcriptional activity. Breeding of Rcan1 trisomic null Ts16 mice with nontrisomic mice restored the disomy of Rcan1 in these mice; however, disomic Rcan1 could not recover craniofacial and cardiac abnormalities (Vega et al., 2003).

Incidence of solid tumors in individuals with DS is remarkably less and considered an age-related disease, though DS children are at increased risk of developing childhood leukemia. The situation has been explained by the increased protective action of some of the trisomic HSA21 genes present in DS. Trisomic RCAN1 is expressed 1.8- and 1.7-times in human DS fetal tissues and Ts65Dn mice, respectively (Vega et al., 2003). Human RCAN1 renders suppression of angiogenic signaling mediated through calcineurin pathway using vascular endothelial growth factors (VEGFs). Testing of two tumor models, such as Lewis lung carcinoma and B16F10 melanoma, demonstrated a considerably higher suppression of tumor growth than wild type, and a reduction in microvessel density. Rcan1 transgenic mice presented a similar result along with a significant decrease in $CD31^+$ and $CD45^-$ (CD31 is an endothelial marker, while CD45 is a hematopoietic one) cells relative to wild-type mice. Co-culture of low numbers of Lewis lung carcinoma cells from transgenic and wild-type mice with a view to generation of slow growth of tumor cells demonstrated an increased expression of Rcan1 that inhibited initial expansion and extended growth of the transplanted tumors. The result indicated an inhibition of neoangiogenesis and cooperation of the existing blood vessels (Vega et al., 2003).

Ts65Dn/$Rcan1^{+/+/-}$ mice generated through breeding exhibited abrogated tumor protection accompanied by enhanced microvessel density indicating the crucial role of overexpressed Rcan1 in these processes. Examination of the contribution of Dyrk1A (trisomic in DS and Ts65Dn and regulates NFAT signaling) described that overexpression of both Dyrk1A and Rcan1 in endothelial cells resulted in higher inhibition of VEGF-mediated endothelial proliferation than the cells with overexpressing Rcan1 alone. The result suggested that Dyrk1A contributes significantly to suppress tumor growth in Ts65Dn/$Rcan1^{+/+/-}$ mice than wild type (Baek et al., 2009). Increased dosage of Rcan1 attenuates VEGF-calcineurin-NFAT signaling.

3.13.3 *SYNJ1*

Synj1 appeared essential for postnatal development in Synj1 null mice. By gene targeting, these null mice for Synj1 were developed using a vector designed for replacement of 103 base pairs from the 3′ region of the first coding exon and 1571 base pairs of the adjoining intron with a neomycin resistance cassette (Cremona et al., 1999). Heterozygous mice for the null allele were fertile and phenotypically normal. Cross breeding of the heterozygotes produced pups in Mendelian ratio with the expected genetic makeup. Homozygous mice (~85%) died within 24 h, while 15% heterozygotes died within 15 days due to severe weakness, ataxia, reduced growth, and generalized convulsions (evoked by the tail flick test). Expression of several neuronal proteins, including intrinsic membrane proteins of synaptic vesicles, synaptojanin 1 interactors, plasma membrane t-SNAREs, were demonstrated to play a role in synaptic vesicle endocytosis and in the synaptic vesicle cycle, and enzymes of the phosphoinositide (PI) metabolism (reviewed in Vacano et al., 2012).

The cultured cortical neurons of the null mice revealed a 1.6-fold enhancement of PtdIns(4,5)P_2 due to reduction in dephosphorylation while PtdInd(4)P remained unaltered. The cells showed an elevated number of clathrin-coated vesicles located around the cluster of synaptic vesicles, along with a number of vesicles isolated from the plasma membrane, in an electron microscopic study. The binding of PI and clathrin coat proteins indicated that enhancement of clathrin-coated vesicles could be influenced, at least in part, by the elevation of PtdIns(4,5)P_2. Consistent result was observed in cell-free assays with incubation of protein-free liposomes from lipid extracts from crude brain in brain cytosol supplemented with ATP and GTP. A 4-fold increase in coated vesicles was observed in cytosol of the null mice compared to wild-type cytosol. An enhancement of clathrin and AP2 binding to liposomes was observed in biochemical analysis of the null cytosol. Addition of purified synatojanin1 to the null cytosol counteracted the outcome (reviewed in Vacano et al., 2012). The hippocampal slices showed decrease in synaptic transmission in hippocampal synapses from null mice and a depression of synaptic response, indicating an effect on actin dynamics due to absence of synaotojanin1, and that result in trapping of clathrin coated vesicles within an actin matrix. Notably, actin cytoskeleton is regulated by PI. In Ts65Dn mice, around 40% enhancement of overexpression of Synj1 caused 33% increased production of phosphatidylinositol monophosphate (PtdInsP) in brain cytosol detected in fluorescent labeled water soluble substrate (NBD-PtdIns(4,5)P_2), and the overproduction of PtdInsP was reduced to control level by making Synj1 disomic in these mice. Overexpression of Synj1 was hypothesized to cause learning deficits in Ts65Dn mice and that was evidenced in Morris water maze test in transgenic mice for either murine Synj1 or human SYNJ1 (Voronov et al., 2008).

3.14 Knockout mice developed for understanding the cancer spectrum in DS

Knockout mice having a specific gene or region knocked out or deleted may extract valuable information for understanding the function of genes, in particular when compared with segmental trisomic models. Several knockout models have been created to delineate the contribution of individual gene on specific DS phenotype.

3.14.1 *ERG*

ERG (the v-ets erythroblastosis virus E26 oncogene homolog; in murine: Erg), is a transcriptional regulator and belongs to the family of the erythroblast transformation-specific (ETS) transcription factors. ERG contributes to chromosomal rearrangements resulting in fusion of Erg with a member of Tet subfamily of DNA-binding proteins leading to development of myeloid leukemia and Ewing sarcoma (5%–10%). Further, deletion of ERG in a subset of acute lymphoid leukemias indicates its potential for oncogenic transformation. Control of ERG expression and associated chromosomal rearrangements by androgen-responsive 5′ elements of TMPRSS2 observed in >50% prostate cancer patients, and its triplication in DS indicates its strong association with hematopoiesis and leukemogenesis mechanism in DS.

Screening of the regulators of hematopoietic stem cell function has detected a germ line mutation represented as Erg^{Mld2} where transition of thymine to cytosine has occurred at exon 12 and that resulted in substitution of proline for serine at residue 329 in the first α-helix of the Ets-DNA binding domain (Loughran et al., 2008). The half-life of the mutant protein appeared similar to that of wild type in pulse chase experiments conducted in human embryonic kidney cell line and that indicated its stability in vivo. Also, the mutation showed wild-type like retention of DNA-binding ability and affinity to E74 enhancer element (an Erg-binding site). However, the mutant protein was unable to activate transcription and promote megakaryocytic differentiation in human erythroleukemic cell line K562.

The trisomic Ts65Dn mice developed progressive thrombocytosis, megakaryocytosis, and megakaryocytic dysplasia in bone marrow, disrupted splenic structure, and extra medullary hematopoiesis in spleen and many other hematopoietic complications (Ng et al., 2010). Breeding of these trisomic mice with the Erg^{Mld2} mutation produced Erg-disomic mice with diminished histopathologic myeloproliferative features relative to wild-type levels. However, another Erg-trisomic mouse strain (Ts1Cje) did not show evidence of myeloproliferative disorder, leading to the hypothesis that Erg is essential but needs cooperation of other genes for expression of the myeloproliferative features in Ts65Dn mice. No transgenic mice have been generated for ERG gene.

Heterozygous Erg^{Mld2} mice were presented with lower platelet counts than wild-type mice and with normal hemoglobin level. Also, histopathologic examination of adult thymus did not reveal any gross abnormality in bone marrow, lymph nodes, spleen, pancreas, liver, kidney, and other organs. Bone marrow and spleen yielded lesser number of colonies and progenitor cells of all lineages in single cell suspension cultures compared to wild type. Erg^{Mld2} heterozygous mice exhibited lower frequency of committed hematopoietic progenitor cells and multipotent stem cells along with a small population of lineage negative Sca-1+c-kit (LSK) cells (cells that represent hematopoietic stem cells and early hematopoietic progenitors). Interestingly, homozygous $Erg^{Mld2/MLD2}$ mice developed from mating of heterozygous Erg^{Mld2} mice demonstrated lethality at embryonic age (viability detected at E10.5 culminated into complete lethality at E13.5). Additionally, developmental delay was grossly accompanied by absence of hematopoietic progenitors for any lineage at E10.5, suggesting failure of hematopoiesis in these homozygous mice for $Erg^{Mld2/MLD2}$ allele (reviewed in Vacano et al., 2012).

3.14.2 *ETS2*

Ets2 gene was manipulated using a vector for replacing all or part of three exons coding for Ets2 DNA binding domain (Ets2db1) with pMC1NeoA and that caused a deletion of a critical region of the gene producing a large fusion transcript and a truncated protein that binds to an Ets2 antibody (Henkel et al., 1996). An abnormality in formation of the extraembryonic tissues did not allow production of homozygous mice from mating of heterozygous mice; however, complementation using tetraploid embryos (that produce active extraembryonic tissues) could have produced homozygous mice for the allele (reviewed in Vacano et al., 2012). The homozygous Ets2db1 mice developed curly whiskers at birth and additionally wavy hairs and slightly rounded forehead at ~ 2 weeks of age. On the whole, skin showed misalignment of hair follicles leading to growth of curly hairs that failed to penetrate the epidermis. Similar phenotype with whisker hair and hair follicle was observed in mice deficient in TGF-α and mice with a point mutation in the EGF receptor. These mice were fertile and unaffected for lymphoid and myeloid development compared to wild type. Gene knockout produced an Ets2 hypomorphic allele (Ets2A72) where threonine residue of Erk-phosphorylation was replaced by alanine (reviewed in Vacano et al., 2012). The homozygous mice with Ets2A72 allele appeared with unaffected fertility and life span without exhibiting abnormalities of hair and hair follicles relative to that of the homozygous mice with Ets2db1. Organ morphologies appeared normal with normal development of mammary glands in females.

Owing to complete lethality of the Ets2db1 homozygous mice caused by placental insufficiency, double-null Ets1/Ets2 was produced with Ets2db1 allele for examining the function of both genes in the ras/Raf/Mek/Erk pathway and preventing the rescue of the individual null phenotypes (Wei et al., 2009). Abnormalities in angiogenesis during development leading to complete lethality by day E14.5 might be caused by mutations in both genes (similar effect was produced by Ets2 nulls (db1 and fl) and mutated

Ets2 (A72)). Both genes are known proto-oncogenes, promote epithelial cell survival in angiogenesis, and may affect tumor angiogenesis in endothelial cells.

While misregulation of ETS2 is associated with cancer, its increased gene dosage lowers the risk of cancer of solid tumors in DS. Examination of Ts65Dn and Ts1Rhr mice having trisomy of Ets2 showed suppression of intestinal tumors in these mice when bred with Apc^{Min} strain having increased susceptibility to intestinal tumors (Sussan et al., 2008), and the trend was consistent with reduced tumor incidence observed in DS. Ts65Dn mice exhibit anomalies in cranial skeleton and thymus. Similar abnormalities were reported in transgenic mice that have constitutive overexpression of Ets2 transcript under the control of metallothionein promoters. These transgenic mice were studied for evaluation of the function of native Ets2 on the craniofacial and thymus features of DS and revealed that reduction of Ets2 expression did not rescue thymus and craniofacial anomalies, except for mesoderm-derived elements. The superoinfero height of the occipital bone appeared decreased by ~16% in Ets2+/− heterozygous mice relative to euploid, and a ~4% reduction was observed in Ts65Dn mice compared to euploid mice. The result indicated that ETS2 might be potential for suppression of tumors in DS; however, Ets2 alone does not cause skeletal or thymus anomalies in trisomic Ts65Dn mice (reviewed in Vaccano et al., 2012).

3.14.3 *RUNX1*

Null mice for Runx1 was developed using a vector for replacement of the splice acceptor and first 20 bp of the exon that encodes the central 52 amino acids of the Runt homology domain (RHD) required for DNA binding with a hygromycin B cassette. The homologous recombination facilitates introduction of stop codons in all three reading frames for producing a truncated protein. The null recombination of the Runx1 in the homozygous embryos caused complete lethality by day E12.5 along with severe hemorrhage within the vertebral canal and ventricle of the central nervous system affecting the origin of the ganglia of the cranial nerves going into the third and lateral ventricles. Acute hemorrhage was also noted in the peritoneal cavity and pericardial space of ~100% null embryos. At day E11.5, ~87% of null embryos appeared viable and indistinguishable from heterozygous or wild-type embryos. However, the liver appeared slightly pallor and showed complete absence of liver-derived hematopoiesis with no trace of erythroid, megakaryocytic, or myeloid cells, while primitive nucleated erythrocytes were present in channels and hepatic sinusoids (reviewed in Vaccano et al., 2012). Stem cells in Runx1 null embryos were able to differentiate into primitive erythroid cells in vitro. Notably, yolk sac or liver of the null embryos demonstrated complete absence of hematopoietic colonies in culture and suggested that Runx1 is necessary for hematopoiesis in liver. Heterozygous null embryos appeared normal with no apparent difference in white blood cell differentials, nucleated blood cell counts, hematocrits, and distribution of peripheral lymphocytes in flow-cytometric FACS scan (reviewed in Vaccano et al., 2012).

In Ts65Dn trisomic mice, reduced proliferation of $CD34^+$ cells was reported in an in vitro hematopoietic study. Development of highly penetrant and progressive myeloproliferative diseases appeared in Ts65Dn mice accompanying with mild anemia, thrombocytosis, bone marrow fibrosis, extramedullary hematopoiesis, and distorted stem, and myeloid progenitor compartments, which were reminiscent to age-related chronic idiopathic myelofibrosis of humans (reviewed in Vaccano et al., 2012). Cross breeding of the trisomic Ts65Dn mice with Runx1 null mice for making mice with disomic Runx1 exhibited that increased Runx1 dosage was not essential for development of extramedullary hematopoiesis, reticulin fibrosis, and megakaryocytic hyperproliferation appeared in Ts65Dn mice with trisomic Runx1 (reviewed in Vaccano et al.,

2012). Studies in Ts1Cje mice having trisomy of a smaller region of MMU16 demonstrated a similar hematopoietic phenotype observed in Ts65Dn mice, except absence of thrombosis or myeloproliferative disease in Ts1Cje mice (Carmichael et al., 2009). This feature observed in Ts1Cje mouse strongly suggested that one more trisomic gene within the trisomic region in Ts65Dn mice is associated with pathomechanism of thrombosis and myeloproliferative disease development. Conversely, other trisomic genes of the trisomic region common in both Ts65Dn and Ts1Cje mice are associated with other hematopoietic anomalies detected in these trisomic animals of DS model.

4. iPSCs with HSA21 as model in DS research

To overcome the uncalled variations in gene expression observed in several DS models, trisomic iPSC lines were developed from individuals with DS (DS-iPSC). A vast range of different cellular models have been developed over the decades for better understanding of the cellular, molecular, and patho-physiological mechanisms involved in neurodevelopment and neurodegeneration in individuals with DS, which includes mouse hybrid cells, transchromosomic mouse embryonic stem cells (ESCs), human ESCs and iPSCs derived from different sources. All of these model systems have provided enormous information in the field of DS research (Table 14.2). The emerging approaches have facilitated study of developmental effects of trisomy 21 in different generations of iPSCs through subsequent differentiation to different cell and tissue types, including cardiac, hematopoietic, neural developments, and many other processes (Antonarakis, 2017; Gough et al., 2020). The iPSC lines were mainly derived from DS patients or amniotic fluid cells of pregnancies with trisomic HSA21 using fibroblasts, lymphoblasts, or neural progenitors. Apart from DS modeling, use of iPSCs also advocated for testing of candidate drugs and chemicals on aneuploidal cells carrying trisomic HSA21, patient-care optimization, and eventually, autologous stem cell replacement therapies. The importance of this DS model has directed development of banking facility and sharing of these cells with global researchers with appropriate regulatory and ethical understanding.

Lymphoblastoid cells derived from DS patients and control individuals were considered for understanding the association of disruptive gene-dosage effect in myriads of studies. Approximately, 29% (22% increase was proportional to gene-dosage effect and 7% were amplified) of HSA21 genes were overexpressed in DS that correspond to open reading frames (ORFs) or genes, while 71% of the expressed sequences were either compensated (~56%, large proportion of predicted genes and antisense transcripts) or due to interindividual variation (possibly due to differences in hereditary genomic differences; 15%) (Graison-Yahya et al., 2007; discussed in previous chapters).

Nevertheless, the variability produced in iPSCs even from a single sample could be overcome by producing more iPSC lines from the same single sample with several replications, which would further be important for understanding the developmental processes of variations in DS pathomechanism. Moreover, the endpoints observed in trisomic iPSCs were always compared with euploid iPSCs derived from normal healthy individuals. Remarkably, DS-iPSC model has facilitated inactivation of one chromosome 21 in vitro, and the approach has created a test system to investigate the changes in genomic expression and cellular pathologies of trisomic HSA21, which is free from genetic and epigenetic noise. Remarkably, silencing of trisomic HSA21 by heterochromatinization surmounts a major first step toward potential development of "chromosome therapy" (Jiang et al., 2013).

TABLE 14.2 Trisomic induced pluripotent stem cells (iPSCs) studied from Down syndrome patients.

Cellular differentiation	Sample sources	Observations
Hematopoietic progenitors	DS patient	Increased tendency of erythropoiesis
	DS patient (isogenic, extra 21 was lost in one iPSC)	Increased propensity of colony formation of erythroid, megakaryocytic and myeloid lineages
	Mosaic DS patient (isogenic)	Faster expansion of hematopoietic lineages
	DS patients (with full and partial trisomy (deletion of 4 Mb) of HSA21), and normal euploids with and without GATA1 mutations	Established as a model for transient myeloproliferative disorder (TMPD); perturbed hematopoietic development through enhanced generation of early hematopoietic progenitors and upregulation of mutant GATA1 expression resulting in accelerated production of abnormally in differentiated cells
Neurons	DS patient	Increased amyloid peptide secretion, tau phosphorylation and cell death in cortical neurons derived from trisomic iPSCs
	DS patients with free (+21) and translocated (+21; t(21q; 21q)); iPSCs reprogrammed from fibroblasts via overexpression of four factors: OCT4, SOX2, KLF4, and c-MYC, by using lentiviral vectors	Similar characteristics of DS-iPSCs , particularly the morphology, surface marker (SSEA4, TRA-1-60, and TRA-1-81) expression, pluripotent-specific transcription-factor expression levels, and methylation status of the OCT4 promoter, to those of human embryonic stem cells; embryoid bodies of Ds-iPSC expressed differentiated markers for three germ layers; DS-iPSCs formed classic teratomas when injected into nonobese diabetic-severe combined immunodeficient (NOD-SCID) mice.
	DS patient	Number of neurons decreased; APP production increased in neuronal progenitor cells.
	DS patient (isogenic)	Perturbation of neurogenesis
	DS patient	Abnormal early glial development
	Mosaic DS patient (isogenic)	Marked deficits in excitatory glutamatergic synapses and inhibitory GABAergic synapses
	DS-iPSC parental line (DS1-iPS4)	XIST-RNA coated HSA21 and HSA21 was transformed into a Barr body by triggering stable heterochromatin modifications, chromosome-wide transcriptional silencing and DNA methylation; rapid reversal of deficits in proliferation and neural rosette formation upon silencing one HSA21.
	Monozygotic twins—DS and normal (isogenic)	Decreased neurogenesis, more astroglia and oligodendroglia; inhibition of DYRK1A reversed these defects
	DS and normal (isogenic and nonisogenic)	Astrocytes with higher RO and lower synaptic molecules; oxidative stress-mediated cell death and abnormal maturation of neurons; minocycline treatment showed partial correction.
	Mosaic DS patient (isogenic)	Reduced neurogenesis; increased mitochondrial dysfunction; elevated secretion of amyloid peptides

Modified from Antonarakis (2017).

5. Organoids (iPSC-based) as chimeric models to study neurogenesis in DS

Promising developments have occurred on use of human iPSCs and their derivatives in mouse models for delineation of the trisomic consequences on development and function relative to characteristics of DS. Indeed, transplantation of differentiated iPSCs from individuals with DS into cells of adult mouse brain with markers of cortical neurons revealed grafting of developing human neurons in mouse cortex consistently organized in large (up to $\sim 100\ mm^3$) vascularized neuron-glia territories with complex cytoarchitecture (Real et al., 2019). Interestingly, cell diversification (human-derived cortical progenitors, neurons, oligodendrocytes, and astrocytes together with host-derived microglia and vessels) was evident in the cellular microenvironment that retained features reminiscent of the human fetal cortex. In vivo longitudinal imaging of >4000 developing human neurons showed development of vascularized neural-glial neighborhoods of cells with human synaptic networks substantially restructured over 4 months, with balanced rates of synapse formation and elimination. Oscillatory population activity mirrored the patterns of fetal neural networks with reduced synaptic plasticity similar to that of DS (Real et al., 2019). However, the extent could not be ascertained to which neurons in human cortical tissue grafts, either generated from human iPSCs (Real et al., 2019) or ESCs, can mimic the maturation, complexity, and functionality of early human fetal cortical networks. Nevertheless, the result demonstrated the potential of human tissue grafts for patient-specific modeling of cortical development, physiology, and pathogenesis.

Keeping the fact that DS patients overproduce $OLIG2^+$ ventral forebrain neural progenitors, brain organoids were produced from iPSCs derived from individuals with DS or a euploid genome (Xu et al., 2019). The DS-iPSC-derived cerebral organoids (COs) showed excessive production of specific subclasses of GABAergic interneurons that caused impaired recognition memory in neuronal chimeric mice reminiscent of DS with elevated expression of the HSA21-encoded transcription factor gene OLIG2. Upon transplantation of the organoids into brains of euploid mice, novel object recognition (NOR) was impaired. Conversely, inhibition of OLIG2 expression in the organoids or shRNA-mediated knockdown of OLIG2 before engraftment largely reverses abnormal gene expression in early-stage DS neural progenitors, reduces interneuron production in DS organoids and chimeric mouse brains, and improves behavioral deficits and NOR in chimeric mice of DS models. The result suggested that overexpression of OLIG2 might underlie the neurodevelopmental abnormalities and cognitive defects in DS patients, and alteration of OLIG2 overexpression could be considered as a potential therapeutic target. Notably, the result showed that $OLIG2^+$ NPCs can generate different subtypes of neuron in 3D culture, while 2D culture can produce only calretinin (CR +) neurons (Xu et al., 2019). Summarily, the report has suggested that OLIG2 could be considered as a potential therapeutic target for DS.

Another study on neuronal differentiation into cortical neurons or brain organoids demonstrated that copy number of APP gene affects amyloid production and deposition but not Tau-pathology, promotes premature neuronal differentiation, and enhances glial differentiation at the expense of neurogenesis. Interestingly, cell type–specific overexpression of different cohorts of HSA21 genes were identified in the single cell RNAseq study (Ovchinnikov et al., 2018; reviewed in Dierssen et al., 2021). COs generated by using CRISPR genome editing technology on isogenic trisomic iPSC lines vary in the gene dosage of APP, including a null line. The RNA and protein expression data appropriately complemented the findings of Ovchinnikov et al. (2018) and provided critical reproducibility in the human stem cell field.

DS patients have increased propensity to AD. Interestingly, organoids derived from DS and familial AD (FAD) patients exhibited spontaneous amyloid plaque deposition and tau hyperphosphorylation, which are more pronounced in DS (Gonzalez et al., 2018). Notably, around 30% of DS patients have delayed onset of dementia, which is demonstrated to be facilitated by triplication of BACE2 (Wiseman et al., 2018). In this direction, the trisomic level of BACE2 might protect the DS-iPSC organoids from early AD-like amyloid plaque pathology. Another study described similar findings of BACE2-mediated suppression of AD neuropathology in iPSC-derived CO of DS models (reviewed in Dierssen et al., 2021). In particular, DS-iPSC organoids secreted increased BACE2 compared to isogenic euploid. Importantly, subtraction of the BACE2 gene copy number from three to two in DS-iPSC organoids by CRISPR/Cas9 gene editing caused decrease in the relative abundance of BACE2-related Aβ cleavage fragments and that strikingly accelerated and triggered early onset of AD-pathologic features, including insoluble amyloid deposits, and pathologically altered Tau in these trisomic organoids (Alić et al., 2021). These results suggested that increase in BACE2 may suppress the development of AD pathology in DS, and thus, BACE2 can be considered as a therapeutic target.

COs derived from DS-iPSCs explored cell-type–specific abnormalities associated with DS of early brain development stage using single-cell RNA-Seq (scRNA-Seq) analysis (Tang et al., 2021). Reduced size of the organoids depicted aberrant neurogenesis with diminished proliferation in the ventricular zone (VZ), decreased neuron distribution in the cortical plate, and reduced expression of layer II and IV markers in cortical neurons in the subcortical regions. These findings provided evidence of impaired neurogenesis of the cortical neurons that may reduce the cortical size in brains of DS individuals. Interestingly, genetic correction using CRISPR/Cas9 and CRISPR interference (CRISPRi) and treatment with small-molecule inhibitor FRAX486 for suppression of the DSCAM/PAK1 pathway demonstrated rescue of NPC proliferation, increased size of these organoids, and reversal of the abnormal neurogenesis. The result indicated the crucial role of DSCAM/PAK1 pathway in regulating proliferation and neurogenesis and that may be responsible for the abnormal development of DS neurons. The DSCAM/PAK1 pathway is linked with enhanced activity in DS. Knocking down DSCAM was demonstrated to increase the complexity of proximal dendritic branching and inhibition in axon growth in mouse cortical neurons (Zhang et al., 2015); however, involvement of DSCAM in cortical neurogenesis in primates was not fully understood. Additionally, its downstream gene PAK1 has been reported to be essential for regulating actin cytoskeleton dynamics, dendritic spine morphogenesis, and cortical neurogenesis in a mouse model (Pan et al., 2015). Thus, the study highlighted the 3D cortical organoids developed in vitro as a valuable model of DS to study the early developmental deficits and provided a direct link between dysregulation of DSCAM/PAK1 pathway and developmental brain defects in DS (Tang et al., 2021). Moreover, the model can be used for identifying therapeutic targets and screening drugs for the treatment of DS, specifically therapeutic efficacy of DSCAM and PAK1 modulators to reverse abnormal neurodevelopment and improve postnatal cognitive function in DS.

DYRK1A mutations are known to cause severe primary microcephaly in human; hence, DYRK1A has been tested as a strong therapeutic target to ameliorate cognitive functions of DS. Interestingly, genetic normalization of Dyrk1a appeared sufficient to normalize early cortical developmental phenotypes in DS mouse models.

In this line, DS-iPSCs were genetically manipulated for generating a single copy DYRK1A-knockout CO model (scDYRK1AKO-COs) where genetic normalization of DYRK1A was expected to result in corrected neurodevelopmental phenotypes reminiscent to normal DYRK1A knock-in (DYRK1AKI) COs (Çağlayan, 2016). Notably, several doses of DYRK1A inhibitors can be tested during neurodevelopment process of DS-COs, and DS-scDYRK1AKO-COs can be considered as optimum models for evaluating phenotypic ameliorations. Dose optimization studies of DYRK1A inhibitors can be targeted for prenatal therapeutic applications in DS, and overdose drug exposure can also be explained in DS-COs by similar defects present in DS-baDYRK1AKO-COs and DYRK1AKO-COs (Çağlayan, 2016). Therefore, further refinement of DS-DYRK1AKO-CO models can be targeted in future as an alternative to generate transplantable and implantable neurons for efficient tissue regeneration in DS individuals.

Brain organoids are derived from iPSCs and ESCs under three-dimensional (3D) culture condition. Gyrencephalic human neocortical development is more complex than that in lissencephalic mice; hence, COs are being increasingly used as model to study the early neurodevelopmental defects of DS. However, there are several limitations in the current CO technology, which can be minimized by the generation of vascularized brain-like organoids giving opportunities to mimic late-stage corticogenesis and complete hippocampal development (Çağlayan, 2016). Interestingly, addition of a few growth factors to iPSCs with or without culture under "organoid forming conditions" precisely recapitulates the developmental process that produces a cortical neuron. While the presence of a few markers does not indicate equivalence to cells expressing those few genes (and thousands of others) in vivo, the production of regions of human-derived cells that reflect some aspects of the physiological impact expected from trisomy demonstrates the potential utility of these systems (reviewed in Moyer et al., 2021).

6. Technical challenges of DS modeling and engineering of genes of interest

6.1 Developing DS models

The trisomic HSA21 in DS is responsible for overexpression of genes contributing to DS characteristics. Interestingly, the short (p) arm of this smallest chromosome encodes largely repetitive DNA, and the long (q) arm spanning ~35 Mb harbors ~500 genes, including 164 protein coding, ~300 transcripts that encode ambiguous ORFs, and ~49 members of the KRTAP gene family. Few of these genes have been experimentally assessed for annotating the DS features. Development of suitable model systems and identification of orthologous genes require fulfillment of several criteria, including statistical considerations for sequence similarity; functional considerations of gaps and coverage, domain composition, gene copy number and splice variants; and technical considerations for analyzing the expression levels, genomic context, stoichiometry of pathways, genomic background of the model organisms, sex dependence of penetrance, and expressivity of DS features (Moyer et al., 2021). Notably, orthologs of HSA21 protein coding genes have been identified with high sequence conservation in mammals, such as chimpanzee (Ptr21q is essentially conserved with HSA21q), mouse (105 protein coding genes in ~28 Mb span of the telomeric region of MMU16; ~20 in an internal segment of ~1.5 Mb span of MMU17; and ~40 genes in an internal segment of ~3 Mb span of MMU10), and rat (RNO11 is syntenic to a segment of MMU16, and contiguous regions of RNO20 are syntenic to MMU17 and MMU10) (Table 14.3) (Moyer et al., 2021).

TABLE 14.3 Orthologs of HSA21q protein coding genes.

Species	Conserved	Syntenic chromosomes	Genetic manipulations	Implied hypothesis of DS pathogenesis	Example phenotypes
Chimpanzee (*Pan troglodytes*)	164	PTR21	Naturally occurring trisomy of Ptr21	Similar to human DS	Not applicable, except observations of captive-born animals
Rat (*Rattus norvegicus*)	158	RNO11, RNO20	Individual gene targeting with CRISPR/Cas9, direct subchromosome duplication, introduction of human chromosomes	Simultaneous overexpression of many genes causes net phenotype	Refined behavior, preferred model for pharmacology, larger animal allows some surgical interventions that are more difficult in mice
Mouse (*Mus musculus*)	158	MMU16, MMU17, MMU10	Segmental duplications of regions with conserved synteny, overexpression or knockout of single genes, introduction of human chromosomes	Overexpression of many genes in segmental duplications or overexpression of single genes causes phenotypes	Diverse phenotypes in many organ systems, diverse behavioral assays
Zebrafish (*Danio rerio*)	137		Addition and knockdown/knockout of genes, many transgenic fish marking key developmental pathways available	Overexpression of one or a few genes is sufficient to cause a phenotype; assumes genes and pathways are well conserved and have conserved roles in development	Developmental phenotypes caused by perturbation of major growth factor pathways (Shh and Wnt), behavioral assays
Fruit fly (*Drosophila melanogaster*)	59		Tissue-specific knockdown and overexpression, temporal control, mutations in most genes available, mosaic analysis	Overexpression of one or a few genes is sufficient to cause a phenotype; assumes genes and pathways are well conserved and have conserved roles in development	Larval crawling, negative geotaxis, phototaxis, flight, activity, rhythmicity, courtship conditioning, habituation, associative learning, cellular morphology, and physiology
Nematode (*Caenorhabditis elegans*)	42		Tissue-specific knockdown and overexpression, temporal control, efficient transgenesis, RNAi	Overexpression of one or a few genes is sufficient to cause a phenotype; assumes genes and pathways are well conserved and have conserved roles in development	Mechanosensation, chemosensation, habituation, thermotaxis, locomotion, feeding, reproduction, cellular morphology, and physiology
Yeast (*Saccharomyces cerevisiae*)	26		Modify expression of single genes, create defined aneuploidies, population selection	General effects of aneuploidy on cell fitness, stress response pathways, etc.	Cell growth, selection under different metabolic conditions, perturbations to overall protein/RNA levels

Reproduced from Moyer et al., 2021.

6.2 Biological differences between mice and humans

Several murine models have been produced for DS research; however, there has not yet been a complete model till date, and thus, search is underway. Mice have their own sets of behaviors and intellectuality, which are significantly different from humans and thus, that cannot be directly extrapolated and correlated. There are genetically conserved orthologous regions on three mouse chromosomes such as MMU10, MMU16, and MMU17 (Table 14.3); however, these are not completely analogous to HSA21. Therefore, biochemical and metabolic differences are expected, which may confer different phenotypic and clinical expression. Of the several variations, metabolic pathways of purine, folic acid, and so on exist between the two creatures. In humans, purine metabolism produces uric acid, which could be relevant to oxidative stress associated with DS; however, otherwise it is an antioxidant. Accumulation of uric acid in blood of DS individuals is significantly high and that could contribute to aging and neurodegeneration in DS. Triplication of the GART gene that encodes the enzyme acts as a catalyst in de novo purine synthesis. Accumulation of uric acid has been demonstrated to be caused by the gene dosage of the GART, or malfunctioning of kidneys for abnormal processing of uric acid in DS individuals, or a collective effect of trisomic genes of HSA21 in presence of other disomic genes. Conversely, mice metabolize the uric acid to allantoin, which is highly soluble and easily excreted. Hence, modeling mice for DS research needs attention for alteration of purine metabolism, such as elimination of uricase (Vacano et al., 2012).

Metabolism of folic acid is somewhat different in mice as the level of folic acid is 10 times higher in murine than human plasma. Alterations in folate metabolism, perhaps due to polymorphisms in genes encoding enzymes of folate metabolism, have been attributed to trisomy 21 in DS individuals, though a direct consequence of the polymorphisms have not been deciphered. It was hypothesized that a specific polymorphism involved in folate metabolism may function only in presence of other polymorphisms (reviewed in Vacano et al., 2012). Importantly, HSA21 harbors several genes involved in folate metabolism, including cystathionine β synthase (CBS), whose mutations are significantly associated with cognitive impairment and cardiovascular disease (Miles and Kraus, 2004). Notably, alteration in folate metabolism has been associated with congenital heart disease (CHD), geriatric biology, and age-related diseases. In addition, the reduced folate carrier (Slc19a1), involved in folate trafficking, is also located on HSA21 and its mutations and/or polymorphisms are linked to sensitivity of methotrexoate. Thus, triplication of Slc19a1 was associated with unusual sensitivity to folate analogs in individuals with DS. CBS and Slc19a1 map to MMU17 and MMU10, respectively, the syntenic regions of HSA21. Several mouse models were created with altered CBS and/or Slc19a1, including strains with inactivation of the endogenous mouse genes and replacement of that with equivalent human genes (reviewed in Vacano et al., 2012); however, much could not be learnt from these mouse strains. Since these genes are present on different mouse chromosomes, production of appropriate mouse models was difficult for studying the consequences of human DS. The fact is that besides the syntenic regions of HSA21 mapped onto MMU16, MMU17, and MMU10, there may be species-specific analogous genes in both human and mouse, and also there may be regulatory sequences and copy number variations (CNVs), which may not code for proteins but may contribute in phenotypic development. Also, the manipulated mouse genome may have untoward consequences, including additional chromosomal rearrangements.

Importantly, production of several mouse models of DS with varying and targeted genetic manipulation has enormously unraveled the understanding of the genotype–phenotype consequences of DS. Although, all mouse models have not yielded expected information reminiscent to DS, knockout models with syntenic genes of HSA21 have contributed significant information on genes involved in causing DS characteristics in human. Creation of more such models with silencing of every single mouse gene orthologous to HSA21 is a continual approach for understanding the genetic expression of triplicated HSA21 as well as the global disomic genes (http://www.mousephenotype.org. accessed May 3, 2021). Additionally, mouse models are becoming extremely important for conducting clinical trials of drugs targeted for treatment of DS. Therefore, comparison of such models, condition of their genetic engineering and phenotypic expression reminiscent to human DS would extract significant information in understanding genetic impact in DS and therapeutic requisites.

6.3 Compendium of DS models

The rapidly evolving genetic engineering techniques has revolutionized development of genetically engineered mouse models for identification and amelioration of the candidate genes responsible for expression of DS characteristics, especially cognitive impairment and AD. A variety of transgenic mouse models have been developed with larger trisomic or duplicated chromosomal segments that mimic the trisomy HSA21 for searching the candidate genes (Dierssen, 2012). The continued effort is evolving new models to better suit the purpose of DS research. However, no model has been proved to be perfect and adequate mainly because of the obvious differences between the mammals, and also technological approaches.

For effective DS studies in model systems, technical consideration for experimenting on targeted single genes appeared important. Construction of cDNA was often considered where normal regulation was absent and expression levels were typically higher than the modest increases of ~50% seen in DS. However, a gross exaggerated result was thought to present a very subtle perturbation with lacking tissue specificity. Another alternative approach of gene knockout and/or knockdown experiments were considered, which appeared convincing for uncovering gene functions but interpretation of results and claims of relevance to DS required serious circumspection. Such approach of investigation was expected to reveal the important contribution of pathways or cellular process in model systems that may reflect the DS phenotype, though such research cannot reflect trisomic gene consequences of DS in total.

Investigations on single gene may miss the important effects of gene–gene interactions and the stoichiometry of pathway components (Moyer et al., 2021). Mouse models having trisomy for the HSA21-encoded DYRK1A protein kinase exhibited divergent phenotypes of DS in the presence of varying numbers and identities of additional HSA21 orthologs (Altafaj et al., 2001; Olson et al., 2007; Jiang et al., 2015). The consequences of elevated expression of DYRK1A kinase and other kinases (or protease, or other protein modifier, of which there are many encoded by HSA21) may well be different from co-overexpression with its targets (as would occur in DS). Another consideration is that several targets of the DYRK1A kinase are encoded by HSA21 (reviewed in Moyer et al., 2021). The development of segmentally trisomic models in organisms other than mammals would indeed be challenging because of short syntenic regions, which are difficult to splice together (Fig. 14.1). Most importantly, no segment of HSA21 contributing to DS features can be excluded for research. Use of DSCR in several cases of partial trisomy HSA21 has been considered in a number of experiments for its association with intellectual disability

(Korbel et al., 2009). High resolution DSCR, a 34-kb segment of HSA21, displays a minimal overlapping region among DS cases due to partial trisomy (reviewed in Moyer et al., 2021). However, the high-resolution DSCR has no stronger contribution for intellectual or phenotypic features than the original DSCR (Korbel et al., 2009). Moreover, no functional genetic elements are annotated in the segment, and no individual trisomic model has been developed exclusively for the DSCR segment. Practically, all HSA21 genes carry importance as candidates for creation of experimental models.

Initial research was mainly focused on the genes located around 21q21–21q22.3 region because the DSCR segment (located proximal to 21q22.3) was thought to confer majority of the features of DS. DS individuals with segmental duplication of HSA21 were considered as the key to build DS phenotypic map and define the DSCR region (Korbel et al., 2009; Korenberg et al., 1994; Lyle et al., 2009). However, limited number of DS individuals with duplications and high incidence of embryonic death could generate the phenotypic map with low resolution. Thus, a precise understanding of the DS genotype–phenotype relationship in human would depend on a systematic analysis of large sample size of DS individuals, including the stillborns. Most of mouse models were developed through irradiation, breeding, and cross-breeding for different HSA21 syntenic regions, which ultimately did not satisfy the targeted goals, though fulfilled largely.

To overcome the technical obstacles, Cre/loxP-mediated chromosome engineering technology was developed to generate chromosomal duplications and/or deletions with predetermined endpoints (Xing et al., 2016). Triple trisomic mouse was developed with MMU16, MMU17, and MMU10 regions homologous to HSA21 (Yu et al., 2010a,b). Technically, three major steps, including 1. Introduction of a loxP site into the first endpoint (upstream of the region of interest) in the genome of the embryonic stem (ES) cells with a positive selectable marker (e.g., neomycin resistant gene *neo*), when the cells are challenged with antibiotic G418, the cells expressing *neo* will survive and be selected; 2. A second loxP site is targeted for a second endpoint (downstream) with a different positive selection marker such as puromycin-resistant gene, and double targeted cloning is done by using a Cre expression vector through electroporation. Thus, a selectable minigene is reconstructed after Cre expression. A deletion (deficiency, Df) is created by locating two targeted *loxP*s on the same chromosome homolog (*cis*) and in the same direction relative to the centromere, while a deletion (Df) and a reciprocal duplication (Dp) occurs when two *loxP*s are located on two homologs and in the same direction relative to the centromere. The desired orientation of the *loxP* in the targeting vector and its genomic sequence may lead to yield a correct recombination of *loxP* on a chromosome. Analysis of sib-selection through positive selection drugs and recombination efficiencies may recognize desired rearrangement in the clones, while southern blotting and/or fluorescence in situ hybridization (FISH) may confirm the desired rearrangements in the genotypes of the engineered minigene in ES cells (Gribble et al., 2013; Yu et al., 2010a). The manipulated ES cells are injected in the blastocytes of mice for generating the chimeric progenies with the targeted rearrangements. Thus, germline transmission of the recombinant genes will produce mutant mouse strains with a desired Df or Dp.

Recombination of HSA21 syntenic regions in mutant mouse with the desired genes can be verified by molecular techniques, including array-based comparative genomic hybridization (a-CGH) or inferring DNA copy number data obtained from next generation sequencing (NGS). Other techniques for generating Cre/loxP-mediated recombination may first consider production of mutant mouse with a single-targeted loxP at either desired endpoints, and then cress-breeding of the mice carrying two targeted loxP sites with a cre transgenic mice, such as Sycp-1-cre or Tg(Pgk-1-cre)1Lni mice (Herault et al., 2017), will produce Cre/loxP-mediated trans-recombination in the compound mutants in the progenies with duplication (Dp) and deletion (Df) rearrangements.

The time required for the Cre/loxP recombination in mouse ESCs is the limiting factor for construction of the targeted vectors by cloning, which have been somewhat resolved by developing pre-made targeted vectors with genomic libraries containing all the required genetic elements (Fig. 14.2) (reviewed in Herault et al., 2017). These libraries were used for making ~153,000 λ-phage clones and subsequently converted in plasmids with the end sequences of their genomic inserts derived from sequencing primers located external to the cloning points. The sequences indicated the genome coordinates of the genomic inserts, and one targeting vector was made for end-point targeting of ~39 kb genome (Adams et al., 2004). These Mutagenic Insertion and Chromosome Engineering Resource (MICER) libraries have replaced the necessity of construction of new targeted vectors with large genomic rearrangements for creating mutant mouse population. The vectors of "Source BioScience" have markedly expedited the procedure of producing desired mouse for studying expression of desired genes and their mutations.

An enormous number of chromosomal rearrangements with HSA21 syntenic regions have been generated in experimental mice (Fig. 14.1 and Table 14.1). Mice with duplication mutants were considered to study the efficacy of the triplication of the entire HSA21 syntenic regions of MMU10, MMU16, and/or MMU17 and also a sub-segment within a HSA21 syntenic region to express DS-features (Li et al., 2007; Pereira et al., 2009; reviewed in Herault et al., 2017). It was speculated that if a small segment is linked to a specific DS phenotype, a larger duplication can be compounded by a deletion of the smaller sub-segment using different combinations of Dp and Df mutants (Olson et al., 2007; Duchon et al., 2010). Additionally, new Dp and/or Df mutants can lead to further dissection of such smaller regions containing ≥10 HSA21 gene orthologs responsible for a specific phenotype. As for example, generation of a compound mutant allele to carry a duplication of the DSCR locus and a null allele of the gene located within the DSCR region can correlate a phenotype in mutant mice carrying the duplication alone. Such subtractive strategy was used for ascertainment of the contribution of individual HSA21 syntenic regions on DS-features (Baek et al., 2009; Salehi et al., 2006; Sussan et al., 2008). Therefore, mice with null alleles, called knockout mice, were created for several genes for examination of individual gene's association with DS phenotypes (reviewed in Herault et al., 2017).

Indeed, 31%–54% of pregnancies with trisomy 21 face spontaneous abortion (illustrated in previous chapters). Owing to variability in expression of DS features in different DS individuals, the long-term goal on DS research has been focused on understanding the subsets of HSA21 genes contributing to specific phenotypic features with a view to developing therapies to control the clinical outcomes, especially cognitive impairment and development of dementia and AD.

Notably, revolutionary development of the CRISPR-mediated rearrangement (CRISMERE) technique (based on the principle of CRISPR/Cas9 genome-editing technology) is increasingly drawing attention for creating targeted models for DS research (Birling et al., 2017). One-cell mouse embryos were injected with two pairs of small guide RNA (sgRNA), each pair selected either upstream or downstream of the targeted region, using Cas9 nuclease and reimplanted. Newborns appeared with chromosomal rearrangements were bred for selection of the carrier

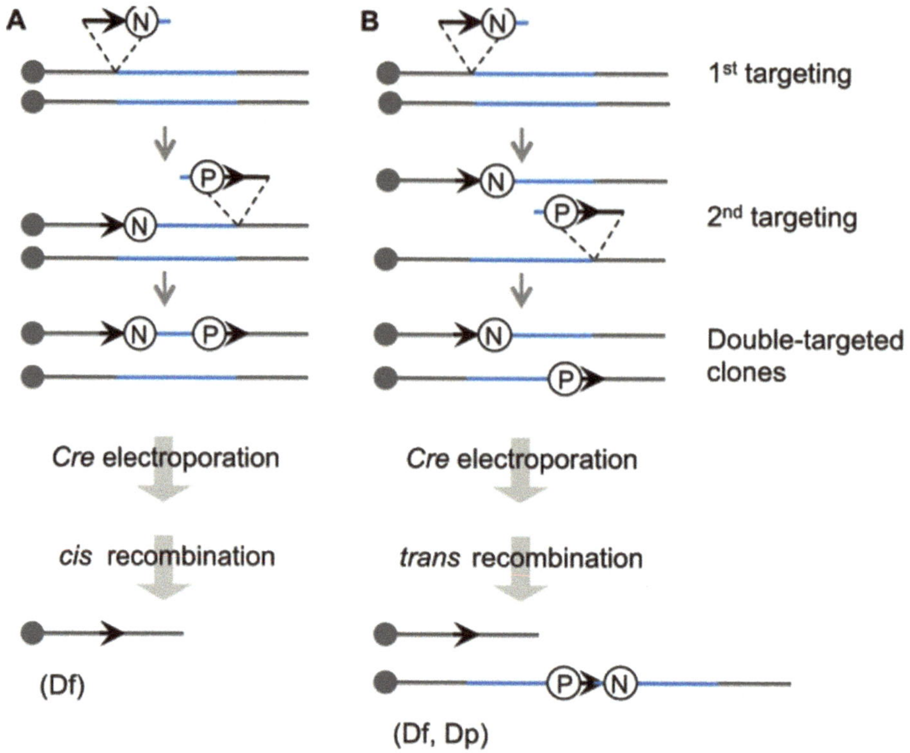

FIGURE 14.2 Cre-/loxP-mediated chromosome engineering technique for creation of deletion/duplication in mouse embryonic stem cells (ESCs): (A) LoxP (*arrow*) site is oriented on the same chromosome and in the same direction and that produced a deletion (Df); (B) LoxP is oriented on two chromosomes and resulted in a Df and Dp (duplication). *Reprinted from Herault et al. (2017). https://doi.org/10.1242/dmm.029728 with permission from Herault Y and under the terms of the Creative Commons Attribution License (http://creativecommons.org/license/by/3.0).*

mice with the new chromosomal complement in the next generation (Herault et al., 2017). Indeed, CRISMERE technique facilitates making chromosomal alterations, such as deletion, duplication, and inversion with 34 Mb genomic segments much faster than germline transmission of the recombinant ES cells. Furthermore, CRISMERE technique has been employed in rats (*Rattus norvegicus*) and other primates. New models of DS have been created with both regions using CRISMERE technique in rats and mice, which are useful for therapeutic testing, validation, and assessment. CRISMERE technique appeared more powerful for manipulating the rodent and probably other genomes in a faster and efficient manner than before. Nowadays, conceptual development of DS-models is no longer restricted to available technology and/or test systems. To meet the challenges of precise delineation of the region/gene of interest for a particular DS phenotype, question arises on whether to generate more models to define smaller regions of interest or make models with larger segments to recapitulate the trisomic HSA21 of human DS. Thus, alongside murine models for DS research in vivo, in vitro models are also being thought off on iPSCs of DS origin with the advent of CRISPR/Cas9 editing facility.

7. HSA21 synteny studied on other models

Synteny, or gene order, is significantly conserved only among mammals. Despite the challenges in experimental design and interpretation, the relative speed and economy of experiments in yeast, nematode, fruit fly, and zebrafish has provided critical roles in DS research. HSA21 orthologs are found entirely on chimpanzee chromosome 21 (PTR21), on segments of rat chromosomes 11 and 20 (RNO11, RNO20), and on mouse chromosomes 10, 16, and 17 (MMU10, MMU16, MMU17). However, in zebrafish, fruit fly, and nematode, HSA21 orthologs are dispersed throughout their genomes. Each colored square represents one HSA21q protein-coding gene or ortholog, and each black square represents ~500 kb of DNA sequence that is not syntenic to HSA21 (Fig. 14.3) (Moyer et al., 2021). The genes and genomic sequences are ordered in a zigzag pattern starting in the upper left-hand corner of each chromosome. Studies using these organisms provide important insights into the functions of individual HSA21 genes and their effects on other target genes and pathways and, in some cases, the development and function of target organs. These results support predictions of how their human orthologs may contribute to the phenotypic features of DS.

7.1 Rat models (*Rattus norvegicus*)

Rats are not considered as the primary choice for genetic engineering, developing embryonic stem (ES) cell lines, and gene editing by CRISPR/Cas9 technology due mainly to requirement of higher consumables compared with mouse. However, several rat lines have been generated through breeding for preclinical pharmacological and other studies, such as metabolic diseases (diabetes), hypertension, and other complex traits, since rats are more closer to

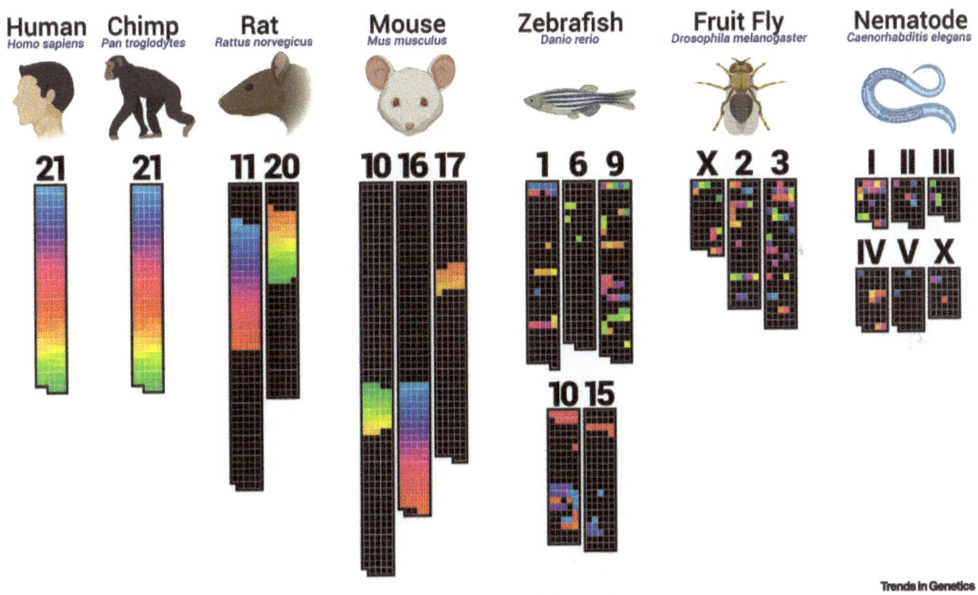

FIGURE 14.3 Locations of syntenic HSA21 and orthologs in select model organisms. *Reprinted from Moyer et al. (2021). https://doi.org/10.1016/j.tig.2020.09.017 with permission from Elsevier and Copyright Clearance Center.*

human in terms of diseases (reviewed in Moyer et al., 2021). Interestingly, the HSA21 homologous regions are located on two chromosomes of rat viz. RNO11 and RNO20. In addition, RNO11 is syntenic to a segment of MMU16, and contiguous regions of RNO20 are syntenic to MMU17 and MMU10. The Lipi-Zbtb21 segment of RNO11 is nearly identical to the homologous region of MMU16, while a unique segment of Umodl1−Prmt2 interval is homologous to RNO20 (Fig. 14.3) (Birling et al., 2017; Herault et al., 2017). Interestingly, direct duplications of regions of the rat genome orthologous to HSA21 have been created. The methods used to transfer the HSA21 for generating the transchromosomic mouse models are currently undergoing further characterization and application (Kazuki et al., 2020; O'Doherty et al., 2005). Nevertheless, number (11 pups for Sprague−Dawley and nine for Wistar rats) and size of litters in rats are larger than mice. In addition, bigger size of rats is useful for handling for surgical interventions and imaging studies via implanted brain electrodes to record activities in real time on larger sample sizes. Notably, although creation of deletions, duplications, and inversions of genomic regions as large as 24.4 Mb in rat founders and germ line transmission of 3.6 Mb fragment size were successful (Birling et al., 2017), a rat model of DS has yet to be described.

7.2 Saccharomyces as a model for understanding the global effects of aneuploidy

In association with the trisomic HSA21 genes with overexpression for specific phenotypes, presence of an extra chromosome may also confer some trisomy-associated features regardless of the specific genes that it encodes, and challenging this hypothesis in vertebrate models is conflicted with nearly nonviable progenies carrying autosomal aneuploidies. In this context, the simple yeast (*Saccharomyces cerevisiae*) with some whole-chromosome aneuploidy appears as an ideal system for studying the general effects of aneuploidy to provide information about the global versus specific effects (some may be relevant to some DS phenotypes), though yeast cells are not considered as suitable model for DS research. *S. cerevisiae* contains an estimated 26 HSA21q orthologs among its 16 chromosomes (Moyer et al., 2021).

Variability in DS phenotypes and pathogenesis is significant among the people with DS owing to increased disruption of homeostatic buffering compared with the general population (discussed in previous chapters). Induction of aneuploidy in yeast by transiently transcribing across the centromere resulted in missegregation and the individual colonies with "acutely aneuploid" cells showed more variation in cell cycle duration and in response to environmental perturbations compared to euploid colonies (Beach et al., 2017; reviewed in Moyer et al., 2021). Such increased variation was observed mostly in haploid strains containing a second copy of a target chromosome (disomes) compared with diploid strains carrying three copies of one chromosome (trisomes). Nevertheless, the results were consistent with increase in phenotypic heterogeneity by aneuploidy.

It was hypothesized that aneuploidies may exert metabolic, mitotic, proteotoxic, and replication stress, which was supported by the evidence that some overexpressed subunits of protein complexes form aggregates to preserve the stoichiometry of subunits within the complex in disomic yeast strains. Furthermore, a transcriptomic profile that is reminiscent of hypo-osmotic stress was reported in heterogeneous populations of aneuploid cells (reviewed in Moyer et al., 2021). In yeast, deletion of some genes affected the fitness of specific disomes, whereas deletion of other genes affected the fitness of multiple disomes. In this line, deletion of components of the conserved oligomeric Golgi

(COG) complex increased the doubling time of disome XVI but had a minimal effect on other disomes. In contrast, deletion of a polyphosphatase encoding PPN1 gene impaired the proliferation of disomes I, VIII, IX, XI, XIII, XIV, XV, and XVI. Disome XVI is a haploid strain with two copies of chromosome XVI. Haploid yeast strains with at least one disome (with one extra copy) showed lower proliferation rate than wild-type strains under standard conditions. In addition, altered expression of a set of genes, similar to that of the yeast environmental stress response, was reported in transcriptional analysis. Even deletion of single gene caused alteration in proliferation of disomic strains (reviewed in Moyer et al., 2021). Taken together, some aneuploidy-associated phenotypes were demonstrated to originate from specific gene effects, while others could be the result of nonspecific impairment of cellular processes.

7.3 Worm orthologs as a model for understanding neurodevelopment

The need for rapid annotation, comparative mapping, and functional characterization of putative genes in the Human Genome Project has led to identify worm orthologs with ~40% of human genes; however, orthologs of HSA21 genes are dispersed throughout the genome of *Caenorhabditis elegans* (*C. elegans*) comprising of six pairs of chromosomes. Thus, creation of aneuploid worms with large number of trisomic HSA21 orthologs appeared impractical and restricted its use to study DS phenotypes. Nevertheless, several studies have assessed the phenotypes caused by knockdown of the orthologs of HSA21 ORFs in *C. elegans* and characterized worm models of APP overexpression and DSAD. In this direction, knockdown of patterning defective 1 (pad-1), the ortholog of DOP1B resulted in mislocalization of differentiated cells and embryonic lethality. Notably, CNV and gain or loss of function mutations in DYRK1A, POFUT2, RCAN1, SYNJ1, ITSN1, and EVA1C orthologs caused alterations in worm phenotypes (reviewed in Moyer et al., 2021). Indeed, knockout mutations in orthologs of 27 HSA21 genes were reported to contribute to the functional annotation of HSA21 genes through analysis of locomotion, exploration, and feeding behavior in worms (Nordquist et al., 2018). Ten HSA21 genes such as SYNJ1, CBS, RUNX1, EVA1C, DONSON, N6AMT1, PDXK1, COL18A1, NCAM2, and POFUT2 were identified to be involved in neurodevelopmental phenotypes. Association of SYNJ1 and EVA1C with neurodevelopment has further supported by these findings. Remarkably, three of the HSA21 orthologs studied had no known link with development of the nervous system. Thus, comparison of the conservation of expression patterns and isoform diversity for each of the candidate genes was essential for validation. In this direction, knockout of dnsn-1, a putative worm homolog of DONSON, caused increase in dwelling time in an exploration assay, suggesting dnsn-1 expression is limited to embryonic development (Nordquist et al., 2018; reviewed in Moyer et al., 2021). However, only 21% amino acid identity was observed in human and worm proteins over 75% of the protein with large gaps in sequence. Expression of Donson protein has been reported in brain of adult mouse (Spellman et al., 2013). Thus, further characterization in worm will decipher the comparable functional properties of dnsn-1 of mouse and human DONSON.

The microscopic (1-mm) nematode *C. elegans* is a model system suitable for genetic analysis of the development of embryonic and postembryonic cell lineages (with 959 adult somatic cells) and the nervous system comprising of 302 neurons. Its transparent body supports fluorescent labeling and is amenable to a variety of genetic manipulations. The simple nervous system and its behavior enable inexpensive large-scale chemical and genetic screening in whole animal. Mutant *C. elegans* have been described with defects in movement and morphology and recapitulated as worm models of human neurological disorders, including AD (reviewed in Moyer et al., 2021). Importantly, further translational studies are necessary to support the contribution of the conserved genes and their interacting partners on disease models with overexpressed human transgenes. Moreover, systematic characterization of HSA21 orthologs in *C. elegans* would be useful for identification of drug targets.

7.4 HSA21 orthologs in Drosophila

Drosophila has the extensive history of use in genetic manipulation for its small size, large number of progeny, and attending economic scale. DS research in Drosophila was thought to be useful to determine the functions of uncharacterized genes, study the genetic interactions between sets of HSA21 orthologs, define the cellular and molecular mechanisms by which protein overexpression impacts on phenotypes, and identify the chemical or genetic modifiers of DS phenotypes. However, creating flies with simultaneous trisomy for large number of HSA21 orthologs appeared conceptually and technically impractical because HSA21 orthologs are dispersed all over the Drosophila genome (Fig. 14.3) (Moyer et al., 2021). Nevertheless, several studies have focused on the HSA21 orthologs of APP, DSCAM, DYRK1A, ITSN1, RCAN1, and SYNJ1 in the fly.

HSA21-encoded DSCAM gene in Drosophila (Dscam, encodes an immunoglobulin-like transmembrane protein) is notable for its capacity of alternative splicing. The gene can theoretically produce up to 38,016 unique mRNAs by alternative splicing of 95 variable exons (reviewed in Moyer et al., 2021). Interestingly, identical Dscam isoforms can bind homophilically, while neuronal self-recognition and self-avoidance is facilitated by the expression of unique sets of isoforms. Notably, defects in dendritic and axonal systems were influenced by the loss of Dscam in flies, but the loss of isoform diversity did not contribute to all of these defects. In particular, the expression level of Dscam (regardless of isoform diversity) controls the presynaptic arbor size, while isoform diversity controls the synaptic targeting in larval class IV dendritic arborization neurons. In addition, Dscam-dependent growth of presynaptic arbors is facilitated by the Abelson kinase (Abl) gene, and thus, inhibition of Abl rescues the morphological defects caused by Dscam cDNA overexpression (Sterne et al., 2015; reviewed in Moyer et al., 2021). The vertebrate DSCAM gene produces only a few protein isoforms, but DSCAM gene contributes to cell type–specific self-avoidance, presynaptic growth, and axon guidance in mammals. Thus, the conservation of a functional property of isoform diversity highlights the importance of experimental validation of protein function, especially in the context of genomic annotations suggesting divergent functions between orthologs (reviewed in Moyer et al., 2021).

Analysis of additive/nonadditive effects and interactions of trisomic subsets of HSA21 on DS phenotypes has been exercised often in vertebrate models; however, it is time consuming and technically challenging. In this regard, Drosophila is ideally suitable for mechanistic studies of protein–protein interactions at the cellular and/or molecular levels. Based on overexpression of RCAN1, ITSN1, and SYNJ1 genes in brains of DS individuals, studies were conducted on overexpressed genes in neurons and systematically examined synaptic transmission and morphology in all combinations of their Drosophila orthologs (variously from genomic clones or cDNA constructs) (reviewed in Moyer et al., 2021). Notably, overexpression of all three transgenes was simultaneously required to perturb the vesicle recycling at the neuromuscular junction. Overexpression of pairwise combinations of all five human and Drosophila cDNAs in the fly heart was also studied. Remarkably, co-overexpression of DSCAM and COL6A2 caused more severe phenotypes than did the overexpression of either cDNA alone. The findings were also studied in mouse. However, approaches with organism-specific protein sequences and alternative splicing are limited, as observed in the fly with lack of orthologs for isoforms of DSCAM, SYNJ1, and ITSN1 found in humans. Altogether, such studies in the fly offer proof-of-principle evidence on interaction of HSA21 proteins and their contribution in causing complex DS phenotypes.

7.5 Zebrafish as a vertebrate system to screen HSA21 effects of DS

Zebrafish (*D. rerio*) has become an established test system for studying the effects of genes on development and several other aspects. The most convenient points of this vertebrate for using in experimental purposes include rapid development (from fertilization stage to free-swimming individual fish in ~5 days) and transparent embryos (allow precise imaging). The facility of analyzing its early development has been considered to assess the gene-dosage effect relevant to DS, specifically the attenuated response of cells with trisomic HSA21 to the effects of Sonic hedgehog growth factor (SHH). The HSA21 orthologs are dispersed throughout the zebrafish genome. A set of 164 cDNAs of conserved protein coding genes representing HSA21 genes were created in vectors, and mRNAs made from each of these genes were injected into early embryos of zebrafish up to 5 days post fertilization. The resulting embryos were studied for assessing the effects on survival, gross morphology, heart development, and four systemic functions that are perturbed in presence of dysfunctional SHH signaling, including pigment abnormalities, cyclopia, rounded somites, and craniofacial structure (Edie et al., 2018). Of the 24 genes produced a gross morphological phenotype, the penetrance of one or more of these phenotypes was significantly enhanced by 11 HSA21-encoded cDNAs. Notably, seven of these genes caused downregulation of the SHH pathway; two showed defects in neural crest migration; one consistently resulted in pericardial edema; and one was embryonic lethal. C21ORF84 (annotated as long intergenic noncoding RNA LINC00313) caused both cyclopia and U-shaped somites. To assess the possible combinatorial gene–dosage effects, C21ORF84 was co-injected with each of six other genes such as RWDD2B, SOD1, YBEY, PCBP3, POFUT2, and RRP1, which contribute to SHH-related phenotypes. Interestingly, the first two did not show a significant change in penetrance when co-expressed, but co-injection with YBEY, PCBP3, or POFUT2 genes significantly increased the number of affected embryos. In particular, co-injection of C21ORF84 with RRP1 resulted in lowering the penetrance than seen with RRP1 alone with 42% penetrance (Edie et al., 2018; reviewed in Moyer et al., 2021). The resultant additive and compensatory effects of combinatorial injections support the concept of complex genetic relationships underlie end phenotypes of trisomy that produce DS. The compensatory effects of overexpression of the trisomic HSA21 genes were likely in DS. Based on this hypothesis, segmentally trisomic mouse models were created, but direct evidence of gene-specific phenomenon was not precisely ascertained. Even though some extent of knowledge has been established on the effects of overexpressing HSA21 candidate genes on alterations of specific pathways related to development in DS, accurate assessment of simultaneous overexpression of two genes (RWDD2B, SOD1) reflecting some DS feature is still lacking. Direct injection of mRNA was thought to yield supraphysiological overexpression of candidate genes. Moreover, partial duplication of the zebrafish genome has created ohnologs of JAM2 and CBR3 (these two are related orthologs of each), which further complicated the dosage effect in this experiment (Edie et al., 2018; reviewed in Moyer et al., 2021).

Moreover, the genome of zebrafish has been sequenced, which shows substantial conservation with mammalian genome, including several growth factor pathways. However, partial duplication makes some genes and their paralogs effectively tetraploid in its genome compared with the diploid condition in mammals, and that status often confuses with the identity of "true orthologs." Indeed, such ambiguity may affect the interpretation of the outcome of experimental genome manipulations. Nevertheless, microinjection can lead to transient overexpression and misinterpretation of the gene expression. Consistent with this, injection of mRNA into the yolk sac of zygotes may result in translation of messages in almost all cells of the early developing embryos of zebrafish. Moreover, gene expression can be altered in zebrafish by transgenesis technique using fluorescent labelled gene transgenics for tracking the specific cell types or tissues. Indeed, the high-throughput direct access to very early stages of development advocates zebrafish system as a more powerful tool for analysis of comparable stages of mammalian development. Furthermore, extensive use of morpholinos and short RNA (shRNA) molecules for target mRNAs that block and down-regulate their translation, and recent use of CRISPR/Cas9 system have enabled targeting specific transcripts and genome manipulation in zebrafish (reviewed in Moyer et al., 2021). Altogether, this test system is useful for expressive dissection of dosage-sensitive genes on early development as single individual locus and/or their combinatorial effects on phenotypes relevant to the DS etiopathology.

8. Models for pharmaceutical developments for DS

Mouse models have been tested for chemicals (Ganguly, 1994) and identification of druggable agents for improvement of learning and memory and control of premature aging and age-related disease consequences (Gardiner, 2010). Testing of memantine, a drug used for treatment of AD in human, showed improvement of learning and memory in Ts65Dn mice (illustrated in previous chapters; reviewed in Vaccano et al., 2012). Use of a γ-secretase rescued learning and memory deficits significantly in young (4-month old) Ts65Dn mice and indicated potential implication of amyloid precursor protein (APP) or its metabolites in learning and memory deficits, even at young age (Netzer et al., 2010). Defects in neurogenesis and brain development of DS has been studied in Ts65Dn model in early postnatal development and that revealed a substantially reduced mitogenic response to Sonic hedgehog protein signaling in granule cells (Roper et al., 2006). Intraperitoneal administration of the smoothened agonist (SAG) for activation of sonic hedgehog pathway (SHH) at postnatal day 2 (P2) showed partial correction of the defects in cerebellar neurogenesis and hippocampal LTP leading to improved spatial memory in Ts65Dn litters (Gutierrez-Castellanos et al., 2013). However, normal cerebellar LTD and working memory could not be restored by SAG therapy, and also, systemic targeting of SHH pathways increases the risk of human cancers of several organs (reviewed in Herault et al., 2017). In this context, another study using γ-secretase inhibitor demonstrated reduced overexpression of the SHH receptor PATCHED1 (PTCH1), that represses the SHH pathway. Overexpression of PTCH1 was thought to be influenced by aggregation of amyloid protein in intracellular domain (AICD) produced by the cleavage of the APP precursor in trisomic neural progenitor cells (NPCs) of the Ts65Dn mice (Trazzi et al., 2013; reviewed in Herault et al., 2017). Collectively, increase in AICD and overexpression of PTCH1 result in impairment in neurogenesis and development of neurites in trisomic NPCs. Interestingly, inhibition of APP cleavage by γ-secretase exhibited restoration of neuronal differentiation in trisomic NPCs, and postnatal treatment could restore the neurogenesis in dentate gyrus (DG) and subventricular zone and normalize APP processing in this mouse model. The result indicated new therapeutic strategies by indirect SHH pathway for restoration of neurogenesis in trisomic brains.

Stem cell–based therapy has drawn major attraction in DS research. Notably, upon injection, neural stem cells (NSCs) can migrate to the sites of damage for implantation and render protection. With this evidence, implantation of NSCs in 12-month old Ts65Dn mice showed reduced extrasomatic granules with TAU and REELIN expression, suggesting neuronal aging (reviewed in Herault et al., 2017). Moreover, similar treatment at P2 exhibited a significant increase in the density of dentate granule cells and long-lasting improvement of cognitive performance. However, the limiting factors of stem cell–based therapeutic strategy, including site-directed short-term efficacy of the injection; direct injection into the site; and transient effects (probably linked to NSC dependent neutrophil production) direct more studies on the targeted outcome.

Another approach of chronic peripheral administration of peptide 6-an11-mer relevant to an active site of ciliary neurotrophic factor was attempted in Ts65Dn mice for enhancing the pool of NPCs and amelioration of learning and memory deficits. Treatment with P7C3, an aminopropyl carbazole for influencing hippocampal neurogenesis, for 3 months restored the neurogenic deficits (reviewed in Herault et al., 2017). Notably, craniofacial phenotypes, including increased cranial vault, were normalized in this mouse model by inhibiting the expression of Dyrk1a through prenatal treatment with epigallocatechine gallate (EGCG; illustrated in previous chapters). In addition, environmental enrichment has been shown to reduce Aβ oligomers and increase sensory-motor stimulation of learning, spatial memory, and motor activity in Ts65Dn mice. Treatment of Ts65Dn mice with JZL184, a selective inhibitor of monoacylglycerol lipase restored hippocampal LTP, long-term memory to normal level, and improved synaptic and locomotor activity. However, deficits in short-term memory and positive thigmotaxis remained unaltered. Imbalance in excitatory/inhibitory neurotransmission was monitored by antagonistic effects of GABA receptors, including RO4938581, CGP55845, and pentylenetetrazol (PTZ) to reduce spatial and long-term memory, hippocampal LTP and expression of several molecular markers, such as BDNF, GAD65/67, VGAT, etc. Consistent with this, antagonistic effects of memantine and MK-801 on N-methyl-D-aspartate receptors (NMDR) improved memory and rescued learning through alterations of APP, vGlut1, and BDNF. Interestingly, clinical trial with memantine is underway on adolescents with DS, though presented limited success (illustrated in Chapter 11). In this context, the Ts65Dn mice were treated with fluoxetine for targeting the serotoninergic system at prenatal and postnatal levels, and that resulted in rescue of long-term memory, synaptic plasticity, and improved neurogenesis; however, similar treatment of the adult mice could enhance neurogenesis, but without improvement of the learning and memory deficits at higher dose, though lower dose could improve short-term memory (reviewed in Herault et al., 2017; illustrated in Chapters 9 and 11). Interestingly, early treatment with another antidepressant drug, lithium, reused LTP, long- and short-term memory and neurogenesis in this mouse model by acting upon the serotonin pathway. Taken together, therapeutic targets included enhancement of neurogenesis and synaptogenesis in trisomic Ts65Dn mice, though the underlying molecular mechanisms are poorly understood. In this direction, treatment with rapamycin, an mTOR-specific inhibitor, restored BDNF-dependent LTP in hippocampal slices and enhanced spatial-memory function in Ts1Cje and Ts65Dn mice (reviewed in Herault et al., 2017).

Use of antioxidant such as α-tocopherol or vitamin E for amelioration of age-related decline in learning and memory significantly reduced the burden of oxidative stress in brains of the Ts65Dn mice (reviewed in Vaccano et al., 2012). Vitamin E appeared useful perinatally and/or throughout the life and also acts upon mTOR pathway. Studies on vitamin E on AD demonstrated conflicting outcome, where treatment after disease onset has been explained as the reason of failure (reviewed in Vaccano et al., 2012) and that indicated the consistence of positive outcome of vitamin E supplementation lifelong in Ts65Dn mice. Another study using choline supplementation of maternal diet during pregnancy and lactation showed improvement of learning and memory and emotional behavior in Ts65Dn offspring at adult age (reviewed in Vaccano et al., 2012; discussed in detail in Chapter 9). The effect in Ts65Dn mice was correlated with association of cholinergic deficits with learning and memory development in mice. Interestingly, perinatal choline supplementation showed beneficial effects on the hippocampal development in mice (reviewed in Vaccano et al., 2012); however, long-term effects of perinatal treatment were not ascertained on behavior or learning and memory in mice. Several other pharmaceutical candidates, including ethosuximide (ETH, an inhibitor of Kcnj6), γ-secretae inhibitor DAPT, anti-Aβ-vaccine, constituents of norepinephrine, xamoterol, estrogen treatment, minocycline (prevents progressive BFCN decline), etc. were considered for testing in Ts65Dn mice for improvement of certain parameters of neurodegeneration, learning, and memory and synaptic plasticity of DS characteristics. Altogether, efficacy of the targeted druggable candidates was preclinically tested in DS mouse models (mostly in Ts65Dn strain); some of which yielded positive results, while several such modulators have failed to present satisfactory outcome. Nevertheless, validation of interlaboratory reproducibility of the outcome and testing of reproducibility in other DS models would be essential for targeted drug development for individuals with DS.

Silencing of the extra HSA21 in DS would be a remarkable postnatal therapy for the individuals with DS. The in vitro attempt to this effect was successful through introduction of the XIST (noncoding X-inactivation gene) into the DYRK1A locus on one of three HSA21 in trisomic iPSCs and silencing of most of the protein coding genes on the XIST-containing HSA21 but without affecting the expression of other genes located elsewhere in the genome (Jiang et al., 2013). The result has indicated the technical feasibility of in utero embryonic manipulation of trisomic HSA21 and other full chromosome aneuploidies. Indeed, such audacious goal of silencing protein coding genes could be exercised for correction of anomalous genes and/or chromosomes therapeutically; however, that remains as a therapeutic dream so far. Notably, the RNA product of the noncoding XIST gene accumulates across the interphase over the inactive X-chromosome (the second X-chromosome in females), and silences transcription of most of the genes through numerous heterochromatin modification processes. Nevertheless, that inactivation of the extra 21 could rescue the DS phenotypes is yet to be tested in DS models in vivo. Moreover, theoretical use of this approach in preimplantation embryos and/or embryonic inactivation or correction of overexpressing genes by CRISPR/Cas9 gene-editing technology is still subject to ethical or societal debates. In addition, there could be significant side effects potentially influenced by such gene manipulation techniques. A new strategy of DOX-inducible XIST after neurogenesis has been suggested more useful than at the pluripotent stem cell stage (Dierssen et al., 2021).

Summarily, therapeutic aspects of DS have been attempted with inhibitors of GABAergic and glutamatergic systems; kinase inhibitors of DYRK1A; activators of SHH pathway; activators of NMDA receptors, serotonergic signaling, neurogenesis, and dendritic spines; degeneration of cholinergic system; inactivation of the extra 21; and so on in DS models in vivo and DS-iPSCs. Future therapeutic targets may consider fetal and/or embryonic stage for evaluating neurogenesis and neuronal migration to the neocortex; however, neurogenesis and neuronal migration in the hippocampal DG and cerebellum continue after birth, while hippocampal neurogenesis continues throughout life at slow rate. Thus, most of the therapeutic testing in mouse has focused on hippocampal circuit–related performance in memory and learning (Dierssen, 2012; Antonarakis, 2017).

9. Conclusion and perspectives

Indeed, advancement in technological development for genetic manipulation and creation of genetically modified animals for testing the expression of genes of choice has enormously contributed in the field of DS-research that has enhanced knowledge on contribution of genes for DS-specific phenotypes. The target is to study the expression of every gene triplicated in HSA21 and its association with phenotypic characteristics of DS of all ages. Mouse models have been produced with chromosomal regions syntenic to HSA21 in parallel with transgenic and/or gene knockout models. Indeed, target is underway to evaluate all mouse genes and their contribution in phenotypic development through manipulation of gene numbers, their structure, and expression for drawing an appropriate interpretation of the consequences of trisomic status and possible interference/interaction of/with the global disomic genome. The approach has enabled unprecedented opportunities of unraveling the pathomechanism in DS patients, extracted crucial information on genotype–phenotype association, and identified therapeutic molecules for pharmaceutical developments, specifically for addressing cognitive impairment and proneness to AD.

Various mouse models have been generated, of which Ts65Dn model has been extensively considered in multiple aspects of DS research, including aging and age-related complications of human DS. These mice are trisomic for 60 MMU17 genes and disomic for 16 MMU16 genes orthologous to HSA21, and remarkable phenotypic similarities with Dp(16)1Yey/+ mice that is exclusively trisomic for the entire MMU16 region syntenic to HSA21. MMU10 and MMU17 regions syntenic to HSA21 might ameliorate some of the expression of the trisomic MMU16 regions. Importantly, more and more studies will unravel the expression of genes independently or in cooperation with others, which may lead to revision of today's knowledge of genotype–phenotype association in DS individuals. Creation of transgenic and knockout mice appeared with characteristics consistent with genes with known functions and also helped in elucidating its other functions contributing to DS characteristics. Interestingly, breeding of the trisomic mice with knockout mice for some alleles have enabled bringing trisomic mice back to disomic state for some targeted genes, which allowed study of expression of genes in disomic as well as trisomic condition, and also facilitated understanding the rescue of the trisomic effect in the cross-bred mice. However, some of the important and crucial points to be taken into account in studies of DS-research, which includes, at least: the expression of trisomic genes against the genetic makeup of different mouse models, which may act differently, and individuals with DS are not inbred. Thus, the expression observed in a noninbred mouse strain or in different genetic backgrounds might be more rational for studying the relevance to the human DS than the effects reported in inbred mouse models. Therefore, association of a single gene and/or a single phenotype shall be considered critical for different mouse models of human DS.

Noticeably, remarkable results have been collected from in vitro iPSC and ESC cell lines and with mouse models of DS on gene expression, DS pathobiology, and therapeutic targets. There are research questions still remain unaddressed relating to various aspects of DS features (Antonarakis, 2017), which includes 1. To understand the function of all HSA21 elements and identify the exact functional genomic elements from three copies that have phenotypic effects on DS; 2. to identify the genomic determinants of the phenotypic variability in DS (DS Genome Project); 3. to identify the influence of trisomic HSA21 on different cell types and tissues in DS (mouse models would be ideal for the purpose); 4. to understand the gene-dosage effect on proteome (there is considerable buffering from the transcriptome to proteome); 5. to intensify research on therapeutic developments, identify molecular targets, and test single or combinatorial therapies on specific targets; 6. to differentiate trisomic iPSCs of different cell types (trisomic iPSCs can complement mouse models to some extent); 7. to develop various iPSC-based in vitro tissues/organoid models for DS research; 8. to study all genomic and chromatin and biochemical and epigenetic markers in order to understand regulation of gene expression in different cell and tissue types; 9. to study brain development process and function for elucidating intellectual disability of DS; and 10. to study metabolomes qualitatively and quantitatively for identification of biomarkers of different DS phenotypes. further, in vitro chromosome correction or inactivation needs further investigations, specifically at the early-embryonic stage of development. Integrative approaches of epigenetic chromosome correction beyond the early stages of embryo development would be ideal because time would be insufficient between the diagnosis and intervention at early embryonic stages. Since CRISPR/Cas9 gene editing in humans has ethical and societal issues, identification and testing of more natural products (such as extract of green tea with EGCG) can reduce the function of overexpressing genes.

References

Adams, D.J., Biggs, P.J., Cox, T., et al., 2004. Mutagenic insertion and chromosome engineering resource (MICER). Nat. Genet. 36, 867–871.

Alić, I., Goh, P.A., Murray, A., Portelius, E., Gkanatsiou, E., Gough, G., et al., 2021. Patient-specific Alzheimer like pathology in trisomy 21 cerebral organoids reveals BACE2 as a gene dose-sensitive AD suppressor in human brain. Mol Psychiatr. 26, 5766–5788. https://doi.org/10.1038/s41380-020-0806-5.

Altafaj, X., Dierssen, M., Baamonde, C., Marti, E., Visa, J., et al., 2001. Neurodevelopmental delay, motor abnormalities and cognitive deficits in transgenic mice overexpressing Dyrk1A (minibrain), a murine model of Down's syndrome. Hum. Mol. Genet. 10 (18), 1915–1923.

Antonarakis, S.E., 2017. Down syndrome and the complexity of genome dosage imbalance. Nat. Rev. Genet. 18, 147–163.

Antonarakis, S.E., Lyle, R., Dermitzakis, E.T., Reymond, A., Deutsch, S., 2004. Chromosome 21 and Down syndrome: from genomics to pathophysiology. Nat. Rev. Genet. 5, 725–738. https://doi.org/10.1038/nrg1448.

Arriagada, C., Bustamante, M., Atwater, I., Rojas, E., Caviedes, R., Caviedes, P., 2010. Apoptosis is directly related to intracellular amyloid accumulation in a cell line derived from the cerebral cortex of a trisomy 16 mouse, an animal model of Down syndrome. Neurosci. Lett. 470 (1), 81–85.

Ashe, K.H., Zahs, K.R., 2010. Probing the biology of Alzheimer's disease in mice. Neuron 66 (5), 631–645.

Baek, K.H., Zaslavsky, A., Lynch, R.C., et al., 2009. Down's syndrome suppression of tumour growth and the role of the calcineurin inhibitor DSCR1. Nature 459 (7250), 1126–1130.

Bambrick, L.L., Fiskum, G., 2008. Mitochondrial dysfunction in mouse trisomy 16 brain. Brain Res. 1188 (1), 9–16.

Beach, R.R., et al., 2017. Aneuploidy causes non-genetic individuality. Cell 169, 229–242 e221.

Belichenko, P.V., Masliah, E., Kleschevnikov, A.M., Villar, A.J., Epstein, C.J., Salehi, A., Mobley, W.C., December 2004. Synaptic structural abnormalities in the Ts65Dn mouse model of Down Syndrome. J. Comp. Neurol. 480 (3), 281–298. https://doi.org/10.1002/cne.20337.

Belichenko, P.V., Kleschevnikov, A.M., Salehi, A., Epstein, C.J., Mobley, W.C., 2007. Synaptic and cognitive abnormalities in mouse models of Down syndrome: exploring genotype-phenotype relationships. J. Comp. Neurol. 504, 329–345. https://doi.org/10.1002/cne.21433.

Belichenko, P.V., Kleschevnikov, A.M., Masliah, E., et al., 2009. Excitatory-inhibitory relationship in the fascia dentata in the Ts65Dn mouse model of Down syndrome. J. Comp. Neurol. 512 (4), 453–466.

Belichenko, P.V., Kleschevnikov, A.M., Becker, A., Wagner, G.E., Lysenko, L.V., Yu, E.Y., et al., 2015. Down syndrome cognitive phenotypes modelled in mice trisomic for all HSA 21 homologues. PLoS One 10, e0134861. https://doi.org/10.1371/journal.pone.0134861.

Besson, V., Brault, V., Duchon, A., Togbe, D., Bizot, J.-C., Quesniaux, V.F.J., Ryffel, B., Hérault, Y., 2007. Modeling the monosomy for the telomeric part of human chromosome 21 reveals haploinsufficient genes modulating the inflammatory and airway responses. Hum. Mol. Genet. 16, 2040–2052.

Birling, M.C., Schaeffer, L., André, P., Linder, L., Maréchal, D., Ayadi, A., et al., 2017. Efficient and rapid generation of large genomic variants in rats and mice using CRISMERE. Sci. Rep. 7, 43331.

Blazek, J.D., Gaddy, A., Meyer, R., Roper, R.J., Li, J., 2011. Disruption of bone development and homeostasis by trisomy in Ts65Dn Down syndrome mice. Bone 48 (2), 275–280.

Brault, V., Duchon, A., Romestaing, C., Sahun, I., Pothion, S., Karout, M., Borel, C., Dembele, D., Bizot, J.-C., Messaddeq, N., et al., 2015. Opposite phenotypes of muscle strength and locomotor function in mouse models of partial trisomy and monosomy 21 for the proximal Hspa13-App region. PLoS Genet. 11, e1005062.

Çağlayan, E.S., December 2016. Generation of improved human cerebral organoids from single copy DYRK1A knockout induced pluripotent stem cells in trisomy 21: hypothetical solutions for neurodevelopmental models and therapeutic alternatives in Down syndrome. Cell Biol. Int. 40 (12), 1256–1270. https://doi.org/10.1002/cbin.10694.

Carmichael, C.L., Majewski, I.J., Alexander, W.S., et al., 2009. Hematopoietic defects in the Ts1Cje mouse model of Down syndrome. Blood 113 (9), 1929–1937.

Choi, J.H.K., Berger, J.D., Mazzella, M.J., et al., 2009. Age dependent dysregulation of brain amyloid precursor protein in the Ts65Dn Down syndrome mouse model. J. Neurochem. 110 (6), 1818–1827.

Cortez, M.A., Shen, L., Wu, Y., Aleem, I.S., Trepanier, C.H., Sadeghnia, H.R., et al., 2009. Infantile spasms and Down syndrome: a new animal model. Pediatr. Res. 65, 499–503. https://doi.org/10.1203/PDR.0b013e31819d9076.

Cremona, O., DiPaolo, G., Wenk, M.R., et al., 1999. Essential role of phosphoinositide metabolism in synaptic vesicle recycling. Cell 99 (2), 179–188.

Davisson, M.T., Schmidt, C., Reeves, R.H., Irving, N.G., Akeson, E.C., Harris, B.S., Bronson, R.T., January 1, 1993. Segmental trisomy as a mouse model for Down syndrome. Prog. Clin. Biol. Res. 384, 117–133.

Deitz, S.L., Roper, R.J., 2011. Trisomic and allelic differences influence phenotypic variability during development of Down syndrome mice. Genetics 189, 1487–1495.

Delabar, J.M., Theophile, D., Rahmani, Z., Chettouh, Z., Blouin, J.L., Prieur, M., et al., 1993. Molecular mapping of twenty-four features of Down syndrome on chromosome 21. Eur. J. Hum. Genet. 1, 114–124.

Dierssen, M., 2012. Down syndrome: the brain in trisomic mode. Nat. Rev. Neurosci. 13, 844–858. https://doi.org/10.1038/nrn3314.

Dierssen, M., Herault, Y., Helguera, P., Martínez de Lagran, M., Anna, V., Bradley, C., et al., 2021. Building the future therapies for Down syndrome: the third international conference of the T21 research society. Mol. Syndromol. 12, 202–218. https://doi.org/10.1159/000514437.

Duchon, A., Besson, V., Pereira, P.L., Magnol, L., Herault, Y., 2008. Inducing segmental aneuploid mosaicism in the mouse through targeted asymmetric sister chromatid event of recombination. Genetics 180, 51–59.

Duchon, A., Pothion, S., Brault, V., Sharp, A.J., Tybulewicz, V.L., Fisher, E.M., Herault, Y., 2010. The telomeric part of the human chromosome 21 from Cstb to Prmt2 is not necessary for the locomotor and short-term memory deficits observed in the Tc1 mouse model of Down syndrome. Behav. Brain Res. 217, 271–281.

Duchon, A., Raveau, M., Chevalier, C., Nalesso, V., Sharp, A.J., Herault, Y., 2011. Identification of the translocation breakpoints in the Ts65Dn and Ts1Cje mouse lines: relevance for modeling Down syndrome. Mamm. Genome 22 (11–12), 674–684. https://doi.org/10.1007/s00335-011-9356-0.

Dwyer, J., Li, H.E., Xu, D., Lui, J.P., 2007. Transcriptional regulation of telomerase activity: roles of the the ets transcription factor family. Ann. N. Y. Acad. Sci. 1114, 36–47.

Edie, S., et al., 2018. Survey of human chromosome 21 gene expression effects on early development in *Danio rerio*. G3 (Bethesda) 8, 2215–2223.

Epstein, C.J., 1981. Animal models for autosomal trisomy. In: de la Cruz, F.F., Gerald, P.S. (Eds.), Trisomy 21 (Down Syndrome). University Park Press, Baltimore, Md, USA, pp. 263–273.

Ermak, G., Morgan, T.E., Davies, K.J.A., 2001. Chronic overexpression of the calcineurin inhibitory gene DSCR1 (adapt78) is associated with Alzheimer's disease. J. Biol. Chem. 276 (42), 38787–38794.

Fernandez, F., Morishita, W., Zuniga, E., Nguyen, J., Blank, M., Malenka, R.C., Garner, C.C., April 2007. Pharmacotherapy for cognitive impairment in a mouse model of Down syndrome. Nat. Neurosci. 10 (4), 411–413. https://doi.org/10.1038/nn1860.

Fernandez, F., Garner, C.C., 2007. Object recognition memory is conserved in Ts1Cje, a mouse model of Down syndrome. Neurosci. Lett. 421, 137–141. https://doi.org/10.1016/j.neulet.2007.04.075.

Galante, M., Jani, H., Vanes, L., Daniel, H., Fisher, E.M., Tybulewicz, V.L., Bliss, T.V., Morice, E., 2009. Impairments in motor coordination without major changes in cerebellar plasticity in the Tc1 mouse model of Down syndrome. Hum. Mol. Genet. 18, 1449–1463.

Ganguly, B.B., 1994. Bone marrow clastogenicity of trimethyltin. Mutat. Res. 312, 9–15.

Ganguly, B.B., Kadam, N.N., Mandal, P.K., 2017. Complexity of chromosomal rearrangements in a Down syndrome leukemia. J. Cancer Res. Therapeut. 13, 381–383.

Ganguly, B.B., Kadam, N.N., 2017. Down syndrome: from the age of characterization to the era of curative approach. Nucleus 60, 197–208. https://doi.org/10.1007/s13237-016-0187-y.

Gardiner, K.J., 2010. Molecular basis of pharmacotherapies for cognition in Down syndrome. Trends Pharmacol. Sci. 31 (2), 66–73.

Gardiner, K.J., 2014. Pharmacological approaches to improving cognitive function in Down syndrome: current status and considerations. Drug Des. Dev. Ther. 9, 103–125. https://doi.org/10.2147/DDDT.S51476.

Gardiner, K., Fortna, A., Bechtel, L., Davisson, M.T., 2003. Mouse models of Down syndrome: how useful can they be? Comparison of the gene content of human chromosome 21 with orthologous mouse genomic regions. Gene 318, 137–147. https://doi.org/10.1016/s0378-1119(03)00769-8.

Glass, T.J., Valmadrid, L.C.V., Connor, N.P., 2019. The adult Ts65Dn mouse model of Down syndrome shows altered swallow function. Front. Neurosci. 13, 906. https://doi.org/10.3389/fnins.2019.00906.

Gonzalez, C., Armijo, E., Bravo-Alegria, J., Becerra-Calixto, A., Mays, C.E., Soto, C., 2018. Modeling amyloid beta and tau pathology in human cerebral organoids. Mol. Psychiatr. 23, 2363–2374. https://doi.org/10.1038/s41380-018-0229-8.

Gough, G., O'Brien, N.L., Alic, I., Goh, P.A., Yeap, Y.J., Groet, J., Nizetic, D., Murray, A., 2020. Modeling Down syndrome in cells: from stem cells to organoids. Prog. Brain Res. 251, 55–90. https://doi.org/10.1016/bs.pbr.2019.10.003.

Graison-Yahya, E.A., Aubert, J., Dauphinot, L., Rivals, I., Prieur, M., Golfier, G., et al., 2007. Classification of human chromosome 21 gene-expression variations in Down syndrome: impact on disease phenotypes. Am. J. Hum. Genet. 81, 475–491. https://doi.org/10.1086/520000.

Gribble, S.M., Wiseman, F.K., Clayton, S., et al., 2013. Massively parallel sequencing reveals the complex structure of an irradiated human chromosome on a mouse background in the Tc1 model of Down syndrome. PLoS One 8, e60482.

Gropp, A., Giers, D., Kolbus, U., 1974. Trisomy in the fetal backcross progeny of male and female metacentric heterozygotes of the mouse. I. Cytogenet. & Cell Genet. 13 (6), 511–535.

Gupta, M., Dhanasekaran, A.R., Gardiner, K.J., 2016. Mouse models of Down syndrome: gene content and consequences. Mamm. Genome 27, 538–555. https://doi.org/10.1007/s00335-016-9661-8.

Gutierrez-Castellanos, N., Winkelman, B.H.J., Tolosa-Rodriguez, L., Devenney, B., Reeves, R.H., De Zeeuw, C.I., 2013. Size does not always matter: Ts65Dn Doen syndrome mice show cerebellum-dependent motor learning deficits that cannot be rescued by postnatal SAG treatment. J. Neurosci. 33, 15408–15413.

Harrington, W., Hillaire, C.S., Zweifel, L.S., et al., 2011. Recruitment of actin modifiers to TrkA endosomes governs retrograde NGF signaling and survival. Cell 146 (3), 421–434.

Hattori, M., Fujiyama, A., Taylor, T., Watanabe, H., Yada, T., Park, H.-S., et al., 2000. The DNA sequence of human chromosome 21. Nature 405 (6784), 311–319. https://doi.org/10.1038/35012518.

Head, E., Lott, I.T., Patterson, D., Doran, E., Haier, R.J., 2007. Possible compensatory events in adult Down syndrome brain prior to the development of Alzheimer disease neuropathology: targets for nonpharmacological intervention. J. Alzheim. Dis. 11 (1), 61–76.

Heber, S., Herms, J., Gajic, V., et al., 2000. Mice with combined gene knock-outs reveal essential and partially redundant functions of amyloid precursor protein family members. J. Neurosci. 20 (21), 7951–7963.

Henkel, G.W., McKercher, S.R., Yamamoto, H., Anderson, K.L., Oshima, R.G., Maki, R.A., 1996. PU.1 but not Ets2 is essential for macrophage development from embryonic stem cells. Blood 88 (8), 2917–2926.

Herault, Y., Delabar, J.M., Fisher, E.M.C., Tybulewicz, V.L.J., Yu, E., Brault, V., October 1, 2017. Rodent models in Down syndrome research: impact and future opportunities. Dis. Model Mech. 10 (10), 1165–1186. https://doi.org/10.1242/dmm.029728.

Holtzman, D.M., Santucci, D., Kilbridge, J., et al., 1996. Developmental abnormalities and age-related neurodegeneration in a mouse model of Down syndrome. Proc. Natl. Acad. Sci. U.S.A. 93 (23), 13333–13338.

Ishihara, K., Amano, K., Takaki, E., et al., 2010. Enlarged brain ventricles and impaired neurogenesis in the Ts1Cje and Ts2Cje mouse models of Down syndrome. Cerebr. Cortex 20 (5), 1131–1143.

Ishihara, K., 2021. Genes associated with disturbed cerebral neurogenesis in the embryonic brain of mouse models of Down syndrome. Genes 12, 1598. https://doi.org/10.3390/genes12101598.

Jiang, J., Jing, Y., Cost, G.J., et al., 2013. Translating dosage compensation to trisomy 21. Nature 500 (7462), 296–300. https://doi.org/10.1038/nature12394.

Jiang, X., et al., 2015. Genetic dissection of the Down syndrome critical region. Hum. Mol. Genet. 24, 6540–6551.

Joshi, K., Shen, L., Michaeli, A., Salter, M., Thibault-Messier, G., Hashmi, S., et al., 2016. Infantile spasms in down syndrome: rescue by knockdown of the GIRK2 channel. Ann. Neurol. 80, 511–521. https://doi.org/10.1002/ana.24749J.

Kazuki, Y., et al., 2020. A non-mosaic transchromosome mouse model of Down syndrome carrying the long arm of human chromosome 21. Elife 9, e56223.

Keating, D.J., Dubach, D., Zanin, M.P., et al., 2008. DSCR1/RCAN1 regulates vesicle exocytosis and fusion pore kinetics: implications for Down syndrome and Alzheimer's disease. Hum. Mol. Genet. 17 (7), 1020–1030.

Korbel, J.O., Tirosh-Wagner, T., Urban, A.E., Chen, X.-N., Kasowski, M., Dai, L., Grubert, F., Erdman, C., Gao, M.C., Lange, K., et al., 2009. The genetic architecture of Down syndrome phenotypes revealed by high-resolution analysis of human segmental trisomies. Proc. Natl. Acad. Sci. U.S.A. 106, 12031–12036.

Korenberg, J.R., Chen, X.N., Schipper, R., Sun, Z., Gonsky, R., Gerwehr, S., Carpenter, N., Daumer, C., Dignan, P., Disteche, C., 1994. Down syndrome phenotypes: the consequences of chromosomal imbalance. Proc. Natl. Acad. Sci. U. S. A. 91, 4997–5001. https://doi.org/10.1073/pnas.91.11.4997.

Lana-Elola, E., Watson-Scales, S., Slender, A., Gibbins, D., Martineau, A., Douglas, C., Mohun, T., Fisher, E.M.C., Tybulewicz, V.L.J., 2016. Genetic dissection of Down syndrome-associated congenital heart defects using a new mouse mapping panel. Elife 5, e11614.

Lee, H.C., Md Yusof, H.H., Leong, M.P., Zainal Abidin, S., Seth, E.A., Hewitt, C.A., et al., September 2019. Gene and protein expression profiles of JAK-STAT signalling pathway in the developing brain of the Ts1Cje down syndrome mouse model. Int. J. Neurosci. 129 (9), 871–881. https://doi.org/10.1080/00207454.2019.1580280.

Letourneau, A., Santoni, F.A., Bonilla, X., Sailani, M.R., Gonzalez, D., Kind, J., Chevalier, C., Thurman, R., Sandstrom, R.S., Hibaoui, Y., Garieri, M., Popadin, K., Falconnet, E., Gagnebin, M., Gehrig, C., Vannier, A., Guipponi, M., Farinelli, L., Robyr, D., Migliavacca, E., Borel, C., Deutsch, S., Feki, A., Stamatoyannopoulos, J.A., Herault, Y., van Steensel, B., Guigo, R., Antonarakis, S.E., 2014. Domains of genome-wide gene expression dysregulation in Down's syndrome. Nature 508 (7496), 345–350. https://doi.org/10.1038/nature13200.

Levine, S., Saltzman, A., Levy, E., Ginsberg, S.D., 2009. Systemic pathology in aged mouse models of Down's syndrome and Alzheimer's disease. Exp. Mol. Pathol. 86 (1), 18–22.

Li, Z., Yu, T., Morishima, M., Pao, A., Laduca, J., Conroy, J., et al., 2007. Duplication of the entire 22.9 Mb human chromosome 21 syntenic region on mouse chromosome 16 causes cardiovascular and gastrointestinal abnormalities. Hum. Mol. Genet. 16, 1359–1366. https://doi.org/10.1093/hmg/ddm086.

Liu, C., Morishima, M., Yu, T., et al., 2011. Genetic analysis of Down syndrome-associated heart defects in mice. Hum. Genet. 130 (5), 623–632.

Liu, C., Morishima, M., Jiang, X., Yu, T., Meng, K., Ray, D., Pao, A., Ye, P., Parmacek, M.S., Yu, Y.E., 2014. Engineered chromosome-based genetic mapping establishes a 3.7 Mb critical genomic region for Down syndrome associated heart defects in mice. Hum. Genet. 133, 743–753.

Lorenzo, L.P., Shatynski, K.E., Clark, S., Yarowsky, P.J., Williams, M.S., August 2013. Defective thymic progenitor development and mature T-cell responses in a mouse model for Down syndrome. Immunology 139 (4), 447–458. https://doi.org/10.1111/imm.12092.

Loughran, S.J., Kruse, E.A., Hacking, D.F., et al., 2008. The transcription factor Erg is essential for definitive hematopoiesis and the function of adult hematopoietic stem cells. Nat. Immunol. 9 (7), 810–819.

Lyle, R., Béna, F., Gagos, S., Gehrig, C., Lopez, G., Schinzel, A., Lespinasse, J., Bottani, A., Dahoun, S., Taine, L., et al., 2009. Genotype-phenotype correlations in Down syndrome identified by array CGH in 30 cases of partial trisomy and partial monosomy chromosome 21. Eur. J. Hum. Genet. 17, 454–466.

Mendioroz, M., Do, C., Jiang, X., Liu, C., Darbary, H.K., Lang, C.F., Lin, J., Thomas, A., Abu-Amero, S., Stanier, P., et al., 2015. Trans effects of chromosome aneuploidies on DNA methylation patterns in human Down syndrome and mouse models. Genome Biol. 16, 263.

Miles, E.W., Kraus, J.P., 2004. Cystathionine β-synthase: structure, function, regulation, and location of homocystinuria causing mutations. J. Biol. Chem. 279 (29), 29871–29874.

Miller, R.A., Harrison, D.E., Astle, C.M., et al., 2011. Rapamycin, but not resveratrol or simvastatin, extends life span of genetically heterogeneous mice. J. Gerontol.—Series A Biol. Sci. & Med. Sci. 66 (2), 191–201.

Moore, S., Hawkins, C., Franca, A., et al., 2011. Increased male reproductive success in Ts65Dn "Down syndrome" mice. Mamm. Genome 21 (11–12), 543–549.

Moore, C.S., Roper, R.J., 2007. The power of comparative and developmental studies for mouse models of Down syndrome. Mamm. Genome 18 (6–7), 431–443. https://doi.org/10.1007/s00335-007-9030-8.

Moyer, A.J., Gardiner, K., Reeves, R.H., May 2021. All creatures great and small: new approaches for understanding Down syndrome genetics. Trends Genet. 37 (5), 444–459. https://doi.org/10.1016/j.tig.2020.09.017.

Netzer, W.J., Powell, C., Nong, Y., et al., 2010. Lowering betaamyloid levels rescues learning and memory in a Down syndrome mouse model. PLoS One 5 (6). Article ID e10943.

Ng, P., Hyland, C.D., Metcalf, D., et al., 2010. Trisomy of Erg is required for myeloproliferation in a mouse model of Down syndrome. Blood 115 (19), 3966–3969.

Nordquist, S.K., et al., 2018. Systematic functional characterization of human 21st chromosome orthologs in *Caenorhabditis elegans*. G3 (Bethesda) 8, 967–979.

O'Doherty, A., Ruf, S., Mulligan, C., Hildreth, V., Errington, M.L., Cooke, S., Sesay, A., Modino, S., Vanes, L., Hernandez, D., et al., 2005. An aneuploid mouse strain carrying human chromosome 21 with Down syndrome phenotypes. Science 309, 2033–2037.

Olson, L.E., Richtsmeier, J.T., Leszl, J., Reeves, R.H., 2004. A chromosome 21 critical region does not cause specific Down syndrome phenotypes. Science 306, 687–690. https://doi.org/10.1126/science.1098992.

Olson, L.E., Roper, R.J., Sengstaken, C.L., et al., 2007. Trisomy for the Down syndrome "critical region" is necessary but not sufficient for brain phenotypes of trisomic mice. Hum. Mol. Genet. 16 (7), 774–782.

Olson, L.E., Mohan, S., 2011. Bone density phenotypes in mice aneuploid for the Down syndrome critical region. Am. J. Med. Genet. 155 (10), 2436–2445.

Ovchinnikov, D.A., Korn, O., Virshup, I., Wells, C.A., Wolvetang, E.J., 2018. The impact of APP on Alzheimer-like pathogenesis and gene expression in Down syndrome iPSC-derived neurons. Stem Cell Rep. 11 (1), 32–42.

Pan, X., et al., 2015. PAK1 regulates cortical development via promoting neuronal migration and progenitor cell proliferation. Mol. Brain 8, 36.

Pawlikowski, B., Betta, N.D., Elston, T., Williams, D.A., Olwin, B.B., March 2018. Muscle stem cell dysfunction impairs muscle regeneration in a mouse model of Down syndrome. Sci. Rep. 8 (1), 4309. https://doi.org/10.1038/s41598-018-22342-5. Bibcode: 2018NatSR … 8.4309P.

Pereira, P.L., Magnol, L., Sahún, I., et al., 2009. A new mouse model for the trisomy of the Abcg1-U2af1 region reveals the complexity of the combinatorial genetic code of Down syndrome. Hum. Mol. Genet. 18 (24), 4756–4769.

Raveau, M., Lignon, J.M., Nalesso, V., Duchon, A., Groner, Y., Sharp, A.J., Dembele, D., Brault, V., Hérault, Y., 2012. The app-runx1 region is critical for birth defects and electrocardiographic dysfunctions observed in a Down syndrome mouse model. PLoS Genet. 8, e1002724.

Real, R., et al., June 11, 2019. Vivo modelling of human neuron dynamics and Down syndrome. Science 362 (6416), eaau1810.

Reeves, R.H., Irving, N.G., Moran, T.H., Wohn, A., Kitt, C., Sisodia, S.S., et al., 1995. A mouse model for Down syndrome exhibits learning and behaviour deficits. Nat. Genet. 11, 177–184. https://doi.org/10.1038/ng1095-177.

Reinholdt, L.G., Czechanski, A., Kamdar, S., King, B.L., Sun, F., Handel, M.A., 2009. Meiotic behavior of aneuploid chromatin in mouse models of Down syndrome. Chromosoma 118, 723–736. https://doi.org/10.1007/s00412-009-0230-8.

Reynolds, L.E., Watson, A.R., Baker, M., Jones, T.A., D'amico, G., Robinson, S.D., Joffre, C., et al., 2010. Tumor angiogenesis is reduced in the Tc1 mouse model Down's syndrome. Nature 465, 813–817.

Richtsmeier, J.T., Baxter, L.L., Reeves, R.H., 2000. Parallels of craniofacial maldevelopment in Down syndrome and Ts65Dn mice. Dev. Dynam. 217 (2), 137–145.

Richtsmeier, J.T., Zumwalt, A., Carlson, E.J., et al., 2002. Craniofacial phenotypes in segmentally trisomic mouse models for Down syndrome. Am. J. Med. Genet. 107, 317–324.

Roper, R.J., Reeves, R.H., 2006. Understanding the basis for Down syndrome phenotypes. PLoS Genet. 2, e50.

Roper, R.J., Goodlett, C.R., Martínez de Lagrán, M., Dierssen, M., 2020. Behavioral phenotyping for Down syndrome in mice. Curr. Protoc. Mouse Biol. 10 (3), e79. https://doi.org/10.1002/cpmo.79.

Roper, R.J., Baxter, L.L., Saran, N.G., Klinedinst, D.K., Beachy, P.A., Reeves, R.H., 2006. Defective cerebellar response to mitogenic Hedgehog signalling in Down syndrome mice. Proc. Natl. Acad. Sci. U.S.A. 103, 1452–1456.

Roubertoux, P.L., Bichler, Z., Pinoteau, W., Jamon, M., Seregaza, Z., et al., 2006. Pre-weaning sensorial and motor development in mice transpolygenic for the critical region of trisomy 21. Behav. Genet. 36 (3), 377–386.

Rueda, N., Flórez, J., Martínez-Cué, C., May 22, 2012. Mouse models of Down syndrome as a tool to unravel the causes of mental disabilities. Neural Plast. 584071. https://doi.org/10.1155/2012/584071.

Sago, H., Carlson, E.J., Smith, D.J., Kilbridge, J., Rubin, E.M., Mobley, W.C., et al., 1998. Ts1Cje, a partial trisomy 16 mouse model for Down syndrome, exhibits learning and behavioral abnormalities. Proc. Natl. Acad. Sci. U.

S. A. 95, 6256–6261. https://doi.org/10.1073/pnas.95.11.6256.

Salehi, A., Delcroix, J.D., Belichenko, P.V., et al., 2006. Increased App expression in a mouse model of Down's syndrome disrupts NGF transport and causes cholinergic neuron degeneration. Neuron 51 (1), 29–42.

Shaw, P.R., Klein, J.A., Aziz, N.M., Haydar, T.F., 2020. Longitudinal neuroanatomical and behavioral analyses show phenotypic drift and variability in the Ts65Dn mouse model of Down syndrome. Dis. Model Mech. 13 (9), dmm046243.

Shukkur, E.A., Shimohata, A., Akagi, T., et al., 2006. Mitochondrial dysfunction and tau hyperphosphorylation in Ts1Cje, a mouse model for Down syndrome. Hum. Mol. Genet. 15 (18), 2752–2762.

Siarey, R.J., Villar, A.J., Epstein, C.J., Galdzicki, Z., 2005. Abnormal synaptic plasticity in the Ts1Cje segmental trisomy 16 mouse model of Down syndrome. Neuropharmacology 49, 122–128. https://doi.org/10.1016/j.neuropharm.2005.02.012.

Simón, M., Schiapparelli, L., Salazar-Colocho, P., et al., 2009. Overexpression of wild-type human APP in mice causes cognitive deficits and pathological features unrelated to Aβ levels. Neurobiol. Dis. 33 (3), 369–378.

Spellman, C., et al., 2013. Expression of trisomic proteins in Down syndrome model systems. Gene 512, 219–225.

Sterne, G.R., et al., 2015. Dysregulated Dscam levels act through Abelson tyrosine kinase to enlarge presynaptic arbors. Elife 4, e05196.

Sussan, T.E., Yang, A., Li, F., Ostrowski, M.C., Reeves, R.H., 2008. Trisomy represses ApcMin-mediated tumours in mouse models of Down's syndrome. Nature 451 (7174), 73–75.

Tang, X.Y., Xu, L., Wang, J., Hong, Y., Wang, Y., Zhu, Q., Wang, D., Zhang, X.Y., Liu, C.Y., Fang, K.H., Han, X., Wang, S., Wang, X., Xu, M., Bhattacharyya, A., Guo, X., Lin, M., Liu, Y., June 15, 2021. DSCAM/PAK1 pathway suppression reverses neurogenesis deficits in iPSC-derived cerebral organoids from patients with Down syndrome. J. Clin. Invest. 131 (12), e135763. https://doi.org/10.1172/JCI135763.

Trazzi, S., Fuchs, C., Valli, E., Perini, G., Bartesaghi, R., Ciani, E., 2013. The amyloid precursor protein (APP) triplicated gene impairs neuronal precursor differentiation and neurite development through two different domains in the Ts65Dn mouse model for Down syndrome. J. Biol. Chem. 288, 20817–20829.

Troca-Marín, J.A., Alves-Sampaio, A., Montesinos, M.L., 2011. An increase in basal BDNF provokes hyperactivation of the Akt-Mammalian target of rapamycin pathway and deregulation of local dendritic translation in a mouse model of Down's syndrome. J. Neurosci. 31 (26), 9445–9455.

Vacano, G.N., Duval, N., Patterson, D., 2012. The use of mouse models for understanding the biology of Down syndrome and aging. Curr. Gerontol. & Geriatr. Res. 20. https://doi.org/10.1155/2012/717315. Article ID 717315.

Vacik, T., Ort, M., Gregorova, S., Strnad, P., Blatny, R., et al., 2005. Segmental trisomy of chromosome 17: a mouse model of human aneuploidy syndromes. Proc. Natl. Acad. Sci. U. S. A. 102 (12), 4500–4505.

Vega, R.B., Rothermel, B.A., Weinheimer, C.J., et al., 2003. Dual roles of modulatory calcineurin-interacting protein 1 in cardiac hypertrophy. Proc. Natl. Acad. Sci. U.S.A. 100 (2), 669–674.

Villar, J., Belichenko, P.V., Gillespie, A.M., Kozy, H.M., Mobley, W.C., Epstein, C.J., 2005. Identification and characterization of a new Down syndrome model, Ts[Rb(12.1716)]2Cje, resulting from a spontaneous Robertsonian fusion between T(1716)65Dn and mouse Chromosome 12. Mamm. Genome 16 (2), 79–90.

Voronov, S.V., Frere, S.G., Giovedi, S., et al., 2008. Synaptojanin 1-linked phosphoinositide dyshomeostasis and cognitive deficits in mouse models of Down's syndrome. Proc. Natl. Acad. Sci. U.S.A. 105 (27), 9415–9420.

Wang, Y., Chang, J., Shao, L., Feng, W., Luo, Y., Chow, M., et al., June 2016. Hematopoietic stem cells from Ts65Dn mice are deficient in the repair of DNA double-strand breaks. Radiat. Res. 185 (6), 630–637. https://doi.org/10.1667/RR14407.1. Bibcode: 2016RadR...185...630W.

Wei, G., Srinivasan, R., Cantemir-Stone, C.Z., et al., 2009. Ets1 and Ets2 are required for endothelial cell survival during embryonic angiogenesis. Blood 114 (5), 1123–1130.

Westmark, C.J., Westmark, P.R., Malter, J.S., 2010. Alzheimer's disease and Down syndrome rodent models exhibit audiogenic seizures. J. Alzheimers Dis. 20, 1009–1013. https://doi.org/10.3233/JAD-2010-100087.

Wiseman, F.K., Pulford, L.J., Barkus, C., Liao, F., Portelius, E., Webb, R., et al., 2018. Trisomy of human chromosome 21 enhances amyloid-β deposition independently of an extra copy of APP. Brain 141 (8), 2457–2474.

Xing, Z., Li, Y., Pao, A., Bennett, A.S., Tycko, B., Mobley, W.C., Yu, Y.E., 2016. Mouse-based genetic modelling and analysis of Down syndrome. Br. Med. Bull. 120, 111–122. https://doi.org/10.1093/bmb/idw040.

Xu, R., et al., 2019. OLIG2 drives abnormal neurodevelopmental phenotypes in human iPSC-based organoid and chimeric mouse models of Down syndrome. Cell Stem Cell 24, 908–926.

Yanagida, M., Osato, M., Yamashita, N., et al., 2005. Increased dosage of Runx1/AML1 acts as a positive modulator of myeloid leukemogenesis in BXH2 mice. Oncogene 24 (28), 4477–4485.

Yu, T., Li, Z., Jia, Z., et al., 2010a. A mouse model of Down syndrome trisomic for all human chromosome 21 syntenic regions. Hum. Mol. Genet. 19 (14), 2780–2791. https://doi.org/10.1093/hmg/ddQ199. ArticleID ddq179.

Yu, T., Liu, C., Belichenko, P., Clapcote, S.J., Li, S., Pao, A., et al., 2010b. Effects of individual segmental trisomies of human chromosome 21 syntenic regions on hippocampal long-term potentiation and cognitive behaviors in mice. Brain Res 1366, 162–171. https://doi.org/10.1016/j.brainres.2010.09.107.

Zhang, L., et al., 2015. DSCAM and DSCAML1 regulate the radial migration and callosal projection in developing cerebral cortex. Brain Res. 1594, 61–70.

Zheng, H., Jiang, M., Trumbauer, M.E., et al., 1996. Mice deficient for the amyloid precursor protein gene. Ann. N. Y. Acad. Sci. 777, 421–426.

Index

Note: 'Page numbers followed by "f" indicate figures and "t" indicate tables.'

L

M

FSC
www.fsc.org
MIX
Papier | Fördert gute Waldnutzung
FSC® C083411

Kolibri 360 GmbH
Ferdinand-Jühlke-Straße 7
99095 Erfurt, Deutschland
produktsicherheit@kolibri360.de